BUSINESS INNOVATION AND DEVELOPMENT IN EMERGING ECONOMIES

PROCEEDINGS OF THE 5TH SEBELAS MARET INTERNATIONAL CONFERENCE ON BUSINESS, ECONOMICS AND SOCIAL SCIENCES (SMICBES 2018), BALI, INDONESIA, 17–19 JULY 2018

Business Innovation and Development in Emerging Economies

Edited by

Irwan Trinugroho
Universitas Sebelas Maret (UNS), Indonesia

Evan Lau
Universiti Malaysia Sarawak, Malaysia

CRC Press
Taylor & Francis Group
Boca Raton London New York

CRC Press is an imprint of the
Taylor & Francis Group, an **informa** business

A BALKEMA BOOK

Published by:
CRCPress/Balkema
P.O. Box 447, 2300 AK Leiden, The Netherlands
e-mail: Pub.NL@taylorandfrancis.com
www.crcpress.com – www.taylorandfrancis.com

First issued in paperback 2020

ISBN 13: 978-0-367-72942-4 (pbk)
ISBN 13: 978-1-138-35996-3 (hbk)

Visit the Taylor & Francis Web site at
http://www.taylorandfrancis.com

and the CRC Press Web site at
http://www.crcpress.com

Library of Congress Cataloging-in-Publication Data

Names: Trinugroho, Irwan, editor. | Lau, Evan, editor.
Title: Business innovation and development in emerging economies / edited by Irwan Trinugroho, Universitas Sebelas Maret (UNS), Indonesia, Evan Lau, Universiti Malaysia Serawak, Malaysia.
Description: Leiden, The Netherlands : CRC Press/Balkema, [2019] | Includes bibliographical references and index.
Identifiers: LCCN 2019015377 (print) | LCCN 2019018078 (ebook) | ISBN 9780429433382 (ebook) | ISBN 9781138359963 (hardcover : alk. paper) Subjects: LCSH: Industries–Developing countries. | Management–Developing countries. | Economic development–Developing countries.
Classification: LCC HC59.7 (ebook) | LCC HC59.7 .B8656 2019 (print) | DDC 338.709172/4–dc23
LC record available at https://lccn.loc.gov/2019015377

Typeset by V Publishing Solutions Pvt Ltd., Chennai, India

Business Innovation and Development in Emerging Economies – Trinugroho & Lau (Eds)
© 2019 Taylor & Francis Group, London, ISBN 978-1-138-35996-3

Table of contents

Development

Entrepreneurship

Financial accounting

Financial reporting and disclosure

HRM & OB

Innovation and strategic management

Intellectual capital

Macro and monetary economics

Marketing

Preface

This publication contains a selection of papers presented at the 5th Sebelas Maret International Conference on Business, Economics and Social Sciences (SMICBES) 2018, held in Bali, Indonesia on July 17–19, 2018. This conference, with the particular theme "Business Innovation and Development in Emerging Economies", is organized by the Faculty of Economics and Business, Universitas Sebelas Maret (FEB UNS) and supported by some other institutions. SMICBES was first organized by FEB UNS in 2013 under a different name. It has been the regular event of FEB UNS since 2016.

The conference objective is to provide a forum for researchers and policymakers to exchange their views about current issues related to business, economic and social sciences and the intersection among those fields. More than 300 papers, either empirical or theoretical, were presented in this conference.

This conference and this publication were made possible by support from many people. We therefore would like to thank the Dean of FEB UNS Dr. Hunik Sri Runing Sawitri, Vice-deans, the organizing committee of the conference led by Mr. Linggar Ikhsan Nugroho, M.Ec.Dev, participants, reviewers and all supporting institutions for their valuable support to the conference.

Editors
Irwan Trinugroho
(Universitas Sebelas Maret)

Evan Lau
(Universiti Malaysia Sarawak)

Auditing

Business Innovation and Development in Emerging Economies – Trinugroho & Lau (Eds)
© *2019 Taylor & Francis Group, London, ISBN 978-1-138-35996-3*

Audit firm selection and voluntary audit firm rotation: Which factors matter?

Sulhani & N.K.E. Putri
Sekolah Tinggi Ekonomi Islam Tazkia, Jawa Barat, Indonesia

ABSTRACT: Investigations related to voluntary audit firm rotation are still rare, whereas voluntary audit firm rotation can be an indication of the particular situation that is being faced by the company. This study was conducted with the aim to understand the influence of a company's financial condition, earnings management and financial statement manipulation to the company's tendency in audit firm selection when perform voluntary audit firm rotation. The sample of this study is all companies listed on the Indonesia Stock Exchange, except the financial industry, which performed voluntary audit firm rotation in the period 2010 to 2015. The data is tested using the ordinal logistic regression method because the dependent variable is ordinal scale data. The results of this study indicate that a company's financial condition have no significant effect on the tendency of audit firm selection when voluntary audit firm rotation is performed. Meanwhile, earnings management has a negative significant effect and the manipulation of financial statements has a positive significant effect on the tendency audit firm selection in the event of voluntary audit firm rotation. Companies that do earnings management tend to switch the audit firm to a lower-quality, while when companies are facing manipulation of financial statements they will switch the audit to better quality audit firm. The implication of this study is that voluntary audit rotation conducted by the company is an indication of certain events that exist in the company that can be taken into consideration by decision making.

Keywords: Voluntary Audit Firm rotation, Financial Condition, Earnings Management, Manipulation of Financial Statement

1 INTRODUCTION

Audit firm rotation began to be an interesting topic since the Enron case in the United States in 2001. To prevent similar cases from recurring, the United States government issued the Sarbanes-Oxley Act (SOX) in 2002 and one of its contents was that corporate must switch their audit firm periodically. According to Kim et al. (2015) audit firm rotation aims to improve the quality of corporate governance and financial reporting. Due to the importance of firm rotation audits in the financial reporting process, the Indonesian government also acted swiftly by issuing KMK RI number 423/KMK.06/2002 which was updated with PMK number 17/PMK.01/2008 on Public Accounting Services, which is providing audit services to financial statements by audit firm into six consecutive yearbooks. This regulation requires companies in Indonesia to replace the audit firm if the audit engagement period has reached the six-year deadline, known as mandatory audit firm rotation (Siregar et al., 2012).

Besides, due to regulatory requirements, audit firm rotation can also come from the client's desire to terminate audit firms before the engagement ends, which is better known as voluntary audit firm rotation. According to Davidson et al. (2005), voluntary audit firm rotation generally occurs due to opportunistic behavior of management. The results of Nasser et al. (2006) stated that one important factor in decision-making audit firm selection is the condition of the company. When the company's financial condition is down, management tends to look for smaller audit firms. This is because small audit firm are reluctant to show

3

their disagreement with the company. While Lennox et al. (2014) states that the voluntary audit firm rotation can also occur when the client lays off the existing audit firm to look for another audit firm at a cheaper cost or look for another non-conservative audit firm.

Conservative auditors that are not in line with management often leads to conflict between client and auditor (Khrisnan, 1994; Defond & Subramanyam, 1998; and Kim et al. 2003). The emergence of such conflicts is due to the auditor not being willing to comply with the wishes of the client. So it can be concluded that the company avoids auditors who behave conservatively to reduce the possibility of detection of manipulation and earnings management by previous auditors (Lou et al., 2009). In addition, low financial condition and firm performance often encourage management to hide the actual financial condition of the company so that the company can avoid reporting the company's losses (Wasiuzzaman et al., 2015). Companies typically also manipulate financial statements when debt levels are high and liquidity levels are low (Dechow et al., 2011). Companies with this cases tend to switch to other non-conservative audit firms or those who can give a better opinion of the existing audit firm. This is done to maintain corporate image (Davidson et al., 2005).

Manipulation of financial statements is often done by the company when the company's financial condition is declining (Dechow et al., 1995). Besides that, the company also performs earnings management for the purpose of avoiding extreme performance degradation (Davidson et al., 2005). In contrast to manipulation, earnings management is allowed to be made while still complying with accounting standards. Earnings management can have a positive or negative impact on the company. Earnings management is done with the aim to affect the profit for the purpose of achieving targeted earnings. Earnings management will have a negative impact if it is done for management to get compensation (Erickson et al., 2015). Under these circumstances, a conservative auditor may create conflict-related incentive reporting between managers and auditors. So that company will replace their existing audit firm that is considered conservative.

Previous research on voluntary audit firm rotation was done by Johnson et al. (1990) and Lennox et al. (2011). Johnson et al. (1990) conducted a study of the financial characteristics and performance of a firm's stock price against the decision to perform voluntary audit firm rotation. The results of this research mentioned that clients tend to replace the audit firm to get an audit firm with lower costs. In addition, Lennox et al. (2011) conducted a study of credit ratings on private companies in the UK as a sample. The results suggest that moving from mandatory to voluntary audit firm rotation has an important role in providing a bad signal for the capital market, where CFOs in private companies tend to manipulate earnings to maintain their corporate credit rating. This is done with the aim of avoiding the possibility of breach of the contract of debt.

This study is different from previous research, because this study wanted to test whether the tendency to perform voluntary audit firm rotation can be a bad signal for the stakeholder, by examining the factors of financial condition, earnings management and financial statement manipulation of voluntary audit firm rotation. This study focuses only on voluntary audit firm rotation and overrides companies that perform mandatory audit firm rotation. In addition, this study will also test the tendency of companies in choosing audit firms if they are not in a good financial condition, have high earnings management and the tendency of manipulation of financial statements at the time of voluntary audit rotation.

2. THEORETICAL FRAMEWORK AND HYPOTHESIS

2.1 *Audit firm rotation*

Independence of the auditor becomes a contentious issue among the accounting profession. This becomes an important issue because independent auditors will effect audit quality. Therefore, the government as a regulator is obliged to intervene in dealing with the problem by establishing rules on the period of audit services that can be provided by the audit firm to its clients. Given this provision, the government may work on the parties concerned, both the company and the auditor. In Indonesia, a mandatory audit rotation became regulated in the Decree of the Minister of Finance of the Republic of Indonesia Number 423/KMK.06/2002 concerning Public Accounting Services. The rules governing that the general audit of the financial statements of the entity become six consecutive yearbooks.

Auditor independence is a foundation for a reliable auditor's report (Public Oversight Board, 2000). Therefore, there have been many studies that discuss mandatory audit firm rotation. A mandatory audit firm rotation is considered to have a positive impact and is likely to improve audit quality (Johnson et al., 2002; Chung et al., 2003; Vanstralen, 2000), whereas there are also studies that find that mandatory audit firm rotation can negatively impact and tend to decrease audit quality (Geiger & Raghunandan, 2002; Gao, 2003; Myers et al., 2003; Jackson et al., 2008). Mandatory audit firm rotation could have positive impact because it will maintain the independence of the auditor but on the other side, when mandatory audit firm rotation is performed, the company will engage with a new auditor that could be have less information about the client.

According to Davidson et al. (2005) and Lennox et al. (2014), the election of an audit firm becomes an important decision for a company. Companies must be careful in deciding which audit firm will give a good opinion for the company. Unlike the mandatory audit firm rotation in Indonesia that is regulated in KMK.06/2002, the decision to adopt voluntary audit firm rotation is the management authority. There are several reasons for the company's voluntary audit firm rotation:

a. Requires more effective and independent audit firm.
The company does a voluntary audit firm rotation when a company wants to improve its image in the public eye. A more independent audit firm can improve the trust of shareholders and creditors. This is an advantage for shareholders (Nasser et al., 2006).
b. Reduce the cost of audit services.
Previous research has suggested that the decrease in audit service costs encourages management to perform voluntary audit firm rotation. There is a decrease in audit service costs if company growth is decreasing (Johnson et al., 1990).
c. Stopping the conservative audit firm.
Khrisnan (1994) finds that voluntary audit firm rotation is more due to the conservative attitude of the auditor when compared with the Exceptional Exclusion by the auditor. In fact, the voluntary audit firm rotation rate is increased when the auditor is skeptical and provides a reasonable opinion to the company. In addition, Davidson et al. (2005) state that the choice of managers to perform voluntary audit firm rotation is motivated by the desire to report better corporate financial condition. Therefore, companies tend to choose a less conservative audit firm. Therefore, earnings management is considered as one of the mechanisms that can be allowed in order to make as if the target of the company has been achieved.
d. Looking for a lower quality audit firm.
DeFond & Subramanyam (1998) state that companies that do not get unqualified opinions are likely to make an audit rotation using the Big 6 audit firms rather than the non-Big 6 audit firms. It is strongly suspected that the company conducts an opinion audit with the non-Big 6 audit firm.

2.2 *Financial condition*

According to Wang et al. (2007), financial conditions are defined as financial performance as measured by cumulative changes of net assets, equity and net cash flows. Financial achievement can be seen from the ability of the company to meet its financial obligations in a timely manner. In conducting its business activities, the company creates financial liabilities in the form of expenses, costs and debts. All these types of obligations require payment at this time, as well as deferred in the future. If a company can pay its liabilities without causing significant financial hardship, then it can be assumed that the company is in sound financial condition.

Summers & Sweeney (1998) revealed that the low performance of the company will worsen the financial condition. The impact of this low financial condition tends to motivate the internal company to take action aimed at improving the appearance of the company's financial position. The insiders of this company are focused on management, where management knows the company's financial information and condition with certainty. This is done so that management does not lose their job.

Financial condition becomes one of the factors that influences the client's decision to maintain its audit firm. Insolvent clients tend to seek audit firms that have a higher level of independence to increase creditor and shareholder confidence. Insolvency is a condition in which the client is unable to meet its short-term obligations due to lack of liquidity. The inadequacy faced by these clients indicates that the lack of management competence in managing corporate assets (Nasser et al., 2006).

In many studies the measurement of corporate finance often uses the Altman Z Score model (1968). This model is a commonly used model as an indicator for measuring financial distress. The Altman model is regarded as the most superior model because of its simplicity, its practicality and its accuracy in measuring the company's financial condition. Altman performs an analysis of financial ratios that can be used to assess profitability and corporate risk. The financial ratios used are the five elements of the ratio that have an important value in describing the ability of management in managing its assets. The five elements of the ratio include working capital to total assets, retained earnings to total assets, earnings before interest and tax to total assets, market value to book value of total debt and total revenue to total assets (Rawi et al., 2008; Ramadhani et al., 2009: Sudiyatno et al., 2010).

2.3 *Earnings management*

Healy & Wahlen (1998) state that earnings management occurs when managers use judgment in financial reporting and change the financial reporting value of the value that should be reported. The value of this financial reporting relates to the fundamental economic performance figures of the company and the contractual results. Thus, the change in figures in these financial statements will outwit the stakeholders in the framework of decision making.

According to Scott (2015), earnings management is included in accounting policies that managers can choose to influence earnings so that specific earnings reporting objectives can be achieved. Earnings management is done for various reasons, among others, to improve management compensation, and affect the performance of stock prices. Thus, the impact of the existence of earnings management will affect the figures and financial statement data used by stakeholders as a source of information in decision making, especially for investors.

Scott (2015) also argues that earnings management can be viewed from two different perspectives, namely opportunistic earning management and efficient earning management. From an opportunistic perspective, earnings management is made by choosing accounting policies appropriate to management interests, although the accounting policy cannot achieve the goals and interests of investors. Earnings management is made to maximize the benefit to be received by management, such as obtaining compensation, maintaining management reputation and meeting investor earnings expectations. Failure of management in achieving earnings of investor expectations affect the decline in stock prices of the company. Furthermore, this will affect the reputation of the management.

Meanwhile, earnings management based on efficient perspective is done by choosing an accounting policy that can control the cost faced by the company. This is done with the aim of improving the quality of profit value information that has not been reflected in the financial statements and convey certain information to stakeholders. Thus, this can reduce the level of agency costs arising from conflicting interests between managers and stakeholders.

2.4 *Manipulation of financial statement*

Fama (1980) argues that goals and interests between principals and agents are often contradictory. This will lead to conflict between the two parties. According to Jensen & Meckling (1976), conflicts between principals and agents are due to information asymmetries, whereas the information asymmetry occurs when the information possessed by the agent is greater than the principal. Thus, the agent is compelled to do intentionally things that should not be done. An example is to manipulate financial reporting. Manipulation is deliberated by preparing financial statements to be materially miss-stated, so the financial statements become unreliable because it contains elements of fraud and also cause losses. The Association of Certified Fraud Examiners (ACFE, 2016) classifies report manipulation as one of three forms of fraud acts. Two other acts considered fraud are asset misappropriation and corruption. The case of asset abuse is the most frequent case, accounting for over 83% of reported cases. Furthermore, corruption is 35% and the manipulation of financial statements is 10%. However, if sorted based on nominal losses generated, the manipulation of financial statements causes the greatest loss compared to the other two categories.

2.5 Financial condition and the tendency of audit firm selection in voluntary audit firm rotation

Financial condition becomes one of the factors that influences the client's decision to maintain its audit firm. Companies that are experiencing growth tend to maintain auditors to maintain audit quality. And so is the case if the company's financial condition is decreasing. Companies tend to retain their existing auditors. This is because the changes of audit firm will further complicate the company's financial condition due to the possibility of increasing the audit service fee. In addition, companies experiencing insolvency circumstances and an unhealthy financial position will be more likely to retain their auditors to maintain the confidence of investors and shareholders, and to prevent litigation risks (Nasser et al., 2006).

This is in line with the study of Chung et al. (2003) that found that the Big 6 audit firms emphasize conservative accounting on clients when clients' financial performance has decreased from expectations, thereby reducing the likelihood that firms will overestimate revenue recording. Meanwhile, Hudaib et al. (2005) found that firms with diminishing financial conditions will be more likely to replace audit firm due to business uncertainty run by the company. The decline in business processes leads to a decrease in the rate of profit gained in a period. So the selection of an audit firm with a cheaper cost is done to reduce the cost of company expenses. So based on the argument, a hypothesis can be drawn as follows:

H1: The financial conditions affect the tendency of audit firm selection when voluntary audit firm rotation is performed.

2.6 Earning management practices and the tendency of audit firm selection when voluntary audit firm rotation is performed

According to Wasiuzzaman et al. (2015), the motivation for earnings management varies across industries. The main motivation for management to do earnings management is to manage earnings in such a way that reported earnings do not decrease from the previous year, not to avoid reporting losses. Earning management is done by distorting the relationship between stock returns and reported earnings, which can degrade the quality of financial information. The impact of earnings management is to present information about the performance of a company that is misleading for creditors and competitors and not to be observed by them.

According to Davidson et al. (2005), by doing earnings management, the chances of managers being laid off will decrease because with earnings management the earnings performance of the company can be managed better. Usually companies do earnings management because the company's operating performance deteriorates. So then the company will conduct audit firm rotation to obtain an audit firm that is not conservative and is therefore willing to approve the level of earnings management conducted by the company. Management tends to shift from the Big 6 audit firms to the non-Big 6 audit firms when it receives a modified audit opinion from previous auditors to obtain a non-conservative audit firm (Davidson et al., 2005). Non-conservative audit firms tend to use less stringent rules in the auditing process. Consistent with Muttakin et al. (2017) that the Big 4 audit firm is more sceptic to client reporting misstatements. This is done on the basis of investor protection.

According to Dechow et al. (2011), when company performance declines, managers make earnings management to maintain their company's stock price. Motivation to maintain or increase the company's stock price arises because the manager wants compensation related to the performance of the stock. In other words, earnings management is done to convince investors that the prospect of the company is going well. Based on the above literature, the authors propose a hypothesis in this study as follows:

H2: Earnings management practice has an effect on the tendency of audit selection when voluntary audit firm rotation is performed.

2.7 Manipulation of financial statements and the tendency of audit firm selection when voluntary audit firm rotation is performed

The Association of Certified Fraud Examiners (ACFE, 2016) states that there are three main categories in the case of fraud: misuse of assets, corruption and manipulation of financial statements. Compared to the case of misuse of assets, the rate of financial manipulation is considered low. However, the nominal losses incurred from the manipulation of these financial statements are much greater than other two categories. In addition, cases of financial report manipulation are reported to increase each year (ACFE, 2016).

According to Carcello & Nagy (2004), cases of manipulation in financial statements tend to occur two to three years early in the audit engagement. Most cases of financial report manipulation are generally done by companies that have been listed on the exchange. This arises because the public company is driven to achieve the profit target, so the greater the incentive to achieve profit, the greater the chance to manipulate.

This is consistent with the results of the research by Lou & Was (2009), which states that 36% of cases of fraud occur in the first two years of the engagement. Thus, in the early engagement, the client performs firm rotation audits to reduce the likelihood of detecting manipulations in the financial statements. This is contrary to the results of research conducted by Hastuti & Gozali (2015) that there is no evidence of manipulation in the financial statements of companies that effect the audit firm rotation.

Research conducted by Francis (2004) and Francis & Wang (2006) states that the Big 4 audit firms have the competence to conduct tighter and better audits than non-Big 4 audit firms, so the Big 4 audit firms probability of finding misstatements is greater. In addition the Big 4 audit firms are considered to provide higher audit quality than non-Big 4 audit firms. This happens because the Big 4 audit firms are required to maintain their reputation, so the Big 4 audit firms tend to report misstatements they find during the audit process.

In addition, Francis & Wang (2006) also stated that the Big 4 audit firms in a law-abiding state are considered to be very protective of the interests of investors. Big 4 audit firms are very conservative on behalf of their clients with the aim of protecting the interests of investors; not infrequently the Big 4 audit firms are demanding a non-conservative auditor. Strict protection against these investors will make the quality of audits higher. This means that when manipulation is high, it will impact on low audit quality and high risk of litigation. It can be concluded that the board of commissioners of independent companies will tend to avoid the risk of manipulation by management to avoid the high risk of litigation. This is in line with Payamta's (2006) research findings that non-Big 4 audit firms will be more often faced with litigation risk than the Big 4 audit firms. Therefore, based on the argument previously described, the proposed hypothesis is as follows:

H3: Financial statements manipulation effect the tendency of audit selection when voluntary audit firm rotation is performed.

3 RESEARCH METHOD

This research is explanatory research which aims to test whether a company's financial condition, earnings management and financial statement manipulation influence the tendency of audit firm selection in the event of voluntary audit firm rotation.

3.1 Population and sample

The population in this study is all non-financial sectors of companies listed in the Indonesia Stock Exchange (IDX) from 2010–2015. Sampling in this study uses the purposive sampling technique and the sampling is based on the criteria that have been set to fit the research objectives. The following are the sample selection criteria in this study.

a. Non-financial public companies listed on the Indonesia Stock Exchange from 2010–2015.
b. Companies whose financial statements have currency Rupiah.
c. The company replaces the public accounting firm prior to the end of the engagement.
d. The company has all the data required for the calculation of the variables in this study.

3.2 Research variables

Table 1. Operationalzation variables.

Variable	Name	Proksi	Scale
Dependent	*Voluntary Audit Firm Rotation*	0 = if the change of audit firm Big 4 to non-Big 4 audit firm 1 = if change audit firm from non-Big 4 or Big 4 to Big 4 audit firm 2 = if the change of audit firm from non-Big 4 to Big 4 audit firm	Ordinal
Independent	Financial Condition	$Z = 0.717 Z1 + 0.847 Z2 + 3.107 Z3 + 0.420 Z4 + 0.998 Z5$	Ratio
	Earnings Management	$TACC_{it-1} / TA_{it-1} = \alpha 1 (TA_{it-1}) + \alpha 2 (\Delta REV_{it} - \Delta REC_{it})/ TA_{it-1} + \alpha 3\, PPE_{it}/TA_{it-1} + \varepsilon_{it}$	Ratio
	Manipulation of financial statement (Benish Model)	M Score = $-4.840 + 0{,}920 DSRI + 0{,}528 GMI + 0{,}0404\, AQI + 0{,}892 SGI + 0{,}115 DEPI - 0{,}172 SGAI + 4{,}679\, TATA - 0{,}327\, LVGI$	Ratio
Control	Year	YEAR	Ratio

4 RESULT AND DISCUSSION

Objects used in this study are all sectors of companies listed in the Indonesia Stock Exchange during the years 2010–2015, except companies listed in the financial sector. There are 380 listed companies in the Indonesia Stock Exchange during the 2010–2015 period consisting of six sectors, namely agriculture, mining, property, real estate and building construction, infrastructure, utility and transportation, trade, service and investment manufacturing sector. The following is the result of company selection based on the criteria established in this research:

4.1 Sample procedures

Based on Table 2 it can be seen that companies listed in the IDX during the 2010–2015 period amounted to 380 companies, then companies that use the dollar currency in the financial statements totaled 68 companies, then companies that do mandatory audit firm rotation amounted to 281 companies, then companies that do not have complete data to be used in this research totaled 7 companies. So the total sample companies that do voluntary audit firm rotation amounted to 24 companies.

4.2 Descriptive statistics

Descriptive statistical tests on research models of a company's financial condition, earnings management and manipulation of financial statement on tendency of KAP selection in the event of voluntary audit firm rotation is used to see the spread of data of variables used in the model. The results of statistical test descriptive of the four variables in this study are as follows.

Table 2. Sample selection procedures.

Sample Selection	Number
Non-financial companies listed in IDX 2010–2015 period continuously	380
Companies that use dollar currency	(68)
Companies that perform mandatory audit firm rotation	(281)
Companies whose financial statements are incomplete	(7)
Total Sample (Company that conduct voluntary audit rotation)	**24**

Table 3. Statistics descriptive.

Variable	N	Mean	Std. deviation	Min.	Max.
VAR	24	1	0.417028	0	2
FCOND	24	1.292	2.186096	0.158	11.147
EAR_MGT	24	0.076875	0.060759	0.01	0.222
MNPLT	24	−0.017791	2.413439	−2.242	8.312

Table 4. Statistic output.

Variable	Odds ratio	Coef.	P > |z|
FCOND	1.248887	0.2222526	0,202
EAR_MGT	2.320006	−12,97539	0,086*
MNPLT	1.538714	0,4309469	0,028**
YEAR	1.637597	0,4932298	0,395

* Significant at the level of 10%.
** Significant at the level of 5%.

4.3 The feasibility of data and hypothesis testing

The Likelihood Ratio is a substitute for F-stat that serves to test whether all the slope of independent variable regression coefficient (company's financial condition, earnings management and financial statement manipulation) and control variable (year) together influence the dependent variable. The probability of a statistical likelihood ratio is 0.0392 <0.05. So the results of this test conclude that H_0 is rejected, meaning that all four variables simultaneously affect the level of change of voluntary audit firm rotation. In addition, the value of Wald chi2 of 10.08 with Prob> chi2 of 0.0392 indicates that simultaneously, the independent variable in this research model can explain the dependent variables in the research model.

Partial test (z statistic test) is performed instead of statistical test t. Partial test is done to find out the influence of the relationship of each independent variable with the dependent variable. This can be done by looking at Prob> chi2 from each of the independent variables as listed in the Table 4. The following is the result of the statistical test z of the four variables.

4.4 Financial condition and the tendency of audit firm selection in voluntary audit firm rotation

In testing the financial condition to the tendency of audit firm selection when performing voluntary audit firm rotation, it was found that the value of P value is greater than the 0.05 (5%) significance level that is 0.202. From these results it can be concluded that this variable has no significant effect on the tendency of audit firm selection when doing voluntary audit firm rotation. Hence hypothesis 1 is rejected in this study.

This study does not support the results of research of Chung et al. (2003) and Hudaib & Cooke (2005). The results of this study indicates that the company's financial condition does not affect the tendency of companies to choose audit firm when doing voluntary audit firm rotation. Voluntary audit firm rotation that done by companies not influenced by the financial condition of the companies. This could be happening because, based on sample research data, it can be seen that 83.33% of samples or as many as 20 companies use non-Big 4 audit firms and companies that experience poor financial condition (Z Score <1.81).

This is contrary to the results of the Hudaib & Cooke (2005) study that states that insolvent corporate circumstances and unhealthy financial conditions allow the company to choose a Big 4 audit firm to keep the trust of shareholders and creditors. But the results of this study support research conducted by Nasser et al. (2006) that poor financial condition is not a factor for the company to conduct audit firm selection in the event of audit rotation. This is because most of the companies sampled in this study use a non-Big 4 audit firm services. Thus, the transfer of audit services to a Big 4 audit firm will further complicate the company's financial condition due to the increase in audit service fees. In addition, companies experiencing unhealthy financial conditions will be more likely to retain their auditors to maintain the confidence of investors and shareholders and to prevent litigation risks.

4.5 *Earning management practices and the tendency of audit firm selection when voluntary audit firm rotation is performed*

In testing the effect of earning management to the tendency of audit firm selection when performing voluntary audit firm rotation, it was found that the value of P value is lower than the significance level of 0.10 (10%), that is equal to 0.086 with the constant −12,97539. From these results it can be concluded that this variable has a negative significant effect on the tendency of voluntary audit firm rotation. Hence hypothesis 2 is accepted in this study. It means that when a company practice earnings management, the company will choose an audit firm with lower quality or a non-Big 4 audit firm.

This is in line with the results of Dechow et al. (2011), Wasiuzzaman et al. (2015) and Muttakin et al. (2017), which examined the effect of earnings management on voluntary audit rotation and the results of their research found that earnings management can significantly influence the company's decision to perform voluntary audit firm rotation. Most of the companies sampled in this study use a non-Big 4 audit firm. According to Davidson et al. (2005), non-Big 4 audit firms tend to adhere to the wishes of the client. This is because non-big 4 audit firms tend to be afraid of losing their clients, so companies will tend to choose another audit firm that has the same level, typically a non-Big 4 audit firm that is willing to agree on the level of earnings management conducted by the company.

Lennox et al. (2014) states that by replacing a non-Big 4 audit firm, the company tends to earn earnings management on the basis of opportunistic behavior of management. The move to a non-Big 4 audit firm was done in order to find a less conservative audit firm. A less conservative audit firm will result in lower audit quality. So it can be concluded that companies that make a change of audit firm to a non-Big 4 audit firm has a value of earnings management that is greater when compared with companies audited by a Big 4 audit firm. In addition, Nasser et al. (2006) also argues that a Big 4 audit firm is more independent than a non-Big 4 audit firm in order to withstand management pressure in the event of a conflict. The Big 4 audit firm has more clients than a non-Big 4 audit firm, so a Big 4 audit firm will be easier to release their troubled clients.

4.6 *Manipulation of financial statements and the tendency of audit firm selection when voluntary audit firm rotation is performed*

In testing the manipulation of financial statement on the tendency of audit firm selection when voluntary audit firm rotation is performed, it was found that the value of P value is smaller than the level of significance 0.05 (5%), that is equal to 0.028 with the constant 0,4309469. From these results it can be concluded that this variable has a positive significant effect on the tendency of voluntary audit firm rotation. From the results of statistical tests, it can be concluded that companies that have manipulation of financial statement tend to choose a Big 4 audit firm compared with a non-Big 4 audit firm, so the hypothesis is accepted.

This result shows that boards of directors and committee audits that have the responsibility to choose audit firms tend to avoid the risk of manipulation of financial statements made by management. So companies prefer to choose a Big 4 audit firm to audit their companies when they perform voluntary audit firm rotation. This is consistent with the results of the research carried out by Francis (2004) and Francis & Wang (2006), which states that a Big 4 audit firm has the competence to conduct tighter and better audits than a non-Big 4 audit firm, so the probability to find misstatements is greater. In addition, Big 4 audit firms are considered to provide higher audit quality than non-Big 4 audit firms. This happens because the Big 4 audit firm has more resources to conduct audit with high quality, and they are also required to maintain their reputation and not get involved with the manipulation of financial statement of a company.

In addition, the Big 4 audit firm in a law-abiding state is considered to be very protective of the interests of investors. The Big 4 audit firm is very conservative when dealing with its clients with the aim to protect the interests of investors. Strict protection against these investors will make the quality of audits higher. This means that when the possibility of manipulation is high, it will impact on low audit quality and high risk of litigation, so the company will choose an audit firm of high quality to avoid the possibility of litigation from the stakeholder. Payamta (2006) also found that a non-Big 4 audit firm will more often be faced with litigation risk when compared to a Big 4 audit firm.

5 CONCLUSION AND LIMITATION

This study aims to determine whether the company's tendency to perform voluntary audit firm rotation can be a bad signal for the stakeholder by examining the company's financial condition, earnings management and financial statement manipulation. The result of this research is that companies with an earnings management practice tend to choose audit firms with less quality, while the companies with an indication of manipulation tend to choose high quality audit firms to avoid litigation risk because of manipulation of their financial statement.

So companies that make earnings management will replace their firm's audit with a low-quality audit firm to sanction their earning management. It is because the earning management is done to maintain the company's performance, and earning management does not damage the image of the company as a whole. Whereas if the company is indicated to manipulate the financial statements, the company will choose to replace their auditor with a better quality auditor. This is done to maintain corporate image and avoid the possibility of lawsuits from their stakeholders. The implication of this study is that voluntary audit rotation conducted by the company is an indication of certain events that exist in the company that can be taken into consideration by decision making.

The weakness of this research is that there are very few samples. This is because, in the period of research taken by researchers, companies that do voluntary audit rotation are very few. Suggestions for further research may develop a measurement of the tendency of firm audits at the time of the company's voluntary audit firm rotation.

REFERENCES

ACFE. (2016). *Report To The Nations On Occupational Fraud and Abuse.* ACFE.
Altman, E.I. (1968, September). Financial ratios, discriminant analysis and the prediction of corporate bankruptcy. *Journal of Finance,* 23(4),589–609.
Beneish, M.D. (1999). The detection of earnings manipulation. *Financial Analysts Journal, 55(5),24–36.*
Carcello, J.V., & Nagy, A.L. (2004). Audit Firm tenure and fraudulent financial reporting. *Auditing: A Journal Oof Practice and Theory,* 23(2), 55–69.
Chung, R., Firth, M., & Kim, J.B. (2003). Auditor conservatism and reported earnings. *Accounting and Business Research*, 23(1), 19–32.
Davidson, W.N., Jiraporn, P., & DaDalt, P. (2005). Causes and consequences of audit shopping: an analysis of auditor opinions, earnings management, and auditor *changes. Quarterly Journal of Business and*

Economics, 45(1/2), 69–87. Dechow, P.M., Larson, C.R., & Sloan, R.G. (2011). Predicting material accounting misstatements. *Contemporary Accounting Research,* 28(1), 17–82.

Dechow, P.M., Sloan, R.G., & Sweeney, A.P. (1995). Detecting earnings management. *The Accounting Review,* 70(2), 193–225.

DeFond, M.L., & Subramanyam, K.R. (1998). Auditor changes and discretionary accruals. *Journal of Accounting and Economics,* 25(1), 35–67.

Erickson, M., Hanlon, M., & Maydew, E.L. (20065). Is there a link between executive equity incentives and accounting fraud?. *Journal of Accounting Research, 44(1),,* 113–143.

Fama, E.F. (1980). Agency problems and the theory of the firm. *The Journal of Political Economy,* 88(2), 288–307.

Francis, J., & Wang, D. (2006). Investor protection and auditor conservatism: are big 4 auditors conservative only in the United States? *Working Paper: University of Miossuri and University of Nebraska.*

Francis, J.R. (2004). What do we know about audit quality? *The British Accounting Review,* 36(4), 345–368.

Geiger, M.A., & Raghunandan, K. (2002). Auditor tenure and audit reporting failures. *Auditing: A Journal of Practice & Theory,* 21(1), 67–78.

Hastuti, T.D., & Gozali, I. (2015). Manipulation detection in financial statements. *International Journal of Humanities and Management Sciences,* 3(4), 222–229.

Healy, P.M., & Wahlen, J.M. (1998). A review of the earning management literature and its implications for standards Setting. American Accounting Association, *Preliminary Draft.*

Hudaib, M., & Cooke, T.E. (2005). The impact of managing director changes and financial distress on audit qualification and auditor switching. *Journal of Business, Finance & Accounting,* 32(9–10), 1703–1739.

Jackson, A.B., Moldrich, M., & Roebuck, P. (2008). Mandatory audit firm rotation and audit quality. *Managerial Auditing Journal,* 23(5), 420–437.

Jensen, M.C., & Meckling, W.H. (1976). Theory of the firm: managerial behavior, agency cost and ownership structure. *Journal of Financial Economics,* 3(4), 305–360.

Johnson, V.E., Khurana, I.K., & Reynolds, J.K. (2002). Audit-firm tenure and the quality of financial reports. *Contemporary Accounting Research,* 19(4), 637–660.

Johnson, W.B., & Lyn, T. (1990). The market for audit services: evidence from voluntary auditor changes. *Journal of Accounting and Economics,* 12(1–3), 281–308.

Khrisnan, J. (1994). Auditor switching and conservatism. *The Accounting Review,* 69(1), 200–215.

Kim, H., Lee, H., & Lee, J.E. (2015). Mandatory audit firm rotation and audit quality. *The Journal of Applied Business Research, 31,* 1089–1106.

Kim, J.B., Chung, R., & Firth, M. (2003). Auditor conservatism, asymmetric monitoring and earning management. *Contemporary Accounting Research,* 20(2). 323–359.

Lennox, C.S., Wu, X., & Zhang, T. (2014). Does mandatory rotation of audit partners improve audit quality? *The Accounting Review,* 89(5), 1775–1803.

Lennox, C.S., & Pittman, J.A. (2011). Voluntary audits versus mandatory audits. *The Accounting Review,* 86(5), 1655–1678.

Lou, Y.-I., & Was, M.-L. (2009). Fraud risk factor of the fraud triangle assessing the likelihood of fraudulent financial reporting. *Journal of Business & Economics Research,* 7(2), 61–78.

Muttakin, M.B., Khan, A., & Mihret, D.G. (2017). Business group affiliation, earnings management and audit quality: evidence from Bangladesh. *Managerial Auditing Journal,* 32(4/5), 427–444.

Myers, J.N., Myers, L.A., & Omer, T.C. (2003). Exploring the term of the auditor-client relationship and the quality of earnings: a case for mandatory auditor rotation? *The Accounting Review,* 78(3), 779–799.

Nasser, A.T., Wahid, E.A., Nazri, S.N., & Hudaib, M. (2006). Auditor-client relationship: the case of audit tenure and auditor switching in Malaysia. *Managerial Auditing Journal,* 21(7), 724–737.

Payamta. (2006). Studi Pengaruh Kualitas Auditor, Independensi, dan Opini Audit Terhadap Kualitas Laporan Keuangan Perusahaan. Jurnal Bisnis & Manajemen 6(1).

Public Oversight Board (POB). (2000). *Panel on audit effectiveness: report and recommendation.* Stamford: CT POB.

Ramadhani, A.S., & Lukviarman, N. (2009). Perbandingan Analisis Prediksi Kebangkrutan Menggunakan Model Altman Pertama, Altman Revisi, Dan Altman Modifikasi Dengan Ukuran Dan Umur Perusahaan Sebagai Variabel Penjelas. *Siasat Bisnis,* 15–28.

Rawi, K.A., Kiani, R., & VeEdd, R.R. (2008). The use of altman equation for bankruptcy prediction in an industrial firm (case study). *International Business & Economics Research Journal,* 7(7), 115–128.

Regulation of Ministry of Finance Number 17/PMK.01/2008 concerning Public Accounting Firm. Jakarta, Indonesia: Kemenkeu.

Scott, W.R. (2015). *Financial accounting theory 7th Edition.* Canada: Pearson.

Siregar, S.V., Amarullah, F., Wibowo, A., & Anggraita, V. (2012). Aaudit tenure, audit rotation, and audit quality: the case of Indonesia. *Asian Journal of Business and Accounting,* 5(1). 55–74.

Summers, S.L., & Sweeney, J.T. (1998). Fraudulently misstated financial statements and insider trading: an empirical analysis. *The Accounting Review*, 73(1), 131–146.

U.S General Accounting Office (GAO). (2003). *Public accounting firms: required study on the potential effects of mandatory audit firm rotation (GAO-04-2016).* retrieved http://www.gao.gov/new.items/d04216.pdf.

Wang, X., Dennis, L., & Tu, Y.S. (2007). Measuring financial condition: a study of U.S. states. *Public Budgeting & Finance*, 27(2), 1–21.

Wasiuzzaman S., Niloufar, Iman S. & Nejad, R. (2015). Prospect theory, industry characteristics and earnings management: a study of Malaysian industries. *Review of Accounting and Finance, 14(3), 324–347.*

Business Innovation and Development in Emerging Economies – Trinugroho & Lau (Eds)
© *2019 Taylor & Francis Group, London, ISBN 978-1-138-35996-3*

Analysis of the effect of public accounting firm size on audit fee in Indonesia

Elwimina Nurjanah &Vera Diyanty

Master's Program in Accounting, Faculty of Economics and Business, Universitas Indonesia, Jakarta, Indonesia

ABSTRACT: This study aims to analyze whether the Big 4 Public Accounting Firms set higher audit fee than non-Big 4 Firms, and middle firms set higher audit fees than the small firms. This study used 345 observations from 115 public companies in Indonesia that were not in the financial sector during the period 2014–2016. The results of this study indicate that Big 4 sets a higher audit fee than Non-Big 4, and middle firms did not set higher audit fee than the small firms. This indicates the existence of premium audit fee received by Big 4. This study also suggests The Indonesian Institute of Certified Public Accountants (IAPI) to evaluate whether non-Big 4 accounting firms has set fair audit fees to prevent unhealthy competition among accounting firms and lower the audit quality.

Keywords: Audit Fee, Public Accounting Firm, Auditing

1 INTRODUCTION

In setting the audit fee, there is a risk of abnormal audit fees, especially public accounting firms that have dominant market shares. To decrease the abnormal audit fees, Indonesian Public Accountant Institute established the Decision Letter of General Head of Indonesian Public Accountant Institute Number KEP 024/IAPI/VII/2008 regarding audit fee setting policy that will be charged to clients. This policy aims to normalize audit fee charged to the client. The policy then updated by the Leader Regulation Number 2 Year 2016 regarding to the setting of audit fee of financial statements fee. This regulation has sanctions for those who don't comply with the regulation. The heaviest sanction is ending the membership of Indonesian Public Accountant Institute.

Even there's a regulation about audit fee setting, there is still some public accounting firms which receive higher audit fee than others. This phenomenon happens in many countries. Wang et al (2014), Krauss et al (2014), Carson et al (2012) found that Big 4 received higher audit fees than non-Big 4. This is what is called by audit fee premium received by the Big 4.

There are two general explanations about audit fee premium. First, audit fee premium is received because of audit quality that has been delivered by the Big 4. Simon (2007) interpreted that audit fee premium is an indication of product differentiation, i.e. because of audit quality. Choi et al (2010) said that bigger public accounting firms can charge higher audit fee because bigger public accounting firm is considered can deliver higher audit quality. Moizer (1997) also said that reputation of public accounting firms is matter in audit fee setting. Thus, top tier of public accounting firms can set higher audit fees than others.

Some arguments explained audit quality is not relevant affecting audit fee. Campa (2013) said that higher audit fee exists due to Big 4 has more powerful position in deciding the audit fee than the client. Krauss et al (2014) said that audit fee discount doesn't reflect the decrease of audit effort, so that the audit quality delivered by the public accounting firms has been standardized.

The second explanation about audit fee premium is the market dominating public accounting firms receive higher audit fee premium than others. Bain (1951, 1956) on Feldman (2006) said that when an entity on a market is more dominant, the entity will have the stronger market power to set the price and receive more profit. Carson et al (2012) found that the changes of Big N premium is related to the changes of Big N period. Audit fee premium in Big 4 and Big 5 period was higher than in Big 6 period. Wang et al (2014) found that audit fee premium received by the Big 4 was lower when there was tight competition with two largest public accounting firms in China.

Basioudis and Fifi (2004) found that in Indonesia, there is no difference in audit fee between Big 5 and non-Big 5. The time frame of research was after Asia financial crisis in 1997 and 1998. The companies had tight budget, so they could not pay higher audit fee to the Big 5. This condition is different with other countries, where the clients pay higher audit fee to the Big N.

The purpose of this research is to analyze whether Big 4 set higher audit fee to their clients in Indonesia with the newest time frame. This research also contribute to test whether the size of public accounting firms set higher audit fee in Indonesia, not only comparing Big 4 to non-Big 4, but also middle public accounting firms to small public accounting firms. The method used in this research is quantitative research, using secondary data published by the public companies. The first section of this paper is introduction. The second section is literature review. The third section is research method and data collection. The fourth section is the result of this research, and the last part is the conclusion.

2 THE MATERIAL AND METHOD

2.1 *Agency theory*

Jensen and Meckling (1976) explained that agency relationship is a contractual relationship between principals and agents, and agents are requested to act as principal interest. In the context of the company, principal is the shareholders, and the agent is the management. The problem is, agent sometimes does not act as principal interest, because agent also has his/her own interest. Agent also has more information, better than the principal. This is because agent operates the business, so agent has more knowledge about the business.

As a response to the responsibility to the principle, agents prepare financial report for the principal. In order to convince the principal that agent has acted as principal interest, agent assigned independent parties, i.e. public accountant, to assess whether the financial statement generated by the agent have reliable information, and provide an opinion whether the financial statements have been fairly and reliable presented and have no material misstatements.

2.2 *Audit pricing*

In determining audit fee, public accounting firms considers three main attributes, namely client attribute, auditor, and engagement (Hay et al., 2006). Client attributes are client-side attributes, such as client size, audit complexity, inherent risk, profitability, leverage, ownership form, internal control, corporate governance, and client industries. On client attributes, the elements that positively affect the audit fee are the client's size (Hay et al., 2006; Wang et al., 2014; Patroomsuwan, 2017; Nagy, 2014; and Zaman & Chayasombat, 2014), audit complexity (Hay et al., 2006; Wang et al., 2014; and Nagy, 2014), inherent risks (Hay et al., 2006 and Doogar et al., 2015), and industrial tpes (Hay et al., 2006). From the element of profitability, companies that suffer losses are subject to higher audit fees (Hay et al, 2006 and Wang et al, 2014). Forms of corporate ownership and corporate governance have diverse influences on audit fees (Hay et al, 2006). While leverage and internal control does not significantly affect the audit fee (Hay et al, 2006).

Auditor attributes consist of auditor quality, auditor tenure, and auditor location. Auditor quality was assessed to be positively associated with audit fees (Hay et al, 2006, Simon,

1997, and Choi et al, 2010). Tenure auditors and auditor locations have no significant effect on audit fees (Hay et al., 2006).

Engagement attributes are attributes that arise from the engagement relationship between the auditor and the client, which consists of report lag (time range between the date of the statement of financial position to the date of audit report), busy period, audit issues, non-audit services, and the complexity of financial reporting. Elements of the engagement attributes that affect audit fees are positively report lags (Hay et al., 2006), non-audit services (Hay et al., 2006), and reporting (Hay et al., 2006; Lin & Yen, 2016; and Goncharov et al, 2014). Elements of attachment attributes that have no effect on audit fees are busy periods (Hay et al., 2006 and Zaman & Chayasombat, 2014) and audit issues (Hay et al., 2006 and Wang et al., 2014).

2.3   *Audit fee premium*

Big N premium audit fee is audit fee variance that clients pay for Big N compared to other public accounting firms (Ireland & Lennox, 2002 and Chaney et al., 2004 in Carson et al., 2012). Various studies explain the phenomenon of audit fee variance between Big N and non-Big N, among them Simon (1997), Moizer (1997), Basioudis and Fifi (2004), Carson et al (2012), Campa (2013), Krauß et al (2014), Wang et al (2014); and Litt et al (2015). These studies explain that premium audit fees occur due to differences in audit quality provided by Big N compared to other public accounting firms (Simon, 1997, Moizer, 1997 and Choi et al, 2010) and market share dominance controlled by Big N (Carson et al, 2012; Campa, 2013; and Wang et al, 2014).

Audit quality by DeAngelo (1981) is defined as the possibility of the auditor finding out what is being violated in the client's financial records and reporting the violations. The violation committed by the client depends on the auditor's ability and the equipment used in the audit, such as the audit procedures performed, including the sampling method used, and the technology used. Audit quality is considered to have an effect on the amount of premium audit fee received by public accounting firms. Simon (1997) interpreted premium audit fees as an indication of product differentiation, one of the reasons is due to auditor quality. Simon (1997), Choi et al (2010) reveals that the size of public accounting firm may affect premium audit fees. Large firms can charge a larger audit fee than smaller one. This is because of the view that a large public accounting firm can provide better audit quality. Apart from public accounting firm sizes, audit quality may also be affected by the public accounting firm's reputation. Moizer (1997) argues that the reputation of a public accounting firm gives a signal to market participants that the public accounting firm has a higher audit quality. Therefore, the public accounting firm included in the top tier ranks will impose a higher audit fee because the price is associated with quality.

However, there are some arguments in which premium audit fees are not affected by audit quality. Campa (2013) disclosed that premium audit fees occurred because of the higher bargaining power positions in Big 4 than the clients. Krauß et al (2014) disclosed that audit fee discounts do not reflect the decreasing of audit effort so the quality of the audit decreases, but rather because of the higher bargaining power position of the client. Low audit fees do not make the auditor lower his business in conducting audits, so the quality of the audit remains standardized.

Another explanation of the phenomenon of the emergence of premium audit fees is due to the control of market share by the public accounting firm. Bain (1951, 1956) in Feldman (2006) said that in the structure-performance hypothesis, increased concentration of the market increases the market power of firms that survive in the industry, so that firms can set higher prices and earn higher returns. Feldman (2006) disclosed that after Arthur Andersen, one of the big public accounting firms, ceased operations in the audit market, the concentration in the audit market increased. Arthur Andersen's clients move on to other large firms in order to increase the audit market concentration. As a result of Arthur Andersen's exit from the audit market, there was a significant increase in audit fees. Carson et al (2012) also revealed similarly, that changes in Big N premium audit fees change with the changes of Big

6, Big 5, and Big 4 period. Audit premium fees increased drastically in the Big 4 and Big 5 period, compared to the Big 6 period. Wang et al (2014) revealed that the premium audit fee received by Big 4 decreased after an increase in competition at two of the largest local public accounting firms in China. The decrease in Big 4 premium audit fee was due to competition from local public accounting firms so that Big 4 was forced to lower the price. Litt et al (2015) revealed that the entry of Big 4 into the Indian audit market led to a price-cutting reaction on local public accounting firms in order for local public accounting firms to be more competitive against new players.

2.4 *Hypothesis*

The audit market in Indonesia and other countries of the world is controlled by Big 4. Previous studies have made it clear that Big 4 obtains a larger audit fee than non-Big 4 (Simon, 1997; Choi et al, 2010; Campa, 2013; Carson et al., 2012; Wang et al., 2014; and Litt et al., 2015). The studies were conducted with place setting outside Indonesia. As far as the authors know, the research on premium fee audit in Indonesia was conducted by Basioudis and Fifi (2004). The result of the research is in the audit market of Indonesia there is no difference between the audit fees is subject to Big 5 with non-Big 5. The research time frame of Basioudis and Fifi (2004) is the year 2000, at which time the large public accounting firms is still called the Big 5 and after the Asian financial crisis of 1997 and 1998. The hypothesis of this research is:

H_1: Big 4 receives higher audit fee compared to non-Big 4.

The size of a public accounting firms also affect the amount of audit fee received by the firms (Simon, 1997 and Choi et al, 2010). This is because the larger public accounting firm is allegedly able to provide better audit quality, so it can charge a higher audit fee. Wang et al (2014) shows that in the Chinese audit market, two local public accounting firms can compete with Big 4 so that Big 4 lowers its audit fee. Although the audit fee received by Big 4 in China is still higher than the two largest local public accounting firms, but the number of audit fees is not as big as before the competition. The second hypothesis of this research is:

H_2: Middle public accounting firm receives higher audit fee compared to small accounting firms.

2.5 *Research methods and data collection*

This research model is designated using regression equation, referring to Wang et al (2014) model with some modifications. In this research, audit fee is scaled by the total assets. Some control variables such as audit tenure and audit problem are not included because they did not have significant effect on Wang et al (2014) research. Subsidiary variable control is also not included because it is hard to calculate the real subsidiaries of a company group. Ownership variable controls such as foreign ownership and state-ownership are changed to family ownership. It refers to report established by Credit Suisse (2011) mentioned that family companies dominated 61% of market capitalization in Indonesia. It points out that the use of family ownership is more relevant for this research.

This research used two models. The first model is as follows:

$$\text{AUDFEE} = \alpha it + \beta 1 \text{BIG4it} + \beta 3 \text{ASSETSit} + \beta 4 \text{INVRECit} + \beta 5 \text{LOSSit} + \beta 6 \text{LEVERAGEit} + \beta 7 \text{FAMILYit} + eit$$

AUDFEE : audit fee scaled by the client size
BIG4 : dummy variable (1 if Big 4, 0 is others)
ASSETS : natural logarithm of total assets
INVREC : (account receivable + inventory)/total asset
LOSS : dummy variable (1if the income statement is reported loss, 0 if others)

DEBT   : total debt scaled by the client size
FAMILY : family ownership percentage
e   : error term

The aim of first model is to answer the first hypothesis of this research. If the BIG4 coefficient have positive sign and statistically significant, it can be concluded that Big 4 accounting firms charges higher audit fee than non-Big 4 accounting firms.

The second model is as follows:

$$AUDFEE = \alpha it + \beta1MIDDLE4it + \beta3ASSETSit + \beta4INVRECit + \beta5LOSSit + \beta6LEVERAGEit + \beta7FAMILYit + eit$$

AUDFEE : audit fee scaled by the client size
MIDDLE : dummy variable (1 if middle accounting firms, 0 is others)
ASSETS : natural logarithm of total assets
INVREC : (account receivable + inventory)/total asset
LOSS   : dummy variable (1if the income statement is reported loss, 0 if others)
DEBT   : total debt scaled by the client size
FAMILY : family ownership percentage
e   : error term

The aim of second model is to answer the second hypothesis of this research. If the MIDDLE coefficient have positive sign and statistically significant, it can be concluded that middle accounting firms charges higher audit fee than small accounting firms. The proxy used to categorize as middle public accounting firms follows the PPPK (2015), that is the public accounting firms which has more than ten partners but not categorize as Big 4. The list of middle public accounting firms is as follows:

Table 1. Middle public accounting firms.

No.	Middle accounting firm name	Number of partners as of 2015
1	Paul Hadiwinata, Hidajat, Arsono, Ade Fatma, dan Rekan	28
2	Hendrawinata Eddy Siddharta dan Tanzil	27
3	Amir Jusuf, Aryanto, Mawar, dan Rekan	23
4	Johan Malonda Mustika dan Rekan	21
5	Tanubrata Sutanto Fahmi dan Rekan	18
6	Doli, Bambang, Sulistiyanto, Dadang dan Ali	17
7	Kosasih, Nurdiyaman, Tjahjo dan Rekan	14
8	Hadori Sugiarto Adi dan Rekan	13
9	Heliantono dan Rekan	13
10	Mulyamin Sensi Suryanto dan Lianny	11

Data used in this research is from annual reports and financial statement reports of non-financial sector public companies in Indonesia within time range 2014–2016 and has completed data needed in this research. There are 115 companies in this research. There are no multicolinearity problem found on this research. Homoskedasticity is found. To handle this problem, this research use *xtreg* regression with *cluster( )* option on Stata application. Autoregression problem is not a big problem in this research.

2.6 *Descriptive analysis*

Below is the table of descriptive analysis.

From the descriptive statistics table (Appendix 1), the average audit fee paid in this study amounted to Rp1.5 billion, with a median of Rp732 million. A total of 83 observational samples or 50.3% of the total observed samples with the higher median on AUDFEE variable

Table 2. Descriptive Statistics.

	Obs.	Average	Median	Dev.standard	Max	Min	Skewness
AUDFEE	345	1,505,718,125	732,348,000	3,548,216,132	36,500,000,000	46,750,000	0.00006364
ASSETS	345	–	–	–	–	–	(0.03838954)
INVREC	345	–	–	–	–	–	–
LEVERAGE	345	–	–	–	–	–	–
FAMILY	345	–	–	–	–	–	–

		Variable Dummy	
	Obs.	% if value 1	% of value 0
BIG4	153	44.35%	55.65%
MIDDLE	144	41.74%	58.26%
LOSS	77	22.32%	77.68%

were audited by Big 4. The remaining 42.42% was audited by the middle accounting firms and 8.48% were audited by small accounting firms. This gives a preliminary conclusion from this study that Big 4 received a higher audit fee than the Non-Big 4.

The largest audit fee is paid by PT Telekomunikasi Indonesia (Persero) Tbk in 2016 amounting to Rp36.5 billion. PT Telekomunikasi Indonesia also pays the largest audit fee in 2015 and 2014, amounting to Rp34.4 billion and Rp31.5 billion, respectively. The largest assets are also owned by PT Telekomunikasi Indonesia (Persero) Tbk, which in 2016, 2015 and 2014 amounted to Rp179.611 trillion, Rp166.173 trillion, and Rp141,822 trillion, respectively.

The smallest fee audit is paid by PT Wahana Pronatural Tbk with audit fee of Rp46.5 million in 2014, 2015 and 2016. Total assets owned by PT Wahana Pronatural Tbk in 2014, 2015 and 2016 amounted to Rp109, 04 billion, Rp107.57 billion and Rp105.89 billion.

When using the measured audit fee scaled by client size (total assets), the largest company paying the audit fee is PT Humpuss Intermoda Transportasi Tbk in 2016, with an audit fee paid of Rp115 million, with total assets of Rp46.7 billion. While the scaled audit fee with the smallest company size is paid by PT Solusi Tunas Pratama in 2014 amounting to Rp170 million, with total assets of Rp12.89 trillion. It can be concluded that the amount of audit fee is not always proportional to the total amount of the asset. There are other variables that affect the amount of audit fee.

This research tried to evaluate clients profile audited by Big 4, middle and small accounting firms from the statistic descriptive analysis. Big 4 inclined to engage with clients which have bigger size, higher leverage, and higher family ownership percentage relative to non-Big 4 clients. Middle accounting firms inclined to engage with clients which have smaller size, higher inherent risk, higher leverage, and lower family ownership percentage. Small accounting firms inclined to engage with clients which have smaller size, smaller inherent risk, smaller leverage, and smaller family ownership percentage.

Public accounting firms inclined to charge higher audit fee to clients with have smaller size, does not report loss on the financial reports, lower leverage, and higher family ownership percentage. Big 4 inclined to charger higher audit fee than non-Big 4.

Based on Pearson correlation analysis for the first model, INVREC and FAMILY are positively correlated to AUDFEE. ASSETS, LOSS, and LEVERAGE are negatively correlated to AUDFEE. In the second model, INVREC and FAMILY are positively correlated to AUDFEE. MIDDLE, ASSETS, LEVERAGE, and LOSS are negatively correlated to AUDFEE. Further relationship analysis is on results and discussion section.

3 RESULTS AND DISCUSSION

3.1 First model

Below is the regression result for the first model:

Table 3. Regression Result for the first model.

	Prediction	Coefficient	P > |t|
Constanta		.0068381	0.000***
Independent Variable			
BIG4	+	.000142	0.022***
Control Variables			
ASSETS	+	−.0002286	0.000***
INVREC	+	1.86e-07	0.875
LOSS	+	−.0000102	0.590
LEVERAGE	+	.0000417	0.636
FAMILY	+/−	1.94e-07	0.798
Observation Number	345		
R²	0.4272		
Prob > F	0.0000		

AUDFEE: audit fee scaled by the client size; *BIG4*: dummy variable (1 if Big 4, 0 is others); *ASSETS*: natural logarithm of total assets; *INVREC*: (account receivable + inventory)/total asset; *LOSS*: dummy variable (1if the income statement is reported loss, 0 if others); *DEBT*: total debt scaled by the client size; *FAMILY*: family ownership percentage.
Note: *** Significant at $\alpha = 1\%$

The research model above has probability value> F equal to 0.0000, significant at $\alpha = 1\%$. This proves that the independent variable (BIG4) and control variables (ASSETS, INVREC, LOSS, LEVERAGE, and FAMILY) jointly influence the dependent variable, i.e. AUDFEE at 99% confidence level. The above model is significant to explain the determinants of audit fees. The value of R-Square is 0.4272, which means that the influence of independent variable in influencing the dependent variable is 42.72%.

From the regression results show that Big 4 has significant positive effect on the audit fee. This is shown from *p* value of 0.022, significant at $\alpha = 5\%$. This indicates that Big 4 receives higher audit fee than with non-Big 4. Therefore, H_1 in this research is accepted. This is in line with research conducted by Simon, 1997; Choi et al, 2010; Campa, 2013; Carson et al, 2012; Wang et al, 2014; and Litt et al., 2015stating that Big N received a larger audit fee than the non-Big N. The results of this study also provide evidence of an indication of premium audit fees received by Big 4 in Indonesia.

There are two main reasons that lead to premium audit fees, namely audit quality and market share audit market. Simon's research (1997), Moizer (1997), Choi et al (2010), and Siddiqui et al (2013) show that Big 4 provides better audit quality, thus Big 4 set higher audit fees than non-Big 4. In addition, larger public accounting firm sizes also provide a signal that the bigger accounting firm size will provide higher audit quality, which will impose a higher audit fee as well (Simon, 2007). Big 4 has better resources, workforce, and job quality than the non-Big 4.

Another thing that supports Big 4 obtains a higher audit fee compared to non-Big 4 is market share dominance performed by Big 4. The positive effect of market share of audit market on audit fee is stated by Feldman (2006), Carson et al (2012), Wang et al (2014), and Litt et al (2015). Feldman (2006) discloses that due to Arthur Andersen's dissolution, the audit market concentration has increased, as Arthur Andersen clients moved to other remaining Big N. This event triggered an increase in audit fees. Carson et al (2012) also calculates changes to premium fee audits in the Big 6 period, Big 5, and Big 4. Premium fee audits become larger when Big N becomes less. Wang et al (2014) and Litt et al (2015) revealed that the participation of public accounting firms in competition in the audit market led to a decrease in audit fees received due to a decrease in audit market concentration. In the Indonesian audit market, there is a positive effect of the market share of the audit market on audit fees. This statement is also supported by the facts that occurred in the field, that Big 4 controls the market share of the audit market, amounting to Rp2.17 trillion or 62% of the total audit market revenue in Indonesia by 2015 (PPPK, 2015). Future research is expected to discuss more deeply whether the quality of audit or market share audit more influences the audit premium fee in Indonesia.

3.2   Second model

Below is the regression result for the first model.

In the second model, the observations used are observations that are audited by middle public accounting firms and small public accounting firms only, so the number of observations used in the second model is not as much as in the first model. The second model use 192 observations.

The research model above has probability value > F equal to 0.0000, significant at $\alpha = 1\%$. This proves that the independent variable (MIDDLE) and control variables (ASSETS, INVREC, LOSS, LEVERAGE, and FAMILY) together influence the dependent variable, i.e. AUDFEE at the 99% confidence level. The above model is significant to explain the determinants of audit fees. The value of R-Square is 0.4413, which means that the influence of independent variable in influencing the dependent variable is 44,13%.

From the result of regression above, it is known that MIDDLE variable has negative coefficient and not statistically significant. It can be interpreted that there is no evidence that middle public accounting firms receives higher audit fee than the small public accounting firms. Thus, H_2 of this study was rejected. The results of this study are in line with research conducted by Wang et al (2009), which states that only Big 4 obtain premium audit fees, and second-tier public accounting firms and other local public accounting firms in China do not receive premium audit fees.

The results of this study also provide evidence that larger public accounting firm measures do not always receive larger audit fee as well, as proposed by Wang et al (2014). In addition, there is the possibility that small public accounting firms improve the audit quality to prevent a decrease in market share audit owned by the small public accounting firms.

3.3   Variable controls on the first and second model

The firm's size control variables (ASSETS) significantly have a negative effect on the audit fee. This proves that audit fee in Indonesia is not always determined based on company size

Table 4.   Regression Result for the second model.

	Prediction	Coefficient	$P > |t|$
Constanta		0.0064059	0.000***
Variabel Independen			
MIDDLE	+	−0.0000366	0.403
Variabel Kontrol			
ASSETS	−	−0.000215	0.000***
INVREC	+	5.82e-07	0.685
LOSS	+	0.0000205	0.417
LEVERAGE	+	−0.000051	0.626
FAMILY	+/−	3.50e-07	0.708
Observation Number	**192**		
R^2	**0.4413**		
Prob > F	**0.0000**		

AUDFEE: audit fee scaled by the client size; MIDDLE: dummy variable (1 if middle accounting firms, 0 is others); ASSETS: natural logarithm of total assets; INVREC: (account receivable + inventory)/total asset; LOSS: dummy variable (1if the income statement is reported loss, 0 if others); DEBT: total debt scaled by the client size; FAMILY: family ownership percentage.
Note: *** Significant at $\alpha = 1\%$
** Significant at $\alpha = 5\%$

variable. Cause of firm size that negatively affect audit fees are explained through the exposure of scale of economies audits. Large firm sizes tend to have smaller scaled audit values. This indicates an audit of scale of economies (Abidin et al, 2010). Audit scale of economies emerges when a company's audit fee falls due to the growing size of the firm. Clients with larger sizes tend to have better internal control systems, thus facilitating the audit process. This will lower the audit fee relative to the total assets.

INVREC representing inherent risk have no significant effect on the audit fee. This is in line with research conducted by Wang et al (2014) and Nelson & Rusdi (2015). The high competition in the audit market makes the public accounting firms when determining audit fees tend to underestimate the inherent risk that a client has. To pursue public accounting firm targets, there is a tendency that auditors will take clients with high inherent risks with audit fees that do not vary greatly with clients with low inherent risk.

LOSS representing the client's profitability has no significant effect on the audit fee. This is in line with research conducted by Nagy (2014) which states that there is no significant evidence that companies that are experiencing losses will be subject to higher audit fees. This is due to the possibility of differences in perception among public accounting firms about profitability effect on audit fees. Some public accounting firms assume that clients who experience losses will have the ability to pay a smaller audit fee, thus the audit fees will be smaller. But on the other hand, public accounting firms with higher bargaining power charge audit fee with no much different from other companies that are not experiencing losses.

LEVERAGE control variable has no significant effect on audit fee. This is in line with research conducted by Hay et al (2006). Leverage describes the audit risk experienced by the auditor. Leverage increases the risk of going concern, and the risk of going concern increases audit risk for public accounting firms in giving an audit opinion. The effect of non-significant leverage on audit fees is likely to be due to high competition among public accounting firms so that public accounting firms are less likely to pay attention to the risks it faces.

The family ownership control variable (FAMILY) does not have a significant effect on the audit fee. This is consistent with research conducted by Ali and Lessage (2013), which states that family firms have agency issues of differentiation between principals (shareholders) and agents (managements), but also have agency problem between majority shareholders and the minority shareholders. This leads to a tendency to lack the demand for better audit quality from family firms, so that family ownership does not have an effect on audit fees.

4 CONCLUSION

This study aims to analyze the tendency of audit fees received by Big 4, middle public accounting firms, and small public accounting firms in Indonesia in the period 2014–2016. The result of this study is Big 4 has significant positive effect on the audit fee. This indicates a premium audit fee received by Big 4 that is not received by non-Big 4. The reason for the premium audit fee is that Big 4 provides better audit quality to its clients and Big 4 dominates the audit market so they have stronger bid to set audit fee than the clients.

Middle public accounting firms have no significant effect on the audit fee. This indicates that there is no evidence that the middle public accounting firms receive more audit fees than the small public accounting firms. The results of this study also provide empirical evidence that larger firm sizes do not always obtain a larger audit fee as well.

This research is expected to make the regulator to review more about the audit fee in Indonesia whether public accounting firms audit fee settings have been in accordance with the regulation. It aims to prevent unhealthy competition among public accounting firms that can trigger a decrease in audit quality. The Indonesian Institute of Certified Public Accountants (IAPI) issued the Management Regulation No. 2 of 2016 on the Determination of Financial Statement Audit Fees. This latest regulation sanctioned IAPI members who did not comply with the regulation, with the most severe sanction is the termination of membership of IAPI by the IAPI Discipline and Investigation Committee. However, there is no further study or analysis on whether the middle public accounting firms and small public accounting firms have established audit fees in accordance with IAPI regulations.

REFERENCES

Adelopo, I. (2009). *Modelling issues in the relationship between audit and non-audit fees.* Journal of Applied Accounting Research, 10(2), 96–108. Lin & Yen, 2014.

Abidin, S., Beattie, V., & Goodacre, A. (2010). *Audit market structure, fees and choice in a period of structural change: Evidence from the UK–1998–2003.* The British Accounting Review, 42(3), 187–206.

Ali, C.B., & Lesage, C. (2014). *Audit fees in family firms: Evidence from US listed companies.* Journal of Applied Business Research, 30(3), 807.

Basioudis, I., & Fifi, F. (2004). *The Market for Professional Services in Indonesia.* International Journal Of Auditing, 8(2), 153–164.

Campa, D. (2013). "Big 4 fee premium" and audit quality: latest evidence from UK listed companies. Managerial Auditing Journal, 28(8), 680–707.

Carson, E., Simnett, R., Soo, B.S., & Wright, A.M. (2012). *Changes in audit market competition and the Big N premium.* Auditing: A Journal of Practice & Theory, 31(3), 47–73.

Choi, J.H., Kim, J.B., & Zang, Y. (2010). *Do abnormally high audit fees impair audit quality?.* Auditing: A Journal of Practice & Theory, 29(2), 115–140.

DeAngelo, L.E. (1981). *Auditor size and audit quality.* Journal of accounting and economics, 3(3), 183–199.

Doogar, R., Sivadasan, P., & Solomon, I. (2015). Audit fee residuals: Costs or rents? Review of Accounting Studies, 20(4), 1247–1286.

Feldman, E.R. (2006). *A basic quantification of the competitive implications of the demise of Arthur Andersen.* Review of Industrial Organization, 29(3), 193–212.

Goncharov, I., Riedl, E.J., & Sellhorn, T. (2014). *Fair value and audit fees.* Review of Accounting Studies, 19(1), 210–241.

Hay, D.C., Knechel, W.R., & Wong, N. (2006). *Audit Fees: A Meta-analysis of the Effect of Supply and Demand Attributes.* Contemporary Accounting Research, 23(1), 141–191.

Jensen, M.C., & Meckling, W.H. (1976). *Theory of the firm: Managerial behavior, agency costs and ownership structure.* Journal of financial economics, 3(4), 305–360.

Krauß, P., Quosigk, B.M., & Zülch, H. (2014). *Effects of initial audit fee discounts on audit quality: evidence from Germany.* International Journal of Auditing, 18(1), 40–56.

Lin, H.L., & Yen, A.R. (2016). *The effects of IFRS experience on audit fees for listed companies in China.* Asian Review of Accounting, 24(1), 43–68.

Litt, B., Desai, V., & Desai, R. (2015). *Incumbent audit firm pricing: a response to entry of the Big Four accounting firms in India.* Journal Of Accounting In Emerging Economies, 5(4), 382–394.

Moizer, P. (1997). *Auditor reputation: the international empirical evidence.* International Journal of Auditing, 1(1), 61–74.

Nagy, A.L. (2014). *Audit partner specialization and audit fees.* Managerial Auditing Journal, 29(6), 513.

Nelson, S.P., & Mohamed-Rusdi, N.F. (2015). *Ownership structures influence on audit fee.* Journal of Accounting in Emerging Economies, 5(4), 457–478.

PPPK. (2015). *Laporan Tahunan Kantor Akuntan Publik Tahun Takwim 2015.* Jakarta: Sekretariat Jenderal Kementerian Keuangan – Pusat Pembinaan Profesi Keuangan.

Pratoomsuwan, T. (2017). *Audit prices and Big 4 fee premiums: further evidence from Thailand.* Journal Of Accounting In Emerging Economies, 7(1), 2–15.

Siddiqui, J., Zaman, M., & Khan, A. (2013). *Do Big-Four affiliates earn audit fee premiums in emerging markets?.* Advances in Accounting, 29(2), 332–342.

Simon, D.T. (1997). *Additional evidence on the large audit-firm fee premium as an indication of auditor quality.* Journal of Applied Business Research, 13(4), 21.

Wang, K., Sewon, O., & Chu, B. (2014). *Auditor Competition, Auditor Market Share, and Audit Pricing-Evidence From a Developing Country.* Journal of Global Business Management, 10(1).

Zaman, M., & Chayasombat, J. (2014). *Audit pricing and product differentiation in small private firms: evidence from Thailand.* Journal of Accounting in Emerging Economies, 4(2), 240–256.

Business Innovation and Development in Emerging Economies – Trinugroho & Lau (Eds)
© *2019 Taylor & Francis Group, London, ISBN 978-1-138-35996-3*

The effects of gender, task complexity, obedience pressure, auditor experience, and knowledge audit on audit judgment

Muammar Khaddafi
Universitas Malikussaleh, Indonesia

St. Dwiarso Utomo, Zaky Machmuddah & Imang Dapit Pamungkas
Universitas Dian Nuswantoro, Indonesia

Feby Milanie
Universitas Pembangunan Panca Budi Medan, Indonesia

ABSTRACT: Judgment is a central activity in conducting audit work. The accuracy of the judgment generated by the auditor in completing the audit work gives a significant influence on the final conclusion or opinion that will be generated. Auditor performance in making audit judgment is influenced by various factors. This study aims to examine the effect of gender, obedience pressure, task complexity, auditor experience and knowledge audit on auditor performance on audit judgment.

The importance of audit judgment to provide an opinion about the fairness of the presentation of financial statements in accordance with accepted accounting standards or accounting principles and the consistent application of such standards or principles. This study aims to determine the effect of gender, obedience pressure, task complexity, and auditor experience to audit judgment at the BPK representative office of Semarang. This study used survey method using primary data obtained from questionnaire. The population in this study is all auditors who work at the BPK representative office semarang. The number of samples in this study as many as 42 respondents taken convinience sampling technique.

Data analysis technique used multiple regression test. The result of the research shows that Gender has an effect on audit judgment. This is shown from the results of t test which shows the significance value of 0.013. Obedience pressure affects the audit judgment, this is shown from the results of t test with a value of 0.003 significance. The task complexity affects audit judgment, this is shown from the t test results, which indicates a significance value of 0.008. The auditor's experience has no effect on audit judgment. This is indicated by a significance value of 0.313.

Keywords: Task complexity, Auditor experience, Knowledge audit, Audit judgment

1 INTRODUCTION

Auditor is a profession that has the duty and function to audit various types of economic activities. In performing its duties and functions, the auditor is required to provide judgment or consideration to an audit evidence. Jamilah et al (2007) argues that judgment is an ongoing process of obtaining information (including feedback from previous actions), the choice to act or not to act, the acceptance of further information. In making a judgment, the auditor will collect various relevant evidence at different times and then integrate information from the evidence (Praditanigrum, 2012).

Audit judgment is a consideration that affects documentation of evidence and opinion judgments made by the auditor (Taylor, 2003) and therefore judgment has a significant effect on the final conclusion, thus indirectly affecting whether or not the decision will be taken by outsiders. When conducting an audit of financial statements and giving an opinion on their fairness often requires judgment. When an auditor performs his duties to make an audit judgment influenced by many factors.

Gender is considered to be one of the individual level factors that influence audit judgment as changes in task complexity and ethics compliance influence. The findings of cognitive and marketing psychological literature research also suggest that women are thought to be more efficient and effective in processing information when there is a task complexity in decision-making compared to men (Jamilah, et al. 2007). In addition, factors that may affect audit judgment are the obedience pressure. In auditing it is not uncommon for auditors to encounter difficulties such as pressure from superiors and clients, but in performing their duties, the auditor must be professional and stick to professional ethics and auditing standards. Any compliance pressure from a superior or client may affect the judgment of an auditor in which the auditor should choose to follow the orders of the superior and the client or follow the applicable accounting profession ethic code.

High task complexity can be a burden if the lack of auditor capability and ability. Raiyani and Saputra (2014) stated that the complexity of the audit assignment can be used as a tool to improve the quality of work. The task complexity can make an auditor inconsistent and unaccountable. Professional judgment is the application of relevant knowledge audit and auditor experience, in the context of accounting auditing and ethical standards, to achieving appropriate decisions in situations or circumstances during the course of audit assignments, and personal qualities, meaning that judgments differ among experienced auditors. An experienced auditor has a better ability to process irrelevant information in the audit judgment process by predicting fraud occurring within a government agency and may find material misstatements. Abdolmohammadi and Wright (1987) which suggests that inexperienced auditors have a significantly higher error rate compared to the more experienced auditors performing with the best. This result is supported by Praditaningrum and Januarti (2012).

Knowledge audit also affects the research or judgment of an auditor. In an auditor who possesses an ever-increasing knowledge, both obtained from both direct and indirect learning and training that can be supportive in his assignment as a auditor. Judgment given an auditor will be better if the auditor has good knowledge and many. Otherwise, if the knowledge possessed by an auditor little or lower then the less qualified judgment given by him (Jamilah, 2007). The results of Chung and Monroe (2001) studies suggest that high task complexity affects judgment taken by the auditor. Abdolmohammadi and Wright (1987) show that there are judgment differences that auditors take on the complexity of high tasks and low task complexity. Zulaikha (2006) found that task complexity does not significantly affect judgment accuracy. The purpose of this research is to know the effect of gender, obedience pressure, task complexity, auditor experience and knowledge audit of audit judgment.

2 LITERATURE REVIEW

2.1 Goal setting theory

The goal-setting theory is part of the motivational theory proposed by Edwin Locke in 1978. This theory asserts that individuals with more specific goals and challenging their performance will be better than obscure goals, such as "doing what is best of ourselves", a specific easy goal or no purpose at all. Locke and Latham, (1990) reveals that: (a) the objectives are quite difficult, it results in higher performance levels than easier goals and (b) specific objectives, quite difficult will result in higher levels of output (Latham, 2004 in Verbeeten, 2008). Edwin Locke concludes that setting a goal affects not only work, but also stimulates employees to seek or use the most effective work methods. Similarly, he argues that involving employees in setting goals can generate work motivation, and achievement of maximum work. The goal-setting theory assumes that there is a direct correlation between the definition of a specific and measurable objective to performance, if managers know what their goals are, they are more motivated to exert effort that can improve their performance in a job or task (Locke and Latham, 1990).

2.2 *Auditor performance on audit judgment*

Performance effectiveness refers to how well a particular task is performed. For the auditor, the quality of the work is assessed by seeing whether or not the answer given by the auditor for each audit task is accurate. In the conduct of detailed audit procedures, the auditor makes many judgments that affect the auditor's evidence documentation and judgment (Pramono, 2007). Audit judgment is required at four stages in the audit process of financial statements, namely engagement acceptance, audit planning, audit testing and audit reporting (Mulyadi, 2002). Judgment auditor is required because the audit is not conducted on all the evidence, because it will take a long time and cost is not small, so inefficient.

This is the evidence used to express an opinion on the audited financial statements. For example, when an auditor wants to accept an audit engagement, it must perform an audit judgment on several aspects of management integrity, independence, objectivity, the ability to use professional proficiency with accuracy and ultimately the decision to accept or not an audit engagement. At the time of the audit planning stage, the auditor should recognize the risks and materiality level of a pre-determined account balance. Judgment at this stage is used to determine audit procedures which are subsequently implemented, since judgment in the early stages of the audit is determined on the basis of consideration at the level of materiality foreseen.

In relation to the financial statements, judgments decided by the auditor will affect an auditor's opinion on the fairness of the financial statements. There are various factors forming an auditor's opinion about the fairness of client's financial statements, namely the reliability of the client's internal control system, the suitability of recording the accounting transactions with generally accepted accounting principles, whether or not the client's audit restrictions, and the consistency of the accounting record. Therefore, it can be said that Judgment is a central activity in conducting audit work. The accuracy of the judgment generated by the auditor in completing the audit work gives a significant influence to the final conclusion or opinion that will be produced.

So indirectly also will affect the right or wrong decision will be taken by outsiders who rely on audited financial statements as a reference. The auditor is assigned and obtains a fee from the company issuing the financial statements, but is responsible to the users of the financial statements. If the auditor fails to perform its role, the detrimental impacts of self-esteem, organizational morale, business relationships, and reputation in society are far greater than the value of money spent in defense of these demands. Ultimately, the performance of the auditor in making the audit judgment will reflect the auditor's independence and objectivity and the quality of the Public Accounting Firm.

2.3 *Hypotheses development*

2.3.1 *Gener on audit judgment*

According to Meyer and Levy (1986) men in information processing usually do not use all the information available so the decisions are less comprehensive. As with women, they tend to process information more carefully by using more complete information and re-evaluate the information and not easily give up. Women are relatively more efficient than men while gaining access to information. In addition, women also have a sharper memory of a new information than men and so the ability to process information more carefully so that in making judgment decisions more precisely than men. This argument is supported by Praditaningrum and Januarti (2012), Siagian, Hardi and Azhar (2014), Yendrawati and Mukti, (2015).

Differences based on gender issues in information processing and decision-making are based on different approaches that men and women use core information processing to solve problems and make core decisions. Men in general solve problems do not use all the information available, and they also do not process information thoroughly, so it is said that men tend to do information processing is limited. Women are seen as more detailed information processors who process information on most core information for decision making or judgment.

H1: Gender affects the audit judgment.

2.3.2 *Obedience pressure affects the audit judgment*

Obedience pressure is a condition in which an auditor is faced with the dilemma of standard auditor application (Tielman, 2012). Pressure is usually given by someone who has more authority or authority over others. Obedience pressure is a state in which a person performs an action or a job by having to obey the commands of a superior or a person who has more power than him. Obedience pressure is given so that what he desires can be achieved. For a person who receives the obedience pressure can lead to conflict because he must be able to work in accordance with the command but also must be in accordance with existing codes of ethics. Obedience pressure can make a person deviate from the code of ethics. The act of deviation is done solely to fulfill the command. Sometimes the boss (the person who gives the order) does not think about the consequences when the subordinate (the person who receives the order) takes a deviant action. Based on McGregor's motivational theory, an individual who is under pressure from obedience from a supervisor or entity who is examined, that they will tend to take a safe and dysfunctional way this will result in the auditor not being able to make good and proper judgment.

H2: Obedience pressure affects the audit judgment

2.3.3 *Task complexity on audit judgment*

The task complexity is an important factor that can affect the performance of audit judgment. An understanding of the complexity of different audit tasks can help managers make better tasks and decision-making exercises (Bonner, 2002). The results of Chung and Monroe (2001) study that high task complexity affects the judgment taken by the auditor. The auditor feels that the audit task he faces is a complex task so that the auditor has difficulty in performing the task and cannot make professional judgment. As a result judgment taken by the auditor becomes incompatible with the evidence obtained.

The difficulty level of the task is always associated with the amount of information about the task, while the structure is related to information clarity. The existence of high task complexity can undermine the judgment made by the auditor. If the auditor is faced with a task with high complexity the auditor will have difficulty in completing the task. As a result the auditor is not able to integrate the information into a good judgment.

H3: Task complexity affects the audit judgment.

2.3.4 *Auditor experience on audit judgment*

Experiences are things that have been done, felt perceived or heard by someone. A person's experience can be seen from how long he worked or how often he did or experienced a similar event. An experienced person can do something quickly and correctly because he or she has done it often. The actions performed by a person are much better than those who are not yet or inexperienced. The auditor is a work done repeatedly. The steps taken by the auditor in auditing for each auditor are the same. Auditors may face similar or similar events or evidence when taking audit judgment in performing audit tasks on different entities. Audit judgment made by auditors with experience of far or better than audit judgment made by auditors who are not or have not experienced. The amount of experience in the field of auditing can assist auditors in completing tasks that tend to have the same pattern (Ariyantini, et al, 2014).

H4: Auditor experience affects the audit judgment.

2.3.5 *Knowledge audit on audit judgment*

The level of knowledge that owned auditor is a thing very important that can be affect the auditor within make decisions. Knowledge is one of the keys work effectiveness (Arleen, 2008). An auditor will be able to take a decision in a manner good if supported with knowledge possessed. Auditor must have that knowledge required in running the task. Raiyani and Suputra (2014) prove that knowledge audit have a positive and significant effect to audit judgment. Increasingly high knowledge owned the auditor will get better in providing deep judgment the tasks it handles. According Fitriani and Daljono (2012) level of knowledge audit owned by auditors is a very important thing that can affect the auditor in making decisions. With the auditor's high level of knowledge, the auditor will not only be able to complete an audit job effectively but will also have a broader view of things.

H5: Knowledge audit affects the audit judgment.

3 METHODOLOGY

3.1 Research design

In this research is explanatory research aims to explain the influence of gender, obedience pressure, task complexity, auditor experience and knowledge audit on audit judgment. method used in this research by using convinience sampling. This study takes samples of auditors who work at the office of the Supervisory Board of Finance and Development (BPKP) located in the representative of Sermarang city.

This research uses data analysis method that is descriptive statistic, data quality test, classical assumption test, multiple linear regression analysis. The analysis was performed using the Stastical Program For Social Science (SPSS) program. The object of this research is at BPKP representative office of Central Java in Semarang. The collected data is primary data, that is the data collected by the researcher directly from the first source or place of research object done (Siregar, 2012). In this study through the submission of questionnaires or questionnaires to employees who run the audit in BPKP, as many as 42 people who made the respondents.

3.2 Operationalization of variables

3.2.1 Dependent variables
Dependent variable is a variable that is explained or influenced by independent variable. The dependent variable of this research is Audit Judgment. Audit judgment in this research is auditor's policy in determining opinion about its audit result which refers to the formation of an idea, opinion or estimate about an object, event, status or other type of event. This variable is measured using indicators developed by Praditaningrum (2012) with some modifications tailored to the government audit environment.

The definition of judgment refers to Jamilah, et al (2007) which is a continuous process in the acquisition of information (including feedback from previous actions), the choice to act or not to act and acceptance of further information. Judgment is measured by a simple case relating to the engineering of the transaction and determining the level of materiality and modified according to the government audit environment. The question uses a five-point likert scale with a choice (1) strongly disagree; (2) Disagree; (3) neutral; (4) agree; and (5) strongly agree. The high scale shows the level of judgment performed by high and low-scaled auditors shows a low level of judgment conducted by the auditor.

3.2.2 Independent variables
Independent variables are the variables that influence other variables either positively or negatively (Sekaran, 2006). The independent variables in this research are gender, pressure of word, task complexity, auditor experience and knowledge audit.

Table 1. Selection of research sample.

No	Sample kreteria	Total
1	Questionnaire distribution	72
2	The questionnaire did not return	(32)
3	Complete questionnaire and can be processed	42
Sample		42

Source: Secondary data procces, 2018.

29

1. Gender

Gender in this study is a cultural concept that seeks to make a difference in terms of the roles, mentality, and emotional characteristics between men and women who develop in Jamilah, et al (2007). Gender variables are measured using Nominal scale. Nominal Scale is a measurement scale that states categories, groups or classifications of constructs measured in variable form (Nur Indriantoro and Bambang Supomo, 2009). Nominal scale is expressed by the numbers: 0 (men) and 1 (women).

2. Obedience Pressure

Obedience pressure is the desire of the auditor to be more obedient to the client and the superior order or the auditor's wish to be more obedient to the audit standards. Obedience pressure received by the auditor in the face boss to client take action deviate from ethical standards. Questionnaire on this variable refers to research previously by Jamilah, et.all (2007). Variable in This study was measured by indicators: (1) Auditor compliance against professionals or obedience to orders boss (2) Client code of ethics and standard. The measurement scale used is Likert scale (score 1–5).

3. Task Complexity

Individual perceptions of task difficulties caused by limited information and clarity of information about the task, limited memory and ability to integrate problems owned by decision makers Jamilah, et al (2007). In measuring Indicators the complexity of the audit is derived in the form of 6 statements on the questionnaire. The measurement scale used is Likert scale (score 1–5).

4. Auditor Experience

Experience is a process learning and accretion potential developments behave behavior that can bring someone to a pattern higher behavior. In-house audit experience this study shows experience possessed by auditor in running his profession as an auditor external government. Question adopted by Melthakasih (2012). The variables in this study were measured by indicators: (1) Duration auditor works (2) Amount an audit assignment ever handled. The measurement scale used is Likert scale (score 1–5).

5. Knowledge Audit

Knowledge is defined with a level of understanding auditor against a work, both conceptually or theoretical. This question adopted from Ika's research Ardiani (2011). Variable in this study was measured by indicators: (1) Understanding standards, accounting standards financial SPAP (2) The entity related as well as the need to be education. The measurement scale used is Likert scale (score 1–5).

3.3 *Technique of analysis*

Multiple Linear Regression Analysis To test the hypothesis used multiple regression analysis tools, on the grounds that this tool can be used as a prediction model of the dependent variable with some independent variables. Hypothesis test is done with SPSS 22.0 program. To formulate multiple regression model as follows:

$$Y = a + b1X1 + b2X2 + b3X3 + b4X4 + b5X5 + e$$

Y   : Audit Judgment
a   : Interception Value (Constants)
b1...b5 : Regression direction coefficient
X1   : Gender
X2   : Obedience Pressure
X3   : Task Complexity
X4   : Auditor Experience
X5   : Knowledge Audit
e   : error

4 RESULT AND DISCUSSION

In this study testing H1, H2, H3, H4 and H5 are done with using multiple linear regression analysis. The results of processing can be seen in Table 2 below.

4.1 Gender on audit judgment

The result of research indicate that gender by using t test is obtained value of significance X1 with value equal to 2,337 with significant level equal to 0,025 because level of significance less than 0,05 ($\alpha = 5\%$). Thus the first hypothesis states that Gender has a significant effect on Audit Judgment. Gender is one of the factors affecting audit judgment. Gender is an analytical concept used to identify differences between men and women from a non-biological point of view, ie from social, cultural, and psychological aspects. Women are thought to be more efficient and more effective in processing information when there is a complexity of tasks in decision making than men. Women are more profound in analyzing the essence of a decision and generally have a higher level of moral judgment than men. Thus based on these results, this study is in line with research conducted by Praditaningrum (2012) which states that Gender affects Audit Judgment.

4.2 Obedience pressure on audit judgment

The result of research indicate that obedience pressure by using t test obtained significant value X2 with value equal to 2,934 with significant level equal to 0,006. because the significance level is less than 0.05 ($\alpha = 5\%$). Thus the second hypothesis states that the Obedience pressure has a significant effect on Judgment Audit. Obedience pressure is a state in which a person performs an action or a job by having to obey the commands of a superior or a person who has more power than him. The pressure of obedience is given so that what he desires can be achieved. For a person who receives the obedience pressure can lead to conflict because he must be able to work in accordance with the command but also must be in accordance with existing codes of ethics). bedience pressure obedience can make a person deviate from the code of ethics. The act of deviation is done solely to fulfill the command. Sometimes the boss (the person who gives the order) does not think about the consequences when the subordinate (the person who receives the order) takes a deviant action. Thus based on these results, this study is in line with research conducted by Reni Yendrawati and Dheane Kurnia Mukti (2015) which states the obedience pressure has an effect on Audit Judgment.

4.3 Task complexity on audit judgment

The result of research indicate that Task Complexity by using t test in earn significant value X3 with value equal to −0,953 with significant level equal to 0,347 because level of significance less than 0,05 ($\alpha = 5\%$). Thus the third hypothesis states that the Task Complexity has no

Table 2. Hypotheses test.

Model	Unstandardized coefficient			Standardized coefficients		
	B	Std. error		Beta	t	Sig.
1	Constant	0.049	0.076	−,047	0.644	0.524
	Gender	0.030	0.013	0.172	2.337	0.025
	Obedience Pressure	0.257	0.087	0.295	2.934	0.006
	Task Complexity	−0.082	0.086	−0.077	−0.953	0.347
	Auditor Experience	0.403	0.144	0.336	2.804	0.008
	Knowledge Audit	0.366	0.104	0.396	3.512	0.001

Source: Results of SPSS version 22.

significant effect on audit judgment. The complexity of the task is an important factor that can affect the performance of audit judgment. An understanding of the complexity of different audit tasks can help managers make better tasks and decision-making exercises. Thus based on these results, this study is in line with research conducted by Siagian, Hardi and Azhar (2014), which states the Complexity of Task has no direct effect on Audit Judgment.

4.4 *Auditor experience on audit judgment*

The result of research indicate that experience by using t test in earn significant value X4 with value equal to 2,804 with significant level equal to 0,008 because level of significance less than 0,05 ($\alpha = 5\%$). Thus the fourth hypothesis states that auditor experience has a significant effect on Audit Judgment. Experiences are things that have been done, felt perceived or heard by someone. A person's experience can be seen from how long he worked or how often he did or experienced a similar event. An experienced person can do something quickly and correctly because he or she has done it often. The actions performed by a person are much better than those who are not yet or inexperienced. Thus based on these results, this research is in line with research conducted by Ariyantini, Sujan and Darmawan (2014), which stated that auditor experience has significant effect on Audit Judgment.

4.5 *Knowledge audit on audit judgment*

The result of research indicate that Task Complexity by using t test in earn significant value X5 with value equal to 3,512 with significant level equal to 0,001 because level of significance less than 0,05 ($\alpha = 5\%$). Thus the third hypothesis states that the Task Complexity has a significant effect on audit judgment. The level of knowledge that the auditor has is very important that can affect the auditor in making decisions. With the auditor's high level of knowledge audit, the auditor will not only be able to complete an audit job effectively but will also have a broader view of things. Thus based on these results, this research is in line with research conducted by Siagian, Hardi and Azhar in 2014, which states that knowledge audit directly affects Audit Judgment.

5 CONCLUSION

Based on the results of data analysis and discussion that has been done, it can be concluded as follows: 1. Gender variable in partial test obtained t value 2,337 with significant level equal to 0,025 because level of significance less than 0,05 ($\alpha = 5\%$), then variable of Gender (X1) have significant influence to Audit Judgment (Y). Gender has a positive effect on Audit Judgment, so the more Gender, the better the Audit Judgment will be. Pressure Variables obedience pressure in the partial test obtained value of t arithmetic of 2.934 with a significant level of 0.006 because the level of significance is smaller than 0.05 ($\alpha = 5\%$), then the obedience pressure (X2) variable has a significant influence on Audit Judgment (Y). Obedience pressure positively affects Audit Judgment, so the higher the Pressure of Obedience level, the better Judgment Audit it performs. 3. Complexity Variables Tasks in the partial test obtained t value of −0.953 with a significant level of 0.0347 because the level of significance is smaller than 0.05 ($\alpha = 5\%$), then the Task Complexity (X3) has no significant effect on Audit Judgment (Y). 4. Experiential variables in partial test obtained value of t arithmetic of 2.804 with a significant level of 0.008 because the level of significance is smaller than 0.05 ($\alpha = 5\%$), then the auditor experience variable (X4) has a significant influence on Audit Judgment (Y). Auditor experience has a positive effect on audit judgment, so the higher the level of auditor experience, the better audit judgment it does. 5. Knowledge audit variable in partial test is obtained by value of t count equal to 3,512 with level of significance equal to 0,001 because signification level smaller than 0,05 ($\alpha = 5\%$), then knowledge audit variable (X5) has significant influence to Audit Judgment (Y). Knowledge audit positively affects Audit Judgment, so the higher the level of knowledge, the better Audit Judgment it does.

6 SUGGESTION

1. This research is limited to only one Office of Representative Agency **BPKP** Central Java Province. For that the researcher suggested for further research to expand the object of research at other representative office of **BPKP** of Province, so the result can be generalized and most likely bring bigger benefit.
2. The auditor shall be professional in order that the audits performed in accordance with his or her capabilities are not affected by personal affairs.
3. The auditor should have an independence over his profession so that whatever evidence is obtained that is the reference to the actions that should be done.
4. In future research, it is necessary to add the variables and other factors affecting audit judgment in the Office of the **BPKP** Representative Office of Central Java Province, such as: Accountability, Leadership.

REFERENCES

Abdolmohammadi, M., & Wright, A. (1987). An examination of the effects of experience and task complexity on audit judgments. *Accounting Review*, 1–13.

Ariyantini, K.E., Edy Sujana, S.E., Msi, A.K., & Darmawan, N.A.S. (2014). The Effect Of Auditor Experience, Pressure And Complexity Task On Auditing Judgement (Empirical Study at BPKP Representative of Bali Province). JIMAT (Scientific Journal of Accounting Students) Undiksha, 2 (1).

Bonner, S.E., & Sprinkle, G.B. (2002). The effects of monetary incentives on effort and task performance: theories, evidence, and a framework for research. *Accounting, Organizations and Society*, 27(4–5), 303–345.

Chung, J., & Monroe, G.S. (2001). A research note on the effects of gender and task complexity on an audit judgment. *Behavioral Research in Accounting*, 13(1), 111–125.

Idris, S.F., & Daljono. (2012). The Influence of Compliance, Complexity of Duties, Knowledge and Ethical Perceptions of Judgment Audit (case study at BPKP representative of DKI Jakarta province) (Doctoral dissertation, Faculty of Economics and Business).

Indriantoro, Nur and Bambang Supomo. (2009), Business Research Methodology. BPFE. Yogyakarta.

Jamilah, S., Fanani, Z., & Chandrarin, G. (2007). Pengaruh gender, tekanan ketaatan, dan kompleksitas tugas terhadap audit judgment. *Simposium Nasional Akuntansi X*, 26–28.

Locke, E.A., & Latham, G.P. (1990). *A theory of goal setting & task performance*. Prentice-Hall, Inc.

Meyers-Levy, Joan (1986), "Gender Differences in Information Processing: A Selectivity Interpretation", unpublished doctoral dissertation, Marketing Department, Northwestern University, Evanston, IL

Praditaningrum, A.S., & Januarti, I. (2012). Analysis of factors influencing audit judgment (Study on BPK RI Representative of Central Java Province) (Doctoral dissertation, Faculty of Economics and Business).

Raiyani, N.L.K.P., & Suputra, I.D. (2014). Effect of competence, task complexity, and locus of control on audit judgment. E-Journal of Accounting, 429–438.

Sekaran, U., & Bougie, R. (2016). *Research methods for business: A skill building approach*. John Wiley & Sons.

Siagian, R.M., & Al-azhar, L. (2014). Factors influencing audit judgment (empirical study on BPK) representative of Riau province). Journal Online Students (JOM) Field of Economics, 1 (2), 1–15.

Taylor, M.H., DeZoort, F.T., Munn, E., & Thomas, M.W. (2003). A proposed framework emphasizing auditor reliability over auditor independence. *Accounting Horizons*, 17(3), 257–266.

Tielman, E.M.A., & Pamudji, S. (2012). Influence of pressure of obedience, time budget pressure, task complexity, knowledge and experience of auditors on audit judgment (Doctoral dissertation, Faculty of Economics and Business).

Yendrawati, R., & Mukti, D.K. (2015). Gender influence, auditor experience, task complexity, compliance pressure, work ability and auditor's knowledge of audit judgment. Asian Journal of Innovation and Entrepreneurship, 4 (01), 1–8.

Verbeeten, F.H. (2008). Performance management practices in public sector organizations: Impact on performance. *Accounting, Auditing & Accountability Journal*, 21(3), 427–454.

Zulaikha. 2006. The Influence of Gender Interaction, Complexity of Duties, and Auditor's Experience on Judgment Audit. National Symposium on Accounting 9. Padang.

Business Innovation and Development in Emerging Economies – Trinugroho & Lau (Eds)
© *2019 Taylor & Francis Group, London, ISBN 978-1-138-35996-3*

The moderating effect of managerial ownership and institutional ownership on auditor opinion, and auditor switching for fraudulent financial statements

Z. Machmuddah, St. Dwiarso Utomo & I.D. Pamungkas
University of Dian Nuswantoro, Semarang, Indonesia

ABSTRACT: A fraudulent financial statement is the type of financial fraud that has the most adverse impact. The purpose of this study was to find empirical evidence of the influence of auditor opinion and auditor switching on fraudulent financial statements, and empirically prove that a good corporate governance mechanism moderate the influence of auditor switching in relation to fraudulent financial statements. The population for this research is public companies in the manufacturing sector in the period 2013 to 2015. A purposive judgment sampling method resulted in a sample of 135 companies. The data used are secondary data from companies' annual financial statements. This study uses PLS (Partial Least Squares) in the context of variance-based Structural Equation Modeling (SEM) to simultaneously test the measurement model and the structural model. WarpPLS 4.0 is used as a data analysis tool. The first test results prove that auditor opinion has a positive but insignificant effect on fraudulent financial statements; however, auditor switching has a positive and significant effect. Second, good corporate governance mechanism proxy by managerial ownership and institutional ownership. The results show that managerial ownership proved able to moderate or weaken the influence of auditors switching on fraudulent financial statements, but institutional ownership has an moderate to insignificant effect on the influence of auditors switching.

Keywords: auditor opinion, auditor switching, fraudulent financial statement, good corporate governance mechanism

1 INTRODUCTION

The Association of Certified Fraud Examiners (ACFE, 2014), based on the frequency of fraudulent acts, states asset misappropriation represents the highest frequency of fraud, followed by corruption and, lastly by fraudulent financial statements. However, the fraudulent financial statement is the type of fraud that has the highest and most adverse impact. A fraudulent financial statement can be triggered by internal and external pressure factors, the opportunity to engage in crime, and rationalization of such action (Romney et al., 2012). Motive trigger fraud problems by Cressey (1953) backed by three factors, namely pressure, opportunity and rationalization, called fraud triangle theory. This is in line with agency theory. Messier et al., (2006) stated that the relationship between principal and agent causes problems of conflict of interest and information asymmetry.

A fraudulent financial statement is something that can happen in any company. In 2002, Enron started the unraveling of a major scandal in America. In 2009, the Satyam computer company engineered a bank balance of about 1.04 billion dollars. Subsequently, Satyam's company has overstated debtors, falsified accrued interest and recognized understated liability. The case of the Toshiba Corporation fraud in 2015 has dragged the CEOs, the motive employed by this company is to fabricate profits by raising a profit rate of almost $1.22 billion over five years. Fraudulent financial statements also occur in Indonesia. Based on OJK records in 2014 and 2015, as many as 20 companies violated rules related to reporting and presenting financial statements or rules VIII.G.7. In addition, some companies violated the rules of information disclosure for financial statements. As illustrated in some of the above cases, proving fraud becomes a very serious problem and has a big impact. ACFE report as much as 77% fraud is by the departments of accounting, purchasing and sales. In 2012, ACFE recorded an increase in the rate of fraudulent financial statements to 9% from 7.6%, with 85.4% of these attributed to the problem of misuse of assets much larger even than accounting fraud. Despite fraudulent financial statements being less frequent, the resulting impact is greater.

Audit results obtained by the company can affect the occurrence of fraud committed by the company. This means the better the audit results of the company the lower the level of fraud that exists in a company, and vice versa, the company's audit results the higher the level of fraud that exists within a company. The auditor may provide some opinions about the company being audited in accordance with the conditions of the company (Mafiana, et al. 2016). With audit opinion, management considers such actions to be general and rational. This is in line with Sukirman and Sari (2013) who examined the effect of rationalization on fraudulent financial statements with the proxy of audit opinion. The main cause is because the audit opinion on the financial statements is needed to provide information and confidence to investors and users of financial statements to take investment decisions.

Companies that commit fraudulent financial statements are more likely to auditor switch, because it reduces the possibility of detecting fraudulent financial statements (Sorenson et al. 1983). Loebbecke et al. (1989) showed evidence that 36% of cheats in their samples were alleged in the initial two years of an auditor's tenure. Further, Krishnan and Krishnan (1997) and Shu (2000), in Lou and Wang (2009), provide evidence that an auditor's resignation is positively associated with the possibility of litigation. Companies that fail to manage, have a greater tendency to auditor switch than healthier companies. DeFond (1992) explains that the presence of auditor switching has an effect on the likelihood of fraudulent financial statements, as management is more likely to auditor switch in anticipation of some agency issues. This is in line with Chen et al. (2007) stating that firms with more frequent auditor switching tend to be more attributed to fraudulent financial statements.

The United States Senate (1976) states that the distinction in the auditor switching is mandatory and voluntary on the basis of which side is the cause of concern over the issue. The main concern of voluntary auditor switching is the client side having the option of auditor switching. Conversely, the mandatory one is rule side that every three years there should be auditor switching. Corporate governance can be used as a way of ensuring and overseeing the governance systems within an organization (Walsh and Schward, 1990 in Sabeni, 2005). In general, the mechanisms that can control the behavior of management, or often called corporate governance mechanism according to Lins and Warnock (2004), in Fala (2007), can be classified into two groups, namely internal mechanisms and external mechanisms. Internal mechanisms are a way of controlling a company using internal structures and processes such as general shareholder meetings, board composition, managerial ownership and meetings with the board of directors. While external mechanisms are a way of influencing companies other than by using internal mechanisms, such as control by the company, institutional ownership and market control.

The implementation of a structure of good corporate governance in a company can be seen through the existence of an independent board of commissioners, an audit committee, audit quality and managerial ownership. Managerial ownership means that management has share ownership. It is used by owner to motivate manager to gain his better performance. Research by Farida et al., (2010) shows that managerial ownership affects fraudulent financial statements. The proportion of share ownership by the managerial party will affect the decision made by the manager, as the decision will affect his position as a manager of the company as well as the a shareholder. Thus, there will be an alignment of interests between management and shareholders. The above mechanism will increase investor confidence that the behavior of managers performing actions to manipulate financial statements can be minimized (Mudjiono, 2010). Similarly, institutional ownership acts as an overseer and controller from the external parties that will make fraudulent financial statements. Institutional owner-ship is the participation of other institutions in the ownership of the company's shares, such as government institutions, banks, insurance companies, investment companies, pension funds and other institutions.

Based on the above background, the purpose of this study is to analyze and obtain empiri-cal evidence about the influence of auditor opinion and auditors switching on fraudulent financial statements. To analyze and obtain empirical evidence on good corporate govern-ance mechanism, managerial ownership and institutional ownership variables are used as proxy to test the moderating role on the relationship between the auditor switching relation-ship and fraudulent financial statement.

2 LITERATURE REVIEW AND HYPOTHESES DEVELOPMENT

2.1 *Fraud triangle theory*

Bologna and Lindquist (1995) define fraud as lies, plagiarism, and theft. Fraud can create significant financial losses very quickly and undermine the trust of customers, employees, investors, partners, and suppliers. Webster's New World Dictionary, cited by Rezaee and Riley (2010) defines fraud as an act being committed deliberately to force others to surrender something to which they are entitled. The Treadway Commission, in Rezaee and Riley (2010), defines fraudulent financial statements as frivolous acts, or omissions that cause materially and misleading financial statements. Well (2007) defines fraud as any kind of criminal form by using deception as the basis of the operation of an individual or criminal group carrying out its criminal plan. Singleton (2006) defines fraud a crime, a generic term, and embraces all the multifarious means which human ingenuity can devise, which are resorted to by one individual, to get an advantage over another by false representations.

Audit Standards (SA) No. 240 defines fraudulent actions to gain an advantage in an unfair manner and unlawfully violates by management, employees or third parties. Fraud is often associated with management, because management is responsible for presenting reliable financial statements. As well as fair presentation, honesty, and the quality of the presentation process of financial statements is the responsibility of management (Rezaee & Riley, 2010). The Statement of Auditing Standard (SAS) No. 99 (2002) defines fraud as a deliberate act which results in and affects the misstatements of material financial statements.

Sutherland introduced the first fraud triangle concept in 1949. It was then developed by the criminologist Donald R. Cressey (1953) in his research on cases of embezzlement (Drew & Drew, 2010). In 2002, American Institute of Certified Public Accountants (AICPA) issued SAS no. 99 based on the fraud triangle theory Cressey (1953). This theory links three factors that always exist in the event of cheating. The cheating-forming factor developed by Donald R. Cressey (1953) is called the fraud triangle theory and explains that fraud is formed by three background factors: pressure, opportunity, and rationalization.

2.2 *Agency theory*

The pioneers of agency theory are Jensen and Meckling (1976). This theory analyzes the rela-tionship between the principal (the owner of the company) and the agent (the manager of the company). These relationships tend to create differences of interests between principals and agents, because, in principle, human beings will seek to maximize utility for their own inter-ests. The position of the agent as the manager of the company is more profitable than the principal, because the agent knows the internal information and prospects of the company in the future. The manager, as an agent, is obliged to provide information about the company's condition to the principal. However, the information conveyed sometimes does not match the actual conditions of the company. This condition is known asymmetry information.

The assumption of the difference in importance between principals and agents built by Eisenhardt (1989) led to the emergence of agency problems. Problems will occur if the principal cannot determine and know what the agent has done. The agency problem can be divided into two categories. First, known as an adverse selection, occurs when an agent fails to provide his or her abilities when a contract has occurred. The second, is known as a moral hazard, that is, the lack of an agent's efforts to carry out the responsibilities afforded, or without the knowledge of the principal acting for his own benefit or otherwise contrary to the principal's interests.

2.3 Auditor opinion on fraudulent financial statements

A fraud conducted by a company can be detected from the audit results. The better the audit results of the company, the lower the level of fraud that exist within a company. Audit opinion is often used to assess the effectiveness of company performance and to assess whether the financial statements presented by management have been accountable and transparent. The auditor may provide some opinions about the company being audited in accordance with the conditions of the company. With audit opinion, management considers such actions to be general and rational because the auditor only provides opinions to be tolerated for corporate error by raising an audit opinion with explanatory language. A study conducted by Sukirman and Sari (2013) examined the effect of rationalization on fraudulent financial statements with the proxy of audit opinion. From the research it can be seen that by using the proxy of audit opinion, rationalization can affect fraudulent financial statements.

Vermeer (2003) found that auditors tolerate more clients' efforts to manage earnings over time. Profit management is a management decision-making process that paves the way for a management push or understanding of terms that might lead to rationalization of fraudulent financial statements (Vermeer 2003). Auditor opinion that uses additional explanatory language is a tolerant form of the auditor for these financial statement reduction. Shelton (2014) says rationalization is how to justify the mind in committing criminal acts. According to Skousen et al. (2009), rationalization is a difficult factor to measure to detect fraud. This is supported by the statement of Sukirman and Sari (2013), namely the external auditors need to identify and consider the risk factors that cause their audit clients to commit fraud. The auditor may provide some opinions on the company being audited in accordance with the circumstances of the company. One of the auditor's opinions is unqualified with clear explanatory language. The opinion is a tolerant form of the auditor for fraud and earnings management. This allows management to be rationalized or assume the mistake it made is not wrong, because it has been tolerated by the auditor through the explanatory language given in his opinion.

H1: Auditor opinion negatively influences fraudulent financial statements.

2.4 Auditor switching on fraudulent financial statements

Management is more likely to change its auditors in anticipation of some agency issues (DeFond 1992). Chen et al. (2007) argue that firms with more frequent auditor turnover tend to be more attributed to fraudulent financial statements. Schwartz and Menon (1985) argue that companies that fail managing it have a greater tendency to change auditors than healthier companies. The form of rationalization of fraud triangle theory according to SAS No.99 and American Institute of Certified Public Accountants (2002) in addition to audit opinion that is the turn of the auditor, states that the effect of frequent auditor switching can be an indication of fraud. According to Lou and Wang (2009), auditor switching is a way to reduce the possibility of detecting fraudulent financial statements by the auditor. The old auditor may be better able to detect any possible fraud committed by the management directly or indirectly. Moreover, they claimed that a failing company replacing its auditor, having a preference for replacing a public accountant's office with a different quality, tended to degrade the quality of the auditor the company used.

A significant factor in influencing auditor switching by the firm is the auditor's opinion, as shown by Chow and Rice (1982). This is supported by Hudaib and Cooke's (2005) research that found that clients have a tendency to auditor switch after receiving a qualified audit opinion. The opinion of a qualified auditor may be due to the detection of non-conformity or fraud in the presentation of financial statements. Defond (1992) states that the determination of income should involve judgment and wisdom, giving managers an opportunity to manipulate high revenues and opportunities to manipulate income that leads to auditor switching. Auditors who have conducted an audit of the company can find out the opportunities for accounts that have the potential for fraud in the presentation of the financial statements. In their research, Sorenson et al., (1983) indicate the possibility of clients auditor switching to reduce the possibility of fraud detection in their financial statements. This is supported by the research of Loebbecke et al., (1989), who found a large number of fraud indications contained within the auditor's sample within the first two years of the auditor's tenure.

The auditor is an important watchdog in the financial statements. Based on the auditor's information it can be seen that there are companies that commit fraud. Companies that conduct fraud do more frequent auditor switching. This is to reduce the possibility of detection of a fraudulent financial statement by the company. Sorenson et al., (1983) state that a company may auditor switch to reduce the likelihood of detecting fraudulent financial statements by auditors (Lou and Wang, 2009). Loebbecke et al., (1989) showed that 36% of the fraud in their samples were alleged within the initial two years of the auditor switching. Further, Krishnan and Krishnan (1997) and Shu (2000) found evidence that the auditor's resignation was positively related to litigation likelihood (Lou and Wang, 2009).

H2: Auditor switching positively influences fraudulent financial statements.

2.5 *Moderating effect of managerial ownership on the relationship between auditor switching and fraudulent financial statements*

Management has a responsibility to shareholders that is related to information disclosure financial statements. According to Lou and Wang (2009), the purpose of the auditor switching is to get maximum audit results. The goal is to remain undetected and reduce the possibility of detection of fraudulent financial statements directly or indirectly by the auditor. This decision was taken because the manager was worried that it was detected by the auditor. These concerns are triggered by management's participation in decision-making. If the decision taken is wrong then it is management who bear the loss. In SAS No.99 (American Institute of Certified Public Accountants, 2002) states that the effect of auditor switching within the company may be an indication of fraud. The old auditor may be better able to detect any possible fraud committed by the management, whether directly or indirectly. However, with the auditor switching, the possibility of cheating will increase. This statement is evidenced from the results of research by Trinanda (2016) and Kurniawati and Raharja (2012) stating that with the resignation or auditor switching, it will affect the possibility of fraudulent financial statements.

Based on agency theory, there is often a problem between the owner of the company and the management of the company caused by the existence of different interests within the company. The issue of interest leads to the importance of a mechanism that is useful for protecting the interests of the shareholders and managers having a high number of shares in the firm (Jensen & Meckling, 1976). Managerial ownership is considered to overcome the problems of the agency that has always happened. Due to managerial ownership, managers will be more energized in increasing the value of the company and can motivate managers to work in accordance with the interests of the principal. The results of Boediono (2005) showed that managerial ownership negatively affects the occurrence of fraud. The greater the share ownership by management, the smaller the possibility of fraud. This is because the manager also has a role as a share owner, therefore he will work in accordance with the interests of the principal.

H3: Managerial ownership weakens the relationship between auditor switching and fraudulent financial statements.

2.6 Moderating effect of institutional ownership on the relationship between auditor switching and fraudulent financial statements

Institutional ownership plays an active role in the managerial oversight process and the reporting process so as to impact on the decline in the opportunity of the company's management to commit fraud. With the declining opportunity of management to commit fraudulent actions that potentially harm the owners of the company, the confidence of the owner of the company in the credibility of the information contained in the financial statements will increase.

Institutional ownership has a very important role in minimizing the agency conflict that occurs between managers and shareholders. The existence of institutional investors is considered capable of being an effective monitoring mechanism in every decision taken by managers, because institutional investors are involved in strategic decision-making so it is not easy to believe in fraudulent acts of financial statements. The percentage of institutional shares is the sum of the percentage of shares of companies owned by institutions or the institutions (insurance companies, banks, investment companies, asset management and other institutional ownership) both within and outside the country. With institutional ownership encouraging more optimal supervision of manager performance (Susiana & Herawaty, 2007), institutional ownership is instrumental in monitoring manager behavior so that financial statements are well maintained. This is because, with supervision, the manager will be more careful in their decision-making. Therefore, institutional ownership plays a role in reducing fraudulent financial statements. Institutional ownership has the advantage of having professionalism in analyzing information to test the reliability of information and have a strong motivation to carry out stricter supervision of activities that occur within the company. Corporate oversight by institutional investors can encourage managers to focus more attention on the company's performance to reduce opportunistic or self-serving behavior. Institutional ownership can reduce the problems that occur as a result of information asymmetry. The existence of institutional ownership will encourage oversight of the performance of management, because the institutional shareholders have the ability and professionals are good at assessing a presented report. High institutional ownership limits managers to manage earnings and can reduce fraudulent financial statements.

H4: Institutional ownership weakens the influence of auditor switching against fraudulent financial statements.

3 RESEARCH METHODOLOGY

The type and source of this research data is secondary data. Secondary data used in this research is the audited annual report of manufacturing companies listed in the Indonesia Stock Exchange. The reason for using a manufacturing company is because there are differences in characteristics between the companies in the manufacturing sector and the financial sector companies in their financial statements. PWC (2010) research results amounted to 83% of fraud perpetrators of financial statements were made by the industries in the field of manufacturing. The data used in this research is obtained from www.idx.co.id, www.sahamok.com, company websites and Indonesian Capital Market Directory (ICMD) in 2013–2015. The sample selection uses a purposive judgment sampling method. The sample is selected based on the criteria that can invert the population. Sampling criteria are as follows: 1) companies in the manufacturing sector listed on the Indonesia Stock Exchange and published an annual report audited during the period 2013–2015; 2) the financial statements are expressed in rupiah (Rp), in order that the value will not be affected by the fluctuation of the rupiah against the dollar; 3) manufacturing companies that have complete data and related to research variables, such as managerial ownership, during the period 2013–2015.

WarpPLS 4.0 is used as a data analysis method. PLS (Partial Least Squares) is a variance-based Structural Equation Modeling (SEM) technique that can simultaneously perform testing of measurement models as well as structural model testing. The measurement model is used for testing validity and reliability, while the structural one is used for testing hypothesis. The advantages of PLS are: 1) PLS is able to model many dependent and independent variables (complex models);, 2) results remain solid despite abnormal data;, 3) it can be used on reflective and formative constructs;, 4) it can be used on a small sample;, 5) it does not require normally distributed data;, 6) it can be used on data with different scale types, namely: nominal, ordinal, and continuing.

3.1 Operationalization variables

3.1.1 Fraudulent financial statements

The dependent variable in this study was measured by the accrual discretionary as a measure of earning management to see the manipulative content in 120 financial statements, as a proxy for fraudulent financial statements. Discretionary accruals in this study were calculated by the Jones model modified by Dechow, et al. (1995), with reasons according to Abdurrahim (2016). This modified Jones model is more accurate in calculating earnings management than other measurements, although it shares some of the weaknesses and advantages of each. To measure discretionary accruals through several stages, the first counts the total accrual (TAC) in year t. Then, the total accrual value is calculated by the accrual value of the total assets of firm i in year t, with the regression equations:

$$TAC_{it} = NI_{it} - CFO_{it} \tag{1}$$

$$TAC_{it}/A_{it-1} = \beta 1(1/A_{it-1}) + \beta 2(\Delta Rev_t/A_{it-1} - \Delta Rec_t/A_{it-1}) + \beta 3(PPE_t/A_{it-1}) + e \tag{2}$$

Using the above regression coefficients, Non-Discretionary Accrual (NDA) values can be calculated with the equation:

$$NDA_{it} = \beta 1(1/A_{it-1}) + \beta 2(\Delta Rev_t/A_{it-1} - \Delta Rec_t/A_{it-1}) + \beta 3(PPE_t/A_{it-1}) + e \tag{3}$$

The next stage of NDA_{it} calculates the discretionary accrual (DACC) with the following equation:

$$DACC_{it} = (TAC_{it}/A_{it-1}) - NDA_{it} \tag{4}$$

where
TAC_{it} = Total accruals of firm i in year t
NI_{it} = Net profit
CFO_{it} = Operating cash flow
A_{it-1} = Total assets of company i in year t-1
ΔRev_t = Changes in corporate income i in year t
ΔRec_t = Change of receivables of company i in year t
PPE_t = The company's fixed assets in year t
NDA_{it} = Non-discretionary accruals of company i in year t
$DACC_{it}$ = Discretionary accruals of company i in year t
TAC_{it} = Total accruals of company i in year t
e = error

3.1.2 Auditor opinion

Auditor opinion is the independent variable in this research. Audit opinion may serve as a benchmark of possible fraud indications. According to Skousen et al. (2009), rationalization is a difficult factor to measure. This study produces rationalization with audit opinions (AO), as measured by dummy variables. A company receiving an unqualified opinion with explanatory language during the period 2013–2015 is coded 1, and a company that obtains the other is coded 0.

3.1.3 *Auditor switching*

The independent variable in this research is that the auditor switching in a company can be assessed as an effort to eliminate a fraud trail which was found by a previous auditor. This tendency encourages companies to replace their independent auditors to cover the fraud within the company. Therefore, this study produces the auditor switching which is a dummy variable, coded as 1 if the company auditor switches within two years before the fraud occurs, and coded as 0 if there is no auditor switching (Lou & Wang, 2009).

3.1.4 *Managerial ownership*

The moderating variable in this research is managerial ownership which is the shareholders from the management who actively participate in corporate decision-making (director and commissioner). The ownership of some shares by insiders can be used as a control in fraudulent financial statement (Skousen et al., 2009). The ratio of ownership of shares by insiders (MNJL) is calculated by the number of shares owned by the management of all the share capital of the company in circulation multiplied by 100% (Skousen et al., 2009).

$$MNJL = \frac{\text{Number of shares owned by management} \times 100\%}{\text{Total shares outstanding}}$$

3.1.5 *Institutional ownership*

Institutional ownership is a moderating variable in this research which is the shareholders from the institutions who actively participate. The ownership of some shares by insiders can be used as a control in fraudulent financial statement (Skousen et al., 2009). The ratio of ownership of shares by an institution (INST) is calculated by the number of shares owned by the institutions of all the share capital of the company in circulation multiplied by 100% (Skousen et al., 2009).

$$INST = \frac{\text{Number of shares owned by institutions} \times 100\%}{\text{Total shares outstanding}}$$

4 RESULTS AND DISCUSSION

In accordance with the predefined criteria, there were 135 manufacturing companies in the sample for the period 2013–2015. The results of data processing are presented in the Table 1.

4.1 *Auditor opinion on fraudulent financial statements*

Based on the information presented in Table 1, it can be seen that the beta value of auditor opinion test against fraud financial statements is −0.014 with a significance of 0.417. Based

Table 1. Hypothesis testing results (Source: PLS processing results, 2018).

Path	Direct effect		Remark
	Coefficient	p-value	
AUOPIN → FRAUD	−0.014	0.417	H1 is rejected
AS → FRAUD	0.095	0.083*	H2 is accepted
AS → MNJLOWN → FRAUD	−0.173	<0.001***	H3 is accepted
AS → INSTOWN → FRAUD	−0.008	0.454	H4 is rejected
Model fit indicators			
Average Path Coefficient (APC)	0.073	0.072	
Average R-square (ARS)	0.041	0.138	
Average Variance Inflation Factor (AVIF)	1.223		

Note: *, **, and *** indicate significance (one-tailed) at the 0.10, 0.05, and 0.01 levels, respectively.

on this, it means that the audit opinion variable with the explanatory language (AO) has a negative and insignificant effect on the possibility of fraud in the financial statements. Thus, the results of this study reject the first hypothesis. The negative beta value means that better audit opinions obtained by the company will further decrease the fraudulent financial statements.

The underlying reason of this research is that the explanatory language in the independent auditor's report is an explanation of certain matters for which translation is required. The addition of this explanatory language does not affect the materiality in the financial statements and does not alter the reasonableness of the financial statements themselves so the addition of this explanatory language does not affect the possibility of fraud by the company's management. In addition, audit opinion is very difficult to observe as it is public data such as financial statements. The rationalization of these frauds can only be accurately demonstrated by interviews with fraudsters (Cressey 1953).

The result of this test is that the audit opinion has no effect on the fraudulent financial statement, due to the addition of the explanatory language in the independent auditor's report. If the explanation of certain matters such as fair opinions given in part is based on other independent reports, additional information is required by the Indonesian Institute of Accountants. This opinion is given when certain circumstances require the auditor to add an explanatory paragraph to the audit report, although it does not affect the unqualified opinion expressed by the auditor. In addition, the addition of the explanatory language does not affect the materiality of the financial statements, so it does not affect the possibility of rationalization of fraud on the financial statements by the management company. It indicates whether or not the audit opinion is obtained. It does not affect the possibility of rationalization of fraud on the financial statements by the management company. The auditor's opinion variable is proxied by looking at whether or not a company has an unqualified opinion. According to Annisya (2016), unqualified opinion with explanatory language is due to three things, namely: a tolerant form of auditors on the existence of earnings management in a company, opinion in the financial statements involving other auditors and opinion of an explanation of certain things.

4.2 *Auditor switching on fraudulent financial statements*

In accordance with the information presented in Table 1, the significance value of the test of the auditor switching influence of 0.083 with a positive beta value of 0.095 means that the hypothesis is accepted. It means that the more frequent auditor switching, the more fraudulent financial statements. SAS No. 99 states that the effect of auditor switching in a company can be an indication of the occurrence of fraud. The old auditor may be better able to detect any possible fraud committed by the management, whether directly or indirectly. However, with auditor switching, the possibility of cheating will increase. The new auditor cannot directly detect fraudulent forms committed by management, as they are not yet familiar with auditing the company.

Auditor switching occurs when the contract of work agreed between the auditor and the assignor has expired and the assignor has decided not to renew it (Gagola, 2011). Auditor turnover occurs for several reasons, including: (i) a client company is a merger between several companies that initially have different auditors; (ii) the need for broader professional services; (iii) the client company is not satisfied with the old Public Accounting Firms (KAP); (iv) a desire to reduce audit revenue; and (v) mergers between several auditors (Boynton, 2001). Another reason that encourages a change of auditors is Finance ministry decree of number 359/KMK.06/2003 Article 2 on "Public Accounting Service". This regulation states that the provision of general audit services to the financial statements of an entity may be conducted by the Public Accounting Firm (KAP) for a maximum of five consecutive years and by a public accountant for the last three consecutive years. The rules are then updated by Finance Ministry rule of 17/PMK.01/2008 with the obligation to replace the KAP after conducting audits for six consecutive years and by a public accountant for a maximum of three consecutive years.

Loebbecke et al., (1989), in Lou and Wang (2009), noted that 36% of the company's accounting fraud samples were performed within the first two years of an auditor's tenure. Several studies shown that the incidence of audit failure after auditor switching happen. Lou & Wang (2009) estimate that auditor switching is positively correlated with possible fraud. The findings of this study are in line with the research conducted by Kurniawati and Raharja (2012), who state that with the resignation or auditor switching, it will affect the possibility of fraudulent financial statements.

4.3 *Moderating effect of managerial ownership on the relationship between auditor switching and fraudulent financial statements*

Based on Figure 1, the research findings prove that the significant value of the moderation test is <0.001 with a negative direction beta value of −0.173 then the hypothesis is accepted. This means that managerial ownership is able to weaken the effect of auditor switching on fraudulent financial statements. The findings of this study are in accordance with the hypotheses and theories. Managerial ownership acts as a watchdog, as the more a manager participates in the company, the more careful he will be in his decision-making because the decisions taken will affect the company and shareholders who are none other than himself. So, when the manager carries out auditor switching with the aim of avoiding detection of fraudulent financial statements then managerial ownership is considered to overcome the problems of the agency. With managerial ownership, managers will be more energized to increase the value of the company and this can motivate managers to work in accordance with the interests of the principal.

Management integrity (attitude) is the main determinant of the quality of financial statements. An independent auditor is an important watchdog for financial reporting. The relationship between managers and auditors shows the rationalization of corporate management. An external auditor is a supervisory mechanism for controlling management behavior related to corporate financial reporting. The Statement of Auditor Standards (PSA) No. 70 indicates that a strained relationship between management and the current auditor/previous auditor

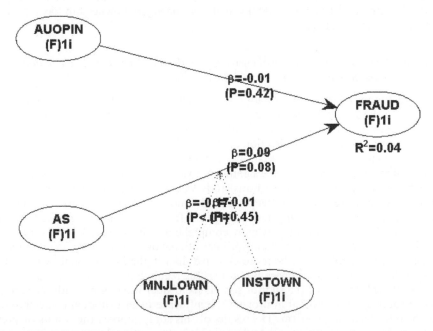

Figure 1.   Direct test results.
Note: AUOPIN: audit opinion; AS: auditor switching; FRAUD: fraudulent financial statement; MNJLOWN: manajerial ownership; INSTOWN: institusional ownership.

is an indication of a fraudulent financial statement. Summers and Sweeny (1998), in Gagola (2011), indicate that clients can use switching auditing mechanisms to reduce the likelihood of detecting fraudulent financial statements by companies. Sorenson et al. (1983) argue that clients may replace auditors to reduce the likelihood of the detection of a fraudulent financial statement. The risk of audit failure and subsequent litigation in initial engagement is higher than in subsequent years (Stice, 1991). Krishnan and Krishnan (1997) and Shu (2000) found that auditor resignation positively affected litigation.

Studies conducted by Stice (1991) and St. Pierre and Anderson (1984) suggest that when auditor switching occur for legitimate reasons, the risk of audit failure and subsequent litigation will be higher than in the following years. Loebbecke et al. (1989) found that a large number of the frauds in their samples were performed within the first two years of an auditor's tenure. SAS No. 99 or Albrecht et al. (2004) suggests auditor switching is associated with fraudulent financial statements. Substitution of auditors conducted by the company can lead to a period of transition and stress hitung the company. The existence of auditor switching in a two year period may be an indication of fraud occurring. Managerial ownership is the ownership of shares by the management company. The proportion of share ownership by the managerial party will affect the decisions made by the manager, as the decision will affect his position as a manager of the company as well as a shareholder. Thus, there is a parallel between the interests of shareholders and managers, because managers directly benefit from the decisions taken and managers who bear the risk if there are losses that arise as a consequence of poor decision-making. This mechanism will add to investor confidence that the behavior of managers to take action to manipulate earnings can be minimized (Mudjiono, 2010).

Jensen and Meckling (1976), in Mudjiono (2010), argue that a convergence of interest between managers and owners can be achieved by providing share ownership to managers. If managers have shares in the company, they will have interests that tend to be equal to other shareholders. With the unification of interests the agency conflict will be reduced so that managers are motivated to improve the company's performance and shareholder wealth. This opinion is in accordance with the results of research conducted by Midiastuty and Machfoedz (2003), who state that managerial ownership with earnings management is negatively related. Ujiyantho and Pramuka (2007) stated that managerial ownership has a significant negative effect on management.

4.4 *Moderating effect of institutional ownership on the relationship between auditor switching and fraudulent financial statements*

Table 1 and Figure 1 provide information that when institutional ownership as a moderating variable in the switching auditor relationship and fraudulent financial statements shows a negative beta value of –0.008 with a significance level of 0.454, the final hypothesis in this study was rejected. Actually, the direction of the beta is properly negative, meaning that institutional ownership will weaken the relationship between the switching auditor and fraudulent financial statements. Nevertheless, the significance value does not support the assumption of the research so the final hypothesis in this study was rejected. Reasons that strengthen the results of this study are that the supervision and control performed by institutional ownership not only focuses on switching auditors against fraudulent financial statements, but the supervision and control performed by more comprehensive institutional ownership is associated with fraudulent financial statements. If institutional ownership exercises comprehensive supervision and control, it will be more effective than if the institutional ownership only focuses on the switching auditor because the existing regulations in Indonesia have regulated the period of replacement of the auditor so that each issuer will comply with the regulation.

This research did not succeed in proving the influence of institutional ownership on the change of public accounting firm. The results of this study support the results of previous research conducted by Suparlan and Andayani (2010). Institutional ownership is also used to create an organizational management with transparency and accountability. The results of this study indicate that the company does not change the old public accounting firm because it is qualified. According to Suparlan and Andayani (2010), high ownership by institutional investors encourages monitoring activities. It is because the magnitude of their voting power will affect management policies including in selecting auditors, where independent auditors play an important role in the monitoring process in order to produce quality and reliable financial reports.

5 CONCLUSION

Based on the data processing and the results of the analysis using PLS (Partial Least Squares) as a variance-based Structural Equation Model (SEM) in this study, it can be concluded that the auditor's opinion has a positive and insignificant effect on fraudulent financial statements. However, the switching auditor has a significant positive effect on fraudulent financial statements. Managerial ownerships weakens the effect of auditors switching on fraudulent financial statements. Institutional ownership did not moderate the effect auditors switching on fraudulent financial statement.

REFERENCES

Abdurahim, A. (2015). Detecting earning management. *Jurnal Akuntansi dan Investasi, 1*(2), 104–111.
Association of Certified Fraud Examiners. (2014). Report to the nation on occupational fraud and abuse. *The ACFE Foundation*. https://www.acfe.com/foundation.aspx.
Albrecht, W.S., Albrecht, C.C., & Albrecht, C.O. (2004). Fraud and corporate executives: Agency, stewardship and broken trust. *Journal of Forensic Accounting, 5*(1), 109–130.
Annisya, Mafiana. (2016). Fraud Detection of Financial Statements Using Fraud Diamond. *Business and Economic Journal*, 4(1), 72–89.
Bologna, J., & Lindquist, R.J. (1995). *Fraud auditing and forensic accounting: New tools and techniques*. Hoboken, NJ: John Wiley & Sons.
Boynton, P.R. (2001). Quality assurance and audit. In *The prevention and treatment of pressure ulcers* (pp. 231–246). Edinburgh, UK: Mosby.
Boediono, G.S. (2005). Quality of profit: Study of the effect of corporate governance mechanisms and the impact of earnings management using path analysis. *Simposium Nasional Akuntansi VIII*, 9, 175–194.
Cressey, D.R. (1953). Other people's money: A study of the social psychology of embezzlement. *American Sociological Review, 19* (3).
Chen, K.Y., Elder, R.J., & Hsieh, Y.M. (2007). Corporate governance and earnings management: The implications of corporate governance best-practice principles for Taiwanese listed companies. *Journal of Contemporary Accounting and Economics, 3*(2), 73–105.
Chow, C.W., & Rice, S.J. (1982). Qualified audit opinions and auditor switching. *Accounting Review*, 57 (2), 326–335.
Dechow, P.M., Sloan, R.G., & Sweeney, A.P. (1995). Detecting earnings management. *Accounting Review*, 70 (2), 193–225.
DeFond, M.L. (1992). The association between changes in client firm agency costs and auditor switching. *Auditing, 11*(1), 16–31.
Drew, J.M., & Drew, M.E. (2010). Establishing additionality: Fraud vulnerabilities in the clean development mechanism. *Accounting Research Journal, 23*(3), 243–253.
Eisenhardt, K.M. (1989). Agency theory: An assessment and review. *Academy of Management Review, 14*(1), 57–74.
Fala, Dwi Yana Amalia S. 2007. The Effect of Accounting Conservatism on Corporate Equity Valuation Moderating by Good Corporate Governance. *Simposium Nasional Akuntansi X*, 1–23.
Farida, Y.N., Prasetyo, Y., & Herwiyanti, E. (2010). The influence of corporate governance implementation on earnings management in assessing financial performance in banking companies in Indonesia. *Jurnal Bisnis dan Akuntansi, 12*(2), 69–81.
Hudaib, M., & Cooke, T.E. (2005). The impact of managing director changes and financial distress on audit qualification and auditor switching. *Journal of Business Finance & Accounting, 32*(9–10), 1703–1739.

Jensen, M.C., & Meckling, W.H. (1976). Theory of the firm: Managerial behavior, agency costs and ownership structure. *Journal of Financial Economics*, *3*(4), 305–360.

Khusnul, Y. (2012). Influence of good corporate governance and compliance reporting mechanism to company's financial performance (empirical study of banking company in Indonesia). (Doctoral dissertation, UPN "Veteran" Yogyakarta, Indonesia).

Krishnan, J., & Krishnan, J. (1997). Litigation risk and auditor resignations. *Accounting Review*, 72 (4), 539–560.

Loebbecke, J.K., Eining, M.M., & Willingham, J.J. (1989). Auditors' experience with material irregularities: Frequency, nature, and detectability. *Auditing A Journal of Practice & Theory*, *9*(1), 1–28.

Messier, W.F., Glover, S.M., & Prawitt, D.F. (2006). Auditd an Assurance Service. Jakarta, Indonesia: Penerbit Salemba Empat.

Lou, Y.I., & Wang, M.L. (2009). Fraud risk factor of the fraud triangle assessing the likelihood of fraudulent financial reporting. *Journal of Business & Economics Research*, 7 (2), 61–78.

Midiastuty, P.P., & Machfoedz, M.U. (2003). Analysis of corporate governance mechanism relationships and earnings management indications. *National Accounting Symposium VI*, 176–186.

Mudjiono. (2010). The effect of profit income actions against market reactions with auditor quality and managerial ownership as a variable moderator. *Eksplanasi, 5 (2)*, 1–11.

Rezaee, Z., & Riley, R. (2010). Financial statement fraud: Prevention and detection. Hoboken, NJ: John Wiley & Sons.

Romney, M.B., & Steinbart, P.J. (2014). *Accounting information systems*. Upper Saddle River, NJ: Pearson Higher Education.

Sabeni, A. (2005). The role of accountants in enforcing the good corporate governance principles in companies in Indonesia (perspective review of agency theory). Proposal speech at Guru Besar Universitas, Diponegoro, Indonesia.

Schwartz, K.B., & Menon, K. (1985). Auditor switches by failing firms. *Accounting Review*, 60 (2), 248–261.

Shelton, Austin. (2014). Analysis of Capabilities Attributed to the Fraud Diamond. Undergraduate Honors These. Paper 21. Available at http://dc.etsu.edu/honors/213, 25 September 2017.

Shu, S.Z. (2000). Auditor resignations: Clientele effects and legal liability. *Journal of Accounting and Economics*, *29*(2), 173–205.

Singleton, T. (2006). Fraud auditing and forensic accounting. Toronto, Canada: John Wiley and Sons.

Skousen, C.J., Smith, K.R., & Wright, C.J. (2009). Detecting and predicting financial statement fraud: The effectiveness of the fraud triangle and SAS No. 99. In *Corporate Governance and Firm Performance*, (pp. 53–81). Bingley, UK Emerald Group.

Sorenson, J.E., Grove, H.D., & Selto, F.H. (1983). Detecting management fraud: An empirical approach. *Symposium on Auditing Research, 5 (11)*, 73–116.

Statement of Auditing Standard (SAS) No. 99: Consideration of Fraud in a Financial Statement Audit, commonly abbreviated as SAS 99, is an auditing statement issued. *The Auditing Standards Board of the American Institute of Certified Public Accountants* (*AICPA*) in October, 2002.

Stice, J. D. (1991). Using financial and market information to identify pre-engagement factors associated with lawsuits against auditors. *Accounting Review, 66 (3)*, 516–533.

St. Pierre, K., & Anderson, J.A. (1984). An analysis of the factors associated with lawsuits against public accountants. *Accounting Review, 59 (2)*, 242–263.

Sukirman & M.P. Sari. (2013). Fraud Triangle Based Fraud Detection Model (Case Study in Public Companies in Indonesia. *Accounting and Auditing Journal, 9 (2)*, 199–225.

Suparlan, & W. Andayani. (2010). Empirical analysis of replacement of public accountant offices after mandatory audit rotation. *National Accounting Symposium XIII*, 1–24.

Ujiyantho, M.A., & Pramuka, B.A. (2007). The mechanisms of corporate governance, earnings management and financial performance. *National Accounting Symposium X*, 1–26.

United States Senate, (1976). The accounting establishment. Prepared by the Subcommittee on Reports, Accounting and Management of the Committee on Governmental Affairs Washington, DC: US Government Printing Office.

Vermeer, T.E. (2003). The Impact of SAS No. 82 on an Auditor's Tolerance of Earnings Management. *Journal of Forensic Accounting,* Vol. 14, 21–34.

Well, D.N. (2007). Accounting for the effect of health on economic growth. *The Quarterly Journal of Economics, 122 (3)*, 1265–1306.

Business Innovation and Development in Emerging Economies – Trinugroho & Lau (Eds)
© 2019 Taylor & Francis Group, London, ISBN 978-1-138-35996-3

The impact of debt covenants violation on audit fee

Jayanti Kania & Fitriany
Universitas Indonesia, Depok, Indonesia

ABSTRACT: This study aimed to prove that debt covenant violations have impact on firm's audit fee. The samples used in this study were loan agreements owned by non-financial public companies listed on the Indonesia Stock Exchange from 2012 to 2015. This study found evidence that firm's audit fee increases when company violates debt covenants. This study found that the debt covenant violations led to an increase in audit fees in the first year after the violation. External auditors considered debt covenant violations as an indicator of financial distress that increases audit risk. Auditor's risk of litigation also increases due to debt covenant violations. Furthermore, company's risk also increases due to debt covenant violation.

Keywords: debt covenant, debt covenant violation, audit fee

1 INTRODUCTION

Debt financing is one of the important activities in business that may result in conflict between shareholders and debtholders. Conflict between shareholders and debtholders is one of agency problems that may occur in a company. Both parties have different interests that may be conflicting. In this type of conflict interest, the management can transfer the company's wealth from debtholders to shareholders by making decision that's beneficial to shareholders. To safeguard debtholders' interest, there are two mechanisms that can be applied: (1) demanding high interest rates as compensation for debtholders' losses during the event of conflict interest or (2) restricting borrower's actions and forcing it to maintain certain financial conditions. The second mechanism is known as protective covenants.

As borrower, company will prefer the use of protective covenants in loan agreement because it allows the company to pay lower interest rate. However, violation of the agreed debt covenants can lead to costly consequences for the company, such as to experience a technical default, to pay the loan faster and to pay fines (Smith, 1993). Debt covenant violations may also result in higher costs to refinance and restructure the loan agreement. Futhermore, companies that violate debt covenant will also get negative market reactions.

Although debt covenant violations may negatively impact the company, previous researches show that debt covenant violations also encourage a positive change in corporate governance practices. Chava, Livdan, & Purnanandam (2009) found that debt covenant violations can be used as an indicator of an increase in company's agency problem so that the violation should be able to trigger improvement in governance practices. Ferreira, Ferreira, & Mariano (2018) found that debtholders were more involved in independent directors' election if the company previously violated the debt covenants. Nini, Smith, & Sufi (2012) also found a significant changes in company's management when debt covenant violations occurred.

The increase in company's agency problem can be solved by improving audit and monitoring activities. Increase in audit and monitoring activities can mitigate the agency problems occurred due to debt covenant violations. The demands for audit and monitoring activities are in line with an increase in company's agency problems (Defond & Zhang, 2014).

The significance of audit and monitoring activities in agency problem mitigation raises the question of how debt covenant violations affect company's audit fee. Previous studies related to debt covenant violations are only focused on their relation with earnings management. Due to the lack of studies on the relationship between debt covenant violations and audit activities, this study is aimed to prove that debt covenant violations have effect on company's audit activities.

Previous studies related to debt covenant violations and audit activities have only been conducted in United States. Jiang & Zhou (2017) examined the role of audit activities in resolving debt covenant violations in public companies. They found an increase in company's needs of audit services after debt covenant violations in order to ensure that there would be no further violations in the future. This increase in needs of audit services led to an increase in audit fees. The study also found that the design and tightness of debt covenant also influenced the increase in audit fees in the event of debt covenant violations. Furthermore, the study also found that increase in audit fees due to debt covenant violations could also decrease the increase in subsequent interest rates. The study by Robin, Wu, & Zhang (2017) also found that the use of qualified auditors reduces the increase in subsequent interest rates after debt covenant violations. Gao, Khan, & Tan (2013) also found an increase in audit fees in the event of debt covenant violations. Furthermore, Bhaskar, Krishnan, & Yu (2017) also found an increase in audit fees and an increase in company's probability to obtain a going concern opinion from the external auditor as well as an increase in external auditor's resignation probability when the company violated debt covenant.

In Indonesia the study related to debt covenant violations and audit activities is still rare. Naturally the impact of debt covenant violations on audit activities in Indonesian companies will be different from American companies. Factors such as regulation, disclosure requirements, and litigation environment have different influence on audit activities in various countries (Taylor & Simon, 1999). Therefore, this study would like to examine how the debt covenant violations influence audit activities in company from a developing country using data from non-financial public companies in Indonesia.

Through the replication of previous study from Jiang & Zhou (2017), this study is aimed to examine how debt covenant violations will affect the audit fees of Indonesian companies. With this research, it is expected that literature related to debt covenant violations in Indonesia can be increased.

This paper is divided into the following sections. **Section 2** describes the theory, prior researches, and strategy development. **Section 3** describes the research design and **Section 4** presents the data. **Section 5** analyzes the test results. Finally, **Section 6** presents the conclusions and limitations of this study.

2 LITERATURE REVIEW AND HYPOTHESIS DEVELOPMENT

2.1 *The impact of covenant violations on audit fees*

Previous studies (Gao et al., 2013; Jiang & Zhou, 2017; Robin et al., 2017; Bhaskar et al., 2017) found debt covenant violations led to an increase in company's audit fees. This occurs because the auditor viewed debt covenant violations as an indicator of financial difficulties that directly influenced auditor risk (Pratt & Stice, 1994). Although the violations were committed by companies who were not experiencing financial difficulties, but such violations also increased company's financial distress risk. In addition, debt covenant violations also increased the auditor risk of litigation. Auditor's litigation risk will increase in the event that the shareholder value of the firm decreases (Lys & Watts, 1994). To decrease the audit risk, auditor had to improve the audit process and activities. Therefore, the auditor would request a higher audit fee

From the company side, debt covenant violations was an indicator of an increase in company's agency problems and would require improvements in corporate governance practices (Nini et al., 2012). Debt covenant violations transferred control of the company to the debtholders (Dewatripont & Tirole, 1994; Aghion & Bolton, 1992) which would lead to an increase in agency problems. To resolve these problems, more levels of oversight were required. The need for more oversight and monitoring would encourage company to request for better audit services quality provided by auditors (Defond & Zhang, 2014). Companies were also willing to pay higher audit fees, if the audit services were high quality (Hay et al., 2006). Therefore, audit fee would rise. Hence the hypothesis can be formulated that:

H1:   There is an increase in audit fees when the company violates debt covenant in the loan agreement.

3   RESEARCH METHODOLOGY

3.1   *Overview*

This study used purposive sampling approach with predetermined criteria. The criteria used in the sample selection are as follows: (1) All non-financial companies listed in Indonesia Stock Exchange in the period 2012 to 2015 and publish annual financial statements, (2) Companies that have reporting period ended on December 31, (3) Companies that disclose information on audit fees (4) Companies that disclose the detailed loan agreements information in annual report, and (5) Company that disclose public informations to collect data for control variables. The data in this study were collected from official sources such as Annual Report or Corporate Website as well as from software such as Datastream or Thomson Reuters Eikon.

3.2   *Research method*

The research model used in this study is a replication previous study by Jiang & Zhou (2017) with modifications by excluding credit rating, foreign, SOX as control variables. The study also didn't examine the effect of the design and tightness of debt covenants that were violated on company's audit fees due to the limited sample being used. In addition, the study also used bankruptcy predictions and the number of subsidiaries as one of the control variables.

To test H1 hypothesis, Model 1 will be used in this study with audit fees as the dependent variable.

$$\text{FEEAUD}_{it} = \alpha_0 + \alpha_1\text{EVENT}_{it} + \alpha_2\text{POST1}_{it} + \alpha_3\text{POST2}_{it} + \alpha_4\text{SIZE}_{it} + \alpha_5\text{SEG}_{it} + \alpha_6\text{SUB}_{it}$$
$$+ \alpha_7\text{ROA}_{it} + \alpha_8\text{LOSS}_{it} + \alpha_9\text{INVREC}_{it} + \alpha_{10}\text{LEVv}_{it} + \alpha_{11}\text{MTB}_{it} + \alpha_{12}\text{CR}_{it}$$
$$+ \alpha_{13}\text{ALTMAN}_{it} + \alpha_{14}\text{BIG}_{it} + \alpha_{15}\text{GC}_{it} + \varepsilon \qquad \textbf{(Model 1)}$$

FEEAUD :   Corporate audit fee in the form of natural logarithm.
EVENT   :   An indicator valued 1 if the company breaches the debt covenant in year t and 0 if no violation occurs.
POST1   :   An indicator valued 1 if the company violates the debt covenant in year t + 1 and 0 if no violation occurs.
POST2   :   An indicator valued 1 if the company violates the debt covenant in year t + 2 and 0 if no violation occurs.
SIZE   :   Total assets of the company in the form of natural logarithm.
SEG   :   The number of business segments owned by the company in the form of natural logarithm.
SUB   :   The number of subsidiaries owned by the company in the form of square root.
ROA   :   Return on Assets (ROA) of the company.
LEV   :   Leverage of the company calculated with Total Liabilities/Total Assets.
LOSS   :   An indicator is worth 1 if the company loses and 0 if it does not lose money.
INVREC :   Total receivables and inventorie divided by total asset values.

49

MTB	:	Market Value-to-Book Value of the company.
CR	:	The current ratio of company.
ALTMAN	:	The corporate bankruptcy prediction indicator valued 1 if the firm has z-score altman <2.073 and 0 if the value of altman z-score >2.073.
BIG	:	An indicator valued 1 if the company's external auditor is Big 4 and 0 if not.
GC	:	An indicator valued 1 if the firm has a going concern issue in the auditor's opinion and 0 if it does not get the going concern issue
α_0 and α_n	:	Cut points and parameters of each research variable
ε	:	The residual value of the research model estimate

3.2.1 *Operationalization of variables*
3.2.1.1 *Dependent variable*
Corporate audit fee (FEEAUD) as the dependent variable which represent the costs paid by the company to external auditors for audit services. The audit fees related data were collected from annual report. In this study audit fees in year t were used in the form of natural logarithm.

3.2.1.2 *Independent variable*
The main independent variable used in this study is an indicator of the debt covenant violations. Consistent Study of Jiang & Zhou (2017), this study specifically used four independent variables as indicators of debt covenant violations. The four independent variables were dummy variable with value of 1 if the company violates its debt covenants or is experiencing a technical default and 0 if the opposite condition occurs. All independent variables used in this study are predicted to have a positive sign.

The first variable is EVENT which is indicator of company's debt covenant violation in year t. The variable POST1 is used as an indicator of potential debt covenant violations in year t + 1 or in the first year after the violation. The variable POST2 is used as an indicator of potential debt covenant violations in year t + 2 or in the second year after the violation. All three variables were used in all models in this study. While variable PRE was used as an indicator of debt covenant violation in year t − 1 or in the year prior to the violation. This variable was used only in model related to the activities and the effectiveness of the audit committee and the board of commissioners.

To obtain an indicator of covenant violation conditions, this study compared the actual results of the ratios used as covenant to the ratios agreed in the loan agreement. The study also examined whether the loan agreements disclosed in Notes of Financial Statement contained the keywords "waiver", "renegotiation", or "default". If there was one of these keywords then the loan agreement was categorized as technical default.

3.3 *Data analysis*

In conducting data analysis, descriptive analysis and regression for panel data are fixed-effect. Outliers are overcome by using the winsorization method at the outmost 3% of each continuous variable used in this study in order to provide the real impact of all variables.

4 RESULTS

By using purposive sampling method with the criteria described in Section 3, Table 1 below presents the summary of sample selection results.

Based on Table 2, it can be seen in the average audit fee paid by the company is Rp1,947,309,633 (FEEAUD variable). The lowest audit fee paid by PT Pelayaran Nelly Dwi Putri Tbk in 2013 was amounted Rp 125,000,000. While the highest audit fees were paid by PT Telekomunikasi Indonesia Tbk in 2015 was amounted Rp 34,400,000,000. For SIZE variables, it can be seen that the average company has a total asset value of Rp15.952.454,531,984, with PT Yanaprima Hastapersada Tbk as the smallest company with total assets of Rp279,189,768,590 in 2015 and PT Telekomunikasi Indonesia Tbk as the largest company with total assets of Rp166,173,000,000,000 in 2015.

Table 1. Summary of sample selection results.

Information	Number of companies			
	2012	2013	2014	2015
Companies listed on the Indonesia Stock Exchange	447	484	498	507
Less companies engaged in the financial services sector (JASICA Code 8)	(71)	(74)	(79)	(87)
Less Companies whose annual report were not found and didn't ended on December 31	(10)	(6)	(18)	(27)
Less companies that didn't disclose audit fees	(272)	(263)	(253)	(232)
Number of Companies that disclose audit fees	**94**	**141**	**148**	**161**
Less Companies that did not have loan agreements or did not disclose information about accounting based covenant	(64)	(101)	(98)	(115)
Total Companies	**30**	**40**	**50**	**46**
Number of Loans	**83**	**93**	**112**	**132**
Total Number of Loans in Sample		**420**		

Table 2. Statistic descriptive.

Variable	Mean	Std. Dev	Min	Max
FEEAUD (Rp)	1,947,309,633	3,945,727,668	125,000,000	34,400,000,000
SIZE (Rp)	15,952,454,531,984	21,972,972,471,172	279,189,768,590	166,173,000,000,000
SEG	53.928	30.556	1	28
SUB	43.976	37.494	0	19
ROA	0.0353	0.0761	−0.2027	0.2789
INVREC	0.2647	0.2164	0.0005	0.8073
CR	15.500	0.9997	0.1933	7.798
LEV	0.5794	0.1679	0.1454	0.9781
MTB	22.744	16.897	0.1263	12.1807

Dummy variable	Dummy = 1	Dummy = 0
EVENT	0.4333	0.5667
POST1	0.4476	0.5523
POST2	0.4738	0.5261
LOSS	0.119	0.8810
ALTMAN	0.6214	0.3786
BIG	0.2786	0.7214
GC	0.0167	0.9833
FIN	0.5738	0.4262

N: 420.

For the SEG control variable, it can be seen that the company has an average of five different business segments. The company that has the most business segments is PT Citra Tubindo Tbk with 28 (twenty eight) business segments. The segment consists of business segments based on business activities and business segments based on geography. Whereas, from the SUB variable, it can be seen that the average company has four subsidiaries. The company that has the most subsidiaries is PT Jaya Agra Wattie Tbk with 19 (nineteen) subsidiaries.

Seeing the independent variable, namely the condition of the violation of the covenant itself, there is an increase from the percentage of companies that commit violations from year to year. There was an increase of 43.33% of the observation samples that committed violations in year t to 44.76% of the observation samples that committed violations in year t + 1. Whereas in year t + 2, there were 47.38% of the observation samples that committed violations. The lowest lending conditions occur in year t – 1 which is equal to 39.52% of the observation sample.

While, from LOSS control variable, it can be seen that 11.90% of the observation sample has a negative net income during the observation period. For GC variables, it can be seen that only 1.67% of the observation sample obtained the going concern opinion from the auditor and for the ALTMAN variable it can be seen that 62.14% of the observation sample was predicted not experiencing financial distress. The use of BIG 4 auditors can be seen from the BIG variable, where it turns out that only 27.86% of the observation samples use BIG KAP 4. The low in use of KAP BIG 4 in this research observation may occur because the companies audited by KAP BIG 4 have loans that less than companies not audited by KAP BIG 4.

4.1 *Analysis of debt covenants demographic*

From the 420 sample loan agreements obtained, there are five ratios that are commonly used as accounting-based covenants in Indonesia, namely Current Ratio, Debt to Equity Ratio, Debt Service Coverage Ratio, Interest Service Coverage Ratio, and Debt/EBITDA ratios. The ratio of Debt to Equity (DER) is the most frequently used as debt covenant in loan agreement. DER ratio is widely used because this ratio indicate the company's solvency risk by directly presenting company's capital structure. Aside from DER, Debt Service Coverage Ratio and Interest Service Coverage Ratio is the second and third most used ratio. Both ratios are used because the debtholders want the company to maintain its ability to pay interest. The debtholders also use the Current Ratio as covenant to force the company to maintain its liquidity. Figure 1 shows the number of accounting-based ratios used as covenants in the loan agreement.

Figure 2 shows that 41% of loan agreements use two ratios as accounting based covenants and 36% of loan agreements use more than two ratios, whereas only 23% of loan agreements use only one ratio. This result indicates that loan agreements held by companies in Indonesia are quite tight because more than 50% of loan agreements use two or more than two financial ratios as covenants.

Figure 3 shows that the debt covenant violations continue to increase from year to year. Most violations occurred in 2015, in which there were 63 violations in one year. This study argues that the increase in debt violations is related to the condition of the Indonesian

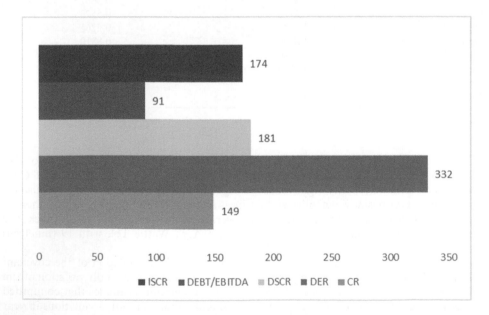

Figure 1.   The number of accounting based ratios used as debt covenant.

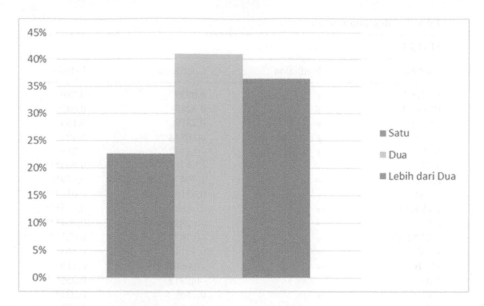

Figure 2.   Financial covenant intensity

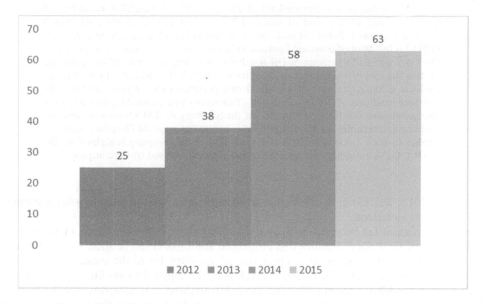

Figure 3.   The number of debt covenant.

economy that is experiencing a slowdown and rupiah exchange rate that tend to weaken in the period 2014–2015.

5   DISCUSSION

Based on the test results in Table 3, Variable POST2 and EVENT have insignificant coefficients and variables POST1 have significant coefficients in 10% level. This result indicates that there is an increase in audit fee in the first year after the debt covenant violations conditions. This result is consistent with the results of previous studies (Jiang & Zhou, 2017;

53

Table 3. Regression result.

Model 1

Variable	Prediction	Coefficient	P-*Value*
EVENT	**+**	**0,0939**	**0,289**
POST 1	**+**	**0,3285**	**0,092***
POST 2	**+**	**0,2310**	**0,158**
SIZE	+	−0,0814	0,406
SEG	+	−0,4577	0,288
SUB	+	12,471	0,000***
ROA	+	1,354	0,079*
LOSS	+	0,6516	0,013***
INVREC	+	23,568	0,013***
CR	−	−0,2818	0,004***
ALTMAN	+	−0,7573	0,001***
LEV	+	0,7914	0,163
MTB	+	−0,0607	0,115
BIG	+	−0,2312	0,056*
GC	+	0,7868	0,033**
CONS		191,966	0,000***
PROB	**0,000***	**R^2**	**67,07%**

***, **, * indicates significance level of 1%, 5%, 10%. FEEAUD is an audit cost model variable (in the form of ln). EVENT is an indicator of covenant breach condition in year t. POST1 is indicator of covenant breach condition in year t + 1. POST2 is indicator of covenant breach condition in year t + 2. SIZE is company size. SEG is a natural logaritman of the number of business segments of the company. SUB is the root of the number of subsidiaries. ROA is the ratio of return on assets LOSS is a dummy variable of 1 if the firm experience a loss in year t and 0 if the opposite condition. INVREC is the sum of inventory plus receivables divided by the total asset value. CR is the current ratio of the company. ALTMAN is an indicator of corporate bankruptey. LEV is the ratio of corporate leverage. MTB is the company's market-to-book value ratio. BIG is an indicator if the company is audited by Big KAP 4. GC is an indicator of going concern opinion received by the company.

Gao et al., 2013; Bhaskar et al., 2017) that also found an increase in audit fees in the event of debt covenant violations.

Increase in audit fee occurred because the auditor viewed the debt covenant violation as an indicator of financial difficulties that influenced audit risk (Pratt & Stice, 1994). Debt covenant violations also increased the auditor risk of litigation due to the greater pressure from the debt holders. The higher the risk of the company, the higher the risks faced by the auditor.

Furthermore, debt covenant violations also led to an increase in company risk. To decrease the risk, more monitoring activities were required in order to prevent the further violations in the future. The need for more supervisions will encourage companies to request increase audit quality provided by the external auditors (Defond & Zhang, 2014). This will result in increase of company's audit fees.

This study also found a lag in the increase of audit fees. The lag implied that the auditor didn't immediately react to the increased risks faced by the company. Furthermore, the lag also occurred because the process of engagement with the auditor has been resolved so that the audit fees have been settled and the company couldn't immediately request improvements in audit services quality.

This study found that control variables, SUB, ROA, LOSS, INVREC, GC proved to have a significant positive effect on audit fees. These results indicate that the more the number of subsidiaries, the higher the value of inventory and receivable, will made audit fee higher. This is because the effort that must be carried out by the auditor is greater if the audited company has more subsidiaries and more accounts receivable and inventory. The higher the return on company assets, the auditor will usually charge a higher audit fee. Audit fees are also higher for companies that experience loss and obtain going concern opinions. this is due to the greater risks faced by the auditor.

6 CONCLUSION

This study is aimed to prove the impact of debt covenant violations on audit fees. This study found that the debt covenant violations led to an increase in audit fees in the first year after the violation. External auditors considered debt covenant violations as an indicator of financial distress that increases audit risk. Auditor's risk of litigation also increases due to debt covenant violations. Furthermore, company's risk also increases due to debt covenant violation. To decrease the risk, more oversight activities are required to prevent the further violations in the future. The need for more oversight encouraged companies to request improvements in audit services quality.

6.1 *Limitations of research*

This study has limitations and can be considered for further research, This study only used companies other than companies engaged in the financial industry during the period 2012–2015 due to the limited disclosure of audit fee information before the year 2012. Further study can examine longer time periods.

REFERENCES

Aghion, P., & Bolton, P. (1992). An incomplete contracts approach to financial contracting. *The review of economic Studies*, 59(3), 473–494.

Bhaskar, L.S., Krishnan, G.V., & Yu, W. (2017). Debt covenant violations, firm financial distress, and auditor actions. Contemporary Accounting Research, 34(1), 186–215.

Chava, S., Livdan, D., and Purnanandam, A. (2009). Do shareholder rights direct the cost of bank loans? Review of Financial Studies, 22(8):2973–3004.

DeFond, M., & Zhang, J. (2014). A review of archival auditing research. Journal of Accounting and Economics, 58(2), 275–326.

Dewatripont, M., & Tirole, J. (1994). A theory of debt and equity: Diversity of securities and manager-shareholder congruence. The quarterly journal of economics, 109(4), 1027–1054.

Ferreira, D., Ferreira, M.A., & Mariano, B. (2018). Creditor control rights and board independence. *The Journal of Finance*, 73(5), 2385–2423.

Gao, Y., Khan, M., & Tan, L. (2017). Further evidence on consequences of debt covenant violations. *Contemporary Accounting Research*, 34(3), 1489–1521.

Hay, D.C., Knechel, W.R., & Wong, N. (2006). Audit fees: A Meta-analysis of the effect of supply and demand attributes. Contemporary accounting research, 23(1), 141–191.

Jiang, L., & Zhou, H. (2017). The role of audit verification in debt contracting: evidence from covenant violations. Review of Accounting Studies, 22(1), 469–501.

Li, H., Meng, L., Wang, Q., & Zhou, L.A. (2008). Political connections, financing and firm performance: Evidence from Chinese private firms. Journal of development economics, 87(2), 283–299.

Lys, T., & Watts, R.L. (1994). Lawsuits against auditors. Journal of Accounting Research, 65–93.

Nini, G., Smith, D.C., & Sufi, A. (2012). Creditor control rights, corporate governance, and firm value. *The Review of Financial Studies*, 25(6), 1713–1761.

Pratt, J., & Stice, J.D. (1994). The effects of client characteristics on auditor litigation risk judgments, required audit evidence, and recommended audit fees. Accounting Review, 639–656.

Robin, A., Wu, Q., & Zhang, H. (2017). Auditor quality and debt covenants. Contemporary Accounting Research, 34(1), 154–185.

Smith Jr, C.W. (1993). A perspective on accounting-based debt covenant violations. *Accounting Review*, 289–303.

Taylor, M.H., & Simon, D.T. (1999). Determinants of audit fees: the importance of litigation, disclosure, and regulatory burdens in audit engagements in 20 countries. *The International Journal of Accounting*, 34(3), 375–388.

Banking

Business Innovation and Development in Emerging Economies – Trinugroho & Lau (Eds)
© *2019 Taylor & Francis Group, London, ISBN 978-1-138-35996-3*

Accounting and capital measurement of banking risk: Evidence from the Indonesian banking industry 2011–2015

Wilson Ruben Lumbantobing
Accounting Study Program, Faculty of Business, Sampoerna University, Surabaya, Indonesia

ABSTRACT: This study examines and provides empirical evidence for the relationship between accounting and capital market risk measurement for a sample of ten Indonesian banks traded on the Indonesia Stock Exchange from 2011 to 2015. By applying panel data regression analysis, the results show that loan loss provision, liquidity, equity to assets, and bank size all appear to be significantly associated with total risk. Equity to assets and size are also significantly related to systematic risk, while size is also significantly related to specific risk within the chosen banks during the period 2011 to 2015. Some suggestions are put forward relating to the banks which were used as samples in this research and the investors who may wish to invest in banking firms, and for facilitating academic research.

Keywords: accounting measures of risk, capital market measures of risk, total risk, systematic risk, specific risk

1 INTRODUCTION

As financial institutions, banks play a very important role in enhancing the distribution of economic growth in Indonesia. The development of the banking industry and its importance in the country's economic development leads banks to face increasingly complex risks. This increases the need for banks to apply appropriate risk management to minimize the risks associated with their banking activities.

Risk in the banking sector is a major issue which is often discussed in terms of financial stability. Recently, the Indonesian banking industry has been faced with more complex risks, as business development in Indonesia progresses and as several external factors have directly impacted on the economy of the country. Risks that should be anticipated are credit risk and inflation. In addition to external factors which affect the Indonesian economic situation, Domestic economic conditions also need to be considered to minimize risks (Bisnis.com, 2015).

To avoid or minimize the risks which face Indonesian banks, a measurement that can accurately represent the risks likely to arise is required. To choose a suitable measurement to evaluate the risk taking of banks, accounting information as well as capital market measures of risk can be used. The appropriate measure for assessment depends on the objective of the assessment and the conditions within which it is applied.

In the case of Indonesian banks, there are few previous studies which have observed the accounting measurement of risk, despite the fact that accounting measurement is particularly important because balance sheet information in Indonesian banking firms is able to reflect banking risk. Therefore, accounting information as presented in the balance sheet is appropriate for the measurement of banks' risk-taking activities.

As there are different types of accounting information presented in the financial statements of each bank, their management of bank risk and the link between accounting and capital market measures of risk may be different. The present study pools data from ten listed banks operating in Indonesia during 2011 to 2015, and finds that market risk measures are significantly related.

After testing the fit of statistical models for the study, the fixed-effects model is preferred. Loan loss provision (LLP), liquidity (LIQ), equity to assets (EQTA), and bank size all appear to be significantly associated with total risk. For systematic risk, equity to assets ratio (EQTA), and bank size are significant, whereas for specific risk as the dependent variable, only size shows a significant relationship.

2 LITERATURE REVIEW

Dhouibi and Mamogli (2009) examined the significant relationships between accounting and market risk measures in the emerging economy of Tunisia. The objective of the researchers was to examine the ability of the capital markets to reflect the risk undertaken by Tunisian listed commercial banks during the ten-year period of 1998–2007.

This research was conducted via a panel data methodology using a feasible generalized least squares estimator. The researcher used this tool to overcome the presence of serial correlation, contemporaneous (spatial) correlation, and heteroscedasticity. To measure capital risk, stock prices were used to indicate the total risk, the systematic risk, and the specific risk of each bank. These measures are the dependent variables, while the independent variables are accounting measures of credit, leverage, and liquidity risks.

The three dependent variables are regressed to several accounting measures of risk used to reflect total risk (SDROA and Z-score), leverage risk (EQTA and DEPEQ), liquidity risk (LIQATA), and credit risk (LLPGL, LLRGL and NPLGL).

The results from the feasible generalized least squares estimation indicate that when total return risk is used as the dependent variable, only LIQATA is significant and has a negative relationship. When systematic risk is used as the dependent variable, only the LLPGL variable is significant, again negatively. When specific risk is used as the dependent variable, EQTA, DEPEQ, and LIQTA show significant relationships. The relationships between LLPGL, LLRGL, and NPLGL and capital market risk measures are not significant and do not display the expected signs. In summary, in Dhouibi and Mamogli's (2009) research, the relationships between the accounting and the capital market measures of risk indicate that capital market risk measures do not accurately reflect the risk taken by Tunisian banks. It shows that bank specific risk is more important than systematic risk and indicates that the market is not able to reflect accurately the risks taken by banks. It can therefore be concluded that the market is not able to reflect the most important sources of risk in Tunisian banks.

Agusman et al. (2008) examined 46 locally incorporated, listed, commercial banks in Asia during the period 1998–2003, using panel data analysis and F-testing. The capital market measures of risk used as the dependent variables in this research were total risk, systematic risk, and non-systematic risk. These variables are regressed to six accounting measures designed to reflect total risk (SDROA), leverage risk (EQTA), credit risk (GLTA and LLRGL), and liquidity risk (LIQATA) and CVCSTF). The researchers also added country dummy variables (DCOUNTRY) to accounting risk measures to control for differences between banks. The results show that when total return risk is used as the dependent variable, SDROA and LLRGL show significant positive relationships. When non-systematic risk is used as the dependent variable, LLRGL and GLTA exhibit significant positive relationships. However, when systematic risk is the dependent variable, none of the coefficients are significant.

The results of this study indicate that firm specific risk is more important than systematic risk in the observed Asian countries. In Asian countries, banks are the primary source of financing, and so the behavior of banks is able to affect other firms in the market index. In this study, the researchers found that the accounting ratios explained a substantial portion of capital market risk.

Salkeld (2011) analyzed 326 publicly traded banks in the U.S. from 1978 to 2010 using accounting ratios and macroeconomic indicators, to investigate the effects of individual bank operations and changes in the economic environment. In this study, the dependent variables of size, equity to assets ratio (EQTA), loan loss provision (LLP), liquidity (LIQ), allowance for loan loss ratio, liquidity ratio, loan to asset ratio, growth in real GDP, growth in the money supply, and the interest rate are regressed against total risk (SDROE and SDROA).

The data are analyzed in four different models: 1) the base model, 2) a model with fixed-effects for time, 3) a model with fixed-effects for each firm, and 4) a model with fixed-effects for both time and each firm. For the interpretation of the results in this research, the researcher uses model 1. In this model, coefficient of size is negative and statistically significant when SDROE is the dependent variable. When using a regression with SDROA as the dependent variable, size has a positive and significant result. The equity to assets ratio is statistically significant in all models and has a negative sign for total risk. For loan loss allowance ratio, the coefficient is statistically significant and positive in all models. The liquidity ratio exhibits a positive relationship with total risk, while loan to asset ratio is positive and statistically significant to total risk for banks. For the dividend payout ratio, in most of the models, the coefficient was negative, meaning it is negatively related to total risk, but the results were highly insignificant. The relationship between changes in the economic environment and total risk is statistically significant, while the growth in the money supply variable is negatively related to total risk. For the interest rate spread variable, the result is statistically significant and has a positive relationship with total risk. These macroeconomic indicators yield the expected and significant results in most models that do not control for time.

As results show that changes in the economic environment are significant to a bank's total risk level, when indicators suggest the country is going through an economic downturn, banks can adjust their operations accordingly to protect against the higher risk level. Based on this study, the statistically significant accounting ratios that increase risk can be identified and bank management can closely monitor these measures. The statistically significant ratios from this study that reduce risk can be used as tools to strengthen the financial health of a bank and help it to be more resilient during a recession.

3 RESEARCH METHOD

This study tests empirically the relationship between capital market risk measures and accounting ratios by using the following general model:

$$\Upsilon_{i,t} = b_0 + b_1 X_1 + b_2 X_2 + b_3 X_3 + b_4 X_4$$

where Υ = total risk, specific risk, and systematic risk, respectively, and the independent variables are as described in Table 1. Table 1 presents the variables employed in the study along with their formulas and definitions. Standard capital market measures of risk are total risk,

Table 1. Variable operationalization.

Variable	Definition
Loan loss provision (LLP)	The amount of loans a bank does not expect to collect from customers
Liquidity (LIQ)	A measure of a bank's ability to absorb unexpected changes in its assets and liabilities
Equity to total assets (EQTA)	Identifies the amount of assets that shareholders contribute
Size	The size of a banking institution
Total risk (TR)	The risk that affects the entire banking industry
Systematic risk (β)	The risk that affects the entire market segment (e.g. economic crisis, recession, inflation)
Specific risk (SR)	Risk that is very specific to a company or small group of companies. (e.g. sudden strike by the employees of a company; a new governmental regulation affecting a particular group of companies)

specific risk, and systematic risk. These variables are regressed for six accounting measures designed to reflect credit risk (LLP), liquidity risk (LIQ), and leverage risk (EQTA).

The data were gathered from the online archives of the Indonesia Stock Exchange (IDX) and from Yahoo Finance. The accounting data collected from the financial statements and the annual reports for each bank are used to analyze and calculate the variables. Also used in this study are data comprising weekly stock market prices and IDX composite values (Joint Stock Price Index) from January 2011 to December 2015, providing 50 observations in total.

Baltagi (2005) states that panel data can minimize bias by providing information on ordering of time and controls for individual heterogeneity, as well as specifying the time-varying relationship between dependent and independent variables. It also consists of simpler effect measurement compared to time-series or cross-sectional data and is more informative, has less collinearity, and more degrees of freedom, allowing the estimation to be more efficient. This study uses a panel data methodology, together with a Hausman's test used to compare the fixed-effects model with the random-effects model.

4 RESULTS AND DISCUSSION

The descriptive statistics for each of the dependent and independent variables analyzed are presented in Table 2. The dependent variables in this research are TR, β, and SR, while the independent variables used are LLP (X1), LIQ (X2), EQTA (X3), and size (X4).β negative means an investment moves in the opposite direction from the stock market. When the market rises, a negative β investment generally falls. The table below shows a negative β in the minimum β, during the period of 2011–2015 both BRI and BNI experienced negative β in year 2014 and 2015 respectively. BTPN and DANAMON experienced negative β 3 years respectively during the same period while DANAMON in experienced negative β in four years.

This study runs the regression using a panel data methodology. From testing the fittest of the models using the Hausman Test, the fixed effect model is used with the ordinary least squared (OLS) method. To study the relationship between accounting measurement and capital market measurement of risk, the data are analyzed using four different models. In Model 1, all independent variables are regressed against TR, β, and SR, while in Model 2, only LLP is regressed against TR, β, and SR. In Model 3, the independent variable regressed against TR, β and SR is LIQ only. Finally, in Model 4, only EQTA is regressed against TR, β, and SR. The results of the four models for each dependent variable are presented in Table 3.

Table 3 shows that when TR is used as the dependent variable, all of the independent variables are significant in relation to it. In Model 2, the result of LLP is similar to Model 1, as the LLP variable is significantly related to TR and exhibits a positive relationship. From the results of this model, it is found that the LIQ variable is also significant toward TR which is consistent with Model 1. However, statistically the coefficient shown by the LIQ variable exhibits a negative relationship with TR in Model 3. The result for EQTA in Model 4 is similar to Model 1, in that the EQTA variable is significantly related to TR and exhibits a negative relationship toward it.

Table 2. Descriptive statistics.

	TR	β	SR	LLP	LIQ	EQTA	SIZE
MEAN	0.0473	0.1681	0.0378	0.0221	0.8225	0.1238	8.129
MEDIAN	0.0408	0.1119	0.0351	0.0197	0.8350	0.1222	8.241
MAXIMUM	0.2265	1.4011	0.0782	0.0559	0.9999	0.2376	8.959
MINIMUM	0.0190	(0.5851)	0.0138	0.0008	0.6042	0.0597	6.706
ST. DEV.	0.0307	0.3734	0.0162	0.0118	0.0951	0.0379	0.6250
OBSERVATION	50	50	50	50	50	50	50

LIQ is also significant, but statistically the coefficient exhibits a negative relationship. The result for EQTA in Model 4 is insignificant toward β which is inconsistent with Model 1. Statistically, the coefficient shown by the EQTA variable exhibits a negative relationship with β in Model 4, similarly to Model 1.

As shown in Table 4, when β is used as the dependent variable, there are two independent variables that show significant results toward it: EQTA and size. In Model 2, the result for LLP is similar to Model 1, in that LLP is significantly related to β and exhibits a positive relationship toward it. From the results in Model 3, it is found that Table 5, shows that for Model 1,

Table 3. Total risk.

Variables	Model 1	Model 2	Model 3	Model 4
LLP	1.643509 (0.0000)***	1.33836 (0.0067)***		
LIQ EQTA	0.098083 0.0085)***		-0.053849 (0.0036)***	
SIZE	-0.170338 (0.0011)***			-0.233298 (0.0010)***
	-0.064 (0.0031)***			
Observations	50	50	50	50
Adjusted R-squared	0.400852	-0.042978	0.277704	-0.041571
F-statistic	0.001379	0.631108	0.008584	0.625466

Table 4. Systematic risk (β).

Variables	Model 1	Model 2	Model 3	Model 4
LLP	6.938076 (0.5600)	8.528172 (0.4117)		
LIQ	-0.238218 (0.8146)		-0.713071 (0.1545)	
EQTA	1.832443 (0.0002)***			-1.437787 (0.4214)
SIZE	-1.311886 (0.0112)**			
Observations	50	50	50	50
Adjusted R-squared	0.227978	0.137213	0.200481	0.136591
F-statistic	0.038313	0.097384	0.036311	0.09825

Table 5. Specific risk.

Variables	Model 1	Model 2	Model 3	Model 4
LLP	0.361255 (0.1199)	0.465802 (0.0148)**		
LIQUIDITY	8.60E-05 (0.9974)		-0.041747 (0.0319)**	
EQTA	-0.01651 (0.8149)			-0.112982 (0.031)**
SIZE	-0.031635 (0.0001)***			
Observations	50	50	50	50
Adjusted R-squared	0.204352	0.178312	0.230101	0.234912
F-statistic	0.026766	0.052316	0.021558	0.019735

only the independent size variable is significant toward SR. The result for Model 2 shows that LLP is significant and is positive. The result for Model 3 is that the LIQ variable has a significant value and negative, in contrast to Model 1. In Model 4, it is found that EQTA has a different result to Model 1 in that it has a significant and negative sign. Based on the results of this research, it is better to focus on accounting measures rather than capital market risk measurements to appropriately assess the risk taking of Indonesian banks.

5 CONCLUSIONS

The focus of this research is to examine whether accounting risk performance effects the capital market measures of bank risk. To measure the relationship between accounting and capital market measurements, this study uses four measurements of accounting: loan loss provision (LLP), liquidity (LIQ), equity to assets ratio (EQTA), and size, together with three measurements of capital markets: total risk (TR), systematic risk (β), and specific risk (SR). The sample of this research is ten banks listed in the IDX for the period 2011 to 2015. The methodology used is panel data using a fixed-effects approach. The relationships between accounting and capital market measures are presented in Table 6. Positive relationship means the increase in independent variable will increase the value of dependent variable while the negative sign means an increase in independent will decrease the value dependent variable and vice versa.

From the results presented in Table 6 it can be concluded that:

1. Loan loss provision (LLP) has significant positive influence on total risk (TR) in the ten listed Indonesian banks over the period 2011 to 2015. LLP is related to TR because when a bank's LLP increases there will be more loans that do not perform as expected. This increases the firm's risk as it will experience a loss.
2. Liquidity (LIQ) has significant positive influence on total risk (TR) in the studied banks over the period 2011 to 2015. LIQ is related to TR because when banking firms have excess liquidity they tend to use their excess capital inefficiently, reducing the resources available for better-performing segments.
3. Equity to assets ratio (EQTA) has significant negative influence on total risk (TR) but positive influence on systematic risk (β) in the studied banks. EQTA is negatively related to TR because banking firms which utilize higher EQTA to finance their operations experience less risk in terms of the cost of interest expenses for borrowed money. EQTA is positively related to β because banking firms which utilize higher EQTA have higher likelihood of experiencing systematic risk.

Size has significant negative influence on total risk (TR), systematic risk (β), and specific risk (SR) for the ten listed Indonesian banks over the period studied. Size is negatively related to all the dependent variables because large banking firms tend to be better able to avoid such risks than smaller firms.

The results indicate that in Indonesian banks, accounting variables extracted from the financial statements of each bank reflect the risks taken by them. From the results, it can be

Table 6.   Independent/Dependent variable relationship.

Independent variable	Dependent cariable		
	TR	B	SR
LLP	+	0	0
LIQ	+	0	0
EQTA	−	+	0
Size	−	−	−

concluded that total risk is able to describe banking risk better than systematic and specific risks. Although market risk and specific risk only indicate a bank's risk related to some variables, these market risk measurements can reflect banks' risks overall. Therefore, based on this research it is better to focus on accounting measures to appropriately assess the risk-taking behavior of Indonesian banks.

The results indicate that in Indonesian banks, accounting variables from the financial statements of each bank reflect the risk taken by it. From the results, it can be concluded that total risk is able to describe bank risk better than systematic risk and specific risk. Even though market risk and specific risk only indicates a bank's risk related to some variables, these market risk measurements can reflect bank risk.

REFERENCES

Agusman, A., Monroe, G.S., Gasbarro, D. & Zumwalt, J. (2008). Accounting and capital market measures of risk: Evidence from Asian banks during 1998–2003. *Journal of Banking & Finance, 32*, 480–488.

Baltagi, B.H. (2005). *Econometrics Analysis of Panel Data* (3rd ed.). Chichester: John Wiley & Sons Ltd.

Dhouibi, R. & Mamoghli, C. (2009). Accounting and capital market measures of banks' risk: Evidence from an emerging market. *Banks and Bank Systems, 4*(4), 1–8.

Salkeld, M. (2011). Determinants of banks' total risk: Accounting ratios and macroeconomic indicators. *Honors Projects*, 24. Retrieved from https://digitalcommons.iwu.edu/busadmin_ho.

Business Innovation and Development in Emerging Economies – Trinugroho & Lau (Eds)
© *2019 Taylor & Francis Group, London, ISBN 978-1-138-35996-3*

The effect of financial ratios to profitability bank BUKU 3 listed on Indonesia Stock Exchange

Reny Fitriana Kaban, Puji Hadiyati & Oktanisa Rahmawati
Perbanas Institute, Indonesia

ABSTRACT: This study was conducted to examine the effect of financial ratios such as Non Performing Loan (NPL), Loan to Deposit Ratio (LDR), Capital Adequacy Ratio (CAR), Net Interest Margin (NIM), and Operating Expense Ratio (OER) to the Profitability, as proxied by Return on Asset (ROA) in Bank BUKU 3. These banks being listed on the Indonesia Stock Exchange (IDX) for period of 201–2016. The population in this study are 11 banks listed on the Indonesia Stock Exchange. Data collection method used is documentation (library research) that is using published bank financial statements as of 31 December period year 2013–206. Data Panel analysis is used for data analysis technique, and the result of data processing from EViews 8.0 application shows the result studies of NPL, LDR, CAR, NIM, and OER simultaneously have a significant influence on ROA. While partially, NPL, LDR, and CAR have no significant effect on ROA. The result of determination coefficient of test shows that Adjusted R-square value is 92.01 percent, which means that NPL, LDR, CAR, NIM and OER variables contribute to bank profitability (ROA) of 92.01 percent, while the rest of 7.99 percent is explained by other factors not included in the model.

Keywords: ROA, NPL, LDR, CAR, NIM and ER

1 INTRODUCTION

Banks are known as financial institutions function of collecting and channeling funds from and to the public. Another function of the Commercial Banks is to provide financial services. Based on this function, it is often referred to a financial intermediary institution, which is an intermediary between parties who have excess funds with the parties who need funds. Therefore banks should be able to maintain public trust by ensuring maintained liquidity levels and operating effectively and efficiently in order to achieve optimal profitability. The fundamental objective of the banking business is to obtain optimal benefits by providing financial services to the public.

Financial performance is one benchmark of measure management success in managing resources optimally. For the bank as the financial institution, it shows what the management orientation is in running the organization and accommodating the management interests (board), shareholders (owners), customers, monetary authorities, and the general public whose activities related to banking. One of the main indicators used as the basis for the assessment of financial performance is the financial statements.

They are one of the financial information sourced from the internal company, showing the past financial performance and current financial position of the company. Based on them will be calculated a number of financial ratios which are commonly used as the basis for assessment bank soundness. Financial ratio analysis allows management to identify key changes in the number trend, and the relationships and reasons for those changes. The result of financial statement analysis will help to interpret the various relationships and trends that provide a basis for consideration of the potential success of the company in the future.

In the banking business, one of the performance measurements can be seen from how the bank uses the owned assets to generate profits (earnings) which reflected in Return on Asset (ROA) ratio. The National banking ROA is currently fluctuating from year to year. This is due to the unstable growth of banking profit in Indonesia. The decline in the profit of Indonesian banks is due to the high level of credit failure and operating expenses of companies that are too large and inefficient. Profitability is the most important indicator to measure bank performance. ROA is important for banks because it is used to measure the company effectiveness in generating profits by utilizing its assets.

Some bank of financial ratios which effect the bank performance are Non Performing Loan (NPL) to assess credit risk, Loan to Deposit Ratio (LDR) to measure bank liquidity level through total credit given, Capital Adequacy Ratio (CAR) to measure the capital adequacy bank, Net Interest Margin (NIM) to determine the ability of bank management in terms of productive asset management to generate net profit, and Operating Expense Ratio (OER) to measure bank efficiency in its operational activities. This evaluation of financial performance can be a benchmark for whether the bank is declared healthy or not.

This research discusses the performance of Banks belonging to BUKU 3. The underlying reason for the selection is because Bank BUKU 3 may conduct business activities in Rupiah and foreign currency. Additionally, it can invest 25% in financial institutions at home and abroad but limited in Asian region. So, the selected topic of this study is *The effect of Financial Ratios to Profitability (ROA) of Bank BUKU 3 Listed on Indonesia Stock Exchange (IDX)*.

2 LITERATURE REVIEW AND HYPOTHESES

2.1 Bank terminology

A Bank is a trust institution that serves as an intermediary institution, facilitates the smoothness of payment systems, as well as institution that assists the government in implementing monetary policy. Bank terminology according to Indonesia Law Number 10 year 1998 about Banking, bank is a business entity that collects funds from the public in form of savings and channels them to the community in form of credit, and or other forms in order to improve the standard of living of many people. According to Dendawijaya (2001), Bank is a business entity whose main task is as a financial intermediary institution that channels funds from parties who excessive funds at the time specified. Kasmir (2007: 11), defines the Bank as a financial institution whose main activity is to raise funds from the community and channel them back to the community and provide other bank services. Meanwhile, according to Hasibuan (2007), Bank is a financial institution, means a bank is a business entity whose main assets in form of financial assets also has profit and social motivation.

Based on the definition of some of the expert above, the bank functions can be summarized as follows:

1. A Bank as an institution that collects or receives funds from the community which then distributes back to them.
2. A Bank as an institution that launches trade transactions and payment traffic within a country.

Bank BUKU 3 is a Bank with a core capital of Rp 5 trillion up to less than Rp 30 trillion, in accordance with Bank Indonesia Regulation 10/15/PBI/2008 about Minimum Capital Requirement of Commercial Banks. It may conduct all business activities in Rupiah and foreign currency and participate in 25% of financial institutions at home and abroad in Asia as mentioned in Bank Indonesia Regulation 15/15/PBI/2013 about Minimum Reserve Requirement of Commercial Banks in Rupiah and Foreign Currency for Conventional Commercial Banks. It is obliged to distribute credit or productive financing, including credit or financing to SME, with a target at least 65%t of total credit or financing. It can open branch offices, representative offices, and other types of offices within and outside the country, limited in Asia.

2.2 Financial ratio

Financial ratio is the calculation result for two kinds of bank financial data, which is used to explain the relationship between the two financial data which are generally expressed numerically, either in percentage or times (Pandia, 2012). Terminology of financial ratio according to Kashmir (2014: 104) is the activity of comparing the figures contained in the financial statements by dividing one number with another number. Comparison can be made between one component with another component in one financial statement or between existed components. Then the comparable numbers can be numbers in periods.

2.3 Profitability

Profitability is the ability of a company to generate profits during a certain period. The profitability of the firm shows the comparison between profit and the asset or capital that generates the profit. Profitability is measured by ROA which measures the ability of bank management in obtaining profit as a whole (Dendawijaya, 2005). Meanwhile, according to Hasibuan (2007: 100), bank profitability is the ability of a bank to earn profits expressed in percentages. Profitability is basically the profit (Rupiah) expressed in profit percentage.

ROA is a multiplication of net profit margin with turnover asset. Net profit margin shows the ability to earn profit from every sale created by the company. While turnover asset shows how far the company is able to create sales from the owned assets. If both factors increase then ROA will also increase. If ROA increases then the profitability of the company increases so the ultimate impact is the increase in profitability enjoyed by shareholders. ROA formula is:

$$ROA = \frac{Net\,profit\,before\,tax}{Total\,assets} \times 100\%$$

2.4 Non Performing Loan (NPL)

Non Performing Loan is part of bank credit management, because the problem loan is faced risk by banking business. According to Pandia (2012) Non Performing Loan is a condition in which the customer is unable to pay part or all of his obligations to the bank as he agrees. Meanwhile, according to Dendawijaya (2005) *"bad loan is the repayment of principal and interest which has been delayed for more than one year since its maturity according to the agreed schedule"*.

From the two opinions above can be concluded that Non Performing Loan is a credit which the maturity can not be repaid by the debtor appropriately as agreement. The definition of maturity is in accordance with the level of bank collectivity. The increase in occurring NPL affects the decline in liquidity for the banking sector, because there is no funds coming in either in the form of principal payments or interest on loans from bad loans. So if this situation is allowed it will affect loss of income from the credit sector and the bank lost trust from the people for not being able to manage customers funds safely. Based on Bank Indonesia Regulation 13/1/PBI/2011 Number 1 Year 2011 about Health Level Assessment of Commercial Banks, the gross NPL ratio is less than 5%. The NPL ratio can be calculated by the following formula:

$$NPL = \frac{Non\,Performing\,Loan}{Total\,credit} \times 100\%$$

2.5 Loan to Deposit Ratio (LDR)

One of the bank liquidity measurement is by Loan to Deposit Ratio. According to Kashmir (2007: 272), the ratio of LDR is the compare ratio between the amount of funds disbursed to the public (credit) with the amount of public funds and own used capital. Dendawijaya (2005) defines the Loan to Deposit Ratio as the ratio between the total amount of credit granted by the bank with funds received by the bank.

Loan to Deposit Ration (LDR) is an independent variable that affects ROA based on the bank risk level that leads to bank profitability (ROA). The LDR ratio is used to measure the ability of the bank to be able to pay its debts and repay the depositors and meet the demand for credit proposed. This ratio can be formulated as follows (Kasmir, 2014):

$$LDR = \frac{Credit}{Third\ party\ funds} \times 100\%$$

Loans given are credits granted by bank that have been withdrawn or disbursed by the bank. Loans provided exclude loans to other banks, whereas according to Kashmir (2012) the definition of third party funds are:

1. Demand Deposit: is a third party deposit in a bank whose withdrawal may be made any time by check, payment order, or by transfer.
2. Time deposit: is a third party deposit in banks whose withdrawal may only be made in a certain period time under agreement between the third party and the bank.
3. Public saving: is a third party deposit in banks whose withdrawals can only be made under certain conditions.

2.6 *Capital Adequacy Ratio (CAR)*

According to Dendawijaya (2005), CAR is a ratio showing how much all bank assets that contain risks (credit, investments, securities, bills with other banks) are financed from the bank own capital funds also obtaining funds from sources, such as public funds, loans (debt) and others. In other words, the Capital Adequacy Ratio is the ratio of bank performance to measure the capital adequacy of the bank to support Risk Weighted Asset (RWA). This ratio can be formulated as follows:

$$CAR = \frac{Capital}{RWA} \times 100\%$$

2.7 *Net Interest Margin (NIM)*

Net Interest Margin is a ratio used to measure the bank management capability in managing its earning assets in generating net interest income, considering that the bank operating income is highly dependent on the interest difference from the distributed loans (Pandia, 2012). According to Dendawijaya (2005) NIM is a ratio that describes the level of profit obtained by the bank compared with income received from its operational activities.

The NIM ratio is used to determine the net interest income in 12 (twelve) months that can be obtained by the bank compared with the average bank earning assets. Net interest income is derived from interest income received from the loan granted subtract interest expenses from the source of the funds granted. Calculated Earning asset is assets that generate interest, such as placements in other banks, securities, investments and loans (Riyadi, 2006). According to Bank Indonesia Regulation no 7/2/PBI/2005 about Asset Quality Assessment of Commercial Banks, the amount of NIM to be achieved by banks is above 6%. The formula used to measure NIM as follows:

$$NIM = \frac{Interest\ income - interest\ expenses}{Average\ bank\ earning\ assets} \times 100\%$$

2.8 *Operating Expense Ratio (OER)*

According to Riyadi (2006), OER is comparison between operating cost and operating income in measuring the efficiency and ability of banks in conducting their operations.

According to Sugiyono (2013), an operating expense is an expenditure whose useful life is not more than one year or expenditure which is directly related to the income in a certain period.

OER ratio shows bank efficiency level with ratio close to 75% means that bank performance shows good efficiency. If the ratio above 90% and close to 100% means low efficiency. According to Bank Indonesia Regulation no. 13/1/PBI/2011, Bank Indonesia (BI) tolerates the OER ratio maximum of 93.25% (Riyadi, 2006). The formula used to measure OER as follows:

$$OER = \frac{\text{Operating cost}}{\text{Operating income}} \times 100\%$$

2.9 Previous research

Some of the previous research result relating to this study are as follows:

Prasanjaya and Ramantha research in 2013 with the research title *"Analysis The Effect of CAR, OER, LDR Ratio and Company Size on Profitability of Banks Listed on BEI"* gives the result partially LDR and OER have significant effect on ROA, while CAR and company size do not have significant effect on ROA. Simultaneously CAR, OER, LDR, and company size variable have a significant effect on ROA.

Restiyana and Mahfud research in 2011, with the title *"Analysis The Effect of CAR, NPL, OER, LDR and NIM on Banking Profitability (Study on Commercial Banks in Indonesia period 2006–2010)"*. It indicates that CAR, LDR, and NIM variables have positive and significant effect on ROA, while NPL and OER have negative and significant effect on ROA.

According to Mouri and Chabachib in 2012, in their research entitled *"Analysis of Capital Adequacy Ratio (CAR), Non Performing Loan (NPL), Net Interest Margin (NIM) OER and Loan to Deposit Ratio (LDR) on Return to Assets (ROA)"* gives result that partially CAR, LDR, and NIM significant affect to ROA, while NPL and OER have negative and significant effect to ROA. Simultaneously CAR, NPL, NIM, OER, and LDR variable have significant effect to ROA.

According to Ponco in 2008 in his research entitled *"Analysis The Effect of CAR, NPL, OER, NIM and LDR to ROA (Case Study on Banking Companies Registered in Indonesia Stock Exchange Period 2004–2007)"*, gives result that partially Variable CAR, NIM, and LDR have positive and significant effect on ROA. The OER variable has negative effect and significant to ROA, and the NPL variable has negative and insignificant effect on ROA.

Study of Perkasa in 2007 with the research title *"Analysis The Effect of Financial Ratios to the Performance of Commercial Banks in Indonesia (Empirical Studies of Commercial Banks that Operate on Year 2005)"* shows the result that partially NPL, NIM, and OER variables effect the ROA, while the CAR and LDR variables do not affect the ROA.

Arimi and Mahfud research in 2012 with the research title *"Analysis The Factors Effect Banking Profitability"* showed that partially CAR has positive and not significant effect to ROA, NPL has negative and not significant to ROA, NIM has positive and significant effect to ROA, LDR has positive and not significant to ROA, and OER has negative and not significant effect to ROA.

2.10 Hypothesis

Based on the previously described background concerning the importance of banking efficiency in Indonesia and there is still research gap of previous research and theoretical mismatch, so the hypothesis of this research is:

Ha1: There is positive effect between NPL on bank profitability level at Bank BUKU 3.
Ha2: There is positive effect between LDR on bank profitability level at Bank BUKU 3.
Ha3: There is positive effect between CAR on bank profitability level at Bank BUKU 3.
Ha4: There is positive effect between NIM on bank profitability level at Bank BUKU 3.
Ha5: There is positive effect between OER on bank profitability level at Bank BUKU 3.

Ha6: NPL, LDR, CAR, NIM, and OER have significant effect simultaneously on bank profitability level at Bank BUKU 3.

3 RESEARCH METHODS

3.1 Variables, research objects, types and data sources

Independent variables in this study are the level of NPL, LDR, CAR, NIM, and OER. The dependent variable in this research is ROA. The object of the research is Bank BUKU 3.

The type of data used in this study is secondary data which is bank financial ratios data derived from the financial statements of bank publication from the year 2013–2016. Sources of data from the publication of annual financial statements of banks on Indonesia Stock Exchange and bank website in the year 2013–2016, without going through calculations.

3.2 Population and sample research

The population in this study are all commercial banks in Indonesia for the period of 2013–2016. Sample selection in this research using purposive sampling method that is sample selection method with certain criterion (Ghozali, 2007). Sample criteria of this research are:

1. Bank BUKU 3 category in Indonesia which submits annual financial report on Indonesia Stock Exchange period 2013–2016.
2. The financial statements which are annual financial statements not quarterly reports.

Based on these criteria, the number of samples used in this study are 11 banking companies in the category of BUKU 3 which can be seen in Table 1 below.

3.3 Data collection and processing method

This study uses two methods of collecting secondary data by means of library studies and documentary studies. Secondary data obtained by using statistical software such as Ms. Excel 2007 and Eviews 8.0. used to perform the significance test of multiple linear regression analysis of panel data.

3.4 Panel data regression analysis

Panel data is a combination of cross section and time series. The panel data regression model can be formulated with Widarjono (2013: 354):

$$Yit = \alpha + \beta1\ X1it + \beta2\ X2it + \beta3\ X3it + \beta4\ X4i\ t + \beta5\ X5it + \varepsilon i$$

Table 1. Research sample bank BUKU 3.

No.	Name of banking company	Code
1	PT Bank Pan Indonesia Tbk	PNBN
2	PT Bank Maybank Indonesia Tbk	BNII
3	PT Bank CIMB Niaga Tbk	BNGA
4	PT Bank Danamon Indonesia Tbk	BDMN
5	PT Bank Permata Tbk	BNLI
6	PT Bank OCBC NISP Tbk	NISP
7	PT Bank Mega Tbk	MEGA
8	PT Bank Bukopin Tbk	BBKP
9	PT Bank Tabungan Pensiun Nasional Tbk	BTPN
10	PT Bank Tabungan Negara (Persero) Tbk	BBTN
11	PT Bank Pembangunan Daerah Jawa Barat dan Banten Tbk	BJBR

where as:
Y = Return on assets (ROA)
A = Constanta
β_1–β_5 = independent variable of regression coefficient
X_1 = Non Performing loan (NPL)
X_2 = Loan to deposit ratio (LDR)
X_3 = Capital Adequacy Ratio (CAR)
X_4 = Net interest margin (NIM)
X_5 = Operating expense ratio (OER)
ε = nuisance variable
i = 11 of the company sample
t = observation period 2013–2016

In estimating the model parameters with panel data, there are three techniques that can be used: ordinary least square (OLS) or Common Effect, Fixed Effect, and Random Effect model (Nachrowi and Usman (2006: 311)). The explanation is as follows:

1. *Ordinary Least Squares (OLS) or common effect*
In analyzing regression with panel data can use Ordinary Least Square (OLS) model analysis or called common effect model. According to Widarjono (2013: 355) common effect approach is the simplest technique to estimate panel data only by combining time series and cross section data. Regression equation with Ordinary Least Square method can be written by:

$$Yit = \alpha + \beta Xit + \varepsilon it$$

2. *Fixed effect model*
According to Nachrowi and Usman (2006: 313) the fixed effect method is the method that may make α changes in each individual (i) and time (t). Mathematically model panel data using fixed effect approach according to Rosadi (2012:272) is as follows:

$$Yit = \beta Xit + ci + dt + \varepsilon it$$

The existence of variables that are not all included in the model equation may make the intercept is not constant or in other words, this intercept may change for each individual and time. This thinking is the basis for the formation of the above model (Nachrowi and Usman, 2006: 311).
3. *Random effect model*
Random effect model shows the difference of individual characteristic and time with error model (Nachrowi and Usman, 2006:316). Since there are two components contribute for potential error, i.e. individual and time, then random error in random effect method needs to be divined as error to individual component, time component and both errors. Equation of random effect model can be formulated as follow:

$$Yit = \alpha + \beta Xit + \varepsilon it$$

3.5 *Technic selection of panel data estimation model*

In order to estimate panel data regression there are three models: common effect, fixed effect, and random effect. From the three models of panel data will be selected one should be used for the panel data regression equation. Therefore, three tests are used to determine the most appropriate model. The explanation of each regression test in technic selection of panel data as follow:

1. *Chow test*
Chow test is used to select the model used whether to use the common effect model or the fixed effect model. The hypothesis of the chow test is as follows:

H0: Model follows common effect
Ha: Model follows fixed effect

The basic in making the decision is as follows:
1. Based on comparison F statistic with F_{table}:
   – If F statistic > F_{table}, then Ho is rejected Ha accepted. This means that the model used using by the research using fixed effect.
   – If F statistic < F_{table}, then Ho accepted, Ha rejected. This means that the model to be used in research is Common Effect.

2. Based on the probability value
   – If the probability value (p-value) < alpha (0.05) then H0 is rejected and Ha is accepted. This means that the model used by the research using Fixed Effect.
   – If the probability value (p-value) > alpha (0.05) then H0 is accepted and Ha is rejected. This means that the model to be used in research is Common Effect.

2. *Hausman test*

Hausman test is performed to choose which model is better, whether using fixed effect model or random effect model. The hypothesis in testing Hausman test is as follows:

H0: Model follows random effect
Ha: Model follows fixed effect

The basic in making the decision is follows:
1. Based on the comparison of chi-square statistic with chi-square$_{table}$
   – If chi-square statistic > chi-square$_{table}$, then H0 is rejected and Ha accepted. This means that the model used by the research using Fixed Effect.
   – If chi-square statistic < chi-square$_{table}$, then H0 is accepted and Ha is rejected. This means that the model to be used in research Random Effect.

2. Based on the probability value
   – If the probability value (p-value) < alpha (0,05) then H0 is rejected and Ha is accepted. This means that the model used by the research using Fixed Effect.
   – If the probability value (p-value) > alpha (0,05) then H0 is accepted and Ha is rejected. This means that model to be used in research is Random Effect.

3. *Lagrange multiplier test (LM test)*

Lagrange Multiplier (LM) test is used to find out which model is better, whether better estimated using the Common Effect model or with the Random Effect model. The hypothesis used in the LM test is as follows:

H0: Model follows common effect
Ha: The model follows the random effect

The LM test is based on chi-square with degree of Freedom (df) for the number of independent variables. If the LM-test is smaller than the chi-square table value, then H0 is accepted and Ha is rejected. So the model used is common effect. However, if the LM-test is greater than the chi-square table value, then H0 is rejected and Ha is accepted. This means that the model used is the random effect.

3.6 *Classic assumption test*

In the use of regression model, hypothesis test should avoid the possibility of deviation of classical assumptions. In the study using panel data then the classical assumption test is considered the most important which do not occur multicollinearity between independent variables and heteroscedasticity or inconstant variance. Some classical assumption tests of regression model as follows:

1. *Multicollinearity test*

According to Ghozali (as cited in Perkasa, 2007: 46) multicollinearity test aims to test whether the regression model is found the correlation between independent variables. A good regression model should not be correlated among independent variables. The correlation test is performed to determine the presence or absence of multicollinearity by using correlation matrix. As a basis of reference can be summarized as follows:

– If the correlation coefficient > 0.8 then there is multicollinearity
– If the correlation coefficient < 0.8 then there is no multicollinearity

2. *Heterocedasticity test*
The heterocedasticity test aims to test whether in the regression model there is inequality variance from residual of an observation to another observation. A good regression model is homocedasticity or no heterocedasticity occurs. This often occurs in cross section data since panel data contains cross section data and it is suspected heterocedasticity (Nachrowi and Usman, 2006: 330). To see if the model has heteroscedasticity problem uses White test. This test is done by comparing chi-square statistic with chi-square table (Widardjono, 2013: 126).

H0: There is no heteroscedasticity problem
Ha: There is a problem of heteroscedasticity

If chi-square statistic > chi-square$_{table}$, then H0 is rejected and Ha accepted. This means there is a problem of heterocedasticity. Conversely, if the chi-square statistic is calculated < chi-square$_{table}$, then H0 is accepted and Ha is rejected. This means there is no problem of heterocedasticity.

3.7 *Hypothesis test*

To test the analyzed hypothesis, it is necessary to do regression analysis through *t*-test and *F*-test. It aims to know the effect of Non Performing Loan (NPL), Loan to Deposit Ratio (LDR), Capital Adequacy Ratio (CAR), Net Interest Margin (NIM) and Operating Expense Ratio (OER) to dependent variable, which is Return On Assets (ROA) partially (*t*-test) and simultaneously (*F*-test).

3.8 *Coefficient determination test (R^2)*

This test is performed to see the effect or contribution of independent variables to the dependent variable. The value of coefficient of determination is 0 to 1, means if the value of coefficient of determination close to 0 indicates the weaker relationship between independent variable to the dependent variable. Conversely, if the coefficient of determination close to 1 then shows a strong relationship between independent variable to the dependent variable. According to Kuncoro (2013: 247), each addition of one independent variable will increase R^2, no matter the variable has a significant effect on the dependent variable or not. Hence in this study the researchers used adjusted R^2 to measure the percentage of the effect of independent variable to the dependent variable.

4 RESULT AND DISCUSSION

4.1 *Result data analysis result*

4.1.1 *Panel data regression estimation*
In estimating the model parameters with panel data, there are three techniques that can be used: Ordinary Least Square (OLS) or common effect, fixed effect model, and random effect model.

1. *Common effect model*
Table 2 shows the estimation using common effect method or OLS method that the adjusted R^2 is 92.01 percent. If the probability of free variable < $\alpha = 0,05$ is significant. Variable NIM

Table 2. Estimation calculation of common effect.

Dependent Variable: ROA
Method: Panel least squares
Date: 07/22/17 Time: 07:39
Sample: 2013–2016
Periods included: 4
Cross-sections included: 11
Total panel (balanced) observations: 44

Variable	Coefficient	Std. error	t-statistic	Prob
NPL	−0.050689	0.073539	−0.689278	0.4948
LDR	−0.004264	0.005362	−0.795069	0.4315
CAR	0.003123	0.024398	0.128004	0.8988
NIM	0.170110	0.032228	5.278352	0.0000
OER	−0.090933	0.005237	−17.36235	0.0000
C	8.853017	0.886290	9.988848	0.0000
R-square	0.929451		Mean dependent var	1.656136
Adjusted R-square	0.920169		S.D. dependent var	1.284279
S.E. of regression	0.362866		Akaike info criterion	0.936556
Sum square resid	5.003519		Schwarz criterion	1.179855
Log likelihood	−14.60424		Hannan-Quinn criter	1.026783
F statistic	100.1272		Durbin-Watson stat	2.387272
Prob. (F statistic)	0.000000			

Table 3. Estimation calculation of fixed effect.

Dependent variable: ROA
Method: Panel least squares
Date: 07/22/17 Time: 07:42
Sample: 2013–2016
Periods included: 4
Cross-sections included: 11
Total panel (balanced) observations: 44

Variable	Coefficient	Std. error	t-statistic	Prob
NPL	−0.080878	0.103537	−0.781155	0.4413
LDR	−0.006267	0.006680	−0.938141	0.3562
CAR	−0.010010	0.038086	−0.262816	0.7946
NIM	0.185155	0.043083	4.297653	0.0002
OER	−0.093149	0.007373	−12.63448	0.0000
C	9.411251	1.417729	6.638259	0.0000
	Effect Specification			

Cross-section fixed (dummy variables)

R-square	0.935611		Mean dependent var	1.656136
Adjusted R-square	0.901116		S.D. dependent var	1.284279
S.E. of regression	0.403851		Akaike info criterion	1.299748
Sum square resid	4.566685		Schwarz criterion	1.948544
Log likelihood	−12.59445		Hannan-Quinn criter	1.540353
F-statistic	27.12366		Durbin-Watson stat	2.562296
Prob (F-statistic)	0.000000			

and OER significant to ROA with probability value from each variable $0,00 <$ from value $\alpha = 0,05$. While NPL, LDR, and CAR variables are not significant to ROA with probability value $> \alpha = 0,05$ that is 0,4948 for NPL, 0,4315 for LDR, and 0,8988 for CAR.

2. *Fixed effect model*

Based on the test result of this fixed effect model, can be seen that the adjusted R^2 is lower than the common effect model that is equal to 90.11 percent. If the probability of free variable $< \alpha = 0,05$ is significant. In Table 3, the result of significance of the five variables gives the same result with the common effect model i.e. the NIM and OER variables are significant to ROA, where the probability value of each variable is $<\alpha = 0,05$, i.e. 0,0002 for NIM and 0,0000 for OER. While NPL, LDR, and CAR are not significant to ROA where the probability value is $> \alpha = 0,05$: that is 0,4413 for NPL, 0,3562 for LDR, and 0,7946 for CAR.

3. *Random effect model*

Based on the test result on this random effect model, it can be seen that the adjusted R^2 is higher than the fixed effect model of 92.02 percent. In Table 4 shows the same thing with the common effect and fixed effect model that the NIM and OER variables are significant to the ROA with the probability value of each variable that is $0.001 <$ of the value of $\alpha = 0.05$. While the NPL, LDR, and CAR variables are not significant to ROA with probability value $> \alpha = 0.05$ i.e, 0,5394 for NPL, 0,4794 for LDR, and 0,9090 for CAR.

Table 4. Estimation calculation random effect.

Dependent variable: ROA

Method: Panel EGLS (Cross-section random effect)
Date: 07/22/17 Time: 07:42
Sample: 2013–2016
Periods included: 4
Cross-sections included: 11
Total panel (balanced) observations: 44
Swamy and Arora estimator of component variances

Variable	Coefficient	Std. Error	*t*-Statistic	Prob
NPL	−0.050689	0.081845	−0.619326	0.5394
LDR	−0.004264	0.005968	−0.714380	0.4794
CAR	0.003123	0.027153	0.115014	0.9090
NIM	0.170110	0.035868	4.742670	0.0000
OER	−0.090933	0.005829	−15.60030	0.0000
C	8.853017	0.986396	8.975112	0.0000

	Effect Specification			
			S.D	Rho
Cross-section random			0.000000	0.0000
Idiosyncratic random			0.403851	1.0000
	Weighted Statistics			
R-square	0.929451		Mean dependent var	1.656136
Adjusted R-square	0.920169		S.D. dependent var	1.284279
S.E. of regression	0.362866		Sum square resid	5.003519
F-statistic	100.1272		Durbin-Watson stat	2.387272
Prob (*F*-statistic)	0.000000			
	Unweighted Statistics			
R-square	0.929451		Mean dependent var	1.656136
Sum square resid	5.003519		Durbin-Watson stat	2.387272

4.2 *Choosing the panel data regression model*

1. *Chow test*
Chow test is used to choose the better used model whether with common effect or fixed effect. The basic in making the decision is as follows:
1. Based on comparison F statistic with F_{table}:
   Given: F statistic = 0,267839 < F_{table} = 5,05
   Conclusion: H0 accepted and Ha rejected

2. Based on the probability Value
   Given: probability Value 0,9835 > α Value = 0,05
   Conclusion: H0 accepted and Ha rejected

From the two basic decision-making result of Chow test above it can be stated that the common effect model is better used than the fixed effect model.

Based on calculations obtained using the Chow test stated that the common effect model is selected then do not have to do Hausman test, because fixed effect model has been declared not feasible so the next stage directly tested Lagrange Multiplier to determine which model will be selected choose between common effect and random effect.

2. *Lagrange multiplier (LM) test*
This test is performed to select which model will be used between common effect model or random effect. LM statistic value will be compared with Chi square$_{table}$ value with degree of freedom as many as independent variable and alpha or significance level of 5%. If the value of

Table 5. Result of chow test.

Redundant fixed effect tests
Equation: FE
Test cross-section fixed effect

Effect Test	Statistic	df	Prob
Cross-section F	0.267839	(10,28)	0.9835
Cross-section Chi-square	4.019578	10	0.9465

Cross-section fixed effect test equation:
Dependent Variable: ROA
Method: panel least squares
Date: 07/22/17 Time: 07:44
Sample: 2013–2016
Periods included: 4
Cross-sections included: 11
Total panel (balanced) observations: 44

Variable	Coefficient	Std. Error	t-Statistic	Prob
NPL	−0.050689	0.073539	−0.689278	0.4948
LDR	−0.004264	0.005362	−0.795069	0.4315
CAR	0.003123	0.024398	0.128004	0.8988
NIM	0.170110	0.032228	5.278352	0.0000
OER	−0.090933	0.005237	−17.36235	0.0000
C	8.853017	0.886290	9.988848	0.0000
R-square	0.929451	Mean dependent var		1.656136
Adjusted R-square	0.920169	S.D. dependent var		1.284279
S.E. of regression	0.362866	Akaike info. criterion		0.936556
Sum square resid	5.003519	Schwarz criterion		1.179855
Log likelihood	−14.60424	Hannan-Quinn criter		1.026783
F-statistic	100.1272	Durbin-Watson stat		2.387272
Prob. (F-statistic)	0.000000			

LM statistic > Chi square$_{table}$ then the selected model is random effect and vice versa if the value of LM statistic < Chi square$_{table}$ then the model selected is the common effect. The value of Chi square$_{table}$ on the 5 degrees of freedom 5 and the 5% alpha value is 11.07 and the LM value is 3.42516, so that LM is bigger than Chi square$_{table}$, then the selected model is random effect.

Furthermore, the classical assumption test should be performed, where for panel data model the classical assumption test only uses multicollinearity and heterocedasticity test.

4.3 Classic assumption test

1. *Multicolinearity test*
Multicolinearity test aims to test whether the regression model found a correlation between independent variables. A good regression model should not be correlated between independent variables. The result of multicollinearity testing are as follows:

Multicolinearity test result can be seen in the Table 6 centered column VIF. Variance inflation factors (VIF) measure how much the variance of the estimated regression coefficients is inflated as compared to when the predictor variables are not linearly related. It is used to explain how much amount multicollinearity (correlation between predictors) exists in a regression analysis. VIF values for NPL, LDR, CAR, NIM, and OER variables are respectively 1, 26094: 1,190394: 2,038171: 2, 044929: and 1,177178 shows a value less than 5, it can be said that there is no multicollinearity in the five independent variables.

2. *Heterocedasticity test*
Heterocedasticity test aims to test whether in the regression model there is an inequality of variance of the residual from one observation to another. A good regression model is homocedastic or no heteroscedasticity occurs. To see if the model has heteroscedasticity problem, do the Glejser test

H0: There is no heteroscedasticity problem
Ha: There is a problem of heteroscedasticity

The result of heteroscedasticity testing can be seen in Table 7 as follow:
The decision is there heteroscedasticity or not on linear regression model is by looking the Prob value of F-statistic. If Prob value of F statistic is greater than alpha level 0.05 (5%) then H0

Table 6. Multicolinearity test result.

Variance inflation factors
Date: 07/22/17 Time: 07:31
Sample: 0001 0044
Included observations: 44

Variable	Coefficient Variance	Uncentered VIF	Centered VIF
NPL	0.005408	5.195046	1.260942
LDR	2.88E-05	79.03297	1.190394
CAR	0.000595	64.00612	2.038171
NIM	0.001039	14.00863	2.044929
OER	2.74E-05	68.48499	1.177178
C	0.785510	262.4898	NA

Table 7. Heterocedasticity result test.

Heterocedasticity Test: Glejser

F-statistic	0.700031	Prob. F (5,38)	0.6268
Obs*R-square	3.710992	Prob. chi-square (5)	0.5917
Scaled explained SS	5.207378	Prob. chi-square (5)	0.3911

is accepted, which means no heteroscedasticity, whereas if the Prob value F-statistic is smaller than the alpha level of 0.05 (5%) then H0 is rejected which means heterocedasticity occurs.

Probability value of F-statistic is equal to 0.6268, that is, greater than alpha level 0.05 (5%), so H0 is accepted, meaning no heteroscedasticity occurs.

4.4 Hypothesis tests

1. *Partial test or t test*

Partial hypothesis test or t test to know whether or not there is influence of each independent variable (NPL, LDR, CAR, NIM, and OER) to dependent variable (ROA).

From the table above the regression equation can be obtained in this research is:

$$ROA = 8,853017 - 0,050689NPL - 0,004264LDR + 0,003123CAR + 0,170110\ NIM - 0,090933OER + e$$

Based on the *t* test result, the effect of independent variables on dependent variable is as follows:

a. The effect of NPL on ROA

Result test showed the value of NPL variable coefficient –0,050689 with probability value 04948 which means smaller than alpha (0.05). At the NPL *t*-statistic –0,689278 which result is smaller than the *t* table of 1,6848 thus Ho accepted and Ha rejected. From the analysis, it can be concluded that the NPL variable has no significant effect on ROA.

b. The effect of LDR on ROA

The result test showed the value of coefficient of LDR variable equal to –0,004264 with probability value equal to 0,4315, which is bigger than alpha (0,05). At *t*-statistic equal to –0,7950, which is smaller than *t* table equal to 1,6848 Ho accepted and Ha rejected. From the analysis, it can be concluded that LDR variable has no significant effect on ROA.

c. The effect of CAR on ROA

The result test showed the value of CAR variable coefficient 0,003123 with a probability value 0,8988 greater than alpha (0.05). At CAR statistic 0,128004 which is smaller than *t* table 1,6848 thus H0 is accepted and Ha is rejected. From the analysis, it can be concluded that CAR variable has no significant effect on ROA.

d. The effect of NIM on ROA

The result test showed the value of NIM variable coefficient 0.170110 with a probability value 0.0000 is smaller than alpha (0.05). At *t*-statistic equal to 5,27835 which result bigger than *t* table equal to 1,6848 H0 rejected and Ha accepted. From the analysis, it can be concluded that the NIM variable has significant effect on ROA.

e. The effect of OER on ROA

The result test showed the value of OER variable coefficient –0,09093 with a probability value 0.0000 smaller than alpha (0,05), thus H0 rejected and Ha accepted. At *t*-statistic equal to –17,36235 which is smaller than the *t* table of 1,848 thus H0 is accepted and Ha is rejected. From the analysis, it can be concluded that OER variable has a significant negative effect on ROA.

2. *Simultaneous test or F test*

The *F*-statistic test shows whether all the independent variables (NPL, LDR, CAR, NIM and OER) included in the model have a mutual influence on the dependent variable (ROA).

Table 8. *t* test result.

Variable	Coefficient	Std. error	t-Statistic	Prob
NPL	–0.050689	0.073539	–0.689278	0.4948
LDR	–0.004264	0.005362	–0.795069	0.4315
CAR	0.003123	0.024398	0.128004	0.8988
NIM	0.170110	0.032228	5.278352	0.0000
OER	–0.090933	0.005237	–17.36235	0.0000
C	8.853017	0.886290	9.988848	0.0000

Table 9.   F test result: Weighted statistics.

R-square	0.929451	Mean dependent var	1.656136
Adjusted R-square	0.920169	S.D. dependent var	1.284279
S.E. of regression	0.362866	Akaike info. criterion	0.936556
Sum square resid	5.003519	Schwarz criterion	1.179855
Log likelihood	−14.60424	Hannan-Quinn criter	1.026783
F-statistic	100.1272	Durbin-Watson stat	2.387272
Prob (F-statistic)	0.000000		

Table 10.   Coefficient determination test: Weighted statistics.

R-square	0.929451	Mean dependent var	1.656136
Adjusted R-square	0.920169	S.D. dependent var	1.284279
S.E. of regression	0.362866	Akaike info. criterion	0.936556
Sum square resid	5.003519	Schwarz criterion	1.179855
Log likelihood	−14.60424	Hannan-Quinn criter	1.026783
F-statistic	100.1272	Durbin-Watson stat	2.387272
Prob (F-statistic)	0.000000		

From Table 10 shows the NPL, LDR, CAR, NIM, and OER variables simultaneously obtained probability value 0.000000 is smaller than the alpha value (0.05), and F-statistic 100.1272 is greater than the F table 5.05. Thus, Ho is rejected and Ha accepted, so it can be concluded that simultaneously variables NPL, LDR, CAR, NIM, and OER have significant effect to ROA.

3. *Coefficient determination test (R2)*

This test is done to see the influence or contribution of independent variable to dependent variable. The coefficient of determination for regression with more than two independent variables is suggested to use Adjusted R2.

From the calculation result through EViews 8.0 program, it is estimated that the value of Adjusted R-square 0.920169 (92.01 percent) indicates that ROA can be explained simultaneously by NPL, LDR, CAR, NIM, and OER variables 92.01 percent, while the rest of 7.99 percent is explained by other factors not included in the research model.

5   DISCUSSION

1. *The effect of non performing loan to profitability (ROA)*

NPL has negative effect on profitability (ROA). If we see from the regression equation that shows –0.05068, it can be interpreted that any 1 percent increase in NPL will result in a decline in profitability of banking companies by 5.068 percent. Based on test results with t test (partially) showed that the variable NPL in this study has a non-significant effect on ROA. This is indicated by a probability value 0.4948, which is greater than alpha (0,05), and t statistic –0.68927, which result is smaller than the t table 1,68488. The result of this study is in line with Ponco (2008) study which states that NPL has negative but not significant effect on ROA. However, the result of this study differs from Restiyana and Mahfud (2011) and Mouri and Chabachib (2012), which states that the NPL variable has a negative and significant effect on ROA.

2. *The effect of loan to deposit ratio on profitability (ROA)*

LDR has a negative effect on profitability of banking companies. If we see from the regression equation that shows –0.00426, it means that every 1 percent increase in LDR it will lead to a decline in the profitability of banking companies by 0,426 percent. Based on test result with t test (partially) shows that the LDR variable in this study did not have significant effect on ROA. The result of this research is shown with probability value 0,4315 bigger than alpha

(0,05) and *t*-statistic equal to –0,79506 smaller than *t* table equal to 1,6848. No effect of **LDR** on **ROA** because the credit disbursed by the bank does not contribute much profit. So, there are banks that do not optimize the third party funds, on the other hand there is an excessive bank in giving credit but the credit is in problems. The result of this study is in line with research Perkasa (2007) which states that the **LDR** variable has no effect on **ROA**. However, the result of this study differs with Mouri and Chabachib (2012) and Restiyana and Mahfud (2011) which states that **LDR** has a significant positive effect on **ROA**. This difference occurs if the bank as the object of this research has different category with the bank in those studies.

3. *The effect of capital adequacy ratio on profitability (ROA)*

From the test result above, it can be seen that **CAR** has effect on profitability of banking companies. If we see from the regression equation that shows 0.00312, the figure can be interpreted that any 1 percent increase in **CAR** will result in an increase in the profitability by 3.12 percent. Based on test result with t test (partially) showed that the **CAR** variable in this study has a non-significant effect on **ROA**. This is indicated by a probability value 0.8988, which is greater than alpha (0,05), and *t*-statistic 0,12800, which result is smaller than *t* table 1.6848. The inability of **CAR** to **ROA** is likely due to the operating banks in that year keep the existing or owned capital strictly. It can also happen because banks have not been able to throw credit as expected or not optimal yet. The result of this research is in line with research of Bilian and Purwanto (2015) which states that **CAR** variable has positive but not significant effect on **ROA**.

4. *The effect of net interest margin on profitability (ROA)*

NIM has a positive effect on profitability. If we view from the regression equation that shows the number 0.17011, the figure can be interpreted that any 1% increase in magnitude of **NIM**, it will result in an increase in profitability by 17.011 percent. Based on the result with *t* test (partially) showed that **NIM** has a significant effect on **ROA**. This is indicated by a probability value 0.00000 which is smaller than alpha (0,05) and *t*-statistic is 5,27835, which is bigger than *t* table equal to 1.6848. The result of this study is in line with the research of Ponco (2008), Hutagalung et al (2011), and Perkasa (2007) stating that **NIM** variable has positive and significant impact on **ROA**.

5. *The effect of operational efficiency ratio (OER) on profitability (ROA)*

From the test result above it can be seen that **OER** has negative effect on bank profitability. If seen from the regression equation that shows the number of –0.09093, the figure can be interpreted that any 1 percent increase in the amount of **OER** it will result in a decline in profitability banking companies 9.093 percent. Based on test result with *t* test (partially) showed that the **OER** variable in this study has a significant effect on **ROA**. This is indicated by a probability value 0.0000, which is smaller than alpha (0,05) and t-statistic –17.36235, which is smaller than *t* table equal to 1.6848. The result of this study is in line with Arimi and Mahfud (2012), Mouri and Chabachib (2012), and Ponco (2008) studies, which state that **OER** variable have a negative and significant effect on **ROA**. This is in accordance with the theory that the greater the operational cost than the operational income means that the bank has not been able to run its business operations efficiently, because any increase in cost means it will lead to decreased profitability.

6 SUMMARY

6.1 *Conclusion*

Based on the result of data analysis on the effect of financial ratios such as **NPL, LDR, CAR, NIM**, and **OER** on Profitability represented by **ROA** at Bank BUKU 3 period 2013–2016, the following conclusions are obtained:

1. NPL has positive but insignificant effect on Profitability (ROA) at Bank BUKU 3.
2. LDR has negative but insignificant effect on Profitability (ROA) at Bank BUKU 3.
3. CAR has positive but insignificant effect on Profitability (ROA) at Bank BUKU 3.
4. NIM has positive and significant effect on Profitability (ROA) at Bank BUKU 3.
5. OER has negative and significant effect on Profitability (ROA) at Bank BUKU 3.

6. NPL, LDR, CAR, NIM and OER simultaneously effect the Profitability (ROA) at Bank BUKU 3.
7. Adjusted R2 value of 0.920169 indicates that NPL, LDR, CAR, NIM, and OER variables simultaneously explain the effect on ROA of 92.01 percent, while the rest of 7.99 percent is explained by other factors not included in the model.

6.2 *Recommendation*

1. The management of the bank should maintain and increase the value of its banking financial ratios in accordance with the provisions set by Bank Indonesia so that its profits increase and public trust is maintained
2. To include other independent variables such as Net Open Position (NOP) and Allowance for Earning Assets Losses (PPAP) in subsequent research in the extent of independent variables used to explain the dependent with a larger percentage rate than the previous research result.

REFERENCES

Arimi, Millatina and Mahfud, Mohammad Kholiq. (2012). Analysis factors effect banking profitability (Study to commercial bank listed in Indonesia Stock Exchange 2007–2010). *Diponegoro Journal of Management*, I (II)), 80–91.
Bilian and Purwanto. (2015). *Analysis the effect of CAR, NIM, OER, and LDR to Profitability in Commercial Bank.* Journal. Bekasi: Faculty of Business President University.
Dendawijaya, Lukman. (2005). *Banking Management.* Jakarta: Ghalia Indonesia.
Ghozali, Imam. (2007). *Aplication of multivariate analysis with SPPS program.* Semarang: UNDIP Publisher.
Ghozali, Imam and Ratmono, Dwi. (2013). *Analysis multivariate & econometrics theory, concept, & application with Eviews 8.* Semarang: UNDIP Publisher.
Hasibuan, Malayu. (2007). *Basics of Banking.* Jakarta: PT. Bumi Aksara.
Hutagalung, E. Novelina. et al. (2013). Analysis financial ratio to performance of commercial banks in Indonesia. *Journal Management Application*, II (1): 123–130.
Ikatan Bankir Indonesia. (2013). Understanding bank business: Sertification Module Level I General Banking. Jakarta: PT Gramedia Pustaka Utama.
Indonesi (2003) *Law Number 10 year 1998: about Banking.* Jakarta: Ghalia Indonesia.
_____. (2005). *Bank Indonesia Regulation 7/2/PBI/2005 about Asset Quality Assessment of Commercial Banks.* Paper of Republic Indonesia Number 12 Year 2005. Extra Paper of Republic Indonesia Number 4471. Jakarta: Bank Indonesia.
_____. (2008) *Bank Indonesia Regulation 10/15/PBI/2008 about Minimum Capital Requirement of Commercial Banks.* Paper of Republic Indonesia Number 135 Year 2008, Extra Paper of Republic Indonesia Number 4895. Jakarta: Bank Indonesia.
_____. (2011). *Bank Indonesia Regulation 13/1/PBI/2011 about Health Level Assessment of Commercial Banks.* Paper of Republic Indonesia Number 1 Year 2011. Extra Paper of Republic Indonesia Number 5184. Jakarta: Bank Indonesia.
_____. (2013). *Bank Indonesia Regulation 15/15/PBI/2013 about Minimum Reserve Requirement of Commercial Banks in Rupiah and Foreign Currency for Conventional Commercial Banks.* Paper of Republic Indonesia Number 235 Year 2013. Extra Paper of Republic Indonesia Number 5478. Jakarta: Bank Indonesia.
Kasmir. (2007). *Bank and other financial institution* (Sixth Edition) Jakarta: PT Raja Grafindo Persada.
_____. (2012). *Banking Management.* Jakarta: PT Raja Grafindo Persada.
_____. (2014). *Financial report analysis.* Jakarta: PT Raja Grafindo Persada.
Kuncoro, Murajad. (2013). *Research method for business and economics* 4th Edition. Jakarta: Indonesia Erlangga Publisher.
Mouri, Tryo Hasnan and Chabachib. (2012). *Analysis the effect of capital adequacy ratio (CAR), non performing loan (NPL), net interest margin (NIM), OER, and loan to deposit ratio (LDR) to return on assets (ROA)* Journal. Semarang, Indonesia: Universitas Diponegoro.
Nachrowi, D. Nachrowi. & Usman, Hardius. (2006). *Econometrics for financial and economic analysis.* Jakarta, Indonesia: Economics Faculty Universitas Indonesia Publisher.
Pandia, Frianto. (2012). *Funding management and bank's health.* Jakarta, Indonesia: Rineka Cipta.

Perkasa, Ponttie. Prasnanugraha. (2007). *Analysis the effect of financial ratios to performance of commercial banks in Indonesia (Empirical study of commercial banks in Indonesia)*. Thesis. Semarang: Post Graduate Program Universitas Diponegoro,

Ponco, Budi. (2008). *Analysis The Effect of CAR, NPL, OER, NIM, and LDR to ROA (Case study to banks listed in Indonesia Stock Exchange Period 2004–2007)*. Thesis. Semarang: Post Graduate Program Universitas Diponegoro,

Prasanjaya, A.A. Yogi. and Ramantha, I. Wayan. (2013). Analysis The Effect of CAR, OER, LDR, and Company Size to Profitability of Bank Listed in IDX. *Accounting E-Journal Udayana University*, 4(1), 230–245.

Restiyana and Mahfud, M. Kholiq. (2011). *Analysis the effect of CAR, NPL, OER, LDR, and NIM to bank profitability (Study to commercial bank in Indonesia period 2006–2010)*. Essay. Semarang: Economy Faculty Diponegoro University.

Riyadi, Selamet. (2006). *Banking assets and liability management*. Jakarta, Indonesia: Economy Faculty Indonesia University Publisher.

Rosadi, Dedi. (2012). *Econometrics and analysis of time series application*. Yogyakarta: Andi Yogyakarta.

Sugiyono. (2005). *Administration research method*. Bandung: CV. Alfabeta.

_____. (2013). *Quantitative, qualitative, and combination research method (mixed method)*. Bandung: CV. Alfabeta.

Veitzal, Rivai, S. Basir, S. Sudarto & A.P. Veithzal. 2013. *Commercial bank management, banking management from theory to practice*. Jakarta: Raja Grafindo Persada.

Widarjono, Agus. (2013). *Econometrics: Introduction and Application*. Yogyakarta: UPP STIM YKPN.

Business Innovation and Development in Emerging Economies – Trinugroho & Lau (Eds)
© *2019 Taylor & Francis Group, London, ISBN 978-1-138-35996-3*

The impact of variables of macroeconomic and bank-specifics on non-performing loan in banking industry in Indonesia

Mahrus Lutfi Adi Kurniawan, Siti Aisyah Tri Rahayu & Agustinus Suryantoro
Department of Economics, Universitas Negri Sebelas Maret (UNS), Surakarta, Indonesia

ABSTRACT: The purpose of this study is to analyze the effect of macroeconomic variables and bank-specific variables to non-performing loans in the banking industry in Indonesia. The data used in this study is quarterly secondary data from 2015:01-2017:04 across 106 commercial banks. The analysis used is dynamic Generalized Method-of-Moment (GMM). The result of this research is that real Gross Domestic Product (GDP) growth, Return on Asset (ROA) and efficiency (BOPO) have a positive impact on the NPL. Inflation and Return on Equity (ROE) have a negative impact on the NPL and interest rate has no effect on the NPL. The Sargan test failure to reject the null hypothesis. It implies that the instruments are valid and the serial correlation test for the second order is higher than alpha by 5%, which means the model has no autocorrelation.

Keywords: macroeconomics variables, bank specifics variables, non-performing loan, dynamic GMM

1   INTRODUCTION

In recent years, banking activities have become very complex. In addition to the intermediary institutions, the stability of the banking system has become very important because the failure to maintain banking stability. It will adversely affect macroeconomic conditions and even social aspects of society and, in the event of a crisis, it will require a very large amount of money to cope with the crisis (Claessens et al., 2013).

Banks always confront credit or Non-Performing Loan (NPL) risks because of its main task as an intermediary institution. However, the NPL issue is not only a measure of bank performance but is also very important for the monetary authority in determining the direction of its policy, as it can dictate the state of the economy of a country.

In Indonesia, there are 116 banks consisting of Bank Persero, Foreign Exchange BUSN Bank, Non-Foreign Exchange BUSN, BPD, Joint Bank and Foreign Bank. This study explores the NPL of each bank. Research on NPLs has been done extensively from various countries (Mondal, 2016) (Karahanoglu, 2015) (Abrebese, 2016) (Parab, 2018) (Anjom, 2015) and (Janvisloo, 2013).

The Basel Committee for Banking Supervision (2001) defines NPL as a probability of losing either part or all of the loan balance due to credit default events. In general, the performance of a bank can be affected by two things: the influences of the internal and the external. Unpredictable global economic conditions leading to the selection of macroeconomic variables becomes very important so the obtained results become more comprehensive.

A high NPL level will let the bank act more selectively toward new debtors and causes it to be difficult to disburse credit. This needs to be done so that the bank is able to maintain the confidence of depositors and that they should always keep the NPL level at a low point to encourage economic growth and maintain the stability of the country's economic system.

As mentioned earlier, there are internal and external factors that affect the NPL. There have been many studies using either internal or external factors and even both. This study sees the importance of using these two factors to test NPL in Indonesia.

2 LITERATURE REVIEW

Non-Performing Loan (NPL) is an anomaly that is always confronted by every bank. Measurements of distributed credit quality will always be linked to any credit risk or NPL that the bank earns. The NPL amount in the bank depends on its ability to assess the credit risk of the loan proposed by an applicant where its normally measured using default probability, loss given default and default exposure (Mileris, 2014).

Many researchers conducted a study to build the factors affecting the NPL. However many obstacles were encountered in the determination of the right variables. Differences in cross-country economic conditions and diverse studies can measure the determinants of different NPLs, and those variables indicate different relationships.

A study by Parab and Patil (2018) employed panel data model approach and specific bank variables and macroeconomic variables in India. The results showed that the use of these two variable factors gave exposure to the relationship with NPL up to 55%. Specific bank variables, such as ROA, ROE, channelled credit, inefficiency operations, CAR and bank size had an inverse relation to NPL while loan growth variable positively affected NPL. The macroeconomic variables used in the research were GDP growth, lending rates, unemployment and inflation, which had significant inverted linkage on NPL in India.

In contrast to research conducted by Mondal (2016), it only used macroeconomic variables such as GDP, inflation, interest rates and unemployment rate against NPL in the banking industry in Bangladesh with least square and Granger causality test methods. The results showed that inflation and interest rates had a negative effect on NPL, but GDP and unemployment rates had a positive relationship with NPL. GDP growth would lead to an increase in people's incomes and ultimately help the economy keep NPL at a low point.

According Ofori-Abebrese et al (2016) with the research using macroeconomic variables (inflation, exchange rate, money supply M2, GDP and treasury bill) on NPL at commercial banks in Ghana with the Autoregressive Distributed Lag (ARDL) method approach showed that inflation and treasury bill significantly influenced the NPL so that uncertainty on macroeconomic conditions would directly affect the performance of banks. It required policymakers whose policies were able to manage the economy well so that NPL would stay fixed at low point.

Research conducted by Anjom & Karim (2016) using macroeconomic variables and specific bank variables examined the relationship between these two factors to NPL in the banking industry in Bangladesh. The correlation approach was used to measure the causality between macroeconomic variables and bank-specific variables to NPL in which the results showed that in Bangladesh, GDP growth and total assets had a positive correlation to NPL, whereas inflation, loan to asset ratio and Loan to Deposit Ratio (LDR) had a negative correlation on NPL. Domestic debt variable as a macroeconomic variable and positively affected the NPL, thus increasing domestic debt would lead to an increase in Bangladesh's NPL.

Janvisloo et al., (2013) investigated the relationship between the macroeconomic and NPL in commercial banks in Malaysia. Observations on 23 commercial banks and more than 250 data observations indicated that FDI-net outflow (% GDP) was a variable that positively affected NPL and there was a negative relationship to the NPL of the GDP variable with the robust of the time influence on lags 2. Inflation and domestic credit growth had a positive and negative relationship and the influence of these two variables that specifically lasted up to two years.

This study examines how macroeconomic variables and specific bank variables affect the NPL of commercial banks in Indonesia. This study employ macroeconomic variables that are GDP, inflation and interest rates, while the specific bank variables used are Return On Assets (ROA), Return On Equity (ROE) and efficiency for operational cost. Dynamic panel data approach is used because it combines cross-section data of 106 commercial banks in Indonesia and time series using quarter data from 2015: 01 – 2017: 04 with total 1272 observation.

3 RESEARCH METHODS

3.1 *Data and data sources*

This research uses a quantitative approach. The quantitative method is a scientific approach that views a reality to be classified, concreted, observable and measurable, and the relationship of variables are causal where the research data is in the form of numbers. This study focuses on the explanation of the relationship between the NPL as the dependent variable and GDP, inflation and interest rates, ROA, ROE and BOPO as independent variables. The type of data used in this study is secondary data as obtained from the Bank Indonesia (BI), the Financial Services Authority (OJK) and the Central Bureau of Statistics (BPS).

3.2 *Definition of operational variables*

These variables are grouped into two, namely the dependent variable (restricted variable) and independent variable (free variable). Dependent variable used in this research is NPL in a commercial bank in Indonesia, while the independent variable used are GDP growth, inflation, interest rate, ROA, ROE and BOPO. The definition of variables used in this study is: first, Gross Domestic Product (GDP), which is the amount of gross value added by all producers in an economy plus taxes and minus subsidies that are not included in the value of a product. This study uses a percentage of quarterly GDP growth rate. GDP growth data on constant prices is accessed from the Bank Indonesia. Second, the Consumer Price Index (CPI) as a proxy of data inflation is accessed from the Indonesian Central Bureau of Statistics (BPS). Third, the interest rate (IR) as a policy tool used by the monetary authorities to influence the amount of credit disbursed and the money supply in a country. All three variables are grouped as macroeconomic variables.

The bank-specific variables in this study are; first ROA, which is a ratio that measures the ability of banks in generating profit or profit, which is measured base on how effective banks in using their assets in generating revenue. Second, ROE, which is the profitability ratio that compares the net profit of the bank with its net assets (equity or capital). This ratio uses the relationship between profit after tax with the used and owned capital. Third, BOPO as a measure of efficiency and operational effectiveness of a company/banking. The lower BOPO means the more efficient the bank is in controlling its operational costs with the efficiency of the cost of the profits obtained by the bank will be greater. The use of both internal and external factors as independent variables is considered important to see the effect on NPL in the banking industry in Indonesia. All specific bank variables are accessed from OJK.

Common models of dynamic data panels are:

$$Y_{it} - Y_{it-1} = (1 - \alpha)Y_{it-1} + \beta_1 X_{it} + \beta_k X_{it} + \eta_i + \varepsilon_{it} \tag{1}$$

Dynamic model panel data is:

$$Y_{it} = \alpha Y_{it-1} + \beta_1 GDP_{it} + \beta_2 CPI_{it} + \beta_3 IR_{it} + \beta_4 ROA_{it} + \beta_5 ROE_{it} + \beta_6 BOPO_{it} + \eta_i + \varepsilon_{it} \tag{2}$$

Explanation:
Y   = Non-performing loan;
GDP   = GDP growth rate;
CPI   = Consumer Price Index;
IR   = Interest Rate;
ROA   = Return On Asset;
ROE   = Return On Equity;
BOPO = Operation Expenses to Operation Income;
η_i = Unobserved banks-specific effect term
ε_{it} = Error term
i   = Cross-section on bank's index
t   = Time index

The econometric method is used by estimation of the GMM dynamic panel, suggested by Arellano & Bond (1991) and developed by Blundell & Bond (1998). This method is applied because of the need to overcome the bias in the models and effects of any banking industry. To eliminate the specific market effect problems, in the Arellano & Bond models (1991), it is suggested transforming the model to first-difference Eq. (2) and use lag at the level in regression to eliminate the bias in the model. Prudence is required in applying the model, as it may lead to erroneous conclusions caused by persistent independent variables. To solve the problem, Blundell & Bond (1998) suggest applying the GMM system estimator at both level and first-difference. This model can reduce the bias and inaccuracy associated with the difference in estimation.

This research utilize the GMM system because the estimation result is consistent and free from bias compared with Ordinary Least Square (OLS), fixed effect, and different GMM. The GMM system is able to overcome the problem of endogeneity because it produces an efficient estimation of the difference of GMM or fixed effect. The GMM system has two estimations, one-step estimation and two-step estimation. Theoretically, the estimation with two-step the GMM system is more efficient than using one-step estimation because it uses optimal weighting. This research model applies the GMM system with two-step estimator to test the effect of macroeconomic variables and bank-specific variables on NPL in the banking industry in Indonesia. The consistency of the GMM system result depends on approximately two specification tests, i.e. the Sargan test to state that the model is valid and correct, and an autocorrelation test (AR2) to eliminate time-series problems in the model.

4 RESULTS AND DISCUSSION

The model estimation in this study uses a data panel combining cross-section data with 106 data industry banks in Indonesia and time series that form with quarter data from 2015: 01 – 2017: 04. Industry banks recorded on OJK is counted to 116. The selection of up to 106 banks is due to several banks not having published reports and occurrence of bank mergers, so there is no recent report validated. Additional information that the sample in the model shows is that cross-section N data is larger than the time series data T (N > T), so it is suggested to use the GMM panel data model analysis.

Based on Table 1, it is known that the variable NPL, GDP growth and ROA have a standard deviation value greater than the average. It indicates that there is heterogeneity in the data in the period of observation. The greater the value of the standard deviation, the greater the average distance of each unit of data is against the average value of the count. Inflation, interest rate, ROE and BOPO variables have a standard deviation value below the average value of the count, not indicating that the data is homogeneous. However, the degree of heterogeneity is no greater than the previous variable. The total observation data is 1,272, and the maximum value of the NPL touches 95.95 as occurred in 2016 in the third quarter of the state pension savings bank. This study focuses after the crisis of 2008, and the application of

Table 1. Statistic descriptive.

Variable	Obs	Mean	Std deviation	Min	Max
NPL	1272	3.00967	3.480428	0	95.95
GDP Growth	1272	1.207459	2.302919	−1.802197	3.852051
Inflation	1272	0.385	0.3110215	−0.05	0.96
Interest Rate	1272	5.916667	1.340355	4.25	7.5
Return On Asset	1272	2.093687	2.561783	0	81.34
Return On Equity	1272	12.6251	10.14594	0.02	89.26
BOPO	1272	85.07987	16.78725	0.86	195.7

Source: Stata 13

banking in maintaining financial stability, one of them by keeping the NPL value at a low point. The next step is the GMM panel data estimation, as follows:

Table 2 shows the result of fixed effect panel data estimation, two-step GMM difference panel and two-step GMM System. The probability value of lagged dependent is significant, which indicates that the GMM panel model is appropriately used. Both results in the table show consistent value and interest rate variable have negative coefficient value to NPL, but no significant effect. This is in accordance with Anjom & Karim's (2015) research that interest rate has no significant effect on NPL. GDP growth positively affects the NPL, indicating that the increase in GDP will affect the increase in NPL. The behaviour of a bank that follows its business cycle will increase the risk as proven by the research of Mondal (2016), which states a correlation between GDP growth and NPL.

The inflation variable shows a probability value smaller than the alpha of 5%, which means that inflation affects the NPL. However, the coefficient of inflation is negative and the rise in inflation will decrease the NPL level in Indonesian banks. The rise in inflation is reflected in rising prices of goods so that people tend to reduce their demand for credit due to the greater risk of default. These findings are supported by Mondal (2016). The three macroeconomic variables show the determination of different NPL.

The specific variable of the bank shows that the three variables ROA, ROE and BOPO have effect on the level of 5% significance to NPL in Indonesia. The ROA and ROE variables have positive and negative coefficients, ROA and ROE demonstrate the level of bank efficiency in the use of equity and assets. However, higher asset utilization rates will have an impact on increasing NPL where, in this case, it indicates that the level of use of assets in the bank should be managed carefully and this applies equally to equity (Parab, 2018). BOPO has a positive effect on NPL so the higher BOPO will have an impact on the increase of NPL.

The consistency of GMM panel data depends on two things. First is the value of the Sargan Test; the results in Table 2 show that the Sargan test value shows a probability value above 5%, meaning that the model is free of excessive restriction problems or the model is said to be valid. Second is free of this autocorrelation problem as stated in Table 2 by identifying Arellano & Bond (1991) or AR2 values above from alpha 5%.

Table 2. Results of difference GMM two-step estimations and GMM two-step system.

Variabel	Difference GMM two-step system	GMM two-step model system
NPLt-1	0.003	0.002
	(0.000)***	(0.000)***
GDP Growth	0.042	0.056
	(0.000)***	(0.000)***
Inflation	−0.098	−0.080
	(0.000)***	(0.000)***
Interest Rate	−0.010	−0.0003
	(0.426)	(0.971)
Return On Asset	1.209	1.232
	(0.000)***	(0.000)***
Return On Equity	−0.130	−0.128
	(0.000)***	(0.000)***
BOPO	0.020	0.044
	(0.000)***	(0.000)***
Constanta	0.113	−1.843
	(0.370)	(0.000)***
Sargan Test	0.143	0.054
Arellano Bond Test (AR2)	0.118	0.594
Observation	1051	1160
Number of Banks	106	106

Sources: Stata 13.

5 CONCLUSION

Using data from 106 banks ranging from 2015: 01 – 2017: 04, this study examines macroeconomic variables and bank-specific variables on Non-Performing Loans (NPL) in the banking industry in Indonesia using the GMM panel data approach. Both factors have a very important role in maintaining banking stability in Indonesia. The role of regulators and stakeholders in actualizing a policy that is able to maintain stable macroeconomic conditions will have a very wide impact in economic conditions, especially in banking.

From the results above, it can be conclude:

1. The bank is procyclical, meaning that the behaviour of the bank follows its business cycle. The increase of GDP growth will affect when the level of demand for credit is increased and the policy of countercyclical and selective demand for credit must be applied accordingly so as not to repeat the crisis of 2008 caused by over-optimism from high economic growth.
2. Emphasizing the importance of bank efficiency in terms of asset, equity and operational efficiency, which will eventually minimize any risk.

REFERENCES

Anjom, W. & Karim, A.M. (2016). Relationship between non-performing loan and macroeconomic factors with bank specific factors: a case study on loan portofolio – SAARC countries, ELK Asia Pacific. Journal of Finance and Risk Management Vol 7 Issues 2.

Arellano, M. & Bond, S. (1991). Some tests of specification for panel data: Monte Carlo evidence and an application to employment equation. *The Review of Economic Studies, 58,* 277–297.

Blundell, R. & Bond, S. (1998). Initial conditions and moment restrictions in dynamic panel data models. *Journal of Econometric, 87,* 115–143.

Claessens, S., Ghosh, S.R. & Mijet, R. (2013). Macro-prudential policies to mitigate financial system vulnerabilities. *Journal of International Money and Finance, 39,* 153–185.

Herring, J. (1999). Credit risk and financial instability. *Oxford Review of Economic Policy, 91,* 401–419.

Janvisloo, A.M. & Muhammad, J. (2013)., Non-performing loans sensitivity to macro variables: panel evidence from Malaysian Commercial Banks. *American Journal of Economics, 3,* 16–21.

Kuncoro, M. (2013). Metode Riset untuk Bisnis dan Ekonomi: Bagaimana Meneliti dan Menulis Tesis? Erlangga, Jakarta.

Mankiw, N.G. (2003). *Teori Makroekonomi.* Jakarta: Penerbit Erlangga.

Minsky, H.P. (1982). The financial instability hypothesis. *Levy Economics Institute Working Paper, 74.*

Parab, C.R. & Patil, R. (2018). Sensitivity of credit risk to bank Specific and macroeconomic determinants: empirical evidence from Indian Banking Industry. *International Journal of Management Studies, 5,* 46–56.

Business strategy

Business Innovation and Development in Emerging Economies – Trinugroho & Lau (Eds)
© *2019 Taylor & Francis Group, London, ISBN 978-1-138-35996-3*

Greengrocers, fish sellers, butchers, and chicken sellers at traditional markets: Model of business governance and business behavior

T.A. Lubis, Firmansyah & R. Savitri
Universitas Jambi, Jambi, Indonesia

ABSTRACT: This research aims to identify a model of business governance and business behavior for greengrocers, fish sellers, butchers, and chicken sellers at traditional markets in Jambi province, Indonesia. This research used a mixed qualitative and quantitative method following a sequential exploratory design. Results of this study indicated that business governance among greengrocer, fish seller, butcher, and chicken seller at traditional markets reflected aspects of business ethics, openness, fairness, and equality. These aspects have significantly positive impacts on business behavior, reflected in traders taking risks and being proactive in facing competition.

Keywords: Business governance, business behavior, traditional markets

1 INTRODUCTION

Traditional market in its original context is defined as a physical place where buyers and sellers meet in order to make exchanges (Kotler, Wong, Saunders & Armstrong, 2005 as cited in Celhasic, Grdic & Ozer, 2008). They also revealed that the tangibility of products in the traditional market makes it possible for the consumer to gain hands on experience and learn about the product by trying it out. Food production and distribution processes present various environmental challenges, especially in relation to waste disposal, energy use, air pollution, water degradation and a decrease in bio-diversity (Werner and Bammert, 2001).

Some research has been carried out in Indonesia into greengrocers, fish sellers, butchers, and chicken sellers at traditional markets. Nabila and Husaini (2017) suggest that butchers at Aceh Besar traditional market use inherited business behaviors, while Ain et al. (2017) find that traders of this type at traditional markets in Rembang use domestic factors such as competitive advantage to make their businesses efficient. Takalamingan et al. (2017) conclude that businesses at traditional markets in the town of Tuna in Bitung, – use two traditional business models.

Bekkers et al. (2017) find that grocers in markets must provide prices to consumers in accordance with local circumstances. Minten et al. (2016) conclude that coffee merchant in markets in Ethiopia are transparent in their activities and so elicit trust from consumers in terms of price determination and quality of product.

Studies into the contexts of business behavior have also been carried out in various regions of the world, and these put forward a variety of findings. Tezel et al. (2017) conclude that suppliers and traders must have mechanisms that allow them to deal with competition in the market. Bogomolova et al. (2017) suggest that traders have to use intuition in making pricing decisions, and that in determining the decisions to take to develop their businesses, merchants need advices from practitioners. Battistella, et al. (2017) put forward the view that businesses should be able to show apply innovative strategies in their activities, and Carlucci and Schiuma (2017) also indicate that entrepreneurs need to innovate within their businesses.

Hernandez and Carra (2016) conclude that if a business is to survive it needs to maintain a relationship and communication with business centers, universities, and public research organizations. Furthermore, Assefa, et al. (2016) indicate that food traders in the EU use strategy management in cultivating diverse business risk. Myskova and Doupalova (2015) conclude that management in small businesses should be able to make appropriate business decisions under all circumstances and should be able to control the risks associated with their businesses. In research related to business governance and business behavior in sub-Saharan Africa, Roba, et al. (2018) conclude that traders of foodstuffs in sub-Saharan Africa need to obtain specific information related to the price ranges on offer from various merchants in the market as a form of proactive behavior in facing business competition.

Greengrocers, fish sellers, butchers, and chicken sellers comprise the largest group of traders in traditional markets. This predominance reflects high and rising levels of consumption of vegetables, fish, meat and chicken. Increased consumption by local communities provides potential for increased business, but if this is not matched by proper governance, poor business behavior will result.

2 METHOD

This research used a mixed qualitative and quantitative method following a sequential exploratory design in which the first stage was qualitative in nature while remaining stages were quantitative. The research was conducted by looking at greengrocers, fish sellers, butchers and chicken sellers at traditional markets in 11 regencies and cities in Jambi province. Quota sampling provided 440 samples, with ten samples from each of the four types of traditional market traders.

3 RESULTS AND DISCUSSION

In terms of ownership, the sample of greengrocers, fish sellers, butchers, and chicken sellers at traditional markets, 82.5% owned their own businesses and the remaining 17.5% were family-owned. In terms of governance type, 86.36% were self governed and the remaining 13.64% were family governed.

Research found that 71.59% of merchants conducted business research while the remaining 28.41% did not. In the sample, 87.5% sourced raw materials from suppliers, while 12.5% supplied raw materials themselves. All of the researched businesses stated that they carried out recording of all transactions, separated business and personal money management, and actively managed financial and operational aspects of their businesses.

The initial qualitative stage of the study involved interviews with some of the research sample. The results of the interviews were tabulated to provide conclusions about which indicators would be used for the two research variables, namely business governance (X) and business behavior (Y). Five business governance indicators were identified: openness (X11), business ethics (X12), responsibility (X13), independence (X14), and fairness and equality (X15). Meanwhile three indicators were chosen for the business behavior variable: taking risk (Y11), continuous improvement (Y12), and proactively meeting competition (Y13).

Following this development of the indicators, the quantitative stage of the research questionnaire gathered data via a questionnaire for subsequent inferential analysis using the partial least squares (PLS) method.

The results of the initial calculation model are presented in Figure 1.

According to Ghozali (2006), convergent validity of the measurement model with reflected indicators can be seen from the correlation between the score and score indicator item. The indicators are considered reliable if the correlation value is above 0.70. However, research on the development scale of 0.5 to 0.6, loading is still acceptable. Based on these guidelines, the indicators X13, X14, and Y12 were removed. Thus the final model used in this research is as presented in Figure 2.

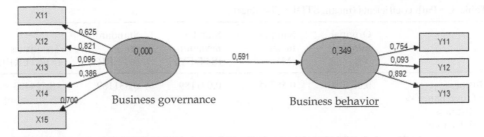

Figure 1. Initial research model results.

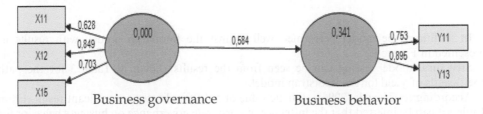

Figure 2. Final research model results.

Table 1. Average Variance Extracted (AVE) and communality.

Variable	AVE	Communality
Business behavior	0.683400	0.683400
Business governance	0.536190	0.536191

Source: Data processed by PLS.

Table 2. Cronbach's alpha and composite reliability.

Variable	Cronbach's alpha	Composite reliability
Business behavior	0.549793	0.810795
Business governance	0.563444	0.773421

Source: Data processed by PLS.

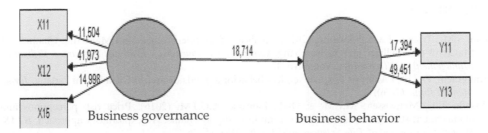

Figure 3. The results of the bootstrapping model.

From Figure 2 it can be seen that all the loading factors have indicators of above 0.6, indicating that the outer model achieves convergent validity.

As the average variance extracted (AVE) and communality values all exceed 0.5, the outer model is valid so there is no need to mention the omission of indicator variables and we can continue with the analysis of the inner models (structural models).

Table 3. Path coefficients (mean, STDEV, T-values).

	Original sample (O)	Sample mean (M)	Standard deviation (STDEV)	Standard error (STERR)	T statistics (|O/STERR|)
Business Governance -> Business Behaviour	0,583680	0,587552	0,031189	0,031189	18,714403

Source: Data processed by PLS.

In addition, the model performed well against the bootstrap model, with results as presented in Figure 3.

Structural model testing can be seen from the results shown in Table 3 for the path coefficients of yield for the bootstrap model.

According to Ghozali (2006), statistic value of > 1.96 indicates a significant result. From Table 3 it can be inferred that the influence of corporate governance on business behavior for traders of vegetables, fish, meat, and chicken showed significant effects.

Minten et al. (2016). In the context of business behaviors that are reflected by taking risks and being proactive in terms of meeting competition, the results of this study are in line with those of Bogomolova et al. (2017), Hernandez and Carra (2016), and Myskova and Doupalova (2015).

However, the research results do not indicate that continuous improvement is reflected in the business behavior of traders in traditional markets and this finding is contrary to the findings of Carlucci and Schiuma (2017).

4 CONCLUSION

The results of this study suggest that the governance of greengrocers, fish sellers, butchers, and chicken sellers has positive significant effect on business behavior. Governance as reflected by openness, business ethics, and fairness and equality has positive significant effects on business behavior that are reflected by the taking of risks and being proactive in facing competition.

REFERENCES

Ain, S.M., Triarso. I & Sardiyatmo. (2017). Analysis of Competitiveness and Marketing Channels of Mackerel (Rastrelliger sp.) In Rembang, Central Java. *Journal of Fisheries Resources Utilization Management and Technology,* Vol. 6. No. 4.

Anggraeni, D. (2017 Analysis of street vendors behavior at roadside area at Citra Pantai Padang. *Jurnal Menara Ilmu, 11*(76).

Assefa, T.T., Meuwissen, M.P.M. & Oude Lansink, A.G.J.M. (2016). Price risk perceptions and management strategies in selected European food supply chains: An exploratory approach. *NJAS-Wageningen Journal of Life Sciences, 80*(March 2017), 15–26.

Battistella, C., De Toni, A.F., De Zan, G. & Pessot, E. (2017). Cultivating business model agility through focused capabilities: A multiple case study. *Journal of Business Research 73*(C), 65–82.

Bekkers, E., Brockmeier, M., Francois, J. & Yang, F. (2017). Local food prices and international price transmission. *World Development, 6,* 216–230.

Bogolomova, S., Szabo, M. & Kennedy, R. (2016). Retailers and manufacturers price-promotion decision: Intuitive or evidence-based? *Journal of Business Research, 76,* doi: 10.1016/j.jbusres.2016.05.020.

Carlucci, D. & Schiuma, G. (2017). The power of the arts in business. *Journal of* Business Research. Volume 85 Pages 342–347 https://doi.org/10.1016/j.jbusres.2017.10.012

Firmansyah., Lubis, T.A. & Zulkifli. (2015). Behavior Development Model of Fisheries Financial Management in Jambi Province. Report of Competitive Grants Research. Research institutions and community service *(unpublish)*.

Guissoni, L.A., Sanchez, J. & Rodrigues, J.M. (2018). Price and in-store promotions in an emerging market. Emerald insight. Vol. 36 Issue: 4, pp. 498–511, https://doi.org/10.1108/MIP-08-2017-0154.

Ghozali, I. 2006. Structural Equation Modeling; Alternative Method with PLS. Undip Publishing Board. Semarang.

Hernandez, R. & Carra, G. (2016). A conceptual approach for business incubator interdependencies and sustainable development. *Agriculture and Agricultural Science Procedia, 8*, 718–724.

Lofsten, H. (2016). New technology–based firms and their survival: The importance of business networks, and entrepreneurial business behaviour and competition. *Local Economy, 31*(3).

Lubis. T., Erwita, Andi, Raja. (2014). The Street Vendor Governance Model in Jambi. Journal of Management Dynamics. Vol. 2. No. 2.

Lubis, Tona Aurora., Junaidi. (2014). Traditional Market Price Monitoring Survey in Jambi. Indonesia Bank of Jambi Area. *(Unpublish)*.

Lubis, Tona Aurora., Kernali, Emi., Ridwan. (2014). Financial Behavior of Cooperatives in Sarolangun District, Jambi Province. Research Report. Center for Research and Development of SMEs and Cooperatives. Institute of Research and Community Service of Jambi University. *(Unpublish)*.

Lubis, Tona Aurora., Zulkifli, Firmansyah. (2016). Financial Behavior Model of Batik Craftsmen in Olak Kemang District. Research Report of Jambi University Management Masters Program. *(Unpublish)*.

Minten B., Assefa, T. & Hirnoven, K. (2016). Can agricultural traders be trusted? Evidence from coffee in Ethiopia. *World Development, 90*(C), 77–88.

Myskova, R. & Doupalova, V. (2015). Approach to risk management decision-making in the small business. *Journal of Business Economics and Management*. Procedia Economics and Finance 34 (2015) 329–336.

Nabila. A. & Husaini, A.Z. (2017).). Prosopography of Beef Traders in Lambaro Main Market, Ingin Jaya Subdistrict, Aceh Besar Regency, 1986–2016. *Student scientific journal. 2*(1). Volume 2, Nomor 1, Januari 2017, hlm. 1–9.

Roba, G.M., Lelea, M.A., Hensel, O. & Kaufmann, B. (2018). Making decisions without reliable information: The struggle of local traders in the pastoral meat supply chain. *Food Policy, 76*, 33–43.

Schmidbauer, E. & Stock, A. (2017). Quality signaling via strikethrough price. (Universiteit Van Pretoria dissertation). *International Journal of Research in Marketing*. Volume 35, Issue 3, September 2018, Pages 524–532. https://doi.org/10.1016/j.ijresmar.2018.03.005.

Takalamingan, M., Longdong, F.V. & Jusuf, A. (2017). Efficiency Analysis of Distribution Channels and Risk of Businessmen in the Cakalang Asap Fish Supply Chain in Girian Atas Village, Bitung City, North Sulawesi Province. *E-journal Unsrat. 5*(9).

Tezel, A., Koskela, L. & Aziz, Z. (2017). Current condition and future directions for lean construction in highway projects: A small and medium-sized enterprises (SMEs) perspective. *International Journal of Project Management, 36*(2).

Werner, F. & Bammert, M. (2001). Environmental protection in the food and consumer goods industry. Greener Management International. No. 34, pp. 27–33.

Corporate finance

Business Innovation and Development in Emerging Economies – Trinugroho & Lau (Eds)
© 2019 Taylor & Francis Group, London, ISBN 978-1-138-35996-3

Bankruptcy prediction of manufacturing companies using Altman and Ohlson model

Amanda Meisa Putri & Imo Gandakusuma
Faculty of Economics and Business Universitas Indonesia, Indonesia

ABSTRACT: Financial distress can be regarded as an early warning of trouble that can lead to bankruptcy. Predicting bankruptcy becomes one thing that companies can do to discover the state of the company's financial health. A total of 585 firm-years of manufacturing companies that listed in Indonesia Stock Exchange are sampled for this research where 113 of them are categorized in financial distress state. Bankruptcy prediction models may be examined to assess a company's economic situation for further purposes. Altman and Ohlson are some of notable researchers to which their models are referred to evaluating the health of companies.

Keywords: Financial Distress, Bankruptcy Prediction, Manufacturing Companies, Altman, Ohlson

1 INTRODUCTION

Based on United Nations Statistics Division (2016), Indonesia ranks 16 out of 212 countries by its manufacturing industry's contribution to Gross Domestic Product (GDP) with 20.51%. Within Southeast Asia Indonesia ranks fourth below Thailand (27,42%), Malaysia (22,84%), dan Myanmar (22,44%). The manufacturing industry is a sector that contributes greatly to Indonesia's GDP. The contribution of the manufacturing industry is fairly large at an average of 21.01% during 2012 to 2016. The following is the contribution of the manufacturing industry in each year. Although it is a significant contribution to GDP, it is noted that the contribution is decreasing every year. In 2012, contributions of 21.45% fell to 20.51% within five years. In fact, this figure is much larger in 2001 that is 27%. In Indonesia there are 198 manufacturing companies listed on the Stock Exchange in 2016. A total of 585 firm-years are sampled from this study where 113 are categorized as financial distress.

Financial distress can be regarded as an early warning of problems by companies (Ross, 2016). Financial distress can lead to bankruptcy of a company so predicting bankruptcy becomes one thing that companies can do to know the state of the company's financial health. It is difficult to define precisely what distress or bankruptcy is due to the variety of accounting procedures or rules in different countries at different times, as well as various events that sent the firms into financial distress (Zhang, 2010). Much of the literature on predicting bankruptcy has emerged and is used in various literatures. Altman and Ohlson models are some examples of popular models that are often used. Altman uses a discriminative multivariate analysis (MDA) in the list of financial ratios to identify ratios that are statistically related to future bankruptcy (Altman, 1968). It is used primarily to classify and/or make predictions in problems where the dependent variable appears in qualitative form, for example, male or female, bankrupt or nonbankrupt (Altman, 2000). Therefore, the first step is to establish explicit group classifications. There are several improvements to the model studied by Altman, for listed manufacturing companies, private companies, and models for emerging markets. Ohlson uses a logit model (Ohlson, 1980), which uses less restrictive assumptions than those adopted by the MDA approach (Y. Wu, 2010).

Based on the above arguments, the studies aims to use Altman and Ohlson model to predict bankruptcy of manufacturing companies in Indonesia.

2 LITERATURE REVIEW

2.1 *Financial distress/bankruptcy*

Financial distress are situations where the operating cash flow of a company is insufficient to meet current liabilities (such as trade credit or interest expense) and the company is forced to take corrective action. Financial distress may cause the company to default on the contract, and may involve financial restructuring between the company, its creditor, and equity investor. Usually companies are forced to take action that will not be taken if it has sufficient cash flow (Ross, 2016). Ross further also said that financial distress can be an early warning for the company matters of a problem. Companies with more debt will experience financial distress earlier than companies with less debt. Companies with low leverage will experience financial distress in the future and, in many instances, are forced to liquidate. Another argument said that companies do not encounter financial difficulties because they have too much debt, but because they are not profitable enough. A heavy debt burden does no more than hasten the onset of financial difficulties (Vernimmen, 2005).

2.2 *Ratio analysis in bankruptcy prediction*

Every department in every business produces some kind of information that can be used by its manager to measure performance. This information may be related to operational considerations within the department, the financial condition of the entire company, or the performance of a company's suppliers and customers. Unfortunately, managers may not be aware of the multitude of measurements that can be used to track these different levels of performance or of the ways that these measurements can yield incorrect or misleading information (Bragg, 2007). There are many relationships between financial accounts and between expected relationships from one point in time to another. Ratios are a useful way of expressing these relationships. Ratios express one quantity in relation to another (usually as a quotient). Several aspects of ratio analysis are important to understand. First, the computed ratio is not "the answer." The ratio is an indicator of some aspect of a company's performance, telling what happened but not why it happened (Robinson, 2009).

Ratio analysis has an impressive ability to predict corporate bankruptcy by analyzing the characteristics of different corporate groups so it is an important tool for financial managers. Financial ratios could not "track the cause" of failures but can only try to gauge the extent to which the company's policies and problems are performing poorly. If company management is able to recognize a potentially bleak future through its ratio analysis, alternatives will then be explored and bankruptcy may be avoided.

"Failure" is defined as the inability of a firm to pay its financial obligations as they mature. Operationally, a firm is said to have failed when any of the following events have occurred: bankruptcy, bond default, an overdrawn bank account, or nonpayment of a preferred stock dividend. A "financial ratio" is a quotient of two numbers, where both numbers consist of financial statement items. A third term, predictive ability, also requires explanation but cannot be defined briefly (Beaver, 1966).

2.3 *Altman model*

Altman's research analyzed 33 inactive companies and 33 active companies. In the end, Altman's Z-score results found that the accuracy of the first and second year prior to failure was 95% and 72%, respectively. Public manufacturing companies were used in the original Altman Z-score model to predict bankruptcy. Five variables were chosen as doing the best overall job together in the prediction of company bankruptcy (Altman, 1968).

The model has limitations that only apply to public manufacturing companies because stock price data is required. So in 1983 Altman revised the original model to be used in private manufacturing companies. For private manufacturing companies, the fourth variable uses the book value of the entity, not the market value of the equity. The two models apply to manufacturing companies. Therefore, to predict the possible bankruptcy of non-manufacturing companies in developing countries, Altman developed the Z" model (Altman, 2005). In this model, the fifth variable has been ignored as it tends to be significantly higher for retail and service companies than with a manufacturing company. Furthermore, the book value of equity is used on the 4th variable.

2.4 Ohlson model

Ohlson introduces an alternative measurement method that is using logit analysis approach. However, as with most logit approach models, Ohlson does not provide the scale for converting values into bankruptcy probabilities (Y. Wu, 2010). There are nine independent variables used to predict bankruptcy in Ohlson Model. Three sets of estimates were computed for the logit model using the predictors previously described. Model 1 predicts bankruptcy within one year, Model 2 predicts bankruptcy within two years, given that the company did not fail within the subsequent year, Model 3 predicts bankruptcy within one or two years (Ohlson, 1980).

3 RESEARCH METHOD

The research method is done by first classifying the data into healthy and distress, the company experiencing financial distress is a company that has negative cumulative earnings for two consecutive years or net asset value (NAV) per share is below par value (Zhang, 2007). Next is to classify the health of the company based on the Altman model, Altman Emerging Market, and Ohlson. Each model has different classification and cutoff point. The three models and their classification are presented in Table 1.

Further classification with these three predictive models is compared with the initial classification to validate the model and find out if there are errors in the categorization of the company's health. And the last step is to test the validity of the model by grouping into two namely Type I error when the model classifies a bankrupt company as a healthy company and Type II error is a healthy company that is classified as a bankrupt company. Models that have a smaller error rate in Type I and Type II lead to higher accuracy so it is concluded that the model is more suited to predict the bankruptcy of manufacturing companies in Indonesia.

4 RESULT AND DISCUSSION

This study begins by classifying firm-years that are considered as financial distress and considered healthy, companies that have negative cumulative earnings for two consecutive years are considered as companies experiencing financial distress.

Table 1. Bankruptcy prediction models.

	Model	Cutoff point
Altman	$Z = 1.2X1 + 1.4X2 + 3.3X3 + 0.6X4 + 0.99X5$	$Z \geq 2,99$: Healthy $1,81 < Z < 2,99$: Gray Area $Z \leq 1,81$: Bankrupt
Altman EMS	$Z" = 6,56X1 + 3,26X2 + 6,72X3 + 1,05X4 + 3,25$	$Z \geq 2,6$: Healthy $1,1 < Z < 2,6$: Gray Area $Z \leq 1,1$: Bankrupt
Ohlson	$0 = -1.32-0.407\ X_1 + 6.03X_1 - 1.43X_1 + 0.0757X. -2.37X_5$ $- 1.83X_6 + 0.285X_7 - 1.72X_8 - 0.521X_9$	$O \geq 0,38$: Bankrupt $O < 0,38$: Healthy

Based on Table 2, of the total sample 585 firm-years, there are 113 firm-years with negative cumulative earning for two consecutive years or more and as many as 472 firms-years considered as healthy companies.

The next step is to classify the company's health using three models discussed in this study (Altman, Altman Emerging Market, and Ohlson) and then analyze it. Table 3 shows the descriptive statistics of each model including the mean values, median values, and standard deviations from the data.

Each model has a different cutoff point so it is reasonable when the value of the descriptive statistics in Table 3 differs considerably. The calculation of the classification of manufacturing firms based on each model is shown in Table 4.

As shown in Table 4, the Altman original model predicts only 166 firm-years in a healthy state, 83 firm-years predicted to be in the gray area, and 336 firm-years are predicted to go bankrupt. The Altman Emerging Market model is so different from Altman that predicts 530 firm-years in a healthy state, 3 firm-years into the gray area, and 52 firm-years predicted to go bankrupt. While the Ohlson model predicts 473 healthy firm-years and 112 bankrupt firm-years. Ohlson categorize company's health only into healthy and bankrupt.

After knowing the health classification of companies based on each model then compared with Zhang's initial classification. This is done to determine the validity of the best model in predicting bankruptcy.

Table 5 shows that Altman Emerging Market has the highest percentage of ability to predict both company's health and bankruptcy in average 73%. Ohlson model is not much different from Altman EMS with an average of 72%. While Altman model has highest level of ability to predict bankruptcy. But Altman model also has the lowest level of ability to predict the health of samples.

From Table 5 also can be determined the Type I Error group and Type II Error group as presented in Table 6.

Based on Table 6, the Altman model has 7% Type I error which categorized bankrupt as healthy, but has a very high Type II error rate of 60% in predicting a healthy company to become a bankrupt company. The Altman EMS model has highest Type I error rate of 54% but Type II error is very small that is 1%. While the Ohlson model has 45% Type I errors and 11% Type II errors.

Table 2. Zhang manufacturing firm classification.

Firm classification based on zhang	
Healthy	471
Distress	113

Table 3. Statistical descriptive.

	Altman	Altman EMS	Ohlson
Mean	2.7342	1.7561	−1.1909
Median	1.3749	1.2672	−1.5743
Std Deviation	5.4026	4.5860	3.3048
Max	47.9817	85.0939	31.0146
Min	−18.7088	−19.3706	−6.5306

Table 4. Altman 1968, Altman EMS, and Ohlson manufacturing companies bankruptcy classification.

	Altman	Altman EMS	Ohlson
Healthy	166	530	489
Grey Area	83	3	0
Bankrupt	336	52	96

Table 5. Comparison to zhang classification.

	Altman		Altman EMS		Ohlson	
	Healthy	Bankrupt	Healthy	Bankrupt	Healthy	Bankrupt
Healthy	158	234	469	3	434	38
Distress	8	102	44	38	55	58
% Correct	40%	93%	99%	46%	92%	51%
	67%		73%		72%	

Table 6. Error grouping.

Altman		Altman EMS		Ohlson	
Type I	Type II	Type I	Type II	Type I	Type II
7%	60%	54%	1%	45%	11%

5 CONCLUSION

Based on sample manufacturing companies in Indonesia during 2012 to 2016, there are 472 healthy firm-years and 113 firm-years considered financial distress. The Altman, Altman EMS, and Ohlson models provide mixed results in predicting sample bankruptcy. Altman EMS has the highest percentage on predicting both bankruptcy and health of manufacturing companies in Indonesia in average with 73%, followed by Ohlson model with 72% and Altman model with 67%. This means that Altman EMS has the higher level of consistency and more appropriate in predicting the bankruptcy of manufacturing companies in Indonesia.

REFERENCES

Altman, E.I. (1968). Financial Ratios, Discriminant Analysis and the Prediction of Corporate Bankruptcy. *Journal of Finance*.

Altman, E.I. (2000). Predicting financial distress of companies: Revisiting the Z-Score and ZETA® models. *Handbook of Research Methods and Applications in Empirical Finance*, 53(July), 428–456.

Altman, E.I. (2005). An emerging market credit scoring system for corporate bonds. *Emerging Markets Review*, 6(4), 311–323.

Beaver. (1966). Financial Ratios As Predictors of Failure Authors (s): William H. Beaver Source : Journal of Accounting Research, Vol. 4, Empirical Research in Accounting : Selected Published by : Wiley on behalf of Accounting Research Center, Booth School of Busi, 4(1966), 71–111.

Bragg, S.M. (2007). *Business ratios and formulas*. New Jersey: John Wiley & Sons, Inc.

Ohlson, J.A. (1980). Financial Ratios and the Probabilistic Prediction of Bankruptcy. *Journal of Accounting Research*, 18(1), 109.

Robinson, T.R., van Greuning, H., Elaine, H., & Broihahn, M. (2009). *International financial statement analysis*. New Jersey: John Wiley & Sons, Inc.

Ross, Westerfield, Jaffe, & Jordan. (2016). Corporate Finance. New Jersey: McGraw-Hill Education.

United Nations Statistics Division. (2016). Retrieved April 8, 2018, from www.unstats.un.org.

Vernimmen, P., Quiry, P., Dallocchio, M., Le Fur, Y., & Salvi, A. (2005). Corporate Finance: Theory and Practice. West Sussex: John Wiley & Sons, Inc.

Wu, Y.C.G. (2010). A Comparison of Alternative Bankruptcy Prediction Model. *Journal of Contemporary Accounting & Economics*.

Zhang, L., Altman, E.I., & Yen, J. (2010). Corporate financial distress diagnosis model and application in credit rating for listing firms in China. *Frontiers of Computer Science in China*, 4(2), 220–236.

Business Innovation and Development in Emerging Economies – Trinugroho & Lau (Eds)
© *2019 Taylor & Francis Group, London, ISBN 978-1-138-35996-3*

Valuation of stock using the discounted cash flow model and Ministry of Finance regulation: Study of PT Indosat Tbk

Astried Minang Nathalia & Zuliani Dalimunthe
Faculty of Economics and Business, Universitas Indonesia, Indonesia

ABSTRACT: This study aims to compare the estimated stock fair value for a telecommunications company in Indonesia—PT Indosat Tbk (Indosat Ooredoo)—based on the Discounted Cash Flow (DCF) model and that based on Ministry of Finance regulation. We conducted valuation by using historical data and financial statement projections. Indosat Ooredoo is a large company in the telecommunications industry that rebranded in 2015 in an effort to increase its growth and market share and become a leader in digital services in Indonesia. The first model applied is the Free Cash Flow to Firm and Free Cash Flow to Equity, based on Damodaran (2012), while the second model is based on Indonesian Ministry of Finance regulation SE-54/PJ/2016. The Discounted Cash Flow (DCF) model estimates that the intrinsic value of Indosat Ooredoo is Rp7,695 per share, while valuation based on the Ministry of Finance regulation estimates that the fair value is Rp7,935 per share. This result indicates that the share price of Indosat Ooredoo on December 31, 2016, at Rp6,450, is undervalued compared to the intrinsic value (fair value) derived from both models.

Keywords: stock valuation, DCF method, regulation-based valuation, Indonesia, financial projection

1 INTRODUCTION

The digitalization of information and communication technology has been acknowledged as a disruptive innovation (Ireland et al., 2003) that encourages the shifting of consumer behavior from voice and short message services into data services. The market participants in the telecommunications industry, especially in Indonesia, have adapted their business model to this industry.

Two factors support the increasing demand for data services in Indonesia: the increased number of cellular customers and the enhancement of internet users (APJII, 2016). PT Indosat Tbk, one of the four major players in the telecommunications industry in Indonesia, has tried to increase its market share to enhance its corporate value. Rebranding in 2015 was one of the efforts to achieve this company goal (Indosat, 2016).

In 2016, Indosat experienced the highest increase of cellular customers among its competitors (APJII, 2016). According to a survey conducted by the Association of Internet Service Providers Indonesia (APJII) in 2016, the penetration of internet users in Indonesia increased by approximately 51.8% compared to internet users in 2014. The data might encourage potential investors to invest in the telecommunications industry, especially Indosat, to gain attractive returns. Investors expect to receive two types of cash flow when they buy the stock: a dividend in the period during which they owned the stock and a capital gain from sales prices at the end of the period (Plenborg, 2002).

Before making an investment decision, investors should determine whether the company can create enough returns and sustainable growth. The analysis conducted by investors can include fundamental analysis and technical analysis (Koller et al., 1994).

Previous studies compare stock valuation based on the Residual Income (RI) and Discounted Cash Flow (DCF) approaches. However, the RI and DCF approaches are theoretically equivalent. This is because both approaches may result in an identical estimate of firm value if implemented properly and consistently. This study shows that simple assumptions affect the estimate of firm value differently. This study also argues that, because the framework for forecasting is based on accrual accounting, while budget control is generally based on accounting numbers rather than cash flow measures, it seems logical to estimate firm value based on the concepts known as accrual accounting and financial statement analysis. According to those criteria, the RI approach seems to be an attractive alternative to the DCF approach (Plenborg, 2002).

The purpose of this study is to investigate the fair value of shares of PT Indosat Tbk by using the discounted cash flow method and Ministry of Finance (MoF) regulation SE-54/PJ/2016. The discounted cash flow model uses Free Cash Flow to Equity (FCFE) and Free Cash Flow to Firm (FCFF), while the MoF regulation uses the income approach and asset approach to estimate a fair value. This study will make projections of Indosat's financial statements to predict the financial condition of Indosat over the next five years with assumptions made according to the fundamental analysis that has been done previously. The discount rates used by the authors are the cost of equity in the FCFE model and the cost of capital in the FCFF model.

This paper's outline is as follows: Section 2 presents a literature review of valuation; Section 3 presents the data and method of the study; Section 4 discusses the results of the study, and Section 5 provides the conclusion and recommendations.

2 LITERATURE REVIEW

The fair value of an asset is the hypothetical price at which a business can sell its assets (exit price), and it is not the price that needs to be paid to buy the asset (entry price) (Subramanyam, 2014). For a liability, the fair value is the price at which a business can transfer the liability to a third party, not the price it will get to assume the liability. The intrinsic value is the value of securities or a business from the perspective of a security analyst (Laro & Pratt, 2005). A previous study uses the real option method and the FCFF method to determine the fair value of the initial share price in the Indonesia Stock Exchange, and to determine the best valuation method (Paramitha et al., 2014).

Other studies that use the Dividend Discount Model (DDM) and Discounted Cash Flow (DCF) method show that the application of the two methods to calculate the fair price per share issued by each sector does not produce very different values. This can be seen through comparison of the fair price of shares with the stock market price

For efficient marketers, valuation is a useful exercise to determine why stocks are sold at the same price. A simple valuation for firms with well-defined assets that generate predictable cash flows is important. The real challenge involves expanding the valuation framework to include varied companies (Damodaran, 2012).

2.1 Fundamental analysis

Fundamental analysis uses information concerning the current and prospective profitability of the company to assess its fair market value. Its purpose is to identify misplaced stock prices relative to some measure of the 'actual value' that can be derived from observable financial data (Bodie et al., 2014). Fundamental analysis can be done by using a top-down approach that consists of macroeconomic analysis, industry analysis, and microanalysis (Bodie et al., 2014).

2.2 Financial forecasting and projected cash flow

Financial forecasting uses a Free Cash Flow (FCF) calculation as the basis of valuation (Koller et al., 1994). We measure the projected cash flow in two ways. The first method begins with net income and adjusts according to the difference between accounting information and cash flow (Kaplan & Ruback, 1995):

Net income
+ Depreciation & amortization
+ New debt issues
– Debt repayment
– Capital expenditure
– Incremental working capital
= Free cash flow to equity

Our second method for measuring capital cash flows begins with Earnings Before Interest and Taxes (EBIT):

EBIT (1-T)
+ Depreciation & amortization
– Capital expenditure
– Incremental working capital
= Free cash flow to firm

2.3 *Discount rate*

The discount rate is used to discount the projected cash flow to become the present value of the company. There are two levels of discount that can be used to determine the fair value of a company: cost of equity and cost of capital (Modigliani & Miller, 1958). The cost of equity is determined by three factors: the risk-free rate of return, the market-wide risk premium (the expected return of the market portfolio less the return of risk-free bonds), and the risk adjustment that reflects each company's riskiness related to the average company (Koller et al., 1994). The cost of capital is an advantage expected by investors (both shareholder and creditors), because it has placed its funds either in the form of equity (shares) or loan funds (debt) in a company (Eitman et al., 2013).

2.4 *Terminal value*

To estimate a company's value, we separate a company's expected cash flow into two periods: the present value of cash flow during the explicit forecast period and the present value of cash flow after the explicit forecast period (Koller et al., 1994). The terminal value assumes that the terminal cash flow will grow at a constant nominal rate for good (Kaplan & Ruback, 1995).

2.5 *Valuation with discounted cash flow method*

A valuation is performed to determine the fair value of a company in which the fair value can be used as a reference to determine whether the intrinsic value of the company is undervalued or overvalued. There are two models of valuation usually used by investors, the equity valuation and the firm valuation (Mielcarz & Mlinarič, 2014). While previous research showed that a comparative analysis of a classical Free Cash Flow to Equity (FCFE) and Economic Value Added (EVA) methodology makes a strong case for Free Cash Flow to Firm (FCFF) as the most efficient approach (Mielcarz & Mlinarič, 2014), this study will use both the FCFE and the FCFF models.

The equity valuation is obtained by discounting the expected cash flows to equity with the cost of equity. In the FCFE technique, the cash flows and the required rate of return are calculated from the equity owner's perspective (Mielcarz & Mlinarič, 2014). The company's valuation is obtained by discounting the cumulative cash flow to all claim holders in the company using the Weighted Average Cost of Capital (WACC). The company's cash flow comes from Free Cash Flow to Firm (FCFF), which is the amount of cash flow to all claims holders in the company, including common shareholders, bondholders, and preferred shareholders

(Mielcarz & Mlinarič, 2014). This calculates the firm's intrinsic value, which is expected to grow faster than stable firms in the initial period and at a stable level thereafter (Damodaran, 2012).

2.6 Valuation with Ministry of Finance regulation (SE-54/PJ/2016)

According to MoF regulation SE-54/PJ/2016 on the Property Appraisal Guidelines, Business Assessment, and Assessment of Intangible Assets for Taxation Purposes, valuation is a series of activities to determine the amount of a particular value at a certain time, which is conducted objectively and professionally based on standards for implementing the provisions of the taxation legislation. Meanwhile, business valuation is a process to estimate the market value of the company including various ownership interests (business ownership interests) as well as transactions and activities that affect corporate value (Pajak, 2016).

There are three valuation approaches to estimate the fair value of an assessment object, consisting of the market approach, income approach, and asset approach. The market approach was completed by comparing companies with other comparable companies that have a known market value. The income approach was conducted by projecting the discounted economic incomes with a certain discount rate to get the market value of the valuation object. Meanwhile, the asset approach can be used to obtain the market value of a company, the value of the invested capital market, the market value of the capital structure, and the market value of the company's net assets (equity). We must choose at least two approaches based on the asset's characteristics and then reach a reconciliation to calculate the fair value of assets (Pajak, 2016).

3 RESEARCH DESIGN

3.1 Data collection

This study uses secondary data consisting of Indosat's financial statements, annual reports, and macroeconomic reports (GDP, inflation, currency exchange rate, interest rate, and unemployment rate) within the five-year period from 2012 to 2016. The data are collected annually from the Indonesia Stock Exchange, Bank Indonesia, and Central Bureau of Statistics reports. Other data, such as the beta industry, market equity risk premium, and country risk premium, are collected from Damodaran. Risk-free rate data are collected from the Thomson Reuters Datastream.

Financial forecasting is conducted for the following five years, from 2017 to 2021. From 2022 onward, we assume that the growth of Indosat is constant. The projection for the first five years is called the Plan Period (PP), and the projection from the sixth year onward (perpetuity) is called the terminal value (Damodaran, 2012). We chose 31 December 2016 as the date of valuation, and the market price of Indosat's share at that time was Rp6,450.

3.2 Discounted Cash Flow (DCF) model

The analysis method for the discounted cash flow method is shown in Figure 1.

The Free Cash Flow to Firm (FCFF) model was carried out through several phases, namely:

1. Calculated Weighted Average Cost of Capital (WACC)
2. Cash flow projection (Free Cash Flow to Firm) for the next five years
3. Calculated the value of the firm (present value) by discounting free cash flow to firm on WACC with this formula:

$$Value\ of\ firm = \sum_{t=1}^{t=n} \frac{FCFFt}{(1+WACC,pp)^t} + \frac{\frac{FCFEn+1}{(WACC,tv-gebit)}}{(1+WACC,tv)^n}$$

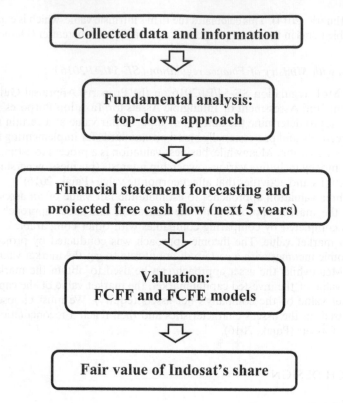

Figure 1. The analysis method.

where:

FCFF$_t$ = *Free cash flow to firm in year t*
WACC$_{,pp}$ = *Cost of capital in plan period*
WACC$_{,tv}$ = *Cost of capital in terminal value*
FCFF$_{n+1}$ = *Free cash flow to firm in year n+1*
G$_{EBIT}$ = *Growth rate after the terminal year forever*

4. Calculated total equity by adding the value of firm to cash and then reduce with debt value

5. To examine the fair value of Indosat's share, we divide total equity by the outstanding shares.

While the Free Cash Flow to Equity (FCFE) model is calculated by:

1. Calculated cost of equity with the formula in Equation 1:

$$Ke = Rf + \beta i \,(\text{implied market US}) + \text{country risk premium} \qquad (1)$$

where:

K$_e$ = *Cost of equity*
R$_f$ = *Risk-free rate*
β$_i$ = *Leveraged beta of Indosat*

We have to convert the cost of equity (Equation 1) by inserting the inflation rate for Indonesia compared to that of the USA.

2. Cash flow projection (Free Cash Flow to Equity) for the next five years

3. Calculated the value of the firm (present value) by discounting free cash flow to equity on cost of equity with this formula:

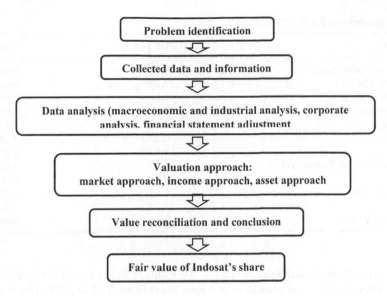

Figure 2. Valuation process (SE-54/PJ/2016).

$$Value = \sum_{t=1}^{t=n} \frac{FCFEt}{(1+Ke,pp)^t} + \frac{\dfrac{FCFEn+1}{(Ke,tv-gNI)}}{(1+Ke,tv)^n}$$

where:
$\quad$ FCFE$_t$ $\quad$ = *Free cash flow to equity in year t*
$\quad$ K$_{e,pp}$ $\quad$ = *Cost of equity in plan period*
$\quad$ K$_{e,tv}$ $\quad$ = *Cost of equity in terminal value*
$\quad$ FCFE$_{n+1}$ = *Free cash flow to equity in year n+1*
$\quad$ G$_{NI}$ $\quad\;$ = *Growth rate after the terminal year forever*
4. To examine the fair value of Indosat's share, we divide total equity value by the outstanding shares.

3.3 Ministry of finance regulation model

The valuation process based on MoF regulation SE-54/PJ/2016 is shown in Figure 2.

According to Figure 2, this study uses two approaches to estimate the fair value of Indosat's share. These approaches are the income approach and the asset approach. The income approach was conducted by the DCF method with adjustment of the projected income statement based on SE-54/PJ/2016 rules while applying the asset approach conducted by the Excess Earning Method (EEM).

The EEM is used to assess the equity of operating companies that have relatively stable growth rates of income and profit. EEM was conducted by calculating the value of collective intangible assets (big pot theory of goodwill) by dividing excess earning by the relevant capitalization rate for intangible assets. The market value of invested capital is calculated by adding the value of intangible assets and then the tangible asset value (Pajak, 2016).

4 RESULTS OF THE STUDY AND DISCUSSION

4.1 Model assumptions (plan period)

This valuation should make some assumptions for the plan period and terminal value as a basis to estimate the fair value of Indosat's share. The plan period assumption is shown in Table 1.

Table 1. Assumptions for the plan period.

Description	Plan period	Source
Growth of Projected Sales	6.82%	assumption
Market Debt Ratio	37.99%	assumption
Market Equity Ratio	62.01%	assumption
Beta Unleverage	0.66	damodaran
Beta Leverage	0.963	calculation
Risk-free rate	6.00%	T-Bill Ina
Cost of Equity	14.82%	calculation
Implied Market Risk Premium (US)	4.5%	damodaran
YTM	9.11%	annual report
Cost of Debt (kd)	6.83%	calculation
WACC	11.79%	calculation
Country Risk Premium	3.13%	damodaran

The growth of projected sales is 6.82%. This figure is generated from the historical data from Indosat's revenue growth in previous years from 2012 to 2016. The market debt and the equity ratio are assumed to be 38% and 62%, respectively. These calculations are in line with Indosat's future strategy, which is to reduce debt in funding its operational activities. Beta leverage of Indosat at the time of valuation is 0.963, which comes from the unleveraged beta of the telecommunications industry (service) in Indonesia, which is formed by the projected capital structure on the market value of debt and equity market value during the plan period. Indostat's cost of equity is 14.82%, which refers to the cost to be paid by Indosat to gain funding from the shareholders. Indosat must pay 6.83% (cost of debt) to gain a source of funding, which can originate from creditors, loans, corporate bonds, *sukuk*, and finance lease obligations. The weighted average cost of capital is 11.79%, calculated from the weighted market equity ratio and market debt ratio with each discount rate.

4.2 Model assumptions (terminal value)

To determine the terminal value of Indosat, we also make some assumptions regarding the discount rate and growth for the perpetuity period. The assumptions are shown in Table 2.

The study assumes that the Return On Equity (ROE) and Return On Invested Capital (ROIC) for the terminal value are 7% and 6%, respectively. The cost of equity terminal value is 6%, while the cost of capital (WACC) is 5.31%. The EBIT growth and net income from the sixth year onward are assumed to be 4.3% and 4.5%, respectively, which indicate lower figures than the projected growth of Indonesia's GDP in the period of terminal value.

We perform financial forecasting, including historical financial statement analysis, projected sales, projected income statement, projected cash flow, projected statements of financial position, and calculated ROIC, to find the free cash flow to equity and free cash flow to firm as proxies to examine the fair value of Indosat.

4.3 Valuation with discounted cash flow method

The Indosat valuation uses a two-stage free cash flow to firm model in which it is assumed that Indosat grows faster than stable conditions during the plan period and grows steadily at the terminal value. The growth in the plan period is 6.82% (see Table 1), and the growth in the terminal term period is stable at 3.68% (see Table 2).

Based on the free cash flow to firm model in Table 3, the corporate value of Indosat is Rp67,354,303 million and the equity value is Rp41,812,583 million. As of December 31, 2016 the number of outstanding Indosat shares is 5,433,933,500 and the intrinsic value (fair value) per share is Rp7,695.

Table 2. Assumptions for terminal value.

Description	Terminal value
ROE	7%
ROIC	6.00%
Ke	6.00%
WACC	5.31%
DPR EBIT	39%
DPR NI	39%
gEBIT	3.68%
gNI	4.30%
gGDP	4.50%

Table 3. Valuation with free cash flow to firm model.

	0	1	2	3	4	5	6
Description		2017	2018	2019	2020	2021	2022
EBIT(1-T)	25%	2,599,105	2,864,220	2,375,502	2,937,267	3,135,106	3,251,280
Depreciation		9,116,210	9,812,976	11,440,052	11,720,437	12,569,603	
Capex		8,842,917	6,470,762	6,462,598	11,072,084	4,351,781	
Incremental WC		(122,271)	123,281	132,396	141,420	151,059	
FCFF		2,994,670	6,083,153	7,220,560	3,444,199	11,201,869	
PVIF	11.79%	0.8945	0.8002	0.7158	0.6403	0.5728	
PV		2,678,877	4,867,839	5,168,709	2,205,480	6,416,662	
Total PP Value	21,337,567						
Terminal Value	44,166,311					77,103,213	
Cash	1,850,425						
Corporate Value	67,354,303						
Debt Value	25,541,720						
Equity Value	41,812,583						
Outstanding Shares	5,433,933,500						
Price per Share (Fair Value of Share)	7,695						

The equity valuation in Indosat uses a two-stage free cash flow to equity model in which it is assumed that Indosat grows by 6.82% (see Table 1) during the plan period and grows steadily by 4.30% (see Table 2) at the terminal value. Another assumption applied is the discount rate, which has a cost of equity of 14.82%, and other assumptions are made on the terminal value, like growth net income and return on equity.

According to Table 4, the valuation with free cash flow to equity model, the equity value of Indosat is Rp41,812,583 million with outstanding shares numbering 5,433,933,500. As of December 31, 2016, the intrinsic value (fair value) per share of Indosat is Rp7,695.

4.4 Valuation with Ministry of Finance regulation (SE-54/PJ/2016)

Based on MoF regulation (SE-54/PJ/2016), the valuation was conducted using two approaches: the income approach and the asset approach. We used the discounted cash flow method with the free cash flow to firm model to estimate the fair value of stock in the income approach. The difference with the previous DCF method lies in the projected income statement in which, if we use SE-54/PJ/2016, we need to adjust some revenue or expense accounts according to the applicable tax rules.

According to Table 5, using the same assumptions as the previous method (Table 1 and Table 2), the valuation with the income approach (FCFF model) based on MoF regulation (SE-54/PJ/2016) estimates the corporate value of Indosat as Rp68,968,038 million. With

Table 4. Valuation with free cash flow to equity (FCFE) model.

Description	0	1	2	3	4	5	6
		2017	2018	2019	2020	2021	2022
NI		1,172,045	1,478,857	1,062,482	1,732,185	2,079,144	2,169,030
Depreciation		9,116,210	9,812,976	11,440,052	11,720,437	12,569,603	
New Debt Issues		1,940,641	3,698,388	5,227,905	6,477,336	7,387,504	
Debt Repayments		2,928,193	3,489,308	4,125,739	4,846,435	5,661,351	
Capex		8,842,917	6,470,762	6,462,598	11,072,084	4,351,781	
Incremental WC		(122,271)	123,281	132,396	141,420	151,059	
FCFE		580,057	4,906,870	7,009,705	3,870,018	11,872,060	
PVIF	14.82%	0.8709	0.7585	0.6605	0.5753	0.5010	
PV		505,169	3,721,649	4.630.161	2,226,254	5,947,758	
Total EP Value	17,030,991						
Terminal Value	24,781,592					49,465,453	
Equity Value	41,812,583						
Outstanding Shares	5,433,933,500						
Price per Share (Fair Value of Share)	7,695						

Table 5. Valuation with the income approach (FCFF model).

Description	0	1	2	3	4	5	6
		2017	2018	2019	2020	2021	2022
EBIT(1-T)	25%	2,680,937	2,951,629	2,468,869	3,036,996	3,241,633	3,361,156
Depreciation		9,116,210	9,812,976	11,440,052	11,720,437	12,569,603	
Capex		8,842,917	6,470,762	6,462,598	11,072,084	4,351,781	
Incremental WC		(116,210)	123,694	132,837	141,891	151,561	
FCFF		**3,070,440**	**6,170,149**	**7,313,485**	**3,543,458**	**11,307,893**	
PVIF	11.79%	0.8945	0.8002	0.7158	0.6403	0.5728	
PV		**2,746,656**	**4,937,454**	**5,235,227**	**2,269,040**	**6,477,394**	
Total PP Value	21,665,772						
Terminal Value	45,451,840					**79,347,421**	
Cash	1,850,425						
Corporate Value	68,968,038						
Debt Value	25,541,720						
Equity Value	43,426,318						
Outstanding Shares	5,433,933,500						
Price per Share (Fair Value of Share)	7,992						

5,433,933,500 outstanding shares at December 31, 2016, the intrinsic value (fair value) per share of Indosat is Rp7,992.

The valuation asset approach with excess earning method assumes that Indosat's rate of return from the Net Tangible Asset Value (NTAV) is 10.75%, calculated using a weighted average of equity cost and cost of net tangible asset capacity in generating loans. Table 6 shows that Indosat's market value invested capital is Rp82,815,390 million.

To find the fair value of Indosat on the basis of MoF regulation, we reconcile the estimated market value of the two prior valuation approaches by weighting each of them. This weighting is based on the presence or absence of data required to obtain the fair value of Indosat's shares. The result of weighting each approach is 71.43% using the income approach and 28.57% using the asset approach.

Table 6. Valuation with the asset approach (excess earning method).

Description	Rate	Value
Net Tangible Asset Value (NTAV)		38,721,849
Rate of Return (Net Asset)	10.75%	
Earning Return on NTAV		4,161,513
EBIT(1-T)		2,955,415
Depreciation		(8,972,570)
Capital Expenditure		(2,743,294)
Net Working Capital		(63,891)
Free Cash Flow (FCF)		**9,120,800**
Excess Earning (EE)		**4,959,287**
Capitalization Rate (Cap Rate)	11.25%	
Capital Excess Earning (GOODWILL)		**44,093,541**
Market Value Invested Capital (MVIC)		**82,815,390**

Table 7. Reconciliation of fair value of Indosat.

Description	Corporate value	Weighted	Value
Income Approach (FCFF)	68,968,038	71.43%	49,262,884
Asset Approach (EEM)	82,815,390	28.57%	23,661,540
Corporate Value (Weighted)			72,924,424
Debt Value			25,541,720
Equity Value			47,382,704
DLOM			9%
Equity Value after DLOM			43,118,261
Outstanding Shares			5,433,933,500
Price per Share (Fair Value of Share)			7,935

Table 8. Comparison of share price with fair value of Indosat.

Share Price (Dec, 31 2016)	Fair Value (Dec, 31 2016)		Result
	DCF Method	Ministry of Finance Regulation	
6,450	7,695	7,935	Undervalued

Table 7 shows that the weighted corporate value is Rp72,924,424 million. The equity value earned by subtracting from the weighted corporate value the market value of the debt is Rp47,382,704 million. According to SE-54/PJ/2016, where the valuation object is stocked, the market value of equity needs to be adjusted by a given discount rate, that is, the Discount for Lack Of Marketability (DLOM). The calculation of DLOM for Indosat is 9%, so the value of equity after DLOM is Rp43,118,261 million. With the outstanding shares totaling 5,433,933,500, it can be deduced that as of December 31, 2016, the intrinsic value per share of Indosat is Rp7,935.

Table 8 shows that the two methods used to calculate the fair price demonstrate less difference in price; both free cash flow to firm and free cash flow to equity models show that the intrinsic value of Indosat on December 31, 2016, is Rp7,695. On the other side, per share, Indosat's fair value is Rp7,935 using MoF regulation SE-54/PJ/2016.

The actual share price of Indosat at the same time is Rp6,450, which indicates that it is undervalued. A company is considered undervalued if its stock price in the market is less than its intrinsic value (fair value). This indicates that Indosat's share price does not reflect all of the information available in the market. It also could be said that our market is not perfectly efficient, which is inconsistent with the efficient market hypothesis (Fama, 1995).

5 CONCLUSION

The valuation with Discounted Cash Flow (DCF) method estimates that the fair value of Indosat's shares is Rp7,695. Meanwhile, the valuation based on the Ministry of Finance regulation (SE-54/PJ/2016) estimates the fair value as Rp7,935. Both of these methods result in a higher value than the market value of the firm's stock on December 31, 2016, at Rp6,450. Investors who own Indosat shares are advised to keep them, while prospective investors are advised to buy Indosat shares as part of their portfolio. Indosat, as one of the major players in the telecommunications industry, still has an appeal for potential investors who want to put their funds into the telecommunications sector.

REFERENCES

APJII. (2016). *Penetrasi dan Perilaku Pengguna Internet Indonesia. Jakarta,* Indonesia.
Bodie, Z., Kane, A. & Marcus, A.J. (2014). *Investment* (10th ed.). New York, NY: McGraw-Hill Education.
Damodaran, A. (2012). *Investment valuation* (3rd ed.). Hoboken, NJ: Wiley.
Eitman, D.K., Stonehill, A.I. & Moffett, M.H. (2013). *Multinational business finance* (13th ed.). Upper Saddle River, NJ: Pearson.
Fama, E.F. (1995). Random walks in stock market prices. *Financial Analysts Journal, 51*(1), 75–80. doi:10.2469/faj.v51.n1.1861.
Indosat. (2016). *Laporan Tahunan 2016.* Jakarta, Indonesia: PT Indosat Tbk.
Ireland, R.D., Hitt, M.A. & Sirmon, D.G. (2003). A model of strategic enterpreneurship: The construct and its dimensions. *Journal of Management, 29*(6), 963–989. doi:10.1016/S0149-2063(03)00086-2.
Kaplan, S.N. & Ruback, R.S. (1995). The valuation of cash flow forecasts: An empirical analysis. *The Journal of Finance, 1*(4), 1059–1093.
Koller, T., Goedhart, M. & Wessels, D. (1994). *Valuation, measuring and managing the value of companies* (3rd ed.). McKinsey & Company.
Laro, D. & Pratt, S.P. (2005). *Business valuation and taxes.* Hoboken, NJ: Wiley.
Mielcarz, P., & Mlinarič, F. (2014). The superiority of FCFF over EVA and FCFE in capital budgeting. *Economic Research-Ekonomska Istraživanja, 27*(1), 1–14. doi:10.1080/1331677X.2014.974916.
Modigliani, F. & Miller, M.H. (1958). The cost of capital, corporation and the theory of investment. *American Economic Review, 48*(3), 261–297.
Pajak, D.J. (2016). SE54PJ2016 ttg Petunjuk Teknis Penilaian Properti, Bisnis, ATB untuk Tujuan Perpajakan.pdf.
Paramitha, A.L., Hartoyo, S. & Maulana, N.A. (2014). The valuation of initial share price using the free cash flow to firm method and the real option method in Indonesia Stock Exchange. *Jurnal Manajemen dan Kewirausahaan, 16*(1), 9–16. doi:10.9744/jmk.16.1.9-16.
Plenborg, T. (2002). Firm valuation: Comparing the residual income and discounted cash flow approaches. *Scandinavian Journal of Management, 18*(3), 303–318. doi:10.1016/S0956-5221(01)00017-3.
Subramanyam, K.R. (2014). *Financial statement analysis* (11th ed.). New York, NY: McGraw-Hill Education.

Business Innovation and Development in Emerging Economies – Trinugroho & Lau (Eds)
© *2019 Taylor & Francis Group, London, ISBN 978-1-138-35996-3*

Do corporate governance, firm characteristics, and financial ratio affect firm performance?

Evita & Silvy Christina
Trisakti School of Management, Jakarta, Indonesia

ABSTRACT: The purpose of this study was to determine the factors that affect the performance of manufacturing companies listed on the Indonesian Stock Exchange. The independent variables consisted of frequency of board meetings, attendance of board members, board size, leverage, firm age, firm size, sales growth and asset turnover. The dependent variable was firm performance. The sample consisted of 186 manufacturing companies listed on the Indonesian Stock Exchange during 2014–2016. The hypothesis test used in this research was multiple linier regression analysis. The results show that firm age, firm size, and asset turnover had a significant effect on the firm performance. Frequency of board meetings, attendance of board members, board size, leverage and sales growth had no significant effect on company performance.

Keywords: firm performance, frequency of board meetings, attendance of board members, board size, leverage, firm age, firm size, sales growth and asset turnover

1   INTRODUCTION

A firm's performance shows its ability to generate profit. Firm performance is a parameter used to identify if the organization's performance is already good enough and has implications for the businesses conducted. To compete globally, firms must maximize their performance (Al-Matari et al. 2014). Other than that, firms also have to be able to present valid and accountable financial reports. Financial reports not only present income statements, but also a firm's financial position and performance, so they can provide complete information to stakeholders. The information can then be used as a consideration to make further decisions. Firms that implement good transparency will be able to present their financial reports transparently and responsibly to their stakeholders. There are several factors that affect a firm's performance, such as frequency of board meetings, board size, firm size, sales growth, etc.

This study was conducted to develop a previous study conducted by Bhatt and Bhattacharya (2015). The purpose of this study was to obtain empirical evidence about the impact of board meetings frequency, attendance of board members, board size, leverage, firm age, size, sales growth and asset turnover on firm's overall performance.

2   THEORETICAL FRAME AND HYPOTHESES DEVELOPMENT

2.1   *Agency theory*

Jensen and Meckling (1976) stated that agency theory describes shareholders as the principal and management as the agent. The management is a party that is trusted by shareholders to work for the interests of the shareholders. For that reason, management is given the authority to make decisions to benefit shareholders.

2.2 Signaling theory

According to Godfrey et al. (2014, p. 375), signaling theory describes a manager's behavior as giving directions based on the firm's financial reports to give an expectation signal and prospect about the future. This signal becomes the communication basis in the assessment of the firm's performance. This theory shows that every action contains information about the firm's conditions that are told to other shareholders.

2.3 Firm performance

The concept and definition of performance is a controversial and multidimensional issue. A good parameter of a firm's performance can be seen from how the firm maximizes its profit, assets and makes the most of funds provided by stakeholders (Zeitun & Tian 2007). Consequently, a firm's performance gets the most attention from investors all over the world. Zeitun and Tian (2007) stated that a firm's performance can affect the prices in stock and capital markets. For example, if the stock market price does not show any improvement, it means that the market reaction is not very good.

2.4 Frequency of board meetings on firm performance

Frequency of board meetings describes how frequent the board members conduct meetings to discuss the progress issues being faced by the firm. For the board to be able to work optimally, it has to conduct meetings to discuss and make decisions responding to policy and issues being faced by the firm. This statement is supported by Garcia-Ramos and Ola-lla (2011) and Arora and Sharma (2016), who found a correlation between board meetings frequency and performance. According to that description, the following hypothesis was formulated:

Ha$_1$: There is a correlation between board meeting frequency and firm performance.

2.5 Attendance of board members on firm performance

Attendance of board members is one of the important factors, other than the board of directors' meetings. This factor emphasizes the purpose of the meetings. If there are not enough board members present at the meetings, the purpose of the meetings is hardly achieved. As the previous section has explained, the frequency of board meetings may correlate with firm performance. This statement is supported by Bhatt and Bhattacharya (2015) and Chou et al. (2013), who found a correlation between board members attendance and firm performance. According to that description, the following hypothesis was formulated:

Ha$_2$: There is a correlation between board members attendance and firm performance.

2.6 Board size on firm performance

Board size is the size of the board of directors in a firm. The function of the board in each country varies, so the results also vary. The board of directors is considered as the party who is able to solve issues and conflicts that arise in the firm. A larger board size can gather better information for the good of the firm. Studies conducted by Arora and Sharma (2016) and Fauzi and Locke (2012) stated that board size has implications for firm performance. According to that description, the following hypothesis was formulated:

Ha$_3$: There is a correlation between board size and firm performance.

2.7 Leverage on firm performance

Leverage is a benchmark ratio of the degree to which firm is funded by debts. The higher the leverage ratio, the worse the perception of the firm's performance. The firm will be perceived to have more funding sourced from creditors than stakeholders. This will decrease potential shareholders and creditors' interest in the firm. On the other side, the lower the leverage value of a firm, the better the perception of its performance. This statement is supported by Sulong et al. (2013) and Khatab et al. (2011), who found a correlation between leverage and firm performance. According to that description, the following hypothesis was formulated:

Ha_4: There is a correlation between leverage and firm performance.

2.8 *Firm age on firm performance*

Firm age is the length of time a firm has been established. Older firms are considered to be more established, experience and have expertise in dealing with various problems and obstacles. Older firms are considered to have better performance because they have survived. This statement is supported by Bhatt and Bhattacharya (2015) and Chada and Sharma (2015) who found a correlation between the age of a firm and its firm performance. They found that the older a firm, the better its performance. According to that description, the following hypothesis was formulated:

Ha_5: There is a correlation between a firm age and firm performance.

2.9 *Firm size on firm performance*

Firm size is how large or how small the firm's scale is, which is usually measured from its assets. The larger the firm's assets, the larger its size. In addition, firms with larger assets are considered to perform better than ones with smaller assets. This statement is supported by Sulong et al. (2013) and also Zeitun and Tian (2007), who found a relationship between firm's size and performance. They found that larger firms are better at maximizing their assets, so the performance is also better. According to that description, the following hypothesis was formulated:

Ha_6: There is a correlation between a firm size and firm performance.

2.10 *Sales growth on firm performance*

Sales growth is a change in the sales value every year. Sales growth is considered to increase profits and investment in a firm. It can be said that sales growth affects a firm's performance. This statement is supported by Chadha and Sharma (2015), Bhatt and Bhattacharya (2015) and Khatab et al. (2011), who found a relationship between sales growth and firm performance. According to that description, the following hypothesis was formulated:

Ha_7: There is a correlation between a firm's sales growth and firm performance.

2.11 *Asset turnover on firm performance*

Asset turnover is one of the parameters that is used as a reference to whether the firm has utilized their assets maximally or not and whether the firm is able to generate profits from the assets or not. If the firm has used the assets well, it means the profits of the firm will increase which will reflect a good performance. Chadha and Sharma (2015) expected to find relationships and influence on firm performance, but they found no significant effect. However, differences in the year and place of study are considered as a factor that contributed to the differing results of the studies. According to that description, the following hypothesis was formulated:

Ha_8: There is a correlation between a firm's asset turnover and firm performance.

3 RESEARCH METHOD

The objects of this study were every constantly registered manufacturing firm on Indonesian Stock Exchange during the 2014–2016 period. Sampling was conducted using a purposive sampling method. The selected samples can represent the population of this research. The total data used in this study comprised 186 data sets, where the sample selection procedure can be seen in Table 1 below.

3.1 Definition of operasional variable and the measurement of the dependent variable

3.1.1 Dependent variable
Firm performance

Firm performance is a measurement of whether the firm's management has done its job effectively or not (Al-matari et al. 2014). In this study, firm performance was measured with a ratio scale. Based on previous research by Bhatt and Bhattacharya (2015), firm performance was given the *TOBINSQ* symbol and can be calculated by the following formula:

$$TOBINSQ = \frac{(\text{Book value of total debt} + \text{Market value of common stock})}{\text{Book value of total asset}}$$

3.1.2 Independent variable
3.1.2.1 Frequency of board meetings
Frequency of board meetings is the frequency of board meetings in a year (Bhatt & Bhattacharya, 2015). In this study, frequency of board meetings was measured with a ratio scale. Based on previous research by Bhatt and Bhattacharya (2015), frequency of board meetings was given the *MEET* symbol and was calculated by the following formula:

$$MEET = Frequency\ of\ board\ meetings\ in\ a\ year$$

3.1.2.2 Attendance of board members
Attendance of board members is the amount of attendance at board meetings that were held in a year (Bhatt & Bhattacarya, 2015). In this study, attendance of board members was measured with a ratio scale. Based on previous research by Bhatt and Bhattacharya (2015), attendance of board members was given the *ATTEN* symbol and was calculated by the following formula:

$$ATTEN = \frac{Attendance\ of\ board\ members}{\text{Total of board members} \times \text{Frequency of board meetings}}$$

3.1.2.3 Board size
Board size is a group of people that were chosen by shareholders to resolve agency conflict between firm and shareholders (Jensen & Meckling 1976). In this study, board size was

Table 1. Sample selection procedure.

Sample criteria	Total firms	Total datas
– Manufacturing firms that were consistently registered on the Indonesian Stock Exchange from 2013–2016	137	411
– Manufacturing firms that published financial reports which end on December 31st	(7)	(21)
– Manufacturing firms that provided financial reports in the form of Rupiah	(28)	(84)
– Manufacturing firms that have complete annual reports from 2013–2016	(7)	(21)
– Manufacturing firms that did not input complete datas from 2014–2016	(33)	(99)
– Total research samples	62	186

Source: IDX (Indonesia Stock Exchange).

120

measured with a ratio scale. Based on previous research by Bhatt and Bhattacharya (2015), board size was given the *BSIZE* symbol and was calculated by this formula:

$$BSIZE = total\ of\ board\ size\ in\ the\ firm$$

3.1.2.4 *Leverage*

Leverage is a ratio that calculates how large the firm's assets were that were funded by debt and became one of the ratios used as a reference for potential lender (Sulong et al. 2103). In this research, leverage was measured with a ratio scale. Based on previous research by Bhatt and Bhattacharya (2015), leverage was given the *LEV* symbol and was calculated by the following formula:

$$LEV = \frac{\text{Total debts}}{\text{Total asset}}$$

3.1.2.5 *Firm age*

The age of a firm shows that the firm's reputation has stood for a long time. The firm's life is calculated from the firm's start of operation until now (Jackling & Johl 2009). In this study, the age of a firm was measured with a ratio scale. Based on previous research by Bhatt and Bhattacharya (2015), the size of the firm was given the *AGE* symbol and was calculated by the following formula:

$$AGE = the\ age\ of\ a\ firm\ counted\ from\ the\ beginning\ of\ the\ firm$$
$$operation\ until\ the\ year\ of\ this\ study.$$

3.1.2.6 *Firm size*

The size of a firm is considered to potentially affect it performance. The size of a firm is a scale that classifies a firm as large or small. Larger firms tend to be perceived as finding it easier to obtain funds from external parties (Keasey, 1999). In this study, the size of a firm was measured with a ratio scale. Based on previous research by Bhatt and Bhattacharya (2015), the size of the firm was given the *SIZE* symbol which was calculated by the following formula:

$$SIZE = Ln\ total\ assets$$

3.1.2.7 *Sales growth*

Sales growth is considered to potentially affect at firm's performance and finance structure. Sales growth is seen as a percentage of sales increase every year (Keasey 1999). In this study, sales growth was measured with a ratio scale. Based on previous research by Ntim and Osei (2011), sales growth was given the *SALES_GROWTH* symbol and was calculated by the following formula:

$$SALES\ GROWTH = \frac{\text{Net sales t} - \text{Net sales t} - 1}{\text{Net sales t} - 1}$$

3.1.2.8 *Asset turnover*

Asset turnover measure how far a firm's utilizes their assets and becomes one of the reference points for stakeholders (Chada & Sharma 2015). In this research, asset turnover was measured with a ratio scale. Based on previous research by Chadha and Sharma (2015), asset turnover was given the *TURNOVER* symbol which was calculated by the following formula:

$$TURNOVER = \frac{Net\ sales}{Total\ assets}$$

4 RESULTS

The variables used in the descriptive tests can be seen in the Table 2 below.

The hypothesis test results of all independent variables can be seen from the significance score (Sig.) in Table 3 below.

The *frequency of board meetings* does not have any effect on firm performance. The frequency of board meetings cannot reduce agency conflict. The frequency of board meetings cannot generate an equal importance between stakeholders, which prevents them achieving their objectives (Jackling & Johl 2009). Attendance of board members does not have any effect on firm performance. Board members only supervise the firm's proceedings and apply good corporate governance according to existing regulations (Chou et al. 2013). *Board size* does not have any effect with firm performance. Board size plays a role as the number of supervisors ensure the operation is in accordance with standard operational procedures (Beiner et al. 2003). *Leverage ratio* does not have any effect on firm performance. This shows that assets that are funded by debt do not have any correlation with a firm's performance. Zeitun and Tian (2007) said that high *leverage* caused agency conflict, which may cause an increase in interest rates, additional costs in supervising and decreased investment. The increased debt is not always for asset additions and it is not proven to increase firm performance.

Firm age has a positive effect on firm performance. Firm age is an important aspect for customer consideration because the firm's products are more publicly known the older the firm is (Chadha & Sharma, 2015). Firm size has a positive effect on firm performance. The size of a firm shows how the public believe in the firm, yet also shows the possibility for prospective shareholders to invest their shares because they think that larger firms have a better performance than smaller firms (Keasey, 1999). Sales growth of a firm do not guarantee that the firm's

Table 2. Descriptive statistic test results.

Variable	N	Minimum	Maximum	Mean	Std. deviation
TOBINSQ	186	0.3041	18.6404	1.956269	2.5730554
MEET	186	3	66	18.322581	13.1725852
ATTEN	186	0.5764	1.0196	0.923719	0.9991125
BSIZE	186	2	15	5.274194	2.7322064
LEV	186	0.0715	3.0291	0.462622	0.3499412
AGE	186	6	98	40.290323	16.6766259
SIZE	186	25.2455	33.1988	28.227487	1.6682120
SALES_GROWTH	186	−0.9001	6.9255	0.123807	0.7027189
TURNOVER	186	0.0267	8.4293	1.076217	0.8431235

Source: SPSS 23.0 data process result.

Table 3. Results of t-test.

Variable	B	T	Sig.
(Constant)	−12.562	−2.982	0.003
MEET	−0.002	−0.172	0.864
ATTEN	1.495	0.836	0.404
BSIZE	−0.024	−0.275	0.784
LEV	0.709	1.344	0.181
AGE	0.052	4.842	0.000
SIZE	0.362	2.479	0.014
SALES_GROWTH	−0.148	−0.614	0.540
TURNOVER	0.626	2.883	0.004

Source: SPSS 23.0 data process result.

performance is good. On the other hand, there are other factors that determine the firm's performance such as resource utilization. Asset turnover has a positive influence on firm performance. This score also shows that firms that utilize assets optimally will be experience an increase in performance. This also shows that the ratio becomes a reference for the shareholders because it is an indicator of whether the firm's manager could utilize assets or not (Bashir et al. 2013).

5 CONCLUSION

From this research, it can be concluded that:

1. Firm age, firm size and asset turnover have a correlation with firm performance.
2. Frequency of board meetings, attendance of board members, board size, leverage and sales growth do not have a correlation with firm performance.

REFERENCES

Al-Matari, E. Mohammed, Abdullah K. Al-Swidi, and Faudziah H.B. & Fadzil. (2014). The measurements of firm performance's dimensions. *Asian Journal of Finance & Accounting. 6* (1), 24–49.

Arora, Akshita and Chanda Sharma. (2016). Corporate governance and firm performance in developing countries: evidence from India. *16* (2), 420–436, Retrieved from http://www.emeraldinsight.com/doi/full/10.1108/CG-01–2016–0018 (accessed February 28, 2017).

Bashir, Taqadus, Anum Riaz, Sabeen Butt and Abida Parveen. (2013). Firm performance: A comparative analysis of ownership structure. *european scientific journal, 9* (31), 1857–7881.

Beiner, S., S. Drobetz, F. Schmid, dan H. Zimmermann. (2003). *Is board size an independent corporate governance mechanism?* Retrieved from https://www.researchgate.net/publication/4993900_Is_Board_Size_an_Independent_Corporate_Governance_Mechanism (accessed March 13,2017).

Bhatt, R, Ratish and Sujoy Bhattacharya. (2015). Board structure and firm performance in Indian IT firms. *Journal of Advances in Management Research, 12* (3), 232–248. Retrieved from www.emeraldinsight.com/0972–7981.htm (accessed February 27, 2017).

Chadha, Saurabh and Anil K. Sharma. (2015). Capital structure and firm performance: empirical evidence from India. Retrieved from http://vision.sagepub.com (accessed March 25, 2017).

Chou, Hsin-I, Huimin Chung and Xiangkang Yin. (2013). Attendance of board meetings and company performance: evidence from Taiwan. *Journal of Banking & Finance, 37*, 4157–4171.

Fauzi, Fitriya and Stuart Locke. (2012). Boad structure, ownership structure and firm performance: A study of New Zealand Listed-Firms. *Asian Academy of Management Journal of Accounting and Finance, 8* (2), 43–67.

Garcia-Ramos, Rebeca and Myriam Garcia Olalla. (2012). Corporate governance, weak investor protection and financial performance in Southern Europe. *International Journal of Business and Management Studies, 4* (2), 129–139.

Godfrey, J., Hodgson, A., Tarca, A., Hamilton, J. and Holmes, J. (2014). Accounting theory. 7 John Wiley & Sons, Inc.

Jackling, Beverley and Shireenjit Johl. (2009). Board structure and firm performance: Evidence from India's top companies. *Corporate Governance: An International Review, 17* (4), 492–509.

Jensen, M.C. & Meckling, W.H. (1976). Theory of the firm: Managerial behavior, agency cost and ownership structure. *Journal of Financial Economics, 3* (4), 305–360.

Keasey, K. (1999). Managerial ownership and the performance of firms: Evidence from the UK. *Journal of Corporate Governance Finance 5* (1), 79–101.

Khatab, Humera, Maryam Masood, Khalid Zaman, Sundas Saleem, and Bilal Saeed. 2011. Corporate governance and firm performance: A case study of Karachi stock market. *International Journal of Trade, Economics and Finance, 2* (1), 39–43.

Ntim, Collins G. and Kofi A. Osei. (2011). The impact of corporate board meetings on corporate performance in South Africa. *Journal Compilation ©2011 African Centre for Economics and Finance, 2* (2), 84–103.

Sulong, Zunaidah, Gardner, J.C., Amariah Hanum Hussin, Zuraidah Mohd Sanusi, and C.B. McGowan Jr. (2013). Managerial ownership, leverage, and audit quality impact on firm performance: Evidence from Malaysian ace market. *Accounting & Taxation, 5* (1), 59–70. Retrieved from https://www.researchgate.net/publication/254964013_Managerial_Ownership_Leverage_and_Audit_Quality_Impact_on_Firm_Performance_Evidence_from_the_Malaysian_Ace_Market (accessed March 25, 2017).

Zeitun, Rami and Gary Gang Tian. (2007). Capital structure and corporate performance: Evidence from Jordan. *The Australian Accounting Business & Finance Journal, 1* (4), 40–61.

Business Innovation and Development in Emerging Economies – Trinugroho & Lau (Eds)
© *2019 Taylor & Francis Group, London, ISBN 978-1-138-35996-3*

The effect of managerial ownership on liquidity, agency cost and performance of credit society banks in West Nusa Tenggara Province of Indonesia

I Nyoman Nugraha Ardana Putra & Siti Sofiyah Abdul Mannan
Mataram University, Mataram, Indonesia

Tatang Ary Gumanti
Jember University, Jawa Timur, Indonesia

Nengah Sukendri
STAHN Gde Pudja Mataram, Mataram, Indonesia

ABSTRACT: This research aims to examine the effect of managerial ownership on liquidity, agency cost, and the profitability of Credit Society Bank (CSB) in West Nusa Tenggara Province, Indonesia. Five consecutive quarterly financial reports are examined using Partial Least Square (PLS). Results show that managerial ownership does not affect the liquidity, agency cost and financial performance. The bank liquidity level does not affect the agency costs. Agency costs affect negatively the financial performance of CSBs, which can be interpreted that the higher the agency costs of the banks, the lower is the financial performance of the banks.

Keywords: Managerial Ownership, liquidity, Agency Cost and Performance

1 INTRODUCTION

One of the eight obstacles faced by Credit Society Bank (CSB), known as the Bank Perkreditan Rakyat (BPR), relates to the efficiency in the form of the operational costs (Putra, 2012). Higher operational costs are associated with the existence of agency costs in the company. Holloh (2001) documents that low efficiency is not uncommon issue among CSBs in East Java, Bali, and West Nusa Tenggara provinces. In addition, a report by Bank Indonesia, the Central Bank of Indonesia, showed that as of September 2015, the operational efficiency of the CSBs in West Nusa Tenggara province, measured as the ratio between operational expenses and operation income (OEOI), was below the national standard of 92%. Table 1 shows some indicators of important measures of CSBs in the province.

As shown in Table 1, the CSBs with managerial ownership has average OEOI of 84.2%, whilst CSBs with institutional ownership and state ownership records an average OEOI of 66.9%. This indicates that CSBs with managerial ownership has a lower efficiency compared to CSBs with institutional and state ownership. Ideally, bank with larger managerial ownership would have better efficiency. Rashid (2015) and Du (2013) show that managerial ownership will improve the efficiency of the company. However, Thepot (2015) argues that this low efficiency could be a result of opportunistic behavior of the management as they are motivated to pursue short-term personal gains.

Interestingly, the financial performance of both types of ownership is larger than the standard set up by Bank Indonesia. The average Return on Assets (ROA) for both of groups of ownership are 3.2% and 7.3%, consecutively. This means that the ability to generate profits is quite good. The phenomenon is similar to Herri *et al.* (2006), who find that CSBs

Table 1. Financial Indicators of CSB in West Nusa Tenggara Province as of September 2015.

Ownership	Efficiency OEOI (%)	Profitability ROA (%)	Liquidity LDR (%)
Managerial	84.2	3.2	77.7
Institutional and State (Regional State)	66.9	7.3	94.5
BI Standard	92.0	1.22	89.0

Source: Bank Indonesia (2016).
Note: OEOI is ratio between operational expenses and operational income, ROA is return on assets, and LDR is loan to deposit ratio.

with Regional Government Ownership have relative more capital and higher profitability (performance) than CSBs with Institutional Ownership and Independent CSBs (Managerial Ownership).

Other data show that the Loan to Deposit Ratio (LDR) for CSBs with managerial ownership is lower compared to those with institutional and state ownership. The average LDR of CSBs with managerial ownership is 75.7% as opposed to 94.5% for CSBs with institutional and state ownership. Thus, the liquidity of CSBs with managerial ownership is lower. Rubin (2007) asserts that banks with institutional ownership would have stronger liquidity than those with insider ownership. However, Chalermchatvichien et al. (2011) report that the higher concentration of insider ownership is positively related the higher liquidity.

Previous studies seem to be inconclusive when examining the relationship between managerial ownership and company performance. For example, Demsetz and Lehn (1985) and Himmelberg et al. (1999) conclude that managerial ownership is not related to company performance. A study of Indonesian Banks shows no relationship between ownership and financial performance (Putra, 2013). Yet, McConnell and Servaes (1990) report that managerial ownership positively affects firm performance. Using market value as a proxy for performance, Randoy and Goel (2003) show relationship with ownership.

Referring to the aforementioned findings, the current study is aimed at examining the relationship between ownership, liquidity, agency costs and the bank financial performance. A total of 14 CSBs in West Nusa Tenggara province are examined. The study has 70 data that are formed by five quarterly financial reports from September 2014 to September 2015, consecutively. Results show that the bank financial performance is negatively affected by their agency costs. The remainder of the paper is organized as follows. Section two presents the review of related literature. Section three provides the research methods, followed by the presentation of the finding and discussion. Final section concludes the paper.

2 REVIEW OF RELATED LITERATURE

Fama (1980) and Fama and Jensen (1983) broadened the view of Jensen and Meckling, (1976) on contractual relationships. They argue that companies characterized by the separation of ownership and control can still survive and function efficiently and effectively as long as there are good internal and external control mechanisms such as managerial ownership, boards of directors, the labor market for managers, and market controls for firms to disciplining agents (especially managers).

The agency problem between managers and shareholders arises when the company generates massive free cash flow, where free cash flow is the net cash flow left in the company after the company finances all projects that have a positive NPV. The manager should maximize shareholder wealth by distributing free cash flow as dividends. In fact, managers prefer to withhold free cash flow so that funding sources under manager authority increase and reduce liquidity risk. In other words, investing with a negative NPV (Jensen, 1986). In addition, managers invest in projects with low risk, resulting in low yields due to low variability of cash flow.

Another cause of conflict between managers and shareholders is the financing decision. Shareholders are only concerned with systematic risks rather than total risk, as shareholders can make investments that minimize unsystematic risks through diversification with the establishment of a portfolio. Instead managers are more concerned about total risk in analyzing main investment projects. According to Fama (1980), there are two underlying reasons. The first is the substantive portion of their wealth in the firm's specific human capital, which makes their wealth non-diversifiable. The second is the manager will be in jeopardy of his reputation, if the company faces bankruptcy due to the failure of its investment.

Ang et al. (2000) contend that there are two measures of efficiency ratios as the proxy for agency costs. The first is in the form of 'cost ratio', calculated as the ratio between operating expenses and sales and the second is asset utilization ratio, calculated as the ratio between sales and total assets. In the case of micro finance institution for which CSBs belong to this type, efficiency can be calculated through two indicators (Arsyad, 2008). The first is the ratio of operational costs, which is the ratio of the operational costs to the average loan disbursed over customers. The second is by using the ratio of salary, that is the ratio between salaries paid to employees.

Efficiency in CSBs is determined by Bank Indonesia (SE BI No 8/31 dated December 12, 2006). It is measured using the same measure of operational efficiency, known as the cost ratio (Ang et al., 2000). This ratio compares the operational cost to operating income (OCOI). It is used to determine the operational efficiency level of the CSBs.

Asymmetric information is a condition where one party has better information than the other. In the case of company, company management knows more about the company than investors in the capital market (Atmaja, 1999). Assumed to acting in the best interest of the shareholders, managers are generally motivated to convey good information about the condition of the company. Submission of information by signaling to investors or public through management decisions becomes very important. Investors will appreciate positive signals by the company, but they will punish for negative signals. The existence of asymmetric information has made investors to be less aware of the prospect of the company than the manager. There is also a tendency that managers try to maximize the value of current shareholders, not new shareholders, (Brigham and Daves, 2004:569). That is, if the company has good prospects, management does not want to issue new shares, but uses retained earnings. Yet, if the prospect of the company is not good, the company offers new shares that will benefit the current shareholders.

Some studies have been carried out to examine the effect of managerial ownership on the liquidity, agency costs, and financial performance of the company. There is strong evidence that managerial ownership positively affect liquidity of the company (Chalermchatvichien et al., 2011; Rubin, 2007). There is also mounting evidence showing that managerial ownership affect the performance of company (Demsetz and Lehn, 1985; McConnel and Servaes, 1990; Moh'd et al., 1998; Himelberg et al., 1999; Randoy and Goel, 2003; Luo et al., 2011; Margaritis and Psilaki, 2010). Studies also show that agency costs are affected by managerial ownership (Ang et al., 2000; Fuentes and Vergara, 2003; Fleming et al., 2005; Chow, 2005; Rashid, 2015; Thepot, 2015). Gorton and Huang (2004) and Petropoulos and Kriyazopoulos (2010) document that liquidity of the company affects significantly the agency costs. There is also strong evidence showing that agency costs affect the financial performance of the company (Tan and Wang, 2010; Dietrich et al., 2011; Athanasoglou et al., 2008).

3 RESEARCH CONCEPTUAL FRAMEWORK

This paper is built using the agency theory framework of Jensen and Meckling (1976). The theory posits that agency costs arise because of the separation role of owners and managers. Agency costs will be zero if there is no separation of ownership and control. Ang et al. (2000) contend that agency cost is significantly higher when in owner-managed companies and agency costs are inversely related to ownership, similar to those disclosed by Jensen and

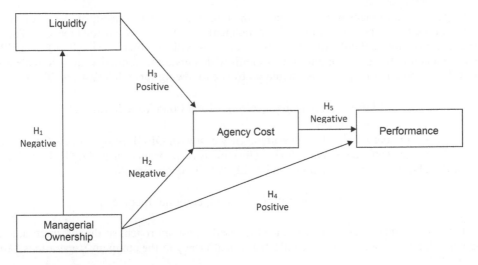

Figure 1. Research conceptual framework.

Meckling (1976). Figure 1 shows the conceptual framework of the study along with the predicted effect of the variables under investigation.

4 RESEARCH METHODS

All CSBs in West Nusa Tenggara Province are the population of the study. The selected CSB must meet two requirements. First, it must have quarterly financial reports as of 30 September 2015 to 30 September 2016. Second, it must have managerial ownership structure. All data can be accessed from the Bank Indonesia. A total of 14 CSBs satisfy the selection criteria and be used as the samples of the study.

Managerial ownership is measured if the top management of the CSB has a stake of ownership. Claessens *et al.* (2000) report that the managers and commissioners dominate the ownership in most of firms of Indonesia and Malaysia companies with the portion of 84.6% and 85% respectively. In small and medium-sized banks, ownership positions can vary greatly. Some banks may have multiple shareholders, who control the majority share of the bank, but others may have multiple shareholders, with no one has dominant ownership position. As a result, the ownership and management structure of the bank may lead to a number of different controls and different agency cost issues (Spong *et al.*, 1995). Measurement of managerial ownership approach used in this study is the percentage of top management ownership, CEO and board of commissioners in the CSB. This measure was first used by Demsetz and Lehn, 1985. Furthermore, the measurement of the managerial ownership structure is then used by Morck *et al.* (1988), Crutchley and Hansen (1989), Jensen *et al.* (1992), Yang *et al.* (2008), and Demsetz and Villalonga (2001). Following Demsetz and Villalonga (2001), ownership ratio is measured using the following formula.

$$LavMH = Log \frac{AvMH}{100 - AvMH}$$

The ownership data approach used in managerial ownership is Corporate Data Exchange (CDE), since research object do not trade their shares in the capital market and they have a bias when the ownership is below 0.2% (Kole, 1995). AvMH stands for average managerial shareholding or the average of fraction or part of ownership of the managing director, director and board of commissioners (Demsetz and Villalonga, 2001). AvMh distribution value to 100−AvMH is then calculated in the log so that obtained LavMH which stands for log of average managerial ownership.

Nonperforming loans represent one of the risks that arise as a result of the inability of the bank's customers to pay the credit installment to the bank for various reasons. Therefore, the measurement indicator of credit risk used is NPL (nonperforming loan). The NPL is to determine the nominal amount of credit with substandard quality, questionable and loss. The nonperforming loan is calculated by the following formula (Taswan, 2010).

$$\text{LDR } (\textit{Loan to Deposit Ratio}) = \text{Total Loan/Total Saving}$$

Efficiency indicator is used to measure agency cost with OEOI ratio to assess the level of CSBs operational efficiency. The following formula is used for measuring OEOI (Bank Indonesia Regulation No. 8/31 dated December 12, 2006 in Sembiring, 2010).

$$\text{OEOI} = \text{Operational Cost/Operational Income}$$

This level of efficiency is an efficiency that used a cost approach, so the interpretation of values will be reversed. This value reflects the inefficiency in the company's operations (Ang *et al.*, 2000).

The company performance demonstrates the company's ability to earn revenues for business operations and as a measure of management effectiveness. It is measured using the following formula (Taswan, 2010; SE BI No 8/31 dated December 12, 2006 in Sembiring, 2010).

$$\text{ROA (Return on Asset)} = \text{Earning before Tax/Total Asset}$$

Partial Least Square is used to test the proposed hypotheses. The structural equations used in this study are:

$$\text{Liquidity} = \beta_0 + \beta_1 \text{ Managerial Ownership} + \varepsilon_1 \tag{1}$$

$$\text{Agency Cost} = \beta_0 + \beta_1 \text{ Managerial Ownership} + \beta_2 \text{ Liquidity} + \varepsilon_2 \tag{2}$$

$$\text{Performance} = \beta_0 + \beta_1 \text{ Managerial Ownership} + \beta_2 \text{ Agency Cost} + \varepsilon_3 \tag{3}$$

5 RESULTS AND DISCUSSION

5.1 *Descriptive statistics analysis*

Descriptive statistics is aimed to provide an overview or description of a data in terms of mean, standard deviation, maximum and minimum value. Descriptive statistics of variables are shown in Table 2 below.

The average fraction of share ownership by managers and board of commissioners in BPR is −0.12. The ownership of shares by the management can be accounted for up to 0.2%

Table 2. Descriptive statistics of variables.

Variable	Minimum	Maximum	Mean	Std. deviation
OWN	−1.69	0.95	−0.12	0.79
OEOI	63.00	192.00	86.26	14.74
LDR	54.00	122.00	74.73	13.25
ROA	−14.00	11.00	2.27	2.63

Note: OWN is AvMh reflects the average of managerial holding, OEOI is operational cost over operating income, LDR is loan to deposit ratio, ROA is return on assets.

ownership or it has a high degree of accuracy. The average value of agency costs is 86.26% which indicated that in general CSBs has worked efficiently since the standard set by Bank Indonesia in this proxy is a maximum of 92%. The average value of non-performing loans during the period of September 2014 to September 2015 is 74.73%. If evaluated from the standard set by Bank Indonesia, the average value of liquidity risk faced by CSBs is below 89% which means liquidity risk faced by CSBs during this period is high because CSBs of managerial ownership are classified as illiquid, which is below the tolerance level allowed by Bank Indonesia. The indication given by this average value implies that CSBs are less able to perform the intermediary function, because the credit that can be distributed is relatively small compared to the value of their deposits, so that if analogous to the manufacturing company, the company is unable to sell their manufactured goods. The average Return on assets (ROA) value of CSBs during the period of September 2014 through September 2015 is 2.27%. If evaluated from the standard set by Bank Indonesia, then the average value is considered healthy, because it is above the standard of 1.22%. This means that the performance of CSBs with managerial ownership during the period fall into good category.

The result with PLS analysis shows the influence of outer model and inner model. Outer model results are not discussed because validity testing is necessary if a latent variable consists of several indicators, and is used to determine whether the indicator is valid and can represent a latent variable. Structural model testing (inner model) is examined by looking at the R^2 value, which is a goodness-fit-model test. R^2 is the ability of proxies in a variable in explaining the proxy of other variables. The results of R^2 test with PLS analysis are shown in Table 3.

As shown in Table 3, there are three variables: liquidity risk, agency cost and performance that have R^2, while managerial ownership variable does not have R^2. This is because there are four equations tested in the proposed research model. In addition, only managerial ownership variables are purely positional as exogenous variables (independent variables).

The liquidity risk variable has an R^2 value of 0.015. This means that the ability to explain managerial ownership variable to credit risk is only 1.5%. Or liquidity risk can be explained by 98.5% by other variables outside the model. In the model, agency cost variables are influenced by variables of managerial ownership and liquidity risk. The ability to explain the effect of these two variables on agency cost variable is seen as 0,5%, in other words as much as 99,5% variable of agency cost is actually influenced by other variables. In the performance variable, R-square shows the number of 0.625. This means that variables such as managerial ownership and agency costs are able to explain (determination) on performance of 62.5%, while as much as 37.5% is explained by other variables outside the model.

The ability of the variables in explaining the overall model is described by *Q-Square Predictive Relevance*. It is a description of *goodness of fit* from *inner model* or a relationship between variables. It can be calculated by the following formula:

$$Q^2 = 1 - (1 - R12)(1 - R22) ... (1 - RP2)$$

Overall, the ability of the variables in explaining the model is:

$$\begin{aligned}
Q^2 &= 1 - (1 - 0.015)(1 - 0.002)(1 - 0625) \\
&= 1 - (0.9985 \times 0.998 \times 0.375) \\
&= 1 - 0.3737 \\
&= 0.6263
\end{aligned}$$

Table 3. R^2 Estimation Results.

No.	Variable	R-square
1	Liquidity (LDR)	0,015
2	Agency Cost (OEOI)	0,002
3	Performance (ROA)	0,625

It means that the ability of the variables in explaining the overall model is 62.63%, or as much as 37.37% can be explained by other variables outside the model. These results indicated that the determination of the variables in the overall model is strong because it has the ability to explain greater than 50% and is relevant as predictive.

Results of hypotheses testing are shown in Table 4. As seen from the 5 tested paths, there are 4 insignificant path, and 1 significant path.

Managerial ownership does not affect the liquidity of CSBs (t = 1.312; p = 0.246). Even the coefficient is negative which is in contrast to the prediction. This means hypothesis is rejected. The lack of influence of ownership of course is different from the results of previous researches. Uno and Kamiyama (2010) stated that the influence of ownership concentration has a negative effect on liquidity, which if observed, the relationship or direction is the same as the result of this research but not significant. In contrast to the results of Chalermchatvichien *et al.* (2011) which explains the positive influence between the concentrations of ownership with liquidity. The results of this study showed no significant and with the LDR value below 89 (74.72), it means that the level of liquidity is relatively low. Company's liquidity is more influenced by other things such as interest rates, because low interest rates will increase the real sector especially related to ease funding from the banking industry to the community.

Managerial ownership is not related to agency costs of CSBs (t = 0.108; p = -0.918). Thus, there is no effect in increasing the agency costs experienced by the company. This means the hypothesis is rejected. The results of this study are similar to the grand theory of agency costs presented by Jensen and Meckling (1976), that managerial ownership reduces agency costs. This is caused by the absence of conflict of interest between the owner and the management of the company that is the owner who is also the manager. This means that the decision makers, executors, and supervisors of the implementation of decisions are conducted by the same party.

Liquidity level does not related to the agency costs (t = 0.283; p = 0.788). This coefficient is in contrast to the prediction. This means hypothesis 3 is rejected. LDR is an indicator of the liquidity level, which is the ratio between the amount of credit channeled with the amount or total deposits of customer funds in the bank. Increasing the amount of credit and savings associated with outsiders or customers in this case is the community. This leads to agency theory not applicable, but there is a relevance to the theory of asymmetric information because one of the causes of non-validity of agency theory because there are cases of information inequality (asymmetric information). This is confirmed by Bebzuck (2003) who stated that there are 3 forms of information asymmetry in relation to the provision of credit to customers namely moral hazard, adverse selection and monitoring costs.

Managerial ownership structure does not affect the level of company's financial performance (t = 0.072; p = 0.945). Thus, hypothesis 4 is rejected. Similar to hypothesis 2, the existence of managerial ownership is not related to performance. It was also mentioned by Jensen and Meckling (1976), who stated that through managerial ownership the company would make a decision not on the basis of efficiency but rather on profits or activities that could enrich or satisfy the owner's wishes in this case the manager or commissioner of the company itself.

Agency costs affect negatively and significantly the company's financial performance (t = 5.659; p = 0.002). This indicates the extent of agency costs reversely affects the financial performance. This means the higher the level of agency costs incurred by the company, the

Table 4. Hypotheses testing results.

Path	Relationship between variable	Path coefficient	t – stat.	p-value	Result
1.	Managerial Ownership to Liquidity	−0.124	1.312	0.246	H1 rejected
2.	Managerial Ownership to Agency Cost	−0.010	0.108	0.918	H2 rejected
3.	Liquidity to Agency Cost	0.040	0.283	0.788	H3 rejected
4.	Managerial Ownership to Performance	0.005	0.072	0.945	H4 rejected
5.	Agency Cost to Performance	−0.791	5.659	0.002	H5 accepted

lower is the financial performance of the company. Thus, hypothesis is accepted. The fifth *path* is the only path that has the influence of the agency costs on the performance of the company. Overall, this description can explain that the actual cost of agency in the company and its effect to the company's performance. However, it is interesting that none of the managerial and liquidity ownership variables affect the agency costs. This means that there are other factors that affect the agency costs which then must be tested again in future research.

6 CONCLUSION

The objectives of the study are to examine the effect of managerial ownership on the liquidity, agency costs, and financial performance of CSBs in West Nusa Tenggara Province of Indonesia. In addition, it also examines of whether agency costs affect financial performance of the company. Results using path analysis generate the following conclusion. Managerial ownership does not affect the liquidity level, the agency costs, as well as the financial performance of the company. The liquidity level does not affect the agency costs. Agency costs affect the financial performance of CSBs, which means the higher the agency costs, the lower is the financial performance, and vice versa.

Based on the results, discussion, and limitations that appear from this study, several things can be suggested as follows. First, future study may improve the CSBs liquidity by lowering the interest rate to attract the interest of the customers to access credit. It is advisable to enlarge the institutional ownership to sharpen the analysis of the decisions made. Further research may re-examine the issues especially the one related to agency cost on CSBs, so it can be known that what is really affects the agency costs on the CSBs managerial ownership.

REFERENCES

Ang, J.S, Rebel A.C, & Lin, J.W. (2000) Agency costs and Ownership Structure. *The Journal of Finance.* [Online] 81–106. Available from: https://doi.org/10.1111/0022-1082.00201.

Arsyad, L. (2008) Lembaga Keuangan Mikro: Institusi, Kinerja dan Sustanabilitas. Yogyakarta, Penerbit Andi.

Athanasoglou, P., Brissimis, S. & Delis, M. (2008) Bank-Specific, Industry-Specific and Macroeconomic Determinants of Bank Profitability. *Journal of International Financial Markets, Institutions and Money*, [Online] No. 18, 121–136. Available from: https://doi.org/10.1016/j.intfin.2006.07.001.

Atmaja, LS. (1999) *Manajemen Keuangan.* Yogyakarta, Penerbit Andi Offset.

Bebczuk, R.N., 2003. *Asymmetric Information In Financial Market, Introduction and Applications.* UK, Cambrige University Press, The Edinburg Building.

Brigham, E.F. & Daves, P.R. (2004) *Intermediate Financial Management.* International Student Edition, 8th Edition, South-Western, Thompson.

Chalermchavichien, P.S., Jiraporn, J.P. & Singh, M. (2011) The Effect of Bank Ownership on Capital Adequacy, Liquidity and Capital Stability (Basel II and Basel III). [Online] 1–45. Available from: https://doi.org/10.2139/ssrn.1888486.

Claessens, S., Djankov, S. & Lang, H.P. (2000) The separation of ownership and control in East Asian Corporations. *Journal of Financial Economics.* [Online] No. 58, 81–112. Available from: https://doi.org/10.2139/ssrn.206448.

Chow, S., & Yuk, J. (2005) Agency Cost and Ownership Structure: Evidence from Small Business Finance Survey Data Base. *Small Business Research Summary.* [Online] No. 268, 1–31.

Crutchley, C.E. & Hansen, R.S. (1989) A Test of the Agency Theory of Managerial Ownership, Corporate Leverage, and Corporate Dividends. *Financial Management.* [Online] Winter, No. 50, 36–46. Available from: https://doi.org/10.2307/3665795.

Demsetz, H. & Lehn, K. (1985) The Structure of Corporate Ownership: Causes and Consequences. *Journal of Political Economy.* [Online] No. 93, 1155–1177. Available from: https://doi.org/10.1086/261354.

Demsetz, H. & Villalonga, B. (2001) Ownership Structure and Corporate Performance. *Journal of Corporate Finance.* [Online] No. 7, 209–233. Available from: https://doi.org/10.1016/s0929-1199(01)00020-7.

Dietrich, A. & Wanzenried, G. (2011) Determinants of Bank Profitability before and during the Crisis: Evidence from Switzerland. *International Finance Market, Institution and Money.* [Online] No. 21, 307–327. Available from: https://doi.org/10.2139/ssrn.1370245.

Du, X. (2013) Does Religion Matter to Owner-Manager Agency Costs? Evidence from China. *Journal Business Ethic*. [Online] No. 118, 319–347. Available from: https://doi.org/10.1007/s10551-012-1569-y.

Fama, E.F. (1980) Agency Problem and The Theory of the Firm. *Journal of Political Economy*. [Online] No. 38, 288–307. Available from: https://doi.org/10.1086/260866.

Fama, E.F. dan M.C. Jensen, 1983, Separations of Ownership and Control, *Journal of Law and Economics*, vol. 27. 301–325. Available from: https://doi.org/10.1086/467037.

Fleming, G., Heaney R. & McCosker R. (2005) Agency Cost and Ownership in Australia. *Pacific-Basin Finance Journal*. [Online] No. 13, 29–52. Available from: https://doi.org/10.1016/j.pacfin.2004.04.001.

Fuentes, R. & Vergara, M. (2003) Explaining Bank Efficiency: Bank Size or Ownership Structure?. *Central Bank Chile*. Chile.

Gorton, G. & Huang, L. (2004) Liquidity, Efficiency, and Bank Bailouts. *American Economic Review*. [Online] No. 94. Available from: https://doi.org/10.3386/w9158.

Herri. T.H., Syarif S., Suhairi E.H. & Ma`ruf. (2006) Studi Peningkatan Peran Bank Perkreditan Rakyat (BPR) Dalam Pembiayaan Usaha Mikro Kecil (UMK) Di Sumatera Barat. *Penelitian Kerjasama Bank Indonesia dan Center for Banking Research*. Padang, Universitas Andalas.

Himmelberg, C.P., Hubbard R.G. & Palia, D. (1999) Understanding The Determinants of Managerial Ownership and The Link Between Ownership and Performance. *Journal Financial Economics*. [Online] No.53, 353–384. Available from: https://doi.org/10.3386/w7209.

Holloh, D. (2001) ProFI Microfinance Institutions Study, Promotion of Small Finance Institutions. Denpasar, Indonesia.

Jensen, M.C., & Meckling, W.H. (1976) Theory of The Firm: Managerial Behavior, Agency Cost and Capital Structure. *Journal of Financial Economics*. [Online] 305–360. Available from: https://doi.org/10.1016/0304-405x(76)90026-x.

Jensen, M.C. (1986) Agency Cost of Free Cash Flow, Corporate Finance and Takeover. *American Economics Review*. [Online] No. 76, 323–329. Available from: https://doi.org/10.2139/ssrn.99580.

Jensen, G., Solberg, D. & Zorn, T. (1992) Simultaneous Determination of Insider Ownership, debt, and Dividend Policies. *Journal of Financial and Quantitative Analysis*. [Online] Vol. 27, 247–263. Available from: https://doi.org/10.2307/2331370.

Kole, S.R. (1995) Measuring Managerial Equity Ownership: A Comparison of Sources of Ownership Data. *Journal of Corporate Finance*. [Online] No. 1, 413–435. Available from: https://doi.org/10.1016/0929-1199(94)00012-j.

Luo, W., Zhang Y. & Zhu N. (2011) Bank Ownership and Executives Perquisites: New Evidence From an Emerging Market. *Journal of Corporate Finance*. [Online] No. 17, 352–370. Available from: https://doi.org/10.1016/j.jcorpfin.2010.09.010.

Margaritis, D. & Psillaki, M. (2010) Capital Structure, Equity Ownership and Firm Performance. *Journal of Banking and Finance*. [Online] No. 34, 621–632. Available from: https://doi.org/10.1016/j.jbankfin.2009.08.023.

Mc, Connell, John J., & Servaes H. (1990) Additional Evidence on Equity Ownership and Corporate Value. *Journal of Financial Economics*. [Online] No. 27, 595–612. Available from: https://doi.org/10.1016/0304-405x(90)90069-c.

Moh'd, MA. Perry, LG. & Rinbey JN. (1998) The Impact of Ownership Structure on Corporation Debt Policy: A Time Series Cross Sectional Analysis. *The Financial Review*. [Online] No.33, 85–98. Available from: https://doi.org/10.1111/j.1540-6288.1998.tb01384.x.

Morck, R., Shleifer, A & Visnhy, R.W. (1988) Management Ownership and Market Valuation, An Empirical Analysis. *Journal of Financial Economic*. [Online] No. 20, 293–315. Available from: https://doi.org/10.1016/0304-405x(88)90048-7.

Petropoulos, D.P., & Kyriazopoulos, G. (2010) Profitability, Efficiency And Liquidity Of The Co-Operative Banks In Greece. *International Conference On Applied Economics*. [Online] ICOAE, 603–607.

Putra, I.N.N.A. (2012) Kepemilikan Manajerial dan Risiko Kredit Sebagai Pemicu Biaya Keagenan Pada Lembaga Keuangan Mikro. *Jurnal Keuangan dan Perbankan* [Online] Vol. 16, no. 3, 437–444.

Putra, I.N.N.A. (2013) Perbedaan Profitabilitas Dan Tingkat Pengawasan Sebelum Dan Sesudah Merger Pada Bank Perkreditan Rakyat. *Jurnal Keuangan dan Perbankan*. [Online] Vol. 17, no. 2, 302–309.

Randoy, T. & Goel, S. (2003) Ownership Structure, Founder Leadership and Performance in Norwegian SMEs: Implication for Financing Entrepreneurial Opportunity. *Journal of Business Venturing*. [Online] No. 18, 619–637. Available from: https://doi.org/10.1016/s0883-9026(03)00013-2.

Rashid, Afzalur. (2015) Revisiting Agency Theory: Evidence of Board Independence and Agency Cost from Bangladesh. *Journal Business Ethic*. [Online] No. 130, 181–198. Available from: https://doi.org/10.1007/s10551-014-2211-y.

Rubin, Amir. (2007) Ownership Level, Ownership Concentration, and Liquidity. *Journal of Financial Markets*. [Online] Vol. 10, Issue 3, August, 219–248. Available from: https://doi.org/10.1016/j.finmar.2007.04.002.

Sembiring, S. (2010) Himpunan Peraturan Perundang-undangan Republik Indonesia "Bank Perkreditan Rakyat" (BPR). Bandung, Penerbit Nuansa Aulia.

Spong, K., Sullivan R. & DeYoung, R. (1995) What Makes a Bank Efficient? A Look at Financial Characteristics and Bank Management and Ownership Structure. *Federal Reserve Bank of Kansas City Financial and Industry Perspectives*. [Online] pp. 1–20.

Tan, J & Wang, L. (2010) Flexibility-Efisiensi Trade Off and Performance Implication among Chines SOEs. *Journal of Business Research*. [Online] No. 63, 356–362. Available from: https://doi.org/10.1016/j.jbusres.2009.04.016.

Taswan. (2010) Manajemen Perbankan, Konsep, Teknik dan Aplikasi. Yogyakarta, UPP STIM YKPN.

Thepot, J. (2015) Negative Agency Cost. *Theory and Decision*. [Online] Vol. 78, issue 3, 411–428. Available from: https://doi.org/10.1007/s11238-014-9427-2.

Uno, J. & Kamiyama, N. (2010) Ownership Structure, Liquidity and Firm Value. *Waseda University Institute of Financial Studies*. [Online] Working Papers Series. Available from: https://doi.org/10.2139/ssrn.1455995.

Yang, C.Y., Hung N.L. & Boon L.T. (2008) Managerial Ownership Structure and Earning Management. *Journal of Financial Reporting and Accounting*. [Online] Vol. 6, 35–53. Available from: https://doi.org/10.1108/19852510880000634.

Business Innovation and Development in Emerging Economies – Trinugroho & Lau (Eds)
© 2019 Taylor & Francis Group, London, ISBN 978-1-138-35996-3

Comparing properties of net income and total comprehensive income: A study on manufacturing industry in Indonesia

Retno Yulianti
Department of Accounting, Faculty of Economics and Business, Universitas Pembangunan Nasional "Veteran" Yogyakarta, Indonesia
Doctorate Program of Economics Science, Faculty of Economics and Business,
Universitas Sebelas Maret, Indonesia

Ari Kuncara Widagdo, Doddy Setiawan & Bambang Sutopo
Department of Accounting, Faculty of Economics and Business, Universitas Sebelas Maret, Indonesia

ABSTRACT: The purpose of this research is to acquire empirical evidence of the benefit of PSAK No. 1 (2009) implementation by testing whether the properties of total comprehensive income are better than those of net income. Four properties of net income and of total comprehensive income were compared. This research also investigated whether the value relevance of total comprehensive income depends on location where the users can obtain this information.

Stock return was measured with average daily stock returns over the fiscal year subtracted with market return. A dummy variable was applied to determine companies that report Other Comprehensive Income (OCI) in their equity statement or in their comprehensive income statement. For companies with OCI, the dummy variable was used to determine the period before and after the implementation of PSAK No. 1 (2009). It represents a place where the users can obtain information about total comprehensive income. This study used a sample of 122 manufacturing firms listed in the Indonesian Stock Exchange from 2008 to 2014. The analytical tool used to test all hypotheses was ordinary least square.

The result of this research demonstrates that net income is more persistence, less variable, and more have predictive ability than total comprehensive income, yet total comprehensive income has more value relevance than net income. In addition, this research reveals that location of OCI is insignificant for the value relevance of total comprehensive income.

Keywords: Indonesia, persistence, variability, predictive ability, value relevance, net income, total comprehensive income, and other comprehensive income

1 INTRODUCTION

PSAK No. 1 (2009) that adopts IAS No.1 of 2009, effective for fiscal year beginning on 1 January 2011 or after, regulates the presentation of comprehensive income statement. The presentation of comprehensive income statement of a period that covers the net income of such period added or subtracted with the changes in other comprehensive income (OCI) items provides information that aid investors and creditors to comprehend more about transformation in equity interest and a company's ability in generating future cash flow (Casabona, 2014). The distinction of income statement presentation after PSAK is only in the disposition of other comprehensive income presentation, which was previously presented in the statement of changes in equity.

The benefits of comprehensive income can be viewed from two perspectives: investor usefulness and contracting usefulness (Black, 2015). Those benefits have been studied as a response to the establishment of SFAS 130 of 1997, with various research results. Some support the appeal for comprehensive income statement presentation, yet some others do not. Those who support the idea claim that clean surplus describing only net income and dividend is internally consistent, less supportive to manipulation, and more suitable with valuation theory. On the other hand, those who do not support such idea state that earnings should be summarized from the influence of transitory to illustrate current operating performance and the manager will exclude items that exceed their capability to control those items (Biddle and Choi, 2006).

Earnings quality is observable from 4 perspectives: persistence, volatility, ability to predict one-year-ahead cash flows from operating activities and net income, and association with contemporaneous stock returns (Dechow and Schrand, 2004). Comprehensive income statement determines which of net income or total comprehensive income (TCI) is more qualified, and whether the location of OCI reporting influences the total comprehensive income in explaining stock return. Kabir and Laswad (2011) have confirmed that net income is more persistent and more capable in explaining stock return than TCI. The variability and predictive ability of both are indifferent. In addition, the location of OCI does not influence the value relevance of TCI. Moreover, they distinguish the location of OCI presentation on the basis of whether the comprehensive income is a single continuous statement of comprehensive income or two separate but consecutive statement. However, nearly all companies in Indonesia report OCI in a single continuous statement of comprehensive income. Therefore, this study aims to retest the research conducted by Kabir and Laswad (2011), yet depending on whether OCI is reported in the statement of changes in equity or in the income statement (comprehensive income) to determine the location of OCI reporting.

Whether the location of OCI presentation about income statement is beneficial becomes a concern for some researchers. Hirst and Hopkins (1998) demonstrate with experiment method that an explicit disclosure of income statement on comprehensive income and its component can effectively improve transparency of earnings management activity and reduce the analyst's valuation judgment. Maines & McDaniel (2000) find out that the participants significantly weigh on their volatility valuation when unrealized gains and losses (URGL) appeared in SFAS 130 comprehensive income, but not in alternative SFAS 130 (URGL reported in changes in equity) and statement of changes in equity SFAS 115. On the contrary, Chambers (2007) confirms otherwise. OCI is valued by investors when it stands on the statement of changes in equity because investors are familiar with the predominant OCI presentation location. While Darsono (2012), who proved that the role of OCI information before the regulation on comprehensive income presentation is effective, verifies that OCI influences company value. Since he acquires OCI information from the statement of changes in equity, his finding then confirms that there is no issue about OCI presentation location.

The experiment by Hirst and Hopkins (1998) divides participants into two groups, namely a group with comprehensive income report (CI) and a group without comprehensive income report (non CI), which that comprehensive income information is presented in the statement of changes in equity. The result points out some differences. Based on that, this research would like to retest the research by Kabir and Laswad (2011) in connection with the location of OCI, because according to the research by Hirst and Hopkins (1998), the result would be different when other comprehensive income is not comprehensively presented together with the income statement. Even in the discussion about the research by Hirst and Hopkins (1998), Lipe (1998) emphasizes that even if another group of participant is added, which is the participant informed with the source of another comprehensive income (CI) out of the income statement, the results is much the same, different from the expected result which processing and judgment would be different from the non CI group. It reveals that CI format apparently has advantages.

Value relevance is the correlation of financial statement numbers with the price or stock return. Many researches on earnings value relevance and its items have been conducted. Lipe (1986) finds that stock return variation of industrial and commercial companies are explained better by earnings component than aggregate earnings. Brown (2001), who compares the quality of earnings by three earning components, namely net income, earning per share from operation, and earning per share before extraordinary item, finds that net income has higher earnings quality compared to the other. Jaggi and Zhao (2002), who specifically research the impact of transformation from SFAS 12 into SFAS 115 of 1993 about available-for-sale financial assets reclassification, find that the information of earnings components is more relevant for investment decisions after the implementation of SFAS 115.

The result of the research shows that net income in compliance with SFAS 130 dominates traditional net income (Biddle and Choi, 2006) and total comprehensive income (Bidlle and Choi, 2006; Kabir and Laswad, 2011) in explaining stock return. Chambers, et al. (2007) who specifically demonstrate the benefit of other comprehensive income presentation prove that two components of other comprehensive income, which are foreign currency translation adjustment and unrealized gains and losses from available securities for sale, have higher value relevance than the other components. It appears so for Lin, et al. (2007) as well, who uses samples from various country, that confirms that total other comprehensive income has value relevance beyond net income in many countries, but comprehensive income has less value relevance than the other measurement of income, which are operating income and net income. On the other hand, the researches that do not promote the use of comprehensive income statement are that by Dhaliwal et al. (1999) and Doukakis (2010). Dhaliwal et al. (1999) conclude that comprehensive income is not closely associated with stock return nor predicts cash flow or earnings better than net income. Doukakis (2010), who employed the adoption of IFRS phenomenon, validates that the adoption of IFRS that requires comprehensive income statement presentation does not improve persistence of earnings and its component systematically for future profitability. On the contrary, operating and non-operating income persistence are lower, so is the explanatory power after the adoption of IFRS.

With a variety of confirmation about the use of comprehensive income statement, this research aims to test the use of comprehensive income statement in Indonesia for the regulation on comprehensive income statement reporting has only been effective since 2011. This research highlights the role of comprehensive income statement presentation with the perspective of investors usefulness in describing investors' response. Since there is an OCI component that becomes the distinctive feature of comprehensive income statement presentation compared to the previous version of presentation, this research therefore highlights the role of OCI in differentiating the presentation format.

From the view of efficient market hypothesis, the market reaction will be indifferent since investors are not affected by where the information is presented. They can still utilize the information to make decision. However, based on the argument that a market cannot be as efficient as expected, and low stock price proportion that can be explained by historical net income (Scott, 2015), measurement approach and clean surplus theory justify the significance of a research on investors' reaction to the transformation of this income statement presentation.

It is substantial to compare the earnings quality between net income and total comprehensive income in order to determine which of the two is better to use for company performance assessment (Black, 1993). Furthermore, evidences are currently required to evaluate the implementation of PSAK No. 1 (2009). Based on information that 11 of the 20 companies with the largest market capitalization in the Indonesia Stock Exchange have business in manufacturing industry sector (www.kemenperin.go.id), this research focuses on the manufacturing industry, by testing whether the net income and total comprehensive income have differences in earnings quality and value relevance.

This research used 384 manufacturing firm-year observations to test the persistence, variability, and predictive ability of NI and TCI, and 377 manufacturing firm-year observations were used to test the value relevance of both. Observations were conducted from 2008 to 2014, 3 years prior to the presentation of comprehensive income, and 4 years after. This research validates that NI is more persistent, less variable, and more capable in predicting operating cash flow than TCI, with slight difference. On the other hand, unlike NI, TCI may explain return, yet the location of OCI does not influence the value relevance of the TCI.

2 LITERATURE REVIEW

2.1 *Comprehensive income*

Accounting information serves a purpose of providing information for investors about the company's performance to make business decision and to evaluate managers' performance. Designing and implementing well-combined concept and standard, the role of accounting information for the investors, and evaluating the managers' performance are the fundamental issues of accounting theory.

Scott (2015) explains that in ideal condition, there isn't any fundamental issues of accounting theory. The application of current value based accounting in financial reporting is second best condition. Although the use of current accounting in financial reporting is more realistic, the fundamental issues of accounting theory such as mentioned above will still persist. Current value reflects more about the value in the mean time so that it will be more preferable by the investors. Yet, it can be less preferable because the investors are accustomed to historical cost. Moreover, current accounting includes unrealized gain or losses of assets and debts changes. It can increase income volatility and does not reflect the actual performance of managers, and it should not be used to assess operating performance of managers because unrealized gain or losses is not the result of operating performance.

Current value of assets and debts is more potential than historical cost in drawing investors' attention because current value provides the finest available indication on company performance and future investment return. However, managers might think that unrealized gains and losses from the adjustment of assets' and debts' registered value in the current value does not reflect their actual performance. Accounting standards board immediately mediates conflict of interest between managers and investors by trying to make standards that can accommodate the interest of both sides.

Comprehensive income statement covers all changes in equity of the shareholders except transactions with the shareholders as in stock purchase or repurchase and dividend distribution (Casabona, 2014). Ohlson (1995) states that all changes in asset/debt that are not related to dividend have to go through income statement. So, comprehensive income is the income that covers all changes of asset/debt that are not related to the transactions with the owner (changes from dirty surplus to clean surplus).

PSAK No. 1 of 2009 demands all companies to present earnings in comprehensive income statement that elaborates such component into gross profit, operating income, current earning of the year which is net income before other comprehensive income, other comprehensive income, and comprehensive income. The fundamental change of PSAK No. 1 of 2009, effective from 2011, compared to the previous PSAK is the separation of unrealized gains and losses into the category of other comprehensive income. Other comprehensive income contains: unrealized gains and losses of available-for-sale financial assets (PSAK 55), revaluation surplus of tangible and intangible assets (PSAK 16 and 19), gains and losses of defined benefit plan actuary, gains and losses of foreign exchange rate changes from overseas business operation (PSAK 10), and gains and losses of hedging instrument (PSAK 55).

The first step of the comprehensive income concept implementation was executed through SFAS No. 130, issued by FASB in 1997. SFAS No. 130 is a request from AIMR (Association for Investment and Research) that recommends several changes in financial reporting model. That standard demands companies to report comprehensive income along with its components in a statement with similar eminence to basic financial statement (Hirst and Hopkins, 1998). Initially, comprehensive income statement for a reporting period covers net income, as reported in income statement, added or subtracted with items of other comprehensive income unreported in income statement. In June 2011, FASB issued Accounting Standar Update (ASU) No. 2011–05 that regulates the manner of presentation of entities to report comprehensive income in financial statement. Previous FASB directive through Accounting Standar Codification (ASC) 220–45–8 allowed three options in reporting comprehensive income as follows (Casabona, 2014): 1) the total of comprehensive income of a period, as well as other comprehensive income, can be reported under net income total in a single combined statement of income and comprehensive income, 2) in separated comprehensive income statement beginning with net income, and 3) in equity statement of the shareholders. Later on, ASU 2011–05 removed the third option, so these two options remain: a single continuous statement of comprehensive income and two separate but consecutive statement. By a single continuous statement of comprehensive income, entities have to include the component of net income, total of net income, component of other comprehensive income, total of other comprehensive income, and total of comprehensive income. Whereas by two-separated-but-consecutive-statement, entities have to report the components of net income and total of net income in net income statement (which is income statement), that has to be followed up with statement of other comprehensive income covering components of other comprehensive income and total of other comprehensive income, and total of comprehensive income. PSAK No. 1 of 2009 also provides 2 options of comprehensive income statement as in ASU 2011–05, which are single and separated.

2.2 *Value relevance of earnings*

Earnings figure is the main concern in a financial statement. By looking at earnings figure, the financial statement information receivers can view a company's overall performance. Earnings are the function of a company's financial performance, and such function demonstrates the accounting system that converts unobservable financial performance into observable earnings figure (Dechow et al. 2010). Therefore, earnings figures cannot completely describe a company's financial performance. Why can't accounting measuring system completely measure the actual performance of a company? Dechow et al. (2010) present three reasons. First is various models of decision. An accounting system that delivers single earnings figure statement cannot provide relevant financial performance representation for all kinds of decisions. The decision maker board creates trade-off to anticipate the users' needs, and eventually there isn't any single decision makers that acquires the image of performance of the company that is perfectly relevant for decision making. Second is the variation in financial performance measurement. There is no single standard that will measure financial performance perfectly for all kinds of company. As an example, the means of COGS (cost of good sold) measurement depends on when a company recognize its revenue. Different companies go with different basis of recognition. Third is implementation. An accounting system that measures unobservable financial performance construct inherently involves estimation and judgment. It can cause either unintentional or intentional mistake (example: earnings management).

How faithful earnings explain actual financial performance is in fact approached with the earnings quality. Earnings quality is a concept that has no general definition in the literature. Schipper and Vincent (2003) define earnings quality from the decision usefulness perspective, that is how faithful an income explains Hicksian income, including assets' net changes other than transactions with the owner. Ayres (1994) states that earnings quality is related to permanent earnings, which is high quality of earnings that portrays sustainable earnings for a long period of time. Bellovary et al. (2005) defines earnings quality as the ability of the reported income to explain a company's actual earning, just like the ability of reported income to predict future earnings. Earnings quality also explains stability, persistence, and low variability of the reported income.

A high quality of earnings figure may deliver the description of current operating performance, a good indicator of future operating performance, and accurate representation of a company's intrinsic value. However, not all earnings figures are created equal due to the dependency of earnings on the compositions of the earnings itself, the stage of the firm's life cycle, the time period, and the industry (Dechow and Shcrand, 2004).

Measuring earnings quality is significant because earnings are normally used to arrange compensation or debt agreement. A contract decision based on low quality earning will result in an unexpected welfare transfer. For example, an overstated income that is used as an indicator of manager performance will create an overstated compensation as well. From the perspective of investment, low earnings quality is unwanted because it is a signal of poor resource allocation, which is inefficient because it distorts potential projects to less profitable projects.

Cornell and Landsman (2003) express that the disagreement on earnings quality is based on more fundamental issue. The key issue that becomes a concern for the regulators and standards board is an efficient capital allocation in a proper capital market function. The precondition for efficient allocation is that market value explains economic value as big as possible. They make two arguments as follows: 1) not even a kind of earnings measurement, including GAAP, satisfyingly represents financial statement information for the purpose of prediction, 2) there is no meaningful criteria to determine whether an earnings measures better than the other components, even for certain companies, than by experience in predicting, and then determining that historical time series is more accurate for predicting.

2.3 *Incomplete revelation hypothesis*

In the efficient market hypothesis, the form of presentation, either simple or detailed, will not influence the judgment of information users because in an efficient market, the users of information can process all kinds of information quickly and properly. However, the information receiver might react differently toward different kind of information presentation form. The theory that explains such possibility is Incomplete Revelation Hypothesis (Bloomfield, 2002) or IRH. IRH predicts that investors use substantial resource to identify mispriced stock on the basis of public data, where manager tries to increase stock price by hiding bad news in the footnotes, and where the regulators want to prevent such effort, because the information that is difficult to be extracted from the financial statement will not be reflected in the stock price.

IRH distinguishes data from statistics. Data is written text or numbers or the ones saved in the computer file. Whereas statistics is useful facts that are extracted from that data, for example is profit figures or financial ratio. So if the data is extracted into meaningful information, then it becomes statistics. Market players have various tolerance towards the risk and funds related to that risk. Statistics can increase trading desire if the sellers collect it more and if those who collect it have big risk tolerance. Statistics with expensive extraction cost from available public information on the contrary will decrease trading desire, and such statistics will not completely reflected in the stock price. So, EMH predicts that stock market price completely describes all available public information, but IRH is otherwise. IRH predicts that stock market price does not completely describe all available public information.

IRH is also supported by psychological researches that prove that an information will not be used if it is unavailable and not processed immediately. Therefore, Hirst and Hopkins (1998) suggest that the judgment of analyst valuation will be influenced with the clarity of relevant information value disclosure. By applying experiment method, they find that an explicit disclosure of income statement in comprehensive income and its items can improve the transparency of management activity and company profit, and diminish the judgment of analyst valuation.

3   HYPOTHSES

As Kabir and Laswad (2011) once performed, this research also observed 6 hypotheses.

3.1 *Persistence and variability of NI and TCI*

PSAK No. 1 of 2009 demands all companies to present earnings in comprehensive income statement that elaborates it into gross profit, operating income, current earning of the year which is net income before OCI, OCI, and TCI. OCI comprises unrealized gains and losses from several items, arising from the fair value change of asset or liability.

The unrealized gain or losses from the value changes of assets and debts can increase income volatility (Scott, 2012) because it is transitory and does not describe core earning (Chambers, et al., 2007), which is income earned from the core activity of the company. Income volatility will complicate the investors to make return estimation in assessing investment. Unrealized gain or losses also does not reflect the actual performance of managers, and it should not be used to assess operating performance of managers because unrealized gain or loss is not the result of operating performance. Unrealized gain or loss is influenced by market factors that cannot be controlled by managers and usually is a result of unpredictable random process (Chambers, et al., 2007). Thus, NI may be more persistent compared with TCI, and TCI is more variable than NI.

H1: NI is more persistent than TCI.
H2: The cross-sectional variation of TCI is more than that of NI.

3.2 *Predictive ability of NI and TCI*

Cash flow of the operating activities is the cash flow utilized and deriving from the primary activities of the company. Unrealized gains and losses from non-primary activities are reported as OCI which is a component of TCI. Therefore, a question is raised concerning which of NI or TCI that is more capable in predicting future cash flows.

Two arguments were then developed. The followers of all-inclusive (clean surplus) believe that TCI contains all changes of economic value resulting from every company activity excluding transactions with the owner. Therefore, investor and creditor will exhaustively comprehend the future prospect of the firm and will be capable to predict future earnings and cash flows in a better manner (Kanagaretnam, 2009). From another point of view, those who promote current operating income (dirty surplus) consider that net income shall define the strength of permanent earnings of a company gained from the primary and repeated company activity that is measured objectively using historical cost based on realization principle. The change of company value resulting from transitory activity may impair its predictive capability.

Kanagaretnam (2009) finds that total comprehensive income correlates more to price and stock return and predicts future cash flows better than net income. Contrarily, Barth et al. (2001) and Dechow and Schrand (2004) confirm that NI predicts future cash flows and NI better than accruals earned from the transitory change of value within TCI.

H3: Predictive ability of NI to predict one-year-ahead CFO is better than TCI.
H4: Predictive ability of NI to predict one-year-ahead NI is better than TCI.

3.3 *Value relevance of NI and TCI*

Value relevance is the ability of financial statement information to capture or summarize information that influences the company value (Collins et al., 1997; Francis and Schipper, 1999) or stock price (Hellstorm, 2005). High earnings quality provides information about the company's financial performance that is relevant to make specific decision. Income is related to capital market performance. An earnings is high in quality if the correlation between income and stock price or market return is strong. By so, the measurement of earnings value relevance is performed by testing the correlation between earnings and stock price.

Lipe (1986) finds that stock return variation of industrial and commercial companies are better explained by earnings component than aggregate earnings. He find that every component provides information that completes aggregate earnings information, and if the information of earnings component is merged into aggregate earnings figure, then that information will disappear. Different from Lipe (1986), Jaggi and Zhao (2002) specifically conduct a research on the impact of the change of SFAS 12 into SFAS 115 of 1993 that reclassifies the presentation of investment securities on banking corporation. Securities reclassification is expected to decrease management discretion in classifying securities and reduce the gap of unrealized gains and losses between companies, and further will increase the content of unrealized loss and profit information. The result of their research verifies that the information of earnings component is more relevant for investment decision after the implementation of SFAS 115.

The information content of income and its components can be measured from its correlation with stock return. Investors will react to an information if that information is relevant for them to make decision and it will be reflected in the stock price or stock return. Prior to the implementation of PSAK No. 1 of 2009, companies present income statement consisting of three components; gross profit, operating income, and net income. After the implementation, companies are demanded to present profit in comprehensive income statement that elaborates it into gross profit, operating income, current earning of the year which is net income before other comprehensive income, other comprehensive income, and comprehensive income. With comprehensive income, that has more income components compared to the previous income statement, a question arises about which income is preferred by investors in making decision. In practice, in fundamental analysis, net income still becomes the top priority in making decision on investment. Further, transitory components, or components that are not the core operating activity of the company, become the following consideration. Bidlle and Choi (2006) confirm that net income that conforms with SFAS 130 dominates traditional net income. So do Kabir and Laswad (2011) that prove that net income can explain stock return better than comprehensive income.

From the point of view of investor, income is the amount of profit for the purpose of investment valuation. Investors assess investment based on the opportunity cost, which is market rate of return (Suwardjono, 2005). Thus, from the investors' point of view, income is the internal rate of return of future cash flow that can be earned if those investors invest their assets somewhere else (opportunity cost), meaning if investors invest their assets somewhere else, then that income becomes opportunity cost for them. Income from the point of view of investor is the economic income (real income), which is income in the form of the increase of economic prosperity.

Net income may be more relevant to investors because it is persistent, less varied, and more able to predict return, then net income may be more relevant. However, total comprehensive income may be more relevant than net income because based on IRH information will be used if it is available and processed immediately.

H5: The value relevance of NI differs from TCI.

The fundamental change of PSAK No. 1 of 2009 compared to previous PSAK lies in the presentation of unrealized gains and losses components into other comprehensive income category previously presented in the changes of equity. The location of OCI is evidently capable to improve the transparency of earnings management activities and reduce the analyst's valuation judgement (Hirst and Hopkins, 1998) as well as influence the use of volatility to assess company performance (Maines & McDaniel, 2000). Although all information in an efficient capital market can be processed by the information users, the users of financial statement prefer to utilize the information when such information is clear and presented with simple method (Johson, et al. 1988). It is the foundation to predict that the location of OCI presentation in comprehensive income statement is more beneficial than the presentation in changes of equity statement.

H6: The value relevance of TCI differs depending on its reporting location.

4 METHODOLOGY

4.1 Data and sample

The accounting data for this research was taken from complete financial statement retrieved from www.idx.co.id and the market data was taken from daily stock price acquired from Indonesia Stock Exchange corner. The population of this research is the companies listed in

Indonesia Stock Exchange (Bursa Efek Indonesia/BEI), and the sample of this research is the manufacturing companies listed in BEI from 2008 to 2014. The sample collection technique applied was purposive sampling method with the criteria that financial statement and market price data were acquired from 2007 to 2015.

4.2 *Variables and operational definition of variables*

The variables in this research are:

4.2.1 *Dependent variable*

The dependent variables for this research are one-year-ahead net income (NI_{t+1}), one-year-ahead total comprehensive income (TCI_{t+1}), one-year-ahead cash flow from operation (CFO_{t+1}), stock return (R_t), and stock price (P_t). The value relevance of earnings in this research was measured with stock return, which is the equity return of company t in year t, that was calculated from the gap of stock price of year t and t−1 added with dividend per share of year t, scaled with stock price of year t−1. Since the companies that divided the dividend is only few from the entire sample, then the numbers of dividend is ignored.

4.2.2 *Independent variable*

The independent variable for this research is net income of the current year, that is net income before other comprehensive income (NI), total comprehensive income (TCI), book value (BV), and the presentation of OCI location (Period), which is the period before the implementation of statement of comprehensive income where OCI is located in the statement of changes in equity and the period after the implementation of comprehensive income reporting where OCI is incorporated within the income statement.

4.3 *Models*

The first and second model to test the persistence of NI and TCI in the first hypothesis proposed in this research are as follows (Dechow and Schrand, 2004):

$$NI_{it+1} = \alpha_0 + \beta NI_{it} + \varepsilon_{it} \tag{1}$$

$$TCI_{it+1} = \alpha_0 + \beta TCI_{it} + \varepsilon_{it} \tag{2}$$

Notes:

NI_{it} = net income of company i in year t are scaled by the weighted average number of shares.

TCI_{it} = total comprehensive income of company i in year t are scaled by the weighted average number of shares.

From those two models, the coefficient (β) of NI or TCI variable is observed to find which is significant. If apparently both are significant, the biggest value of adjusted Rsquare between those models will be used to determine whether hypothesis 1 is supported.

The variability of NI and TCI were compared with the basis of standard deviation. Higher standard deviation reflects higher variability.

Predictive ability of NI and TCI on the third hypothesis proposed in this research was tested using third model and fourth model are as follows (Dechow and Schrand, 2004):

$$CFO_{it+1} = \alpha_0 + \beta NI_{it} + \varepsilon_{it} \tag{3}$$

$$CFO_{it+1} = \alpha_0 + \beta TCI_{it} + \varepsilon_{it} \tag{4}$$

Notes:

CFO_{it+1} = one-year-ahead cash flow from operating activities are scaled by average number of shares.

From those two models, the coefficient (β) of NI or TCI was then observed to find which is significant. If apparently both are significant, the highest value of adjusted Rsquare between those models will be used to determine whether hypothesis 1 is supported.

The predictive ability of NI and TCI was also tested by comparing which of NI or TCI is capable to predict NI in the following year. The test applied model 5 below (Dechow and Schrand, 2004):

$$NI_{it+1} = \alpha_0 + \beta TCI_{it} + \varepsilon_{it} \tag{5}$$

If the coefficient (β) of NI in model 1 is higher than the coefficient (β) of TCI in model 5, hypothesis 4 is supported.

The value relevance of NI and TCI was tested using model 6 to 11. Model 6 to 9 were applied to test the fifth hypothesis proposed in this research. The test of value relevance was carried out using return model (Dechow, 1994 and Easton, 1999) and price model (Dhaliwal, 1999) as follows:

$$R_{it} = \alpha_0 + \beta NI_{it-} P_{t-1} + \varepsilon_{it} \tag{6}$$

$$R_{it} = \alpha_0 + \beta TCI_{it-} P_{t-1} + \varepsilon_{it} \tag{7}$$

$$P_{it} = \alpha_0 + \beta_1 BV_{it} + \beta_2 NI_{it} + \varepsilon_{it} \tag{8}$$

$$P_{it} = \alpha_0 + \beta_1 BV_{it} + \beta_2 TCI_{it} + \varepsilon_{it} \tag{9}$$

Notes:

R_{it} = annual average of equity return of the company i in year t, calculated from the stock price gap of year t and t-1, scaled with the stock price of year t-1.

P_{it} = price per share of the company i at the end of the fiscal year t

BV_{it} = book value per share of the company i at the end of the fiscal year t

$NI_{it-} P_{t-1}$ = net income to common per share of company i in year t are scaled beginning-of-year stock price

$TCI_{it} P_{Pt-1}$ = total comprehensive income to common per share of company i in year t are scaled beginning-of-year stock price

If the coefficient (β) of NI in model 6 and or 8 is higher than the coefficient (β) of TCI in model 7 and or 9, hypothesis 5 is supported.

The research by Hirst and Hopkins (1998) results in different conclusion when other comprehensive income is not comprehensively presented collaboratively with income statement. Hence, OCI location as the determinant of the difference between NI and TCI shall be a notice to the user. By modifying the model used by Kabir and Laswad (2011), the following model 10 and 11 were applied:

$$R_{it} = \alpha_0 + \beta_1 TCI_{it-} P_{t-1} + \beta_2 TCI_{it-} P_{t-1} Period_t + \varepsilon_{it} \tag{10}$$

$$R_{it} = \alpha_0 + \beta_1 BV_{it} + \beta_2 TCI_{it} + \beta_3 TCI_{it} * Period_t + \varepsilon_{it} \tag{11}$$

$Period_t$ = dummy variable for year t, 1 if the company reported OCI in its income statement, 0 if the company reported OCI in the statement of changes in equity.

From the models above, if the coefficient β_2 and or β_3 of TCI interaction and the dummy variable, i.e. $\beta_2 TCI_{it-} P_{t-1} * Period_t$ and or $\beta_3 TCI_{it} * Period_t$ is positive significant, hypothesis 6 is then supported.

5 FINDINGS

5.1 Descriptive statistic

The data of manufacturing firms listed in Indonesia stock exchange during the observation period was acquired from Fact Book published annually by Indonesia Stock Exchange. The following is the sampling result for the hypothesis testing:

Table 1. Sample.

Manufacturing firms in BEI from 2008–2014	122 companies
Observation	854 firm-year observations
Financial statements are not available	45 firm-year observations
NI and TCI are the same (have no OCI items)	425 firm-year observations
Total observation for model 1 to 5 examinations	384 firm-year observations
Closing Prices are not available	7 firm-year observations
Total observation for model 6 to 11 examinations	377 firm-year observations

The descriptive statistics for the 384 firm-year observations and 377 firm-year observations is as follows:

Table 2. Descriptive statistics.

Variables	N	Mean	SD
NI_t	384	0,000389	0,002911
NI_{t+1}	384	0,000201	0,000546
TCI_t	384	0,000411	0,002966
TCI_{t+1}	384	0,000223	0,000551
CFO_t	384	0,000431	0,002980
CFO_{t+1}	384	0,000244	0,000587
NI_P_{t-1}	377	0,238057	1,344062
$TCI_t\_P_{t-1}$	377	0,410249	2,417801
$TCI_t\_P_{t-1}$*Period	377	0,132658	0,918423
TCI_t*Period	377	336,663993	2979,028206
Return	377	0,002869	0,009169
BV	377	1805,083269	3820,573128
Pt	377	4539,917800	22857,704920

From the table above, the total comprehensive income (TCI) of manufacturing firms is more diverse than net income (NI) as represented by the higher standard deviation. It indicates that TCI is more volatile due to OCI. However, both the standard deviation and average of NI and TCI are nearly similar, most likely due to small percentage of OCI compared to total asset, which is at 1.6%. The return between firms is less diverse as presented by low standard deviation.

5.2 Result of hypothesis testing

The testing result of all hypotheses is displayed in Tables 3 and 4 below:

Table 3. Result of hypothesis testing.

	Model				
	1	2	3	4	5
	Dependent variables				
	NI_{it+1}	TCI_{it+1}	CFO_{it+1}	CFO_{it+1}	NI_{it+1}
Independent variables					
Constant	0,000	0,000	0,000	0,000	0,000
	(6,751***)	(7,466***)	(7,769***)	(7,766***)	(6,737***)
NI_{it}	0,036 (3,835***)		0,030 (2,912***)		
TCI_{it}		0,035		0,027	0,034
		(3,721***)		(2,727***)	(3,725***)
N	384	384	384	384	384
Adjusted R^2	0,035	0,032	0,019	0,017	0,033
F-statistic	14,708***	13,847***	8,478***	7,435***	13,879***

Notes: Statistically significant at *10, **5, and ***1 percent.

144

Table 4. Result of hypothesis testing.

	Model					
	6	7	8	9	10	11
Dependent variables	Dependent variables					
	R_{it}	R_{it}	P_{it}	P_{it}	R_{it}	R_{it}
Independent variables						
Constant	0,003	0,003	1694,322	1671,622	0,003	1781,276
	(5,932***)	(5,603***)	(1,287)	(1,273)	(5,697***)	(1,352)
$NI_{it\_}P_{Pt-1}$	0,000087					
	(0,249)					
$TCI_{it\_}P_{Pt-1}$		0,001			0,001	
		(2,580***)			(2,867***)	
$TCI_{it\_}P_{Pt-1}$ * PERIOD					-0,001	
					(-1.268)	
BV_{it}			1,757	1, 790		1,877
			(4,045***)	(4,130***)		(4,244***)
NI_{it}			-0,825			
			(-1,461)			
TCI_{it}				-0,869		-4,005
				(-1,572)		(-1,240)
TCI_{it} * PERIOD						3,096 (0,986)
N	377	377	377	377	377	377
Adjusted R^2	-0,003	0,015	0,048	0,048	0,016	0,048
F-statistic	0,062	6,659***	10,385***	10,560***	4,138**	7,364***

Notes: Statistically significant at *10, **5, and ***1 percent.

Model 1 and 2 were applied in the test of persistence of NI and TCI in hypothesis 1. From the test result in Table 3, both models are significant, either NI or TCI influences one-year-ahead NI and one-year-ahead TCI. However, if compared to adjusted R^2 of both equations, model 1 has higher adjusted R^2, thus hypothesis 1 proposing that NI is more persistent than TCI is supported. This corresponds to the finding of Kabir and Laswad (2011).

The variability of NI and TCI in the second hypothesis was tested by the comparison of standard deviation. From the result in Table 2, standard deviation of TCI is higher than that of NI, indicating the higher variability of TCI than NI. Thus, hypothesis 2 is supported.

Table 3, specifically in model 3 and 4, illustrates the test result of predictive ability of NI and TCI as the third hypothesis. Both models imply the significancy of NI and TCI in predicting one-year-ahead CFO, yet the adjusted R^2 of model 3 (0.019) is higher compared to model 4 (0.017). Therefore, hypothesis 3 is supported, denoting that the predictive ability of NI to predict one-year-ahead CFO is better than TCI. The predictive ability of NI and TCI to predict one-year-ahead NI was also tested using model 1 and 5. In model 1 and 5 at Table 3, both NI and TCI are significant to predict one-year-ahead NI. However, the adjusted R^2 of model 1 is higher than that of model 5. Thus, hypothesis 4 is supported, clarifying that the predictive ability of NI to predict one-year-ahead NI is better than TCI.

The value relevance of NI and TCI was tested using model 6 to 9. As clarified in Table 4, only model 7 that exhibits the capability of TCI in describing stock return significantly at 1%, whereas NI is on the contrary. In the price model, either NI or TCI is incapable of describing stock price. Thus, hypothesis 5 is supported that the value relevance of NI differs from TCI.

The response of the users of income statement towards TCI may also be influenced by OCI location. Model 10 and 11 were used to test such assumption. As the result, Table 4 presents how model 10 and 11 clarify that OCI location is insignificant for the users in responding to TCI. Hypothesis 6 is then unsupported, so the TCI reporting location does not affect the relevance of TCI values.

Although the result of this research signifies that NI is more persistent and has a better predictive capability than TCI, the comparison of adjusted R2 of both items only shows slight difference. NI is even less variable than TCI, and the gap of standard deviation between the two is tight. The possible cause concerns with the sample of this research that comprises manufacturing companies. Manufacturing companies have small amount and low frequency of unrealized gain or losses compared to service and financial companies (Dhaliwal et al., 1999) so that the influence of unrealized gain or losses to net income or comprehensive income is relatively low.

6 CONCLUSION, LIMITATION, AND FUTURE RESEARCH

The purpose of this research is to acquire empirical evidence on the comparison between properties of net income and total comprehensive income. The result of this research demonstrates that net income is more persistence, less volatile, and greater in predicting one-year-ahead cash flows from operation, than total comprehensive income. In contrast, total comprehensive income is more capable in describing stock return than net income and the location of OCI is insignificant towards the value relevance of TCI.

The finding of this research verifies the research by Kanagaretnam (2009) affirming that total comprehensive income may describe stock return. Thus, this research does not corroborate the researches by Bidlle and Choi (2006), Dhaliwal et al (1999), and Kabir and Laswad (2011) concluding that net income in comprehensive income reporting is more capable in describing return than total comprehensive income. This research also promotes the research by Kabir and Laswad (2011) about persistence, variability, and predictive ability that suggests net income is more capable in those three properties, yet only has subtle difference if compared with total comprehensive income. This research may demonstrate how total comprehensive income excells over net income, yet still being under net income in connection with the foregoing properties. By this conclusion, this result may not present full endorsement towards the implementation of PSAK No. 1 of 2009.

This research does not support the hypothesis formulating that the location of OCI influences the value relevance of TCI. The inability of this research in supporting such hypothesis may be due to the low frequency and amount of unrealized gains and losses of the manufacturing companies compared to those of finance companies. In accordance with the aforementioned, further research is expected to conduct retest on companies with high other comprehensive income, e.g. companies engaging in service and financial industries.

REFERENCES

Ayres, F.L. 1994. Perception of earnings quality: What managers need to know. *Management Accounting* 75 (9): 27–29.

Barth, M.E., D.P. Cram, and K.K. Nelson. 2001. Accruals and the prediction of future cash flows. The Accounting Review 76 (1): 27–58.

Bellovary, J.L, D.E Giacomino, and M.D. Akers. 2005. Earnings quality: it's time to measure and report. *The CPA Journal* 75 (11): 32–37.

Bidlle, G.C. and Jong-Hang Choi. 2006. Is comprehensive income useful? *Journal of Contemporary Accounting & Economics* 2 (1): 1–32.

Black, D.E. 2015. Other comprehensive income: A review and directions for future research. *Working Paper.*

Black, F. 1993. Choosing accounting rules. Accounting Horizons 7 (4): 1–17.

Bloomfield, R.J. 2002. The "incomplete revelation hypothesis" and financial reporting. *Accounting Horizons* 16 (3): 233–243.

Brown, L.D and Sivakumar. 2001. *Comparing the quality of three earnings measures.* http://www.ssrn.com/ abstract = 272180.

Casabona, P.A., and T. Coville. 2004. Statement of comprehensive income: new reporting and disclosure requirements. *Review of Business* 35 (1): 23–34.

Chambers, D., T.J. Linsmeier, C. Shakespeasre, and T. Sougiannis. 2007. An evaluation of SFAS No. 130 comprehensive income disclosure. *Review Accounting Studied* 12: 557–593.

Collins, D.W., E.L. Maydew, and I.S. Weiss. 1997. Changes in the value-relevance of earnings and book values over the past forty years. *Journal of Accounting and Economics* 24: 39–67.

Cornell, B., and W. R. Landsman. 2003. Accounting valuation: Is earning quality an issue? *Financial Analyst Journal* 59 (6): 20–28.

Darsono. 2013. Dampak Konservatisma Terhadap Relevansi Nilai Informasi Akuntansi di Indonesia. *Disertasi*. Universitas Gadjah Mada, Yogyakarta.

Dechow, P., and C. Schrand. 2004. Earnings Quality. Research Foundation for CPA Institute, Charlottesville.

Dechow, P., W. Ge, and C. Schrand. 2010. Understanding earnings quality: A review of the proxies, their determinant and their consequences. *Journal of Accounting and Economics* 50: 344–401.

Dhaliwal, D., K.R. Subramanyam, and R. Trezevant. 1999. Is Comprehensive income superior to net income as a measure of firm performance? Journal of Accounting and Economics 26: 43–67.

Doukakis, L.C. 2010. The persistence of earnings and earnings components after the adoption of IFRS. *Managerial Finance* 36 (11): 969–980.

Easton, P.D. 1999. Security returns and the value relevance of accounting data. Accounting Horizons 13 (4):399–412.

Feltham, G.A, and J.A. Ohlson. 1995. Valuation and clean surplus accounting for operating and financial activities. *Contemporary Accounting Research* 11 (2): 689–731.

Francis, J. and K. Schipper. 1999. Have financial statements lost their relevance? *Journal of Accounting Research* 37 (2): 319–352.

Hellstorm, K. 2005. The value relevance of financial accounting information in a transitional economic: The case of Czech Republic. *Working Paper.*

Hirst, D. Eric, and Patrick E. Hopkins. 1998. Comprehensive income disclosures and analysts' valuation judgments. *Working Paper.*

Ikatan Akuntan Indonesia. 2012. Standar Akuntansi Keuangan. Jakarta: Ikatan Akuntan Indonesia.

Jaggi, B. and R. Zhao. 2002. Information Content of Earnings Components of Commercial Banks: Impact of SFAS No. 115. *Review of Quantitative Finance and Accounting* 18: 405–421.

Johson, E.J., J.W. Payne, and J.R. Bettman. 1988. Information displays and preference reversal. *Organizational Behavior and Human Decision Processes* 42: 1–21.

Kabir, M.H. and F. Laswad. 2011. Properties of net income and total comprehensive income: New zealand evidence. *Accounting Research Journal* 24 (3): 268–289.

Kanagaretnam, K., R. Mathieu, and M. Shehata. 2009. Usefulness of comprehensive income reporting in Canada. *J. Account. Public Policy* 28: 349–365.

Lin, S.W., O.J. Ramond, and Jean-Francois Casta. 2007. Value relevance of Comprehensive Income and its components: Evidence from Major European Capital Markets. Working Paper. Florida International University.

Lipe, G.M. 1998. Discussion of reporting and Analysts' valuation judgments. *Journal of Accounting Research* 36 (Supplement): 77–83.

Maines, L.A, and L.S. McDaniel. 2000. Effect of comprehensive-income characteristics on nonprofessional investors' judgment: the role of financial-statement presentation format. *The Accounting Review* 75 (2): 179–207.

Scott, R. William. 2015. *Financial Accounting Theory*. Toronto: Pearson Canada Inc.

Shipper, K., and L. Vincent. 2003. Earning quality. *Accounting Horizons* 17:97.

Suwardjono. 2005. *Teori Akuntansi Perekayasaan Pelaporan Keuangan*. Yogyakarat: BPFE. http://www.kemenperin.go.id/artikel/16062/Emiten-Manufaktur-Diminta-Ekspansi.

Business Innovation and Development in Emerging Economies – Trinugroho & Lau (Eds)
© 2019 Taylor & Francis Group, London, ISBN 978-1-138-35996-3

Firm characteristics, financial leverage, corporate governance, and earnings management in Indonesia

Y.K. Susanto & V. Agness
Trisakti School of Management, Jakarta, Indonesia

ABSTRACT: The objective of this research is to obtain empirical evidence about the influence of firm size, financial leverage, firm age, audit quality, ownership concentration, managerial ownership, institutional ownership, board size and board activity as independent variables on earnings management as a dependent variable in manufacturing companies listed in the Indonesia Stock Exchange. The population in this research is all listed as non-financial companies in the Indonesia Stock Exchange from 2013 to 2016. The sample was obtained through a purposive sampling method, in which 78 listed manufacturing companies in the Indonesia Stock Exchange met the sampling criteria resulting in 305 data. Multiple linear regression was used as the data analysis method. The results of this research show that firm size, firm age, audit quality, ownership concentration, managerial ownership, and board activity does not, statistically, influence earnings management, while financial leverage, institutional ownership, and board size does. Leverage has a negative influence as high leverage can reduce cash available for non-optimal spending and it can trigger the lender to induce spending restrictions.

Keywords: Earnings management, financial leverage, firm characteristics, corporate governance

1 INTRODUCTION

Almost all firm owners do not operate their own companies. They prefer to hire professional people to manage their companies. This separation of ownership and control causes a divergence of interest between shareholder and management, and leads to suboptimal management decisions (Alves, 2012). These decisions provide opportunities for managers to manage earnings for their own interest (Gulzar & Wang, 2011). Owners control these opportunities by using financial statements.

For owners, the financial statement is used as an indicator to measure a company's performance and is analyzed to provide the results of a company's operation for a specific period. Owners can monitor their managers by seeing the results of the financial statement. This will help owners to make a decision whether their managers are good at managing their company or not. If the manager has not achieved the expected result, the owners can hire other professional people to replace him/her. This shows that the financial statement is important.

The managers are also affected by the financial statement because if it shows an increase in profit, the owners are satisfied with the results, and the managers will receive a bonus. Instead, if the financial statement shows a decrease in profit and the owners are not satisfied with the results, the managers will not receive any bonus. As the manager is affected by the results of the financial statement, this encourages managers to commit fraud by manipulating information in the financial statement which is not permitted by the financial standard. This aim to make a company's profit increase where the managers stand to receive bonuses, or more, is known as 'earnings management'.

Corporate accounting scandals started happening at the beginning of the 21st century for example, Enron, WorldCom and Xerox (Bassiouny et al., 2016). Scandals like this creates a phenomenon where earnings management start to escalate. This issue makes earnings management connected to the manipulation of the financial statement. But many accountants do not accept that all earnings management are fraudulent and the managers are hired to manage the earnings.

Earnings management in PT Kimia Farma Tbk was viewed as financially fraudulent and mislead the public about its performance. The public thought that PT Kimia Farma Tbk had IDR132 billion of net income instead of IDR99 billion. This had an effect on public decisions.

This topic is very interesting to research as the earnings which can be manipulated by management can influence a financial user's decision. This problem formed the motivation for this research because earnings management can influence the public which are owners, other people connected to the company and also the economics in one country.

The number of incidences of earnings management keeps increasing and this also makes it an interesting topic to be researched. This research paper's objective is to obtain empirical evidence about the influence of firm size, financial leverage, firm age, audit quality, ownership concentration, managerial ownership, institutional ownership, board size and board activity on earnings management. The result of this research is expected to help create preventive action to avoid earnings management for management, avoid misinterpretation of financial reports for financial statement users, and give additional information about factors affecting earnings management. Overall, it will help auditors to detect fraud which is caused by earnings management.

1.1 *Agency theory*

The definition of the agency theory explains the relationship between shareholders or owners as the principals, and the managers as agents. The owners usually do not manage the business directly; they prefer to hire professional people to manage their companies. When the managers are given control, they also have rights to do tasks and make a decision. This relationship, which the principals give authority to manage the firm to the managers (agents), is called an agency relationship. This relationship can create a condition which is called asymmetric information and can be happen when one party has more information than the other party. This asymmetric information is used by the managers to fulfill their interests which are not the same interests as the owners (principals). When the managers use the asymmetric information to satisfy their interests, the situation will result in agency conflicts.

According to Jensen and Meckling (1976), the agency conflict can happen when there is separation of ownerships, and control and conflict of interest between principals and agents. Therefore, the agents tend to make a decision which is to satisfy their own interest rather than the principals' interests.

The source of the conflict of interest is asymmetric information and this has developed into a theory called the asymmetric information theory. This theory explains the gap in information which exists when the managers have more information than shareholders (as the shareholders are not involved in the firm's operation). This situation makes the owners unable to observe a firm's performance fully while the managers can manipulate the earnings easily (Lobo & Zhou, 2001).

1.2 *Earnings management*

Managers have many ways to manipulate earnings because they are motivated by agency benefit. One way is to use the accrual accounting method. This earnings manipulation to satisfy one party's interest is known as earnings management. Earnings management influences the earnings quality in the financial statements (Subramanyam, 2014). The opportunity of doing earnings management exists because the managers have the flexibility to decide which accounting method will be used (Sulistyanto, 2008).

Earnings managements can be divided into two: cosmetic and real. Cosmetic earnings management by managers is the manipulation of accrual so that the cash flow is not affected. Real management is earnings management by actions which will affect the cash flow.

According to Subramanyam (2014), there are several motivations for doing earnings management: (1) contracting incentives; (2) stock price effects; and (3) other incentives. There are three strategies to earnings management based on Subramanyam (2014) which are:

149

(1) increasing income; managers try to increase the income to show that the company is in good condition; (2) big bath; management activities to include all current and future period costs in the current period; and (3) income smoothing; a common type of earnings management where managers try to reduce volatility by increasing or decreasing income.

1.3 *Firm size and earnings management*

Firms can be classified into three size classes: small, medium, and large. Factors which can affect the firm size are total assets, total sales, average sales and market value of stock (Yuliana & Trisnawati, 2015).

Based on Susanto (2013), a small company has a higher chance of using earnings management in order to attract an investor. Meanwhile based on Agustia (2013), large companies need more funds than small companies. Motivation to get more funds can influence managers to increase the profit through the use of earnings management so that they fulfill owners' expectations, and attract investors and creditors.

Based on Bassiouny et al. (2016), the relationship of firm size and earnings management is negative. There are several reasons that explain this result. Earnings management is small in large-sized firms because their internal control system is effective and has more competent internal auditors. Small-sized firms have ineffective internal control and auditors. An effective internal control helps a firm give reliable financial information, so it can reduce the ability of management to manipulate earnings. The large-sized firms are audited by one of the 'Big Four' auditing firms and this also helps to prevent earnings management.

On the other hand, Barton and Simko (2002) argue that firm size and earnings management has a positive relationship. The reason is large-sized firms have many pressures to meet an analyst's expectations. The other reason is that large-sized firms have greater bargaining power with auditors; they can bargain with the auditors to reduce an effective and efficient audit.

In research by Llukani (2013), it concluded that firm size and earnings management has no relationship. The reason for this result is that a company which is big or small has no differences in motive for earnings management. Based on the descriptions above, the hypothesis is as follows:

H_1: Firm size has an effect on earnings management.

1.4 *Financial leverage and earnings management*

The relationship between debt level and accounting policies is because of debt covenants. The latter will influence a company when the company relies on debt. They need to maintain the leverage so that they will not violate the debt covenants imposed by creditors or bondholders. Bassiouny et al. (2016) argued that financial leverage has a negative relationship on earnings management. There are two reasons for this argument. The first reason is leverage requires debt repayment, thus reducing cash available for non-optimal spending. The second reason is that if a firm employs debt financing, it is often subject to lender induced spending restrictions (Jensen & Meckling, 1976). Research by Selahudin et al. (2014), shows the relationship between financial leverage and earnings management is positive. It means high financial leverage still has the possibility of managers undertaking earnings management. Based on the descriptions above, the hypothesis is as follows:

H_2: Financial leverage has an effect on earnings management.

1.5 *Firm age and earnings management*

Firms discover what they specialize at and learn new techniques to speed up their production processes, to minimize costs and improve the quality of products after many years (Arrow, 1962; Ericson and Pakes, 1995). Based on Bassiouny et al. (2016), firm age has a negative relationship with earnings management. The reason for this is a large-sized firm has a reputation to protect. So, the firms are aware of rules which they need to obey. Savitri (2014) found that firm age had no influence because neither new nor old firms are aggressive in managing earnings to avoid earning loss. Based on the descriptions above, the hypothesis is as follows:

H_3: Firm age has an effect on earnings management.

1.6 *Audit quality*

High quality audit is one of the most effective barriers to earnings management as it has a high chance of detecting and thus reporting errors and irregularities (DeAngelo, 1981). Asymmetries information between managers and shareholders can be reduced using auditing (Watts & Zimmerman, 1990). In auditing, the auditors verify the validity of financial statements and it can reduce the agency costs. The Big Four audit firms. Based on research conducted by Bassiouny et al. (2016), there is no relationship between audit quality and earnings management. This result is supported by a study conducted by Yasar (2013).

On the other hand, Lenard and Yu (2012) argue that firms audited with the 'Non-Big Four' firms have greater discretionary accruals. They concluded that the audit quality has a negative relationship with earnings management. This can happen because the higher quality auditors will not accept any manipulations and always report any error. This result is also supported by Memis and Cetanak (2012). Based on the descriptions above, the hypothesis is as follows:

H_4: Audit quality has an effect on earnings management.

1.7 *Ownership concentration and earnings management*

Shareholders can be divided in two groups based on its ownership: majority shareholders and minority shareholders. Minority shareholders seldom monitor management because the monitoring cost is too high and the benefit is not big for them. Instead, in an efficient monitoring hypothesis, majority shareholders would monitor and influence management actively (Shleifer & Vishny, 1986). Majority shareholders monitor the firm to protect their investments and assume to get the expected return from their investments. Thus, majority shareholders can reduce the agency costs by monitoring actively and alleviating the activity of minority shareholders. Gonzalez and Garcia-Meca (2014) and Alves (2012) concluded that ownership concentration reduces the possibility of managers undertaking earnings management.

However, conflict of interest between majority and minority shareholders may arise from this concentration ownership. Majority shareholders can use their rights to create benefits for themselves and sometimes do not give the same benefits to minority shareholders. This result is supported by Kim and Yoon (2008). The relationship between ownership concentration and earnings management is negative. This result is supported by Iraya et al. (2015). Fatmawati & Sabeni, (2013) concluded that ownership concentration has a relationship with earnings management. The reason is that the majority shareholders do not drive the firms to undertake earnings management. Based on descriptions above, the hypothesis is as follows:

H_5: Ownership concentration has an effect on earnings management.

1.8 *Managerial ownership and earnings management*

The agency theory explains how the separation of control and ownership can inflict agency conflict. When management do not have ownership in a firm, they are most likely to deviate from the goal of maximization of the owner's wealth (Jensen & Meckling. 1976). This will increase agency conflict and encourage management to carry out earnings management. In the agency theory, managerial ownership can encourage management to improve firm value as the managers can enjoy the effect of increasing firm value as shareholders. So, management's stock ownership can reduce agency conflict between managers and owners. The research by Susanto (2013) explained that managerial ownership has no influence. This is due to the small numbers of firms which have managerial ownership (Susanto, 2013). When managerial ownership increases, it is expected that the possibility of managers undertaking earnings management will decrease. This expectation is supported by research carried out by Alves (2012). On the other hand, Ruan et al., (2011) argued that managerial ownership encourages managers toward earnings management or it can be said managerial ownership and earnings management have a positive relationship. When managers own stocks in a firm, they will use their authority to increase the value of the shares. Therefore, higher managerial ownership encourages managers to manipulate earnings which improves the share's price. Based on the descriptions above, the hypothesis as is follows:

H_6: Managerial ownership has an effect on earnings management.

1.9 *Institutional ownership and earnings management*

Institutional investors pool a part of their profit and use the money to invest in securities, real property, and other investment assets, or operating firms. Monitoring of institutional ownership is one of the most important governance mechanisms. Institutional shareholders can more easily monitor management than smaller investors. Thus, the institutional ownership which is better in monitoring management's action can decrease the ability of management to manipulate earnings. Kamran and Shah (2014) concluded that institutional ownership is negatively related to earnings management.

Claessens and Fan (2002) argue that institutional investors do not actively monitor management's action. Institutional investors are more likely to sell their shares than monitoring and improving the performance of a firm. This can happen because the institutional investors do not want monitoring activities to affect their business relationship. Based on descriptions above, the hypothesis as is follows:

H_7: Institutional ownership has an effect on earnings management.

1.10 *Board size and earnings management*

Board size is the number of directors on a board. According to Iraya et al. (2015) and Daghsni (2016), a large board size will decrease the incidence of earnings management because an increase in board size increases a board's monitoring capacity. So, it gives less opportunities for management to undertake earnings management through the use of accounting policies. According to Liu et al. (2013) board size has a positive relationship with earnings management. It explains that the larger the firm, the larger its board size, but with any large firm the political cost is higher than a smaller company. The political cost happens when the company is large and attracts the attention from government, media and customers. This encourages them to decrease net profit to reduce political costs or maybe avoid all political costs even though the monitoring capacity of the board is large. According to Susanto (2013), board size has no relationship with earnings management. This is due to one reason: the size of the directors has no effect to make the directors more optimal in detecting earning management (Susanto, 2013). Based on the descriptions above, the hypothesis is as follows:

H_8: Board size has an effect on earnings management.

1.11 *Board activity and earnings management*

Board activity is an indicator of corporate governance and is measured by the number of board meetings held during a year. The more active the board, the better for the shareholders, as the board spends more energy on the company (Daghsni et al., 2016). In previous research carried out by Iraya (2015) and Gulzar and Wang (2011), the relationship between board activity and earnings management is positive. Daghsni et al. (2016) also supported this result. Based on the descriptions above, the hypothesis is as follows:

H_9: Board activity has an effect on earnings management.

Table 1. Sample selection procedure.

No.	Description	Total number of companies	Total number of data analyzed
1	Manufacturing companies listed consistently in Bursa Efek, Indonesia from 2012 until 2016	131	524
2	Manufacturing firms that did not use IDR in their financial statements from 2012 until 2016	29	116
3	Manufacturing firms that published their financial statements and annual reports as of 31st December from 2012 until 2016	1	4
4	Manufacturing firms that t disclosed the numbers of directors meeting from 2013 until 2016	23	92
5	Total data used for sample	78	312
6	Outlier data		7
7	Total data		305

Source: Data was obtained and processed through IDX's data.

2 RESEARCH METHOD

The sample studied in this research paper was firms chosen using a purposive sampling technique, which means that the sample will be used only if it meets intended criteria. The criteria for the research sample was: (1) manufacturing firms that were consistently listed in the Indonesia Stock Exchange from 2012 until 2016; (2) manufacturing firms that used IDR in their financial statements from 2012 until 2016; (3) manufacturing firms that published their financial statements and annual reports as of 31st December from 2012 until 2016; and (4) manufacturing firms that disclosed the numbers of director board meetings from 2013 until 2016.

The measurement of discretionary accruals was calculated through steps. Firstly, by calculating Total Accruals (TA) for each firm in Year t as shown below:

$$TA_t = NI_t - CFO_t$$

where:

TA_t = total accruals in year t
NI_t = net income in year t
$CFOit$ = cash flow from operating activities in year t

$$TAC_{jt}/A_{jt-1} = \beta_1[1/A_{jt-1}] + \beta_2[(\Delta REV_{jt} - \Delta AR_{jt})]/A_{jt-1} + \beta_3[PPE_{jt}A_{jt-1}] + \varepsilon_t$$

where:

TAC_{jt} = total accruals for Firm j in year t
A_{jt-1} = total assets for firm j in year t–1
ΔREV_{jt} = change in the revenues (sales) for firm j in year t less revenue in year t–1
ΔAR_{jt} = change in accounts receivables for firm j in year t less receivable in year t–1
PPE_{jt} = gross properties, plants and equipment for firm j in year t
β_1 = firm specific parameters
β_2 = firm specific parameters
β_3 = firm specific parameters

Next, non-discretionary accruals were measured for each year and fiscal year combination using the equation:

$$NDA_{jt} = \beta_1[1/A_{jt-1}] + \beta_2[\Delta REV_{jt} - \Delta AR_{jt}/A_{jt-1}] + \beta_3[PPE_{jt}/A_{jt-1}]$$

where:

NDA_{jt} = non-discretionary accruals for firm j in year t
A_{jt-1} = total assets for firm j in year t–1

Table 2. Variables measurements.

Variable	Measurement
Firm size	ln (total assets)
Financial leverage	Total debt ÷ total assets
Firm age	Numbers of years since the firm's foundation
Audit quality	1 if the auditor is a 'Big Four Firm'; 0 if not
Ownership concentration	Fraction of stocks owned by major shareholders
Managerial ownership	1 if there is proportion of stock held by management; 0 if none
Institutional ownership	Percentage of common stock held by institutions
Board size	Total number of directors on the board
Board activity	Number of board meetings held in one year

ΔREV_{jt} = change in the revenues (sales) for firm j in year t less revenue in year t–1
ΔAR_{jt} = change in accounts receivables for firm j in year t less receivable in year t–1
PPE_{jt} = gross properties, plants and equipment for firm j in year t
β_1 = firm specific parameters
β_2 = firm specific parameters
β_3 = firm specific parameters

Then the difference between total accruals and the non-discretionary components of accruals was considered as Discretionary Accruals (DA), as stated in the equation below:

$$DA_{jt} = TAC_{jt}/A_{jt-1} - NDA_{jt}$$

where:
DA_{jt} = discretionary accruals for firm j in year t
TAC_{jt}/A_{jt-1} = total accruals for firm j in year t
NDA_{jt} = non-discretionary accruals for firm j in year t

The independent variables measurements conducted in this research is shown in Table 2.

3 RESEARCH RESULTS

This paper's research examined 78 manufacturing firms listed in the Indonesia Stock Exchange from the year 2013 until 2016. Thus 305 data were used in this research. The statistical results are shown in Table 3.

The results show that firm size has a significance level of 0.370, which is above 0.05. This means that firm size has no influence on earnings management. A big or small company has no differences in motives to do earnings management (Llukani, 2013). This result is consistent with Llukani (2013), but not consistent with Bassiouny et al. (2016) and Barton and Simko (2002).

The results show that financial leverage has a significance level of 0.000, which is below 0.05. This means that financial leverage has a negative influence on earnings management. This can happen for two reasons. The first reason is leverage requires debt repayment, thus reducing the cash available for non-optimal spending. The second reason is that a firm that employs debt financing, is often subject to lender induced spending restriction (Bassiouny et al., 2016). This result is consistent with Bassiouny et al. (2016), but not consistent with Selahudin et al. (2014).

The results show that firm age has a significance level of 0.625, which is above 0.05. This means that firm age has no influence on earnings management. Neither new nor old firms are aggressive in managing earnings to avoid earning loss reporting (Savitri, 2014). This result is consistent with Savitri (2014), but not consistent with Bassiouny et al. (2016).

The results show that audit quality has a significance level of 0.123, which is above 0.05. This means that audit quality has no influence on earnings management. There is no difference in audit quality between the Big Four and Non-Big Four audit firms in constraining the practice

Table 3. Descriptive statistics.

Variable	N	Minimum	Maximum	Mean	Std. Deviation
EM	305	−0.32468	0.25641	−0.0006642	0.07837436
SIZE	305	25.2455	33.1988	28.117186	1.6507663
FLEV	305	00.0005038	4.0951867	0.497161931	0.4579512411
FAGE	305	4	98	38.30	16.562
AUQUL	305	0	1	0.37	0.483
Concentration	305	0.10170	48.83000	0.6586684	2.77709272
MO	305	0	1	0.48	0.500
IO	305	0.00000000	0.98240000	0.69199774356	0.194256499871
BSIZE	305	2	16	5.17	2.643
BACT	305	2	66	16.12	12.558

Source: Data output from SPSS 19.0.

Table 4. Hypothesis test.

Variable	B	Sig.
(Constant)	0.132	0.204
SIZE	−0.003	0.370
FLEV	−0.041	0.000***
FAGE	0.000	0.625
AUQUL	−0.020	0.123
Concentration	0.000	0.861
MO	−0.008	0.381
IO	−0.044	0.069*
BSIZE	−0.004	0.077*
BACT	0.000	0.576

Notes: F 2.976, Sig. 0.002, *10% **5% ***1%.
Source: Data output from SPSS 19.0.

of earnings management (Yasar, 2013). This result is consistent with Bassiouny et al. (2016) and Yasar (2013), but not consistent with Lenard and Yu (2012), and Memis and Cetanak (2012).

The results show that ownership concentration has a significance level of 0.861, which is above 0.05. This means ownership concentration has no influence on earnings management. The majority shareholders do not drive the firms to do earnings management (Fatmawati & Sabeni, 2013). This result is consistent with Fatmawati and Sabeni (2013), but not consistent with Iraya et al. (2015), Gonzalez and Garcia-Meca (2014), and Alves (2012).

The results show that managerial ownership has a significance level of 0.381, which is above 0.05. This means that managerial ownership has no influence on earnings management. This is due to the small number of firms which have managerial ownership (Susanto, 2013). This result is consistent with Susanto (2013), but not consistent with Alves (2012), and Ruan et al. (2011).

The results show that institutional ownership has a significance level of 0.069, which is below 0.1. This means institutional ownership has a negative influence on earnings management. Institutional investors are likely to be monitoring and improving performance in a firm. This result is consistent with Kamran and Shah (2014), but not consistent with Claessens and Fan (2002).

The results show that board size has a significance level of 0.077, which is below 0.1. This means that board size has a negative influence on earnings management. The number of directors makes the board optimal in detecting earnings management. This result is consistent with Daghsni (2016), Iraya (2015), and Liu et al. (2013), but not consistent with Susanto (2013).

The t test result shows that board activity has a significance level of 0.576, which is above 0.05. This means that board activity has no influence on earnings management. According to Sukeecheep et al. (2013), meetings do not provide any constraint to reducing or increasing earnings management. This result is not consistent with Daghsni et al. (2016), Iraya (2015) and Gulzar (2011).

4 CONCLUSIONS

Financial leverage, institutional ownership and board size have an influence on earnings management, and the influence is negative. Firm size, firm age, audit quality, ownership concentration, managerial ownership, and board activity have no influence on earnings management. There were some limitations that existed during this research: (1) the research period used was relatively short (only four years); and (2) the research population was relatively small (only listed manufacturing firms). Based on the limitations above, recommendations that can be used in future research are: (1) to undertake a longer period of research; and (2) to enlarge the research population.

REFERENCES

Agustia, D. (2013). The effect of good corporate governance, free cash flow, and leverage on earnings management. *Jurnal Akuntansi dan Keuangan, 15*(1), 27–42.

Alves, S. (2012). Ownership structure and earnings management: Evidence from Portugal. *Australasian Accounting, Business and Finance Journal, 6*(1), 57–74.

Arrow, K.J. (1962). The economic implication of learning by doing. *The Review of Economic Studies, 29*(3), 155–173.

Barton, J. & Simko, P.J. (2002). The balance sheet as an earnings management constraint. *The Accounting Review, 77*, 1–27.

Bassiouny, S.W., Soliman, M.M. & Ragab, A. (2016). The impact of firm characteristics on earnings management: An empirical study on the listed firms in Egypt. *The Business and Management Review, 7*(2), 91–101.

Claessens, S. & Fan, J.P.H. (2002). Corporate governance in Asia: A survey. *International Review of Finance, 2*(2), 71–103.

Daghsni, O., Zouhayer, M. & Mbarek, K.B.H. (2016). Earning management and board characteristics: Evidence from Frenchlisted firm. *Arabian Journal of Business and Management Review, 6*(5), 1–9.

DeAngelo, L.E. (1981). Auditor size and audit quality. *Journal of Accounting and Economics, 3*, 183–199.

Ericson, R. & Pakes, A. (1995). Markov-perfect industry dynamics: A framework for empirical work. *The Review of Economic Studies, 62*(1), 53–82.

Fatmawati, D. & Sabeni, A. (2013). The effects of geographic diversification, industry diversification, concentration of company ownership, and period of audit engagement on earnings management. *Diponegoro Journal of Accounting, 2*(2).

Gonzalez, J.S. & Garcia-Meca, E. (2014). Does corporate governance influence earnings management in Latin American markets? *Journal of Business Ethics, 121*, 419–440.

Gulzar, M.A. & Wang, Z. (2011). Corporate governance characteristics and earnings management: Empirical evidence from Chinese listed firms. *International Journal of Accounting and Financial Reporting, 1*(1), 133–157.

Iraya, C., Mwangi, M. & Muchoki, G.W. (2015). The effect of corporate governance practices on earnings management of companies listed at the Nairobi Securities Exchange. *European Scientific Journal, 11*, 169–178.

Jensen, M.C. & Meckling, W.H. (1976). Theory of the firm: Managerial behavior, agency costs and ownership structure. *Journal of Financial Economics, 3*(4), 305–360.

Lenard, M.J & Bing, Y. (2012). Do earnings management an audit quality influence over-investment by Chinese companies? *International Journal of Economics and Finance, 4*(2), 21–30.

Kamran. & Shah, A. (2014). The impact of corporate governance and ownership structure on earnings management practices: Evidence from listed companies in Pakistan. *The Lahore Journal of Economics, 19*(2), 27–70.

Kim, H.J. & Yoon, S.S. (2008). The impact of corporate governance on earnings management in Korea. *Malaysian Accounting Review, 7*(1), 43–59.

Liu, J., Harris, K. & Omar, N. (2013). Board committees and earnings management. *Corporate Board: Role, Duties & Composition, 9*(1), 6–17.

Llukani, T. (2013). Earnings management and firm size: An empirical analyze (sic) in Albanian market. *European Scientific Journal, 9*(16), 135–143.

Lobo, G.J. & Zhou, J. (2001). Disclosure quality and earnings management. *Asia-Pacific Journal of Accounting and Economics, 8*(1), 1–20.

Memiş, M.Ü. & Çetenak, E.H. (2012). Earnings Management, Audit Quality and Legal Environment. *International Journal of Economics and Financial Issues, 2*(4), 460–469.

Ruan, W., Tian, G. & Ma, S. (2011). Managerial ownership, Capital Structure and firm value: Evidence from China's civilian-run firms, *Australasian Accounting, Business and Finance Journal, 5*(3), 73–92.

Savitri, E. (2014). Analysis of the influence of leverage and life cycle on earnings management in real estate and property companies listed on the Indonesia stock exchange. *Jurnal Akuntansi, 3*(1), 72–89.

Selahudin, N.F. Zakaria, N.B. & Sanusi, Z.M. (2014). Remodelling the earnings management with the appearance of leverage, financial distress and free cash flow: Malaysia and Thailand evidences (sic). *Journal of Applied Sciences, 14*(21), 2644–2661.

Shleifer, A. & Vishny, R.W. (1986). Large shareholders and corporate control. *Journal of Political Economy, 94*(3), 461–488.

Subramanyam, K.R. (2014). *Financial statement analysis* (11th ed.).: McGraw-Hill Education.

Sukeecheep, S., Yarram, S.R., Farooque, O.A. (2013). Earnings management and board characteristics in Thai listed companies. *The 2013 IBEA, International Conference on Business, Economics, and Accounting.* Bangkok, Thailand.

Sulistyanto, H.S. (2008). Earnings management (theory and empirical model). Jakarta, Indonesia: PT Gramedia Widiasarana.

Susanto, Y.K. (2013). The effect of corporate governance mechanism on earnings management practice (case study on Indonesia manufacturing industry). *Jurnal Bisnis dan Akuntansi, 15*(2), 157–167.

Watts, R.L. & Zimmerman, J.L. (1990). Positive accounting theory: A ten year perspective. *The Accounting Review, 65*(1), 131–156.

Yaşar, Alpaslan. (2013). Big Four Auditors' Audit Quality and Earnings Management: Evidence from Turkish Stock Market. *International Journal of Business and Social Science, 4*(7), 153–163.

Yuliana, A. & Trisnawati, I. (2015). The effect of auditor and financial ratio on earnings management. *Jurnal Bisnis dan Akuntansi, 17*(1), 33–45.

Business Innovation and Development in Emerging Economies – Trinugroho & Lau (Eds)
© *2019 Taylor & Francis Group, London, ISBN 978-1-138-35996-3*

Profile analysis of the importance of equity financing decisions on Package IPO (PIPO) and Shares-only IPO (SIPO) in Indonesia

C. Dhevi
Universitas Negeri Malang, Malang, Indonesia

Y. Soesetio & D.Q. Octavio
Management Department, State University of Malang, Malang, Indonesia

ABSTRACT: This study aims to analyze the association between a firm's financial and non-financial information and their Initial Public Offering (IPO) strategy. Specifically, our focus is on the package IPO and shares-only IPO strategies. The purpose of this study is to analyze the impact of auditor reputation and underwriter reputation on a firm's IPO strategy. Our sample is the IPOs listed on the Indonesian Stock Exchange from 2000 to 2016. We have excluded the year 2008 because of the US economic crisis that affected the Indonesian Stock Market. A total of 178 companies are used in this study. Our results show that a firm's financial risk, a firm's profitability and auditor reputation play an important role in IPO strategy.

Keywords: Package IPO (PIPO), Shares-only IPO (SIPO), unit offer

1 INTRODUCTION

The use of a package offering in equity issuance has become the subject of considerable debate in the corporate finance world. Many IPO bundles include other securities as a sweetener in order to attract investors. Usually, a warrant is a financial security that is bundled into an IPO. This action is popularly called a 'unit offer' or 'unit offering'. How and Howe (2001) introduced other appellations to describe these phenomena. They refer to them as a Package Initial Public Offering (PIPO), for an IPO bundled with other securities, and a Shares-only Initial Public Offering (SIPO) for a normal IPO.

There are various reasons why PIPOs are conducted. First, by using a warrant in an IPO bundle, firms signal that they have the commitment to keep the stock price at a minimum at the same level as the exercise price of the warrant in a seasonal offering. From another perspective, bundling an IPO with a warrant can be seen as a signal to the public that the IPO offer price is underpriced (James & Wier, 1990). In other words, the price of the share has a greater potential to increase. Schultz (1993) also states that companies that sell shares with warrants aim to reduce agency costs. Howe and Olsen (2009) argue that the issuance of a package IPO is a corporate governance mechanism for companies conducting an IPO. They also said that the option or warrant is one of the substitutes for corporate governance of companies conducting an IPO.

This paper analyzes the importance of auditor reputation on the choice of IPO method. To the best of the authors' knowledge, studies relating to the use of PIPO and SIPO in developing countries are very limited. Most of the existing IPO literature focuses on the underpricing of IPOs. Thus, this research will expand on the literature relating to an initial public offering strategy, especially in developing countries such as Indonesia.

Our sample is 178 IPOs issued from 2000 to 2016. Due to the economic crisis, we have excluded IPOs issued in 2008. Nezky (2013) showed that the US crisis in 2008 affected the Indonesian

Stock Market significantly. Many researchers (El-Banany, 2013; Manolopoulou & Tzelepis, 2014) also suggested excluding this crisis period in analysis, since a data outlier can create bias..

To systematize the report, we have divided this study into five sections. The first section reveals the importance of this study. The second section explains the relationship between auditor reputation and IPO strategy. The third section explains the data, variables, definitions and methods. In the fourth section, the empirical results are reviewed. The fifth section summarizes the results of this study.

2 LITERATURE REVIEW

In this section we explain the reasons why each variable has to be considered with regards to the choice of IPO strategy. There are four main independent variables that we believe have the ability to explain the choice of IPO strategy. They are total debt to total asset ratio, return on equity, auditor reputation and underwriter reputation.

2.1 *Debt to total asset ratio*

This variable shows how a firm has grown and acquired its assets over time. This ratio also reflects the ability of a firm to meet all of its liabilities. In other words, this variable can be described as a firm's financial risk. In line with the explanation offered by Schultz (1993), we expect that firms that have a large debt to total asset ratio are more likely to choose PIPO over SIPO, because they are at greater risk.

2.2 *Return on equity*

This variable reflects the ability of a firm to create a profit. Firms that have good profitability may also have a high level of public confidence. Thus, they do not need a 'sweetener' in order to attract the public to buy their IPO. In addition, PIPO firms are smaller, have less income and fewer assets than SIPO firms (Schultz, 1993).

2.3 *Auditor reputation*

IPOs are typically used to fund a firm. The increase to cash flow in the firm can be an opportunity for a manager to undertake other projects, even though these projects may not be efficient or, in the worst case, may have a negative present value. In other words, the function of the manager, which is to maximize the value of shareholders, is not fulfilled. This phenomenon occurs because there is asymmetric information between managers and owners, giving the manager the opportunity to maximize their interests through violating the owner's best interests (Jensen & Meckling, 1976).

A reputable auditor can reduce agency costs significantly (Watts & Zimmerman, 1983). The amount of agency costs in a firm will affect the firm's decision as to whether to choose a PIPO or a SIPO. Issuing an IPO bundled with a warrant can reduce agency costs. This warrant becomes an instrument that can be used to control the activities of a manager, because managers have to please the public in the seasoned offering. Thus, they have a responsibility to increase the price of the stock to more than the exercise price (Howe & Olsen, 2009). Thus, a firm that is suffering agency problems will choose a PIPO as their IPO strategy. In contrast, firms that have few agency costs will have no motivation to choose a PIPO. Moreover, additional administrative fees will be incurred by firms if they issue a warrant and this cost is not relevant for their business.

2.4 *Underwriter reputation*

Underwriters play a significant role in IPO activities. In IPO pricing creation, underwriters and issuers make an agreement to determine the initial price. Indeed, underwriters play a significant role in the decisions regarding an IPO. However, the direct effect of the underwriter remains inconclusive. Some researchers found that an underwriter has a positive effect on underpricing (Cooney et al., 2001; Bates & Dunbar, 2002). However, Carter and Manaster (1990) found that a prestigious underwriter reduces the level of underpricing. Beatty and Welch (1996) argue that the difference in underwriter behavior is attributable to economic conditions.

Indeed, to the best of the authors' knowledge, there is no research that explains the impact of the reputation of the underwriter on the choice of IPO method. However, we propose that underwriters with a good reputation tend to avoid PIPO. This is because having a warrant as a sweetener suggests that the issuer is not confident. This action may damage the underwriter's reputation, because it can be interpreted as meaning that the underwriter does not have the capability to buy all of the shares if they do not sell well in the market.

3 RESEARCH METHOD

Our sample consists of firms that conducted an IPO in the Indonesian Stock Market from 2000 to 2016. In this study, we have excluded the year 2008 because of the significant impact of the US financial crisis on the Indonesian Stock Market (Nezky, 2013). Thus, the number of samples is reduced from 219 to 178 firms. This exclusion is necessary in order to reduce the impact of data outliers. Logit and Probit regressions are chosen as our main method of analysis, because our dependent variable is binary.

3.1 Dependent variable

Our dependent variable type is binary. This variable will be valued as 1 if the firms chose PIPO as their IPO strategy, while it will be valued as 0 if SIPO is chosen as their IPO strategy. The code of this variable is IPO_STRAT.

3.2 Independent variables

There are four main independent variables in our study, which are Auditor Reputation (AUD_REP), Underwriter Reputation (UND_REP), Return on Equity (ROE) and Debt to Total Asset (DAR). Auditor reputation (AUD_REP) is measured using a dummy variable, which will be assigned a value of 1 for a reputable auditor and a value of 0 otherwise. A reputable auditor is a company that is included in the category of 'The Big Four'. Similarly, underwriter reputation (UND_REP) is also calculated using a dummy variable. It is assigned a value of 1 for a reputable underwriter and a value of 0 for an unreputable underwriter. A reputable underwriter is an underwriter included in 'The Top 5' underwriters, based on how much the nominal share value was guaranteed by the underwriter in this observation period, which is from 2000 to 2016. In contrast with the other two variables, return on equity (ROE) and debt to total asset (DAR) are not binary variables and are calculated as ratios. Return on equity is computed as a firm's total profit divided by shareholder equity value, and debt to total asset is calculated as total debt divided by total assets.

3.3 Control variables

In addition to the variables described above, we have also included two control variables, namely firm age and firm size. These variables are less important than our main variables because their definitions are not as specific as those of the main variables. In this study, we transform these two variables into a natural logarithm, because the gap value of both of them compared to the other variables is otherwise too large. Thereby, heteroscedasticity can be reduced.

4 EMPIRICAL RESULTS

4.1 Descriptive statistics and correlation

Our samples are the IPOs of 178 firms listed on the Indonesian Stock Market from 2000 to 2016; 43 firms for PIPO and 135 for SIPO. We have excluded the year 2008 because the economic crisis in that year will affect the data. Table 1 shows the descriptive statistics of the variables that we have analyzed. Our descriptive statistics show the number of observations, mean, standard deviation, minimum value, maximum value and Wilcoxon rank-sum test.

The purpose of this analysis was to reveal the profile of the data that we have analyzed and the differences between PIPO and SIPO. Based on this analysis, it can be seen that there are significant differences between the firms that conducted PIPO and SIPO in all of the aspects. Based on the financial characteristics, the average value of total debt to total asset ratio of firms that conducted PIPO is significantly lower than that of firms that conducted SIPO. Surprisingly, this indicates that firms that conducted PIPO may have less financial risk than firms that conducted SIPO. However, based on the average ROE value, firms that conducted SIPO are significantly more profitable than firms that conducted PIPO. The average value of the assets of firms that conducted PIPO is smaller than that of firms that conducted SIPO. Based on the non-financial characteristics, firms that conducted PIPO have, on average, hired a less reputable external auditor than firms that conducted SIPO. Similar to the previous profile, firms that conducted PIPO have, on average, hired a less reputable underwriter than firms that conducted SIPO. However, the difference is only slightly significant, being only significant at 10%.

Before we conducted the main analysis, we also needed to ensure that there was no multicollinearity in each independent variable. To analyze whether or not any multicollinearity existed, we conducted a correlation matrix using the Pearson correlation as our method. The result is shown in Table 2, and indicates that there is no value above 0.800. Thus, we can conclude that there is multicollinearity in each independent variable.

4.2 Regression analysis

In the regression analysis we used two methods, Logit and Probit. These methods were used because our dependent variable was binary. In Table 3 we show the results of the regression based on these two methods and hierarchical regression in order to deduce the stability of the independent variable in explaining the dependent variable.

Our findings show that DAR has a negative association on the choice of the firm to conduct PIPO. Indeed, the significance value is slightly reduced when control variables are

Table 1. Descriptive statistics.

| Variable | PIPO | | | | | SIPO | | | | | Wilcoxon rank-sum (z) |
	Obs	Mean	Std. Dev.	Min	Max	Obs	Mean	Std. Dev.	Min	Max	
DAR	43	0.463	0.258	0.004	0.950	135	0.600	0.227	0.033	0.970	3.152***
ROE	43	0.061	0.127	−0.176	0.526	135	0.212	0.607	−0.781	6.361	3.971***
AUD_REP	43	0.140	0.351	0.000	1.000	135	0.319	0.468	0.000	1.000	2.282**
UND_REP	43	0.023	0.152	0.000	1.000	135	0.104	0.306	0.000	1.000	1.649*
ln_asset	43	26.174	1.743	23.076	29.864	135	27.081	2.577	17.552	33.154	3.318***
ln_age	43	2.308	0.826	0.693	4.489	135	2.614	0.854	0.693	4.970	2.403**

PIPO denotes package IPO; SIPO denotes shares-only IPO; DAR denotes total debt to total asset ratio; ROE denotes return on equity; AUD_REP denotes auditor reputation, valued as 1 if the firm hired a reputable auditor and 0 otherwise; UND_REP denotes underwriter reputation, valued as 1 if the firm hired a reputable underwriter and 0 otherwise; ln_asset denotes the natural logarithm of the total assets; ln_age denotes the natural logarithm of the firm age. Asterisks (***), (**) and (*) indicate statistical significance at 1%, 5%, and 10% levels, respectively.

Table 2. Correlation matrix.

	DAR	ROE	AUD_REP	UND_REP	ln_asset	ln_age
DAR	1					
ROE	0.0888	1				
AUD_REP	0.1095	−0.0700	1			
UND_REP	0.1181	0.2855	−0.0059	1		
ln_asset	0.2801	0.0850	0.1205	0.1524	1	
ln_age	0.1844	0.0487	0.0361	−0.0532	0.2023	1

DAR denotes total debt to total asset ratio; ROE denotes return on equity; AUD_REP denotes auditor reputation, valued as 1 if the firm hired a reputable auditor and 0 otherwise; UND_REP denotes underwriter reputation, valued as 1 if the firm hired a reputable underwriter and 0 otherwise; ln_asset denotes the natural logarithm of the total assets; ln_age denotes the natural logarithm of the firm age.

Table 3. Regression results.

	Logit	Logit	Probit	Probit
Variable	IPO_STRAT	IPO_STRAT	IPO_STRAT	IPO_STRAT
DAR	−2.071***	−1.764**	−1.178***	−0.988**
	(0.791)	(0.827)	(0.455)	(0.476)
ROE	−2.719**	−2.334*	−1.488**	−1.318*
	−1.220	−1.191	(0.685)	(0.677)
AUD_REP	−1.098**	−1.098**	−0.662**	−0.658**
	(0.503)	(0.507)	(0.280)	(0.282)
UND_REP	−1.141	−1.296	−0.703	−0.772
	−1.079	−1.097	(0.577)	(0.591)
ln_asset		−0.047		−0.034
		(0.080)		(0.048)
ln_age		−0.320		−0.176
		(0.244)		(0.141)
Constant	0.536	2.396	0.280	1.500
	(0.460)	−2.074	(0.274)	−1.237
Observations	178	178	178	178
Pseudo R-squared	0.1217	0.1335	0.1220	0.1339

IPO_STRAT is valued as 1 if the firm conducted package IPO and 0 if the firm conducted shares-only IPO; DAR denotes total debt to total asset ratio; ROE denotes return on equity; AUD_REP denotes auditor reputation, valued as 1 if the firm hired a reputable auditor and 0 otherwise; UND_REP denotes underwriter reputation, valued as 1 if the firm hired a reputable underwriter and 0 otherwise; ln_asset denotes the natural logarithm of the total assets; ln_age denotes the natural logarithm of the firm age. Asterisks (***), (**) and (*) indicate statistically significance at 1%, 5%, and 10% levels, respectively. Figures in round (.) brackets are standard errors while the values above them are the coefficient values.

included, from being statistically significant at 1% to 5%. However, we think that statistical significance at 5% is still quite strong. In addition, this result does not change even when the method is changed. Another finding in the financial variables shows that ROE has a negative association with package IPO. However, this relationship is weak, because the significance value in both methods is reduced when control variables are included, changing from being statistically significant at 5% to 10%.

Looking at the non-financial variables, only auditor reputation has a significant association with the choice of package IPO: our result shows that AUD_REP has a negative association with the choice of package IPO. This result does not change, even if the regression method is changed and control variables are included. The p-value also remains significant at 5%.

5 CONCLUSION

Our results show that a firm's financial risk, represented by total debt to total asset ratio, has a negative association with PIPO. Our regression finding is also strengthened by the result of the descriptive statistics. This shows that, on average, the amount of total debt to total assets of firms that choose SIPO is larger than that of firms that choose PIPO. However, this finding does not fully align with previous research, which found that PIPO issues are associated with higher-risk firms (Schultz, 1993). We assume that this action occurred because firms did not want to give a signal to the public that they really needed the funds. They preferred to maintain their reputation.

Firm profitability, represented by return on equity, has a negative association with PIPO. The reason behind this finding is quite similar to that described in the previous paragraph. Firms that have good performance do not want their reputation to deteriorate if they use a package IPO to fund their activities. Our findings, as shown in the descriptive statistics, also support the result estimated by the Logit and Probit analyses. We find that firms that conduct SIPO have relatively more profit than firms that choose PIPO.

Our third finding shows that firms that have a reputable auditor tend to avoid package IPO. Researchers hypothesized that those firms that chose to bundle their IPO with a warrant were the firms that suffered from agency problems (Schultz, 1993). Package IPO is considered to be a good solution for reducing agency costs (Howe & Olsen, 2009; How & Howe, 2001). Nevertheless, firms that have a reputable auditor tend to have fewer agency costs, because a reputable auditor may be able to reduce agency costs significantly. Thus, these firms do not need to issue a PIPO.

REFERENCES

Bates, T. & Dunbar, C. (2002). *Investment bank reputation, market power, and the pricing and performance of IPOs*. University of Western Ontario.

Beatty, R. & Welch, I. (1996). Issuer expenses and legal liability in initial public offerings. *Journal of Law and Economics, 39*(2), 545–602.

Carter, R. & Manaster, S. (1990). Initial public offerings and under-writer reputation. *The Journal of Finance, 45*, 1045–1067.

Cooney, J., Singh, A., Carter, R. & Dark, F. (2001). *The IPO partial-adjustment phenomenon and underwriter reputation: Has the inverse relationship flipped in the 1990's?* University of Kentucky, Case Western Reserve and Iowa State University.

El-Banany, M. (2013). Impact of global financial crisis and other determinants on intellectual capital disclosure in UEA banks. *University of Sharjah Journal for Humanities & Social Sciences, 10*(1), 23–43.

How, J.C.Y & Howe, J.S. (2001). Warrants in initial public offerings: Empirical evidence. *The Journal of Business, 74*(3), 433–457.

Howe, J.S. & Olsen, B.C. (2009). Security choice and corporate governance. *European Financial Management, 15*(4), 814–843.

James, C. & Wier, P. (1990). Borrowing relationships, intermediation, and the cost of issuing public securities. *Journal of Financial Economics, 28*, 149–171.

Jensen, M.C. & Meckling, W.H. (1976). Theory of the firm: Managerial behavior, agency cost and ownership structure. *Journal of Financial Economics, 4*, 305–360.

Manolopoulou, E. & Tzelepis, D. (2014). Intellectual capital disclosure: The Greek case. *International Journal of Learning and Intellectual Capital, 11*(1), 33–51.

Nezky, M. (2013). The impact of US crisis on trade and stock market in Indonesia. *Bulletin of Monetary, Economics and Banking, 15*, 83–96. doi:10.21098/bemp.v15i3.428.

Schultz, P. (1993). Calls of warrant: Timing and market reaction. *The Journal of Finance, 48*(2), 681–696.

Watts, R.L. & Zimmerman, J.L. (1983). Agency problems, auditing, and the theory of the firm: Some empirical evidence. *The Journal of Law and Economics, 26*(3), 613–633.

Business Innovation and Development in Emerging Economies – Trinugroho & Lau (Eds)
© 2019 Taylor & Francis Group, London, ISBN 978-1-138-35996-3

Factors influencing income smoothing in manufacturing companies listed on the Indonesia Stock Exchange

B.K.D. Cahyo & N. Alexander
Trisakti School of Management, West Jakarta, Indonesia

ABSTRACT: Income smoothing is defined as a way used by management to reduce fluctuations in reported income. The purpose of this study is to examine factors that influence the practice of income smoothing in manufacturing companies listed on the Indonesia Stock Exchange. The data were selected using a purposive sampling method. The sample used in this study was 47 firms listed on the Indonesia Stock Exchange and the observation period was the three-year period from 2014 until 2016. The hypothesis was tested using binary logistic regression. The result of the binary logistic regression showed that only financial leverage influences income-smoothing practices, while firm size, profitability, public ownership, the value of the firm, the deviation standard, institutional ownership and net profit margin do not influence income smoothing.

Keywords: income smoothing, firm size, profitability, financial leverage, public ownership, value of firm, stock risk, institutional ownership, net profit margin

1 INTRODUCTION

Every company wants to be in a good condition. A company in good condition, shown by the performance of the company, will give it a good reputation and allow it to attract potential investors (Peranasari & Dharmadiaksa, 2014). Thus, many companies are doing various things to attract the attention of potential investors, but many investors ignore how the company gets good performance, whether in the right or wrong ways to get the good performance.

According to Zuhriya and Wahidahwati (2015), one of the indicators when assessing a company's performance are financial statements, which can act as a measure of the condition of a company, and usually serve as a means to account for the manager's performance in utilizing the owner's resources. Financial statements themselves consist of several sections, including an income statement. Profit information is used for various purposes and by various stakeholders, including investors whose purpose is to measure risk when they want to make investment decisions (Zuhriya & Wahidahwati, 2015).

Investors have a tendency to pay attention to the figures presented in financial statements without looking deeper into the information. This will make management do deviant behavior by manipulating the profits generated by the company (Zuhriya & Wahidahwati, 2015). Management should undertake income smoothing to maintain profits and provide profitable returns as a finance-friendly company, as well as and conducting transactions with other companies that are in a stable and efficient financial condition, so that the company has high sales (Koch, 1981). Consequently, the company will choose accounting procedures that generate net profits in accordance with the target they want (Mambraku & Hadiprajitno, 2014).

According to Zuhriya and Wahidahwati (2015), the practice of income smoothing is the subject of considerable debate. Some consider the practice of income smoothing as an adverse action because it does not describe the actual financial condition and position of a company. On the other hand, the practice of income smoothing is considered to be a reasonable action because it does not violate accounting standards while still giving permissible signals, although the practice of income smoothing can reduce the reliability of a financial report.

2 FRAMEWORK OF THEORY AND HYPOTHESES

2.1 *Agency theory*

Agency theory was created because a company expanding its business, even if initially managed by family members, must eventually be administered by non-family people to keep business activities running, and in this case non-family members are management (Godfrey et al., 2010). According to Jensen and Meckling (1976), the owner of a company submits all of the company's activities to management under contract. Management then has the authority to manage the company as well as possible to provide for the welfare of the owners of the company. In this way, agency theory describes the relationship between shareholders and management.

2.2 *Signaling theory*

Signaling Theory was developed by Ross (1977) and is where the company's executives, who have better information about their company, will be encouraged to convey that information to potential investors in order to increase the company's stock price. This signal is information about what has been achieved by the agent to satisfy the principal's will in the company. Signals may take the form of other information that suggests that the company is better than other companies. Financial statements are a signal to users of financial information about the company. Managers are generally motivated to convey positive information about their company to the public.

2.3 *Income smoothing*

Beidleman (1973) defines earnings-smoothing behavior as a deliberate action by an agent to smooth out profits, so that profitability is regarded as normal for a company. According to Koch (1981), income smoothing is an approach used by agents to reduce reported profit fluctuations to artificially fit desired targets and accounting methods. In this study, the definition of income smoothing is anything in the business done deliberately by agents to maintain the stability of earnings, either by adding or reducing the amount of profits earned by the company in a given period.

According to Watts and Zimmerman (1986), in Suryandari (2012), income smoothing was motivated by the bonus plan, debt covenant and political cost hypotheses. The hypothesis of bonus plans is that agents tend to choose accounting procedures that can generate greater profits. Because agents, like everyone else, want high rewards, the chances are that they can increase their bonus in that period by reporting the highest net income possible. One way to do this is to choose an accounting policy that increases the profit reported in that period, in accordance with the applicable provisions in accounting, of course. The debt covenant hypothesis is that the closer a firm is to a breach in accounting based on a debt agreement, the more likely it is that the agent of the company will carry out income smoothing. This is to maintain their reputation in the eyes of external creditors.

The political cost hypothesis involves a firm of very large size having higher performance standards, with an appreciation of environmental responsibility. Such large companies have the ability to achieve higher profits, but run the risk of greater political costs. This is because government considers large companies are able to assist in the development of the state through the obligations undertaken by large companies, such as Corporate Social Responsibility (CSR), infrastructure development, income tax, and so on. Such a situation may encourage a company to carry out income smoothing.

2.4   Firm size and income smoothing

Suryandari (2012) reported that firm size has a positive effect on income smoothing. Big companies usually do more income smoothing than small companies, because a big company will smooth its income to maintain its performance, while a small company will undertake earnings maximization to obtain investor trust. Peranasari and Dharmadiaksa (2014) reported that firm size has a positive effect on income smoothing. Moses (1987) indicated that the big companies are expected to minimize income fluctuation, and found smoothing behavior to be subject to management motivation.

The big companies are usually required to bear higher costs because they are expected to have the ability to produce greater profits. Therefore, a company will avoid a drastic increase in profits in order to avoid an increased charge from government and society. Ginantra and Putra (2015) said that larger-sized companies tend to be more concerned about receiving positive attention from analysts, investors and government. Large companies will avoid drastic fluctuations in profits by carrying out income-smoothing actions, because the company will otherwise be burdened with large taxes and will want to minimize the risks that are likely to occur. On the basis of the explanation above, our hypothesis is:

H_1: Firm size has positive effects on income smoothing

2.5   Profitability and income smoothing

Widana and Yasa (2013) reported that profitability has a positive effect on income smoothing. The higher the profitability, the more the company will carry out income smoothing, because the company will want to retain stable profitability from year to year. Wijoyo (2014) reported that profitability has a positive effect on income smoothing. Relatively stable profitability shows the good performance of a company's management in generating profit and, of course, this will give confidence to potential investors. On the basis of the explanation above, our second hypothesis is:

H_2: Profitability has positive effects on income smoothing.

2.6   Financial leverage and income smoothing

Widhyawan and Dharmadiaksa (2015) reported that financial leverage has a positive effect on income smoothing. Companies with high levels of leverage are expected to practice income smoothing because the company wants to retain a good name in front of its creditors. Wulandari et al. (2013) reported that financial leverage has a positive effect on income smoothing. Ginantra and Putra (2015) said financial leverage is an important thing in companies, based on the use of financial resources that have a fixed expense with the goal of generating greater profits. If a company has a relatively large debt it will certainly have increased risk, which is likely to trigger the company to take income-smoothing action to stabilize its financial position. On the basis of the explanation above, our third hypothesis is:

H_3: Financial leverage has positive effects on income smoothing.

2.7   Public ownership and income smoothing

Husaini and Sayunita (2016) reported that public ownership has a negative effect on income smoothing. The higher the level of public ownership, the more a company tends to carry out income smoothing in order to produce a low profit, which means the company has a low risk level. Public ownership of a company always makes management want to show credibility in front of investors by way of performance and good financial statements, such as stabilizing financial ratios to influence the investment decisions of potential investors. This is done so that investors will compete to invest funds in the company. On the basis of the explanation above, our fourth hypothesis is:

H_4: Public ownership has negative effects on income smoothing.

2.8 *Value of firm and income smoothing*

Pernanasari and Dharmadiaksa (2014) reported that firm value has a positive effect on income smoothing. According to Aji and Mita, in Peranasari and Dharmadiaksa (2014), the higher the value of the company the more it tends to carry out income smoothing, because the value of the company is good, the profits generated by the company must be stable so that management does income smoothing. Good corporate value means the company's image is considered good by investors so that they are willing to buy shares. Husaini and Sayunita (2016) reported that firm value has a positive effect on income smoothing. According to Pernanasari and Dharmadiaksa (2014), a high corporate value reflects a stable profit, so a company tends to practice income smoothing to attract investors. On the basis of the explanation above, our fifth hypothesis is:

H_5: Value of firm has positive effects on income smoothing.

2.9 *Institutional ownership and income smoothing*

Kharisma and Agustina (2015) show that institutional ownership negatively affects earnings-smoothing practice. The ownership of shares by institutions means the information asymmetry between the principal and the agent will be lower, so the management will re-think the practice of income smoothing. Husaini and Sayunita (2016) reported that institutional ownership has a negative effect on income smoothing. On the basis of the explanation above, our sixth hypothesis is:

H_6: Institutional ownership has negative effects on income smoothing.

2.10 *Net profit margin and income smoothing*

Widana and Yasa (2013) reported that net profit margin has a positive effect on income smoothing. Manuari and Yasa (2014) reported that net profit margin has a positive effect on income smoothing. According to Manuari and Yasa (2014), when the resulting Net Profit Margin (NPM) ratio for a company turns out to be higher than the NPM level considered normal by management, then management tends to undertake the practice of flattening profit to lower the ratio. Similarly, if the NPM ration is lower than the NPM level considered normal by management, then it will practice income smoothing to increase NPM to the level considered normal.

Ginantra and Putra (2015) reported that net profit margin has a positive effect on income smoothing. As a result, net profit margin has a positive effect on income smoothing because investors tend to see only after-tax profits when taking a decision on investments. On the basis of the explanation above, our final hypothesis is:

H_7: Net profit margin has positive effects on income smoothing.

3 RESEARCH METHODS

This research uses a purposive sampling technique. The population of this study involves data from manufacturing companies in Indonesia in the period 2014–2016. The sample selection is presented in Table 1.

3.1 *Income smoothing*

Income smoothing was measured by the Eckel index created in 1981, using variable earnings after tax (net profit) and net sales variables (CV). Companies with less than one flattening

Table 1. Sample selection procedure.

Criteria	Companies	Sample
Manufacturing companies consistently listed during 2014–2016	366	122
Companies with no periods ending 31 December	(4)	(12)
Companies with no Rupiah currency	(20)	(60)
Companies with no positive profits during 2014–2016	(51)	(153)
Companies selected at last sample	47	141

index are identified as income smoothing, while firms with more than one uneven index are not identified as income smoothing.

$$\text{Income smoothing} = \frac{CV\,\Delta I}{CV\,\Delta S}$$

3.2 Firm size

Company size is calculated by using the natural logarithm formula for total assets, from the research of Iskandar and Suardana (2016).

$$\text{Size} = \text{Ln TA (Natural logarithm of Total Assets)}$$

3.3 Profitability

According to Widana and Yasa (2013), the Return On Asset (ROA) ratio is one of the profitability ratios that can predict future profits. The profitability variable can be measured by the return on asset formula from the research of Widana and Yasa (2013).

$$\text{ROA} = \frac{\text{Earnings after tax}}{\text{average total assets}}$$

3.4 Financial leverage

Financial leverage is measured using the Debt-to-Asset Ratio (DAR) formula from Sherlita and Kurniawan's research (2013).

$$\text{DAR} = \frac{\text{Total debt}}{\text{Total assets}}$$

3.5 Public ownership

According to Ginantra and Putra (2015), public ownership (KP) describes the number of shares circulating in the community, and is measured by the number of shares held publicly, divided by the number of shares outstanding.

$$\text{KP} = \frac{\text{Shares owned by public}}{\text{Outstandind shares}} :$$

3.6 Value of firm

The ratio of stock price to the book value of the company, or Price-to-Book Value (PBV), shows the level of the company's ability to create value relative to the amount of capital invested. Company value is measured using the PBV formula from Peranasari and Dharma-diaksa's research (2014).

$$PBV = \frac{Market\,price\,per\,share}{Book\,value\,per\,share}$$

3.7 *Institutional ownership*

From the research of Kharisma and Agustina (2015), institutional ownership (KI) is measured by the number of shares owned by institutions divided by the total shares outstanding.

$$Institutional\,ownership = \frac{Shares\,owned\,by\,institutions}{Outstanding\,shares} \times 100$$

3.8 *Net profit margin*

According to Ginantra and Putra (2015), Net Profit Margin (NPM) is a measure of efficiency, administration, production, pricing, marketing, funding and tax management. Management will try to show the best performance to increase the company's NPM in order to increase confidence in investing in the company.

$$NPM = \frac{Earing\,after\,tax}{Total\,sales}$$

4 RESULTS

This study uses the financial statements of manufacturing companies listed on the Indonesia Stock Exchange in the period 2014–2016.

The value shown in Table 3 for the firm size variable is 0.072, which is greater than 0.05. This means that firm size has no effect on income smoothing and it can be concluded that H_1 is rejected. This result is consistent with the research of Suryani and Damayanti (2015), Zuhriya and Wahidahwati (2015), and Ginantra and Putra (2015). According to Indarti and Fitria (2015), income smoothing is not only performed by large companies, but also by small companies, so that investors will want to invest in the companies.

The profitability variable (ROA) is 0.129, which is greater than 0.05. This means profitability has no effect on income smoothing. Thus it can be concluded that H_2 is rejected. This result is consistent with the research of Sherlita and Kurniawan (2013), Suryandari (2012), and Ginantra and Putra (2015); Suryani and Damayanti's (2015) research shows that profitability has no effect on income smoothing.

The value of the financial leverage variable (DAR) is 0.038, which is less than 0.05. This means financial leverage has an effect on income smoothing, and it can be concluded that

Table 2. Descriptive statistics.

Variable	N	Maximum	Minimum	Mean	Std. deviation
Size	141	33.1988	25.6195	28.715619	1.7203088
ROA	141	0.4154	0.0008	0.100778	0.0866871
DAR	141	0.8375	0.1331	0.391896	0.1628300
KP	141	0.6693	0.0182	0.271501	0.1547858
PBV	141	62.9311	0.0018	3.632620	8.6092512
KI	141	0.9818	0.3222	0.689058	0.1659410
NPM	141	0.2637	0.0012	0.077752	0.0564479

Table 3. Hypotheses test results.

Variable	B	Sig.	
Size	−0.072	0.601	H_1 rejected
ROA	−9.083	0.129	H_2 rejected
DAR	−3.210	0.038	H_3 accepted
KP	2.739	0.385	H_4 rejected
PBV	−0.016	0.854	H_5 rejected
KI	2.953	0.290	H_6 rejected
NPM	−1.751	0.790	H_7 rejected

H_3 is accepted. This result is consistent with the research of Husaini and Sayunita (2016) but inconsistent with that of Sherlita and Kurniawan (2013), Widana and Yasa (2013), Suryandari (2012), and Ginantra and Putra (2015), which indicate that financial leverage has no effect on income smoothing.

The value of the public ownership variable (KP) is 0.385, which is greater than 0.05. This means that public ownership has no effect on income smoothing, and it can be concluded that H_4 is rejected. This result is consistent with research conducted by Manuari and Yasa (2014), Ramanuja and Mertha (2015), and Ginantra and Putra (2015). According to Ramanuja and Mertha (2015), public ownership is not a reason for a company to conduct income-smoothing practices because the amount of public ownership reflects the actual performance of the company's management and the amount of confidence in the company.

The variable for value of firm (PBV) has a value of 0.854, which is greater than 0.05. This means that the value of a firm has no effect on income smoothing, and it can be concluded that H_5 is rejected. This result is consistent with the research of Setyaningsih and Marisan (2010), and Zuhriya and Waihdahwati (2015). However, the results of this study are inconsistent with the research of Husaini and Sayunita (2016), Handayani and Fuad (2015), and Pernanasari and Dharmadiaksa (2014). According to Setyaningsih and Marisan's (2010) research, the value of stocks that are too high also does not invite investment in the company because Indonesian investors tend to prioritize capital gains, which also do not depend on company performance.

The value of the institutional ownership variable (KI) is 0.290, which is greater than 0.05. This means that institutional ownership has no effect on income smoothing, and it can be concluded that H_6 is rejected. The result of this study is consistent with Suryani and Damayanti's (2015) research. However, the results of this study are inconsistent with Husaini and Sayunita (2016), and Kharisma and Agustina (2015), which indicate that institutional ownership has an influence on income smoothing. According to Suryani and Damayanti (2015), institutional ownership cannot be relied upon to monitor company performance and whether to give authorization to income smoothing.

The value of the net profit margin (NPM) variable is 0.790, which is greater than 0.05. This means the net profit margin has no effect on income smoothing, and it can be concluded that H_7 is also rejected. This result is consistent with the research results of Suryandari (2012) and Pramono (2013). According to Pramono (2013), changes in the net profit margin can affect the tax burden incurred by a company. Then, according to Pramono (2013), economic growth in Indonesia automatically increases the company's current year earnings so that the company doesn't have to perform income smoothing. However, the results of this study are inconsistent with the results of research from Widana and Yasa (2013), Manuari and Yasa (2014), and Ginantra and Putra (2014), which indicate that net profit margin does influence income smoothing.

5 CONCLUSION

Based on the results of this research, it can be concluded that only financial leverage has an effect on income smoothing. The remainder—firm size, profitability, public ownership, value of firm, institutional ownership, and net profit margin—have no effect on income smoothing. There are limitations in this study, because this research only examines a three-year period and the variables used are limited to seven. The recommendation from this research is to increase the research time and conduct it over a longer period, and to include additional variables that may have an effect on income smoothing, such as auditor quality, managerial ownership, independent commissioners, company size, free cash flow, other comprehensive income, bonus plan, industry sector, and cash holdings.

REFERENCES

Beidleman, C. (1973). Income smoothing: The role of management. *The Accounting Review,* 8(4), 653–667.

Godfrey, J., Hodgson, A., Tarca, A., Hamilton, J. & Holmes, S. (2010). *Accounting theory* (7th ed.). Milton, Australia: John Wiley & Sons.

Ginantra, I.K.G.I., Nyoman W.A.P. (2015). The Effect of Profitability, Leverage, Firm Size, Public Ownership, Dividend Payout Ratio, and Net Profit Margin on Income Smoothing. *Jurnal Akuntansi Universitas Udayana, 10*(2), 602–617.

Handayani, F. and Fuad. (2015). Factors Effect on Income Smoothing on Automotive companies Listed on Indonesia Stock Exchange Period 2009–2012. *Diponegoro Journal of Accounting.* Vol. 4(2), 1–12.

Husaini and Sayunita. (2016). Determinant of income smoothing at manufacturing firm listed on Indonesia Stock Exchange. *International Journal of Business and Management Invention,* 9(6) 1–4.

Indarti, T.S. and Astri F. (2015). Factors Effect Income Smoothing Practices on Manufacturing Companies. *Jurnal Ilmu & Riset Akuntansi, 4*(6), 1–20.

Iskandar, A.F. and. Suardana K.A. (2016). The Effect of Firm Size, Return on Asset, dan Winner/Losser Stock on Income Smoothing Practices. *Jurnal Akuntansi Universitas Udayana* Vol. 14(2), 805–834.

Jensen, M.C. & Meckling, W.H. (1976). Theory of the firm: Managerial behaviour, agency costs, and ownership structure. *Journal of Financial Economics,* 3(4), 305–360.

Kharisma, A. & Agustina, L. (2015). The Effect of Corporate Governance and Firm Size on Income Smoothing. *Accounting Analysis Journal, 4*(2),1–10.

Koch, B.S. (1981). Income smoothing: An experiment. *The Accounting Review,* 56(3), 574–586.

Mambraku, P., Milka E dan Basuki H. (2014). The Effect of Cash Holding and Managerial Ownership on Income Smoothing. *Diponegoro Journal of Accounting* Vol. 3(2), 1–9.

Manuari, I.A.R. Gerianta W.Y. (2014). Income Smoothing and The Affecting Factors. *Jurnal Akuntansi Universitas Udayana,* 7((3), 614–629.

Moses, O.D. (1987). Income smoothing and incentives: Empirical test using accounting changes. *The Accounting Review,* 62(2), 358–377.

Peranasari, I.A., Ayu I., Ida B.D. (2014). Income Smoothing and The Affecting Factors. *E-Jurnal Akuntansi Universitas Udayana,* 8(1), 140–153.

Putra, R.A.S., And Ketut A.S. (2016). Influence of Variants in Stock Value, Public Ownership, and Debt to Equity Ratio on the Income Smoothing. *E-Jurnal Akuntansi Universitas Udayana,* Vol. 15(3), 2188–2221.

Pramono, O. (2013). The Effect of ROA, NPM, DER, and Size on Income Smoothing. (Study on Manufacturing Companies Listed in Indonesia Stock Exchange Period 2007–2011). *Jurnal Ilmiah Mahasiswa Universitas Surabaya,* 2(2), 1–16.

Ramanuja, I.G. V., Made. I.M. (2015). The Effect of Variances in Stock Value, Publick Ownership, DER and Profitability on Income Smoothing. *Jurnal Akuntansi Universitas Udayana,* 10(2), 398–416.

Ross, S.A. (1977). The determination of financial structure: The incentive-signalling approach. *Bell Journal of Economics,* 8(1), 23–40.

Setyaningsih, I., Ichwan. M. (2010). Factors Affect Income Smoothing on Manufacturing Companies Listed in Indonesia Stock Exchange. *Jurnal Dinamika Ekonomi & Bisnis,* 7(1), 23–35.

Sherlita, E. & Kurniawan, P. (2013). Analysis of factors affecting income smoothing among listed companies in Indonesia. *Jurnal Teknologi,* 64(3), 17–23.

Suryandari. (2012). Factors Effect Income Smoothing. *Media Komunikasi FIS,* 11(1), 196–205.

Suryani, A.D., Gusti. I.A.E.D. (2015). The Effect of Firm Size, Debt to Equity Ratio, Profitability, and Institutional Ownership on Income Smoothing. *Jurnal Akuntansi Universitas Udayana,* 13(1), 208–223.

Widana N., Nyoman. I.A., Gerianta. W.Y. (2013). Income Smoothing and the Affecting Factors on the Indonesia Stock Exchange. *Jurnal Akuntansi Universitas Udayana,* 3(2), 297–317.

Widhyawan, I.M.I, Ida, B.D. (2015). The Effect of Financial Leverage, Dividend Payout Ratio, and Corporate Governance on Income Smoothing. *Jurnal Akuntansi Universitas Udayana, 13(*2), 157–172.

Wijoyo, D.S. (2014). Variables that Affect Profit Smoothing Practices in Public Manufacturing Companies. *Jurnal Bisnis dan Akuntansi,* 16(1), 37–45.

Zuhriya, S. and Wahidahwati. (2015). Income Smoothing and Factors Affecting Manufacturing Companies on the IDX. *Jurnal Ilmu & Riset Akuntansi,* 4(7), 1–22.

Development

Business Innovation and Development in Emerging Economies – Trinugroho & Lau (Eds)
© 2019 Taylor & Francis Group, London, ISBN 978-1-138-35996-3

Model of Human Resource Development (HRD) in the context of Indonesian food security

M. Ali
Universitas Tamansiswa Palembang, Palembang, Indonesia

J. Siswanto
Universitas Sriwijaya, Palembang, Indonesia

Sardiyo
STIE Musi Rawas, Palembang, Indonesia

F.C. Utomo
Universitas Krisnadwipayana, Jawa Barat, Indonesia

ABSTRACT: Food security is an issue that will never be discussed because food is a sensitive issue and related to the livelihood of many people. Under Article 33 of the Constitution, the Government of Indonesia essentially has the legal basis to use its natural resources to the greatest prosperity of the people by regulating and issuing related policies. In the category of food security, Indonesian policy generally emphasizes food self-sufficiency programs more to achieve food security, which can be inferred from Law no. 18 of 2012 on Food. The Food Law stipulates that food imports can only be undertaken if local production is not sufficient to meet the needs of the Indonesian people's consumption. Therefore, this article will discuss: (a) What are the direct influences of education, training and accompaniment on food security? (b) Can funding moderate education, training, and accompaniment on food security? Based on the results of the influence between variables, it can be seen that the total influence of total accompaniment and training variables are smaller compared to the direct influence between each variable but not for the education variable because it has a greater total value. Thus, it can be concluded that funding is not proven to be an intervening variable capable of mediating the effect of accompaniment and training on food security, but funding is capable of mediating the educational variables. Finally, this article offers two solutions that are more effective and efficient than the idea of food self-sufficiency to achieve the objectives of food security policy.

Keywords: education, training, accompaniment, funding of food security

1 INTRODUCTION

The problem of food security is a global issue that is closely related to human survival. This paper seeks to address emerging issues related to global food security. In the World Food Summit 1996 in Rome, Italy, it is stated emphatically that what is meant by food security is when everyone, at all times, has access to adequate food and security, as well as a nutritious food source to fulfill the needs of life and health.

Note that adequate food for every individual is a human right, as set forth in the International Covenant on Economic, Social and Cultural Rights and the Universal Declaration on the Eradication of Hunger and Malnutrition. In addition, the effort to achieve food security is also included in the agenda of the Sustainable Development Goals (SDGs), especially the second goal, namely to overcome the problem of hunger, achieve food security and increase nutrition, and promote a sustainable agriculture sector. (Dinas Ketahanan Pangan, 2017).

The problem of food security is rapidly changing as a result of population growth, inadequate agricultural infrastructure, and the diminishing number of farm households, even the process of structural transformation is not working. By 2045 the population of Indonesia is estimated to reach 400 million people (Dain, 2015).

Recent food price hikes and the global economic crisis have left their mark, as the number of hungry and malnourished people increased worldwide, particularly in developing countries. Evidence shows that about 902 million people in the developing world were malnourished in 2008, reflecting an increase of about 65 million since 2000–2002 (FAO, 2010). The impact of the decline in household incomes from the global economic downturn has been compounded by relatively high food prices in many developing countries, resulting in further increases in the number of undernourished households in these countries. Preliminary estimates for 2009 indicate that the cutbacks in food expenditure have resulted in the number of undernourished people rising above one billion in developing countries (FAO, 2010). This development makes it increasingly difficult to achieve the first Millennium Development Goal (MDG) of halving the number of hungry people by 2015. Fanzo et al. (2010) identify the lack of political will at both global and national levels as the major cause of the growing divergence from this important MDG. Although food insecurity had attracted little attention in the media and political agendas of developed countries during the last decades, the situation changed in 2008 as riots over higher food prices occurred throughout the developing world (Falcon & Naylor, 2005; Fanzo et al., 2010). The rising numbers of food-insecure persons and the clearly established linkages between food security, national security, and global security have contributed to renewing international interest in the food security policies of developing countries.

Food security involves ensuring both an adequate supply of food and access of the population to that supply, mostly through generating adequate levels of effective demand via income growth or transfers. Food security in developing countries therefore tends to be influenced by both micro and macro factors that include the adoption of new technologies, support for institutions available to farmers, and food price policy, as well as monetary, fiscal, and exchange rate policies that affect overall economic growth and income distribution. The policies that are normally associated with food security usually involve structural changes in relative prices, the general economic environment, as well as other measures such as targeted food subsidies, improving technologies, and institutions available to farmers and consumers (Weber et al., 1988). Policymakers are often confronted with the dilemma of higher food prices to induce increased food production and the food security of low-income consumers, as higher prices impose a heavy cost on this group of consumers. A variety of short- and long-term policy options have been used by governments to promote food security in the developing world. Some measures affect food availability on local markets, others individuals' entitlements to obtain food, while others tend to influence food utilization, i.e., how many nutrients an individual obtains from a given supply of food.

1.1 *Research problem*

1. What are the direct influences of education, training, and accompaniment on food security?
2. Can funding moderate education, training, and accompaniment toward food security?

2 THEORY REVIEW

Indonesia as an archipelagic country with abundant resources has an interest to attain food security in order to improve the living standard of its people. Food security has been defined in various ways, but the most widely accepted definition is the one provided in the 1996 World Food Summit that defines food security as existing "when all people at all times have physical and economic access to sufficient, safe and nutritious food to meet their dietary needs and food preferences for an active and healthy life." Thus, food insecurity exists when people do not have sufficient physical, social or economic access to food. This widely accepted definition highlights four elements of food security: availability, access, utilization and stability. These terms can be defined as follows: food availability is the amount of food that exists in sufficient quantities on a consistent basis; food access is defined as having sufficient resources for acquiring appropriate food for a nutritious diet; food utilization is through adequate diet, clean water, sanitation and health care; and stability is having access to adequate food on a regular basis. In contrast, food self-sufficiency is defined as the ability to meet consumption needs, particularly for staple foods, from a country's own domestic production rather than having to rely on importing or buying from non-domestic sources (minimizing dependence on international trade). People often tend to equate food security and self-sufficiency. It is often believed that increasing self-sufficiency is the best way to attain the goal of food security. The idea of self-sufficiency is politically appealing. A government has more control over its food supply and it is not dependent on imported food. The former World Trade Organization (WTO) Director General Supachai Panitchpakdi, in his speech at the FAO in 2005, stated that "self-sufficiency" and "food security" are not the same and argued that food security is best achieved in an economically integrated and politically interdependent world (FAO, 2006).

Education, on the other hand, is usually more broadly defined as a more general, less specialized or hands-on approach to enhancing knowledge. The Manpower Services Commission (1981), which was superseded by the now defunct Training Commission, UK) defined education as follows: "Activities which aim at developing the knowledge, skills, moral values and understanding required in all aspects of life rather than knowledge and skill relating to only a limited field of activity."

The benefits of employee training are numerous and widely documented (Wilson, 1999; Jensen, 2001; Sommerville, 2007) with organizations, as well as workers, reaping the rewards in terms of improved employee skills, knowledge, attitudes, and behaviors (Treven, 2003) and results like enhanced staff performance (Brown, 1994), job satisfaction, productivity and profitability (Hughey & Mussnug, 1997).

Training has been defined in various ways, including the following: "A planned process to modify attitude, knowledge or skill behavior through a learning experience to achieve effective performance in any activity or range of activities. Its purpose, in the work situation, is to develop the abilities of the individual and to satisfy current and future manpower needs of the organisation" (Manpower Services Commission, 1981); "Training endeavors to impart the knowledge, skills and attitudes necessary to perform job-related tasks. It aims to improve job performance in a direct way" (Truelove, 1992). "Training is characterised as an instructor-led, content-based intervention leading to desired changes in behaviour" (Sloman, 2005)

3   METHOD

This research is an explanatory research, which aims to solve the problem. The framework for problem solving in this study is illustrated in the causal relationships that occur between education, training, accompaniment, funding and food security to be tested in this study. Systematically, the conceptual framework of the study can be seen in Figure 1.

Dimension of variables:

1. **Education variable (X_1)**
   Indicators: (1) formal education; (2) non-formal education
2. **Training variable (X_2)**
   Indicators: (1) kinds of training; (2) materials for training; (3) time of training
3. **Accompaniment variable (X_3)**
   Indicators: (1) coordinate the implementation of the counseling; (2) assessment of companions as technicians; (3) developing activities; (4) monitoring and evaluating farmer groups; (5) assessment of companion in assisting and guarding assistance; (6) assessment

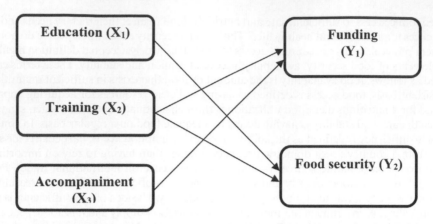

Figure 1. Conceptual framework.

of companion in inventorying and identifying potential of the region; (7) assessment of companion in making an activity implementation report

4. Funding variable (Y₁)
Indicators: (1) price subsidy scheme; (2) social assistance schemes; (3) credit interest subsidy scheme; (4) program/activity scheme

5. Food security variable (Y₂)
Indicators: (1) adequacy of food availability; (2) the stability of food supply without fluctuation from season to season or from year to year; (3) accessibility/affordability of food; (4) quality/food safety

Based on the research objectives and theoretical reviews, the research hypotheses are:

H1: Education, training, and accompaniment together and partially affect food security;
H2: Funding can moderate education, training, and accompaniment toward food security.

4 RESULTS

4.1 Validity and reliability test

The validity test is done by calculating the correlation coefficient (r) between the score of the measurement item and its total score. All of the questions are valid with values greater than 0.201. Testing reliability is obtained from the value of Cronbach's alpha total coefficient reliability and the test results obtained with the values of each variable are above 0.60. Thus, all indicators proved reliable in measuring their respective latent variables.

4.2 Structural model

The structural model in this study presented in Figure 2 combines exogenous variables and endogenous variables. The structural equation model generated in this study is formulated as follows:

$$Y_1 = 0.057\,X_1 + 0.587\,X_2 + 0.283\,X_3$$
$$Y_2 = 0.587\,X_1 + 0.497\,X_2 + 0.263X_3$$
$$Y_2 = 0.587\,X_1 + 0.497\,X_2 + 0.263X_4 - 0.367Y_1$$

4.3 Confirmatory factor analysis

The required loading factor value must meet the criteria ≥ 0.50. A factor loading value < 0.60 gives the view that one variable is dimensionless with another variable in order

178

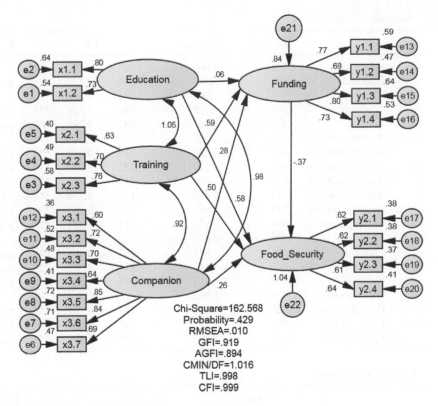

Figure 2. Full standardized model of structural equations.

to describe a formation variable. Table 1 shows that each indicator on education, train-ing, accompaniment, funding and food security variables has a loading factor loading fac-tor > 0.50. Therefore, it can be concluded that all indicators are able to form their latent variables (education, training, accompaniment, funding, and food security).

4.4 *Assumption tests*

4.4.1 *Normality*
Based on the results of data processing, the known Condensation Resistance (CR) value of the multivariate is 13.992. Thus, based on the multivariate normality test the distribution of the data is not normally distributed.

4.4.2 *Outlier test*
The outlier test in this research is done by observing the Mahalanobis distance value at a significant level which is $P < 0.001$ to Chi-squared ($\chi 2$) on the free degree of the indicator number. Based on the tests that have been done then the results of the test states there is no outlier on these test results.

4.5 *Goodness of fit index*

Test results on the goodness of fit index in Table 2 show that the structural model generated in the study fulfill the criteria. The goodness of fit index is good except for **AGFI** whose value still does not fulfill the criteria set. In other words, the value can be said to be marginal and the proposed model is acceptable.

Table 1. Variables of regression weight.

			Std. Est	Estimate	SE	CR	P	Label
x1.2	←	Education	.734	1.000				
x1.1	←	Education	.800	.943	.087	10.777	***	par_3
x2.3	←	Training	.763	1.000				
x2.2	←	Training	.698	.758	.081	9.334	***	par_4
x2.1	←	Training	.630	.761	.092	8.298	***	par_5
x3.7	←	Accompaniment	.688	1.000				
x3.6	←	Accompaniment	.844	1.286	.128	10.042	***	par_6
x3.5	←	Accompaniment	.847	1.229	.122	10.042	***	par_7
x3.4	←	Accompaniment	.637	.909	.117	7.751	***	par_8
x3.3	←	Accompaniment	.695	1.436	.171	8.379	***	par_9
x3.2	←	Accompaniment	.718	.887	.103	8.602	***	par_10
x3.1	←	Accompaniment	.604	.832	.114	7.321	***	par_11
y1.1	←	Funding	.770	1.000				
y1.2	←	Funding	.683	.851	.097	8.788	***	par_12
y1.3	←	Funding	.801	1.172	.110	10.609	***	par_13
y1.4	←	Funding	.725	.917	.098	9.350	***	par_14
y2.1	←	Food security	.618	1.000				
y2.2	←	Food_Security	.620	.927	.134	6.923	***	par_15
y2.3	←	Food security	.607	.961	.141	6.839	***	par_16
y2.4	←	Food security	.638	.906	.128	7.091	***	par_17

Table 2. Goodness of fit index.

GOF	Value	Criteria	Status
χ^2 – Chi-squared (df = 419)	162.568	Small (<467.725)	Good
Probability	0.429	≥ 0.05	Good
RMSEA	0.010	≤ 0.08	Good
GFI	0.919	≥ 0.90	Good
AGFI	0.894	≥ 0.90	Marginal
CMIN/DF	1.016	$\leq 2,00$	Good
TLI	0.998	≥ 0.95	Good
CFI	0.999	≥ 0.95	Good

4.6 Hypotheses test results

The recapitulation of the hypotheses' testing results presented in Table 2 shows that the direct effects of education on funding, accompaniment to funding, and training on funding have influence values of 0.057, 0.283 and 0.587. Meanwhile, the magnitude of the influences of education on food security, accompaniment to food security, training on food security, and funding to food security have values for each influences of 0.585, 0.263, 0.497 and –0.367. The indirect effects of education, accompaniment, and training on food security through funding have the values for each influences of –0.021, –0.104 and –0.216. Furthermore, the total effect of education on food security either directly or indirectly is 0.487. The total influence between accompaniments toward food security either directly or indirectly is 0.147 and the effect of training on food security either directly or indirectly is 0.219.

Based on the results of the influence between variables, it can be seen that the total influence of total accompaniment and training variables is smaller compared to the direct influence between each variable but not for the education variable because it has a greater total value. Thus, it can be concluded that funding is not proven as an intervening variable capable of mediating the effect of accompaniment and training on food security, but funding is capable of mediating the variable of education.

Table 3. Standardized direct effects, indirect effects and total effects.

	Education	Accompaniment	Training	Funding	Food security
Direct effects					
Funding	0.057	0.283	0.587	0.000	0.000
Food security	0.585	0.263	0.497	−0.367	0.000
Indirect effects					
Funding	0.000	0.000	0.000	0.000	0.000
Food security	−0.021	−0.104	−0.216	0.000	0.000
Total effects					
Funding	0.057	0.302	0.527	0.000	0.000
Food security	0.487	0.147	0.219	−0.318	0.000

5 DISCUSSION

Indonesia is the world's fourth most populous country and was the sixth largest agricultural producer in 2013. The contribution of agriculture to Indonesia's GDP has fallen from 17% in 1995 to 13.5% in 2015. However, its share of total employment has fallen faster, from around 44% in 1995 to 34% in 2015. While food crop production is based on small family farms, large commercial farms specialize in perennial crops, in particular palm oil. Palm oil and rubber account for around 60% of total agri-food exports and contribute to a significant surplus in Indonesia's agri-food trade. Indonesia has achieved significant progress in poverty eradication and food security. However, these issues remain important with around 11% of the population continuing to live below the nationally-defined poverty line. The prevalence of undernourishment was 7.6% of the population in 2014–16, half of what it was only a decade ago. (Developments in Agricultural Policy and Support by Country—of the report entitled Agricultural Policy Monitoring and Evaluation 2017).

Based on the analysis of the results above, socialization toward prospective business actors in the sector of the role must be aggressively carried out. Improving farmers' capacity through education and training programs and mentoring should be done continuously.

In addition, to improve the ability of farmers, especially young farmers or future generations in the agricultural sector, supporting capital for agriculture business, the use of appropriate technology, and the use of seeds with superior quality will be strategies for the sustainability of this food security program as an effort to improve the quality of human resources in context as human capital investment (Maulana, 2017).

6 CONCLUSIONS AND SUGGESTIONS

6.1 Conclusions

The conclusion of this study is that the variables of education, training and mentoring together affect the food security program. Educational variables have the most dominant influence either directly or indirectly on food security. Funding variables are only able to moderate education variables on food security.

6.2 Suggestions

The government must be properly focused on creating a new generation in the agricultural sector as well as making strict regulations for land conversion issues. The government is expected to act as a facilitator in providing capital assistance and technological assistance. Policies in relation to the strategy to achieve food security must be carried out continuously. The government also needs to have good cooperation with related institutions and the community on land transfer functions that pose a threat to sustainable food security programs.

ACKNOWLEDGMENTS

The researchers would like to thank the rector of Universitas Tamansiswa Palembang for the motivation and financial support for the implementation of this research. The researchers are also grateful to the parties involved in the process of data collection and data processing such that this paper could be completed.

REFERENCES

Brown, A. (1994). TQM: Implications for training. *Training for Quality, 2*(3), 4–10.

Dain, D. (2015). *The root of the problem of national food security in the future (Akar masalah ketahanan pangan nasional di masa depan)*. Retrieved from https://www.kompasiana.com/dainsyah/akar-masalah-ketahanan-pangan-nasional-di-masa-depan_5535ac686ea834731dda42e2.

Dinas Ketahanan Pangan. (2017). *The problem of global food security (Problem ketahanan pangan global)*. Retrieved from https://dkp.bulelengkab.go.id/artikel/problem-ketahanan-pangan-global-global-food-security-45.

Falcon, W.P. & Naylor, R.L. (2005). Rethinking food security for the twenty-first century. *American Journal of Agricultural Economics, 85*(5), 1113–1127.

Fanzo, J., Pronyk, P., Dasgupta, A., Towle, M., Menon, V., Denning, G., Roth, G. (2010). *An evaluation of progress toward the Millennium Development Goal one hunger target: A country-level, food and nutrition security perspective.* New York, NY: United Nations Development Group.

Food and Agricultural Organization. (2006). *Food security, FAO Policy Brief, Issue 2, June 2006.* Retrieved from http://www.fao.org/forestry/13128-0e6f36f27e0091055bec28ebe830f46b3.pdf.

Food and Agricultural Organization. (2010). Global hunger declining, but still unacceptably high. Retrieved from http://www.fao.org/3/al390e/al390e00.pdf.

Hughey, A.W. & Mussnug, K.J. (1997). Designing effective employee training programmes. *Training for Quality, 5*(2), 52–57.

Jensen, J. (2001). *Improving training in order to upgrade skills in the tourism industry. Tourism and Employment,* Final Report of Working Group B. Copenhagen: European Commission.

Maulana. (2017). Analysis of human capital investment on economic growth in South Sumatra. *International Journal of Economic Research, 14*(8), 177–183.

Sloman, M. (2005). *Training to learning.* Retrieved from http://www.cipd.co.uk/NR/rdonlyres/52AF1484-AA29-4325-8964-0A7A1AEE0B8B/0/train2lrn0405.pdf.

Sommerville, K.L. (2007). *Hospitality employee management and supervision: Concepts and practical applications.* New York, NY: John Wiley & Sons, Inc.

The Manpower Services Commission. (1981). *Glossary of Training Terms.* London: HMSO.

Treven, S.J. (2003). International training: The training of managers for assignment abroad. *Education + Training, 45*(8/9), 550–557.

Truelove, S. (1992). *Handbook of training and development.* Oxford, UK: Blackwell.

Weber, M.T., Staatz, J.M., Holtzman, J.S., Crawford, E.W. & Bernsten, R.H. (1988). Informing food security decisions in Africa: Empirical analysis and policy dialogue. *American Journal of Agricultural Economics, 70*(5), 1044–1052.

Wilson, J.P. (1999). *Human resource development: Learning and training for individuals and organisations.* London, UK: Kogan Page.

Business Innovation and Development in Emerging Economies – Trinugroho & Lau (Eds)
© 2019 Taylor & Francis Group, London, ISBN 978-1-138-35996-3

Does the tourism sector effectively stimulate economic growth in ASEAN member countries?

A.A.A. Islam & L.I. Nugroho
Universitas Sebelas Maret, Surakarta, Indonesia

ABSTRACT: The tourism sector has become one of the rapidly growing services sectors in the world. This sector has increasingly become an important driver for economic and social development for many countries. The methodology used in this study is regression and pooling data. Sample data is taken from member countries of the Association of Southeast Asian Nations (ASEAN) in the period 2002–2016. Secondary data is taken from several sources, including the World Bank, the World Travel and Tourism Council and the United Nations World Tourism Organization. This study was conducted to find out whether the tourism sector can stimulate economic growth in ASEAN member countries. It was found that the contribution of the tourism sector to gross domestic product is significant.

Keywords: Tourism, economy growth, unemployment, ASEAN member countries

1 INTRODUCTION

Recently, the tourism sector has become one of the rapidly growing services sectors. This sector has increasingly become an important driver for economic and social development for many countries in the world (Athanasopoulou, 2013). In line with the United Nations World Tourism Organization (UNWTO, 2017), international tourist arrivals have increased from 674 million in 2000 to 1,235 million in 2016. Likewise, international tourism receipts earned from departures worldwide have surged from US$495 billion in 2000 to US$1,220 billion in 2016. In other words, it can be said that the tourism sector represents 10% of the world's Gross Domestic Product (GDP). Tourism is a major category of international trade in services. In addition to receipts earned from departures, international tourism also generated US$216 billion in exports through international passenger transport services rendered to non-residents in 2016, bringing the total value of tourism exports up to US$1.4 trillion, or US$4 billion a day on average. International tourism represents 7% of the world's exports in goods and services. As a worldwide export category, tourism ranks third after chemicals and fuels, and ahead of automotive products and food. In many developing countries, tourism is the top export category.

The impact of the tourism sector on economic and social development can be enormous. The tourism sector stimulates economic growth through jobs and enterprise creation; furthermore, it provides significant foreign exchange revenues for many countries. The sector generates opportunities for reducing unemployment, at a time when many countries are suffering from high unemployment. In addition, this sector can help to reduce the poverty rate and inequality, preserving natural and cultural heritage, and also upgrading infrastructure.

Such tourism is highly important in many countries and has the potential to be a driver of employment growth. However, there is not much research regarding this topic, and little research in the Association of Southeast Asian Nations (ASEAN) region regarding the impact of the tourism sector on economic growth, or the relationship between the tourism sector and unemployment. Therefore, it is difficult for a government to decide upon regulations for the development of tourism in the future.

On the basis of this background, this study therefore seeks to establish whether the tourism sector effectively stimulates economic growth and reduces unemployment in ASEAN member countries. Thus, the researcher hopes that this study can be a reference for the government to make the best regulation for the Development of the tourism.

The paper is organized as follows. In the next section some literature related to tourism trends in ASEAN member countries and the tourism contribution to economic growth are discussed. In Section 3 the method used in our analysis is presented. Section 4 elaborates the results, Section 5 contains the conclusion, and the final section contains recommendations.

2 LITERATURE REVIEW

2.1 *Tourism*

According to UNWTO (2015), tourism is the social, cultural, and economic phenomena that involve the movements of people from or to another country, or a place outside their environment, to achieve personal or business professional goals.

The Australian Department of Tourism & Recreation (1975), in Jiyanto (2009), defines tourism as an identifiable nationally important industry, which involves a wide cross section of component activities, including the provision of transportation, accommodation, recreation, food, and related services. Meanwhile, UNWTO in 1963, in Jiyanto (2009), explained that a visitor/tourist is a person who visits a country other than their usual place of residence, for any reason other than taking up a remunerated job within the country visited. Moreover, according to Hunziker and Kraph (in Burkart & Medik, 1974), in Jiyanto (2009), tourism is the sum of the phenomena and relationships arising from the travel and stay of non-residents.

Therefore, it can be said that tourism is the social, cultural, and economic phenomena that involve the movements of people who visit a country other than their usual place of residence and have component activities, including the provision of transportation, accommodation, recreation, food, and related services.

2.2 *Tourism trends in ASEAN member countries*

The current ASEAN member countries are Brunei Darussalam, Cambodia, Indonesia, Laos, Malaysia, Myanmar, Philippines, Singapore, Thailand, and Vietnam. The increasing number of tourists will directly increase the output of GDP, and also the foreign exchange reserves that can influence economic growth. There are three aspects to the focus of economic growth as being the increasing output per capita in the long term. The first one is the process of economic growth, which is the dynamic aspect over time. The second one is the output per capita aspect; that economic growth will be increased if there is increasing output per capita. The third is the relationship between aspects of the economy; that economic growth is the result of the whole interaction that happens in the economy.

2.3 *Contribution of tourism*

The positive impacts of tourism activities on the economy are usually captured by the concept of the 'tourism multiplier' (jobs and income), which is the sum of direct, indirect and induced impacts. The direct contribution of the tourism sector to GDP is reflected in the 'internal' spending on the tourism sector (total money spent by residents and non-residents for business and leisure purposes), as well as 'individual' spending by the government on tourism services directly linked to visitors, such as cultural (e.g. museums) or recreational (e.g. national parks) (WTTC, 2015). Indirect impacts happen when, for instance, a hotel buys inputs (goods and services) from other businesses in the economy (Meyer, 2006). The 'induced' contribution measures the GDP, investments and jobs supported by the expenditure of those who are directly or indirectly employed by the tourism Industry.

Table 1. Variables used in this study.

No	Variable	Notes
1	GDP	Is a monetary measure of the market value of all the final goods and services produced in a period of time, often annually or quarterly
2	Expenditure (EKSP)	The total amount of money that a government or person spends.
3	International Tourist, Number of Arrivals (NOA)	International inbound tourists (overnight visitors) are the number of tourists who travel to a country other than that in which they have their usual residence, but outside their usual environment, for a period not exceeding 12 months and whose main purpose in visiting is other than an activity remunerated from within the country visited.

In other words, induced impacts include all of the economic impacts that will result from the paying out of salaries and wages to people who are employed in the tourism sector and/or tourism-related businesses. These additional salaries and wages lead to an increased demand for various consumable goods that need to be supplied by other sectors of the economy. Thus, the total contribution of the tourism sector includes its 'wider impacts' (i.e. the indirect and induced impacts) on the economy.

3 METHODOLOGY

The methodology used in this research is regression and pooling data. This research took 11 ASEAN member countries as the sample from the year 2002 to 2016. The countries are Brunei Darussalam, Cambodia, Indonesia, Lao PDR, Malaysia, Myanmar, Philippines, Singapore, Thailand, and Vietnam. The data obtained are from several sources, including the World Bank, the World Travel and Tourism Council (WTCC), the United Nations World Tourism Organization, and other literature.

3.1 *Model specification*

The model used in this study can be expressed in the Estimated Generalized Least Square (EGLS) Model. EGLS method is used in case of Heteroskedasticity or auto correlation problems.

4 RESULTS AND DISCUSSION

Based on the results shown in Table 2, we can derive the following formula:

$$GDP = 7.52E + 14 - 2965.635EKSP - 15138685 \, NOA$$

The result of hypothesis testing in this study shows that the coefficient variable expenditure is –2,965.635 and is not significant at the 5% rate. Furthermore, the coefficient for the Number of Arrivals (NOA) variable is 15138685, showing that this variable is significant at the 5% rate.

4.1 *Tourism contribution to GDP*

Table 3 shows 11 years of trend data for tourism's contribution to ASEAN economies as measured by the contribution to GDP. In value terms, the sector contributed a total of US$154.3 billion in 2004 and US$260.7 billion in 2014 to regional GDP, demonstrating a significant increase in GDP between 2004 and 2014.

Table 2. Estimation results.

Dependent Variable: GDP
Method: Panel EGLS (cross-section weights)
Date: 10/10/18 Time: 19:04
Sample: 2002 2016
Periods included: 15
Cross sections included: 11
Total panel (unbalanced) observations: 154
Linear estimation after one-step weighting matrix

Variable	Coefficient	Std. error	t-statistic	Prob.
C	7.52E+14	1.37E+13	54.83244	0.0000
EKSP	−2,965.635	3,525.957	−0.841087	0.4017
NOA	15138685	1319623	11.47198	0.0000

Effects specification
Cross-section fixed (dummy variables)
Weighted statistics

R-squared	0.968086	Mean dependent var	8.95E+14
Adjusted R-squared	0.965370	S.D. dependent var	1.28E+15
S.E. of regression	2.88E+14	Sum squared resid	1.17E+31
F-statistic	356.4228	Durbin-Watson stat	0.200112
Prob (F-statistic)	0.000000		

Unweighted statistics

R-squared	0.942425	Mean dependent var	8.41E+14
Sum squared resid	3.81E+31	Durbin-Watson stat	0.056487

Table 3. Total tourism contribution to GDP (US$ in bn, real prices) (Source: www.wtcc.org).

Country		2004	2005	2006	2007	2008	2009	2010	2011	2012	2013	2014
Brunei	% share	8.6	7.3	6.4	7.0	6.9	8.4	7.2	6.8	7.0	7.4	7.6
	US$	1.1	0.9	0.8	0,.9	0.9	1.1	0.9	0.9	1.0	1.1	1.1
Cambodia	% share	22.4	24.6	24.9	24.1	22.7	24.4	27.3	30.6	32.8	32.5	33.3
	US$	2.0	2.4	2.7	2.9	2.9	3.2	3.7	4.5	5.2	5.5	6.0
Indonesia	% share	6.8	6.2	5.7	5.8	6.2	6.2	5.8	5.7	5.6	5.5	5.8
	US$	34.0	32.3	31.5	34.4	38.9	40.7	41.0	42.5	44.4	46.2	51.1
Laos	% share	12.3	13.6	11.5	12.7	13.1	13.7	13.9	13.4	13.0	14.2	13.0
	US$	0.7	0.8	0.8	0.9	1.0	1.1	1.2	1.3	1.3	1.6	1.6
Malaysia	% share	13.1	13.1	13.7	16.4	12.7	13.3	13.1	12.8	12.9	13.7	14.2
	US$	21.9	23.1	25.4	32.5	26.3	27.2	28.7	29.4	31.3	35.0	38.3
Myanmar	% share	4.6	4.9	3.7	3.9	3.7	3.2	2.4	2.6	4.2	4.2	5.5
	US$	1.1	1.3	1.1	1.3	1.4	1.4	1.1	1.3	2.2	2.4	3.3
Philippines	% share	11.4	12.5	13.2	14.9	10.4	10.9	11.9	14.0	14.5	15.0	16.6
	US$	17.6	20.4	22.5	27.0	19.7	21.0	24.6	29.9	33.0	36.7	43.0
Singapore	% share	9.3	8.9	8.3	9.1	8.8	8.8	9.8	9.8	10.2	9.7	9.8
	US$	15.0	15.5	15.7	18.7	18.5	18.4	23.5	25.1	26.9	26.9	28.2
Thailand	% share	17.2	15.9	16.8	17.7	17.0	15.9	14.0	15.2	16.4	18.3	17.6
	US$	49.9	47.8	53.3	59.1	57.7	53.5	50.8	55.7	64.5	73.5	71.5
Vietnam	% share	11.2	10.7	12.9	9.8	11.6	9.2	10.0	9.5	9.7	9.3	9.1
	US$	11.2	11.5	14.8	12.1	15.1	12.6	14.5	14.7	15.9	16.1	16.6

4.1.1 Tourism contribution toward employment

Table 4 shows 11 years of trend data for tourism's contribution to ASEAN economies as measured by contribution to employment. In the ASEAN region, there were 255,199,000 jobs in 2004, which accounted for around 10.62% of total employment.

Table 4. Total tourism contributions to employment (Source: www.wtcc.org).

Country		2004	2005	2006	2007	2008	2009	2010	2011	2012	2013	2014
Brunei	% share	9.1	8.4	7.5	8.1	8.0	9.7	8.2	7.6	7.8	8.1	8.2
	000	15.3	14.5	13.1	14.3	14.4	17.7	15.3	14.4	14.9	15.7	16.1
Cambodia	% share	19.6	21.3	21.4	20.6	19.6	20.8	23.0	26.1	28.9	29.1	29.7
	000	1248.5	1402.8	1451.1	1441.1	1380.2	1548.0	1808.2	2113.3	2297.8	2333.5	2432.3
Indonesia	% share	10.1	9.0	8.3	8.6	9.1	9.0	8.6	8.3	8.3	8.3	9.0
	000	9715.5	8479.1	7960.8	8556.1	9291.0	9500.0	9272.2	9113.4	9351.8	9600.0	10540.9
Laos	% share	10.9	11.6	10.3	11.5	11.8	12.4	13.0	12.5	11.9	12.8	11.8
	000	259.1	282.5	257.8	295.6	310.5	335.2	359.7	353.9	343.9	377.3	356.2
Malaysia	% share	11.3	11.5	11.9	14.7	11.3	12.1	11.8	11.1	11.2	12.0	12.1
	000	1163.5	1181.4	1255.2	1599.4	1242.5	1355.1	1389.6	1389.5	1429.6	1628.6	1686.8
Myanmar	% share	3.8	4.1	3.0	3.3	3.3	3.0	2.4	3.0	4.5	4.4	5.7
	000	736.8	804.4	607.6	660.5	669.5	617.9	499.6	631.2	956.3	934.1	1240.2
Philippines	% share	10.0	10.5	11.0	12.4	8.6	9.0	10.2	12.3	13.0	13.4	15.1
	000	3156.2	3392.4	3584.1	4165.7	2920.4	3166.2	3669.5	4564.5	4892.8	5113.5	5756.4
Singapore	% share	7.3	7.3	6.7	7.5	7.0	6.9	7.8	8.4	8.8	8.6	8.5
	000	158.9	166.5	163.4	198.4	201.8	203.8	238.3	267.9	291.0	295.6	302.9
Thailand	% share	14.5	13.0	14.0	13.9	14.2	13.5	11.5	11.3	13.1	14.9	12.9
	000	5037.7	4565.7	4989.0	5058.6	5238.2	5064.2	4381.8	4348.9	5095.8	5788.4	4910.9
Vietnam	% share	9.7	9.9	12.0	9.2	10.7	8.3	9.1	7.8	8.0	7.7	7.6
	000	4028.3	4236.2	5278.2	4138.9	4965.2	3946.9	4446.0	3918.7	4094.7	4042.2	3993.6

5 CONCLUSIONS

Tourism is an important driver in the economic development and growth strategy, and should be treated well. The contribution to GDP from the tourism sector may be small in the actual amount of income generated in a country or region, but if there is little alternative industry then its relative importance and the percentage of the country's GDP may be large. In the ASEAN member countries in this study, the tourism sector is one of the most important industries in terms of absolute size of employment and output, and industrial linkages are strong and widely dispersed. The tourism industry can also drive reductions in unemployment, generate income, and provide additional quality-of-life benefits to local residents.

6 RECOMMENDATIONS

From the analysis results and conclusion above, there are some recommendations suitable to stimulate economic growth and reduce unemployment in ASEAN member countries through the tourism sector, which can be summarized as follows:

1. ASEAN member countries should invest more in tourism infrastructure to further realize and maximize their market potential. In this case, it is better for them to facilitate the disabled, senior citizens, women, and children.
2. ASEAN member countries should consider the budget to support the tourism sector. In this case, if necessary they should increase the budget for the tourism sector.

REFERENCES

Athanasopoulou, A. (2013). Tourism as a driver of economic growth and development in the EU-27 and ASEAN regions.
Balli, F., Curry, J. & Balli, H.O. (2015). Inter-regional spillover effects in New Zealand international tourism demand. Tourism Geographies, 17(2), 262–278.

Broadhurst R. (2008). *Managing environments for leisure and recreation. London,* UK: Routledge.

Cahyaningrum, D. (2017). Community empowerment based local wisdom in tourism of Bajo community, Wakatobi. *International Journal of Scientific & Technology Research*, *6*(11), 196–201.

Makochekanwa, A. (2013). An analysis of tourism contribution to economic growth in SADC countries. *Botswana Journal of Economics*, *11*(15), 42–56.

Meyer, D, 2006. Caribbean Tourism, local sourcing and enterprise development: Review of the literature. Pro-Poor tourism working paper 18, Pro-Poor Tourism Partnership, London.

Pigram, J.J. & Jenkins, J.M. (2006). *Outdoor recreation management*. New York, NY: Routledge.

Torkildsen, G. (1999). *Leisure and recreation management*. London, UK: Routledge.

Tugcu, C.T. (2014). Tourism and economic growth nexus revisited: A panel causality analysis for the case of the Mediterranean Region. *Tourism Management*, *42*, 207–212.

United Nations World Tourism Organization (UNWTO). (2015). Asean Statistical Leaflet.

United Nations World Tourism Organization (UNWTO). (2017). Asean Statistical Leaflet.

Williams, S. (2003). *Tourism and recreation*. Harlow, UK: Prentice Hall.

World Travel Tourism Council (WTCC). (2015). Travel and Tourism: Economy Impact 2015 World.

Entrepreneurship

Business Innovation and Development in Emerging Economies – Trinugroho & Lau (Eds)
© 2019 Taylor & Francis Group, London, ISBN 978-1-138-35996-3

Intrapreneurship of handwoven crafts toward economic democracy in the border area of SajinganBesar, Indonesia

Elyta
Faculty of Social and Political Sciences, Tanjungpura University, Indonesia

ABSTRACT; One of the sources of income and economic growth forweavers in SajinganBesar, Sambas Regency, Indonesia comes from the production of handwoven crafts. These crafts are in the form of chairs, furniture and household handicrafts. Although there are characteristics that distinguish woven crafts from the Indonesian and Malaysian border, there is a claim for patent rights. This study aims to explain the development and marketing potential of woven crafts through Intrapreneurship and realize economic democracy based on two models. Data was obtained through two sources: secondary data and in-depth interviews. Secondary data was obtained from related institutions, while in-depth interviews were conducted by interviewing plaiters, the Head of Chemical and Multifarious Industrial Crafts of Sambas District, the Trade and Industry Office of Sambas Regency, the Village Chief of Kaliau, weavers and community leaders of SajinganBesar. Qualitative descriptive analysis was performed on the data. The results of this study found that intrapreneurship plays a role in striving to improve competitiveness, develop production and develop capabilities of weavers. Through intrapreneurship, weavers are given the means to develop productivity of their woven crafts throught implementation of the first model, (intrapreneurship model of the creative industry for the woven craft agribusiness sector) and the second model (government development of woven crafts by providing easy access to intrapreneurship funding, patent protection, and improvement in the ability of human resources) as an effort toward economic democracy.

Keywords: hand woven crafts, intrapreneurship, marketing, eco democracy economy

1 INTRODUCTION

Development at the Indonesian borders is a manifestation of state sovereignty that has strategic roles and values that are important in national development. The success of such a development has an important influence on state sovereignty, state welfare, social and economic improvement for the weavers; providing employment and strengthening their influence on the country's security. Even so, until now development in the border area has not shown tangible results; the area is still experiencing poverty. However, contrast to other border area, handicrafts from weavers in Sajingan Besar as the border area of Sambas Regency in Indonesia has the potential to solve this problem.

Their handwoven crafts are one of the creative industries that are characteristic of this area, encouraging many foreign and local tourists to visit. Their handcraft is a tradition inherited from their ancestors. The sale of Sambas Regency's hand woven handicrafts was less attractive to buyers, even though the prices were cheaply pegged, so some weavers marketed their woven products to neighboring Malaysia to reap profits. One craftsman and plaitor said that it was more profitable to sell the woven goods to Malaysia than sell them to his own country.

Intrapreneurship leads to an individual process in a group to obtain opportunities for resources controlled by a group (Stevenson & Jarillo, 1990, p. 23). This attitude in entrepreneurship is a desire and an ability to create something new and innovative that can be useful for others and also themselves, and can help achieve the desired profit. Entrepreneurship is the value shown through the behavior and resilience of business actors to maintain their business.

Based on survey research that conducted by us, the research team found problems with plaited handicrafts in the border area of SajinganBesar, namely weavers have limited understanding of the capital and technology needed to run an intrapreneurship business. This leads to weavers preferring to sell their products to Malaysia and causing woven craft from SajinganBesar patented by the Malaysian State. So far, woven marketing in the West Kalimantan region has only been targeted at domestic buyers, while marketing in Malaysia has been exported to foreign countries. The market and handicraft buyers consider the product to be produced from a Malaysian country, even though it is clear that the manufacture of the products and raw materials comes from Indonesia. Therefore, the issue is that the woven craft has been patented by the Malaysian State. As a result of these issues, a number of officials from the Sambas Regency Industry and Trade Service, the Provincial Government of West Kalimantan and the Ministry of Trade have intervened to review the work of weavers by providing support and creating an intrapreneurship model as an effort towards economic democracy. My research team conducted a study of hand woven intrapreneurship models towards economic democracy on the border of the SajinganBesar.

2 METHOD

The method of analysis used in this study a was the qualitative descriptive method. The study was located in the border area of SajinganBesar, Sambas Regency, Indonesia, and used secondary data from the Ministry of Foreign Affairs of the Republic of Indonesia, books and journals. In addition, information was obtained directly from the field by interviewing several informants, namely the Head of the Border Management Division of the Regional Secretariat of Sambas district, the Head of the Chemical and Multifarious Industrial Crafts Section of Sambas Regency, the Trade and Industry Office of Sambas Regency, the Village Head of Kaliau weavers and SajinganBesar community leaders.

3 RESULTS AND DISCUSSION

The real nature of economic democracy is that it aims to create prosperity for all crafters and involve them in both the production process and the management of the results. The activity of woven crafts is important in realizing economic democracy for the weavers, even though the management which is not yet fully independent is still pursued through the support of guidance from the government.

The results of this study found that the hand waving intrapreneurship model led to economic democracy on the border of the Sambas Regency and gained the right potential skills for the weavers through the development of an intrapreneurship economic institution of creative industries that fostered the competence of the craftsmen and traders.

More clearly, the intrapreneurship model of hand woven crafts towards economic democracy on the border of the Sambas Regency can be seen in Figure 1.

3.1 *Intrapreneurship model for the creative industry based on the woven craft agribusiness sector*

SajinganBesar region has one of the most creative industries in the form of woven crafts, where the method of manufacture is still using a simple process that is undertaken by the local people. There is a very high cultural value contained in hand woven crafts. This is because the craft has a characteristic element in its design, shape and material (it uses natural products in

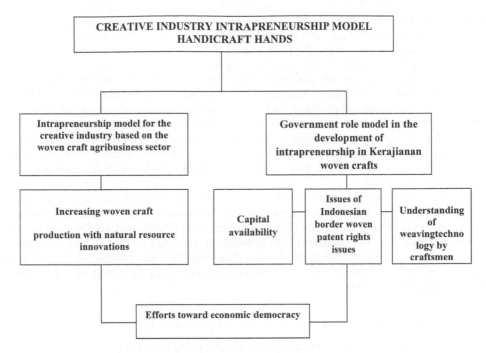

Figure 1.   Creative industry intrapreneurship model handicraft hands.

the form of bamboo). So, it is not a strange thing to know that neighboring countries really want it, besides because of its low price, it is of distinctive cultural value.

Looking at the West Kalimantan Gross Regional Domestic Product (GRDP) growth data, based on this field of business in the manufacturing industry sector from 2013 to 2015, it was seen that the GRDP processing industry sub-sector in 2015 was the largest in the food and beverage industry sub-sector with a total GRDP of IDR17.91 trillion. Whereas, the second position was occupied by the sub-industry of wood, articles made from wood and cork and woven goods from bamboo, rattan and the like for a total of IDR1.4 trillion. The rubber industry, goods made from rubber and plastics were third with a GRDP in 2015 amounting to IDR1.24 trillion. Therefore, it can be seen, hand woven crafts occupy the second position of growth in West Kalimantan GRDP based on business fields in the manufacturing industry sector from 2013 to2015 (Badan Pusat Statistik Provinsi Kalimantan Barat, 2015, p. 21). Thus, it can be said that hand woven crafts have the potential to be developed so that it can be estimated that with intrapreneurship, the market share of handwoven crafts on the border of Sajingan Besarwill expand.

Intrapreneurship teaches individuals in an organization to have motivation in creating business (Antoncic & Hisrich, 2003, p. 20). Business and entrepreneurship are opportunities pursued by individuals to take various initiatives related to the acquisition of resources and organizations in building a new business (Stevenson & Jarillo, 1990, p. 23). Opportunities referred to in entrepreneurship is a recombination of resources that are alleged to be individual benefits (Shane, 2003, p. 18). Then every new business or entrepreneur has different information or manages it with a different perspective (Shane, 2003, p. 41). Businesses and entrepreneurs in the creative industry of hand woven crafts from each of the border regions of SajinganBesar have the potential based on the results of previous work. Intrapreneurship has the authority to promote production, by coordinating and supporting the Office of Industry and Trade, starting with calculating the minimum stock of woven craft demand from the previous year. Hand-woven craft has decreased when compared to machine woven craft.

The results of this research show that border development is an inseparable part of the interests of weavers toward economic democracy, especially in producing and marketing hand woven crafts. Intrapreneurship in the creative economy is applied to marketing the most in demand hand woven products such as Sambas Regency's woven fabric with various innovations. Their handwoven craft has distinctive characteristics for example, from the Iban tribal craftsmen and the Dayaks. The difference in question lies in the weaving and woven motifs.

Weaving handicrafts from the Iban tribe craftsmen have similarities with the DayakSalako tribe in Malaysia, so it is alleged that the Malaysian State may make a claim on Indonesian hand-woven handicraft products. The constraints faced by the weavers in the firstwoven function have not been coordinated until now so that sales have not been optimal. Facing these problems the role of innovation in the Sambas Regency cloth plays an important role. These innovations can be combined with other fabrics and better added with foam so that outsiders are more interested in Sambas Regency woven fabrics compared to fabrics produced from outside regions.

Woven craft is a form of creative industry, namely the creation in terms of distribution of handicraft products. One area consisting that includes several actors with the same business is cooperation that based on Product Operating Standards (POS), which is the same production with high quality will be easier to sell hand woven products. So that an area with several business actors who want to cooperate in accordance with the same production POS with the same quality, will facilitate the sale of woven products with a target capacity of one month to produce thousands of products.

3.2 Government role model in the development of intrapreneurship of woven crafts

The planned economic democracy encompasses democracy from the production location of woven crafts to the access of the weavers to the means of production or important facilities. The planning refers to socially controlled economic aspects in accordance with the provisions set by the government, meaning the involvement of the government with the weavers cannot be ignored if planning a comprehensive goal for the common interest such as efforts to improve the welfare of weavers through woven crafts are continuously developed and supported from various aspects.

3.3 Capital availability

One component of intrapreneurship support is capital. Capital has an important role in running a business. This component can affect the socio-economic aspects of the weavers and minimize the number of weavers who are still too dependenton the neighboring countries. Collaboration between the role of government and the private sector will provide opportunities for weavers to increase their productivity. Craftsmen can get capital through financial institutions from the state and private parties for example, by applying for a loan from the Bank Rakyat Indonesia or through Credit Union.

The results showed that weavers felt there was a lack of funds in running an intrapreneurial business. The existence of intrapreneurship was not followed by an understanding of weavers, especially the border craftsmen of SajinganBesar. These weavers objected to managing their intrapreneurship because they thought it required large capital and they also considered that the capital must be spent with the conditions of the welfare of the weavers. The role of intrapreneurship for weavers continues to be emphasized. Potential development is carried out by identifying and reviewing the work of weavers of Sambas Regency in order to develop productivity.

Furthermore, the results of this study indicated that the Sambas Regency has export potential, especially to Malaysia. This potential is able to strive and improve the welfare standards of the SajinganBesar weavers toward economic democracy. Most of the products produced are processed from the local area then shipped and exported. Industry and trade offices together with the Ministry of Cooperatives and Entrepreneurship Institutions, and weavers have sought the export of international hand woven craft products, namely by aiming at the target of consumers who need to make products according to the desires and needs of consumers abroad. Demand for handwoven crafts from abroad have a standard based on the function of goods that can be useful and a size that suits the wishes of consumers based on product and comfort needs, as well as being environmentally friendly.

3.4 *Issues of Indonesian border woven craft patent rights*

The results of the study found that this government through the Industry and Trade Office of the Cooperative and Small and Medium Enterprises of Sambas Regency, tried to control the issue of Indonesian border woven product being patented by Malaysia. So that hand woven with many types of products can be protected by patent rights. However, patenting the motive is more difficult because most of the motifs have been used during the turn of the generation which also affects changes in the specificity of the motive. Such motives are often not specifically created by an area so it cannot be recognized as their full ownership rights. In addition, the motive is not only created by one person, the dimensions can be reduced in size or can change shape. Therefore, changes in these motives can be recognized as crafts from other regions, especially Malaysia.

3.5 *Understanding of weaving technology by craftsmen*

In addition to constraints on limited capital to run an intrapreneurship, hand woven crafts are also constrained by sophisticated technology used by other countries, especially Malaysia and Taiwan. This is due to a quality gap and lack of human resource skills. The results of this study showed that plaiters in the Sajingan Besar lacked the technology to market their products. In this modern era, the mastery of technology for marketing is very important to supporting the economy. Globalization requires the ability of entrepreneurs to serve consumers quickly and comfortably through technological sophistication. In fact, the plaiters of SajinganBesar are not able to master technology. This is due to several factors including the lack of human resources who can use technology, especially the internet. Limitations of communication technology for handicraft workers is due to lack of equitable development from the government and limited signal and internet access in various regions of the SajinganBesar. Therefore, there is not sufficient marketing for consumers to meet their needs to access online sales.

Intrapreneurship is inseparable from technology in the development of a craft. In order for woven crafts to survive, it is necessary to improve the skill of the weavers and address the existence of simple technology, but also to make it easier for the weavers to produce their craft. The results of this study found that intrapreneurship in handicrafts in the border areas was also constrained by technology to produce woven materials derived from rattan, pandanus and other raw materials in making hand-woven crafts. The need for processing raw materials is adjusted to the form of goods that produced, followed by collaboration with additional materials such as plastic to produce more attractive type of goods for the buyers.

In addition to discussing the capital development, the results of this study indicate that the steps taken to face competitive marketing are empowerment, especially human resources for the border area of Sambas Regency because of the competition with outside countries that are more skilled and have high quality technology produced products. The Indonesian state border is not ready and it is feared that it will be slowly crushed. The weavers produce handwoven and the results are not as many as woven crafts tahat are produced with the help of technology. Therefore, there have been plans from the Department of Industry and Trade to increase the competitiveness of the quality of handicrafts woven by border weavers of SajinganBesar by providing facilities and technology.

Wrighy (as cited in Bin, 2015, p. 253) found that the inability of democratic institutions to control the movement of capital reduces the ability of a democracy to organize collective priorities over the use of social resources. In managing the weaving craft resources, government institutions have not controlled capital movements maximally and it show many shortcomings in the collective regulations of the government. Therefore, this study assesses that efforts to overcome this problem are through the involvement of weavers, synergy with the government, support in the form of guidelines and providing facilities that will strengthen democracy.

4 CONCLUSION

Firstly, the intrapreneurship model of the creative industry of woven craft agribusiness sector. Intrapreneurship is carried out as a tool for developing woven crafts that seeks to minimize the dependence of weavers on Malaysian countries in for its broad marketing. Therefore, the Malaysian government cannot state claims that handwoven crafts originating from the Indonesian border region, especially in SajinganBesar, are its state patents. The results of this study found that the intrapreneurship model of the creative industry in the agribusiness sector used the natural resources of SajinganBesar as its own attraction and innovation in the production of woven crafts to maximize access to international markets. Weaving crafts have the potential to increase economic growth. This potential can be developed through intrapreneurship as an effort to increase the production of woven crafts and become characteristic of the SajinganBesar area.

Secondly, the government's role model in developing intrapreneurship of woven crafts. A spirit of entrepreneurship in various sectors of agriculture, plantations, the craft sector and the service sector is needed to generate competition globally and as a support in the effort to realize economic democracy for the Sajingan Besar weavers. That embodiment also can be supported by increasing cooperation in the sense of holding meetings and trade between countries is needed; especially in trade such as staple food needs which are dominated by the import process, which is easy to access and have agreements. This is to reduce the level of dependence on imports from abroad. In addition to the implementation of intrapreneurship, in this second model, my research team indicates that there is an active role by the government in the development of woven crafts toward economic democracy, namely: (1) availability of capital—the government provides for intrapreneurship development as a means to produce woven crafts compete in the international market who are worried about lack of funding can be provided with loans by Bank Rakyat Indonesia or through Credit Union; (2) issues of Indonesian border products patent rights, and cultural/generational motifs and shapes of woven crafts are cultures that are inherited from generation to generation causing the similarity of the use of motifs and the manufacture of woven craft forms indicated by claims of production. The government through intrapreneurship provides patent protection on the results of woven crafts causing cross-country similarities and claims of production, are managed by the government through intrapreneurship patent protection; and (3) understanding craft technology—one of the steps in competitive marketing is the availability of human resources that have capability to use technology; intrapreneurship is inseparable from technology as a support for the creativity of woven crafts in improving the quality and quantity of their production.

REFERENCES

Antoncic, B. & Hisrich, R.D. (2003). *Clarifying the concept of intrapreneurship. Journal of Small Business and Enterprise Development*. 10 (1), 7–24.
Badan Pusat Statistik Provinsi Kalimantan Barat. (2015). *Profil industri makro dan kecil kalimantan barat*. Kalimantan Barat, Indonesia: Badan Pusat Statistik.
Bin, Daniel. (2014). Macroeconomic policies and economic democracy in neoliberal Brazil. *Economia e Sociedade, Campinas, 24*, (3 (55)), 513–539.
Stevenson, H.H. & Jarillo, J.C.(1990). A paradigm of entrepreneurship: Entrepreneurial management. *Strategic Management Journal, 11*, 17–27.
Shane, S. (2003). *a general theory of entrepreneurship; Nexus individual opportunity*, Cheltenham, United Kingdom: Edward Elgar.

Financial accounting

Business Innovation and Development in Emerging Economies – Trinugroho & Lau (Eds)
© *2019 Taylor & Francis Group, London, ISBN 978-1-138-35996-3*

Quality analysis of accounting information after adoption of International Financial Reporting Standards

K.A. Rahayu, O.S. Heningtyas & Payamta
Faculty of Economics and Business, Sebelas Maret University, Surakarta, Indonesia

ABSTRACT: The purpose of this study is to provide empirical evidence as to whether the adoption of International Financial Reporting Standards (IFRS), which had mandatory implementation in 2012, can improve the quality of accounting information. The quality of accounting information is measured by the value relevance proxy. The population is taken from automotive companies listed on the Indonesia Stock Exchange (IDX) in the period 2006–2011 prior to adoption, and 2012–2016 after the adoption of IFRS. This research uses multiple regression statistical tests. The results of the tests show that there is an increase in the value relevance of accounting information after the adoption of IFRS. These tests also show that an increase in value relevance occurs in net earnings per share and book value per share.

Keywords: Adoption of IFRS, quality of accounting information, price models, automotive companies

1 INTRODUCTION

Since the beginning of 2000, cases of accounting scandals such as Enron and WorldCom reflect the failure of existing accounting standards (such as US GAAP) (Liu et al., 2014). The standardization of accounting on the basis of International Financial Reporting Standards (IFRS) in some countries allows for companies to voluntarily adopt international standards (Kim & Shi, 2012). As of 1 January 2012, companies listed on the Indonesia Stock Exchange (IDX) must prepare their financial statements in accordance with IFRS.

All IFRS standards are principle-based and independent of the nature and specificity of certain industries, so the same principles will apply. However, accounting history has shown that the general principle cannot take account of the accounting nuances of each and every industry. There may be interpretation problems when adopting the same principles in all industries. One of the industries affected by IFRS is the automotive industry because it is a capital-intensive industry that has a higher level of complexity than many other industries (Lavi, 2016).

The adoption of IFRS will have an impact through the high disclosure that companies must make and will improve the quality of information, such as enhancing the appeal and transparency of financial reporting worldwide, which is expected to reduce the capital costs for companies (Armstrong et al., 2010). The quality of accounting information is one of the most fundamental elements that can influence economic decision-making by stakeholders (Deaconu et al., 2010). The information content of the accounting numbers should reflect a true and fair view of the company's financial situation (Chandrapala, 2013).

Research conducted by Zeghal et al. (2012), Chua et al. (2012), Chang et al. (2013), Kargin (2013), Rohmah and Yuni (2013), Edvandini et al. (2014), and Wulandari and Adiati (2015) also provides support for IFRS adoption by stating that the use of IFRS may improve the quality of accounting information when compared to that prior to its use. The results of research conducted by Outa (2011), Karampinis and Hevas (2011), Cahyonowati and Ratmono (2012), Karyadan and Irwanto (2017), and Kouki (2018) show contradictory empirical evidence, indicating no significant improvement in the quality of accounting information after IFRS adoption.

Previous studies have almost all used samples from many industries and show inconsistent results. However, each industry has different traits and characteristics that will have different effects. This research seeks to explore the impact of IFRS adoption on one type of industry, that is, the IDX-listed automotive industry. This research therefore aims to test whether the adoption of IFRS, which had mandatory implementation in 2012, can improve the quality of accounting information for IDX-listed automotive companies. In addition, this study uses the same proxy as that of Van der Meulen et al. (2007), Barth et al. (2008), Kousenidis et al. (2010), Kwong (2010), Outa (2011), Karampinis and Hevas (2011), Agostino et al. (2011), Alali and Foote (2012), Kargin (2013), and Wulandari and Adiati (2015), where the quality of accounting information is measured by value relevance proxies. Such proxies are used in accordance with the results of a study conducted by Barth et al. (2008), which states that high-quality accounting information is information with a high degree of value relevance. Consistent with these studies, this study analyzes the effect of overall IFRS adoption and not the influence of any adopted standards.

2 LITERATURE REVIEW

2.1 *Signaling theory*

Signaling theory is applied by corporate managers who have better information about their companies and are encouraged to convey this information to potential investors, where the aim is to increase the value relevance through a report by sending a market signal through financial statements (Scott, 2012). Complete, accurate and timely information is needed by investors in the capital market, for use as an analytical tool to make investment decisions. The relationship between signaling theory and value relevance of accounting information can be seen on the relevance of the value of accounting information company where accounting figures obtained from the financial statements and stock prices company. The value relevance of the accounting figures in the financial statements is the net income per share and the book value of equity per share, both of which are summarized in the main measurements of a company's financial statements.

2.2 *Quality of accounting information*

According to the Institute of Indonesia Chartered Accountants (Ikatan Akuntan Indonesia, 2017), the qualitative characteristics of financial statements are the typical characteristics that make information in the financial statements useful to users. There are four basic qualitative characteristics:

2.2.1 *Understandable*

The quality of the information contained in the financial statements should be readily understood by the user; for this purpose, the user is assumed to have adequate knowledge of economic and business activities, accounting, and the willingness to acquire information with reasonable diligence.

2.2.2 *Relevant*

To be useful, information should be relevant to user needs in the decision-making process. Information is of relevant quality if it can inform the users' economic decisions and help them evaluate past, present or future events, as well as affirm or correct past user evaluations.

2.2.3 *Reliable*

Information will be useful if it is reliable. Information has reliable qualities if it is free from misleading notions, manifest faults, and reliable for users as a presentation or reasonably expected to be served. Information may be relevant but if the nature or presentation is unreliable then the use of such information can potentially be misleading.

2.2.4 *Comparable*

Users should be able to compare an entity's financial statements between periods to identify tendencies or positioning trends and financial performance. The user should also be able to compare the financial statements between entities to evaluate their financial position, performance, and relative changes in financial position. Therefore, the measurement and presentation of the financial impacts of similar transactions and other events should be consistently carried out for such entities, over the same periods.

2.3 *IFRS adoption*

IFRS is an international accounting standard issued by the International Accounting Standards Board (IASB). The IFRS is composed of four major international organizations: the International Accounting Standards Board, the European Community (EC), the International Organisation of Securities Commissions (IOSOC), and the International Federation of Accountants (IFAC). The IASB, formerly known as the International Accounting Standards Committee (IASC), is an independent institution for the development of accounting standards. The organization has the goal of developing and encouraging the use of high-quality, understandable and comparable global accounting standards (Choi et al., 1999). Indonesia has adopted this international accounting standard, established by the IAI (Institute of Indonesia Chartered Accountants) and hopes that IFRS's use can improve the value of a company in terms of comparability, transparency, and quality of financial statements. With its principality-based approach and fair value measurement, IFRS is considered to have a positive impact on the relevance of the value of accounting information in the presentation of financial statements. The use of IFRS standards can be useful in harmonizing accounting practices, lowering transaction costs, increasing international investment, and the beneficial effectiveness of communications with investors (Iatridis, 2010). The increase in comparability, transparency, and quality of financial statements, proxied by assessing the relevance of the value of accounting information, can provide better and more understandable information that is also relevant and reliable, and can be used by stock market participants, in this case potential investors.

2.4 *Relevance of value information*

Consistent with previous IFRS research (Van der Meulen et al., 2007; Barth et al., 2008; Karampinis & Hevas, 2011; Alali & Foote, 2012), the quality of accounting information in this study is proxied with value relevance. Value relevance reflects the ability of financial information to represent the value of a company, that is, the relationship between accounting numbers and stock prices. The main summary of the relationship between accounting rates and stock quotes can be seen from the accounting figures in the financial statements, by looking at the value of net earnings per share and the book value of equity per sheet, both of which are summarized in key measurements of financial statements that can determine the value of a company's stock price.

Francis and Schipper (1999) define the value relevance of accounting information as the ability of accounting figures to summarize the information underlying stock prices, so that value relevance is indicated by a statistical relationship between financial information and stock prices or returns. A high quality of accounting information is indicated by a strong relationship between stock prices and earnings and book value of equity, because both types of accounting information reflect the company's economic condition (Barth et al., 2008). Ohlson's model or technique (1995) basically connects the firm's market value (stock price) with the profit and book value, as well as to other information that may affect the relevance of the value of accounting information.

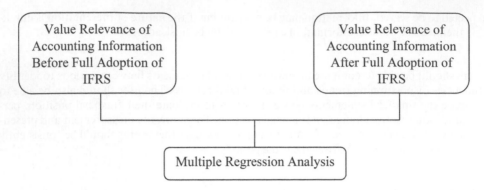

Figure 1. Framework.

2.5 *Development of hypotheses*

IFRS convergence in the years before and after full implementation is expected to provide significant results as to the value relevance of accounting information from firms listed on the Indonesia Stock Exchange (BEI). This particularly relates to automotive firms in the years 2007–2016, which will increase the interest of foreign investors in menanamkan shares in global capital, especially in Indonesia.

The presence of International Financial Reporting Standards will be easier to understand for foreign investors, and the desired information can be obtained by reading financial statements from various countries in the world. The relevance of earnings value and book value of equity increases after the application of IFRS due to the use in IFRS of measurements that may better reflect the economic condition of a company (Barth et al., 2008).

Research conducted by Zeghal et al. (2012), Chua et al. (2012), Chang et al. (2013), Kargin (2013), Rohmah and Yuni (2013), Edvandini et al. (2014), and Wulandari and Adiati (2015) also provides support for IFRS adoption by stating that the use of IFRS may improve the quality of accounting information compared to that prior to use of IFRS.

H1: There is an increase in the value relevance of accounting information after the full adoption of IFRS.

3 RESEARCH METHODS

3.1 *Sample classification*

The population in this study used all automotive companies listed on the Indonesia Stock Exchange or those that can be accessed at http://idx.co.id. Based on data from financial statements that have been obtained from 2007–2016, there are 13 automotive companies listed. The automotive industry listed on the IDX as a population of the research sample is due to the fact that the automotive industry is affected by the impact of IFRS adoption. The observation years of this research are the period before IFRS adoption (2007–2011), and the period after full adoption of IFRS (2012–2016).

The sampling method in this study is purposive sampling, which is the selection of sample groups based on the characteristics of the population set. There are also criteria in this study that is the period before the adoption in 2007–2011 and after the full adoption of IFRS in 2012–2016.

The criteria used in determining the sample are as follows:

1. An automotive company listed on the Indonesia Stock Exchange (IDX) that publishes its financial report/listing within the ten consecutive years 2007–2016.
2. A manufacturing company presenting financial statements, using the rupiah currency in its financial statements during the period 2007–2016. It is used to use the same or equivalent currency.

3. Automotive companies listed on the Indonesia Stock Exchange consistently publishing annual reports in the period 2007–2016 and having the required research data (stock price, book value per share and earnings per share); this is because it is more effective to obtain data fluctuating in one period.
4. An automotive company meeting the above criteria and not having a negative Earning Per Share (EPS) or being in a loss condition for each annual report for the period 2007–2016. This criterion is used so as not to affect the calculation of profit as being relatively very low for value relevance.

3.2 Research data

The data used in the documentation data and secondary data is taken from information in the financial statements of automotive companies listed on IDX for the years 2007–2011, as the period before adoption of IFRS, and also from years 2012–2016 when the companies had fully implemented and adopted IFRS. The data collected from the Indonesia Stock Exchange (BEI) and the Indonesian Capital Market Directory (ICMD) are the annual financial statements and quarterly corporate financial statements, which meet the criteria for the sampling of the automotive industry. Stages in data collection is by reading previous research that has been published in reference journals.

3.3 Research variable

Variables are anything that can differentiate or show variation in values. In this research there are two variables, namely dependent variables and independent variables.

a. The dependent variable is Value Relevance of Accounting Information.
b. The independent variable is the variable that affects the dependent variable. The independent variables of this study relate to data prior to IFRS adoption, and to data after the full adoption of IFRS.

3.4 Operational definition of variables

The operational definition is an indication of how a variable is measured, so that the researcher can know the quality of the measurement. The operational definition of this research consists of:

1. The dependent variable is the variable that is influenced or caused by other variables, where the variables in this study are the value relevance of accounting information from quarterly stock price measurements of closing price, book value per share and earnings per share.
2. The independent variable is the adoption of IFRS, which is tested to prove its influence in terms of the periods before and after full adoption of IFRS in manufacturing companies listed on the Indonesia Stock Exchange in the 2007–2011 period before the adoption, and after full adoption in the period 2012–2016.

3.5 Analysis tool

This study is consistent with previous IFRS studies such as Barth et al. (2008), testing the value relevance of accounting information using the pricing model developed by Ohlson (1995). The Ohlson model equation is estimated by means of statistical test with multiple regression for period data before and after IFRS adoption, separately. The value-relevance test using the adjusted R^2 value is used because more general or R^2 adjustments do not increase in value as the independent variables increase. The adjusted R^2 value is between 0 and 1; if the adjusted R^2 takes a negative value then it means that the independent variable is totally unable to explain the variance of the dependent variable. However, if the value of adjusted R^2 approaches 1, then it is able to explain almost all the information needed to predict the variance of the dependent variable. Thus, if the value of adjusted R^2 is significantly greater for the data period after IFRS adoption than before, then this indicates an increase in the value relevance of the accounting information. Conversely, if there is no significant increase or an actual decrease in adjusted R^2 after IFRS adoption then this shows no uplift in the value relevance of the accounting information upon the adoption of IFRS.

This study uses the following equation:

$$P_{it+} = \alpha_0 + \beta_1 EPS_{it} + \beta_2 BVPS_{it} + e_{it}$$

where
P_{it+} : stock price at end of month $t + 1$;
EPS_{it} : net income at end of each share period;
$BVPS_{it}$: book value of equity at the end of each share period;
e_{it} : error.

4 RESEARCH RESULT AND DISCUSSION

Descriptive statistics of the research variables used in the value-relevance testing model are presented in Table 1. The model used in this study is the same model used in the study of Barth et al. (2008). The variables used are stock market price (P), net earnings per share (EPS), and equity book value per share (BVPS).

The descriptive statistical results in Table 1 show an increase in average stock price of from 2,242.40 before the adoption of IFRS (2006–2011) to 3,666.27 in the period after IFRS adoption (2012–2016). The average net profit per share decreased from 1,176.12 (before adoption) to 252.60 (after adoption). The average equity book value of shares increased from 653.13 before adoption to 1,662.93 after the adoption of IFRS. The increase in the average values of both the stock price and the equity book value of the shares indicates an improvement in accounting performance in 2012–2016 (after the adoption of IFRS) compared to that in 2007–2011 (before IFRS adoption). The net profit per share decreased, which indicates a decline in accounting performance after adoption.

A study examining the impact of IFRS adoption projected with value relevance is presented in Table 2. The method used in this study is similar to the method developed by Barth et al. (2008), by using the price model estimated by multiple regression for each period. The results of the test in Table 2 show that this model is feasible to use because the F value for each period is significantly greater than 0.05 (i.e. 12.929 for 2006–2011, and 113.751 for 2012–2016).

Testing the improved quality of accounting information proxied by value relevance focused on changes in adjusted R^2 values before and after IFRS adoption. If the value of adjusted R^2 increases then the quality of accounting information also increases after IFRS adoption

Table 1. Descriptive statistics.

Period adoption of IFRS		N	Min	Max	Mean	SD
Before IFRS	LnP	15	88.00	7,890.00	2,242.40	2,246.87
	LnEPS	15	56.00	4,393.00	1,176.13	1,392.88
	LnBVPS	15	84.00	1,873.00	653.13	534.022
	LnValid N	15				
After IFRS	LnP	15	600.00	7,950.00	3,666.27	2,798.10
	LnEPS	15	66.00	480.00	252.60	151.28
	LnBVPS	15	142.00	3,456.00	1,662.93	1,193.53
	LnValid N	15				

Description: LnP = the natural logarithm of the stock price on March 31 in $t + 1$; LnEPS = natural logarithm of the per-share earnings; LnBVPS = the natural logarithm of the book value per share.
(Source: Data processed).

Table 2. Results of multiple linear regression analysis.

Variables	Prior to adoption of IFRS (Years 2007–2011)		After adoption of IFRS (Years 2012–2016)	
	Coefficient	p value	Coefficient	p value
Constants	0.089	0930	−2.890	0014
LnEPS	0.428	0676	7.026	0000
LnBVPS	2.085	0061	6.895	0000
F value	12.929	0000	113.751	0000
Adjusted R^2	0.630		0.942	

Description: LnP = natural logarithm of the stock price on March 31 in t + 1; LnEPS = natural logarithm of the per-share earnings; LnBVPS = the natural logarithm of the book value per share.
(Source: Data processed).

(Barth et al., 2008; Outa, 2011; Chua et al., 2012; Zeghal et al., 2012). Table 2 indicates that there is an increase in adjusted R^2. For 2006–2011 (before adoption), the value of adjusted R^2 was 0.630, and it was 0.942 for 2012–2016 (after adoption). These results show that IFRS adoption in automotive companies in Indonesia had an influence on value relevance as measured by net earnings per share and equity book value per share. This empirical evidence shows that there is a change in the value relevance for automotive companies in Indonesia following the adoption of IFRS. The results of this study are consistent with previous research results (Liu & Liu, 2007; Barth et al., 2008; Paananen & Lin, 2009; Chua et al., 2012; Zeghal et al., 2012; Kargin, 2013; Rohmah & Yuni, 2013; Edvandini et al., 2014; Wulandari & Adiati, 2015), which suggest an increase in value relevance after the adoption of IFRS.

The analysis in Table 2 shows that the share earnings coefficient increased from 0.428 to 7.026 after the adoption of IFRS. The coefficient of book value per share also increased from 2.085 to 6.895. These results indicate that there is an increase in the value of relevance for the earnings per share net and book value per share.

5 CONCLUSION

The purpose of this study is to provide empirical evidence on whether the adoption of IFRS implemented in 2012 improved the quality of financial information from companies on the Indonesia Stock Exchange (IDX). The tests were conducted using a model developed by Barth et al. (2008), that is, information that corresponds to the value-relevance measure in the period before and after adoption of IFRS.

This research used a multiple regression statistical test. The results of the tests show the increased value relevance of financial information after IFRS adoption. This indicator also shows the level of relevance that occurs in net earning per share and book value per share.

The research will be expanded by looking at the impact of IFRS adoption on information quality, with a focus on other industries and individual adopted standards. In addition, the study may also consider variables as moderating variables of the relationship between IFRS adoption and value relevance. Subsequent research may also include different programs of accounting quality services, such as earnings quality and earnings management. This research only uses the data of companies listed on the IDX, with the aim of focusing on one type of industry. By using only one type of data, the variation of data available is not too large the data is very sensitive to changes.

REFERENCES

Agostino, M., Drago, D. & Silipo, D. (2011). The value relevance of IFRS in the European banking industry. *Review Quantity Finance Accounting, 36*(3), 437–457.

Alali, F.A. & Foote, P.S. (2012). The value relevance of international financial reporting standards: Empirical evidence in an emerging market. *The International Journal of Accounting, 47*(1), 85–108.

Armstrong, C.S., Barth, M.E., Jagolinzer, A.D. & Riedl, E.J. (2010). Market reaction to the adoption of IFRS in Europe. *Accounting Review, 85*(1), 31–61.

Barth, M., Landsman, W. & Lang, M. (2008). International accounting standards and accounting quality. *Journal of Accounting Research, 46*(3), 467–498.

Cahyonowati, N. & Ratmono, D. (2012). Adopsi IFRS dan relevansi nilai informasi akuntansi. *Jurnal Akuntansi dan Keuangan, 11*(2), 105–115.

Chandrapala, P. (2013). The value relevance of earning and book value: The importance of ownership concentration and firm size. *Journal of Competitiveness, 5*(2), 98–107.

Chang, Y.L., Liou, C.H. & Chen, Y.H. (2013). The effect of IFRS and the institutional environment on accounting quality in Chinese listed firms. *Journal of American Business Review, 1*(2), 122–127.

Choi, F.D.S., Frost, C.A. & Meek, G.K. (1999). *International accounting* (3rd ed.). Upper Saddle River, NJ: Prentice Hall International.

Chua, Y., Cheong, C. & Gould, G. (2012). The impact of mandatory IFRS adoption on accounting quality: Evidence from Australia. *Journal of International Accounting Research, 11*(1), 119–146.

Deaconu, A., Buiga, A. & Nistor, C.S. (2010). The value relevance of fair value: Evidence for tangible assets on the Romanian market, *Transition Studies Review, 17*(1), 151–169.

Edvandini, L., Subroto, B. & Saraswati, E. (2014). Telaah Kualitas Informasi Laporan Keuangan dan Asimetri Informasi Sebelum dan Setelah Adopsi IFRS. *Jurnal Akuntansi Multiparadigma, 5*(1), 88–95.

Francis, J. & Schipper, K. (1999). Have financial statements lost their relevance? *Journal of Accounting Research, 37*(2), 319–352.

Iatridis, G. (2010). International financial reporting standards and the quality of financial statement information. *International Review of Financial Analysis, 19*(3), 193–204.

Ikatan Akuntan Indonesia. (2017). *Standar Akuntansi Keuangan* per 1 Juni 2017.

Karampinis, N. & Hevas, D. (2011). Mandating IFRS in an unfavorable environment; The Greek experience. *The International Journal of Accounting, 46*(3), 304–332.

Kargin, S. (2013). The impact of IFRS on the value relevance of accounting information: Evidence from Turkish firms. *International Journal of Economics and Finance, 5*(4), 71–80.

Karyadan, I.P.F. & Irwanto, A. (2017). Kualitas Informasi Akuntansi pada Tahap Konvergensi International Financial Reporting Standard. *Jurnal Akuntansi Multiparadigma, 8*(2), 227–429.

Kim, J. & Shi, H. (2012). Voluntary IFRS adoption, analyst coverage and information quality: International evidence. *Journal of International Accounting Research, 11*(1), 19–42.

Kouki, A. (2018). IFRS and value relevance: A comparison approach before and after IFRS conversion in the European countries. *Journal of Applied Accounting Research, 19*(1), 60–80.

Kousenidis, D., Ladas, A. & Negakis, C. (2010). Value relevance of accounting information in the pre- and post-IFRS accounting periods. *European Research Studies, 13*(1), 145–154.

Kwong, L. (2010). The value relevance of financial reporting in Malaysia: Evidence from three different financial reporting periods. *International Journal of Business and Accountancy, 1*(1), 505–527.

Lavi, M.R. (2016). *The impact of IFRS on industry*. Chichester, UK: John Wiley & Sons.

Liu, J. & Liu, C. (2007). Value relevance of accounting information in different stock market segments: The case of Chinese A-, B-, and H-shares. *Journal of International Accounting Research, 6*(2), 55–81.

Ohlson, J.A. (1995). Earning, book value, and dividends in equity valuation. *Contemporary Accounting Research, 11*(2), 661–687.

Outa, E. (2011). The impact of International Financial Reporting Standards (IFRS) adoption on the accounting quality of listed companies in Kenya. *International Journal of Accounting and Financial Reporting, 1*(1), 212–241.

Rohmah, A. & Yuni, R. (2013). Dampak Penerapan Standar Akuntansi Keuangan (SAK) Pasca Adopsi IFRS terhadap Relevansi Nilai dan Asimteri Informasi. *Prosiding Simposium Nasional Akuntansi (SNA) XVI*, Manado.

Scott, W.R. (2012). *Financial accounting theory* (6th ed.). Toronto, Canada: Pearson Education.

Van der Meulen, S., Gaermynck, A. & Willekens, M. (2007). Attribute differences between US GAPP and IFRS earnings: An exploratory study. *The International Journal of Accounting, 42*(2), 123–142.

Wulandari, T.R. & Adiati, A.K. (2015). Perubahan Relevansi Nilai dalam Informasi Akuntansi Setelah Adopsi IFRS. *Jurnal Akuntansi Multiparadigma, 6*(3), 341–511.

Zeghal, D., Chtourou, S. & Fourati, Y.M. (2012). The effect of mandatory adoption of IFRS on earning quality: Evidence from the European Union. *Journal of International Accounting Research, 11*(2), 1–25.

Financial reporting and disclosure

Business Innovation and Development in Emerging Economies – Trinugroho & Lau (Eds)
© 2019 Taylor & Francis Group, London, ISBN 978-1-138-35996-3

Information content of forward-looking disclosure

E. Gantyowati, Payamta, J. Winarna & A. Wijayanto
Economic and Business Faculty, Universitas Sebelas Maret, Surakarta, Indonesia

ABSTRACT: The study examines the relationship between forward-looking disclosures and the cost of equity capital for the banking sector. This research is important because financial and non-financial forward-looking disclosures demonstrate bank capability. Financing online service competition conditions influences the role of forward-looking disclosures for the assessment of bank capability. Three years of data and multiple regression are used to test the relationship. The result shows that forward-looking disclosures (non-financial) can significantly decrease the cost of equity capital. There is an interesting finding that forward-looking disclosures are still useful, although other information is available in online media, and they have a role in relation to bank capability.

Keywords: forward-looking disclosure, cost of equity capital, concentrated ownership. JEL-Code: M41 accounting- accounting

1 INTRODUCTION

The rapid development of information technology has led to the increased availability of information for users. This condition is very profitable for investors. Their investment may be better because of this information support. For example, information about a firm can be searched for on the firm's website. Both good news and bad news can be obtained from online media.

On the other hand, policymakers face major challenges, as to whether the information presented can be trusted by the users or not. The information from regulators is usually based on high standards and therefore investors or potential investors believe it. The weakness of information from the regulators is that it is usually slower than information generally available to the public.

In 2018, the public was shocked, when one of the banks revised its financial statements of the previous three years: 2015, 2016 and 2017 (Sugianto, 2018). The financial statements we are audited by one of the big four affiliates. The bank revised its 2016 net profit to Rp183.56 billion, down from a previous figure of Rp1.08 trillion. The biggest decrease was in the proportion of provision and commission income, which is income from credit cards. This revenue decreased from Rp1.06 trillion to Rp317.88 billion (Rachman, 2018).

In addition to credit card issues, a revision also occurred in financing a subsidiary related to the addition of reserve balances for certain debtor impairment losses. As a result, the allowance for impairment losses on financial assets was revised and increased from Rp 649.05 billion to Rp797.65 billion. This bank was actually "punished" for this incident. It has revised down its equity at the end of 2016 by: Rp2.62 trillion, from Rp9.53 trillion to Rp6.91 trillion. Profit balance was downward from Rp2.62 trillion to Rp5.52 trillion, because the profit reported in the previous period is incorrect.

The Capital Adequacy Ratio (CAR) revision was left at 11.62%. Another thing that affected the decline in CAR was its Non-Performing Loan (NPL) ratio. As a result OJK (the Financial Services Authority) began conducting an inspection of this case (Rachman, 2018).

This case shows that information from other media is quickly rapidly available, but the trueth information will be submitted in a formal report. This research tries to examine non-financial forward-looking disclosure information content by seeing its effect on the cost of equity capital. The annual banking report becomes information that represents high standards because it is governed by a basic reporting rule. We suspect that the adoption of these high standards means that the disclosures presented in the annual report remain as the information that users rely on. Therefore forward-looking disclosure as part of the annual report is predicted to have information content that can lower the cost of capital. Investors will ask for a lower return if future corporate performance can be estimated on the basis of reliable information.

Studies linking voluntary disclosure to costs of capital have inconsistency in their results. Studies that indicates a negative correlation between voluntary disclosure and the cost of capital include Lang & Lundholm (2000), Botosan & Plumlee (2002), Francis et al., (2005, 2008), Zhao et al. (2009), Gao (2010), Lopes & De Alencar (2010), and Cheynel (2013), While studies that shows a positive or insignificant relationship are those by Botosan (1997), Botosan and Plumlee (2002), Armitage and Marston (2008), Francis et al. (2008), Gao (2010), Lopez (2011), and Cheynel (2013). Thus, within these studies inconsistent results still occur.

The studies mentioned above combine historical disclosure, current disclosure, and forward-looking disclosure. The study of Dima et al. (2013) finds that financial and non-financial disclosures are positively associated with the Price-to-Earnings Ratio (PER). As Gietzsman (2006) suggests, information on strategy management, including plans, strategies and performance, new products, and research and development, are considered to be in a forward-looking scope. Future information, according to Gietzmann (2006), is important information for investors because investors are more concerned with information about the future than information about the past. Beretta and Bozzolan (2006) presented three kinds of disclosure, namely: future events, decisions, opportunities and risks that can have a likely effect on future results; vision, strategies and objectives expressed by management; and explanation of past events, decisions, facts and results can have a significant impact on future results. These are categorized as forward-looking disclosures.

Specific research on forward-looking disclosure is still not yet comprehensive. Bravo (2016) states that forward-looking disclosure involves the current plans and future forecasts that allow users to assess company performance. Forward-looking disclosure include financial and non-financial information. Some studies only take some part of the forward-looking disclosure. Johnson et al. (2001), for example, only uses forecasting profit and sales forecasting as forward-looking disclosure. Bravo (2016) uses only financial information, by measuring the number of sentences containing future content. Bravo (2016) also judges from the content of financial information and counts its number of its sentences.

Therefore, this research reinforces previous research using non-financial information forward-looking disclosure items. Forward-looking disclosure is related to the cost of equity capital because, as mentioned earlier, forward-looking disclosure is more beneficial to investors than historical data. The cost of equity capital is the return demanded by investors as a return for their investment. Non-financial forward disclosure is the focus of this study, based on the finding of Gantyowati et al. (2017). The result is that a non-financial disclosure is more significant than a financial disclosure.

Mandatory disclosure is regulated in the current financial accounting standards, which have adopted the International Financial Reporting Standards (IFRS). Disclosure is also stipulated in the Monetary Services Authority's decree on the preparation of financial statements. Therefore, the importance of forward-looking disclosure will provide inputs to the compilers of the Financial Accounting Standards, as well as the Capital Market Supervisory Agency.

Forward-looking disclosure is considered as important private information because it shows a firm's prospects and beliefs about the firm's performance in the future. This information is important to reduce the information asymmetry between investors, where it can also be used as the basis for determining the investors' rate of required return over their investments.

Share ownership in developing countries tends to be concentrated (Claessens dan Fan, 2002). This leads to higher information asymmetry. The purpose of this study is to obtain empirical evidence about the role of non-financial forward-looking disclosure in reducing the cost of equity capital in the capital market. Shareholders use information as a part of return parameter determination.

This study is expected to obtain findings regarding the usefulness or content of forward-looking disclosure information in relation to decreasing investors' expected return or the cost of equity capital. The contribution of this study includes, first, that this research makes an academic contribution to the development of accounting study, especially in presenting the type of private information for users in the form of forward-looking disclosure. Second, it makes a practical contribution to showings the benefits of disclosure information from an annual report as a parameter for determining returns.

2 LITERATURE REVIEW

The agency problem is enlarged through a condition of information asymmetry. This happens when one party has more information than another party, or one party has additional private information that others do not, as described by Cheng et al., (2006) and Chang et al., (2008). Information imperfection due to information asymmetry can cause a malfunction in the market. Agents as managers of the company know more internal information and more about the prospects of the company in the future than do its shareholders (Jensen & Meckling, 1976).

The agent will provide information on the future prospects of the company to demonstrate the agent's confidence in the firm's performance. Therefore, the investor needs the information to assess whether the agent's signals are appropriate to what should have been delivered. Agency theory (Jensen & Meckling, 1976) and signaling theories (Spence, 1973) underlie the relationship between independent variables and dependent variables.

Studies that show the relationship between non-financial forward-looking disclosure and the cost of equity capital include the research of Gao (2010), Lopes and De Alencar (2010), Cheynel (2013), Abed et al., (2016), and Bravo (2016). Based on these previous studies, we formulate a hypothesis as follows:

H_1: Non-financial forward-looking disclosure negatively relates to the cost of equity capital.

3 RESEARCH METHOD

3.1 *Samples*

The population of this study includes all banking firms in Indonesia. Banking firms are selected because the specificity of the overall firms' characteristics is closely related to disclosure, and the banking firms are Indonesia's largest investment after gold (Murdaningsih, 2018). Intake of data took three years, due to considerations of data updating: determining the value of the cost of equity capital using the Price-Earnings Growth (PEG). Model requires two years of data, so it needs data from the 2016 annual report that is published in 2017. The sampling technique is purposive sampling. The earnings per share calculations are obtained from annual report banks. Disclosure data is obtained from the firm's annual reports listed on the Indonesia Stock Exchange (IDX), and the firms' annual reports are obtained from the IDX website (www.idx.co.id) and the firms' websites.

4 VARIABLES

4.1 *Cost of equity capital*

The cost of capital uses only the cost of equity capital. This study uses a cost of equity capital measurement based on Francis et al. (2005), Muino & Trombetta (2009), Lopez (2011),

Baginski & Rakow (2012), Gray et al. (2013), Mazzotta & Veltri (2014), and Kim et al. (2014), which is the *ex-ante* cost of equity capital or PEG model.

There are two types of cost of equity measurement, namely implied cost of equity measure (*exante* cost of equity capital), and *expost* realized return or measurement of expected return based on realized return/realized based return measure (Bhattacharya et al., 2012). The PEG model is an implied cost of equity measure model that has a high degree of validity (Botosan & Plumlee, 2002. The determination cost of equity capital, with PEG ratio model, is performed using the following equation:

$$r = \sqrt{(eps_2 - eps_1)/P_o} \qquad (1)$$

5 DISCLOSURE

Hasan and Marston (2010) wrote that the disclosure index uses a widely used self-constructed disclosure index because there is no theory of the type and amount of disclosure included in the index. The index also assumes the quantity of disclosure as being a proxy for the quality of disclosure. The advantage of using an unweighted index is that each of the items is considered to have an equal importance; thus it is becomes more objective because it does not use considerations which may lead to subjectivity (Chavent et al., 2006). Abed et al. (2016) state that both the unweighted index and the weighted index give the same results.

Items of forward-looking disclosure of non-financial information are based on Bravo (2016); Gietzmann (2006). The items of disclosure are broken down into vision/missions, long-term goals, future strategy, research commitments and development, and new products.

The forward-looking disclosure measure is corresponds to Bravo (2016). Bravo makes judgments from the financial information content, counting the number of sentences containing future content. The financial information presented in table form is considered as one sentence. Abed et al. (2016) states that volume measurements are better than measurements in the existing and non-disclosure modes. This study uses manual analysis based on the study of Abed et al. (2016). One of their result showed that manual analysis gave the highest R-squared value, and is more significant as an independent variable than computer analysis. They said that differences in sentence structures may be the source of the discrepancy.

6 DATA ANALYSIS

Hypothesis testing examines the direct relationship between forward-looking disclosures and the cost of equity capital. Hypothesis testing is performed with multiple regression (Ghozali, 2011). Variable relationships are illustrated in the following equation:

$$COEC = b_1 NFFL + b_2 X_1 + b_3 X_2 + e_2 \qquad (2)$$

where
b_1 = coefficient of forward-looking disclosure
$b_2\, b_3$ = coefficient of control variable
COEC = cost of equity capital
NFFL = non-financial forward-looking disclosure
X_1 = control variable—total assets
X_2 = control variable—leverage

The control variables includes total company assets (X2), and leverage (X3). The variable of total assets is used as a proxy for the variable of firm size. This is based on the research of Lang and Lundhlom (1996), Kanagaretnam et al. (2007), and Akhtaruddin et al. (2009). Leverage is calculated from the ratio of total debt total assets: it derives from the studies of Cheng, Collins, & Huang (2006), and Brown and Hillegeist (2007).

Table 1. Statistic descriptive of dependent, independent, control variable and cost of equity capital variable with the equation of PEG model as COEC = b + b_1 NFFL + b_4 TA + b_5 Lev + e.

Variable	Mean	Std. deviation	N
Cost of Equity Capital (COEC)	0.1665	0.1982	135
Non-Financial-Forward-Looking Disclosure (NFFL)	27.563	13.026	135
Total Assets (TA)	8.3054	2.119	135
Leverage (Lev)	0.8312	0.1732	135

Table 2. The hypothesis-testing result, cost of equity with the equation of PEG model as COEC = b + b_1 NFFL + b_4 TA + b_5 Lev + e; COEC PEG = $\sqrt{(EPS_2 - EPS_1)/P_0}$.

Relationship between variables	Coefficient	t-statistic	Prob.	Direction of relationship	Description
Constant	−0.038	−0.400	0.690		
Non-Forward-Looking Disclosure (NFFL)	−0.004	−3.508	0.001	negative	H1 accepted
Total Assets (TA)	0.032	4.222	0.000		
Leverage (Lev)	0.054	0.715	0.476		

7 RESEARCH RESULT

7.1 *Descriptive statistics*

The results of classic assumptions of regression testing showed that there are no multicollinearity, heteroscedasticity, and autocorrelation issues occurring, as well as that the data is normally distributed. Each is tested with tolerance and variance inflation factor (VIF) value, scatter plot, the Durbin-Watson statistic, and the Kolmogorov-Smirnov test.

8 HYPOTHESIS-TESTING RESULT

Table 2 shows that with a t-value of −3.508 and a p-value of 0.001 non-forward-looking disclosure has a negative relationship to the cost of equity capital. The research model is also sufficient according to the F-test value of 16.501 and p-value of 0.000. The R-squared value is 25.8%, which means that non-forward-looking disclosure only explains 25.8% of the relationship to cost of equity.

9 ADDITIONAL ANALYSIS

We also use additional analysis with a one-year examination. The results of the return test using a one-year examination for the cost of the equity capital variable shows a result consistent with the PEG model, as presented in Table 3. Therefore, these test results strengthen the outcome of the main test results.

10 DISCUSSION OF RESEARCH RESULT

10.1 *The relationship between non-financial-forward-looking disclosure and cost of equity capital*

The results of the hypothesis testing are shown below. The slope coefficient or intercept of independent variables of non-financial-forward-looking disclosure indicates a negative

Table 3. Testing result of H_1 a year examination; COEC = b + b_1 Disc + b_2 TA + b_3 Lev + e.

Year	Disclosure coefficient (Disc)	Prob.	Direction of relationship	Description
t1	−0.005	0.010	negative	H_1 accepted
t2	−0.006	0.002	negative	H_1 accepted
t3	−0.0002	0.050	negative	H_1 accepted

significant value. It shows how much influence one independent variable has on the dependent variable by assuming another variable is constant, in comparison to a p-value of 0.001. The error rate received is 5%, and thus H_1 was accepted. It means that the independent variable has an influence on the dependent variable, or the disclosure of forward-looking information has a negatively influence on the cost of equity capital.

The coefficient value of R-squared is 0.27 and R^2 adjusted is 0.258. The coefficient of determination measures the extent of the model's ability to explain variations in the dependent variable. The coefficient of determination ranges between zero and one. The greater the value of the coefficient of determination then the greater the ability of independent variables to explain the variation of the dependent variable. The test results show that the ability of independent variables explains the variation of the dependent variable by 25%, thus the value of this coefficient is quite low, and many other variables influence the cost of equity capital besides that of disclosure.

Forward-looking disclosures serve as information that is used by the controlling shareholders to influence the investors' investment decision-making in determining the returns demanded. Therefore, future disclosure is one of the important information types to be presented by issuers to influence investors in making investment decisions. The test of one hypothesis was re-examined using a year-to-year test. The results are consistent with the model, where the results are also supported by the average disclosures that have been generally presented, which has gained a mean score of 28. There are12 items; therefore one item contains more than two sentences.

The results of the research support the existing literature (Bravo, 2015). The results of the research also strengthen the study of agency theory and signal theory. The presentation of forward-looking financial information will reduce the conflict of interest between the controlling shareholders and the minority shareholders because the information is able can help investors in obtaining information that can be used to predict future returns and reduce the risks that will be faced by the investors.

The statistical F-test, often called the significance test, denotes whether all independent variables can explain the dependent variable or not. If the result is found to be significant then the equation model fits. Test results show the F-value as 16.501 and the value of probability F of 0.000. Therefore, it can be explained that all dependent variables are able to influence the independent variables. This means that forward-looking disclosure variables, earnings volatility, total assets, and leverage together affect the variable cost of equity capital.

Briefly, it can be explained that the first hypothesis which states that the non-financial forward-looking disclosure is negatively related to the cost of equity capital is accepted.

11 DISCUSSION OF CONTROL VARIABLES

Two control variables were used in this study. These control variables are firm size, and leverage. According to the theoretical framework of signal theory, the control variable are related to the cost of equity capital. Each variable other than forward-looking disclosures which theoretically able to influence the cost of equity capital must be separated, because it can be disrupt the result of testing of forward-looking disclosures on the cost of equity capital. The separated variables are treated as control variables.

Total assets, as the first control variable, show that firms that are categorized as large firms have a tendency to reveal more information than small firms, so that the investors ask for smaller returns. The second control variable is leverage. Leverage represents a corporate liability to a third party. The bigger the leverage, the greater an investor's demand because the investor faces greater risks, therefore investors will require higher returns.

12 CONCLUSION

This study aims to test the information content of forward-looking information disclosure for decreasing the cost of equity capital. The result shows that forward-looking disclosure correlates significantly with the cost of equity capital. The greater the disclosure, the lower the cost of equity capital for the investors. These findings strengthen the previous studies of Gao (2010), Lopes and de Alencar (2010), Cheynel (2013), and Bravo (2016).

The Contribution to the literature of accounting and finance is to confirm previous research, which states that forward-looking information is needed by investors to help in predicting future performance or reducing the uncertainty of risks. Furthermore, its findings will provide more valuable information in assisting investors in their investment decision-making. Nevertheless, there are still further challenges to analyze each type of future information and establish which most influences the decrease in the cost of equity capital.

REFERENCES

Abed, S., Al-Najjar, B & Roberts, C. (2016). Measuring annual report narratives disclosure: Empirical evidence from forward-looking information in the UK prior to the financial crisis. *Managerial Auditing Journal, 31*(4/5), 338–361.

Akhtaruddin, M., Hossain, M.A., Hossain, M. & Yao, L. (2009). Corporate governance and voluntary disclosure in corporate annual reports of Malaysian listed firms. *Journal of Applied Management Accounting Research, 7*(1), 1–19.

Armitage, S. & Marston, C. (2008). Corporate disclosure, cost of capital and reputation: Evidence from finance directors. *The British Accounting Review, 40*(4), 314–336.

Baginski, S.P. & Rakow, K.C., Jr. (2012). Management earnings forecast disclosure policy and the cost of equity capital. *Review of Accounting Studies, 17*(2), 279–321.

Beretta, S. & Bozzolan, S. (2006). Quality vs quantity: The case of forward-looking disclosure (Working paper). *In Cardiff Business School Financing Reporting and Business Communication Conference,* (pp.1–49).

Bhattacharya, N., Ecker, F., Olsson, P.M. & Schipper, K. (2012). Direct and mediated associations among earnings quality, information asymmetry, and the cost of equity. *The Accounting Review, 87*(2), 449–482.

Botosan, C.A. (1997). Disclosure Level and the Cost of Equity Capital, *The Accounting Review, 73*(2), 323–349.

Botosan, C.A. & Plumlee, M.A. (2002). Re-examination of disclosure level and the expected cost of equity capital. *Journal of Accounting Research, 40*(1), 21–42.

Bravo, F. (2016. Forward-looking disclosure and corporate reputation as mechanisms to reduce stock return volatility. *Spanish Accounting Review, 19*(1), 122–131.

Brown, S. & Hillegeist, S.A. (2007). How disclosure quality affects the level of information asymmetry. *Review Accounting Study, 12*(2–3), 443–447.

Chang, M., D'Anna, G., Watson, I. & Wee, M. (2008). Does disclosure quality via investor relations affect information asymmetry? *Australian Journal of Management, 33*(2), 375–390.

Chavent, M., Ding, Y., Fu, L., Stolowy, H. & Wang H. (2006). Disclosure and determinants studies: An extension using the Divisive Clustering Method (DIV). *European Accounting Review, 15*(2), 181–218.

Cheng, E.C., Courtenay, S.M. & Krishnamurti, C. (2006). The impact of increased voluntary disclosure on market information asymmetry, informed and uninformed trading. *Journal of Contemporary Accounting & Economics, 2*(1), 33–72.

Cheng, C.S.A., Collins, D. & Huang, H.H. (2006). Shareholder rights, financial disclosure and the cost of equity capital. *Review of Quantitative Finance and Accounting, 27*(2), 175–204.

Cheynel, E. (2013). A theory of voluntary disclosure and cost of capital. Review of Accounting Studies, *18*(4), 987–1020.

Claessens, S. & Fan, J.P.H. (2002). Corporate governance in Asia: A survey. *International Review of Finance*, *3*(2),71–103.

Dima, B., Cuzman, L., Dima, S. & Saramat, O. (2013). Effects of financial and non-financial information disclosure on prices' mechanisms for emergent markets: The case of Bucharest Stock Exchange. *Accounting and Management Information Systems*, *12*(1), 76–100.

Francis, J.R., Khuana, I.K. & Pereira, R. (2005). Disclosure incentives and effects on cost of capital around the world. *The Accounting Review*, *80*(4),1125–1162.

Francis, J., Nanda, D. & Olsson, P. (2008). Voluntary disclosure, earnings quality, and cost of capital. *Journal of Accounting Research*, *46*(1), 53–99.

Gantyowati, E., Payamto, Winarna, J. & Wijayanto, A (2017). *Forward-looking disclosure and cost of equity capital*. PUPT report, (unpublished).

Gao, P. (2010). Disclosure quality, cost of capital and investor welfare. *The Accounting Review*, *85*(1), 1–29.

Ghozali, I. (2011). *Model Persamaan Struktural, Konsep dan Aplikasi (Structural Equation Model, Concepts and Applications)*, Semarang, Indonesia: UNDIP.

Gietzmann, M. (2006). Disclosure of timely and forward-looking statements and strategic management of major institutional ownership. *Long Range Planning*, *39*(4), 409–427.

Gray, S.J., Kang, T. & Yoo, Y.K. (2013). National culture and international differences in the cost of equity capital. *Management International Review*, *53*(6), 899–916.

Hasan, O. & Marston, C. (2010). Disclosure measurement in the empirical accounting literature- A review article. Retrieved from http://ssrn.com/abstract = 1640598.

Hoque, Z. (Ed.). (2006). *Methodological issues in accounting research: Theories and methods*, London, UK: Spiramus.

Jensen, M.C. & Meckling, W.H. (1976). Theory of the firm: Managerial behavior agency costs and ownership structure. *Journal of Financial Economics*, *3*(4), 305–360.

Johnson, M.F., Konznik, R. & Nelson, K.K. (2001). The impact of securities litigation reform on the disclosure of forward-looking information by high technology firms. *Journal of Accounting Research*, *39*(2), 297–327.

Kanagaretnam, K., Lobo, G.J. & Whalen, D.J. (2007). Does good corporate governance reduce information asymmetry around quarterly earnings announcements? *Journal of Accounting and Public Policy*, *26*(4), 497–522.

Kim, JB., Shi, H. and Zhou. J. (2014). International Financial Reporting Standards, institutional infrastructures, and implied cost of equity capital around the world. *Review of Quantitative Finance and Accounting*, *42*(3), 469–507.

Lang, M.H. & Lungdlom, R.J. 1996. Corporate disclosure policy and analyst behavior. *The Accounting Review*, 71(4), 467–492.

Lang, M.H. & Lungdlom, R.J. (2000). Voluntary Disclosure and Equity Offerings: Reducing Information Asymmetry or Hyping The Stock? *Contemporary Accounting Research*, *17*(4), 623–663.

Lopes, A.B. & de Alencar, R.C.D. (2010). Disclosure and cost of equity capital in emerging markets: The Brazilian case. *International Journal of Accounting*, *45*(4), 443–464.

Lopez, E. (2011). *Voluntary disclosure and the cost of equity capital in the Netherlands* (Thesis, Erasmus University).

Mazzotta, R. & Veltri, S. (2014). The relationship between corporate governance and the cost of equity capital. Evidence from the Italian stock exchange. *Journal of Management & Governance*, *18*(2), 419–448.

Muino, F. & Trombetta, M. (2009). Does graph disclosure bias reduce the cost of equity capital? *Accounting and Business Research*, *39*(2), 83–102.

Murdaningsih, D. (2018). *Orang Indonesia lebih Suka Menabung Dibanding Investasi*. Retrieved from https://republika.co.id/berita/ekonomi/keuangan/18/02/11.

Rachman, F.F. (2018). *Bank Bukopin Permak Laporan Keuangan, Ini Kata BI dan OJK (Bukopin make restatement financial statements, Indonesia central bank and Financial Service Authority responses)*, Retrieved from https://finance.detik.com/moneter/d-4002904/ojk-mulai-periksa-laporan-keuangan-bank-bukopin-yang-dipermak.

Spence, M. (1973). Job market signaling. *Quarterly Journal of Economics*, *87*(3), 355–374.

Sugianto, D. (2018). *OJK Mulai Periksa Laporan Keuangan Bank Bukopin yang Dipermak. (Financial Service Authority begin check restatement of Bukopin financial statements)*, Retrieved from https://finance.detik.com/moneter/d-3994551/bank-bukopin-permak-laporan-ini-kata-bi-dan-ojk.

Zhao, Y., Davis, M. & Berry, K.T. (2009). Disclosure channel and cost of capital: Evidence from open vs closed conference calls. *Review of Accounting and Finance*, 8(3), 253–278.

Business Innovation and Development in Emerging Economies – Trinugroho & Lau (Eds)
© *2019 Taylor & Francis Group, London, ISBN 978-1-138-35996-3*

Sustainability report, financial performance and investor reaction: A partial least square perspective of Indonesian public companies

I. Maulidya, I. Ulum, T. Nur & E.D. Wahyuni
Accounting Department, University of Muhammadiyah Malang, Indonesia

ABSTRACT: The purpose of this study was to examine the effect of financial performance on investor reaction with a sustainability report as the mediating variable. The sample
was drawn from 34 Indonesian companies that disclosed a corporate sustainability report.
Partial Least Square (PLS) was used to analyze direct and indirect effects. The results showed
that financial performance (measured by profitability and leverage) have positive and negative effects on investor reaction. In addition, the sustainability report, which is corporate
social responsibility disclosure, mediates the relationship of financial performance and investor reaction. This means that investors in the Indonesian capital market pay attention to a
company's actions concerning social activities disclosed in a sustainability report. Otherwise,
the greatest attention is still on the financial performance.

Keywords: corporate social responsibility disclosure, Indonesian public companies, investor
reactions; sustainability report

1 INTRODUCTION

The availability of funds and access to funding sources will affect the sustainability and
growth opportunities of a company. In the modern economy, one of the sources of external
funding is the capital market (Apriliastuti and Andayani 2015). In 2011, the issuers listed on
the Indonesia Stock Exchange included 442 companies, this number increased to 526 issuers in 2015. In an equity market, investors (lenders) require information about the company
(borrower) to assess the ability and companies' performance. The information published by
the companies would lead to changes in the price of securities and stock trading transactions
(Hartono 2008). The information used by investors before making a decision in general are
the financial statements, but this time of economic decision makers not only see the financial
performance of an entity, because the conclusion of good or bad performance of a company
is not enough to just be seen from the profits generated (Cheng and Christiawan 2011).

Corporate Social Responsibility (CSR) is disclosed in the corporate sustainability report,
often called a sustainability report and based on the guidelines of the Global Reporting Initiative
(GRI). Disclosure of CSR is believed to improve the performance of a company and can be an
added value for companies because the company conducting the CSR will prioritize an aspect of
sustainability into the strategy and operations of the company. Tjiasmanto and Juniarti (2015),
stated that the disclosure of CSR has a significant role in giving a company competitive advantage
for investors that no longer just to look at the financial performance in order to make decisions.

This study examined the effect of company financial performance on investor reaction by
the disclosure of CSR as a mediator for companies listed on the Indonesian Stock Exchange
(BEI) who made a sustainability report in 2015. The companies listed on the BEI that have
a sustainability report were selected as a sample because the sustainability report contains
the company's performance in three aspects: economic, environmental and social. Implementation of a CSR program is a form of implementation of Good Corporate Governance
(GCG). From the explanation provided above, the researcher is interested in addressing this

problem by taking the title 'Effect of Financial Performance for Investor Reaction with Corporate Social Responsibility Disclosure as a Mediator'.

2 HYPOTHESIS DEVELOPMENT

2.1 *Effect of financial performance on investor reaction*

Profitability, proxied by the ROA, shows a company's ability to exploit assets to generate profits. In a study by Tjiasmanto and Juniarti (2015), the higher the ROA of a company, the more efficient the company in utilizing its assets to generate income so that it provides a high return. Based on signaling theory is the information published as an announcement would give a signal for investors in taking investment decision (Hartono 2007). The higher ROA information which contains a positive influence and that information are considered as a good signal (good news). Contrary to profitability, leverage proxied with DER, which is the proportion of total debt to total liabilities, can show how great the company's dependence on debt to finance its operations is. A lower DER indicates good financial performance, while a higher DER indicates poor financial performance, but high or low DER still affect investors' reactions (Rafik and Asyik 2013).

H1: Financial performance affects investor reaction.

2.2 *Effect of financial performance on CSR disclosure*

Financial performance is proxied by profitability and leverage. In this study, it is suspected to affect disclosure of CSR. Results of research conducted by Januarti and Wardani (2013) showed that profitability, as measured by ROA, had a significant effect on the disclosure of social responsibility. Companies that have greater profitability are likely to have wider social responsibility disclosure. Leverage is used to provide an overview of the capital structure of the company, so the level of risk of non-collection of debt can be seen. Therefore, companies with high leverage ratios have an obligation to disclose more social responsibility. Leverage can be considered as additional information needed to dispel the doubts of investors. Based on stakeholder theory, the 'viability and success of the company' depend on how the company meets the needs of stakeholders in economic aspects to generate profit and non-economically to the social performance of companies. Disclosure of CSR is seen as a dialog between companies and stakeholders.

H2: Financial performance has an effect on corporate social responsibility disclosure.

2.3 *Corporate social responsibility disclosure in mediating relationships between financial performance and investor reaction*

Disclosure of CSR is a thing that must be done by a business entity, as it provides additional information for potential investors before making an investment decision. Munawaroh and Priyadi (2014) state that the relationship between the profitability of a company with a CSR disclosure has been postulated (basic assumptions) to reflect the view that the social reaction requires a managerial style, the greater benefits will be increase the company's ability to pay dividend. Thus, profitability is an important consideration for investors in investment decisions. For a company with a lot of debt, increased profits will strengthen the position and security of bondholders more than shareholders (Dewi and Sitinjak 2009). However, leverage can be considered as additional information needed to dispel doubtful investors. Corporate Social Responsibility Disclosure (CSRD) can be seen as a dialog between companies and stakeholders. Because the leverage could be considered as additional information needed to dispel the doubts of investors, companies with a high leverage ratio have an obligation to make a wider disclosure than companies with low leverage ratios. Profitability and leverage can generate a positive and/or negative response for investors, but it is expected that additional information (such as corporate sustainability report) can strengthen the influence of financial performance and investor reaction.

H3: Corporate social responsibility disclosure is able to mediate the relationship between financial performance and investor reaction.

3 METHOD

This research is associative research which is a type of research that aims to identify and analyze the influence of the relationship between a variable with other variables (Ulum and Juanda 2016). This study aims to determine and analyze the influence of financial perform-ance, that is profitability and leverage, on investor reaction through CSRD as a mediator variable. The sample selection for this study was by means of purposive sampling.

The dependent variable in this study is the abnormal return that measured with expected return and market-adjusted model. While the independent variable was the financial per-formance proxied by profitability and leverage. CSRD was used as a mediator variable. The data analysis technique was by means of Structural Equation Modeling (SEM) with a SEM-based variance approach, better known as Partial Least Squares (PLS).

4 RESULTS AND DISCUSSION

The first hypothesis was examined by testing the direct effect between financial performance and investor reaction. This stage is also a requirement of the first step in research on the effects of mediation based on a two-stage regression (Baron and Kenny 1986).

Figure 1 shows that the value of the coefficient is −0365 with $p < .01$ (***). Financial performance was proxied by profitability and leverage jointly affecting investor reaction. Long-term investors would be very concerned about the level of profitability of a company, because remaining profitable can predict the level of profits in the form of dividends (Aini 2015). Investors are more likely to get a higher profit with less risk. Based on signaling theory, such information can be used by investors to consider and determine whether to invest or not, so it can be concluded that financial performance has an effect on investor reaction.

To answer the second and third hypotheses, indirect effect testing was used. Testing the effect of financial performance for the CSRD and testing the effect of the financial perform-ance for the investor reaction by entering the intervening variables can be seen in Figure 2.

Figure 2 indicates that the coefficient value of the indirect effect of Kin_Keu-CSRD is significant at p 0.051 0.356 (**). Companies with a high level of profitability have more

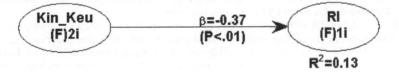

Figure 1. Output warpPLS—direct effect.

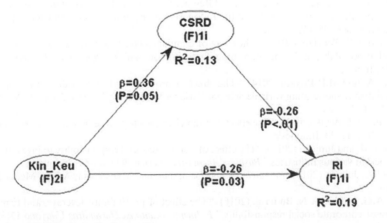

Figure 2. warpPLS Output—indirect effect.

resources for a broader CSRD (Januarti and Wardani 2013). Disclosure of CSRD is seen as a dialog between companies and stakeholders, so that the corporate sustainability report of leverage can be considered as additional information required to convince investors and creditors with more detailed disclosure (Yanti and Budiasih 2016).

Finally, the estimation of the indirect effects of financial performance showed investors reaction as -0.262 coefficient values with significant values on ρ 0.029 (**) and estimated direct effect on the financial performance of investors reaction showed significant coefficient on ρ 0.003 -0.365 (***). Thus, CSRD can become a mediator between financial performance and investor reaction. CSR can be one of the company's strategies and company's tools to show its responsibility to protect the environment, company to its overall responsibility for environmental and social. With the disclosure of CSR, companies can gain social legitimacy and maximize their long-term profit (Herry and Ariyanto 2012).

5 CONCLUSION

Financial performance proxied with profitability and leverage, all affect investor reaction. However, profitability, which is illustrated by ROA, has a positive effect, while the leverage proxy, which is illustrated by the DER, has a negative effect on investor reaction. The positive effect on the financial performance of CSRD was proven to be a variable that can mediate the relationship between financial performance and investor reaction.

REFERENCES

Aini, A.K. (2015). "The effect of company's characteristics to corporate social responsibility at Indonesian companies in LQ-45 index." Kinerja **12**(1), 221–232, Indonesia.
Apriliastuti and Andayani (2015). "The effect of financial performance and firm's size to investor reaction." *Jurnal Ilmu & Riset Akuntansi* **4**(12), 89–101, Indonesia.
Baron, R.M. and D.A. Kenny (1986). "The moderator–mediator variable distinction in social psychological research: Conceptual, strategic, and statistical considerations." *Journal of personality and social psychology* **51**(6): 1173–1182.
Cheng, M. and Y.J. Christiawan (2011). "The effect of corporate social responsibility to abnormal return." *Jurnal Akuntansi dan Keuangan* **13**(1), 132–143, Indonesia.
Dewi, R.R. and M. Sitinjak (2009). "The effect of companies characteristics to earning response coefficient with corporate social responsibility as a intervening variable at the Indonesia manufacturing companies." *Jurnal Informasi, Perpajakan, Akuntansi Dan Keuangan Publik* **4**(2), 76–85, Indonesia.
Hartono, J. (2007). Research method for business: misunderstanding and experiences. Yogyakarta, BPFE, Indonesia.
Hartono, J. (2008). Portfolio theory and stock investment analysis. Yogyakarta, BPFE, Indonesia.
Herry, Y. and S. Ariyanto (2012). "Analysis the differences of profitability level before and after the disclosure of corporate social responsibility at the Indonesia mining and pharmaceutical companies." *Binus Busines Review* **3**(2), 55–67, Indonesia.
Januarti, I. and N.K. Wardani (2013). "The effect of companies characteristics to the disclosure of corporate social responsibility at the Indonesia manufacturing companies." *Diponegoro Journal of Accounting* **2**(2): 1–15, Indonesia.
Munawaroh, A. and M.P. Priyadi (2014). "The effect of profitability to firm value with corporate social responsibility as a moderating variable Sebagai Variabel Moderating." *Jurnal Ilmu dan Riset Akuntansi* **3**(4), 64–72, Indonesia.
Rafik, D.P. and N.F. Asyik (2013). "The effect of financial ratios to market reaction." *Jurnal Ilmu & Riset Akuntansi* **1**(1), 11–20, Indonesia.
Tjiasmanto, V.M. and Juniarti (2015). "The effect of corporate social responsibility to investor response at the Indonesian various industries." *Business Accounting Review* **3**(1): 66–67, Indonesia.
Ulum, I. and A. Juanda (2016). *Research method for accounting.* Yogyakarta, Aditya Media Publishing, Indonesia.
Yanti, N.K.A.G. and I.G.A.N. Budiasih (2016). "The effect of profitability, leverage, and firm size to the disclosure of corporate social responsibility." *E-Jurnal Akuntansi Universitas Udayana* **17**(3), 233–241, Indonesia.

Business Innovation and Development in Emerging Economies – Trinugroho & Lau (Eds)
© *2019 Taylor & Francis Group, London, ISBN 978-1-138-35996-3*

Corporate governance supervision aspect and corporate characteristics on environment disclosure (case study in manufacturing companies in IDX 2012–2016 period)

R. Mustafa Zahri & Muthmainah
Faculty of Economics and Business, Universitas Sebelas Maret, Indonesia

ABSTRACT: This research aims to obtain empirical evidence about the role of corporate governance, and its aspects and company characteristics on environmental disclosure. The independent variables tested in this research consisted of corporate governance aspects (proportion of independent commissioners, ethnicity of the president of the commissioners, educational background of the president of the commissioners, frequency of commissioners' meetings, proportion of independent members of an audit committee, frequency of audit committee meetings, and gender diversity of the commissioners), and company characteristics (profitability, leverage, and size), while the dependent variable of environmental disclosure was measured using Global Reporting Index 4.0. The sample researched consisted of 89 manufacturing companies listed on the Indonesia Stock Exchange from 2012 to 2016. This paper is a quantitative study and the sampling method used was purposive. The data on corporate governance aspects was collected from annual reports and the analysis technique was multiple regression. The results of this research showed that the ethnicity of the president of the commissioners had a negative significant effect, the proportion of independent audit committee members had a positive significant effect, and the frequency of audit committee meetings had a positive significant effect in environmental disclosure. For company characteristics, only leverage had a negative significant effect while firm size had a positive significant effect on environmental disclosure. The proportion of independent commissioners, the educational background of the president of the commissioners, the frequency of the commissioners' meetings, the gender diversity of the commissioners, and profitability had an insignificant effect on environmental disclosure. The results of this research provide an interesting probability for further discussion and research.

Keywords: corporate governance, company aspects, company characteristics, environmental disclosure

1 INTRODUCTION

In general, a company acts as a supplier of goods and services demanded of by the community, but its activities frequently have an adverse impact on the community (Almilia & Wijayanto, 2007). Recently, the impact has worsened as indicated by the increase in environmental pollution issues in Indonesia (Suratno et al., 2007). It has been mainly caused by the negligence of most companies in carrying out environmental disclosures (Anggraini & Reni, 2006).

Environmental pollution has become a common issue in Indonesia. (Wahana Lingkungan Hidup, 2017) reported approximately 30 factories in Dayeuhkolot, Bandung Regency, discharge their waste into the Citepus river, causing odor and water pollution. Such pollutions bring effects on local communities for example, skin diseases. Meanwhile, in Sidoarjo, East Java, a continuous mudflow of sulfur-sludge gas, is evidence of the company's poor attention to the environmental impacts of its industrial activities (Ja'far, 2006).

Uncontrolled environmental pollution is detrimental to the local community, which may lead to claims or demands against the presence of a company (Pratama & Rahardja, 2013. To maintain the company's reputation and to prevent any negative sentiments from the community, the company must confirm its environmental responsibilities through environmental disclosure (Pratama et al., 2013. Moreover, Paramitha and Rohman (2014) suggested environmental disclosure as an instrument of corporate financial statements.

An attempt to improve environmental management and reduce adverse impacts has been undertaken by the Indonesian government through the enactment of the Regulation No. 47 of 2012 on Social and Environmental Responsibility of Limited Liability Companies; specifically, in article 2 and article 3, paragraph 1. Despite this, developing countries such as Indonesia, have relatively low levels of environmental disclosure compared to other countries (Sufian & Zahan, 2013). Indonesia ranked 11th among Asia Pacific countries where the level of implementation of Good Corporate Governance (GCG) is very weak in terms of the Corporate Governance Watch Market Score (Lukviarman, 2016). Consequently, many companies have a propensity to overlook the prevailing rules and eventually refuse to perform an environmental disclosure (Anggraini & Reni, 2006).

Therefore, corporate governance is an important element to increase transparency and accountability of companies so that environmental disclosure continues to be disclosed (Suhardjanto, 2010). Furthermore, Suhardjanto (2010) also stated that companies with GCG are mostly supported by appropriate supervision by aboard of commissioners. It is in accordance with the conditions of the organizational structure of Indonesian companies that use the Two Board System that the board of commissioners functions to oversee the performance of the board of directors (Komite Nasional Kebijakan Governance, 2011).

In carrying out its function, the board of commissioners is assisted by an independent board of commissioners as anunaffiliated agency (Komite Nasional Kebijakan Governance 2006). The effectiveness of corporate supervision is also supported by the characteristics of the president and the board of commissioners which are substantially influenced by ethnic, education and gender diversity (Carter et al., 2003). These characteristics will minimize possible fraud committed by directors, enhance the value of the company from a stakeholders' perspective and improve the effectiveness of supervision (Chtourou et al., 2001; Said et al., 2009).

Furthermore, Suhardjanto (2010) reasserted that an audit committee is another supervisory aspect that supports GCG. This committee serves to provide assistance for the board of commissioners in promoting a company's transparency and accountability with an independent audit (Badan Pengawas Pasar Modal, 2004). Thus, the supervision aspects of corporate governance will augment environmental responsibility and reduce asymmetry information simultaneously (Sulistyowati, 2014).

Additionally, the characteristics of a company will affect the level of environmental disclosure, while the environmental impact depends on the type, characteristics and size (Mirfazli & Nurdiono, 2007; Suhardjanto, 2010). The characteristics are seen from the size, profitability, leverage, number of shareholders, liquidity, industry type and profile (Marwata, 2001). Previous studies related to "aspects of corporate governance supervision of environmental disclosure" are: Akbas (2014); Suhardjanto (2010); Khan et al. (2013); Fortunella and Hadiprajitno (2015); Lie et al. (2008); Kharis and Suhardjanto (2012); Liao et al. (2015); Setyawan and Kamila (2015); and Supatminingsih and Wicaksono (2016). Meanwhile, previous studies that discuss "The characteristics of company toward environmental disclosure" include: Giannarkis (2014); Burgwal and Viera (2014); Ohidoa et al., (2016); Aghdam (2015); Nur and Priantinah (2012); and Oktarina and Mimba (2014).

2 THEORETICAL REVIEW AND HYPOTHESIS DEVELOPMENT

2.1 *Proportion of independent commissioners and environmental disclosure*

The independent commissioners are an essential part whose function is to diminish the possibility of fraud in financial reporting as well as environmental disclosure (Cho & Pattern, 2007). Based on this statement, a hypothesis is formulated as follows:

H_1: The proportion of independent commissioners has a positive significant effect on environmental disclosure.

2.2 Ethnicity of the president of the commissioners on environmental disclosure

This study will examine ethnicityon environmental disclosure. It is presumed that people of Chinese descent havea resolute work ethic, are efficient, disciplined and material-oriented compared to indigenous people (Sugiyono, 2007). For example, with the philosophy of a work ethic possessed by people of Chinese descent, the president of the commissioners is expected to improve the company's environmental disclosure (Setyawan, 2005). The hypothesis is as follows:

H_2: The ethnicity of the president of the commissioners has a positive significant effect on environmental disclosure.

2.3 Educational background of the president of the commissioners and environmental disclosure

This study will determine the educational background that influences the level of environmental disclosure. The assumption is that the president of the commissioners with aneducational background of economics or business, will have more profound business knowledge to manage the company and to comply with regulations related to environmental disclosure (Kharis & Suhardjanto, 2012). The hypothesis is as follows:

H_3: The educational background of the president of the commissioners has a positive significant effect on environmental disclosure.

2.4 Frequency of commissioners' meetings and environmental disclosure

Theoretically, the higher the frequency of commissioners' meetings, the better the communication among the members to achieve GCG, including environmental disclosures (Suryono & Prastiwi, 2011). The hypothesis is as follows:

H_4: The frequency of commissioners' meetings has a positive significant effect on environmental disclosure.

2.5 Proportion of independent audit committee and environmental disclosure

Independent audit committee members not affiliated to any company or committee will make their performance accountable and can improve the quality of corporate control as well as the quality of environmental disclosure (Forker, 1992; McMullen, 1996). The hypothesis is as follows:

H_5: The proportion of independent audit committee has a positive significant effect on environmental disclosure

2.6 Frequency of audit committee meetings and environmental disclosure

The Financial Services Authority issued Regulation No. 55/POJK.04/2015 that stipulated that the audit committee shall hold a meeting periodically and at least once every three months. The more frequent the meetings, the more effective the supervision function of a committee is. It will ultimately improve the frequency of environmental disclosures (Corporate Governance Guidelines, 2007). The hypothesis is as follows:

H_6: The frequency of audit committee meetings has a positive significant effect on environmental disclosure.

2.7 Gender diversity of commissioners and environmental disclosure

The gender diversity of commissioners is an imperative ornament in corporate governance that affects environmental disclosure (Liao et al., 2015). The higher the concern for gender diversity, the wider the perspective or ideas that will generate innovation and creativity as well as minimize conflicts that will potentially affect the performance of commissioners in relation to environmental disclosure (Carter et al., 2003; Lückerath, 2013). The hypothesis is as follows:

H_7: Gender diversity of commissioners has a positive significant effect on environmental disclosure.

2.8 *Profitability and environmental disclosure*

Brammer and Pavelin (2006) argued that corporations with high profitability have excessive funds to support environmental disclosure. Hence, the environmental disclosure is deemed as an approach to lessen social pressure and respond to social needs (Hackston & Milne, 1996). The hypothesis is as follows:

H_8: Profitability has a positive significant effect on environmental disclosure.

2.9 *Leverage and environmental disclosure*

Nasser et al., (2006) revealed the positive effect of leverage on environmental disclosure. This was due to the risks endured by corporations in convincing investors and creditors by providing more specific disclosure. The hypothesis is as follows:

H_9: Leverage has a positive significant effect on environmental disclosure.

2.10 *Firm size and environmental disclosure*

Environmental disclosure is mostly conducted by companies with a high turnover (Despina et al., 2011). Therefore, large companies tend to be under pressure to divulge their business activities due to the impact of numerous corporate activities that affect people, shareholders and investors (Suhardjanto et al., 2008). The hypothesis is as follows:

H_{10}: Firm size has a positive significant effect on environmental disclosure.

3 RESEARCH METHOD

3.1 *Population and sample*

This paper is a quantitative research study. The sample was obtained through purposive sampling in which the sample selected was based on certain criteria. As many as 445 financial statements of manufacturing companies listed on the IDX from 2012 to 2016 were used. Data analysis was undertaken withthe multiple regression method using the Eviews 8 statistical package.

The manufacturing companies were selected for their dominance on the IDX and that their industrial activities were often linked to environmental degradation (Kartadjumena, Hadi and Budiana, 2011). The data in 2012 was specifically chosen by considering the Government Regulation No. 47 of 2012 on Social and Environmental Responsibility of Limited Liability Companies.

4 MEASUREMENT OF INDEPENDENT VARIABLES

4.1 *Proportion of independent commissioners*

The indicators used in this variable were adopted from previous studies, such as Suhardjanto (2010), Khan et al., (2013), and Akbas (2014); namely the percentage of external commissioners compared to internal company commissioners. This indicator is commonly used by researchers to estimate the proportion of independent commissioners; hence the formula is expressed as below:

$$\text{Proportion of independent commissioners} = \frac{\sum \text{Independent commissioner}}{\sum \text{Members of the board of commissioners}}$$

4.2 Ethnicity of the president of the commissioners

The ethnicity of the president of the commissioners was measured using dummy variables. Indicators of this variable were adopted from studies carried out by Kusumastuti et al. (2007) and Suhardjanto (2010). The president of the commissioners that came from an indigenous background was coded 1, from a Chinese background coded 2, and from another country coded 3.

4.3 Educational background of the president of the commissioners

The indicators used for the educational background of the president of the board of comissionerere adopted from studies undertaken by Haniffa and Cooke (2005), Effendi et al., (2012) and Setyawan et al. (2015). Those with a financial or business education background were coded 1, while other educational backgrounds were coded 0.

4.4 Frequency of commissioners' meetings

The frequency of commissioners' meetings indicator was adopted from research by Suhardjanto (2010), Brick and Chidambaran (2007), and Allegrini and Greco (2011); namely the number of meetings conducted by commissioners within a year.

$$\text{Total meetings} = \sum \text{The BoC meetings}$$

4.5 Proportion of independent audit committee members

The indicators used in this research were adopted from Forker (1992), Hoand Wong (2001), and Akbas (2014). This was taken as the percentage of independent audit committee members compared to the entire audit committee members.

$$\text{Proportion of independent audit committee members} = \frac{\sum \text{Independent audit committee members}}{\sum \text{Company audit committee members}}$$

4.6 Frequency of audit committee meetings

The indicator used was the number of audit committee meetings held within one year, and those that complied with the Suhardjanto and Permatasari (2010), and Corporate Governance Guidelines (2007).

$$\text{Total number of meetings} = \sum \text{Audit committee meetings}$$

4.7 Gender diversity of commissioners

In this study, the gender diversity of commissioners' indicator was adopted from Sudiartana (2013) and Akbas (2014), Which is the number of female members in the BoC compared to the total members.

$$\text{Gender diversity} = \frac{\sum \text{Female commisionners}}{\sum \text{Total commisioners}}$$

4.8 *Profitability*

In this study, profitability was measured with an indicator based on Return On Asset (ROA). It is a ratio of net profit after tax to total assets as developed by Akrout and Othman (2013). It is stated as follows:

$$ROA = \frac{\text{Earning After Tax}}{\text{Total Asset}}$$

4.9 *Leverage*

Leverage is measured with indicators of Debt to Asset Ratio (DAR) because it is presumed that assets are used to guarantee all liabilities or corporate debt (Brealey, 2007, p. 76).

$$\text{Leverage} = \frac{\text{Total Debt}}{\text{Total Asset}}$$

4.10 *Firm size*

In this study, the indicator used to measure a firm's size was its assets, which were measured by the logarithm of the total assets of the company (Jogiyanto, 2007, p. 282). Log value is used to prevent any problems in natural data that is not distributed normally (Chen et al., 2005).

$$\text{Size} = \log (\text{book value of total assets})$$

5 MEASUREMENT OF DEPENDENT VARIABLES FOR ENVIRONMENTAL DISCLOSURES

The measurement technique used in this study was Global Reporting Index 4, which encompassed 46 items of environmental aspects. Each of the disclosed environmental issues were scored 1 and those that did not disclose got a value of 0 or called disclosure scoring Al Tuwaijri et al., 2004).

$$ED = \frac{\sum \text{the disclosed item}}{\sum \text{GRI environmetal disclosure}}$$

Environmental disclosure variables used GRI indicators because several countries commonly use GRI as the guideline for social and environmental responsibility disclosure (Isaksson & Steimle, 2009). GRI users were approximately 1000 organizations from 60 countries and they already used the GRI report framework (Visser, 2008).

Regression model in this research is as follows:

$$Y = \beta_0 + \beta_1 X_1 + \beta_2 X_2 + \beta_3 X_3 + \beta_4 X_4 + \beta_5 X_5 + \beta_6 X_6 + \beta_7 X_7 + \beta_8 X_8 + \beta_9 X_9 + \beta_{10} X_{10} + \varepsilon$$

where:

Y	= Environmental disclosure;
β_0	= Constant;
$\beta_{1,2,3,4,5,6,7,8,9,10}$	= Multiple regression coefficients, respectively;
X_1	= Proportion of independent commissioners;
X_2	= Ethnicity of the president of the commissioners;
X_3	= Educational background of the president of the commissioners;
X_4	= Frequency of the commissioners' meetings;
X_5	= Proportion of independent audit committee members;

X_6 = Frequency of audit committee meetings;
X_7 = Gender diversity of the commissioners;
X_8 = Profitability;
X_9 = Leverage;
X_{10} = Firm size;
ε = Randomerror.

6. RESULTS AND DISCUSSION

6.1 Description of research objective

The frequency of research data is described in Table 1.

Table 1 shows atotal sample of 445 annual reports from manufacturing companies listed on the IDX from 2012 to 2016.

6.2 Regression analysis

Regression analysis in this study used a fix effect model as the final output in accordance with the data processing stages through Eviews 8. The results are shown in Table 2.

Table 2 shows that the variable of the Proportion of Independent Commissioners (PIC) was 1.230 with a p-value > 0.05 of 0.63 and a t ratio equal to 0.479. Based on the results of the data testing, for Hypothesis 1, the proportion of independent commissioners has an insignificant effect on the environmental disclosure. It confirms the findings of Akbas (2014) and Supatminingsih and Wicaksono (2016). It is allegedly linked to the absence of a direct relationship between the independent commissioners and the company's daily activities or operations so as to weaken the corporate governance of the company leading to unattained environmental disclosure (Effendi et al., 2012). Moreover, there is an inclination that

Table 1. Sample selection result.

Description	Total
Manufacturing companies listed on the IDX 2012–2016	565
Manufacturing companies that did not publish annual reports	115
Manufacturing companies that did not provide environmental disclosure information	5
Total sample number	445

Source: Data processing.

Table 2. Results of data regression test.

Variable	Coefficient	t-statistic	Prob.	Adjusted r^2
				0.233
PIC	1.230097	0.478858	0.6323	
ETNIC*	−1.142716	−3.028498	0.0026	
EB	0.444133	0.765671	0.4443	
CM	−0.051345	−0.904954	0.3660	
IAC*	4.642455	2.545086	0.0113	
TCM*	0.111730	2.012492	0.0448	
DG	−3.186909	−1.851846	0.0647	
PRO	0.012485	0.508666	0.6112	
LEV*	−1.262728	−2.020362	0.0440	
FS*	4.910008	5.145451	0.0000	

Notes: *significant at 5%. Prob. = probability (p-value).

the appointment of independent commissioners is less effective and not transparent hence their independence is doubted, leading to poor supervisory function and low environmental disclosure (Fortunella and Hadiprajitno, 2015).

For Hypothesis 2, the regression Coefficient of Ethnicity (ETNIC) was −1.142 with a p-value < 0.05 of 0.003 and a t ratio of −3.028. Based on these results, it can be declared that the ethnicity of the president of the commissioners has a negative significant impact on environmental disclosure. It confirms the findings of Katmon et al. (2017) but opposes Suhardjanto (2010) and Setyawan et al. (2015). It indicates that indigenous presidents are more concerned with and care about their corporate social environment than Chinese presidents who prioritize profitability and corporate continuity, rather than environmental and social impact.

For Hypothesis 3, the regression coefficient of the variable Educational Background (EB) was 0.444 with a p-value > 0.05 of 0.444 and a t ratio of 0.766. Based on this result, the educational background of the president of the commissioners has an in significant effect on environmental disclosure. It corroborates the findings of Setyawan et al. (2015) that the number of environmental disclosures does not rely on a president with an economics or business education, but more on their management's awareness and willingness to comply with rules and regulation (Rahmawati & Subardjo, 2017). Nevertheless, management has more knowledge and information about corporate operations compared to the commissioners which are indirectly involved in corporate daily activities (Lukviarman, 2016).

For Hypothesis 4, the regression coefficient of the variable Total Commissioners Meetings (CM) was −0.051 with a p-value > 0.05 of 0.366 and a t ratio of −0.905. Based on this result, the number of commissioners meetings has an insignificant effect on environmental disclosure. This confirms the findings of Suhardjanto (2010). It indicates the frequency of the commissioners' meetings is not effective since the meetings will only expend time, energy and additional cost, but fail to divulge fundamental information, leading to company loss (Muntoro, 2005).

For Hypothesis 5, the regression coefficient of the variable Proportion of Independent Audit Committee Members (IAC) was 4.642 with a p-value < 0.05 of 0.011 and a t ratio of 2.545. Based on this result, the proportion of independent audit committee members has a positive significant effect on environmental disclosure. Thisis in line with Supatminingsih and Wicaksono, (2016). It implies the principle that independent audit committee members have control, transparency, honesty and responsibility toward a company, and hence will provide more reliable information and improve the quality of environmental disclosures (Forum For Corporate Governance In Indonesia, 2009). With this principle, the performance of independent audit committee members is trustworthy and is capable of enhancing a company's quality and environmentaldisclosure level (Forker, 1992; McMullen, 1996).

For Hypothesis 6, the regression coefficient of the variable Total Number of Audit Committee Meetings (TCM) was 0.112 with a p-value < 0.05 of 0.045 and a t ratio of 2.012. Based on this result, the number of audit committee meetings has a positive significant effect on environmental disclosure. It is consistent with Suhardjanto and Anggitarani (2010). It implies that the more frequent audit committee meetings will improve the oversight function effectively in managing and running the company so as to improve the company's environmental disclosure (Corporate Governance Guidelines, 2007).

For Hypothesis 7, the regression coefficient of the variable Gender Diversity (GD) was-3.187 with a p-value > 0.05 of 0.065 and a t ratio of −1.852. It can be declared from the results that gender diversity in the commissioners has an insignificant effect on environmental disclosure. It confirms the findings of Akbas (2014) and implies existing discrimination in the commissioners, mainly due to the presumption that women are incompetent and inferior in the business sector (Hogg & Vaughan, 2008). This inhibits the career development of women and becomes an obstacle for them to accelerate a company's performance and information disclosure (Santrock, 2006).

For Hypothesis 8, the regression coefficient of the variable Profitability (PRO) was 0.012 with a p-value > 0.05 of 0.611 and a t ratio of 0.508. Based on this result, it can be declared that profitability has an insignificant effect on environmental disclosure. It reaffirms the findings of Nur and Priantinah, (2012). In corporations with either high or low profitability, the management assume it is not necessary to report information that will restrain or obstruct the financial advantage of the company (Brugwal et al., 2014). Nevertheless, the financial factor is perceived as an indicator or information for investors to reflect on and determine their investment on a company for its continuity (Suwardjono, 2010).

For Hypothesis 9, the regression coefficient of the variable Leverage (LEV) was −1.263 with a p-value < 0.05 of 0.044 and a t ratio of −2.020. Based on this result, leverage has a negative significant effect on environmental disclosure. This is similar to the findings of Giannarkis (2014) but dissimilar to Ohidoa et al. (2016). It indicates that the lower the level of leverage of the company, the environmental disclosures made will increase disclosure and vice versa, if the company's leverage ratio is high then the disclosures made will be lower because it requires more costs to report environmental information (Andrikopoulus & Kriklani, 2013).

Hypothesis 10 shows the regression coefficient of the variable Firm Size (FS) was 4.910008 with a p-value < 0.05 of 0.000 and a t ratio of 5.145. Based on this result, firm size has a positive significant effect on environmental disclosure. It confirms the findings reported by Giannarkis (2014) and Brugwal et al. (2014). It implies that a large company has many investors and gets attention from numerous stakeholders, thus it is important for it to perform sufficient social and environmental disclosure in order to prevent any public criticismand to maintain its image.

7 CONCLUSIONS, IMPLICATION, AND LIMITATIONS

7.1 *Conclusions*

The proportion of independent commissioners has an insignificant effect, the ethnicity of the president of the commissioners has a negative significant effect, and the educational background of the president of the commissioners has an insignificant effect. The frequency of the commissioners' meetings has an insignificant effect, the proportion of independent audit committee members has a positive significant effect, and the frequency of audit committee meetings has a positive significant effect. Gender diversity has an insignificant effect. Profitability has an insignificant effect, leverage has a negative significant effect, and firm size hada positive significant effect.

7.2 *Implication*

The implication of this research paper's results is that the managers of manufacturing companies listed on the Indonesian stock exchange need to always strive to improve environmental disclosure in terms of corporate governance factors and company characteristics, in order to improve company performance and value for stakeholders and investors. This is important to reduce information asymmetry within the company and can strengthen relationships between stakeholders.

7.3 *Limitations*

The limitation of this study is the use of GRI 4 indexes in the measurement of environmental disclosure, in which it has not implemented entirely on companies in Indonesia. In addition, it is difficult to determine the ethnicity of the president of the commissioners due to the lack of literature that discuss this matter.

It is suggested that for future research, the Indonesian enviromental reporting ndexes modified by Suhardjanto et al. (2008) are used to estimate environmental disclosure, since it is adjusted to the condition of corporate environmental management in Indonesia. Moreover, the source of data should not merely be in the form of annual reports.

REFERENCES

Aghdam, S.A. (2015). Determinants of environmental disclosure: The case of Iran. *International Journal of Basic Sciences and Applied Research*, 4(6), 343–349.

Akbas, H.E. (2014). Company characteristics and environmental disclosure: An empirical investigation on companies listed on Borsa Istanbul 100 index. *The Journal of Accounting and Finance*, 145–164.

Akrout, M.M. & Othman, H.B. (2013). A study of the determinants of corporate environmental disclosure in MENA emerging markets. *Journal of Reviews on Global Economics*, 2, 46–59.

Allegrini, M. & Greco, G. (2011). Corporate boards, audit committees and voluntary disclosure: Evidence from Italian listed companies. *Journal of Management & Governance*, 26, 208–209.

Almilia, L.S. & Wijayanto, D. (2007). Effect of environmental performance and environmental disclosure on economic performance. *Proceedings of the First Accounting Conference*, Depok 7–9 November.

Al-Tuwaijri, S.A., Christensen, T.E. & Hughes II, K.E. (2004). The relations among environmental disclosure, environmental performance, and economic performance: A simultaneous equations approach. *Accounting, Organizations and Society*, 29, 447–471.

Andrikopoulos, A. & Kriklani, N. (2013). Environmental disclosure and financial characteristics of the firm: The case of Denmark. *Corporate Social Responsibility and Environmental Management*, 20(1), 55–64.

Anggraini, F.R. & Reni, R. (2006). Pengungkapan Informasi Sosial Dan Faktor-Faktor Yang Mempengaruhi Pengungkapan Informasi Sosial Dalam Laporan Keuangan Tahunan: Studi Empiris Pada Perusahaan Yang Terdaftar di Bursa Efek Jakarta. *Simposium Nasional Akuntansi*, 9, 1–21.

Badan Pengawas Pasar Modal. (2004). *Keputusan Ketua Badan Pengawas Pasar Modal No. KEP-29/PM/2004 (peraturan No. IX.I.5) Tentang Pembentukan Dan Pedoman Pelaksanaan Kerja Komite Audit.*

Brammer, S. & Pavelin, S. (2006). Voluntary environmental disclosures by large UK companies. *Journal of Business Finance & Accounting*, 33(7–8), 1168–1188.

Brick, E. & Chidambaran, N.K. (2007). *Board meetings, committee structure, and firm performance.* Retrieved from https://papers.ssrn.com.

Brealey, M.M. (2007). *Dasar-Dasar Manajemen Keuangan Perusahaan.* Buku 2 Edisi 5. Jakarta, Indonesia: Erlangga.

Burgwal, D.V.D. & Vieira, R.J.O. (2014). Environmental disclosure determinants in Dutch listed companies. *Revista Contabilidade & Financas*, 25(64), 60–78.

Carter, D.A., Simkins, J. & Simpson, W.G. (2003). Corporate governance, board diversity, and firm value. *The Financial Review*, 38, 33–53.

Chen, K.Y., Kuen L.L. & Jian, Z. (2005). Audit quality and earnings management for Taiwan IPO firms. *Managerial Auditing Journal*, 20(1), 86–104.

Cho, H.C. & Patten, M.D. (2007). The role of environmental disclosure as tools of legitimacy: A research note. *Accounting, Organizations and Society*, 32, 7–8.

Chtourou, S.M., Bedard, J. & Courteau, L. (2001). *Corporate governance and earnings management.* Working Paper, Universite Laval, Quebec City, Canada.

Corporate Governance Guidelines. (2007). *Guidelines on corporate governance.* Retrieved from http://phx.corporate-ir.net/External.File?item=UGFyZW50SUQ9MTQ0NjIwfENoaWxkSUQ9LTF8VHlwZT0z&t=1.

Despina, G., Efthymios, G. & Stavropoulos, A. (2011). The relation between firm size and environmental disclosure. *International Conference on Applied Economics*, 179–186.

Effendi, B., Uzliawati, L. & Yulianto, A.S. (2012). Pengaruh dewan komisaris terhadap environmental disclosure pada perusahaan manufaktur yang listing di IDX tahun 2008–2011). *Makalah Simposium Nasional Akuntansi XV,* Banjarmasin, Indonesia.

Forum For Corporate Governance In Indonesia (2009). *Peranan Dewan Komisaris dan Komite Audit dalam Pelaksanaan Corporate Governance* (Tata Kelola Perusahaan). Jakarta, Indonesia. Jilid 2.

Forker, J.J. (1992). Corporate governance and disclosure quality. *Accounting and Business Research*, 22(86), 111–124.

Fortunella, A.P. & Hadiprajitno, B. (2015). The effects of corporate governance structure and firm characteristic towards environmental disclosure. *Diponegoro Journal of Accounting*, 4(2), 1–11. ISSN: 2337–3806.

Giannarkis, G. (2014). Corporate governance and financial characteristic effect on the extent of corporate social responsibility disclosure. *Social Responsibility Journal*, 10(4), 569–590.

Hackston, D. & Milne, M.J. (1996). Some determinants of social and environmental disclosures in New Zealand companies. *Accounting, Auditing and Accountability Journal*, 9(1), 77–108.

Haniffa, R.M. & Cooke, T.E. (2005). The impact of culture and governance on corporate social reporting. *Journal of Accounting and Public Policy*, 24, 391–430.

230

Hogg, M.A. & Vaughan, G.M. (2008). *Social psychology* (5thed.). London, England: Pearson Education Limited.

Ho, S.S.M. & Wong, K.S. (2001). A study of the relationship between corporate governance structures and the extent of voluntary disclosure. *Journal of International Accounting Auditing and Taxation*, *10*(2), 139–156.

Isaksson, R. & Steimle, U. (2009). What does GRI reporting tell us about corporate sustainability? *The TQM Journal*, *21*(2), 168–181.

Ja'far, M. (2006). Pengaruh Dorongan Manajemen Lingkungan, Manajemen Lingkungan Proaktif Dan Kinerja Lingkungan Terhadap Public Environmental Reporting. *Simposiun Nasional Akuntansi IX*, Padang, Indonesia.

Jogiyanto, (2007). *Teori Portofolio Dan Analisis Investasi* (1st ed.). BPFE Yogyakarta, Indonesia.

Kartadjumena, E., Hadi, D. A. & Budiana, N. (2011). The relationship of profit and corporate social responsibility disclosure (survey on manufacture industry in Indonesia). *2nd International Conference on Business and Economic Research*.

Katmon, N., Mohamad, Z.Z., Norwani, N.M. & Al Farooque, O. (2017). Comprehensive board diversity and quality of corporate social responsibility disclosure: Evidence from an emerging market. *Journal of Business Ethics*, 1, 1–35.

Khan, A., Muttakin, M.D. & Siddiqui, J. (2013). Corporate governance and corporate social responsibility disclosure: Evidence from an emerging economy. *Journal of Business Ethics*, *114*, 207–223.

Kharis, A. & Suhardjanto, D. (2012). Corporate Governance Dan Ketaatan Pada Badan Umum Milik Negara. *Jurnal Keuangan dan Perbankan*, *16*, 37–44.

Komite Nasional Kebijakan Governance (KNKG). (2006). *Pedoman Umum Good Corporate Governance Indonesia*. Jakarta, Indonesia.

Komite Nasional Kebijakan Governance (KNKG). (2011). *Pedoman Umum Good Corporate Governance Indonesia*. Jakarta, Indonesia.

Kusumastuti, S., Supatmi, S. & Perdana, S. (2007). Pengaruh Board Diversity Terhadap Nilai Perusahaan Dalam Perspektif Corporate Governance. *Jurnal Akuntansi Dan Keuangan*, *9*(2), 88–98.

Liao, L., Luo, L. & Tang, Q. (2015). Gender Diversity, Board Independence, Environmental Committee And Greenhouse Gas Disclosure. *The British Accounting Review*, *47*(4), 409–424.

Lie, J., Pike, R. & Haniffa, R. (2008). Intellectual capital disclosure and corporate governance structure in UK firms. *Accounting and Business Research*, *38*(2), 137–159.

Lückerath, R.M. (2011). Women on boards and firm performance. *Journal of Management & Governance*, *17*(2), 492–509.

Lukviarman, N. (2016). *Corporate governance*. Solo, Indonesia: PT Era Adicitra Intermedia.

McMullen, D.A. (1996). Audit committee performance: An investigation of the consequences associated with audit committee. *Auditing: A Journal of Theory and Practice*, *15*(1), 87–103.

Marwata, (2001). Hubungan Antara Karakteristik Perusahaan Dan Kualitas Pengungkapan Sukarela Dalam Laporan tahunan perusahaan publik di Indonesia. *SNA IV*. 155–172.

Mirfazli, E. & Nurdiono. (2007). Evaluasi Pengungkapan Informasi Pertanggungjawaban Sosial Pada Laporan Tahunan Perusahaan Dalam Kelompok Aneka Industri Yang Go Publik di BEJ. *Jurnal Akuntansi dan keuangan*, *12*, 1–45.

Muntoro, R.K. (2005). *Membangun Dewan Komisaris Yang Efektif*. Majalah Usahawan Indonesia No. 11 Tahun XXXVI.

Nasser, A.T., Wahid, E.A., Mustapha, N. & Hudaib, M. (2006). Auditor–client relationship: The case of audit tenure and auditor switching in Malaysia. *Managerial Auditor Journal*, *21*(7), 724–737.

Nur, M. & Priantinah, D. (2012). Analisis Faktor-Faktor Yang Mempengaruhi Pengungkapan Corporate Social Responsibility di Indonesia (Studi Empiris Pada Perusahaan Berkategori High Profile Yang Listing di IDX). *Jurnal Nominal*, *1*(1), 22–34.

Ohidoa, T., Omokhudu, O.O. & Oserogho, I.A.F. (2016). Determinants of environmental disclosure. *International Journal of Advanced Academic Research/Social and Management Sciences*, *2*(8), 49–58.

Oktarina, N.W. & Mimba, N.P.S.H. (2014). Pengaruh Karakteristik Perusahaan Dan Tanggung Jawab Lingkungan Pada Pengungkapan Tanggung Jawab Sosial Perusahaan. *Jurnal Akuntansi*, *6*(3), 402–418.

Otorisasi Jasa Keuangan Republik Indonesia. (2015). *Peraturan No. 55 Tentang Pembentukan Pedoman Pelaksanaan Kerja Komite Audit*. Jakarta, Indonesia.

Paramitha, B.W. & Rohman, A. (2014). Pengaruh Karakteristik Perusahaan Terhadap Environmental Disclosure. *Diponegoro Journal of Accounting*, 3, 188–198.

Pratama, Agny Gallus, dan Rahardja. 2013. Pengaruh Good Corporate Governance Dan Kinerja Lingkungan Terhadap Pengungkapan Lingkungan (Studi Empiris Pada Perusahaan Manufaktur Dan Tambang Yang Terdaftar Pada BEI Dan Termasuk Dalam PROPER Tahun 2009–2011). *Diponegoro Journal of Accounting*, 2 (3), 1–14.

Rahmawati, M.I. & Subardjo, A. (2017). Pengaruh Pengungkapan Lingkungan Dan Kinerja Lingkungan Terhadap Kinerja Ekonomi Yang Dimoderasi Good Corporate Governance. *Jurnal Buletin Studi Ekonomi, 22*(2).

Republik Indonesia. (2012). *Peraturan Pemerintah No. 47 Tentang Tanggung Jawab Lingkungan dan Sosial Perseroan Terbatas.*

Said, R., Zainuddin, Y.H.J. & Haron, H. (2009). The Relationship Between Corporate Governance Characteristics In Malaysian Public Listed Companies. *Social Responsibility Journal, 5*(2), 212–226.

Santrock, J.W. (2006). *Human adjustment.* New York, New York: McGraw-Hill.

Setyawan, S. (2005). Konteks Budaya Etnis Tionghoa Dalam Manajemen Sumber Daya Manusia. *Jurnal Manajemen dan Bisnis BENEFIT, 9*(2), 164–170.

Setyawan, H. & Kamilla, P. (2015). Impact of corporate governance on corporate environmental disclosure: Indonesian evidence. *Proceedings of International Conference on Trends in Economics, Humanities and Management 15th,* Pattaya, Thailand.

Sudiartana, I.M. (2013). Pengaruh Diversitas Gender Dan Latar Belakang Pendidikan Dewan Direksi Terhadap Luas Pengungkapan Sukarela. *Jurnal Riset Akuntansi,* UNUD.

Sufian, M.A. & Zahan, M. (2013). Owner structure and corporate social responsibility disclosure in Bangladesh. *International Journal Economics and Financial, 3*(4).

Sugiyono. (2007). *Menjawab Stigma, Mewariskan Tradisi.* Retrieved from http://www.kabarejogja.com/new/canthing2.html.

Suhardjanto, D. (2010). Corporate Governance, Karakteristik Perusahaan Dan Environmental Disclosure. *Prestasi, 6*(1) ISSN 1411–1497.

Suhardjanto, D. & Anggitarani, A. (2010). Karakteristik Dewan Komisaris Dan Komite Audit Serta Pengaruhnya Terhadap Kinerja Keuangan Perusahaan. *Jurnal Akuntansi, 2,* 125–245.

Suhardjanto, D. & Permatasari, D.N. (2010). Pengaruh Corporate Governance, Etnis, Dan Latar Belakang Pendidikan Terhadap Environmental Reporting Index. *Kinerja, 14*(1), 151–164.

Suhardjanto, D., Tower, G. & Brown, A.M. (2008). Indonesian stakeholders' perceptions on environmental information. *Journal of the Asia-Pacific Centre for Environmental Accountability, 14*(4), 2–11.

Sulistyowati. (2014). *Pengaruh Corporate Governance Terhadap Environmental Disclosure (Studi Empiris Pada Perusahaan Manufaktur Dan Pertambangan Yang Listing di IDX Tahun 2010–2012).* UNEJ, Artikel Mahasiswa, Fakultas Ekonomi.

Supatminingsih, S. & Wicaksono, M. (2016). Pengaruh Corporate Governance Terhadap Pengungkapan Lingkungan Perusahaan Bersertifikasi ISO 14001 di Indonesia. *Jurnal Akuntansi dan Pajak, 17*(1), 32.

Suratno, Darsono & Mutmainah, S. (2007). Pengaruh Environmental Performance Terhadap Environmental Disclosure Dan Economic Performance: Studi Empiris Pada Perusahaan Manufaktur Yang Terdaftar di Bursa Efek Jakarta Periode 2001–2004). *The Indonesian Journal of Accaounting Research, 10*(2).

Suryono, H. & Prastiwi, A. (2011). Pengaruh Karakteristik Perusahaan Dan Corporate Governance Terhadap Praktik Pengungkapan Sustainability Report. *SNA XIV,*Aceh, Indonesia.

Suwardjono. (2010). *Teori Akuntansi: Pengungkapan Dan Sarana interpretatif.* edisi ketiga. BPFE Yogyakarta, Indonesia.

Visser, W. (2008). *The new era of sustainability and responsibility.* CSR International Inspiration Series, 1.

Wahana Lingkungan Hidup (2017). Retrieved from https://daerah.sindonews.com/read/1193865/21/miris-30-pabrik-di-dayeuhkolot-buang-limbah-ke-sungai-1491212988.

HRM & OB

Business Innovation and Development in Emerging Economies – Trinugroho & Lau (Eds)
© 2019 Taylor & Francis Group, London, ISBN 978-1-138-35996-3

A study of employee engagement at Bisnis Indonesia

H.B. Winarko & S. Sihombing
Sampoerna University, Jakarta, Indonesia

ABSTRACT: The aim of this study was to identify and understand the level of employee engagement in a news media company, Bisnis Indonesia, by analyzing its organizational commitment and the job satisfaction levels of its employees. This project was used by its Human Resource (HR) department to design its future plan to better improve organizational commitment and job satisfaction. A model of employee engagement measurement was then developed and the questionnaires distributed. Based on the analysis using the GML performance benchmark method, the study found that the engagement level of the employees in Bisnis Indonesia was considered high. On the other hand, the company needs to consider improving the satisfaction level in order to retain and develop its best employees.

Keywords: Bisnis Indonesia, employee engagement, job satisfaction, news media, organizational behavior, organizational commitment

1 INTRODUCTION

1.1 *Background*

An organization needs to build a special relationship with its employees in order to remain connected or engaged, especially to build connections among the employees with a shared vision and mission of the company. The purpose of this is to make the employees feel that they belong to the company, thus, they may be reluctant to leave the company when other competing companies try to attract them with better offers. Moreover, employees who are engaged tend to contribute their utmost efforts to reach the company's goals and objectives. Such close relationships are called employee engagement.

Dicke et al. (2007) reports that a company should think about its employee engagement first. Further, according to Jack and Suzy Welch, quoted from this report, stated that 'no company, small or large can win over the long run without energized employees who believe in the organizations' mission and understand to achieve it'. That is why every company needs to take a measure of employee engagement at least once a year, through anonymous surveys in which people may feel completely safe to speak their minds. This clearly shows that employee engagement may contribute much to the sustainability of a company.

Employee engagement can be measured regularly once or twice a year, depending on the company's need. However, although it is a crucial thing to do, some companies may not measure its own employee engagement, and that was happened in most companies, including Bisnis Indonesia. There are many Human Resource consultants offering their services to help those companies to conduct surveys and measure employee engagement rate, one of them is GML Performance Consulting. This consulting company specializes in services related to general management consulting, public and in-house training, balanced scorecards, individual performance assessments, e-learning and other HR related services.

The GML Performance Consulting company was giving an employee engagement seminar and workshop to the HR and managers at Bisnis Indonesia in 2013. Based on the method learned from that workshop, it raised the eagerness of the company's management to measure employee engagement in Bisnis Indonesia. While working as an internee at Bisnis Indonesia, the co-author was assigned by the company to conduct an employee engagement survey during the first semester of 2014 and applied the GML Performance Consulting benchmark method as the basis of this consulting research.

1.2 Research problem

In general it is the HR department's task to develop plans, activities, staffing systems, policies and regulations just for one main reason, which is to motivate their employees in reaching the organization's goals according to its vision and mission. Hence, the employees are expected to stay (loyal) and strive (commit) to the organization in which they work because the employees who are engaging are those who love what they are doing (Rutledge, 2009). This means that those employees who tend to have a positive impact on the company, are those who are productive, profitable to the company, create stronger customer relationships and stay longer with the company (Gallup Daily, 2017).

The level of employee engagement really matters and is very important for the HR department as by knowing the engagement rate, HR knows the conditions and feelings of their employees at the time: Do the employees work happily and love the organization or are they just working without really thinking about the organization's future? Do all the things the company offer satisfy their employees? The engagement of the employees is needed, especially when the organization wants to increase its productivity and retain good employees to stay and strive for the future of the organization, either in conditions of recession, stagnation or rapid growth (Hewitt, 2012). Hence, each organization should know its own engagement rate in order to know how many employees really fight for the success of the organization.

1.3 Limitations

The survey used to generate this research was conducted in one group of companies which is located within the same organization. Thus, this survey may have some limitations; but still, is a good reference to be used in any in-house employee engagement survey. Since the survey was used for the first time in the company, this research uses two main variables as the research model. These are the organizational commitment and job satisfaction of the employees. These two main variables have several complex dimensions to measure, but this research simplifies them by taking three dimensions to measure commitment level. These are commitment to the organization (C1), commitment to the department (C2) and commitment to the task (C3). Meanwhile, the other dimensions used to measure the satisfaction level are satisfaction with facilities (S1), satisfaction with safety and comfort (S2), satisfaction with togetherness (S3), satisfaction with opportunities (S4) and satisfaction with actualization (S5). Hence, this might be a good opportunity for future researchers to add new variables to the proposed model.

In addition, the other limitation of this research is due to the survey used to measure the rate of engagement, commitment and satisfaction of Bisnis Indonesia's employees without any comparisons from other companies within the same industry. Comparisons can be measured by conducting similar research in other related companies (i.e. Kompas, Kontan, Tempo, etc.) and then comparing each level of engagement, commitment and satisfaction by using t-scores or z-scores (depending on the sample size and purpose of the future research).

1.4 Research objectives

As mentioned by Williams (2010), 'to avoid platitudes, every organization should define employee engagement to ensure that the information it is gathering from the workforce on can be put into practice'. Further, with that kind of thought, the consulting research's objectives in this report can be developed as follows:

1. Understanding the employee engagement level in Bisnis Indonesia in the first semester of 2014.

2. Developing an employee engagement matrix based on four quadrants: engaged, uncommitted, unsatisfied and disengaged in Bisnis Indonesia.
3. Comparing each department's employee commitment toward the organization, the department and the task in Bisnis Indonesia.
4. Comparing each department's employee satisfaction with its facilities, safety and comfort, togetherness, opportunities and actualization in Bisnis Indonesia.
5. Comparing each position of the employees for its commitment toward the organization, department and task in Bisnis Indonesia.
6. Comparing each position of the employees for its satisfaction with the facilities, safety and comfort, togetherness, opportunities and actualization in Bisnis Indonesia.

2 LITERATURE REVIEW

2.1 *Media management*

According to the basic word in communications science, *'pers'* (in the Indonesian language) or 'press', is a word taken from *'persuratkabaran'* (in the Indonesian language, or 'correspondence'), which means an organization or a place where information about facts and opinions are managed. Therefore, Bisnis Indonesia as a business news media organization or press in general, may also be similar to other organizations that have management and administrations which are managed by professionals. Therefore, press or media companies create added values that can be seen as business entities.

The professionals who manage the companies are those from the top managerial level with support provided by the Human Resources Department (HRD). While the regulations are formulized by the top management, the HRD may directly implement these polices or regulations with the cooperation of the managers. In general, a media company like Bisnis Indonesia typically implements 'the 5M' as its foundation of resources in order to manage the company. These 5M resources are: (1). Man, such as the reporters, editors, staff workers; (2). Money, such as the stockholders, the owners; (3). Materials, such as the data/information sources; (4). Machines, such as the printing equipment, computers; (5). Methods, such as reporting and publishing systems.

2.1.1 *Corporate culture*
Bisnis Indonesia is considered as having a strong corporate culture which is embedded in its organizational DNA and reflected in their Corporate Culture Values which describes the abbreviated letters of 'B-I-S-N-I-S' as follows:

1. Balance (B): the basic principal for all aspects.
   The employees are expected to:
   - Be fair and wise in doing any actions independently and objectively
   - Be balanced in their private and working lives
   - Be balanced in taking profit-orientation and social responsibilities
   - Work harder but in a fun way
2. Integrity (I): behavior and attitudes toward its ethics and regulations.
   The employees are expected to:
   - Be highly committed to the organization
   - Be honest, disciplined, consistent and trusted
   - Be responsible, transparent and accountable
   - Have a sense of belonging
   - Care and keep the goodwill of the organization and its confidentiality
3. Service excellence (S): provide solutions and beyond-expected services.
   The employees are expected to:
   - Understand and proactively fulfill the wants of the client's interests
   - Professionally give the best services based on standards
   - Create added value in providing solutions

- Keep the customer satisfied, both external and internal customers
4. Networking (N): build productive partnerships.
   The employees are expected to:
   - Create a synergy among employees and departments
   - Respect the partner in working
   - Create an open dialog to create togetherness and solid teamwork
   - Be proactive in building and broadening partnerships
5. Innovation (I): be creative to create added value.
   The employees are expected to:
   - Be open-minded
   - Keep learning
   - Undertake evaluation and continuously correct errors
   - Innovate or invent new things in order to create added value
   - Think positively about challenges and changes
6. Strive (S) for success: high commitment to be the best.
   The employees are expected to:
   - Implement a high quality of working
   - Optimize productivity
   - Be competitive and ready to face any challenges
   - Develop self-skills and capability

2.1.2 *Human resources development*

There were two kinds of HR strategic initiatives in Bisnis Indonesia (see Figure 1). Firstly, 'What Strategies' which focus on the business plan, such as talking about targets, individual tasks, hard and soft skills and other aspects related to the job. The objectives of these strategies are to create excellent results at the end of the performance appraisal. Secondly, 'How Strategies' which describe the corporate culture or values that employees must have as the source of employee attitudes in Bisnis Indonesia.

Employee development is created based on a Training Need Analysis (TNA). The training will be received by the employees if it supports the firm's business plan, related with the job description, if the employee's latest Key Performance Indicator (KPI) is good, recommended by the supervisors, or in order to support the employee's career path. In order to evaluate and maintain the employees' quality/performance, the employees may receive an assessment of performance each semester (twice a year).

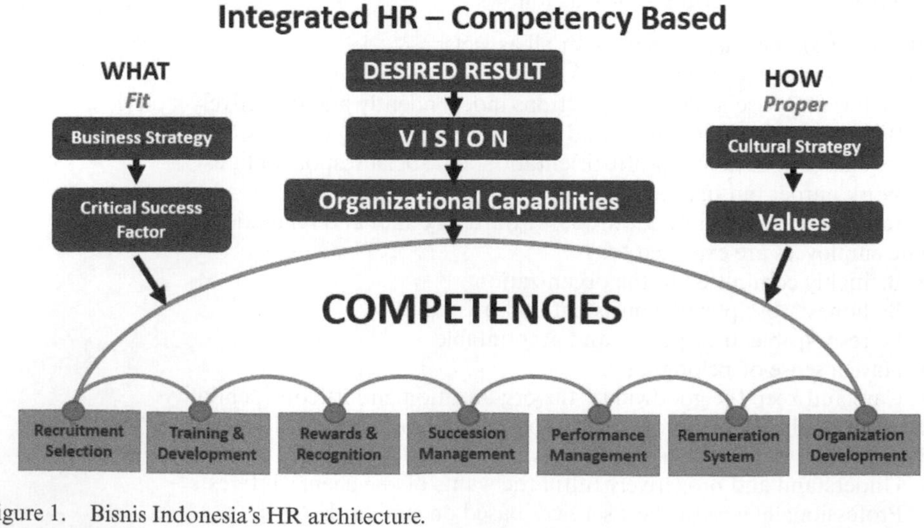

Figure 1. Bisnis Indonesia's HR architecture.
(Source: Author).

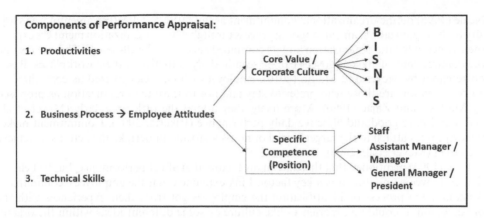

Figure 2.   Bisnis Indonesia's performance appraisal model.
(Source: Author).

Performance appraisals are conducted online twice a year, every June (first semester) and December (second semester). These performance appraisals may affect the income that each employee will get at the end of the month. If the employees were justified as passing the test and able to reach good scores, which are 'very good' or 'excellent' marks, they were eligible to receive a financial reward/bonus according to the performance.

2.2   *Employee engagement*

The employee engagement concept used in this research was adopted based on employee engagement research conducted by Swarnalatha and Prasana (2013), which defines engagement as 'an employee's involvement with, commitment to, and satisfaction with work'. In this sense, the employee engagement itself can be seen as a part of employee retention. Further, Robbins and Judge (2011) defined employee engagement as concerning an individual's 'involvement with, satisfaction with, and enthusiasm for', the work he or she does. Employees who are engaged, will give themselves 'over and over above the demands of the job' (according to Peter Drucker as quoted from Rutledge, 2009). After learning and doing some research, Rutledge (2009) devised the idea that employees who are engaged are those who are 'attracted to, committed to, and fascinated with' their work, which usually they got from direct reflection of their relationship with their boss, supervisor, or manager. This engagement will become a cornerstone of achieving a sustainable competitive advantage of the company. This is because the bottom line of a business is that people buy a service or product, which means that people and their performance are the organization's defining competitive advantage. That is why the measurement of employee engagement is very important.

In addition, when an employee is categorized as being engaged, that employee will speak positively about the organization (say), not want to leave the organization (stay) and always trying to give their best performance (strive) to the organization (Hewitt, 2012). The other theory by Rutledge (2009) says that 'the truly engaged employees are attracted to, and inspired by, their work (I want to do this), committed (I am dedicated to the success of what I am doing) and fascinated (I love what I am doing)'. In other words, if we want to make our employees become our word-of-mouth tools, giving their best efforts to reach the company's goals and loyal to the company, then companies need to create engagement by fulfilling their satisfaction first and selecting the right person at recruitment. Then, the company will not only be satisfying them but the company will also be satisfied with their performance because they are committed and responsible in their jobs.

2.2.1   *Employee commitment*

According to GML Performance Consulting's model of engagement measurement, one of the variables to measure engagement is the employee commitment variable. Northcraft and

Neale (1996) define organizational commitment as 'an attitude reflecting an employee's loyalty to the organization, and an ongoing process through which organization members express their concern for the organization and its continued success and well-being'. This means that the commitment of the employees can be seen in daily activities in their workplaces. Based on research by Meyer and Maltin (2010), employees who are categorized as committed to an organization are those who prefer to stay and not to leave the organization as proposed by Mathieu and Zajac (1990). More likely, they attend the office regularly (Meyer et al., 2002) and have good and effective daily performance (Riketta, 2002). Commitment makes the employees attach to the target (social or non-social) and undertake the actions to achieve the target (Meyer et al., 2006).

Fink (1992) mentioned that there are many factors that affect performance, but he believes that employee commitment is a key factor. Fink explained that the employees' commitment is actually the process of identification the employees got from their experiences with the organization; it could be experiences like culture or some different ideas within the organization and experiences that related to someone like supervisors, co-workers, colleagues or customers. All those experiences will be identified by the employees and in the end, will either make them feel that they belong to the organization or vice versa. In this research, based on organizational effectiveness, commitment was measured by using several dimensions, as follows:

- Commitment to organization (C1)
- Commitment to department (C2)
- Commitment to task (C3)

2.2.2 Employee satisfaction

The other variable used by GML Performance Consulting to measure engagement is employee satisfaction. Employee satisfaction is not the same as employee engagement. According to CustomInsight (2012) employee satisfaction indicates how happy or content the employees are. For some employees, being satisfied means collecting a paycheck while doing as little work as possible. This means that employee satisfaction represents the feelings of the employees; they could be happy (satisfied) or not (dissatisfied) with everything that they get in return for what they do for the company. Locke (1976) gave a comprehensive definition of job satisfaction as a 'positive emotional state resulting from the appraisal of job experience'. Hence, it is like a happy-ending result that employees feel after evaluating their work. Job satisfaction is generally recognized in the organizational field because job satisfaction is the most common and frequent type of attitude studied.

Luthans (1998) found that there are three important dimensions to job satisfaction: (1) job satisfaction as an emotional response to a job situation; (2) job satisfaction defined by how employee expectations and company expectations are met; and (3) job satisfaction based on workplace conditions and the management system. The Oxford Dictionary defines satisfaction as 'the act of fulfilling the need or desire'. This definition is just in line with a research center that says that satisfaction is a fulfillment (Leimbach, 2006). Finally, employee satisfaction is the good feeling employees have because they assume they have fulfilled their appetites and pleasures. Thus, this satisfaction can be associated with the happiness and comfort feelings of the employee. The satisfaction variable in this research also has several dimensions to measure it. The indicators were developed based on the returns for employee which focus on the extrinsic financial and non-financial issues. The satisfaction indicators developed were as follows:

- Satisfaction with facilities (S1)
- Satisfaction with comfort and safety (S2)
- Satisfaction with togetherness with other employees (S3)
- Satisfaction with opportunities: career and self-development (S4)
- Satisfaction with employee actualization (S5)

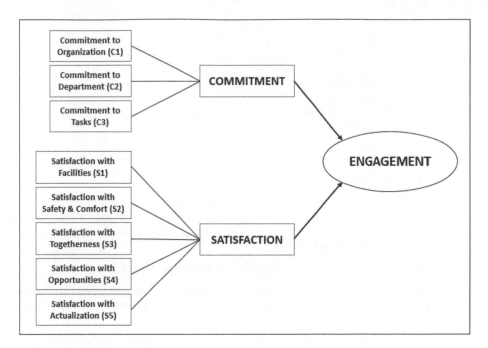

Figure 3. Research model.

3 RESEARCH METHODOLOGY

In order to understand the employee engagement level in an organization, GML utilizes questionnaires as a tool to gather specific data of the sample from respondent's. The distribution of the questionnaires can be either online (Internet-based) or offline (paper-based). After receiving all the information needed, GML analyzed it and compared it to the benchmark they have. The questionnaire design (see Appendix A) consisted of 40 closed-questions using a Likert Scale (scale 1 = Strongly agree, 2 = Agree, 3 = Disagree, 4 = Strongly disagree). Additionally, three open-ended questions were included with the purpose of allowing employees to freely share their thoughts about what encourages them to remain with the organization, what is the most satisfactory thing they received from the organization and what are their suggestions to make the organization better in the future. The questionnaires were distributed to the permanent employees only. The benchmarking technique was established based on previous surveys conducted by GML. According to its past experience, GML found that an average mean score of 2.8 is the best practice industry benchmark level to measure employee engagement to justify whether the employee is engaged or not. If the score is less than 2.8, it is considered low and if the score is equal or more than 2.8, it is considered high.

Further, the engagement matrix of this research was developed by adopting the GML method which classifies the level of employee engagement as 'Engaged', 'Disengaged', 'Unsatisfied' and 'Uncommitted', with the following assumptions:

- Employees grouped as 'engaged' have mean scores of commitment ≥ 2.8 and satisfaction ≥ 2.8.
- Employees grouped as 'uncommitted' are those who have a mean score of commitment <2.8 and satisfaction ≥ 2.8.
- Employees grouped as 'unsatisfied' are those who have a mean score of commitment ≥ 2.8 and satisfaction <2.8.
- Employees grouped 'disengaged' are those who have a mean score of commitment <2.8 and satisfaction <2.8.

Table 1. Sample distribution.

Department	Number of samples
Editor office	56
Operational	43
Marketing	52
Finance & admin	40
Subsidiaries	47
Total	238

3.1 Hypotheses development

Hypotheses were developed to fulfill the needs of this research project. Moreover, the hypotheses are temporary answers to prove which statement/thought may still be consistent when compared to the findings. Applying the GML benchmark score, the hypotheses of the research can be developed as follows:

H1: The employee engagement rate for Bisnis Indonesia is greater than 2.8.

This research may want to create a matrix of engagement, to categorize the employee who has lower degree of engagement, satisfaction, and commitment, therefore the hypothesis can be developed as follows:

H2: There are more engaged employees than disengaged, unsatisfied and uncommitted employees in Bisnis Indonesia.

By calculating the mean score, this research also wanted to measure the commitment and satisfaction rate for each of the four main departments/directorates in Bisnis Indonesia (editor office, operations, finance and administration, and marketing), plus those who worked at its subsidiaries, therefore the following hypotheses were developed:

H3: The employee commitment rate for each department in Bisnis Indonesia is greater than 2.8.
H4: The employee satisfaction rate for each department in Bisnis Indonesia is greater than 2.8.

The mean scores can also be used to understand the commitment and satisfaction levels of employee in each position, focusing on four group of positions: general manager, manager, assistant manager and staff. The following hypotheses were developed:

H5: The employee commitment rate in each position of Bisnis Indonesia is greater than 2.8.
H6: The employee satisfaction rate in each position of Bisnis Indonesia is greater than 2.8.

3.2 Sample and data

The population of this research were Bisnis Indonesia Group's permanent employees and the samples collected were Bisnis Indonesia's employees who were still active employees in April 2014. The sample/respondent profiles were employees who were grouped based on their position, tenure, age and department (see Table 1) and consisted of 238 respondents determined by using a purposive sampling (non-random sampling) technique.

The questionnaires, consisting of 40 closed-questions and three open-questions, were distributed to check validity and reliability. However, the authors found that three of the closed-questions had to be omitted due to invalidity and unreliability (questions numbered 16, 17 and 19).

4 RESULTS AND DISCUSSION

4.1 Respondents' profile

Based on the gathered data, the respondents' profiles based on age, tenure, position and department can be seen in Tables 2–5.

Table 2.   Respondents' age.

Age	Respondents
21–30 years old	111
31–40 years old	53
41–50 years old	59
>50 years old	15

Table 3.   Respondents' tenure.

Tenure	Respondents
<2 years	81
2–5 years	45
5–10 years	35
>10 years	77

Table 4.   Respondents' position.

Position	Respondents
General managers	9
Managers	31
Assistant managers	14
Staff	184

Table 5.   Respondents' department.

Department	Respondents
Editor office	56
Operational	43
Marketing	52
Finance, accounting, administration & business development	40
Subsidiaries	47

4.2   *Employee engagement matrix*

The employee engagement matrix developed from the collected data (see Figure 4) shows that 40.76% of the total respondents were in the 'Engaged' category. This means that almost half of all employees in Bisnis Indonesia were engaged employees (committed and satisfied). The next questions is how to keep these employee engaged. What can the company do to keep a good relationship with its employees, be a good friend and fair to all employees? By staying closer to its employees, managers and the HR department will know the latest information about the employees: are they happy to stay in the company? How about their KPIs, do they show a positive trend or not? If not, what could possibly contribute to a low performance?. Thus, the relationship is key to a deeper understanding of the employees.

At the other hand, the 'Unsatisfied' category was also relatively high (34.03%). Based on the data and analysis, the major contributing factors to this arose from employees who were unsatisfied with the facilities and remuneration aspects. So, this is a challenge for the HR department to determine how to move the employees from the 'Unsatisfied' quadrant to the 'Engaged' quadrant. Employees in this category were actually those who have high commitment to the organization, department and tasks. Thus, they were passionate in working

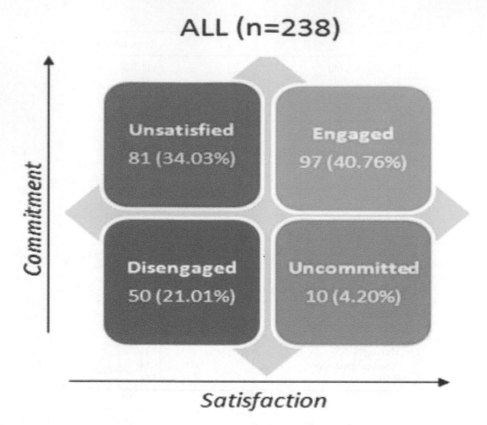

ALL (n=238)

Commitment (vertical axis)

Unsatisfied	Engaged
81 (34.03%)	97 (40.76%)
Disengaged	Uncommitted
50 (21.01%)	10 (4.20%)

Satisfaction (horizontal axis)

Figure 4.   Engagement matrix.

but they felt the rewards they received in return was not worthy compared to what they have done for the company; hence, they were not feeling satisfied. In order to shift the 'Unsatisfied' employee to the 'Engaged' category, if it is possible, the HR can make a benchmarking survey toward other media companies (similar to Bisnis Indonesia) concerning their facilities and remuneration as specific criteria, based on the position, tenure and job description and then compare it with facilities and remuneration offered in Bisnis Indonesia. After that, the HR department create a standard for facilities and remuneration scheme, then communicate the results in a townhall meeting. If the HR professionals agree to apply a new standard, let them talk to the employees in their department to determine whether they may agree or resist. Thus, everyone knows that the company is fair and competitive in providing facilities or remunerations. If some of them disagree with the new standard, ask for feedback or evaluation before the standard is announced as a new regulation.

As for the 'Uncommitted' category, the rate was only 4.20%. Although the rate was considered as low (less than 5%), but this part might be important. The employees in this category were those who are satisfied with all the benefits that the company provides but are not committed to the organization, department or tasks. As a result, those employees may not perform their jobs optimally and do not have the desire to achieve any target. This condition possibly reduces the company's overall achievement.

The last quadrant of the matrix shows the 21.01% of 'Disengaged' category employees. Inside this disengaged quadrant, there were some employees who were less committed and very unsatisfied, and surprisingly the rate was relatively high. These were followed by employees who had a medium scale of commitment and were unsatisfied. Finally, there were those who had a low scale of commitment and were unsatisfied. This latter type may be worth retaining and accepting but with several notes: try to look deeper into why they feel uncommitted and unsatisfied and then try to fix it if possible. However, the employees with medium to higher uncommitment and unsatisfaction level, might possibly considered by company to be laid-off to keep the working environment healthy

4.3   Employee commitment and satisfaction in general

4.3.1   Employee commitment in general

- **Commitment to the organization**

Twenty-eight respondents scored below 2.8 for their average mean. This means only 12% of 238 respondents were uncommitted to the organization.

- **Commitment to the department**

The survey found only seven respondents from 238 respondents uncommitted to their own department, which was only 3%.

- **Commitment to the task**

Only 8% of 238 respondents were uncommitted. Reaching up to 92% of the commitment to the task was not an easy process. However, Bisnis Indonesia's employees could be considered as highly responsible to their individual tasks.

4.3.2   Employee satisfaction in general

- **Satisfaction with facilities**

The satisfaction with facilities was the lowest compared to other satisfaction variables. However, the rate was still good, which was that 61% of all respondents were satisfied with all facilities. The respondents satisfied with the facilities in Bisnis Indonesia were 146, with 92 dissatisfied respondents.

- **Satisfaction with safety and comfort**

From the total respondents of 238, 81% were satisfied, which means that most of the employees feel safe and comfortable working with Bisnis Indonesia.

- **Satisfaction with togetherness**

The satisfaction with togetherness survey results in 88% of the total respondents were satisfied.

- **Satisfaction with opportunities**

The dissatisfied in this variable were higher compared with the others above, as dissatisfied with opportunities reached 32%.

- **Satisfaction with actualization**

The satisfaction in employee actualization was higher in Bisnis Indonesia, with 87% of respondents satisfied with actualization.

4.4   Employee commitment and satisfaction by position

4.4.1   Employee commitment by position

When the comparisons by position were made in more detail (see Figure 5), the study found that two positions (staff and general manager) were below the benchmark value of 2.80, which only occurred in question number 1: 'I am willing to use my own properties to support my work in the office'.

Slightly different from previous results, Figure 6 showed that all positions resulted in better scores in commitment to their department (above 3.00 for each position).

Finally, commitment to tasks (Figure 7) also scored above 3.00. Instead of showing a moderate score level, the general manager position had the highest score for commitment to tasks, which reached more than 3.60 and leveled out at 4.00.

4.4.2   Employee satisfaction by position

Surprisingly, only one position passed the benchmark of 2.8. This was assistant manager with the score of 2.86 (moderate) in question number 21: 'The company is giving enough facilities for me to do my job'. The other questions in this satisfaction with facilities dimension

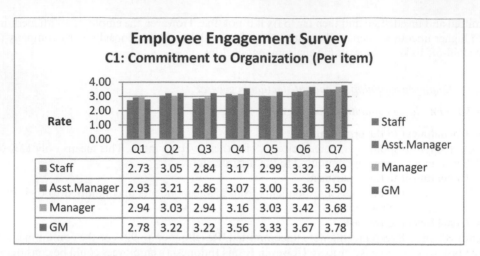

Figure 5. Commitment to organization by position.
*Q means question, i.e. Q1 = question number 1.

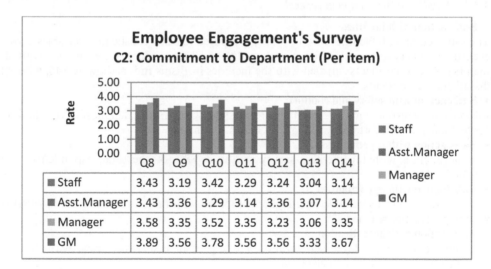

Figure 6. Commitment to department by position.

generated unpredictable results (unsatisfied) from all positions such as staff, assistant manager, manager and general manager (see Figure 8).

As shown Figure 9, questions numbered 25 (information transparency), 26 (information access) and 27 (focused work environment), only one position (assistant manager) who agreed that they were satisfied enough toward the safety and comfort working as employees in Bisnis Indonesia.

Figure 10 shows that all positions were in good shape in terms of scores, which passed the benchmark level of 2.8. This means that every position felt that they were satisfied with the opportunities given by the company.

There were many unsatisfactory indications in the results shown in Figure 11 in terms of togetherness. For instance, in question number 33 (professional development), only the general manager and manager positions reported satisfaction. In question number 34 (clear career path), staff and assistant managers were dissatisfied. Finally, in question number 35 (job rotation), the staff and assistant managers all had scores under the benchmark value of 2.8 (although they were close).

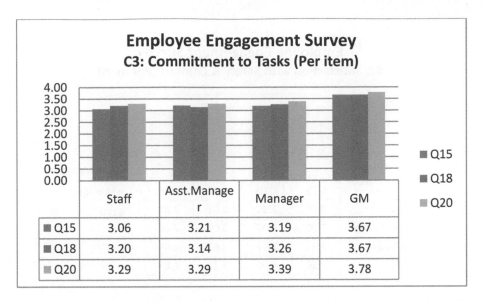

Figure 7.   Commitment to task by position.

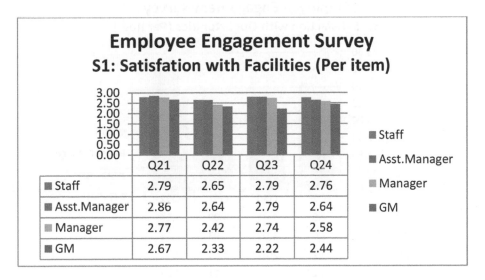

Figure 8.   Satisfaction with facilities by position.

Regarding satisfaction with actualization, all positions were relatively satisfied with their actualization as employees of Bisnis Indonesia as shown in Figure 12.

4.5   *Employee commitment and satisfaction by department*

Furthermore, when looking at each department/directorate, it seems that generally the results showed that overall commitment scores were above the benchmark score of 2.8. Meanwhile, the editor office was less satisfied with the facilities (2.64), safety and comfort (2.76), and opportunities (2.73) given by the company. Besides that, in the subsidiaries, the employees were also less satisfied with the facilities, safety and comfort, and opportunities.

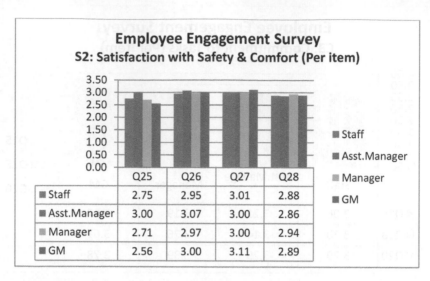

Figure 9.   Satisfaction with safety and comfort by position.

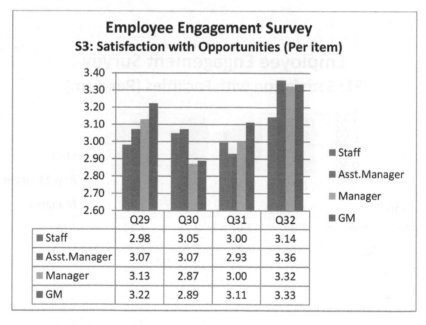

Figure 10.   Satisfaction with opportunities by position.

4.6   *Overall employee engagement evaluation*

Based on Table 6, the overall employee engagement score can be calculated by using the following formula:

Employee Engagement = (Employee Commitment + Employee Satisfaction)/8
$$= (3.10 + 3.28 + 3.21 + 2.73 + 2.91 + 3.05 + 2.80 + 3.00)/8$$
$$= \mathbf{3.01}$$

Further, based on the above discussion, the following hypothesis can be confirmed:

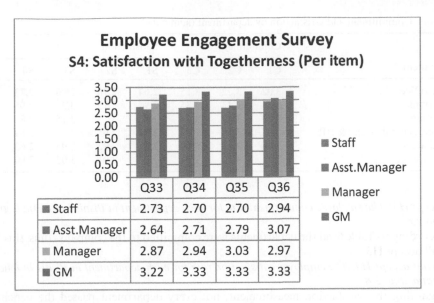

	Q33	Q34	Q35	Q36
■ Staff	2.73	2.70	2.70	2.94
■ Asst.Manager	2.64	2.71	2.79	3.07
■ Manager	2.87	2.94	3.03	2.97
■ GM	3.22	3.33	3.33	3.33

Figure 11.   Satisfaction with togetherness by position.

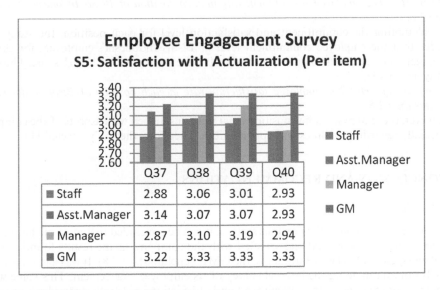

	Q37	Q38	Q39	Q40
■ Staff	2.88	3.06	3.01	2.93
■ Asst.Manager	3.14	3.07	3.07	2.93
■ Manager	2.87	3.10	3.19	2.94
■ GM	3.22	3.33	3.33	3.33

Figure 12.   Satisfaction with actualization by position.

- *Accept H1: The employee engagement rate for Bisnis Indonesia is greater than 2.8.*
  The findings show that the overall employee engagement score of Bisnis Indonesia is greater than the benchmark of 2.8 (3.01).
- *Accept H2: There are more engaged employees than disengaged, unsatisfied and uncommitted employees in Bisnis Indonesia.*
  From the findings of the employee engagement matrix, the employees of Bisnis Indonesia who were categorized as 'Engaged' were 40.76%, 'Uncommitted' were 4.20%, 'Unsatisfied' were 34.04% and 'Disengaged' were 21.01%.

Table 6. Commitment and satisfaction by department (summary).

Department	Commitment			Satisfaction				
	C1	C2	C3	S1	S2	S3	S4	S5
Editor office	3.02	3.22	3.15	**2.64**	**2.76**	2.96	**2.73**	2.88
Operations	3.25	3.43	3.35	2.94	3.15	3.2	2.95	3.1
Marketing	3.19	3.32	3.27	2.82	3	3.15	2.8	3.03
Finance, general affairs & HR	3.02	3.18	3.09	**2.73**	2.9	3	2.89	3.03
Subsidiaries	3.06	3.23	3.19	**2.5**	**2.72**	2.95	**2.63**	2.97
Average score	3.11	3.28	3.21	2.73	2.91	3.05	2.80	3.00

- *Accept H3: The employee commitment rate for each department in Bisnis Indonesia is greater than 2.8.*
  According to Table 6, all the departments were committed to the company; thus, this study shall accept H3.
- *Do not accept H4: The employee satisfaction rate for each department in Bisnis Indonesia is greater than 2.8.*
  Regarding the satisfaction measurement, not every department passed the benchmark score, especially the editor office, finance and administration, and subsidiaries which have scores below the benchmark for satisfaction with facilities, as well as safety and comfort in working in the company. Therefore, the study rejected H4.
- *Accept H5: The employee commitment rate in each position of Bisnis Indonesia is greater than 2.8.*
  By calculating the commitment and satisfaction level for each position, the study concludes that the employee commitment level of Bisnis Indonesia's employees for general manager, manager, assistant manager and staff were above the benchmark score. Thus, the study accepts H5.
- *Do not accept H6: The employee satisfaction rate in each position of Bisnis Indonesia is greater than 2.8.*
  However, there were still some positions that were unsatisfied with aspects of the company, especially regarding facilities and opportunities. Therefore, the study rejected H6.

5 CONCLUSION AND RECOMMENDATIONS

5.1 Conclusions

Based on the discussion in the previous section, it can be concluded that most of the employees were engaged, committed and satisfied with the company, as the results found that the overall engagement level was 3.01 (above the GML benchmark of 2.8). It was a good start for Bisnis Indonesia in doing the internal survey of its employee engagement. This value was a combined average value of the commitment and satisfaction level of all respondents, whereby the commitment level of the employees contributed a greater portion (3.08) compared to the satisfaction level (2.94).

The percentage of employees who were engaged in Bisnis Indonesia was also high. This can be shown from the employee engagement matrix, which reached 40.76%. That percentage was generally acceptable, and there was concern about the 'Unsatisfied' category where the percentage reached 34.03%. It means that HR may need to do more challenging works on how to solve the unsatisfied problems by developing the proper action plan in the future.

Based on the GML benchmark, no matter the position or the department in which the employees belonged, they were considered as having high commitment toward their jobs, as the average mean values were above 2.8. This finding was consistent with the feedback or comments acquired from the open-ended questionnaire. It was mentioned that most of the employees liked and were committed to Bisnis Indonesia, because of its warm environment, friendly co-workers, solid team, helping each other and they perceive Bisnis Indonesia as their second home.

On the other hand, the satisfaction measurement found in each position was not as good as the commitment scores. Some positions had average scores below the benchmark. According to the respondents' feedback, most of the employees were asking for an improvement in facilities, such as good office equipment, opportunities to develop and improve skills, a clear career path and remuneration aspects (rewards, bonus, appreciations, subsidy, etc.). The research found that the employees felt that facilities that the company provide was below from what they expected, and it made most of them became unsatisfied.

All departments perceived that they had high commitment toward Bisnis Indonesia, especially toward their own department and tasks. However, not every department was satisfied. This raised question of why specific departments felt unsatisfied. One of the proposed reasons was because the employees believe there was no specific and clear standardization for them in terms of career path, salary, rewards, bonus and so on. Especially, there were gaps between employees in the headquarters and its subsidiaries, between newly hired and existing employees, between contract and permanent employees, and other aspects that might create inequality. In brief, these findings were consistent with respondents' answers to open-ended questions.

5.2 *Recommendations*

Based on the findings of this research, the core problem in Bisnis Indonesia concerns satisfaction, especially satisfaction with facilities, safety and comfort, and opportunities. In order to improve the satisfaction level with its facilities, the company can create a standardized program about what facilities employees require based on each position. This scheme can be developed based on industry best practices, especially from other reputable media companies similar to Bisnis Indonesia. Bisnis Indonesia must recheck the facilities provided to employees at each level and determine the differences or gaps that may occur. If it is possible, the proposed new facilities for employees in Bisnis Indonesia should be at the same level or a little bit higher than the average media company. The idea of creating a scheme of new facilities will show employees that the proposed facilities in Bisnis Indonesia are competitive and even better than other media companies. Hopefully, this will motivate the employees of Bisnis Indonesia to remain and perform well at the company.

Safety and comfort concern feeling safe in working time and in pension time. It is recommended that Bisnis Indonesia develop and arrange a pension plan for employees. If Bisnis Indonesia already had a pension plan program previously, which did not work as expected, program evaluation would be useful to understand what happened and how to fix the problem. Thus, the employee may get better job security entering their pension time. This also relates to opportunities to grow within the company, especially long-term career paths and employee development. A clear career path should be identified and determined for each employee: how long an employee can take to get to the next position; what kind of criteria shall be considered to get the new position; and how the company will assess each employee. Besides making employees more satisfied, this may also motivate every employee to always improve their job performance. In order to improve the self-development of employees, both for skills and knowledge, the company may need to provide equal opportunities for all employees to attend seminars, workshops or other learning programs as long as it supports the company's objectives. This could be done periodically, once or twice a year per department. Hopefully, this effort may empower employees to become experts in their field, or at least there will be an improvement of their skills and knowledge.

When all of the above recommendations are put into the company regulations, what actually requires more attention is communication. Bisnis Indonesia have already developed existing regulations which manage all aspects related to remuneration, pension plans and so on. However, the communication process regarding regulations and standardization is not implemented well, thus, the employees may possibly not be well informed. Moreover, what the company needs to do is improve communication between the management and employees, as well as keeping the regulations up to date for all newly hired and existing employees.

REFERENCES

AON Hewitt Consulting Talent and Rewards. (2012). *2012 total rewards survey: Transforming potential into value.* Retrieved from: https://www.aon.com/human-capital-consulting/thought-leadership/talent_mgmt/2012_aonhewitt_total_rewards_survey.pdf.

CustomInsight. (2012). *What is employee satisfaction?* Retrieved from https://www.custominsight.com/employee-engagement-survey/what-is-employee-satisfaction.asp.

Dicke, C., Holwerda, J. & Kontakos, A-M. (2007). *Employee engagement: What do we really know? What do we need to know to take action?* White paper published by the Center for Advanced Human Resource Studies (CAHRS), Paris: CAHRS.

Fink, S.L. (1992). *High commitment workplaces.* New York: Greenwood Publishing Group.

Gallup Daily. (2017). *U.S. Employee Engagement.* Retrieved from: https://news.gallup.com/poll/180404/gallup-daily-employee-engagement.aspx.

Gandhi. (2012). Calculating and interpreting reliability estimates for achievement test in graph theory—(A modern branch of mathematics). Retrieved from http://www.ssmrae.com/admin/images/e63e1aa472c19ac-95c417168ef9d6b2 8.pdf.

Hornby, A.S. (2010). *Oxford advanced learner's dictionary* (8th ed.). New York: Oxford University Press

Landy, F. (1989). *The psychology of work behavior.* California: Brooks/Cole.

Leimbach, M. (2006). *Redefining employee satisfaction: Business performance, employee fulfilment and leadership practices.* Retrieved from: https://www.wilsonlearning.com/wlw/research-paper/l/employee-satisfaction

Locke, E.A. (1976). The nature and causes of job satisfaction. In M.D. Dunnette (Ed.), *Handbook of industrial and organizational psychology.* Chicago, IL: Rand McNally.

Luthans, F. (1998). *Organizational behaviour* (8th ed.). Boston: Irwin McGraw-Hill.

Meyer, J.P. & Maltin, E.R. (2010). Employee commitment and well-being: A critical review, theoretical framework and research agenda. *Journal of Vocational Behavior, 77,* 323–337.

Mathieu, J.E. & Zajac, D.M. (1990). A review and meta analysis of the antecedents correlates and consequences of organizational commitment. *Psychological Bulletin, 108,* 171–199.

Meyer, J.P. & Allen, N.J. (1991). A three-component conceptualization of organizational commitment. *Human Resource Management Review, 1,* 61–89.

Meyer, J.P., Allen, N.J. & Smith, C. (1993). Commitment to organizations and occupations: Extension and test of a three-component conceptualization. *Journal of Applied Psychology, 78,* 538–551.

Meyer, J.P., Becker, T.E. & van Dick, R. (2006). Social identities and commitments at work: Toward an integrative model. *Journal of Organizational Behavior, 27,* 665–683.

Meyer, J.P, Stanley, D.J., Herscovitch, L. & Topolnytsky, L. (2002). Affective, continuance, and normative commitment to the organization: A meta-analysis of antecedents, correlates, and consequences. *Journal of Vocational Behavior, 61,* 20–52.

Northcraft, G.B. & Neale, M.A. (1996). *Organizational behavior: A management challenge.* Chicago: Dryden Press.

Riketta, M. (2002). Attitudinal organizational commitment and job performance: A meta-analysis. *Journal of Organizational Behavior, 23,* 257–266.

Robbins, S.P. & Judge, T.A. (2011). *Organizational behavior* (13th Ed.). New Jersey: Prentice Hall.

Rutledge, T. (2009). *Getting engaged: The new workplace loyalty.* Toronto, Ontario, Canada: Mattanie Press.

Swarnalatha, C. & Prasana, T.S. (2013). Employee engagement: An overview. *International Journal of Management Research and Development, 3*(1), 52–61.

Smith, P.C., Kendall, L.M. & Hulin, C.L. (1969). *The measurement of satisfaction in work and retirement.* Chicago: Rand McNally. Retrieved from: www.siop.org/tip/july10/06 jdi.aspx.

Supranto, J. (2003). *Statistics theories and applications* (5th Ed.) [In Indonesia language]. Jakarta: Erlangga.

Ulrich, D. (1996). *Human resource champions: The next agenda for adding value and delivering results.* Boston, Massachusetts: Harvard Business Press.

Vance, R.J. (2006). *Employee commitment and engagement: A guide to understanding, measuring and increasing engagement in your organization.* Alexandria, Virginia: SHRM Foundation.

Williams, R. (January 4, 2010). *Employee engagement: Define it, measure it and put it to work in your organization.* Retrieved from http://www.workforce.com/2010/01/04/employee-engagement-define-it-measure-it-and-put-it-to-work-in-your-organization/.

APPENDIX A. QUESTIONNAIRE DESIGN

Closed-Ended Questions:

1. I am willing to use my own property to support my work in the office.

2. I am trying to always minimize the office budget.
3. I want to do any projects that benefit the office, although not related to my job description.
4. I still want to continue my career in this company for the next two years.
5. I love the culture of my company; thus, I will stay here for a further 3–5 years.
6. I am proud to promote the products or services of my company to others.
7. I will keep the confidentiality of all data in my company.
8. I am always trying to create a comfortable and conducive environment in the office.
9. When we have a meeting, I am proactive by giving suggestions or opinions to my department.
10. I will help my partners in the same department, so my department goals and targets will be achieved immediately.
11. The information in my department is transparent and easy to get.
12. The task coordination Either in teamwork or for independent work, is very clear and good in my department.
13. I'm willing to execute my department's plan, although it is conflicting with my point of views.
14. For me, the target in my department is my own target too.
    Questions 15 to 40: the following questions reflect my personal commitment in accomplishing the job tasks.
15. I want to do overtime work, if I should finish my job according to the plan.
16. If it is needed, I will do the job at home.
17. I will use all of my skills and knowledge to reach the target.
18. I always finish the job well and even beyond expectation.
19. I develop a new method to finish the tasks in efficient way.
20. Innovation is important for me.
21. The company provide enough facilities for me to do my job.
22. I think the salary I get is worth the complexity of my tasks.
23. I get a bonus or incentives based on my performance.
24. My company is providing a subsidy package.
25. All the transformation and change in the organization is very transparent for me as an employee.
26. It is easy for me to get access to information related to my work.
27. I feel the working environment surrounding me is supporting me to have good concentration.
28. I feel my job description is good and fair.
29. My company is trusting me to handle the projects.
30. My supervisor is caring enough about me.
31. The coordination among departments is good in my company.
32. I feel that I am part of my team.
33. I get seminars and workshops to develop my skills.
34. The company has a clear career path for me.
35. I have the chance of job rotation in another division.
36. Information exchange is running well in my company.
37. In my point of view, the culture of my company supports my innovation.
38. The company give me freedom to use my way in finishing my job.
39. I get the chance to show my best performance.
40. In developing the company, my ideas and suggestions are being considered.

Open-Ended Questions:

1. What kind of things (the specific one) have made you stay in this organization? Please give your reason(s).
2. What kind of things (the specific one) have made you satisfied as an employee here? Please give your reason(s).
3. Please give your comments about Bisnis Indonesia.

Thank you!

Business Innovation and Development in Emerging Economies – Trinugroho & Lau (Eds)
© 2019 Taylor & Francis Group, London, ISBN 978-1-138-35996-3

Analysis of leadership and work discipline in improving the performance of employees at the general bureau, staffing and organization of the Ministry of Tourism

B. Hasmanto, U. Rusilowati & L. Herawaty
Pamulang University, Indonesia

ABSTRACT: This study aims to determine the role of leadership and discipline in improving the performance of employees at the General Bureau, Human Resources and Organization of the Ministry of Tourism of Indonesia for Aceh province. Qualitative research methods are used in this study. The scope of this study is limited only to the General Bureau, Human Resources and Organization of the Ministry of Tourism of Indonesia. The data collection techniques used in this study are interviews with informants who are relevant to the research.

The results showed that there are four indicators of the importance of the leadership role in improving employee performance. First, the leader's ability to solve problems by utilizing a variety of existing resources in the Ministry, involving all parties in helping to jointly overcome these problems and providing the opportunity for employees to participate in training and development in order to improve employee performance. Second, the strategy undertaken to achieve the organization's vision and mission should provide an insight to the various parties that they will be jointly responsible for the increase in tourism in the Ministry. Also, they should involve them in the formulation of the Ministry's vision and mission, as well as making them jointly responsible for realizing these through the support of various program activities. Third, the ability to make decisions by involving all of the constituents who express concerns, and ensuring that a mutual decision is taken and that the decision is not just made by the General Bureau Chief, Personnel and Organization alone. The decision made by the head of the General Bureau certainly has an important meaning, both to others and to ourselves. Fourth, the ability to communicate by realizing effective communication. Communication occurs in both directions and the head of the General Bureau needs to establish communications with the whole sector when it is necessary.

In addition to leadership style, the improvement in employee performance due to work discipline is shown by the following indicators. First, the presence of employees shows a high level of discipline, however, there are still employees who arrive late and do not use their hours at work to the best advantage, as they often leave the office during working hours. Second, adherence to labor regulations can be viewed by the ability of employees to complete their work on time, although there are also employees whose work does not fit the mandate of leadership and who do not follow government procedures. Third, adherence to labor standards, although there are still employees who lack discipline. Fourth, high vigilance, which can be seen when employees carrying out the work, but cannot be seen in the use of office facilities and in the assignment of work to these employees. Fifth, work ethic, which involves all employees being friendly to fellow employees, mutual respect for fellow employees in the workplace, and honesty in carrying out the work. However, there are also employees who are not honest when performing their job.

Keywords: Leadership, work discipline, employee performance

1 INTRODUCTION

Organizations and companies have a formidable resource. That resource is humans, who in a healthy organization must be arranged in such a way as to support the achievement of the organization's strategic plan. If the human resources are not managed properly, then it will be difficult to achieve the success of the organization's strategic plan.

As part of these achievements, the organization should strengthen the basis of human resource development through an approach that aims to improve the organizational environment and which puts job satisfaction as the main focus for development and employee awards. The success of an organization relies heavily on either the good or poor performance of the organization's employees. The performance of an organization depends on its employees, with every employee actively involved in the operation of the organization. The performance of its employees will have a direct impact on either the progress or regression of the organization. According to Hanggraeni (2012, p. 3) 'The human is a vital part of the survival and success of an organization. Like fuel, humans form a source of energy for the operation of an organization to achieve its objectives.'

Wirawan (2009, p. 5) states that 'performance is the output generated by functions or indicators of a job or a profession in a certain time and performance of employees is the result of the synergy of a number of factors: internal factors of employees one of which is the discipline of work, while on the organization's internal environment includes leadership, and environmental factors external to the organization'. Meanwhile, Simanjuntak, in Wiratama and Sintaasih (2013), states that 'the factors that affect performance are the individual competence, organizational support, and management support (linked to the managerial capacities of management or leadership)'.

According to Wiranata (2011), "the fundamental issue that bothers leaders is how to undertake development and organizational change in order to improve the utilization of the resources in the organization". This is better known as 5M, which stands for *man, machines, methods, materials, money*. In an era of modern organization in the execution of work and organizational leadership, the style of an organization's leadership is important. One possible consequence of a bad leadership style, or of leadership that is more oriented to the task than to the employee, is a decrease in performance among employees, which will have the effect of decreasing the organization's productivity.

According to Sutrisno (2011, p. 213), "leadership plays an important role, as a leader is someone who will move and steer the organization to achieve results". A corporate leader should have the ability to influence and motivate the employees, which will have an impact on performance improvement. Furthermore, Timpe (2000, p. 127) said that 'leadership effectiveness depends on the relationship with a subordinate leader, and a variety of leadership styles that are used in certain circumstances'.

Meanwhile, Yukl (2005, p. 8) states that "leadership is a process of influencing others to understand and agree with what needs to be done and ensuring that the task is done effectively, as well as the process of facilitating individual and collective efforts to achieve common goals". Leadership is also interpreted as an initiative to act that produces a consistent pattern in order to find a way of solving a problem together. Leadership is the key to the survival of an organization, because the leader is the originator of the planning, organization and mobilization of all of the resources owned in order that the desired goal can be achieved.

Discipline within an organization needs to be applied so that organizational goals can be achieved. There are several types of indiscipline committed by employees of the Ministry of Tourism, such as attendance matters in the workplace, work that does not match the established procedures, a workspace that is not neat and clean, ineffective and inefficient use of work equipment, taking excessive breaks, and other matters, which are all against the rules.

Furthermore, work discipline is very important for both the employee and for the organization, as it will affect employee productivity. Good working discipline reflects a person's sense of responsibility toward their assigned tasks. Especially for a company or organization, obedience to the rules and conditions set out by the organization is very important. In other words, an organization is in dire need of labor discipline, because by applying work discipline they will be able to achieve the needs and objectives of the organization. But conversely, if work discipline is not enforced or not followed by all of the employees of the organization, it will be difficult to meet the demands.

If all employees comply with work discipline then there will be a benefit both for the organization and for the employees. Therefore, the employees themselves need to be aware of the need to comply with the regulations that have been imposed by the organization and representatives of the workers. With regards to the organization, they should make regulations that are clear and easy to understand and that are fair. In other words, "the regulations should apply to all walks of life, from the highest member of the organization to the employees under them" (Mulyadi, 2015, p. 47).

The influence of the leadership and the discipline applied by the Ministry of Tourism and its influence on the performance of members of Aceh province is expected to help to achieve its organizational goals. Both the employees and the Ministry of Tourism will receive advantages and it will also improve the welfare of the organization. It will improve morale and help the organization itself to grow again. Employee performance is a contributing factor in every organization.

Shown here is the implementation of the Government Performance Accountability System, which occurs because of changes in the weighting system in determining the value of the performance evaluation results.

2 THEORETICAL PROBLEMS

1. How can a leadership role improve the performance of employees at the General Bureau, Human Resources and Organization of the Ministry of Tourism?
2. How does work discipline improve the performance of employees at the General Bureau, Human Resources and Organization of the Ministry of Tourism?

3 LITERATURE

3.1 Human Resource Management

Human Resource Management (HRM), according to Hasibuan (2012, p. 1), "is the science and art of regulating the process of the utilization of both human resources and other resources effectively in order to achieve a certain goal".

Furthermore, Werther and Davis, in Hanggraeni (2012, p. 4), "define human resource management as activities that are trying to facilitate the people in the organization to contribute to the achievement of the organization's strategic plan. In brief, human resource management relates to how the formal system of an organization is designed and whether or not it ensures the utilization of human resources effectively and efficiently in order to support the achievement of the objectives and the strategic plan of the organization".

3.2 Leadership

Leadership is the process of a leader influencing or setting an example to their followers in an effort to achieve organizational goals.

According to Sutrisno (2014, p. 213), "leadership is a process whereby someone who is charismatic is motivating others by leading, guiding and influencing them to do something in order to achieve the expected results". Yuniarsih and Suwatno (2008, p. 265) argue that "leadership is the ability and power of one person, or a leader with the 'stance to affect the mind (mindset) of others to be willing and able to follow his will, and to inspire others to design something more meaningful".

From the definitions above, it can be concluded that leadership is the ability to influence others, both subordinates and groups. It is the ability to direct behavior in order to achieve the goals of the organization or group.

3.3 Work discipline

Labor discipline is obedience to the laws and regulations. Discipline is also closely related to the sanctions that need to be imposed on the offending party. Discipline is a management action that is needed in order to encourage members of the organization to meet the demands of various conditions (Siagian, 2008, p. 305). Meanwhile, according to Mulyadi (2015, p. 48), "work discipline is the willingness and attitude needed to abide by and comply with the prevailing regulatory norms". Good employee discipline will accelerate the company's goals, while declining discipline would be prohibitive and inhibit the company in achieving its goals.

In the event that an employee violates the rules of the organization, the employee concerned should be able to accept the punishment that has been agreed upon. Participants in disciplinary problems, both superiors and subordinates in the organization, will provide patterns of organizational culture. In most modern views of work, it is said that work discipline is a fundamental part of the human condition, that is, essential. At its most basic, it might affect the reputation of the organization as well as bind the employees to their responsibilities. "Discipline is a form of training that ensures that the rules of the organization are followed" (Mathis, 2006, p. 511). There are two approaches to discipline, namely:

1. Positive discipline approach
The positive discipline approach relates to the philosophy that the offensive action can usually be corrected constructively without penalty. In this approach, managers are focused on finding out the facts and issuing guidance to encourage the desired behavior, rather than using punishment to prevent undesirable behavior.
2. Progressive discipline approach
Progressive discipline combines a series of steps, where each step becomes progressively harder. This is designed to change employees' inappropriate behaviors. A common system of progressive discipline, and most of the progressive discipline procedures, use verbal and written reprimands, as well as suspension, before dismissal.

3.4 Employee performance

A company or organization is established in order to achieve a particular purpose, and whether or not the organization will be able to achieve its goals is influenced by its behavior. One of the most common activities undertaken within the organization is the work of the employees, which relates to how they do everything associated with any work or role within the organization. According to Mangkunagara, in Mulyadi (2015, p. 63), 'The term is derived from the performance of *Job Performance* or *Actual Permormanse* (actual job performance or achievements attained by someone). So the meaning is the performance, the result of the quality and quantity of work accomplished by an employee in performing their duties in accordance with the responsibilities given to him. Understanding the performance or *the performance of* an overview of the level of achievement of the implementation of a program of activities or policies in realizing the goals, objectives and mission of the organization's vision that pour through the strategic planning of an organization'. Moeheriono (2012, p. 69) "points out that the meaning of the word performance comes from the term *job performance* and it is also called *actual performance* or work performance, or the achievements of an employee".

4 RESEARCH METHODOLOGY

4.1 Place and time of research

In conducting this study, the authors took the General Bureau, Human Resources and Organization of the Ministry of Tourism as the research object, which is located in Sapta Pesona Building, Jalan Medan Merdeka Barat No. 17, Central Jakarta 10110.

4.2   Research methods

In this study, the researchers used qualitative methods. The qualitative research method, according to Lodico, Spaulding, and Voegtle, in Emzir (2014, p. 2), is also called "interpretive research or field research, and is a methodology that has been borrowed from disciplines such as sociology and anthropology and adapted to the educational setting". A qualitative researcher uses inductive reasoning and strongly believes that there are many perspectives that will be disclosed. Qualitative research focuses on social phenomena and capturing the feelings and perceptions of the participants being studied. It is based on the belief that knowledge is generated from social settings and that the understanding of social knowledge is a legitimate scientific process. The researchers will record, manage, present, interpret and analyze the data in order to describe the Analysis of Leadership Roles and Discipline Level Work in Improving Employee Performance at the General Bureau, Human Resources and Organization of the Ministry of Tourism.

4.3   Research focus

This study is focused on analyzing the role of leadership and the level of work discipline needed to improve the performance of the employees of the General Bureau, Human Resources and Organization of the Ministry of Tourism. The approach used concerns legislation, which is to see how the formulation of norms, rules and related legislation form the basis of the analysis of leadership roles and the level of labor discipline needed to improve employee performance at the General Bureau, Human Resources and Organization of the Ministry of Tourism.

4.4   Data types

Two kinds of data were used in this study, namely:

a. Primary data
According to Sujarweni (2014, p. 73–74), primary data is data obtained from respondents through questionnaires, focus groups and panels, or data obtained from interviews with informants and researchers. The data sources directly provide data to the data collectors. In this study, the primary data was gathered through interviews with the informants, namely the General Bureau, Human Resources and Organization of the Ministry of Tourism.
b. Secondary data
According to Sujarweni (2014, p. 74), secondary data is data obtained from records, books, and magazine publishing companies in the form of financial reports, government reports, articles, magazines, and so forth. Data obtained from secondary sources does not need to be processed again, as the sources do not directly provide data to the data collector.

4.5   Data collection technique

According to Sujarweni (2014, p. 74), the data collection techniques used can be explained as follows:

a. Interview
According to Sujarweni (2014, p. 31), this 'is the process of obtaining clarification interviews to collect information using a question and answer which can be done either face-to-face or not, namely through the medium of telecommunications between the interviewer with the interviewee, with or without the use of guidelines'. In other words, the interview is an activity undertaken to obtain in-depth information about an issue or a theme raised in the study.
b. The literature study (library research)
This research was conducted by studying, researching, assessing and interpreting the literature, which was in the form of books, legislation, magazines, newspapers, articles, websites and previous studies that have a relationship to the problems examined. The aim of the literature study was to obtain as many theories as possible that could be expected to support the data collected and further the process of the study. Researchers do not just rely

on their ability to analyze the research object in order to analyze qualitative data. It also needs other capabilities, such as sensitivity to all of the aspects studied, which will allow researchers to analyze the data.

4.6 *Data analysis technique*

According to Kriyantono (2009, p. 63), "the analysis of the data is the process of organizing and sorting data into patterns, categories and basic descriptive units that have a theme". A working hypothesis is then formulated from the analysis of the data. Data obtained from the research results will be analyzed using a descriptive qualitative technique, and will then be interpreted and analyzed in order to provide information for solving problems.

4.7 *Data validation techniques*

According to Tresiana (2013, p. 142), "in order to maintain the correct level of validity, reliable media research is needed to eliminate a degree of culpability and avoid bias or false validity". The technical examinations of the validity of the data used in this study are:

a. Proof by extending the participation of researchers researcher regularly visited to the site for one month.
b. Proof through persistent observation
To demonstrate the problems that could arise in this study, the authors continually observed and provided analysis based on a defined research focus.
c. Proof through triangulation
1. Triangulation of data, which utilizes a variety of sources. In conducting this research, the authors utilized various data sources, such as documents, journals and legislation. Moreover, the authors compared this data with the results of interviews obtained in the field.
2. Triangulation of researchers, which involves a variety of different investigators with a scientific background. In this case, the authors studied other research that has been done and made comparisons.
3. Triangulation theory, in which the authors interpret the data obtained using several theoretical perspectives.
4. Triangulation method, in which the authors do a cross check to verify the data and the information obtained in the field.

5 RESEARCH RESULTS

5.1 *Leadership role in improving employee performance in the General Bureau, Human Resources and Organization of the Ministry of Tourism*

A person in the role of leader in the General Bureau, Human Resources and Organization of the Ministry of Tourism has a part to play in communicating, directing and providing guidance. Their role is to motivate the formation of a conducive working environment, provide supervision and punishment, understand the constraints inherent in motivating others and find ways to overcome problems arising in the provision of motivation. The leadership role at the General Bureau, Human Resources and Organization of the Ministry of Tourism involves improving the performance of employees, which is a very important role, because when there is good leadership, employees tend to undertake their job with enthusiasm and excitement, so optimal results will be achieved that will support the efficient and effective achievement of the desired objectives.

Rules can improve employee performance in the General Bureau, Human Resources and Organization of the Ministry of Tourism, as shown by how employees are disciplined, how punctual they are, and how firmly their leader upholds the rules. On the other hand, it can be said that if the leader has a gentle nature, shows courtesy toward employees, nurtures subordinates, shows no favoritism and treats everyone as being equal, then a sense of togetherness unites the employees of the General Bureau, Human Resources and Organization of the Ministry of Tourism, which ultimately makes the employees very enthusiastic about their performance, although sometimes they feel reluctant to criticize their leader, because the leader is regarded as very nice.

a. Leader's ability in solving problems

In the current era, leaders in the General Bureau, Human Resources and Organization of the Ministry of Tourism have wider powers and are faced with complex issues. There are several problems that occur in the General Bureau, Human Resources and Organization of the Ministry of Tourism, such as the degradation of the performance of an employee, which certainly requires effort from those in leadership roles in order to overcome this problem and produce quality improvements.

The leadership role in the General Bureau, Human Resources and Organization of the Ministry of Tourism involves resolving some of the problems that exist, such as managing the various resources. For example, by working together with employees to tackle the downturn in employee performance by providing the space and opportunity for the Bureau of Public Employees, Personnel and Organization of the Ministry of Tourism to further improve their competence in order to improve employee performance.

Thus there is a pattern of co-operation that can help to produce the desired improvement in performance. The leadership role is important in the pursuit of the desired performance in every organization. Organizations will only prosper if they are led by a visionary leader, who has the managerial skills, as well as the personal integrity, to implement an improvement in performance. Leadership General Bureau, Human Resources and Organization of the Ministry of Tourism would follow the directions given by the organization leader.

b. Realizing vision and mission strategy

The vision and mission that has been formulated by the General Bureau, Human Resources and Organization of the Ministry of Tourism requires a leader to make it happen. The head of the General Bureau, Human Resources and Organization of the Ministry of Tourism must have the right strategy in order to achieve this vision and mission. Leaders should continuously disseminate the vision, mission and objectives of the General Bureau, Human Resources and Organization of the Ministry of Tourism to all employees through various forms of activities that can eventually help them to develop a sense of awareness and understand the importance of the vision, mission and goals of the Ministry of Tourism.

One of the important aspects of leadership roles in realizing the vision and mission of the Ministry is to create an awareness of the need to support the implementation of the activity programs as a form of collective responsibility, which could lead to an improvement in the quality of tourism in Indonesia. A leader must have the ability to ensure the organization is able to work together to realize the vision and mission of the Ministry of Tourism.

c. Decision-making ability

In relation with decision-making, which normally follows a problem-solving stage, leadership on General Bureau, Human Resources and Organization of the Ministry of Tourism functions as the basis of all human activity and is targeted individually and as a group both institutionally and organizationally. In order to support the process of making the right decisions, every branch of General Bureau, Human Resources and Organization of the Ministry of Tourism needs to have a good information management system, because every decision needs to be supported by information that is fast, precise, and accurate. The need for that type of system is felt when we are faced with the increasingly fierce competition that occurs today.

The leaders of the General Bureau, Human Resources and Organization of the Ministry of Tourism need to have the ability to take decisions that involve all of the affected constituents, and to make sure that there is a mutual decision, which is not the decision of the General Bureau, Human Resources and Organization of the Ministry of Tourism alone. Whatever decision the General Bureau, Human Resources and Organization of the Ministry of Tourism makes is certainly important, both for themselves and for others. As we often see, every decision made by the leader of the General Bureau, Human Resources and Organization of the Ministry of Tourism is eagerly awaited by various groups, who each have their own intentions and interests. Decisions made by leaders at the General Bureau, Human Resources and Organization of the Ministry of Tourism are very important, serious and influential and impactful in the organization, even when only very few people are involved. This fact gives an indication that any decision taken by the leader of the General Bureau, Human Resources and Organization of the Ministry of Tourism should not only be taken carefully, but also firmly and boldly. Therefore, a good leader should be able to both solve problems and make decisions. The decisions that are made by the leaders of the General Bureau, Human Resources and Organization of the Ministry of Tourism will have a more meaningful quality when they are made jointly and involve all members of the group, rather than just making individual decisions.

d. Ability to communicate

The communication undertaken by the leader of the General Bureau, Human Resources and Organization of the Ministry of Tourism plays an important role in improving employee performance. Without communication, the organization will come to a standstill. The Bureau of Public Communications, Human Resources and Organization of the Ministry of Tourism aims to provide and receive information, influence others, solve problems, make decisions, and evaluate effective behavior. Without communication, some of these objectives will not be achieved. In order to improve the effectiveness of the organization in achieving their targets, the role of communication is very important and strategic. Distortion, ambiguity, uncertainty and conflict will increasingly disadvantage the organization, until all members of the organization realize the need for effective communication. This means that, through communication, the development of the organization will become a reality.

There are efforts to achieve effective communication among the clerks and other parties at the General Bureau, Human Resources and Organization of the Ministry of Tourism, as its leaders seek to understand the various information deliveries as either input or criticism. Communication occurs in both directions. In order for the employees of the General Bureau, Human Resources and Organization of the Ministry of Tourism to be able to perform their functions, General Affairs, Human Resources and Organization of the Ministry of Tourism should have a good competence in these functions. The level of competence needed by the Chairman of the General Bureau, Human Resources and Organization of the Ministry of Tourism includes the ability to put together a program for the Ministry of Tourism, organize personnel, empower the employees of the Ministry, and utilize the resources optimally. Based on the results of the statements about leadership, it can be concluded that the leadership role can improve the performance of the employees of the General Bureau, Human Resources and Organization of the Ministry of Tourism, as shown by the fact that if the employees are disciplined and their leader upholds the rules firmly, then they are punctual at work. On the other hand, it can be said that if the leaders have a gentle nature, are polite to employees, are protective and nurturing to subordinates, show no favoritism and consider everyone to be equal, then a sense of togetherness becomes embedded in the organization, which ultimately makes the employees very enthusiastic about their performance. However, when a leader shows those characteristics, the employees might doubt his credibility. So, if the employment situation is very supportive, this may result in the desired performance being achieved.

5.2 *Work discipline in improving performance at the General Bureau, Human Resources and Organization of the Ministry of Tourism*

Based on the research that has been done regarding work discipline, it is shown that employees must be:

a. Members of The Bureau of Public Employees, Personnel and Organization of the Ministry of Tourism who are always present during working hours, which are set in accordance with Government Regulation No. 53 of 2010 regarding civil servants' working hours.

b. Employees who obey the rules of the General Bureau, Human Resources and Organization of the Ministry of Tourism, especially those concerning disciplining employees. In particular, those who obey the Government Regulation No. 53 of 2010, regarding the hours of work, how to dress, ethics, and procedures for carrying out their duties and functions, as well as their level of achievement.

c. Employees who adhere to the specified standards of work. For example, employees who are not only always present during working hours, but who also implement their duties and functions in accordance with the standard operating procedures that were established in order to achieve the expected performance.

d. Employees who have a high level of vigilance, which means that, while carrying out their duties and functions, they are always careful to stick to the rules and are able to regulate the use of work equipment properly according to the needs of the organization. This is especially the case in financial processing, which should always be guided by the Ministry of Finance and other applicable rules.

e. Employees who can work ethically, which means carrying out their duties and functions while maintaining order and good relations with fellow employees and the general public. This is particularly important as the Civil Service is a servant of the state.

A high level of discipline shows that an employee respects the rules and regulations in an organization, which is, in this case, the General Bureau, Human Resources and Organization of the Ministry of Tourism. Thus, if there are rules and regulations that are ignored, or frequently violated, this means that the employee has poor discipline and vice versa.

Based on the research that has been conducted, there are five indicators that affect employee discipline. The objective of discipline is to improve efficiency as much as possible in order to prevent the waste of time and energy. In addition, discipline is applied in order to try to prevent damage to or loss of property, equipment and working equipment caused by carelessness. Discipline is applied as an attempt to try to overcome mistakes and negligence caused by inattention, incompetence and delay. Discipline prevents the onset of slow work and tries to overcome differences of opinion among employees. Here are five indicators of labor discipline:

5.2.1 *Attendance*

Attendance is one of the most important indicators when assessing employee discipline, as employees should always be present during the predetermined working hours. The statement explained that the employees were disciplined so that they were not only present, but present in accordance with a predetermined schedule and were at work until it was time to go home.

Employee discipline was seen to be good if they were present, and could also be demonstrated by employees who arrived on time, worked during the appointed hours, were at work during working hours, did not leave the workplace except when there were very important circumstances and always used their working hours to do what needed to be done.

5.2.2 *Obedience to work regulations*

Regulations are also a major factor of concern when assessing discipline, because discipline occurs when a person obeys the work rules. Every organization has rules regarding how duties and functions should be carried out. The regulations are guidelines that must be followed, and this is one measurement tool for the level of employee discipline.

One of the rules for disciplining employees, which is based on the Government Regulation No. 53 of 2010 on Discipline and other regulations, such as guidelines for the implementation of tasks and functions, such as procedures for correspondence and archivists, should be used as guidelines by the employees when implementing mailings discipline so that they can produce letters based on the guidelines and the rules of correspondence that have been set.

The level of obedience to work rules affects employee discipline at the General Bureau, Human Resources and Organization of the Ministry of Tourism. If employees do not follow the work rules this will influence the achievement of organizational goals, because the goals are less likely to be achieved. For example, there are the rules on the distribution of scholarships. If the guidelines set by the General Bureau, Human Resources and Organization of the Ministry of Tourism are not followed, the results or objectives distributed to recipients will not reach the desired target. This is because other elements are being ignored and there are other duties that should really follow the guidelines, in particular with regards to financial problems.

The level of discipline regarding obedience to the work rules at the General Bureau, Human Resources and Organization of the Ministry of Tourism can be demonstrated by the ability of employees to complete the work mandated by their leader on time and in accordance with the procedures. There are still employees who break the rules, but not grave violations, such as reporting late, not typing letters in accordance with the rules of correspondence, and other issues. In the last year, no employee was sanctioned for violations of the labor legislation, which can also be seen as a sign of a good level of leadership.

From the above, it is clear that obedience to the work regulations obviously affects the discipline of an employee, because a violation of work rules is a violation of labor discipline. This has consequences for human resource audit issues, which matters because they are a form of violation of the labor laws.

5.2.3 *Obedience to labor standards*

The standard of work is also a factor of concern when assessing the discipline of employees at the General Bureau, Human Resources and Organization of the Ministry of Tourism, as good discipline is shown when an employee is able to produce work in accordance with the required standard. Every job must be based on a required standard, which provides a measurement of an employee's success in carrying out their duties and functions. An employee at the General Bureau, Human Resources and Organization of the Ministry of Tourism is seen to have good discipline if they are subject to and adhere to specified standards of work.

Employees who are able to produce work according to the standard output show that they are disciplined employees, and vice versa, because when the output is not in accordance with the required standard, the results will not be good. For example, if there are procurements, bonuses or rewards from the head of the General Bureau, Human Resources and Organization of the Ministry of Tourism to an employee, then these bonuses or rewards must be in accordance with the prescribed rules.

5.2.4 *High level of vigilance*

A high level of vigilance is also an area of concern when assessing discipline at the General Bureau, Human Resources and Organization of the Ministry of Tourism, as an employee who is always careful and rigorous in their actions, and who uses the facilities at the General Bureau, Human Resources and Organization of the Ministry of Tourism efficiently, implements the job description of a disciplined employee. A high level of vigilance needs to be considered, particularly with regards to job description, so that output is in accordance with the applicable standards and regulations.

At the General Bureau, Human Resources and Organization of the Ministry of Tourism, a high level of vigilance needs to be applied, especially in the line of duty, because a lack of caution will cause a breach of discipline that will have an impact on the achievement of their objectives and the employees' obligations as executor.

5.2.5 *Work ethic*

Work ethic is also an area of concern when assessing discipline at the General Bureau, Human Resources and Organization of the Ministry of Tourism, because employees who are friendly, have mutual appreciation and respect, and are honest are the embodiment of a disciplined employee. The organization's goals cannot be achieved if there is poor co-operation among employees or no clear division of labor. In order to achieve this, every employee is required to show a good work ethic, which reflects the level of employee discipline.

A good work ethic will create an atmosphere and habits that support the establishment of discipline. At the General Bureau, Human Resources and Organization of the Ministry of Tourism, it is required that every employee should be friendly, have mutual appreciation and respect, and also be honest, particularly when providing services to fellow employees, and especially the community. Each employee is expected to have good work ethic and discipline.

Based on the results of the assessment of labor discipline, it can be concluded that in order to improve the discipline of the employees working at the General Bureau, Human Resources and Organization of the Ministry of Tourism, special attention should be paid to supervision in order to obtain a high level of discipline. The indicators of labor discipline greatly affect the future performance of employees at the General Bureau. The expected qualities of an employee who consciously abides by the disciplinary rules and regulations can be used to educate employees to obey the legislation procedures and policies so as to generate a good performance.

Discipline helps to improve employee performance, which is one key to organizational success. Work discipline is crucial for improving employee performance as it helps to drive attitudes and personal behavior, enabling someone to be able to act and behave in accordance with the rules established to support the achievement of organizational goals. Therefore, as human resources within an organization, employees should promote, direct and improve their work discipline in order to facilitate their tasks and work as an employee.

REFERENCES

Bangun, W. (2011). *Management of Human Resources*. Bandung, Indonesia: Erlangga.
Emzir. (2014). *Qualitative Research Methodology for Data Analysis*. Cet. 4. Jakarta: Rajawali Press.
Feriyanto, A. & Triana, E.S. (2015). *Introduction to Management (3 in 1)*. Kebumen, Indonesia: MEDIATERA.
Ghozali, I. (2005). *Multivariative Analysis Application using SPSS program, Body*. Semarang, Indonesia: Diponegoro University.
Hanggraeni, D. (2012). *Management of Human Resources*. Jakarta, Indonesia: Faculty of Economy, University of Indonesia.
Hasibuan, M.S.P. (2005). *Management of Human Resources* (Revised ed.). Jakarta, Indonesia: Bumi Aksara.
Hasibuan, M.S.P. (2012). *Management of Human Resources*. Jakarta, Indonesia: Bumi Aksara.
Kementerian Pariwisata 2015
Kriyantono, Rachmat. (2009). *Practical Techniques of Communication Research*. Malang: Prenada Media Group.
Law No. 5. (2014). Regarding the Nation's Civil Institutions.
Mangkunegara, A.A.A.P. (2010). *Management of Company's Human Resources*. Bandung, Indonesia: PT Remaja Rosdakarya.
Mangkunegara, A.A.A.P. (2011). *Management of Company's Human Resources*. Bandung, Indonesia: PT Remaja Rosda Karya.
Mathis, R.L. & Jackson, J.H. (2006). *Human resource management (Translated by Dian Angelia)*. Jakarta, Indonesia: Salemba Empat.
Moeheriono. (2012). "Competency Based Performance Measurement". Jakarta: Raja Grafindo Persada.
Mulyadi. (2015). *Management of Human Resources*. Jakarta, Indonesia: In Media.
Priyatno, D. (2011). *SPSS Statistical Data Analysis Pocket Book*. Yogyakarta, Indonesia: MediaKom.
Rivai, V. & Jauvani, E. (2011). *Management of Human Resources for Companies*. Jakarta, Indonesia: Raja Grafindo Persada.
Rivai, V. (2010). *Management of Human Resources for Companies From Theory to Practice*. Jakarta, Indonesia: Raja Grafindo Persada.
Siagian, S.P. (2008). *Management of Human Resources*. Jakarta, Indonesia: PT Bumi Aksara.
Siagian, S.P. (2014). *Management of Human Resources*. Jakarta, Indonesia: Bumi Aksara.
Simanjuntak, P.J. (2011). *Management & Evaluation of Performance*. Jakarta, Indonesia: Fakultas Ekonomi Universitas Indonesia.
Simanjuntak. (2005). *The Performance of Human Resources*. Jakarta, Indonesia: PT Raja Grafindo Persada.
Sugiyono. (2012). *Quantitative, Qualitative and R&B Research Methods*. Bandung, Indonesia: Alfabeta.
Sugiyono. (2014). *Quantitative, Qualitative and R&D Research Methods*. Bandung, Indonesia: Alfabeta.
Sujarweni, V.W. (2014). *Research Methodology*. Yogyakarta, Indonesia: Pustaka Baru Press.
Sutrisno, E. (2010). *Management of Human Resources*. Jakarta, Indonesia: Kencana. Prenada Media Group.
Sutrisno, E. (2011). *Management of Human Resources*. Jakarta, Indonesia: Kencana.
Sutrisno, E. (2014). *Management of Human Resources*. Jakarta, Indonesia: Pranada Media Group.
Suwatno & Priansa, D.J. (2011). *Management of HR in Public Organization and Business Management*. Bandung, Indonesia: Alfabeta.
Timpe, A.D. (2000). *Leadership*. Jakarta, Indonesia: PT Elex Mediaa Komputindo.
Tresiana, Novita (2013). *Qualitative Research Methods*, Research Institute University of Lampung.
Wahyudi, B. (2002). *Management of Human Resources*. Bandung, Indonesia: Sulita.
Wiranata, A.A. (2011). The Influence of Employees' Performance and Stress: A Case Study in CV Mertanadi. *Scientific Journal of Civil Engineering*, *15*(2), Juli 2011. Denpasar, Indonesia: Udayana University.

Wiratama & Sintaasih. (2013). The Influence of Leadership, Trainings, and Work Discipline on Employee Performance at PDAM Tirta Mangutama in Badung. *Journal of Management, Business Strategy and Entrepreneurship.* 7(2), 126. Udayana University.

Wirawan. (2009). *Evaluation of the Performance of the Human Resources Management.* Jakarta, Indonesia: Salemba Empat.

Wirawan. (2009). *Evaluation of the Performance of the Human Resources.* Jakarta, Indonesia: Salemba Empat.

Yani, A.T. (2011). *Management of Human Resources A Strategic Approach.* Bandung, Indonesia: Humaniora.

Yukl, G. (2005). *Leadership on Organizations* (5th ed.). Jakarta, Indonesia: PT Indeks.

Yuniarsih, T. & Suwatno. (2008). *Management of Human Resources.* Bandung, Indonesia: Alfabeta.

Innovation and strategic management

Business Innovation and Development in Emerging Economies – Trinugroho & Lau (Eds)
© 2019 Taylor & Francis Group, London, ISBN 978-1-138-35996-3

Redesigning business model on small and medium-sized enterprises

Faisal
Magister of Management, Faculty of Economics and Business, Universitas Indonesia, Indonesia

Sisdjiatmo K. Widhaningrat
Lecturer at Magister of Management, Faculty of Economics and Business, Universitas Indonesia, Indonesia

ABSTRACT: Small and medium-sized enterprises need to recognize which business model they employ to maintain their sustainability. Business model canvas is one of the instruments used to represent it. This research aims to analyze business models employed by small and medium-sized enterprises using the qualitative method. This study recommends business model reconstruction so that the small and medium-sized enterprises have the ability to compete in the industry.

Keywords: business model, business model canvas, small and medium-size enterprises

1 INTRODUCTION

Despite their size, small and medium sized enterprises (SMEs) have become the leading economic unit in creating employment. In OECD regions, small and medium-sized enterprises play a major role in economic growth. They provide 60%–70% employment and become a source for most new jobs in most countries (OECD, 2000).

In Indonesia, 99.99% of enterprises are SMEs and provide 97% of employment. They also contribute 57% in Gross Domestic Products (GDP) (Sarwono, 2015). On the other hand, there are many things that they have to do to be sustainable in the industry. According to the data, less than one-half of small start-ups survive for more than five years (OECD, 2000).

SMEs have less attention about long-term development and strategies. They are more concerned about how to survive in the industry (Frick & Ali, 2013). However to develop sustainable strategies, SMEs need to understand their competitiveness (Dudin, Lyasnikov, Leont'eva, Reshetov, & Sidorenko, 2015). Possessing knowledge about types of industries and several external and internal factors can help SMEs to survive. Creating a business model can help SMEs to understand, describe, and predict how things work by exploring a simplified representation (Frick & Ali, 2013).

The optimal business model is required to set up and develop business (Dudin et al., 2015). Osterwalder had developed a model that can be understood by everyone, namely the Business Model Canvas. For example, by utilizing Business Model Canvas, Morten Lund tells a story for companies on what could be their strategy (Lund, 2013).

The object of this paper is CV Azka Syahrani—one of the small and medium-sized enterprises in the Muslim fashion industry. This research analyzes the nine building blocks of Business Model Canvas to examine which part can be developed by enterprises to yield better returns and how to reconstruct it. The results of this study are expected to contribute to the enterprises' sustainability.

2 LITERATURE REVIEW

The business model describes the rationale of how an organization creates, delivers, and captures value (Osterwalder, Pigneur, Smith, & Movement, 2010). The business model is like a blueprint to implement strategy through organizational structures, processes, and systems. Osterwalder and Pigneur developed the Business Model Canvas to simplify the concepts of the business model that everybody can understand (Osterwalder et al., 2010). They believe that the business model can be described through nine basic building blocks that explain the logic of how a company intends to make money. Those nine blocks consist of customer segments, value propositions, channels, customer relationships, key resources, key activities, key partnerships, revenue streams, and cost structure. Osterwalder advises to use graphical icons in one page (Frick & Ali, 2013) to make the Business Model Canvas visually appealing (Fallis, 2013).

2.1 Key partners

Key partners can be defined as the network of suppliers and partners that make the business model work (Osterwalder et al., 2010). Network activities have essential roles because the firm does not have all resources and activities to create and deliver value by itself (Coes, 2014). The firm can create alliance to optimize its business model, reduce risk, or acquire the resources (Osterwalder et al., 2010).

2.2 Key activities

The key activities building block represents the actions that a firm performs in order to create and offer value propositions to its customers and make profit out of them (Muhtaro, 2013). These are most important actions the firm must take to operate successfully, or in other words, this is important to make the business model work (Osterwalder et al., 2010).

2.3 Key resources

Key resources are inputs and capabilities that the firm needs in order to deliver value to its customer (Muhtaro, 2013). These resources allow the firm to create and deliver value propositions, reach the market, maintain relationships with customers, and earn revenue (Osterwalder et al., 2010). It includes the physical and intellectual assets, human resources, and financial resources.

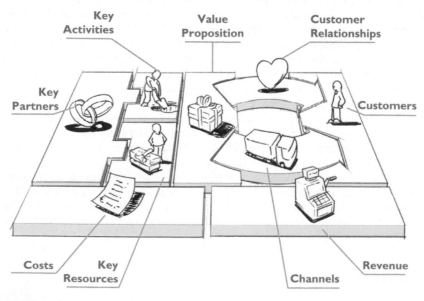

Figure 1. Osterwalder's business model canvas.
Source: Business model generation (Osterwalder et al., 2010).

2.4 Value propositions

Customer value propositions have a critical role in communicating how the firm aims to provide value to customers (Payne, Frow, & Eggert, 2017). Osterwalder defines value propositions as the bundle of products and services that create value for specific customer segments (Osterwalder et al., 2010). If the value propositions do not meet the needs of customer, the firm will not sustain in the future (Coes, 2014).

2.5 Customer relationships

Customer relationship describes the types of relationships a company builds with specific customer segments (Osterwalder et al., 2010). It can be considered as the lifeblood of a firm's business activity (Muhtaro, 2013). The aim of customer relationship is to sell more products or services by improving customer loyalty and finding new customers (Coes, 2014).

2.6 Customer segments

Customer segments can be defined as the different groups of people or organizations an enterprise intends to reach and serve (Osterwalder et al., 2010). A company needs to identify which segments the firm can serve effectively (Kotler & Keller, 2009). Once the segments are selected, the business model can be designed around a strong understanding of specific customer needs (Osterwalder et al., 2010).

2.7 Channels

Channels can be described as how the firm gets in touch with its customers and delivers the value propositions (Muhtaro, 2013). Channels are customer touch points that take important roles in the customer experience (Osterwalder et al., 2010). The way of communication, sales, and distribution strategy need to be accommodated to create good customer awareness about the proposed value (Coes, 2014).

2.8 Cost structures

Cost structure describes all costs incurred by the firm for delivering value propositions to its customers and doing all other business activities (Muhtaro, 2013). It can be characterized depending on business model as cost driven, value driven, fixed costs, variable costs, economies of scales, and economies of scope. Costs can be calculated easily after defining key resources, key activities and key partnerships (Osterwalder et al., 2010).

2.9 Revenue stream

Revenue stream describes the cash a company generates from each customer segment by delivering its value propositions (Osterwalder et al., 2010). It outlines the activities and pricing of the offered values with which a company improves its revenues (Muhtaro, 2013).

3 METHOD

This paper is adapted from the Business Model Canvas—a theoretical and practical concept developed by Alexander Osterwalder and Yves Pigneur. This research is based on a case study of a small and medium-sized enterprise in Muslim fashion industry in Indonesia. The data has been collected through observation at the company and unstructured interviews with 8 participants including Owner, Chief Executive Officer, and Head of Division in the company out of 20 employees. The participants were chosen based on their knowledge about the whole business process in the company.

The observation and interview guides consist of 9 main issues based on the research aims. The issues explored in the interviews and observations are concerned with who the key partners of the firm are, what the key activities of the firm are, what the key resources the firm needs, what value propositions are provided by the firm, what type of relationships the firm builds with its customers, what customer segments the firm aims towards, what channels the firm uses, what costs are incurred in firm activities, and where the revenue of the firm is generated from. After doing the interview and observation, verbal transcripts were made and analyzed with content analysis.

4 RESULTS AND DISCUSSION

In this section, the results of this study are described. The data was translated into the nine building blocks of the business model canvas. Each building block was analyzed to recognize whether there is any part that requires improvement. Then, some elements were changed and a new business model canvas was formed. The analysis in this paper was conducted by comparing the previous and new business model canvas.

4.1 *Key partners*

To create value to its customers, partners are needed by CV Azka Syahrani to supply resources and some activities. Fabric suppliers, tailors, and embroiders are the key partners for main production activities in CV Azka Syahrani. Agents are the key partners for distribution and selling channels. However, after evaluating the performance of key partners,

Table 1. Comparison of previous and new business model canvas.

Building block	Previous	New
Key partners	Fabric Suppliers, Tailors, Embroiders, and Agents.	Fabric Suppliers, Tailors, and Embroiders.
Key activities	Design, Production, Sales	Design, Production, Sales
Key resources	Human Resources (Management, Employees, Tailors, Embroiders, Agents) Production Assets (Factory, Gallery, and Machines)	Human Resources (Management, Employees, Tailors, Embroiders) Production Assets (Factory, Gallery, and Machines) Information Technology Systems
Value propositions	Quality of Products, Hand Made Embroidery, Sharia Model, ISO 9001:2015	Quality of Products, Hand Made Embroidery, Sharia Model, ISO 9001:2015
Customer segments	Men and Women, Young Family, 25–45 Years Old, Muslim, Indonesian Citizenship	Men and Women, Young Family, 25–45 Years Old, Muslim, Indonesian Citizenship
Customer relationships	Customer Service (WhatsApp, Social Media, Website) After Sales Service	Customer Service (WhatsApp, Social Media, Website) After Sales Service
Channels	Offline (Agents, Gallery, Bazaar, Door to Door, Word of Mouth) Online (Website, Facebook, Instagram)	Offline (Gallery, Bazaar, Door to Door, Word of Mouth) Online (Website, Facebook, Instagram)
Cost structures	Material, Labor, Overhead Cost, Salary Expenses, Marketing Expenses, Transportation, Utility Expenses, Agent's Discount, Withdrawal	Material, Labor, Overhead Cost, Salary Expenses, Marketing Expenses, Transportation, Utility Expenses, Withdrawal
Revenue stream	Direct Selling (Online and Offline Channels), Sales via Agents, Sales from Rejected Products, Special Order	Direct Selling (Online and Offline Channels), Sales via Agents, Sales from Rejected Products, Special Order

we found that agents contribute in high quantity of selling products, but they also cause loss for the company. Since the price they requested is far below the retail price, the agents demand 40%–42% discount on every product they bought. The owner and management of CV Azka Syahrani agreed to eliminate the agents and change their selling scheme to direct selling.

4.2 Key activities

The activities of CV Azka Syahrani start from deciding the theme of their products in a season, then the firm designs the model of its products. After the design is finished, the firm starts the mass production of its products. In one season, usually CV Azka Syahrani can produce 17.000–20.000 pcs. CV Azka Syahrani also does the promotion and selling activities. In this building block, there is no significant change.

4.3 Key resources

Production assets are one of the key resources for CV Azka Syahrani, such as machinery, factory, warehouse, and gallery. Another one is human resources which includes management, employees, tailors, and embroiders. Agents have been eliminated from the key resources building block as explained previously in the key partner's building block. The information technology system is added as one of the key resources, because CV Azka Syahrani wants to exploit online channels and strengthen their digital marketing.

4.4 Value propositions

The value that CV Azka Syahrani offers to customers is the quality of products. The product has been checked in the quality control division to make sure that the product is consistent with the standard. CV Azka Syahrani has been certified with ISO 9001:2015. Furthermore, CV Azka Syahrani has established the identity of their products which have handmade embroidery and are in accordance with Islamic Sharia Law. There is no revision in the value propositions building block.

4.5 Customer relationships

To maintain their relationship with customers, CV Azka Syahrani allocates 4–5 persons in charge in customer service activities. Customers can ask about the products or give their complaints on the company's social media, on the website, or by phone. Moreover, CV Azka Syahrani serves customers with after sales service such as changing or customizing the size of the product. This building block also remains unchanged.

4.6 Customer segments

Customer segments that CV Azka Syahrani intends to reach are young families with 1–2 kids and men and women between 25–45 years old. Because the price the company offers are relatively expensive, the company's target customers are those who are in middle and upper economic class. Furthermore, the company aims to reach Muslim Indonesian customers. The customer segment the firm intends to reach remains unchanged.

4.7 Channels

CV Azka Syahrani uses offline and online methods in its sales channel. The company uses social media such as Facebook, Instagram, website, and WhatsApp for its online channel. For their offline channel, CV Azka Syahrani uses direct selling in its gallery, bazaar and exhibition, and door to door selling. Before the agents were eliminated, the company also utilized the agents to make direct selling in broader areas.

4.8 Cost structures

In running its business, certain costs are involved in CV Azka Syahrani activities. The costs incurred are cost of goods sold such as direct material cost, direct labor cost (tailor and embroiderer), and overhead cost. Moreover, operational cost such as salary expenses, utility expenses, marketing expenses, and transport. Withdrawal by the owner is also incurred as the company's cost. Agent's discount has been eliminated from cost structures because there are no more agents in CV Azka Syahrani's schemes.

4.9 Revenue stream

For the revenue stream, CV Azka Syahrani generates cash from customer by selling its product in its gallery, bazaar and exhibition, and order via online channels. The company also receives special order for producing uniform. Moreover, company sells its rejected product in lower price. Since agents have been eliminated from partnership, there is no more revenue from agents.

5 CONCLUSION

The Business Model Canvas can be used to explain the strategies and operations that a company applies. In CV Azka Syahrani's case, by using the Business Model Canvas, the owner and management can get information about which area should be improved and how to improve it. This tool can be used by CV Azka Syahrani to plan new strategies to maintain the sustainability of the company.

The cost structure building block provides information to the management that all this time the costs that they have spent are higher than their revenue. It was caused by the agents who demanded too much discounts. After unsuccessful negotiation with the agents, the management decided to eliminate the agent scheme. On the other hand, to increase quantity of sales, the company uses online channels to promote and sell its products. The application of the Business Model Canvas in CV Azka Syahrani shows that the Business Model Canvas can be used for small and medium-sized enterprises and be useful for the management to set what strategies they can apply.

REFERENCES

Coes, B. (2014). Critically assessing the strenghts and limitations of the business model canvas. *Master Thesis*, 99.

Dudin, M.N., Lyasnikov, N.V. Evich, Leont'eva, L.S., Reshetov, K.J. evich, & Sidorenko, V.N. (2015). Business model canvas as a basis for the competitive advantage of enterprise structures in the industrial agriculture. *Biosciences Biotechnology Research Asia*, 12(1), 887–894.

Fallis, A. (2013). Criticisms, variations and experiences with business model canvas. *European Journal of Agriculture and Forestry Research*, 1(2), 26–37.

Frick, J., & Ali, M.M. (2013). Business Model Canvas as Tool for SME, 142–149.

Kotler, P., & Keller, K.L. (2009). *Marketing Management. Organization* (Vol. 22).

Lund, M. (2013). Innovating a business model for services with storytelling. In *IFIP Advances in Information and Communication Technology* (Vol. 397, pp. 677–684).

Muhtaro, F.C.P. (2013). Business Model Canvas Perspective on Big Data Applications.

OECD. (2000). Small and Medium-sized Enterprises: Local Strength, Global Reach. *Policy Brief*, 1–8.

Osterwalder, A., Pigneur, Y., Smith, A., & Movement, T. (2010). *Business Model Generation. Booksgooglecom* (Vol. 30).

Payne, A., Frow, P., & Eggert, A. (2017). The customer value proposition: evolution, development, and application in marketing. *Journal of the Academy of Marketing Science*, 45(4), 467–489.

Sarwono, H.A. (2015). Profil Bisnis Usaha Mikro, Kecil Dan Menengah (UMKM). *Bank Indonesia Dan LPPI*, 5–57.

Business Innovation and Development in Emerging Economies – Trinugroho & Lau (Eds)
© *2019 Taylor & Francis Group, London, ISBN 978-1-138-35996-3*

Exploring the empirical framework of sustainable innovation-supporting resources and their role in the control of market targets (An empirical study of Indonesian batik SMEs)

Sudarwati
Universitas Sebelas Maret and Universitas Islam Batik, Surakarta, Indonesia

A.I. Setiawan
Universitas Sebelas Maret, Surakarta, Indonesia

ABSTRACT: This research aims to build a model of sustainable innovation-supporting resources, and their role in controlling market targets. This model involves a number of other supporting variables such as market literacy, consumer-preference influence, and technology accessibility. The view of this research framework is relevant to implementation in batik Small-to-Medium Enterprises (SMEs). From 320 SMEs of batik in Solo, 127 are selected as the sample by using a purposive sampling technique and the data is taken from the Cooperatives Department of Surakarta. Data analysis uses the partial least square technique with *WarpPLS 5.0* software.

The results show that market literacy has a positive effect on consumer-preference influence. Market literacy has no effect on sustainable innovation. Sustainable innovation has a positive effect on niche market control. Consumer-preference influence has a positive effect on sustainable innovation, as does technology accessibility. Consumer-preference influence has a positive effect on niche market control.

Sustainable innovation, as the novelty in this research, contributes to market target control.

Keywords: sustainable innovation, market literacy, consumer-preference influence, technology accessibility, niche market control

1 INTRODUCTION

Research on product innovation is still very interesting because it involves some variables which cannot be solved partially, such as in the research conducted by Weiss (2003), Salavou (2008), and Najib and Kiminami (2011). Weiss (2003) states that in the more competitive era, a company should be able to take decisions to create innovation in very diverse market conditions. One company effort to beat the competition is through product innovation. Product innovation can be the best strategy to support a company in maintaining its market share Salavou *et al.* (2004).

In their research, Lin and Chen (2007) state that innovation can be seen from the perspectives of an individual, organization and nation that focus on personal traits, management innovation and the nation's competitiveness. Scholars from various disciplines have been exploring innovation from different perspectives. The difference in research area allows other researchers to obtain better results in understanding the nature of product innovation. Holtzman (2008) also emphasizes the importance of organic and innovative growth as being a company's opportunity to achieve success in the future. A sustained acquisition strategy and process improvement has proven to be successful, yet it is hard to maintain, costly and risky. Other researches also explain that many companies conduct innovation, both technological and marketing (Lin & Chen, 2007; Chen *et al.*, 2012; Meroño-Cerdáns *et al.*, 2008).

Innovation is also said to be a successful key to creating new products, new markets and industry, so that innovation is conducted so companies can survive and sustain themselves (Alpay *et al.*, 2012). This research framework supports innovation of Small-to-Medium Enterprises (SMEs), which gives outstanding contribution in economic development and the effectiveness of company's competitiveness. In their research, Eshlaghy *et al.* (2011) examined 82 small and medium companies in Teheran, showing the importance of the innovation role, which can make positive contributions toward company performance. The company's role in dealing with the fluctuating environment definitely needs innovation. Innovation holds the major role in competitive advantage and the ultimate performance.

Innovation has an important function in management because it relates to a company's performance. This is shown by Zhou (2006), Salavou (2008), Oke *et al.* (2007),Willison and Buisman-Pijlman (2016), and Lin and Chen (2007). Some researchers also explain that innovation is the main key for a company to achieve superior performance (Zhou, 2006; Najib & Kiminami, 2011; Hian *et al.*, 2007; Keskin, 2006).The gap in this research is the absence of a similar understanding of the factors causing the emergence of product innovation and its results. Batik SMEs, as one part of the SMEs in Indonesia, represent one of the industrial strategies in the Indonesian economy. The opportunity for creative economic development of batik has gained increasing momentum after the inauguration of batik as a world heritage by UNESCO in 2009 (Isvandiary, 2016).

This recognition at least obtains brand awareness for batik that is more real in the eyes of the world and strategic to batik's existence in Indonesia. On the other side, it also strengthens the batik industry in Indonesia in penetrating the international market. Batik is a part of Indonesian tourism, and emerges as a comparative advantage and provides economic benefit.

Batik industrial development in Indonesia increased by more than 300% in the 3.5 years to 2013 (Isvandiary, 2016). Based on data from the Ministry of Industry and Commerce (Kemenperindag), there are around 21,600 batik businesses in Indonesia. The batik creative economy has also been contributing toward the national economy with an export value of US$69 million. Furthermore, 99.39% of the 55,912 business units operating in the batik industry are small and micro businesses, with domestic consumers representing more than 72.86 million people. Currently, the batik workforce involves around 3.5 million people spread over various regions, and of course it is very significant in contributing to ward job creation and increasing citizens' incomes.

The batik industry growth is known to have risen significantly. This has resulted in an economic increase for citizens because batik is an industry that involves millions of artisans. Nevertheless, the batik industry is facing challenges when encountering the modern textile industry, which is more decorative, secular and very dynamic in its refreshment cycles. In general, the modern textile industry takes many varied forms, with many and free styles, controlled by the development of various concepts in art and external market control, according to the needs and demands of today's modern society (in fashion). Thus, the national batik industry should act carefully and intelligently in dealing with the conditions associated with the high-fashion modern textile industry. The batik SMEs must create and innovate in order to produce batiks able to enter the high-end global fashion world without sacrificing their batik characteristics.

For Indonesia, the creative economic industry becomes an opportunity as well as a challenge for economic enhancement. The development of the batik creative industry is expected to give two benefits at once: first as leverage for economic growth, and second in strengthening the nation's cultural identity. Thus, batik creative economic development, as one of 14 creative economic components, should continue to be developed and enhanced.

Batik SMEs are also facing tighter competition with the opening of the domestic market. The phenomenon of free trade with China has triggered an influx of Chinese products into Indonesia, including batik. The Central Bureau of Statistics (BPS) data shows that in 2013 almost 56.3 tons of batik from China had been imported into Indonesia, worth Rp14.5 billion, either in the form of clothing or fabric at relatively low prices. This situation means that Indonesian batik products face severe competition. Chinese batik is predicted to dominate market share through low prices and interesting patterns. However, batik from China is printed batik, not handmade batik. Therefore, batik entrepreneurs in Indonesia need to increase their creativity and maintain traditional patterns, and be able to meet market demand in order to promote healthy competition and mutual benefit.

Based on these studies, the presenter search aims to explore the empirical framework of sustainable innovation-supporting resources and their role in controlling market targets, and focuses on empirical study of the effort to create innovative products by batik SMEs.

2 LITERATURE REVIEW

2.1 *Market literacy*

Literacy is one's ability to manage and understand information while reading and writing. A company that has the ability to figure out the cause of an increased offer and demand for its product, will be able to know a competitor's strategy. Literacy has a broad meaning. It can mean technologically literate, politically literate, critical thinking, and being sensitive to the surrounding environment. Literacy is one's ability in using written and printed information to enhance knowledge, so it provides benefits to society.

According to (Narver & Slater, 1990), the attention and response to competitors are the seller's understanding of the strength and weakness of competitors in the short-term and the capability and strategy of main and potential competitors in the long-term. The company that is able to control a consumer's demand is the company that has the capability to direct the consumer's preference.

A company's capability in fostering a commitment influences the business performance and the hope of business partners in establishing long-term relationships (Izquierdo & Gutiérrez Cillán, 2004). A company that has the capability to meet the consumer's needs and understand consumer behavior on trends, tastes, designs and models will be able to provide influence or persuasion to the consumer, so that it can increase sales volumes. A company that has the capability to manage and understand information and read the current situation will be able to produce the new products being demanded by the consumer. Based on these thoughts, two hypotheses are determined as follows:

H1: Literacy market has a positive effect on customer-preferences
H2: Literacy market has a positive effect on sustainable innovation

2.2 *Sustainable innovation*

Innovation is an organization's ability to adopt and implement new ideas, processes or products. The combination of resources and a company's organization creates the innovation strategy for the company, which influences its competitive advantage. In order to be able to compete, the company should be able to meet consumer demands by being able to direct the consumer's preferences. Innovation is one of the basic tools to be considered for a company's strategic growth in entering a new market, which supports market share and equips the company to compete. The ability for product innovation is the ability owned by the company to create new products that confer competitive advantage over its competitors. Production innovation is an activity to make a change in production in order to reduce production and operational costs, increase production numbers, and/or increase product quality, ensuring that production can run as efficiently and effectively as possible. Product innovation is a process that attempts to provide a solution to current problems. The most common problem in business is the quality product but expensive, or the cheap product but has a bad quality.

The company which has an ability to create new products, and to have high competitiveness in the end, will be able to control the market in various lines and even control a market that has not been explored by its competitors. The effect of product innovation on business performance shows that competitive advantage and business performance are influenced by product innovation, as mentioned in the research conducted by (Weiss, 2003; Helen Salavou & Avlonitis, 2008; Najib and Kiminami, 2011). According to the research of Weiss (2003), a company should be able to create innovation under a variety of market conditions. One of the focuses of a company in beating the competition is in creating product innovation. Product innovation can also be the most appropriate strategy to support a company in maintaining its market share (Salavou *et al.*, 2004).

According to Verhees (2005), product innovation in a small company is controlled by its owner and the small company usually innovates products by modifying its own products. Product innovation in Verhees's research uses experimental indicators as something new and accepts changes in the product as a form of the newness. To establish its newness, a company should be able to adapt to changes in the future. Based on that thought, the third hypothesis is determined as follows:

H3: A sustainable innovation has a positive effect on market niche control

2.3 *Technology accessibility*

Accessibility is a broad and flexible concept. Kevin Lynch says that accessibility is a matter of timing and also depends on the attraction and identity of the travel route (Aslam et al., 2017). New technology accessibility is the degree of ease with which a company attains new technologies. John Black says that technology accessibility is a measure of the convenience or ease of acquiring new technologies and relationships with each other, and the accessibility in achieving location is based on a transportation (Yulianto, 2017). The ease with which a company obtains new technology for new product development triggers the company to continually create new products according to consumer demand. Based on that thought, the next hypothesis is determined as follows:

H4: Technology accessibility has a positive effect on sustainable innovation.

2.4 *Influencing consumer preference*

Consumer preference can also be defined as being the likes, choices or things that consumers prefer. This preference is formed by the consumer's perception of the products Munandar & Udin (2013). According to Kotler (2000), consumer preference is someone's like or dislike of the products (or service) consumed. Influencing consumer preference gives an influence or persuasion to the consumer to like the product the company produces. If a company can direct consumer preference, it can control market share, which can eventually increase sales performance.

Market niche is a small group with small market share that has a huge power compared to overall market shares. This group attempts to create new innovations which will attract consumers of the market share and has very high competitiveness. Controlling market niche is the company's ability to control its consumers from various existing lines, or controlling market shares which have not been explored by other competitors. By controlling this market niche, the company can increase its sales performance. Bharadwaj et al. (2012) state that a

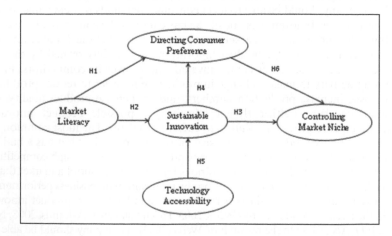

Figure 1. Research model.

278

focus on customers is the basis of competitive advantage. Based on that thought, two further hypotheses are determined as follows:

H5: Consumer preference has a positive effect on sustainable innovation.
H6: Consumer preference has a positive effect on the control of market niche.

Based on the above explanation, a frame work for sustainable innovation in the control of market niche can be created, as shown in Figure 1.

3 RESEARCH METHOD

3.1 *Data collection*

This is quantitative research with a survey using a questionnaire. The research objects are small-to-medium-sized batik enterprises. The data is taken from the Cooperatives Department of Surakarta. There are 320 batik SMEs in Solo, and 127 are selected as the sample. The sampling technique uses simple random sampling in which the sampling is from the random population members, regardless of the stratum in that population.

This research used a partial least square technique with *Warp PLS 5.0* software. This research aims to examine the validity and reliability of the research instrument, confirm the model accuracy, and examine the effect of one variable on the other variables.

3.2 *Operational definition and indicator identification*

Operational definition is a description of the variable boundaries in question or about what is measured by the relevant variable. There are five indicators on each variable so that it can clearly describe each variable. Sustainable innovation is a process that takes into account environmental, social and financial sustainability, integrated in the company from initial idea to Research and Development (R&D) and commercialization. Each variable can be measured by several indicators such as the strength to produce new products, being able to produce new designs, being able to develop new users, being able to modify products with other materials, and having a unique value of product with a quality and price different from the competitors (Verhees, 2005; Weiss, 2003; Boons & Lüdeke-Freund, 2013).

Technology accessibility is a measure of convenience or ease of acquiring new technologies and relationships with each other, and accessibility in achieving location is based on through transportation. Some important indicators are being able to access new technologies for new product development, producing products which have advantages over competitors, being able to get services fast and effectively, being able to develop the facilities which support the comfort and interest of users, and knowing the consumer condition related to their residents (Aslam et al., 2017).

Market literacy is an ability to understand the knowledge and skills needed to manage the market, product, price, sales, and resources to increase profits. The variable has indicators such as understanding the trend changes, knowing the consumer's needs, being able to explain the products disliked by the market, being able to explain the causes of increased demand, and being able to understand consumer behavior (Narver & Slater, 1990).

Influencing consumer preference gives an influence to or persuades the consumer to like the product being produced. The variable could be defined as being able to increase the loyalty, being able to control consumer behaviors, and being able to adjust consumer tastes (Munandar & Udin, 2013).

Controlling market niche is the company's ability to control its consumers in terms of various existing lines or controlling market shares that have not been explored by other competitors. The indicators are things such as being able to develop the market overall (finding new users, new usages, and more users), being able to make the specifications of products, quality, services and price, being able to protect market share, being able to expand market share, and being able to become the cloner, adaptor and imitator for the market follower (Tangendjaja, 2014; Hamel & Prahalad, 2000).

4 DATA ANALYSIS

4.1 *Validity test*

A validity test is a test that is used to show the extent to which the measuring instrument used is measuring what is required. Ghozali (2012) states that validity testing is used to measure validity, or to check whether a questionnaire is valid or not. The validity test consists of a loading factor, Average Variable Extract (AVE), and communality. An indicator is valid when it has a correlation value above 0.7, the average variance extracted >0.5 and determinant validity > 0.5. The construct validity test consists of 25 existing indicators, however two indicators are outlayer. The two indicators are variables of consumer-preference.

Table 1 shows that by eliminating two indicators of the consumer-preference influence variable, the target values of *average variance extract, outer loading* and *communality* are fulfilled for all variables, that is, > 0.50. The determinant validity test matrix also shows that there are very strong significant positive relationships between one variable and the other variables.

4.2 *Reliability test*

The reliability test is used to measure the internal consistency in the research. There are two ways: by using composite reliability, and through Cronbach's alpha.

Table 1. Variables and indicator validity (Source: *WarpPLS 5.0* data processing output).

Variable	Indicator	Outer loading	AVE	Communality
Sustainable innovation	The strength to produce new products	0.821	0.580	0.761
	Able to produce new designs	0.834		
	Able to develop new users	0.792		
	Able to modify product with other materials	0.739		
	Have unique value of product, quality and price different from the competitors	0.796		
Technology accessibility	Able to access new technologies for new product development	0.846	0.601	0.775
	Produce products which have advantages over competitors	0.784		
	Able to get services fast and effectively	0.797		
	Able to get the facilities which support the comfort and the interest of the users	0.745		
	Know the consumer condition related to their residents	0.794		
Market literacy	Understand the trend changes	0.804	0.675	0.822
	Know the consumer's needs	0.785		
	Able to explain the products disliked by the market	0.841		
	Able to explain the cause of increased demand	0.839		
	Able to understand consumer behavior	0.838		
Consumer-preference influence	Able to increase loyalty	0.797	0.677	0.823
	Able to control consumer behaviors	0.854		
	Able to adjust consumer tastes	0.816		
Control of market niche	Able to develop the market overall (finding new users, new usages, and more users)	0.794	0.673	0.820
	Able to make the specifications of products, quality, services and price	0.805		
	Able to protect market share	0.780		
	Able to expand market share	0.871		
	Able to become the cloner, adaptor and imitator for market follower	0.847		

Table 2.   Reliability test (Source: *WarpPLS 5.0* data processing output).

Variable	Composite reliability	Cronbach's alpha
Market literacy	0.912	0.880
Technology accessibility	0.882	0.832
Sustainable innovation	0.872	0.815
Consumer-preference influence	0.862	0.761
Control of market niche	0.911	0.878

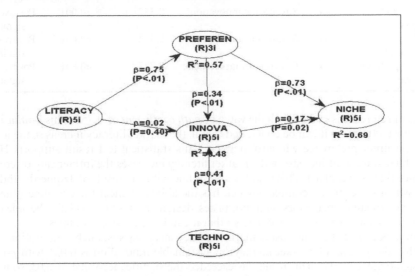

Figure 2.   Inner model.

The results of the reliability test in Table 2 show that the values of *composite reliability* and *Cronbach's alpha* are both > 0.70, such that the value has exceeded *rule of thumb,* that is, > 0.60. This result can also be interpreted in terms of whether the respondent is consistent in answering the questionnaire questions.

4.3   *Structural model analysis*

Table 3 shows that market literacy on consumer-preference influence has a *path coefficient* value of 0.7555 and *p-value* of <0.001 (significance <0.05). This means that market literacy has a positive significant effect on consumer-preference influence. Market literacy on sustainable innovation has *path coefficient* value of 0.021 and *p-value* of 0.404 (significance <0.05). This result means that market literacy has no effect on sustainable innovation.

Sustainable innovation on the control of market niche has *path coefficient* value of 0.174 and *p-value* of 0.022 (significance <0.05). This shows that sustainable innovation has a positive significant effect on the control of market niche. The consumer-preference influence on sustainable innovation has *path coefficient* value of 0.337 and *p-value* of <0.001 (significance <0.05). This shows that the influencing of consumer preference has a positive significant effect on sustainable innovation.

New technology accessibility on sustainable innovation has *path coefficient* value of 0.412 and *p-value* of <0.001 (significance <0.05). This means that new technology accessibility has a positive significant effect on sustainable innovation. The influencing of consumer preference on the control of market niche has *path coefficient* value of 0.727 and *p-value* of <0.001 (significance <0.05). This means that the influencing of consumer preference has a positive significant effect on the control of market niche.

Hypothesis 1 states that an increase of market literacy increases the influencing of consumer preference. Table 3 shows that the *p-value* of market literacy on the influencing of

Table 3. Path coefficients (Source: *WarpPLS 5.0* data processing output).

Variable			Path coefficients	p-values	Description
Market literacy	→	Consumer-preference influence	0.755	<0.001	Positive significant
Market literacy	→	Sustainable innovation	0.021	0.404	Not significant
Sustainable innovation	→	Control of market niche	0.174	0.022	Positive significant
Consumer-preference influence	→	Sustainable innovation	0.337	<0.001	Positive significant
Technology accessibility	→	Sustainable innovation	0.412	<0.001	Positive significant
Consumer-preference influence	→	Control of market niche	0.727	<0.001	Positive significant

consumer preference is <0.001 with the value of *path coefficient* of 0.755. This *p-value* is <0.05 (significance <0.05), so it can be concluded that if the market literacy increases, the influencing of consumer preference will also increase. This statistical test result supports Hypothesis 1, which states that increasing the market literacy increases the influencing of consumer preference (H1 is accepted). This result is also in line with the result of Izquierdo and Cillán (2004), which states that a company which has the ability to meet the consumer's need and understand consumer behaviors on trend, tastes, design and models, will also be able to exert influence or persuasion over consumers so that it can increase sales volumes.

Hypothesis 2 states that increasing market literacy increases sustainable innovation. Table 3 shows that the *p-value* of market literacy on sustainable innovation is 0.021 with *p-value* of 0.404 (significance <0.05). This result has exceeded the significance value of 0.05, so it can be concluded that market literacy has no effect on sustainable innovation. Thus, this statistical test result does not support Hypothesis 2 (H2 is rejected). This research is contrary to the research of Izquierdo and Cillán (2004), which states that a company that has the ability to meet the consumer's need and understand the consumer's behaviors on trend, tastes, designs and models, will be able to give influence or persuasion to consumers such that it will increase sales volumes. Fallah and Lechler (2008) also state that if a company wants to achieve sustainable competitive advantage, it should be able to introduce its products and services on a regular and ongoing basis. The reason market literacy does not increase sustainable innovation may be because SMEs have not yet fully understood innovation. Innovation itself is defined as an activity to make a change in production in order to reduce the production and operational costs, increase the volume of production, and/or improve product quality. Innovation also aims to ensure that production runs more effectively and efficiently. Therefore, product innovation is a process that attempts to provide a solution for the most common problem in business, that is, the quality product but expensive or the cheap product but has a bad quality. This research result is also not in line with the research of Oke et al. (2007), which states that SMEs tend to focus on product innovation rather than consumer services.

Hypothesis 3 states that increasing the sustainable innovation increases the control of market niche. Table 3 shows that the *p-value* of sustainable innovation on the control of market niche is 0.022 (significance <0.05) and has *path coefficient* value of 0.174. This result has exceeded the significance value of 0.05, so it can be concluded that sustainable innovation has an effect on the control of market niche. Therefore, this statistical test result succeeds in supporting Hypothesis 3 (H3 is accepted). This research result is in line with the research of Weiss (2003), Salavou (2008), and Najib and Kiminami (2011), which stated that product innovation has an effect on competitiveness advantages and business performance. Weiss (2003) added in his research that a company should be able to create innovation in a variety of market conditions. Salavou et al. (2004) also stated that product innovation is the most appropriate strategy to support a company in maintaining its market share. Eshlaghy et al. (2011) explained in their research that the innovation role can make a positive contribution to a company's performance.

Hypothesis 4 states that increasing the consumer-preference influence increases the sustainable innovation. Table 3 shows that the *p-value* of consumer-preference influence on sustainable innovation is <0.001 and has a *path coefficient* value of 0.337. This result has exceeded the significance value of 0.05, so it can be concluded that the consumer-preference influence has a positive effect on sustainable innovation. Thus, this test result supports Hypothesis 4 (H4 is accepted).This research result is in line with the research of Oke et al. (2007), which states that SMEs tend to focus on product innovation instead of consumer services. Julienti, Bakar, & Ahmad (2010) also stated in their research that a company's reputation has the highest and cheapest impact on product innovation. Najib and Kiminami (2011) stated that cooperation among all parties, including consumers, has a positive effect on product innovation.

Hypothesis 5 states that increasing technology accessibility increases sustainable innovation. Table 3 shows that the *p-value* of new technology accessibility on sustainable innovation is <0.001 (significance <0.05) and the value of *path coefficient* is 0.412. This test result has exceeded the significance limit of 0.05, so it can be concluded that new technology accessibility has a positive significant effect on sustainable innovation. Based on this statistical test, this research supports Hypothesis 5 (H5 is accepted). This result is in line with the research of Julienti et al. (2010), which stated that all company resources such as physical, financial and human, including intellectual ability, organization ability, and technology accessibility, have a massive impact on product innovation.

Hypothesis 6 states that increasing the consumer-preference influence increases the control of market niche. The *p-value* of consumer-preference influence on the control of market niche is <0.001 (significance <0.05), with *path coefficient* value of 0.727. This test result has exceeded the significance limit of 0.05, so it can be concluded that the consumer-preference influence has a positive significant effect on the control of market niche. Thus, the statistical test result supports Hypothesis 6 (H6 is accepted). This result is in line with the research of Salavou et al. (2004), which states that product innovation is the most appropriate strategy for a company to support the maintenance of its market share. It was also stated by Eshlaghy et al. (2011) that the innovation role can make a positive contribution to a company's performance.

5 CONCLUSION

This research model presents the sustainable innovation-supporting resources and their role in the control of market targets. The process part of this model consists of four variables: market literacy, consumer-preference influence, technology accessibility, and the control of market niche. This model defines optimization of sustainable innovation in influencing the control of market targets. Sustainable innovation should be able to control the market targets as a whole, in this case the control of market niche. Sustainable innovation is also affected by consumer-preference influence and technology accessibility.

Not all hypotheses in this research are supported empirically. The second hypothesis is not proven, in which market literacy is not a significant influence on sustainable innovation because batik SMEs have not fully understood the concept of sustainable innovation, and only focus on current innovation.

However, other hypotheses in this research are supported empirically. Sustainable innovation has a significant effect on two antecedent variables: consumer-preference influence and technology accessibility. Sustainable innovation needs to understand the consumers so it will be easier to direct consumer preference. In addition, sustainable innovation should also follow technology developments so that it can access them. The control of market niche also has a significant effect on two antecedent variables: the influencing of consumer preference and sustainable innovation.

Sustainable innovation has a significant effect on the control of market niche. Sustainable innovation has a terminal role on the control of market target. As a terminal variable, it plays an important role because it is able to influence the intermediary variable that supports the control of market target.

Sustainable innovation has an effect on the control of market niche. This means that increasing sustainable innovation increases the control of market niche. This result is in line with the research of (Weiss, 2003; Salavou & Avlonitis, 2008; Najib & Kiminami, 2011), which stated that product innovation has an effect on competitiveness advantages and business performance. In his research, Weiss (2003) added that a company should be able to create innovation in a variety of market conditions. Salavou et al. (2004) also stated that product innovation is the most appropriate strategy to support a company in maintaining its market share. Eshlaghy et al. (2011) stated in their research that the role of innovation can make a positive contribution to a company's performance.

6 MANAGERIAL IMPLICATIONS AND FUTURE RESEARCH

Based on the discussion and conclusion of the research data processing, managerial implications are determined as input for batik SMEs, specifically in terms of the improvement of marketing activities related to sustainable innovation.

Sustainable innovation is the variable which plays the most important role in the control of market targets. In this research, sustainable innovation is reflected in the strength to produce new products, the ability to produce new designs, develop new users, modify products with other materials, give unique value to products, and generate quality and pricing different from that of competitors.

The control of market niche can be achieved through influencing consumer preference and sustainable innovation. The novelty in this research is the supporting resources of sustainable innovation advantage, which are able to control the market targets.

This research suggests future researches based on its limitation, which is the abnormal data of this research, which requires it to be analyzed by using the partial least square technique. Therefore, the suggestion for future researches is to test the level of data normality by considering the indicators used in the research. Future research should add indicators and research variables in order to satisfy the data normality test.

REFERENCES

Alpay, G., Bodur, M., Yilmaz, C., & Büyükbalci, P. (2012). How does innovativeness yield superior firm performance? The role of marketing effectiveness. *Innovation*, *14*(1), 107–128. https://doi.org/10.5172/impp.2012.14.1.107.

Aslam, A.K., Teknik, J., Wilayah, P., Kota, D.A.N., Sains, F., & Teknologi, D.A.N. (2017). Pengaruh pertumbuhan minimarket terhadap minat dan kebiasaan belanja masyarakat di kelurahan tamamaung kota makassar.

Bharadwaj, N., Nevin, J.R., & Wallman, J.P. (2012). Explicating hearing the voice of the customer as a manifestation of customer focus and assessing its consequences. *Journal of Product Innovation Management*, *29*(6), 1012–1030. https://doi.org/10.1111/j.1540-5885.2012.00954.x.

Boons, F., & Lüdeke-Freund, F. (2013). Business models for sustainable innovation: State-of-the-art and steps towards a research agenda. *Journal of Cleaner Production*, *45*, 9–19. https://doi.org/10.1016/j.jclepro.2012.07.007.

Chen, Y., Yeh, S., & Huang, H. (2012). Does knowledge management "fit" matter to business performance?, *16*(5), 671–687. https://doi.org/10.1108/13673271211262745.

Eshlaghy, A.T., Maatofi, A., & Branch, G. (2011). Learning Orientation, Innovation and Performance: Evidence from Small-Sized Business Firms in Iran, *19*(1), 114–122.

Fallah, M.H., & Lechler, T.G. (2008). Global innovation performance: Strategic challenges for multinational corporations. *Journal of Engineering and Technology Management - JET-M*, *25*(1–2), 58–74. https://doi.org/10.1016/j.jengtecman.2008.01.008.

Ghozali, I. (2012). *Aplikasi Analisis Multivariate Dengan Program IBM SPSS 20.* Semarang: Badan Penerbit Universitas Diponegoro.

Hamel, G., & Prahalad, C.K. (2000). Regime shifts to sustainability through processes of niche formation: the approach of strategic niche management, *10*(2), 127–142.

Hian, C.K., Woo, E.-S., Cohen, A., Sayag, G., Chen, K.Y., Elder, R.J., ... O'Leary, C. (2007). Deakin Research Online. *Managerial Auditing Journal*, *18*(SUPPL.), 68–101. https://doi.org/10.1016/j.jemp.

Holtzman, Y. (2008). Innovation in research and development: Tool of strategic growth. *Journal of Management Development*, *27*(10), 1037–1052. https://doi.org/10.1108/02621710810916295.

Isvandiary, R. (2016). 7 tahun pengukuhan batik Indonesia oleh UNESCO. *Kainusa Jurnal.* Retrived from http://jurnal.kainusa.id/2016/10/11/7-tahun-pengukuhan-batik-indonesia-oleh-unesco.

Izquierdo, C.., & Gutiérrez Cillán, J. (2004). The interaction of dependence and trust in long-term industrial relationships. *European Journal of Marketing, 38*(8), 974–994. https://doi.org/10.1108/03090560410539122.

Julienti, L., Bakar, A., & Ahmad, H. (2010). Assessing the relationship between firm resources and product innovation performance A resource-based view. *Emerald Insight, 16*(No. 3), 420–435.

Keskin, H. (2006). Market orientation, learning orientation, and innovation capabilities in SMEs: An extended model. *European Journal of Innovation Management, 9*(4), 396–417. https://doi.org/10.1108/14601060610707849.

Kotler, P. (2000). Marketing Management, Millenium Edition. *Marketing Management, 23*(6), 188–193. https://doi.org/10.1016/0024-6301(90)90145-T.

Meroño-Cerdán, A.L., Soto-Acosta, P., & López-Nicolás, C. (2008). How do Collaborative Technologies Affect Innovation in SMEs? *International Journal of E-Collaboration, 4*(4), 33–50. https://doi.org/10.4018/jec.2008100103.

Munandar, J.M., & Udin, F. (2013). Analisis Faktor Yang Mempengaruhi Preferensi Konsumen Produk Air Minum Dalam Kemasan Di Bogor. *Speed, 13*, 1–11.

Najib, M., & Kiminami, A. (2011). Innovation, cooperation and business performance Some evidence from Indonesian small food, *1*(1), 75–96.

Narver, J.C., & Slater, S.F. (1990). The Effect of a Market Orientation on Business Profitability. *Journal of Marketing, 54*(4), 20. https://doi.org/10.2307/1251757.

Oke, A., Burke, G., & Myers, A. (2007). Innovation types and performance in growing UK SMEs. *International Journal of Operations & Production Management, 27*(7), 735–753. https://doi.org/10.1108/01443570710756974.

Salavou, H., & Avlonitis, G. (2008). Product innovativeness and performance a focus on SMEs. *European Journal of Innovation Management, 7*(1), 33–44. https://doi.org/10.1108/14601060410515628.

Salavou, H., Baltas, G., & Lioukas, S. (2004). Organisational innovation in SMEs. *European Journal of Marketing, 38*(9/10), 1091–1112. https://doi.org/10.1108/03090560410548889.

Tangendjaja, B. (2014). Daya Saing Produk Peternakan: Ceruk Pasar. *Memperkuat Daya Saing Produksi Pertanian, 287–305.*

Verhees, F.J.H.M. (2005). *Market-Oriented Product Innovation in Small Firms.* Wageningen University Dutch.

Weiss, P. (2003). Adoption of product and process innovations in differentiated markets: The impact of competition. *Review of Industrial Organization, 23*(3–4), 301–314. https://doi.org/10.1023/B:REIO.0000031372.79077.fc.

Willison, J., & Buisman-Pijlman, F. (2016). Article information: *International Journal for Researcher Development, 7*(1), 63–83.

Yeh-Yun Lin, C., & Yi-Ching Chen, M. (2007). Does innovation lead to performance? An empirical study of SMEs in Taiwan. *Management Research News, 30*(2), 115–132. https://doi.org/10.1108/01409170710722955.

Yulianto, R. (2017). Masyarakat Kecamatan Ngronggot, Kabupaten Nganjuk Dampak Beroperasinya Jembatan Papar terhadap Kondisi Sosial dan Ekonomi Masyarakat. *Swara Bhumi., V nomer 4*, 90–98.

Zheng Zhou, K. (2006). Innovation, imitation, and new product performance: The case of China. *Industrial Marketing Management, 35*(3), 394–402. https://doi.org/10.1016/j.indmarman.2005.10.006.

Intellectual capital

Business Innovation and Development in Emerging Economies – Trinugroho & Lau (Eds)
© *2019 Taylor & Francis Group, London, ISBN 978-1-138-35996-3*

Intellectual capital disclosure and post-issue financial performance

Wahyu Widarjo, Rahmawati, Ari Kuncara Widagdo & Eko Arief Sudaryono
Accounting Study Program, Faculty of Economics and Business, Universitas Sebelas Maret, Indonesia

ABSTRACT: This research analyzes the relationship of intellectual capital disclosure with post-issue financial performance of the company. The analysis results on 85 companies which did Initial Public Offering (IPO) in Indonesian Stock Exchange period 2010–2014 show that intellectual capital disclosure affect positively on post-issue financial performance of the company. The results of this research indicate the usefulness of intellectual capital disclosure as a signal of company's performance quality in the future. The company which is more extensive in disclosing intellectual capital in the IPO prospectus is proved to have better future performance.

Keywords: Intellectual capital disclosure, initial public offering, post-issue performance

1   INTRODUCTION

Intellectual capital is a very valuable resource for company in modern business (Williams, 2001). Business practitioners, investors, creditors, regulators, and academics have agreed that intellectual capital is a resource which can create added value and competitive advantage for the company (Bontis, 2001; Chen, 2008; Singh and Zahn, 2009). In addition, the literatures also show intellectual capital importance in improving financial performance and stock performance (see Chen et al., 2005; Tan et al., 2007; Sihotang and Winata, 2008; Zeghal and Maaloul, 2010; Ousama and Fatima, 2015; Saeed et al., 2016). But in practice, intellectual capital has not been fully recognized as a company asset. Only a few elements which can be recognized as assets and listed in the financial report, such as patents, trademarks and copyrights. These conditions result in outside parties difficulty in analyzing and assessing the company resources and prospects, especially at IPO, where there is high asymmetry information between internal parties with external parties (potential investors).

One of the alternatives which are suggested by researchers in intellectual capital field to show the intellectual capital ownership is by disclosing it in company's financial reporting or in financial reporting supplements (Abeysekera, 2013). Disclosure is the delivery of information in financial statements, including the financial statements themselves, financial statements notes, and additional disclosure which related to financial statements. Conceptually, disclosure is an integral part of financial reporting (Evans, 2003). The literatures show that voluntary disclosure may reduce asymmetry information and cost of capital (Lev, 1992; Bottosan, 1997; Leuz and Verrecchia, 2000; Ritter and Welch, 2002; Hanley et al., 2010; Botazzi and Da Rin, 2016). In intellectual capital disclosure context, previous researches also show that intellectual capital disclosure can reduce asymmetry information during initial public offering, thus can decrease cost of capital and increase company value (Widarjo, 2011; Too et al., 2015; Widarjo et al., 2017; Widarjo and Bandi, 2018). Although there have been many researches on the usefulness of intellectual capital disclosure, but analysis of the relationship between intellectual capital disclosure and future performance is still a little (especially post-issue financial performance). Gelb and Zarowin (2002) and Lundholm and Myers (2002) state that better disclosure will help investors in better understanding company's future performance.

Singh and Van der Zahn (2009) have conducted an analysis of intellectual capital disclosure relationship with post-issue stock performance. These research results indicated a negative relationship. They explained that the condition is likely due to investors being overly optimistic with the company's future performance which is more extensive in disclosing its intellectual capital. Therefore, when investor expectations are not immediately fullfilled, investors are aggressively discounting the company stocks, which will have a negative impact on post-issue stock performance of the company.

In contrast to previous research that focus on stock performance, we focus on financial performance. This is based on several arguments: 1) financial performance more represents the company's ability to generate profits through the company's operating activities, 2) the stock performance is more influenced by external factors which are difficult to control (eg, economic conditions, regulatory changes, capital markets, etc.). Based on the explanation, this research aims to provide empirical evidence of intellectual capital disclosure usefulness in predicting the company's financial performance in the future. The next section explains the theoretical framework and hypothesis as well as discussion on the research method in the next section. The results and conclusions will be described at the end of this paper.

2 THEORITICAL BACKGROUND AND HYPOTHESIS

Intellectual capital is one of non-financial information which is needed by stakeholders (especially potential investors) in making investment decision. Therefore, intellectual capital disclosure financial reporting will extensively and systematically reduce asymmetry information and cost of capital during initial public offering (Too et al., 2015; Widarjo et al., 2017). In addition, intellectual capital disclosure will also provide comprehensive information to potential investors about company quality and prospects in the future. Theoretically, Diamond and Verrecchia (1991) explained that increased disclosure can decrease transaction costs, so it can increase stock liquidity and reduce adverse selection problems.

The previous literatures show that intellectual capital performance which is measured by Value Added Intellectual Coefficient (VAIC) is correlated positively on stock performance and financial performance (Firer and Williams, 2003; Riahi-Belkaoui, 2003; Chen et al., 2005; Abdolmohammadi, 2005; Shiu, 2006; Zeghal and Maaloul, 2010), although it is not in the IPO context. Furthermore, Tan et al. (2007) indicated a positive relationship between intellectual capital and current and future corporate performance. In Indonesia, Ullum et al. (2008) supported the research result of Tan et al. (2007). Based on the study of theories and the research results, it can be assumed that companies with high intellectual capital intensity are more extensive in disclosing intellectual capital in financial reporting and IPO prospectus. Furthermore, it can be assumed that companies which are more extensive in disclosing intellectual capital will have future performance which are better than other companies. Therefore, it can be formulated the research hypothesis as follows:

H1: There is a positive relationship between intellectual capital disclosure extent and post-issue financial performance.

3 RESEARCH METHODS

3.1 *Sample and data*

The sample in this research is 85 companies which did IPO in Indonesian Stock Exchange period 2010–2014. Data of intellectual capital disclosure and post-issue financial performance of the company is obtained from the Initial Public Offering prospectus and Indonesian Capital Market Directory (ICMD).

3.2 Variable measurement and statistical analysis

3.2.1 Dependent variables

In this research, the post-issue financial performance is measured by return on equity one year after IPO (ROE_{t+1}). Return on equity is one of the financial performance indicators that measures how well a company uses its investment funds to generate profits. Return on equity is more important for shareholders than return on investment (ROI) because it provides information to investors how effectively their capital is reinvested. Companies with high return on equity are more successful in money making internally. Therefore, investors are always looking for companies with high and growing equity return as an investing place. The return on equity variable after the IPO (ROE_{t+1}) is calculated using the following formula.

$$ROE_{t+1} = Net\ income_{t+1}/Total\ equity_{t+1}$$

3.2.2 Independent variables

Intellectual capital disclosure is measured by disclosure index which has developed by Widarjo et al. (2017). Intellectual capital disclosure level is calculated by the following formula.

$$ICD = \frac{\sum_{ij} DItem}{\sum_{ij} ADItem}$$

Remarks:

ICD : The level of intellectual capital disclosure,
D_{item} : Total score of intellectual capital disclosure in the prospectus,
AD_{item}: Numbers of items in intellectual capital disclosure index.

3.3 Control variables

The ownership structure and the ownership concentration are used as control variables in this research. The ownership structure is proxied with ownership management and foreign ownership. The literatures show that ownership management is related positively on performance (Demsetz and Lehn, 1985; Hermalin and Weisbach, 1991). In addition, foreign ownership has also been shown to have a positive effect on performance (see Chibber and Majumdar, 1999; Douma et al., 2006). Managerial ownership is measured by assigning a value of 1 if management owns company stocks, and 0 for others. Foreign ownership is measured by assigning a value of 1 if there are individuals or foreign institutions holding company stocks, and 0 for others.

High concentrations of ownership will increase the benefits of cost monitoring is also higher. The ownership concentration is an instrument which is used by the involved parties in the agency conflict in achieving a resolution or alignment process so that the parties obtain the optimal utility (Syafruddin, 2006). The previous researches provide evidence of a positive relationship between ownership concentration and performance (Leung, Richardson, and Jaggi, 2014; Vintila and Gherghina, 2014). The ownership concentration is measured by assigning a value of 1 if there is more than fifty percent (>50%) of the company stocks and 0 for others.

Testing of research hypothesis use multiple linear regression analysis. Hypothesis testing use the following research model.

$$ROE_{t+1} = \alpha_0 + \beta_1 ICD + e \tag{1}$$

$$ROE_{t+1} = \alpha_0 + \beta_1 ICD + \beta_2 MO + \beta_3 FO + \beta_4 OC + e \tag{2}$$

Remark:

ROE_{t+1} : Return on equity one year after IPO,
ICD : Intellectual capital disclosure,
MO : Managerial ownership,

FO   : Foreign ownership,
OC   : Ownership concentration,
e   : Error term.

4 RESEARCH RESULT AND DISCUSSION

Descriptive analysis in Table 1 shows the average of ROE_{t+1} is 8 percent and mean of ICD is 46 percent. The average value of intellectual capital disclosure level in this research is relatively higher than researches from outside Indonesia. The research results of Singh and Van der Zahn (2007) showed the average intellectual capital disclosure in Singapore is 27.5 percent, while Singh and Van der Zahn (2008) showed a value of 28 percent. The research of Rashid et al. (2012) in Malaysia showed a value of 34.9 percent, while Too et al. (2015) which also examined companies which did IPO in Malaysia showed an average value of intellectual capital disclosure in amount of 19 percent. The research results of Branswijk and Everaert (2012) in Belgium and Netherlands showed that intellectual capital disclosure level is relatively higher than those researches, the amount of percentage is 34.5 percent. The differences in intellectual capital disclosure level in IPO prospectuses in these countries may be caused by the differences of used index in intellectual capital disclosure. It may caused by intellectual capital disclosure which is still voluntary and there is no standard that regulates the categories and items which should be disclosed in the IPO prospectus. The correlation analysis in Table 1 shows a positive correlation between performance variables (ROE_{t+1}) with intellectual capital disclosure (ICD) and ownership concentration (OC). These results are an early indication of the relationship between intellectual capital disclosure and post-issue financial performance. In addition, correlation analysis results do not show the existence of multicollinearity among independent variables.

The ICD data descriptions per category in Table 2 indicate that the most dominant intellectual capital categories which are disclosed in the prospectus are strategies and processes, while the category of human resources, customers, research and development (R&D) and information technology are relatively less disclosed. The results indicate that the company places greater emphasis on extending business strategy information and business processes (internal and external) to convince potential investors about the potencies and prospects of strategic resources which is owned by the company.

We suspect that intellectual capital disclosure in the initial public offering prospectus is correlated positively on the future financial performance. The regression analysis result in Table 3 shows that the regression coefficient of intellectual capital disclosure variable is marked positive and significant at 5 percent level. The result is consistent after structure

Table 1. Statistic descriptive and correlation.

	ROE_{t+1}	ICD	MO	FO	OC
Min	−0.66	0.27	0.00	0.00	0.00
Max	0.50	0.60	1.00	1.00	1.00
Mean	0.08	0.46	0.44	0.18	0.33
SD	0.16	0.08	0.49	0.38	0.47
ROE_{t+1}	1.000				
ICD	0.226*	1.000			
MO	0.049	−0.034	1.000		
FO	0.100	−0.079	−0.033	1.000	
OC	0.223*	0.106	−0.010	−0.259*	1.000

Notes: *Correlation is significant at the 0.05 level (2-tailed). ROE_{t+1} = Return on equity one year after the IPO; ICD = Intellectual capital disclosure; MO = Managerial ownership; FO = Foreign ownership; OC = Ownership concentration.

Table 2. Description of ICD per category.

	Minimum	Maximum	Mean	Std. deviation
Human Resource	0.20	0.67	0.45	0.09
Customer	0.05	0.63	0.33	0.12
Information Technology	0.00	1.00	0.29	0.38
Process	0.33	0.89	0.59	0.14
Research and Development	0.00	0.80	0.31	0.29
Strategy	0.18	0.88	0.62	0.13

Table 3. The analysis result of regression.

Variable	Equation 1		Equation 2	
	Coeff.	t-value	Coeff.	t-value
Constant	−0.113	−1.188	−0.154	−1.602
Main Variable				
ICD	0.434	2.116**	0.415	2.056**
Control Variables				
MO			0.020	0.621
FO			0.075	1.692*
OC			0.082	2.281**
R^2		0.051		0.126
Adj. R^2		0.040		0.082
F-value		4.477		2.878
Sig		0.037		0.028
N		85		85

Notes: * and ** indicate significance at the 10% and 5% levels, respectively. ICD = Intellectual capital disclosure; MO = Managerial Ownership; FO = Foreign Ownership; OC = Ownership Concentration.

control variable and ownership concentration are added. Therefore, the hypothesis in this research is supported. The company which is more extensive in disclosing intellectual capital in the IPO prospectus is having better future performance than the others.

The results of this research reinforce the arguments and findings of the previous researches which showed that intellectual capital is a resource which can create added value and competitive advantage of the company (Bontis, 2001; Chen, 2008; Singh and Zahn, 2009). Intellectual capital performance can improve efficiency and effectiveness of the company's business processes, so the performance is increased (Riahi-Belkaoui, 2003; Chen et al., 2005; Abdolmohammadi, 2005; Tan et al., 2007; Zeghal and Maaloul, 2010). In addition, the company with high intellectual capital intensity are more competitive and more successful than the others (Youndt et al., 2004).

The analysis results in Table 3 also show that variable of foreign ownership and ownership concentration are correlated positively with post-issue financial performance of the company. These results indicate the effectiveness of monitoring role which is performed by foreign owners. Foreign investors are judged to have better capacity and capability of resources in monitoring business activities of company management. Research findings which indicate a positive correlation between ownership concentration and post-issue performance indicate that majority shareholders obtain greater benefit of cost monitoring than minority shareholders. This means that majority shareholders have a higher chance and authority to monitor the activities of company management. Thus, the company management will work optimally to increase the utility of capital owner.

Table 4. The analysis result of alternative measurement.

Variable	Equation 2		Equation 2	
	Coeff.	t-value	Coeff.	t-value
Constant	−0.154	−1.602	−0.160	−1.547
Main Variable				
ICD	0.415	2.056**		
ICD_Ater			0.396	1.959*
Control Variables				
MO	0.020	0.621	0.021	0.625
FO	0.075	1.692*	0.072	1.629
OC	0.082	2.281**	0.083	2.299**
R^2		0.126		0.122
Adj. R^2		0.082		0.078
F-value		2.878		2.772
Sig		0.028		0.033
N		85		85

Notes: * and ** indicate significance at the 10% and 5% levels, respectively. ICD = Intellectual capital disclosure; MO = Managerial Ownership; FO = Foreign Ownership; OC = Ownership Concentration.

4.1 Robustness checks

We do additional testing to ensure that the results are robust and consistent. This test is focused on the use of alternative measurement of intellectual capital disclosure variables. It aims to anticipate the existence of measurement error which is caused by the disclosure index development which were did by researchers. Therefore, it is necessary to retest the research model using the disclosure index which has developed by Bukh et al. (2005) as an alternative measurement of intellectual capital disclosure variables.

The underlying arguments of the use of Bukh et al. (2005) index as an alternative measurement in this research is the index which has often been used as a reference by previous researchers in several different countries (see Rimmel et al., 2009; Branswijck and Everaert, 2012). In addition, there is a common legal system which is used in this country (Danish country) with the legal system in Indonesia, the code law system.

The analysis result in Table 4 shows the regression coefficient of intellectual capital disclosure variable is marked positive and significant at 10 percent level. The results are relatively consistent with the previous test. Therefore, it can be concluded that there is a positive correlation between intellectual capital disclosure and post-issue financial performance of the company.

5 CONCLUSION

The literature has showed the usefulness of intellectual capital disclosure in reducing asymmetry information and cost of capital. This research extends the previous literatures by analyzing the relationship of intellectual capital disclosure with post-issue financial performance, especially in developing countries. The results support the hypothesis that intellectual capital disclosure is correlated positively with post-issue financial performance of the company. The findings of this research have important implications in literature and practice. The results of this research indicate the usefulness of intellectual capital disclosure as a medium in analyzing and predicting the company's performance in the future. Companies which have high intellectual capital intensity are relatively more efficient and effective in running the business, so it will produce better performance than other companies. Furthermore, the findings of this research support and extend the application of signalling theory which states that the disclosure extent can reduce asymmetry information level and help the potential investors in investment analysis and decision making (see Welker, 1995; Jog and McConomy, 2003; Guo et al., 2004; Yosano, 2015; Widarjo et al., 2017).

This research has several limitations. First, the measurement of intellectual capital disclosure is only from the quantity aspect of disclosure, it has not representated the disclosure quality. It is important to consider because high quantity of disclosure may not necessarily represent a high quality of disclosure. Therefore, further research needs to consider aspects of intellectual capital disclosure quality as an alternative to intellectual capital disclosure measurement (see Cerbioni and Parbonetti, 2007; Yi and Davey, 2010). Second, this research only focuses on financial performance one year after the IPO, it has not analyzed long-run financial performance (three or five years after IPO). Future research can develop the research by analyzing long-run financial performance. In addition, further researchers can also perform intellectual capital disclosure analysis on long-run stock performance such as Singh and Zahn research (2009), especially in developing countries for comparison analysis with research in developed countries.

REFERENCES

Abdolmohammadi, M.J. (2005). Intellectual Capital Disclosure and Market Capitalization. *Journal of Intellectual Capital*, 6(3): 397–416.

Abeysekera, I. (2013). A Template for Integrated Reporting. *Journal of Intellectual Capital*, 14(2): 227–245.

Bontis, N. (2000). Intellectual Capital and Business Performance in Malaysian Industries. *Journal of Intellectual Capital*, 1(1), 85–100.

Bottazi, L. and Da Rin, M. (2016). *Voluntary Information Disclosure at IPO*. Available on line at: http://papers.ssrn.com/sol3/papers.cfm?abstract_id=2810847 (Accessed 10 September 2016).

Botosan, C.A. (1997). Disclosure Level and the Cost of Equity Capital. *The Accounting Review*, 72(3): 323–349.

Branswijck, D. and Everaert, P. (2012). Intellectual Capital Disclosure Commitment: Myth or Reality? *Journal of Intellectual Capital*, 13(1): 3956.

Bukh, P.N., Nielsen, C., Gormsen, P. and Mouritsen, J. (2005). Disclosure on Information Intellectual Capital in Danish IPO Prospectuses. *Accounting, Auditing and Accountability Journal*, 18(6): 713–732.

Cerbioni, F. and Parboneti, A. (2007). Exploring the Effect of Corporate Governance on Intellectual Capital Disclosure: An Analysis of European Biotechnology Companies. European Accounting Review, 16(4): 791–826.

Chen, Y.S. (2008). The Positive Effect of Green Intellectual Capital on Competitive Advantages of Firms. *Journal of Business Ethics*, 77: 271–286.

Chen, M.C., Cheng, S.J. and Hwang, Y. (2005). An Empirical Investigation of The Relationship Between Intellectual Capital and Firms' Market Value and Financial Performance. *Journal of Intellectual Capital*, 6(2): 159–76.

Chibber, P.K. and Majumdar, S.K. (1999). Foreign Ownership and Profitability: Property Rights, Control and The Performance of Firms in Indian Industry. *Journal of Law and Economics*, 46(3): 209–238.

Demsetz, H., and Lehn, K. (1985). The Structure of Corporate Ownership Causes and Consequences. *Journal of Political Economy*, 93: 1155–1177.

Diamond, D. and Verrecchia, R. (1991). Disclosure, Liquidity and The Cost of Capital. *Journal of Finance*, 46(4): 1325–1359.

Douma S., George, R. and Kabir, R. (2006). Foreign and Domestic Ownership, Business Groups, and Firm Performance: Evidence from a Large Emerging Market. *Strategic Management Journal*, 27(7): 637–657.

Evans, T.G. (2003). *Accounting Theory: Contemporary Accounting Issues*. Australia: Thomson, South-Western.

Gelb, D., and Zarowin. P. (2002). Corporate Disclosure Policy and The Informativeness of Stock Prices. *Review of Accounting Studies*, 7: 33–52.

Guo, R.J., Lev, B. and Zhou, N. (2004). Competitive Costs of Disclosure by Biotech IPOs. *Journal of Accounting Research*, 42(2): 319–355.

Hanley W., Kathleen, and Hoberg, G. (2010). The Information Content of IPO Prospectuses. *Review of Financial Studies*, 23(7): 2821–2864.

Hermalin, B. and Weisbach, M.S. (1991). The Effects of Board Composition and Direct Incentives on Firm Performance. *Financial Management*, 20(4): 101–112.

Jog, V. and McConomy, B.J. (2003). Voluntary Disclosure of Management Earnings Forecasts in IPO Prospectuses. *Journal of Business, Finance and Accounting*, 30(1/2): 125–67.

Leung, S., Ricahrdson, G., and Jaggi, B. (2014). Corporate Board and Board Committee Independence, Firm Performance, & Family Ownership Concentration: An Analysis Based on Hong Kong Firm. *Journal of Contemporary Accounting & Economic*, 10(1): 16–31.

Lev, B. (1992). Information disclosure strategy. *California Management Review*, 34(4): 9–32.

Leuz, C. and Verrechia, R.E. (2000). The Economic Consequence of Increased Disclosure. *Journal of Accounting Research*, 38 (Supplement): 91–135.

Lundholm, R., and Myers, L.M. (2002). Bringing The Future Forward: The Effect of Disclosure on The Returnsearnings Relation. *Journal of Accounting Research*, 40(3): 809–839.

Ousama, A.A. and Fatima, A.H. (2015). Intellectual Capital and Financial Performance of Islamic Banks. *International Journal of Learning and Intellectual Capital*, 12(1): 1–15.

Riahi-Belkaoui, A. (2003). Intellectual Capital and Firm Performance of US Multinational Firms: A Study of the Resource-Based and Stakeholder Views. *Journal of Intellectual Capital*, 4(2): 215–226.

Rimmel, G., Nielsen, C. and Yosano, T. (2009). Intellectual Capital Disclosure in Japanese IPO Prospectuses. *Journal of Human Resource Costing & Accounting*, 13(4): 316–337.

Ritter, J. and Welch, I. (2002). A Review of IPO Activity, Pricing, and Allocations. *Journal of Finance*, 57(4): 1795–1828.

Saeed, S., Rasid, S.Z.A. and Basiruddin, R. (2016). Relationship Between Intellectual Capital and Corporate Performance of Top Pakistani Companies: An Empirical Evidence. *International Journal of Learning and Intellectual Capital*, 13(4): 376–396.

Shiu, H.-J. (2006). The Application of the Value Added Intellectual Capital Coefficient to Measure Corporate Performance: Evidence from Technological Firms. *International Journal of Management*, 23(2): 356–365.

Sihotang, P., and Winata, A. (2008). The Intellectual Capital Disclosures of Technology-Driven Companies: Evidence from Indonesia. *International Journal Learning and Intellectual Capital*, 5(1), 63–82.

Singh, I. and J-L.W.M. Van der Zahn. (2007). Does Intellectual Capital Disclosure Reduce an IPOs Cost of Capital: The Case of Underpricing. *Journal of Intellectual Capital*, 8(3): 494–516.

Singh, I. and J-L.W.M. Van der Zahn. (2008). Determinants of Intellectual Capital Disclosure in Prospectuses of Initial Public Offerings. Accounting and Business Research, 38(5): 409–431.

Singh, I. and Van der Zahn, J-L.W.M. (2009). Intellectual Capital Prospectus Disclosure And Post-Issue Stock Performance. *Journal of Intellectual Capital*, 10(3): 425–450.

Syafruddin, M. 2006. Pengaruh Struktur Kepemilikan Perusahaan Pada Kinerja: Faktor Ketidakpastian Lingkungan Sebagai Pemoderasi. *Jurnal Akuntansi dan Auditing Indonesia*, 10(1): 85–99.

Tan, H.P., Plowman, D. and Hancock, P. (2007). Intellectual Capital and Financial Returns of Companies. *Journal of Intellectual Capital*, 8(1): 76–95.

Too, S.W., Fadzilah, W. and Yusoff, W. (2015). Exploring Intellectual Capital Disclosure as a Mediator for the Relationship Between IPO Firm-Specific Characteristics and Underpricing. *Journal of Intellectual Capital*, 16(3): 1–26.

Ulum, I., Ghozali, I., dan Chariri, A. (2008). *Intellectual capital* dan kinerja keuangan perusahaan: suatu analisis dengan pendekatan *partial least squares*. *Simposium Nasional Akuntansi XI*. Pontianak, 23–26 Juli.

Vintila, G. and Gherghina, S.C. (2014). The Impact of Ownership Concentration on Firm Value: Empirical Study of the Bucharest Stock Exchange Listed Companies. *Procedia Economics and Finance*, 15, 271–279.

Welker, M. 1995. Disclosure Policy, Information Asymmetry, and Liquidity in Equity Markets. Contemporary Accounting Research, 11 (2): 801–827.

Widarjo, W. 2011. Pengaruh Modal Intelektual dan Pengungkapan Modal Intelektual pada Nilai Perusahaan yang Melakukan *Initial Public Offering*. *Jurnal Akuntansi dan Keuangan Indonesia*, 8(2): 157–170.

Widarjo, W., Rahmawati, Bandi and Widagdo, A.K. (2017). Underwriter Reputation, Intellectual Capital Disclosure, and Underpricing. *International Journal of Business and Society*, 18(2), 227–244.

Widarjo, W. and Bandi. (2018). Determinants of Intellectual Capital Disclosure in the IPOs and Its Impact on Underpricing: Evidence from Indonesia. *International Journal of Learning and Intellectual Capital*, 15(1), 1–19.

Williams, S.M. 2001. Is Intellectual Capital Performance and Disclosure Practices Related?, *Journal of Intellectual Capital*, 2(3): 192–203.

Yosano, T., Nielsen, C., and Rimmel, G. (2015). The Effects of Disclosing Intellectual Capital Information on the Long-Term Stock Price Performance of Japanese IPO's. *Accounting Forum*, 39(2), 83–96.

Yi, An and Davey, H. (2010). Intellectual Capital Disclosure in Chinese (Mainland) Companies. *Journal of Intellectual Capital*, 11(3): 326–347.

Youndt, M.A., Subramaniam, M. and Snell, S.A. (2004). Intellectual Capital Profiles: An Examination of Investments and Returns. *Journal of Management Studies*, 41(2): 335–361.

Zeghal, D. and Maaloul, A. (2010). Analysing Value Added as An Indicator of Intellectual Capital and Its Consequences on Company Performance. *Journal of Intellectual Capital*, 11(1): 39–60.

Macro and monetary economics

Business Innovation and Development in Emerging Economies – Trinugroho & Lau (Eds)
© *2019 Taylor & Francis Group, London, ISBN 978-1-138-35996-3*

What do you want from me? Demand for skilled labor in the global value chain: The case of Indonesia's apparel industry

M.A.P. Prabowo
Department of Economics, Universitas Indonesia, Depok, Indonesia

P. Wicaksono
Labor Economics at the Department of Economics, Universitas Indonesia,
Depok, Indonesia

T. Bakhtiar
Applied Mathematics at the Department of Mathematics, Bogor Agricultural University,
Bogor, Indonesia

ABSTRACT: Indonesia's apparel industry is well known as being one of the top apparel exporters that make a significant contribution to the global apparel industry, meaning that the industry is already involved in the global value chain. One of the issues related to labor economics is whether the global value chain demands low-skilled or high-skilled labor. This paper analyzes the effect of the global value chain on the demand for skilled labor. The result is that the global value chain demands lower-skilled rather than high-skilled labor, because expanding business globally needs more low-skilled labor than high-skilled labor. Moreover, the decrease in high-skilled labor causes labor productivity to increase during involvement in the global value chain.

Keywords: Global value chain, vertical specialization, skilled labor, non-production worker, productivity

1 INTRODUCTION

Indonesia's apparel industry is already well known in the world nowadays. Indonesia is one of the top ten apparel exporters. Moreover, Indonesia's apparel industry has made a significant contribution by producing apparel products for 80% of global clothing brands in the world, according to the Indonesian Textile Association (API) in 2013. These contributions indicate that Indonesia's apparel industry is involved in the Global Value Chain (GVC) phenomenon, which has recently been booming.

The involvement of Indonesia's apparel industry is not separate from the size of Indonesia's population. As the fourth most populous country in the world, Indonesia has a lot of labor, which is one of the basic factors of production. This results in the average nominal wage in Indonesia being relatively lower than in other Asian countries.

The International Labor Organization reported that Indonesia has the fourth-lowest nominal average wage among the 13 Asian countries, which attracts foreign firms to locate in Indonesia, particularly in labor-intensive industries (ILO, 2014). If we conclude that a huge population is to Indonesia's comparative advantage, then relative wages and demand for unskilled labor in Indonesia are predicted to increase.

However, the involvement in the global value chain has affected the demand for labor. The demand for skilled labor in developing countries surprisingly increases because of this involvement (Kasahara et al., 2016; Goldberg & Pavcnik, 2007; Sánchez-Páramo & Schady, 2003). Manoto et al. (2017) have reported that during the global value chain era, the share of skilled workers in Indonesia's apparel industry increases, but the productivity surprisingly decreases. This violates the common sense of the relationship between the proportion of skilled labor and productivity.

The fact is that the study did not reveal causality between participation in the global value chain and skilled labor, and so did not establish whether or not the change in the share of skilled labor is caused by participation in the global value chain. The same also applied to the correlation between the share of skilled labor and productivity, so that further research is needed to scrutinize both issues.

In the study, moreover, they only used educational attainment to categorize skilled labor. This can be misleading. The skills acquired from a formal education are constant after exiting the educational system, whereas skills can vary over time due to informal education or work experience, and even the quality of formal education can be different for each person (Vera-Toscano et al., 2017).

To measure the skill level of labor, it is more precise to use the type of occupation rather than educational attainment. The current occupation of the labor force describes the current skill they have more accurately, as a result of experience, than does the formal and informal education that they have. Therefore, this paper is going to examine what the impact is of involvement in the GVC of Indonesia's apparel industry on the demand for skilled labor, as well as the correlation between skilled labor and productivity in the apparel industry, by using current occupation to categorize high-skilled and low-skilled labor.

2 ANALYTICAL FRAMEWORK

2.1 Global value chain

In this era, hardly any of the economic activities within a country can be separated from the GVC. Gereffi and Fernandez-Stark (2011) firstly defined a value chain as being a 'full range of activities that firms and workers do to bring a product from its conception to its end use and beyond'. Then, Timmer et al. (2014) defined GVC as the dispersion and fragmentation of economic activities in which different countries undertake specialization of one part or task in the production of a final good.

The concept of a global value chain is more likely similar to vertical specialization. Vertical specialization is an interconnectedness of the production process with a trading chain stretching across many countries vertically, each of which is specialized in a particular part of the production process (Yi, 2003).

The main reason why firms decide to fragment the production process of their product, or participate in the GVC, is simply because trading cost among countries has significantly decreased (OECD, 2012), so that firms can be more efficient by putting their production in other countries to obtain cheaper factory production costs.

Memedovic (2004) has explained that the GVC has two classifications: buyer-driven and supply-driven. The former is associated with 'easy and simple' technologies and labor-intensive industries, such as apparel and footwear, which means the value chain is mainly located in developing countries. Adversely, the latter is strongly associated with 'difficult technologies' and medium- and high-tech industries, such as automotive and electronics.

2.2 Measuring participation in the global value chain

According to the definition of the global value chain, each country would undertake a part of the production process of the final product which the country specializes in. It means there will be a trading activity, both export and import, which is done by countries that are involved in the global value chain. Then, a country can be categorized as being involved in the global value chain in two activities: the export product uses an import content (downstream producer) or the country supplies an intermediate good to the export destination (upstream producer) (European Central Bank, 2013).

To measure the participation in the global value chain, Esmaile (2018) used other variables to measure the degree of global engagement. Those variables are export and foreign ownership/foreign direct investment. Furthermore, we can also use a measurement of vertical specialization, because Yi (2003) has explained that vertical specialization is similar to the global value chain. Hummels et al. (2001) have developed a measurement of vertical specialization and both measurements have been tried in the United States—it is:

$$VS_{ki} = \left(\frac{Imported\,Intermediaries_{ki}}{Gross\,Output_{ki}} \right) Exports_{ki}$$

2.3 Measuring skilled labor

Skilled labor is labor which has a specialization of work supported by an expertise, training, and both formal and informal education. We can also say that skilled labor is a result of investment in human capital. Firms prefer to hire skilled labor because increases in productivity of skilled labor will be higher than that of unskilled labor in every task (Katz & Murphy, 1992). To categorize whether or not labor is skilled labor, researchers usually use educational attainment as a measurement. As the European Union (2016) explained, the level of educational attainment, either general or vocational, allows for comparative analyses of different skills levels, for instance in geographic and socioeconomic groups. Sometimes a researcher uses more detailed information such as a field of educational attainment to analyze the issues in depth.

However, measuring the skill of labor using educational attainment can become misleading. The skill received from an informal education is constant after exiting the educational system, while skills can vary over time due to informal education or work experience; even the quality of formal education can be different for each person (Vera-Toscano et al., 2017). Therefore, we need another measurement for categorizing skilled labor.

An alternative measurement is occupational categories. Such categories are commonly divided into two: production and non-production workers. A production worker is a worker who is in charge of a task such as fabricating, processing, assembling, packing, or warehousing. This also includes the worker who is in charge of repair, janitorial, record-keeping, and other services related to production operations. Thus, a non-production worker is a worker who is in charge of an executive, financial, sales, professional, technical, or other office-routine work (Bureau of Labor Statistics, 1957). Moreover, a production worker, or blue-collar worker, is often categorized as unskilled labor, while a non-production worker, or white-collar worker, is categorized as skilled labor by many researchers in their papers (Kasahara et al., 2016).

2.4 Demand for skilled labor

Labor, as one of the production factors needed to produce output, is demanded by the firm. Because the production factors are not just labor itself, the firm must consider other variables to decide how many laborers they should hire. The basic theory of labor demand explains that there are three variables considered by a firm in their decision-making about hiring labor. Those variables are wages, labor productivity, and capital. Wages have a negative correlation with demand for labor. Capital, another production factor, also has a negative correlation with the demand for labor (Ehrenberg & Smith, 2012). Furthermore, the firm would be willing to hire additional labor if labor productivity remains high, or at least remains higher than wages (Pyndick & Rubinfeld, 2009). However, if we divide labor into skilled and unskilled labor, skilled labor is more complementary to physical capital than unskilled labor, which means that a firm is more likely to substitute unskilled labor with physical capital than skilled labor (Griliches, 1969).

2.5 Linking demand for skilled labor to the global value chain

Because the global value chain is considered as being an international trade activity, we also need to take into account the effect of trade on the demand for skilled labor. Kasahara et al.

(2016) have reported that importing intermediate goods greatly increases the demand for skilled labor among firms in Indonesia, because importing activities require the workers to deal with customs and excise departments, and arrange agreements and shipments with trading partners. Moreover, Sánchez-Páramo and Schady (2003) argued that being involved in global engagement, which is a global value chain, increases demand for skilled labor in developing countries. In their argument, they explained that when a country achieves their import, including importing capital goods, it will demand high-skilled labor.

Exports also affect the demand for skilled labor. An exporting firm will increase the skill upgrading or demand for high-skilled labor more quickly than a non-exporter (Bustos, 2005). This argument is supported by Bernard and Jensen (1997); their paper reported that export is fundamental to the changes in the labor market, particularly in the increase of demand for high-skilled labor. Other arguments are that exporting activities require a quality upgrade for output (Verhoogen, 2008; Brambilla et al., 2015) and also involve an operational service (Matsuyama, 2007; Brambilla et al., 2015). Moreover, Esmaile (2018) has reported that exporting activities will cause a firm to need more labor, and that foreign ownership also has a positive and significant correlation with labor demand.

3 INDONESIA'S APPAREL INDUSTRY: AN OVERVIEW AND RECENT DEVELOPMENT IN THE GLOBAL VALUE CHAIN

Due to the involvement in the global value chain, Indonesia's apparel industry has developed into an 'open' industry. As a characteristic of the GVC, 'open' means that foreign ownership in the apparel industry has recently increased.

Figure 1 shows the trend of foreign ownership in Indonesia's apparel industry from 2001 to 2015. As we can see, the trend line slopes upward, which means that the number of firms owned by foreigners in the industry is increasing over time. According to Esmaile (2018), foreign ownership is one of the variables used to measure global engagement. The more foreign ownership in an industry, the more the industry is involved in the GVC. Therefore, Indonesia's apparel industry has become more involved in the global value chain over time because the number of firms owned by foreigners is increasing.

Besides foreign ownership, we can also use imported raw materials and export value as measurements of participation in the global value chain (European Central Bank, 2013). Figure 2 shows the trend of these variables from 2001 to 2015 in Indonesia's apparel industry. Both are increasing over time, but the export value is steeper in its increase than is the imported raw materials value.

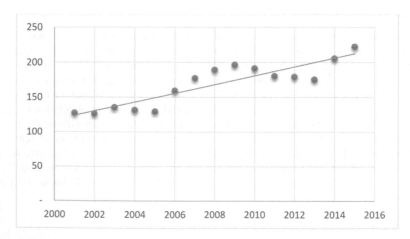

Figure 1.   The trend of foreign ownership in Indonesia's apparel industry, 2001–2015. (Source: Calculated from industrial statistics, Indonesia's Central Agency on Statistics (BPS)).

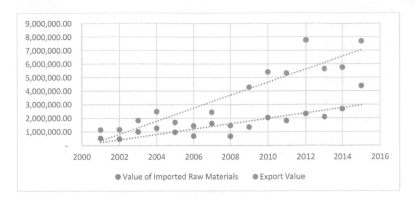

Figure 2.　The trend of import and export in Indonesia's apparel industry in million Rupiah, 2001–2015. (Source: Calculated from industrial statistics, Indonesia's Central Agency on Statistics (BPS)).

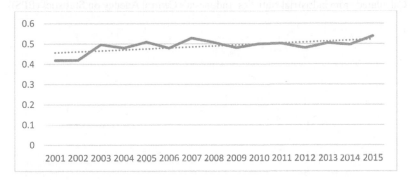

Figure 3.　The trend of average share of imported raw materials in Indonesia's apparel industry, 2001–2015. (Source: Calculated from medium-large enterprises' industrial statistics, Indonesia's Central Agency on Statistics (BPS)).

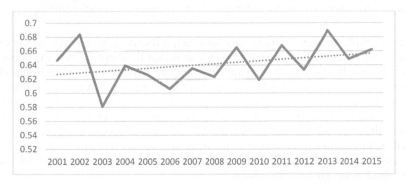

Figure 4.　Trend of average share of exported product in Indonesia's apparel industry, 2001–2015. (Source: Calculated from medium-large enterprises industrial statistics, Indonesia's Central Agency on Statistics (BPS)).

Moreover, in order to strengthen our analysis, we are looking at the share of imported raw materials to overall inputs. Figure 3 shows the trend of the average share of imported raw materials in Indonesia's apparel industry in rupiah from 2001 to 2015. As we can see, there is an increasing trend in the share of imported raw materials. This means that most of the raw materials used in the apparel industry are imported raw materials, so that the firms in Indonesia's apparel industry are not separated from the global value chain scheme.

A similar trend occurs in the average share of exported products. Figure 4 shows the trend of average share of exported products in Indonesia's apparel industry in rupiah from 2001 to 2015.

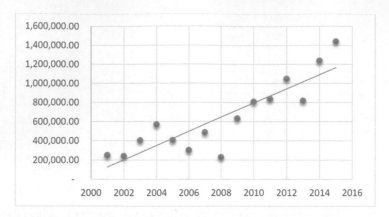

Figure 5. The trend of value of vertical specialization in Indonesia's apparel industry, 2001–2015. (Source: Calculated from industrial statistics, Indonesia's Central Agency on Statistics (BPS)).

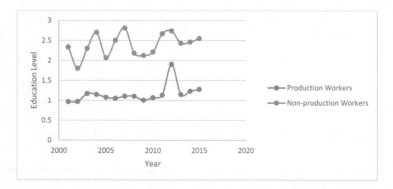

Figure 6. Educational attainment of Indonesia's apparel industry workers, 2001–2015. (Source: Calculated from national labor force survey, Indonesia's Central Agency on Statistics (BPS)).

Based on the graph, the trend is increasing, although it fluctuates more than the import trend. This means that firms in the industry have also increased their export over the period, which fully confirms that Indonesia's apparel industry is highly involved in the global value chain.

Furthermore, we can also use the value of vertical specialization as a measurement of the global value chain because vertical specialization is a similar concept to the global value chain. The value is calculated by using a formula developed by Hummels et al. (2001). Figure 5 shows the trend of value of vertical specialization in Indonesia's apparel industry from 2001 to 2015. The trend is very similar to the trend of foreign ownership, which has increased over the same period of time. This trend fully confirms that Indonesia's apparel industry is highly involved in the global value chain.

4 THE EFFECT OF GLOBAL VALUE CHAIN PARTICIPATION ON DEMAND FOR SKILLED LABOR

In the previous part of the analysis, we have clearly demonstrated that Indonesia's apparel industry is highly involved in the global value chain. Now we are going to analyze the effect of the involvement of demand for skilled labor in Indonesia's apparel industry. According to the Bureau of Labor Statistics (1957), we can divide workers in the industry into two categories: high-skilled labor (white-collar workers) and low-skilled labor (blue-collar workers).

Firstly, we need to justify our definitions of high-skilled and low-skilled. Figure 6 shows the educational attainment of Indonesia's apparel industry workers. As we can see, non-production

Table 1.   Classification of educational level.

Classification of education level	
0	Not Participated
1	Elementary School
2	Junior High School
3	Senior/Vocational High School
4	Diploma
5	Bachelor/Master/PhD

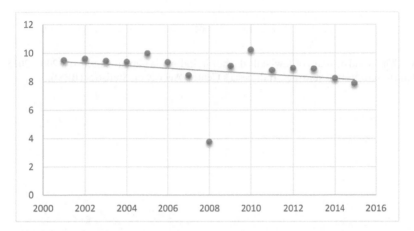

Figure 7.   The trend of share of high-skilled labor in Indonesia's apparel industry, 2001–2015. (Source: Calculated from industrial statistics, Indonesia's Central Agency on Statistics (BPS)).

workers in the industry have a higher educational background (see Table 1) than production workers. A non-production worker is more likely to have attended at junior high school level than a production worker. This satisfies the 'first order necessary condition' of high-skilled labor, which has a higher educational attainment level. Therefore, we could confidently argue that the non-production worker represents high-skilled labor and the production worker low-skilled labor.

Furthermore, we can jump to the main analysis of this paper. Figure 7 shows the trend of the share of high-skilled labor in Indonesia's apparel industry from 2001 to 2015. As we can see, the trend is decreasing over time. This violates the argument of Kasahara et al. (2016), who argued that when an industry in a developing country is involved in the global value chain, the demand for skilled labor increases, particularly high-skilled labor. There are several reasons that can lead to such a decrease.

As we can see in Figure 8, the share of low-skilled labor in Indonesia's apparel industry is obviously increasing. These opposing trends for the two categories of workers in Indonesia's apparel industry occurs because firms want to expand their business through export. Increasing export means firms need to increase production. Therefore, firms will prefer to hire labor associated with the production process, which is low-skilled labor. This is one of the reasons why the global value chain phenomenon demands more low-skilled labor.

Figure 9 shows the trend in the real wage of labor in Indonesia's apparel industry between 2001 and 2015. Both labor types, low- and high-skilled labor, are experiencing real wage increases. However, the real wage of high-skilled labor is always above that of low-skilled labor. This causes firms, particularly those in foreign ownership, to decide to hire more low-skilled than high-skilled labor because high-skilled labor is not needed as much in the production process as low-skilled labor. This makes sense because many firms in Indonesia only take on a simple part of the production process, such as assembling, which does not need much high-

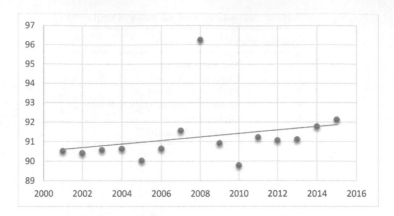

Figure 8. The trend of share of low-skilled labor in Indonesia's apparel industry, 2001–2015. (Source: Calculated from industrial statistics, Indonesia's Central Agency on Statistics (BPS)).

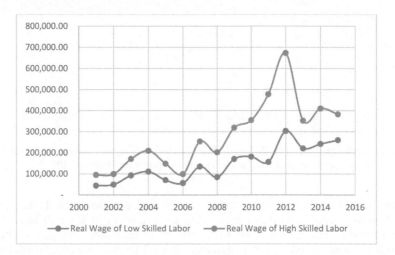

Figure 9. The trend of real wages of labor in Indonesia's apparel industry, 2001–2015. (Source: Calculated from industrial statistics, Indonesia's Central Agency on Statistics (BPS)).

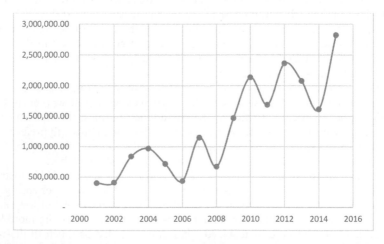

Figure 10. The trend of labor productivity in Indonesia's apparel industry, 2001–2015. (Source: Calculated from industrial statistics, Indonesia's Central Agency on Statistics (BPS)).

skilled labor. Firms will lose their profit if they hire higher-skilled labor, which increases the cost of production but does not make as high a contribution as low-skilled labor to the production process.

Because of an increasing proportion of low-skilled labor, labor productivity in Indonesia's apparel industry increased during 2001–2015. Figure 10 shows the trend of labor productivity in Indonesia's apparel industry. Increasing production causes firms to hire more low-skilled labor, which then leads to higher labor productivity. Therefore, although the proportion of high-skilled labor is decreasing, labor productivity is increasing. However, this result is probably biased because we defined high-skilled labor as non-production workers, who are not involved in the production process.

5 CONCLUSIONS AND REMAINING PROBLEMS

Indonesia's apparel industry has been involved in the global apparel value chain for many years. In fact, the industry has become more deeply involved over time because key indicators of the global value chain (which include foreign ownership, export value, imported raw materials value, and vertical specialization) are increasing every year. However, the share of high-skilled labor is decreasing in the course of this involvement. This occurs because the wages of high-skilled labor are higher than those of low-skilled labor, such that firms prefer to hire low-skilled labor because Indonesia's apparel industry only takes on a 'simple and easy' part of the production process. Moreover, the decrease in high-skilled labor causes labor productivity to increase in the course of involvement in the global value chain. The reason is that of low-skilled labor increases so that labor that takes a part in the production process increases, and then production tends to increase. Nevertheless, there are research gaps because this paper only takes wages into account as a determinant of demand for labor. In addition, the correlation between high-skilled labor and labor productivity might be biased due to the definition of high-skilled labor itself, if we only use descriptive analysis to explain the phenomenon. Further research is needed to complete the analysis of this phenomenon.

REFERENCES

Bair, J. (2005). Global capitalism and commodity chains: Looking back, going forward. *Competition and Change*, 9(2), 153–180.

Bernard, A.B. & Jensen, J.B. (1997). Exporters, skill upgrading, and the wage gap. *Journal of International Economics*, 42(1–2), 3–31.

Brambilla, I., Chauvin, N.D. & Porto, G. (2015). *Wage and employment gains from exports: Evidence from developing countries*. Working paper 2015–28. Paris, France: CEPII Research Center.

Bureau of Labor Statistics. (1957). Nonproduction Workers in Factories, 1919–1956. *Monthly Labor Review*, 80(4), 435–440. Retrieved from http://www.jstor.org/stable/41833721

Bustos, P. (2005). *The impact of trade liberalization on skill upgrading: Evidence from Argentina*. Working paper 1189, Department of Economics and Business. Barcelona, Spain: Universitat Pompeu Fabra.

Dean, J., Fung, K. & Wang, Z. (2008). *Measuring the vertical specialization in Chinese trade*. Office of Economics working paper EC-2008-09-D. Washington, DC: US International Trade Commission.

Dickerson, K.G. (1995). *Textiles and apparel in the global economy*. Englewood Cliffs, NJ: Prentice Hall.

Ehrenberg, R.G. & Smith, R.S. (2012). *Modern labor economics: Theory and public policy* (11th ed.). Pearson Education.

Esmaile, A.Y. (2018). Employment quality and wages effect from global engagement: Evidence from Ethiopian manufacturing firms. In *Proceedings of EconWorld 2018, 23–25 January 2018, Lisbon, Portugal*. Retrieved from https://lisbon2018.econworld.org/papers/Esmaile_Employment.pdf

European Central Bank. (2013). How have global value chains affected world trade patterns? *ECB Monthly Bulletin* (pp. 10–14). Frankfurt, Germany: European Central Bank.

European Union. (2016). *Statistical approaches to the measurement of skills*. Luxembourg: Publications Office of the European Union.

Gereffi, G. & Fernandez-Stark, K. (2011). *Global value chain analysis: A primer*. Durham, NC: Duke Center on Globalization, Governance, and Competitiveness.

Gereffi, G. (1994). The organization of buyer-driven global commodity chains: How US retailers shape overseas production networks. In G. Gereffi & M. Korzeniewicz (Eds.), *Commodity chains and global capitalism* (pp. 95–122). Westport, CT: Praeger.

Gereffi, G. (2013). Economic and social upgrading and workforce development in the apparel global value chain. In *International Labor Brief* (pp. 18–28). Sejong City, South Korea: Korea Labor Institute. Retrieved from https://dukespace.lib.duke.edu/dspace/handle/10161/10946

Goldberg, P.K. & Pavcnik, N. (2007). Distributional effects of globalization in developing countries. *Journal of Economic Literature*, *45*(1), 39–82.

Griliches, Z. (1969). Capital-skill complementary. *Review of Economics and Statistics*, *51*, 465–468.

Hummels, D., Ishii, J. & Yi, K.-M. (2001). The nature and growth of vertical specialization in world trade. *Journal of International Economics*, *54*(1), 75–96.

International Labor Organization. (2014). *Wages in Asia and the Pacific: Dynamic but uneven progress*. Regional Office for Asia and the Pacific.

International Labour Organization. (2010). *A skilled workforce for strong, sustainable and balanced growth: A G20 training strategy*. Geneva, Switzerland: International Labour Organization.

Kasahara, H., Liang, Y. & Rodrigue, J. (2016). Does importing intermediates increase the demand for skilled workers? Plant-level evidence from Indonesia. *Journal of International Economics*, *102*, 242–261.

Katz, L.F. & Murphy, K.M. (1992). Changes in relative wages, 1963–1987: Supply and demand factors. *Quarterly Journal of Economics*, *107*(1), 35–78.

Kuroiwa, I. & Heng, T.M. (2008). *Production networks and industrial clusters: Integrating economies in Southeast Asia*. Singapore: Institute of Southeast Asian Studies.

Matsuyama, K. (2007). Beyond icebergs: Towards a theory of biased globalization. *Review of Economic Studies*, *74*(1), 237–253.

Memedovic, O. (2004). *Inserting local industries into global value chains and global production networks: Opportunities and challenges for upgrading*. Vienna, Austria: United Nations Industrial Development Organization. Retrieved from https://www.unido.org/sites/default/files/2009-12/Inserting_local_industries_into_global_value_chains_and_global_production_networks_0.pdf

OECD. (2012). *Mapping global value chains*. Paris, France: Organisation for Economic Cooperation & Development.

Pyndick, R.S. & Rubinfeld, D.L. (2009). *Microeconomics* (7th ed.). Pearson Prentice Hall.

Sánchez-Páramo, C. & Schady, N.R. (2003). *Off and running? Technology, trade, and the rising demand for skilled workers in Latin America*. Policy Research Working Paper No. 3015. Washington, DC: World Bank.

Timmer, M.P., Erumban, A.A., Los, B., Stehrer, R. & de Vries, G.J. (2014). Slicing up global value chains. *Journal of Economic Perspectives*, *28*(2), 99–118.

Vera-Toscano, E., Rodrigues, M. & Costa, P. (2017). Beyond educational attainment: The importance of skills and lifelong learning for social outcomes. Evidence for Europe from PIAAC. *European Journal of Education*, *52*(2), 217–231.

Verhoogen, E. (2008). Trade, quality upgrading, and wage inequality in the Mexican manufacturing sector. *Quarterly Journal of Economics*, *123*(2), 489–530.

Yi, K.-M. (2003). Can vertical specialization explain the growth of world trade? *Journal of Political Economy*, *111*(1), 52–102.

Business Innovation and Development in Emerging Economies – Trinugroho & Lau (Eds)
© *2019 Taylor & Francis Group, London, ISBN 978-1-138-35996-3*

Fair treatment in employment of global production networks: Lessons from the Indonesian footwear industry

Rieka Evy Mulyanti
Population and Labour Economics, Faculty of Economics and Business, Universitas Indonesia, Depok, Indonesia

Padang Wicaksono
Labour Economics at the Department of Economics, Demographic Institute, Faculty of Economics and Business, Universitas Indonesia, Depok, Indonesia

Toni Bakhtiar
Department of Mathematics, Institut Pertanian Bogor, Bogor, Indonesia

ABSTRACT: The participation of the state in global production networks generate a positive impact on industries. The footwear industry is undergoing rapid growth and created millions of jobs in developing countries especially for women, therefore called labor-intensive industries. Economic developments trigger economic upgrading that is not always followed by social upgrading. Economic upgrading and social upgrading (twin upgrading) are at once experienced by the footwear industry in Indonesia. Global production networks research has recently focused on workers, social upgrading and economic upgrading. However, the research does not specifically clarify the mechanism behind global production networks growth and economic implications on working conditions and social welfare of workers. One of the decent work indicators is fair treatment, this article makes an analysis of fair treatment in employment and global production networks in the Indonesian footwear industry. The fair treatment indicators used in this article are occupational segregation by sex and female share of employment in managerial and administrative occupations. Both indicators are used to measure the equality of opportunity in employment. This article aims to clarify the consequences that workers face in global production networks and focus on the relationship between twin upgrading and the challenge of achieving decent work. In addition, this article also explores the implications of decent work as a result of the emergence of global production networks in the footwear industry of Indonesia.

Keywords: fair treatment, employment, decent work, global production networks, global value chains, footwear industry

1 INTRODUCTION

The participation of the state in global production networks generate a positive impact on industries. The footwear industry is undergoing rapid growth and created millions of jobs in developing countries especially for women, therefore called labor-intensive industries. Women represent an average of 46 percent in the leather and footwear industries (UNIDO, 2013). Indonesia is 6th largest footwear producer and 10th largest consumer in the world according to World Footwear Yearbook 2016. Indonesia's footwear industry is a stable sector in terms of generating job opportunities and increasing gross domestic product. Based on data from the Central Bureau of Statistics (BPS), the trend of export value from Indonesian footwear industry in 2002–2015 tended to increase. Economic developments trigger economic upgrading that is not always followed by social upgrading (Barrientos, Gereffi & Rossi, 2011;

Milberg & Winkler, 2011, 2013; Rossi, 2013; Salido & Bellhouse, 2016). Economic upgrading and social upgrading (twin upgrading) are at once experienced by the footwear industry in Indonesia (Hardini, Wicaksono & Bakhtiar, 2016).

One in five jobs worldwide is expected linked to global production networks (ILO, 2015). However, many global production networks workers are insecured and unprotected but to ensure decent work for more vulnerable workers will cause significant problems. Global production networks research has recently focused on workers, social upgrading and economic upgrading (Barrientos, Gereffi & Rossi, 2011; Coe & Hess, 2013; Rossi, Luinstra & Pickles, 2014). However, the research does not specifically clarify the mechanism behind global production networks growth and economic implications on working conditions and social welfare of workers. One of the decent work indicators is fair treatment, this article makes an analysis of fair treatment in employment and global production networks in the Indonesian footwear industry. The fair treatment indicators used in this article are occupational segregation by sex and female share of employment in managerial and administrative occupations. Both indicators are used to measure the equality of opportunity in employment. This article aims to clarify the consequences that workers face in global production networks and focus on the relationship between twin upgrading and the challenge of achieving decent work. In addition, this article also explores the implications of decent work as a result of the emergence of global production networks in the footwear industry of Indonesia.

This article is organized into four sections as follows. Section 1 consist of the introduction and background of this article. Section 2 consist of an analytical framework comprising fair treatment in decent work (linkage between twin upgrading) and global production networks and global value chains. Section 3 consist of the fair treatment in employment of global footwear production networks in Indonesia comprising occupational segregation by sex and female share of employment in managerial and administrative occupations. Section 4 is the conclusion and some remaining problems.

2 ANALYTICAL FRAMEWORK

2.1 *Fair Treatment in Decent Work (Linkage Between Twin Upgrading)*

Analysis of economic upgrading and social upgrading (twin upgrading) measures is complex because it involves different data sets depending on the scope and understanding of these concepts. Three major methodological challenges in this measurement include the level of analysis and comparison of recent studies (comparability), the quality of available data and the conceptualization of social enhancement (Salido & Bellhouse, 2016). Facing tough competition, companies are looking for ways to meet customer needs for higher quality goods while maintaining flexibility and reducing costs. In this context, supplier companies can use employment strategies that include temporary employment, lower wages, extend working hours, work subcontracting and minimize investments in health and safety at work. Although such employment practices are appropriate to the needs of workers seeking temporary employment situations, it can be considered to increase the social burden on workers engaged in this chain who prefer full-time formal employment relations (ILO, 2016).

Economic upgrading is a process whereby economic actors (firms & workers) move from low-value activities to relatively high-value relatives in the global supply chains and can be analyzed using country-level data export indicators (Gereffi, 2005). Economic upgrading is a multi-faceted & complex process, involving changes in business strategy, production & technology structure, market policy & organization (Bernhardt & Millberg, 2013). Types of economic upgrading include process upgrading, product upgrading, functional upgrading and chains upgrading. Each of these types of upgrading has different implications for skills and job development (Humphrey & Schmitz, 2002).

Social upgrading is generally defined as changes in the working conditions or rights of workers which improve the quality of their employment (Rossi, 2013). The social upgrading in the global value chains concept are part of the profits from economic improvements captured by workers in a particular company, industry or economy in a global production networks. This advantage can come in the form of monetary remuneration or in terms of increased welfare. Social enhancement can be described as a social impact felt by workers engaged in production networks (ILO, 2015). Social upgrading is generally defined as a change in the quality of work of workers. The concept of social improvement is framed by the Decent Work Agenda of the International Labor Organization that includes employment, standards and rights at work, social protection and social dialogue.

Social upgrading can be divided into two components: measurable standards and enabling rights (Elliott & Freeman, 2003; Barrientos & Smith, 2007). Measurable standards include wages, physical wellbeing (e.g. health and safety, working hours), job security (e.g. contract type, social protection) and gender (e.g. women in managerial). While enabling rights include the rights of major trade unions such as freedom of association and collective bargaining, the right to free choice of employment, non-discrimination and vote. Measurable standards are often the result of a complex bargaining process, framed by the spread of workers' rights. It is less easily measured, such as association, collective bargaining rights, non-discrimination, vote, and empowerment. Anker, et al. (2003) proposed 11 groups of indicators, each containing a measurable variable. These 11 groups are employment opportunities, unacceptable work, adequate earnings and productive work, decent hours, stability and security of work, balancing work and family life, fair treatment in employment, safe work, social protection, social dialogue and workplace relations, economic and social context of decent work.

Fair treatment in work that is an intrinsic human hope, which is one of the indicators of decent work (Anker, et al., 2003). At the international level, this has been expressed in equal opportunity in employment and employment, and equal pay for equal value work. The International Labor Organization (Employment and Occupation) Discrimination Convention, 1958 (No. 111) defines discrimination as "any distinction, exclusion or preference made on the basis of race, color, sex, religion, political opinion, national extraction or social origin which has the effect of nullifying or impairing equality of opportunity or treatment in work or work". The Equal Remuneration Convention, 1951 (No. 100) deals with discrimination in equal pay and remuneration for work of equal value. In addition to the absence of discrimination at work and in access to the workplace, fair treatment means working without harassment or exposure to violence, certain levels of autonomy, and fair handling of complaints and conflicts. The latter is closely related to the presence or absence of workplace mechanisms for social dialogue.

Indicators for fair treatment in employment according to suggestion of Anker's, et al. (2002) are different treatments in men and women. Other indicators that can be used include discrimination such as race, ethnicity, religion and migrant status. Segregation of occupations by sex and female share of employment in managerial and administrative occupations are two indicators that measure the equality of opportunity in employment. The main reason for focusing on gender is that gender discrimination is universal and has received the most attention, despite many other sources of discrimination in addition to much better data availability than other aspects such as autonomy, grievance procedures, and absence of harassment or violence.

2.2 *Global Production Networks (GPN) and Global Value Chains (GVCs)*

Global production networks are made up of leading companies with local suppliers participating in different value chains in business activities while global value chains consist of two or more production networks that are compatible with the production process, therefore both are complementary concepts (Kuroiwa & Toh, 2008). Global production networks are a set of inter-firm relationships that bind a group of companies into larger economic units and have proven to be powerful tools for studying economic upgrading because they have been found to form a context in which companies have improved their productive capabilities, especially through learning from relationships with buyers, as these suppliers seek to produce internationally competitive goods and services (Millberg & Winkler, 2011). Companies that participate in global production networks in various hierarchical layers are under the supervision of principal company or lead company so that the growth, strategic direction and position of the company/supplier network is highly dependent on key corporate strategy. Firm-led production networks participate in various value chains and compete with each other (under the leadership of a flagship firm) in a product-specific value chain (Wicaksono & Priyadi, 2016). The global production networks approach stresses the importance of non-firm actors and the embeddedness of production networks in social, institutional and policy contexts, stemming from the early recognition that firms behaviour is strongly influenced by the context in which they operate (Bair, 2009; Henderson et al., 2002).

The literature on global value chains and production networks analyses how production, distribution and consumption are globally interconnected along organizationally fragmented and geographically dispersed chains or networks of activities and actors. Considerable attention has been dedicated to economic upgrading, defined as a firm's, region's or country's trajectory from low- to higher-value activities (Bair and Gereffi, 2003). Many companies looked for ways to minimize production costs to remain competitive by moving their production sites to developing countries. The revival of the global value chain paradigm makes integration into international production networks an opportunity for low- and middle-income countries to improve economic competitiveness through increased access to global markets (Salido & Bellhouse, 2016). It can also increase opportunities and challenges to improve living standards (Bernhardt & Millberg, 2013).

Companies involved in global production networks have the opportunity to improve the economy by producing higher value goods or reposition themselves in the value chain. Global brand-name companies and retailers retain high value-added activities from products but no longer own factories. Global production has moved from commercial plant to subcontractor (Barrientos, Gereffi & Rossi, 2011). The labels that resellers will also face smaller margins and increased costs. Many retailers will go out of business or have to reduce their operations, labels and retailers looking for more flexibility in design, quality, delivery and speed to market in order to be competitive (Kenneally, 2014). Footwear industry is a buyer-driven production chain that produces labor-intensive goods manufactured in accordance with the specifications and designs developed by "manufacturers without a factory" (Coe, Kelly, & Yeung, 2013). The growth competition in the footwear industry is increasingly divided into products and brands of high and low class (called "value"). High-end production consists of factories that use better technology and more skilled workers. These factories have a higher level of multi-party involvement. Conversely, low-end production or value is dominated by a considerable price focus and often poor working conditions (Hess, 2013; Jones, 2013).

Yi (2003) defines vertical specialization as an increase in production process linkages in the sequential vertical chain of trade that spans across many countries, with each country specializing at a certain stage of the order of production of goods. Vertical specialization occurs when the goods are produced in successive stages, two or more countries add value in a good production sequence, and at least one country must use import input at the stage of its production process as well as some of the resulting output must be exported. Vertical specialization has an import side that is a subset of intermediate goods (used to make goods to be exported) and export side that includes end goods and semi-finished goods (Hummels, Ishii & Yi, 2001). Ideally, vertical specialization is calculated on the level of individual goods then combined. This article used vertical specialization as a measure of a firm's involvement in global production networks by declaring the contribution of imported inputs to gross output on exports in every footwear firm. The equation of vertical specialization used in this paper is as follows:

$$VS_{ki} = \left(\frac{Imported\ Intermediaries_{ki}}{Gross\ Output_{ki}} \right) Exports_{ki}$$

where the notation k denotes state, i is goods, and j is country export destination country k.

3 THE FAIR TREATMENT IN EMPLOYMENT OF GLOBAL FOOTWEAR PRODUCTION NETWORKS IN INDONESIA

Indonesia's footwear industry is well integrated in global production networks, from component makers to finished footwear products. One of world's leading footwear manufacturers such as Adidas brand (Reebok is one of the sub brands of Adidas) where the headquarter is located in Herzogenaurach, Germany. Shoes production countries (manufacturing) spread around the world, among others 1% in Europe; 3% in Americas and 96% and in Asia (Vietnam, Indonesia, China, others). Number of supplier factories (shoes & leather) more than 1,100 and employing more than 60,000 people in over 160 countries. Adidas also produce more than 850 million product units every year with turnover generating sales of around € 19 billion. Supplier factories of Adidas in Indonesia consists of several companies by employing thousands of workers, especially female workers.

The linkage of Indonesia in global production networks is also evidenced in the data of ownership structure. Collective investment is the main ownership structure for footwear manufacturers as seen in Table 1. Since the early 2000s with the abolition of rigorous local content policy, the global relations of Indonesian production networks have been significantly strengthened. Foreign investment in the form of full foreign ownership has increased, and many domestic companies have shifted their ownership structure to mutual investment. However, the linkages of global production networks vary at the sub-industry level. The ownership structure trends have moved towards full foreign ownership in major footwear manufacturers as a result of the brand leader's characteristic in the industry. Meanwhile, joint investment is the preferred form of ownership for manufacturers of footwear components. This is because multinational companies can access local resources and comply with local Indonesian content policies; local companies can become original equipment manufacturers (OEMs), which is very profitable for the profitability of foreign and local companies (Wicaksono & Priyadi, 2016). Indonesia as a producer and consumer of footwear is quite large in the world, so tend to be a footwear manufacturer than footwear components because brand manufacturers are closer to consumers to minimize distribution costs.

The number of companies is relatively stable in fifteen years if we only look at firms that do the GPN than framework than non-GPN firms as seen in Figure 1. The number of non-GPN firms tends to increase over the period of 2001–2015, but the decline tends to occur in GPN companies. This may be due to GPN-related footwear manufacturers being overwhelmed by large companies with well-known brands with large capital so that companies are relatively medium capital that is not too big enough to face the footwear market that tends to be dynamic and following the fashion. A significant increase occurred in the number of non-GPNs companies in 2006. in that year indicated a resurgence of the local footwear industry after the occurrence of production sector slump due to quite a lot of unilateral contract termination by foreign investors and termination of employment. In addition, during the period 1995–2005 Indonesia is also one of the countries that became the target of alleged dumping practices for the footwear industry, it is affecting the footwear industry. According to data from the Ministry of Industry, in that year there was an increase in the number of footwear companies in Banten, West Java and East Java provinces.

The economic upgrading shown by the improvement of export performance of a country occurs in Indonesian footwear industry, as seen in Figure 2. Despite fluctuations, export trends show an increasing trend as the market for export of footwear products in Indonesia is able to compete internationally (Bernhardt & Milberg, 2013). Indonesia's footwear products have advantages over Vietnamese and Thai products because their products show better skill, that is, more neatly sewn (Hardini, Wicaksono & Bakhtiar, 2016). A significant increase

Table 1. Footwear industry ownership structure by number of establishments.

Ownership	2001	2002	2003	2004	2005	2006	2007	2008	2009	2010	2011	2012	2013	2014	2015
Domestic Firm	329	319	303	286	297	525	488	431	407	394	401	401	399	409	429
Foreign Investment	11	15	25	20	16	25	31	28	31	35	32	34	31	33	44
Joint Investment	24	22	22	21	14	19	16	14	14	18	18	19	18	19	16
Total	364	356	350	327	327	569	535	473	452	447	451	454	448	461	489

Source: Calculated from Medium and Large Enterprises Industrial Statistics, Badan Pusat Statistik.

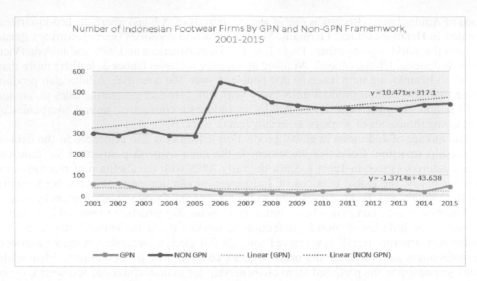

Figure 1.   Number of Indonesian footwear firms by GPN and non-GPN framework, 2001–2015.
Source: Calculated from Medium and Large Enterprises Industrial Statistics, Badan Pusat Statistik.

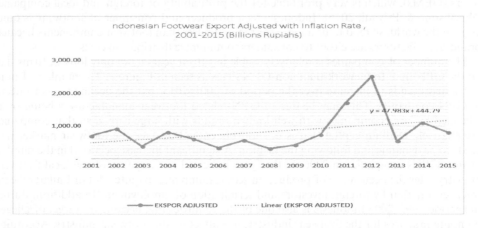

Figure 2.   Indonesian footwear export adjusted with inflation rate, 2001–2015 (Billion Rupiahs).
Source: Calculated from Medium and Large Enterprises Industrial Statistics, Badan Pusat Statistik.

occurred in the export value of footwear in Indonesia in 2012, as shown in Figure 2. Despite the global crisis, the market share of footwear in Indonesia abroad, especially the countries in America and Europe did not decrease, even tended to increase. This is due to the increasing state of export market share that is the countries in the Middle East especially the United Arab Emirates so that the export value of footwear in Indonesia tends to increase. Indonesia's footwear products have advantages over Vietnamese and Thai products because their products show better skill, that is, more neatly sewn (Hardini, Wicaksono & Bakhtiar, 2016).

As for social upgrading which is indicated by the increase of wages of both production and non-production workers in the footwear industry in Indonesia within the period of 2001–2015 tends to increase as shown in Figure 3. Real wage is an indicator of how many workers benefit from the value created in their sector and indicated the labor bargaining power (Bernhardt & Milberg, 2013). The relationship between twin upgrading described by Millberg & Winkler (2011) used economic theory that sees wage growth closely linked to productivity growth. Economic upgrading in this case is measured by productivity growth (e.g. increased output per worker) as a proxy, while social upgrading uses wage growth as a proxy. The marginal theory of productivity uses the basic assumption that perfectly competitive

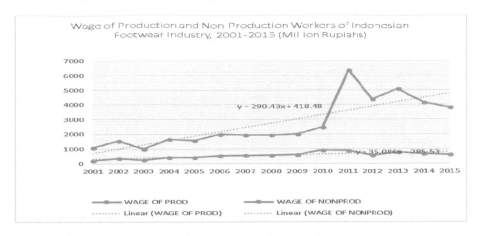

Figure 3. Wage of production and non-production workers of Indonesian footwear industry, 2001–2015 (Million Rupiahs).
Source: Calculated from Medium and Large Enterprises Industrial Statistics, Badan Pusat Statistik.

markets, ruling out technological processes and risks in the economy, wages are made up of competitive job markets where labor demand and supply are the deciding factor. The company is also assumed to always maximize profits. Wages will shift if there is a shift in demand curve or supply of labor force (Borjas, 2010).

3.1 Occupational segregation by sex

Anker (1998) states that about half of all workers in the world are in sex-dominated jobs where at least 80 percent of workers have the same gender, so the worldwide labor market is highly segmented by sex. The rigidity of the labor market reduces employment opportunities especially for women and disrupts economic efficiency. Goldin (2002) in "A Pollution Theory of Discrimination" also states that increased feminization in work negatively affects wage levels in the occupation. The number of Indonesian footwear production workers by sex in the 2001–2015 period also tends to increase and fluctuate mainly in women production workers compared to men production workers as shown in Figure 4. The number of women production workers increases every year although the increase is not as large as the men production workers. In absolute terms the number of women production workers is still greater than that of men production workers. In this case there are improvements in the formalization of women workers and little or no gender discrimination in the footwear industry. Women workers tend to be preferred by employers in the footwear production process because they are more careful and conscientious.

3.2 Women share of employment in managerial and administrative occupations

The extent to which women are in decision-making positions (authority) such as managers and administrative workers. This is because the distribution of women and men at different levels of responsibility is an important measure for equal treatment in employment (Anker's, et al., 2002). In this article consists of two types of workers namely production and non-production. Production workers are used as a proxy for non-skilled workers while non-production workers are used as proxies of skilled workers such as managerial level or administrative occupations. The number of Indonesian footwear non-production workers by sex in the period 2001–2015 tends to increase and fluctuate especially in women workers compared to men workers as shown in Figure 5. Increased exports and real wages at once (twin upgrading) could lead to an increase of share of employment in managerial and administrative occupations. The extent to which women are in decision-making positions

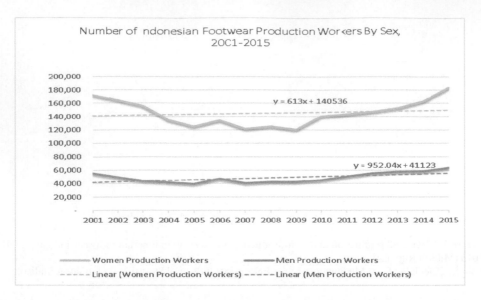

Figure 4.   Number of Indonesian footwear production workers by sex, 2001–2015.
Source: Calculated from Medium and Large Enterprises Industrial Statistics, Badan Pusat Statistik.

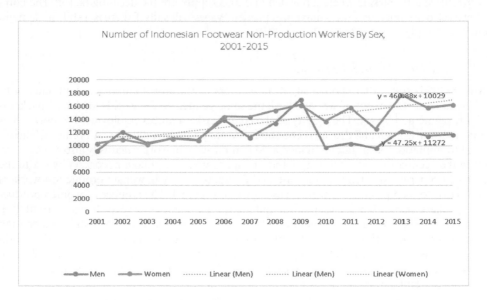

Figure 5.   Number of Indonesian footwear non-production workers by sex, 2001–2015.
Source: Calculated from Medium and Large Enterprises Industrial Statistics, Badan Pusat Statistik.

(authority) such as managers and administrative workers. This is because the distribution of women and men at different levels of responsibility is an important measure of
equal treatment in employment (Anker, et al., 2002). In this article consists of two types of
workers namely production and non-production. Production workers are used as a proxy
for unskilled workers while non-production workers are used as representative of skilled
workers such as managerial or administrative work. The number of Indonesian footwear
non-production workers by sex in the 2001–2015 period tends to increase and fluctuate
mainly in women workers compared to male workers as shown in Figure 5. In absolute
terms the number of women's non-production work is still greater than that of men non-
production workers. Increased exports and real wages at once (twin increase) are likely to

lead to an increase in the share of employment in managerial and administrative jobs especially for women workers. In this case further studies are needed to examine the magnitude of causality.

4 CONCLUSION AND SOME REMAINING PROBLEMS

This article has examined fair treatment in employment to measure equality of employment opportunities by focusing on two elements of decent work: segregation of employment by sex and the division of labor by women in managerial and administrative work. Both indicators are used to measure equality of opportunity in employment. Based on the analysis it was found that although the number of GPN companies in the footwear industry in Indonesia has tended to decline over the last 15 years, twin increases have occurred along with improved employment indicators.

However, there are some limitations in this article. Measurements of twin increase consist of economic improvement and social improvement each using only one indicator. Millberg & Winkler (2011) provides several indicators to improve twin measurements for country, sector/industry and enterprise level. Another limitation is the use of proxies of production workers for non-skilled workers and the proxy of non-production workers for skilled workers. Based on the definition of the Central Statistics Agency, non-production workers in this case are workers who are not directly related to the production process and usually as support workers of the company. Not just managers (not production), heads of personnel, secretaries, typists and managerial or other administrative levels, but also including security, company drivers, etc. However, the number of non-skilled non-production workers tends to be less than that of skilled non-production workers.

The unit of analysis and data sources in this article is a medium and large-scale footwear industry company in Indonesia during the period 2001–2015 and the Annual Survey of Large and Medium Manufacturing Statistics conducted by the Central Bureau of Statistics in 2001–2015. Classification of Footwear Industry by ISIC Rev.4 (Indonesia Standard Industrial Classification) for all economic activities published by UNSD (United Nations of Statistical Division) in 2008 consisting of: footwear industry for daily purposes (15201); sports shoes industry (15202); engineering in the shoe industry / industry (15203), and other footwear industries (15209).

REFERENCES

Anker, R. (1998). *Gender and jobs: Sex segregation of occupations in the world.* International Labour Organization.

Anker, R., Chernyshev, I., Egger, P., Mehran, F., & Ritter, J.A. (2003). Measuring decent work with statistical indicators. *International Labour Review, 142*(2).

Bair, Jennifer, and Gary Gereffi. "Upgrading, uneven development, and jobs in the North American apparel industry." *Global Networks* 3.2 (2003): 143–169.

Barrientos, Stephanie & Sally Smith. (2007). "Do Workers Benefit From Ethical Trade? Assessing Codes of Labour Practice in Global Production Systems", in Third World Quarterly, Vol. 28, No. 4.

Barrientos, Stephanie; Gary Gereffi & Arianna Rossi. (2011). *Economic and Social Upgrading in Global Production Networks: A New Paradigm for a Changing World. International Labour Review* Vol. 150, No. 3–4.

Bernhardt, T., & Milberg, W. (2013). Does industrial upgrading generate employment and wage gains?. *The Oxford Handbook of Offshoring and Global Employment*, 490–533.

Borjas, George J. (2016). *Labour Economics. Seventh Edition.* New York: McGraw Hill Education.

Change Your Shoes (2016).

Coe, N.M., & Hess, M. (2013). Global production networks, labour and development. *Geoforum, 44*, 4–9.

Coe, Neil, Philip Kelly & Henry W. C. Yeung. (2013). Economic Geography: A Contemporary Introduction 2nd Edition. Wiley.

Elliott, Kimberley Ann & Richard B. Freeman. (2003). "The Role Global Labour Standards Could Play in Addressing Basic Needs", in Jody Heymann (ed): Global Inequalities at Work: Work's Impact on The Health of Individuals, Families, and Societies. New York, NY, Oxford University Press, pp. 299–327.

Gereffi, Gary. (1999). *International Trade and Industrial Upgrading in The Apparel Commodity Chain. Journal of International Economics.* Elsevier.

Goldin, C. (2002). A Pollution Theory of Discrimination: Male and Female Earnings in Occupations and Industries. *NBER Working Paper, 4985.*

Hardini, N., Wicaksono, P. & Bakhtiar, Toni (2016). Economic and social development in a Global Network: Production lessons from the Indonesian footwear industry.

Hess, M. (2013). Global production networks and variegated capitalism: (Self-)regulating labour in Cambodian garment factories, Better Work Discussion Paper Series, No. 9. Geneva, ILO, 2013. https://www.adidas-group.com/

Hummels, D., Ishii, J., & Yi, K.M. (2001). The nature and growth of vertical specialization in world trade. *Journal of international Economics, 54*(1), 75–96.

Humphrey, J. (2004). *Upgrading in Global Value Chains.* ILO Policy Integration Department, Working Paper No. 28.

Humphrey, J. & H. Schmitz. (2002). *How Does Insertion in Global Value Chains Affect Upgrading in Industrial Clusters? Regional Studies,* 9, 1017–1027.

ILO Promoting Decent Work in Global Supply Chains in Latin America and the Caribbean. key issues, good practices, lessons learned and policy insights. Lima: ILO Regional Office for Latin America and the Caribbean, 2016. 120 p. (ILO Technical Reports, 2016/1).

ILO (International Labour Organization) (2015), World Employment Social Outlook: The Changing Nature of Jobs (Geneva: ILO).

Jones, R.W. (2013). Interpreting (non-)compliance: The role of variegated capitalism in Vietnam's garment sector, Better Work Discussion Paper Series, No. 11. Geneva, ILO, 2013.

Kenneally, I. (2014). "New Sourcing Survey: Expect Rising Costs and Narrowing Margins", in (online) Sourcing Journal, 24 Mar. 2014, https://www.sourcingjournalonline.com/tag/merchandise-margin.

Kuroiwa, Ikuo & Toh Mun Heng. (2008). *Production Networks and Industrial Clusters: Integrating Economies in Southeast Asia.* Singapore: Institute of Southeast Asian Studies.

Lee, Joonkoo, Gary Gereffi & Dev Nathan. (2013). Mobile Phones: Who Benefits in Shifting Global Value Chains? Capturing The Gains, Revised Summit Briefing, No. 6.1.

Milberg, W. & Winkler, D. (2011). *Economic and Social Upgrading in Global Production Networks Problems of Theory and Measurement.*

Morrison, Andrea, Carlo Pietrobelli & Roberta Rabelloti. (2006). *Global Value Chains and Technological Capabilities: A Framework to Study Industrial Innovation in Developing Countries.* Milano: Universita Commerciale Luigi Bacconi.

Rossi, A., Luinstra, A., & Pickles, J. (Eds.). (2014). *Towards better work: Understanding labour in apparel global value chains.* Springer.

Rossi, Arianna (2013), "Does Economic Upgrading Lead to Social Upgrading in Global Production Networks? Evidence from Morocco," World Development, vol. 46, June pp. 223–233.

Rossi, Arianna. (2011). *Economic and Social Upgrading in Global Production Networks: The Case of The Garment Industry in Morocco.* DPhil Dissertation. Brighton, Institute of Development Studies, University of Sussex.

Salido, J. & Bellhouse, T. (2016). *Economic and Social Upgrading: Definitions, Connections and Exploring Means of Measurement.* United Nations: Economic Commission for Latin America and The Caribbean.

UNIDO. (2013). UNIDO: International Yearbook of Industrial Statistics, Vienna, 2013.

Wicaksono, P. & Priyadi, L. (2016). Decent Work in Global Production Network Lessons Learnt From Indonesian Automotive Sector. *Journal of Southeast Asian Economies (JSEAE)*, 33(1), 95–110.

World Footwear Yearbook 2016 (www.worldfootwear.com).

Yi, K.M. (2003). Can vertical specialization explain the growth of world trade?. *Journal of political Economy, 111*(1), 52–102.

Business Innovation and Development in Emerging Economies – Trinugroho & Lau (Eds)
© 2019 Taylor & Francis Group, London, ISBN 978-1-138-35996-3

Decent work in global apparel production networks: Evidences from Indonesia

Fitria Nur Diana
Graduate School for Population and Labor Economics, Universitas Indonesia, Depok, Indonesia

Padang Wicaksono
Labor Economics at the Department of Economics, Universitas Indonesia, Depok, Indonesia

Toni Bakhtiar
Applied Mathematics at the Department of Mathematics, Bogor Agricultural University, Bogor, Indonesia

ABSTRACT: The engagement in GPNs looks like a double-edged sword since it can offer opportunities and pose threats to workers as well. This study makes a particular attempt to clarify the close links between enterprises' engagement in GPNs and decent works in the Indonesian apparel industry. More specifically, this paper examines the decent wages and employment opportunities for local workers within the decent work framework in the country's apparel sector with the expansion of GPNs. It finds that although the engagement of Indonesia's apparel industry in GPNs is growing, but the local workers are less likely to benefit from it. Furthermore, despite employment opportunities rising, the wages for local employees remain non decent.

Keywords: decent wages, real wages, adequate wages, global production networks, apparel industry

1 INTRODUCTION

The global production network (GPNs) is related to the phenomenon where multinational companies outsource activities to developing countries on the grounds of lower production costs, while maintaining higher value-added intangible activities as their core business (Rossi, 2011). The apparel industry is one of the most popular industries in the GPNs that has an important role in terms of employment, investment, and trade, especially for Asian countries. More than 65% of world garment production is produced by Asian countries, including Indonesia (Appendix 1). Whilst, the country became the 14th largest exporter of apparel in the world, with a positive average annual growth of 4.83 percent (Appendix 2).

Benefiting from the abundant working-age population, Indonesia plays a role in the production process with low economic value added in the global apparel value chain. This corresponds to the pattern of activity in the global value chain that the developed and developing countries tend to concentrate on higher-value activities and lower value activities, respectively (Fernandez-Stark, Frederick and Gereffi, 2011). This position provides opportunities for employment creation, especially for production workers. By 2015, Indonesia's apparel industry[1] was able to employ 684,023 workers in large and medium enterprises which 92 percent of them are production workers. While this figure is the second largest after food industry

---

1. The definition of apparel industry uses International Standard Industrial Classification of all Economic Activities (ISIC) rev. 3 for 2001–2009 period and ISIC rev. 4 for 2010–2015 (see Appendix 3).

(BPS, 2017a), trend in production workers from 2001 to 2015 have increased. The industry has proved substantial in employment creation while stay competitive as a labor-intensive manufactured export.

However, employment creation itself are not enough since it doesn't necessarily guarantee decent works for local workers. Needless to say, the decent works are prerequisite to achieve workers' welfare. Therefore, it is essential to take a look at changes in decent works over time to capture the dynamics of workers' welfare as a consequence of GPNs engagement. The challenge faced in this problem is that engagement in the GPNs has two side effects because it can offer opportunities and pose a threat to workers as well. The quality of work in lower value-added activities on GPNs is often characterized by high degree of flexibility, uncertainty and precariousness (Barrientos, Gereffi, & Rossi, 2011). The apparel industry is one of the job characteristics relevant to the activities.

Research on GPNs is often associated with economic benefits, only a few studies have tried to link GPNs to social aspects (e.g. Wicaksono & Priyadi, 2016). Indeed, some previous literatures examined the implication of GPNs engagement on workers' premium wages (e.g. Kohpaiboon & Jongwanich, 2013; Amiti & Cameron, 2012; Amiti & Davis, 2011; Goldberg & Pavcnik, 2007; Bigsten & Durevall, 2006; Mishra & Kumar, 2005; Attanasio, Goldberg & Pavcnik, 2004; Galiani & Sanguinetti, 2003; Feenstra & Hanson, 2001; Harrison & Hanson, 1999; Currie & Harrison, 1997). Nonetheless, the literatures has not particularly evaluated its implication on decent works issue.

This study aims to specifically close the research gap by explaining the conditions of workers as a consequence of GPNs engagement. More specifically, the authors focus on the implication for decent works as a result of the rise of GPNs in the Indonesian apparel industry. Indicators of decent work discussed in this study are limited to employment opportunities, adequate income and productive work due to limited data[2].

This paper is organized as follows. Section 2 will depict the analytical framework, which comprises the GPNs concept and the mechanism of the relationship between GPNs engagement and decent wages. A brief discussion on real wages and adequate wages in the Indonesian apparel industry will be presented in Section 3. Section 4 will figure out the consequences of GPNs engagement on real wages and adequate wages. The final section 5 concludes and outlines the possible related issues for further research.

2 ANALYTICAL FRAMEWORK

2.1 *GPNs: Vertical specialization*

According to Kuroiwa and Toh (2008), GPNs consists of the participation of local suppliers in the value chain at different levels of hierarchy in business activities under the supervision of the flagship firm. Thus, the growth, strategic direction, and network position of the local firms depend heavily on the firm's flagship strategy. Different from Global Value Chain (GVC) framework that focuses solely on the commercial dynamics between firms in different segments of the production chain, the GPNs focus also emphasizes the social and institutional embeddedness of production, and the varying power relationships between actors as a source spread across multiple developing countries (Barrientos, Gereffi & Rossi, 2010).

GPNs provide opportunities for created new job, particularly in developing and newly emerging economies (Kawakami & Sturgeon, 2011). However, they also face challenges in meeting the commercial demands and buyer quality standards that are sometimes difficult to satisfy (Barrientos, Gereffi, & Rossi, 2011). The risks of global pressure on producer prices, delivery times and tight competition between suppliers can put pressure on wages, working conditions and the rights of workers who participate in the chain (ILO, 2016).

---

2. The definition of apparel industry uses International Standard Industrial Classification of all Economic Activities (ISIC) rev. 3 for 2001–2009 period and ISIC rev. 4 for 2010–2015 (see Appendix 3).

Ernst and Kim (2002) classify principal GPN companies into two types: brand leaders and contract manufacturers. Brand leaders allow suppliers to be independent but demand high performance, whereas contract manufacturers establish an autonomous production network and create an integrated supply chain available to brand leaders.

The apparel industry is one of the industries that implement the GPNs framework for its development. This industry is export-oriented, labor-intensive and low-cost (Gereffi & Memedovic, 2003). The strategy that is widely used by brand leaders are shifting their activities to the developing countries to reduce labor costs. This phenomenon encourages the expansion of production and employment related to GPNs (Barrientos, Gereffi & Rossi, 2010). Production is generally carried out by tiered networks from contract manufacturers in developing countries that make finished goods for foreign buyers. Specifications are provided by brand leaders who order goods (Dicken & Hassler, 2000).

The low cost of the industry is due to the capital, technology, and skills needed by this industry are relatively simple. Since apparel production is deemed not to require a high level of expertise, the industry has become one of the most fragmented industries in the world, with the largest apparel companies deciding to run their production processes in countries where production costs can be kept lower. In the apparel industry, GPNs has become the norm, brands and retailers are in a competitive cycle to lower production costs in order to remain profitable as their competitors (Tager, 2016).

The engagement in GPNs can be explained using the concept of vertical specialization[3]. Vertical specialization is a measure of the engagement of production processes in a vertical chain involving multiple countries, with each country specializing at a certain stage of the order of production of goods. Vertical specialization occurs when goods are produced sequentially, two or more countries add value in the order of goods production, and at least one country must use imported inputs in the production process phase and some of the resulting output must be exported (Hummels, Ishii, & Yi, 2001).

Vertical specialization has import and export sides. On the import side, vertical specialization is an intermediate goods used to make goods for export. While on the export side, the vertical specialization may take the form of final goods and intermediate goods. Vertical specialization is foreign added value embodied in the export, or equivalent, imported input from export. Ideally, vertical specialization is calculated at the individual level goods in a country. Therefore, vertical specialization can be calculated for the apparel's firms level to expresses the contribution of imported inputs into gross output on exports in every apparel's firm. Vertical specialization describes how much engagement the apparel's firm in GPN.

2.2 Linkage between GPNs and decent works

This study specifically looks at the decent works in the form of employment opportunities and decent wages. Employment opportunities illustrate how much work can be created and the quality of workers demanded. It can be used to show our position in the global apparel production networks. While decent wages are illustrated by real wages and adequate wages are used to measure how many workers benefit from the value created by their economic activity in the Indonesian apparel industry to ensure the economic well-being of themselves and their households.

The globalization of production is characterized by the division of production processes in separate stages and allocated economically in many countries according to competitiveness. This encourages the development of international trade. The neo-classical trade theory states that international trade will lead to the specialization of a country according to its comparative

---

3. The term vertical specialization (VS) used in this study refers to the concept formulated by Hummels, Ishii and Yi (2001) as follows:

$$VS = \left(\frac{Imported\ intermediates}{Gross\ output} \right) \cdot Exports$$

Thus the concept of the firm engaged in GPN is if the firm has import activities in intermediate goods and exported the resulting output.

advantage. For developing countries, which have a comparative advantage in the abundance of unskilled workers, international trade will increase demand for workers due to increased export opportunities. Increased demand for workers is expected to also increase the wages of workers, both in quantity (nominal wages) and quality (real wages and adequate wages). Thus, engagement in GPNs is expected to provide benefits to workers in the form of employment opportunities and decent wages.

Indonesia's engagement in global apparel production networks can pose a threat to production workers, as well as opportunities. The apparel industry has a business model based on increasing fashion styles with short product lifecycles, necessitating bringing new products to market with increasing speed and in a shorter span of time (Plank, Rossi & Staritz, 2012). Suppliers are required to have high responsiveness in a flexible way and are able to produce high quality goods at low cost. Supplier companies struggle to accommodate these requirements and can pass that pressure on to their workers. The phenomenon that occurs is that the characteristics of workers in the Indonesian apparel industry are dominated by female workers, with low union rate. Wages of apparel industry workers in Indonesia are very low and fluctuated greatly during the period 2000–2014 (Manoto, Wicaksono & Bakhtiar, 2017). More than one-third of workers are paid below the minimum wage (Cowgill & Huynh, 2016). The impact of Indonesia's engagement in a global apparel production network on decent wages of workers still requires further investigation.

3 RECENT DEVELOPMENT AND STRUCTURE OF THE INDONESIAN APPAREL INDUSTRY

Despite the slowdown, Indonesia's apparel industry is well integrated within GPNs. Since the abolition of apparel export quotas through the Textile and Apparel Agreement (ATC) in 2005, the apparel industry competition in developing countries has become tighter and new countries are emerging as exporters of apparel products in the world (Fernandez-Stark, Frederick & Gereffi, 2011). Indonesia's competitiveness declined, marked by the transfer of production to China and Vietnam by some major international buyers in Indonesia. This is due to several factors such as the high cost of inputs in Indonesia compared to other countries, the weakening of the rupiah, the use of old machines, and the high minimum wage of workers in Indonesia. Nevertheless, the abundance of low-skilled workers is still a comparative advantage of Indonesia to engage in the production chain in the global apparel network. This is indicated by an upward trend of the firms engaged in GPNs, which is described as importing raw materials to be processed into final goods which are then re-exported (Figure 1). Indonesia's apparel industry is able to support the production of several popular and high-end brands such as

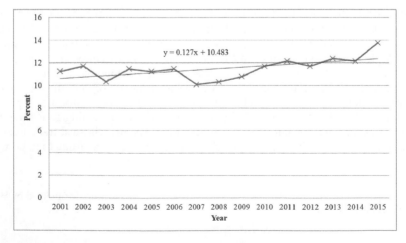

Figure 1. Trend of Apparel Firms Number Engaged in GPNs, 2001–2015.
Source: IBS (2001–2015), author's calculation.

Zara, Hugo Boss, Guess, Mark and Spencer, Mango, and so on. Most of Indonesia's apparel products are exported to developed countries such as USA, Japan, Germany, UK, and others (BPS, 2017b). The value of imports of intermediate goods and exports of Indonesia's apparel industry continued to increase during the period 2001–2015 (Appendix 4).

Participation of Indonesian apparel industry in GPNs can be measured from the development of imported intermediate goods to gross output ratio in the export value of apparel or called vertical specialization. In the period 2001–2015, the value of vertical specialization of the Indonesian apparel industry has an increasing trend. This means an increase in intermediate goods imports is focused to meet the needs of global markets, in addition to the domestic market. This condition illustrates that Indonesia's participation is increasing in GPNs. Increased participation is accompanied by increased productivity of labor (Figure 2). Increased productivity can be attributed to increased worker skills and the use of more advanced technologies. However, since 2010 the trend of increasing the value of productivity tends to slow down. This is linked to a global recession that causes a decrease in world apparel demand, especially from Europe, the United States and Japan. Because productivity

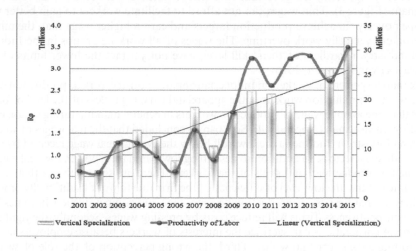

Figure 2. Vertical Specialization and Productivity of Labor in Indonesian Apparel Industry, 2001–2015. Source: IBS (2001–2015), author's calculation.

Table 1. Firm characteristics of Indonesian apparel industry by status of GPNs in selected years.

		Share of firm by size (%)			Share of firm by ownership (%)			
Year	Firm category by status of GPNs	Medium	Large	Total	Domestic	Foreign	Joint investment	Total
2001	Not-engage in GPNs	61.24	38.76	100.00	92.76	3.62	3.62	100.00
	Engage in GPNs	0.00	100.00	100.00	63.27	12.24	24.49	100.00
	All	54.36	45.64	100.00	89.45	5.96	4.59	100.00
2005	Not-engage in GPNs	62.53	37.47	100.00	91.99	2.84	5.17	100.00
	Engage in GPNs	2.04	97.96	100.00	53.06	18.37	28.57	100.00
	All	55.73	44.27	100.00	87.61	7.80	4.59	100.00
2009	Not-engage in GPNs	64.27	35.73	100.00	90.75	3.34	5.91	100.00
	Engage in GPNs	2.13	97.87	100.00	40.43	21.28	38.30	100.00
	All	57.57	42.43	100.00	85.32	9.40	5.28	100.00
2015	Not-engage in GPNs	65.69	28.99	100.00	92.55	2.13	5.32	100.00
	Engage in GPNs	11.67	88.33	100.00	48.33	16.67	35.00	100.00
	All	58.26	37.16	100.00	86.47	9.40	4.13	100.00

Source: IBS (2001–2015), author's calculation.

depends on the value of the output, then this condition may cause productivity growth in the Indonesian apparel industry to slow down.

The linkage of Indonesia in GPNs is also shown by the structure of apparel firms based on size and ownership (Table 1). GPNs firms are now not only limited to large-scale enterprises, but also some medium-scale enterprises. The trends of foreign ownership also increased, illustrated the increased of foreign investment. This increase in foreign investment provides an opportunity for increased economic development as it can bring in foreign capital, foreign technology, management knowledge, jobs and access to new markets (Boly, Coniglio, Prota & Seric, 2014).

4 DECENT WORK CONSEQUENCES DUE TO THE EXPANSION OF GPNS IN THE INDONESIAN APPAREL INDUSTRY

Wage in Indonesia's apparel industry tend to be lower than the average wage of workers in the manufacturing sector. Although the nominal wage trend increases with the increase in the minimum wage set by the government to ensure workers' welfare, the real wage tends to be stagnant (Figure 3). That wages can not afford to facilitate workers to gain better welfare or escape poverty. To achieve production targets and earn a higher wage than the minimum wage, workers generally work overtime. The wages of all working groups in the Indonesian apparel industry, regardless of their skill level, have not yet met the lower limits of family living wages (Van Klaveren, 2016).

More than 90 percent of workers in Indonesian apparel industry work as production workers. The real wage of production workers in the apparel industry, regardless of the level of position, is lower than the average wage of production workers in general manufacturing sector, even for levels below the foreman (Appendix 4). So it can be concluded that the real wages of workers in the apparel industry in Indonesia are low, still below the average real wage received by workers in other industrial sectors. In addition to describing low purchasing power, low real wages also reflect the low productivity and skills of the workers. There are several things that can cause this condition. First, the low level of efficiency of the firm due to work traditionally or with low technology so that the cost of production becomes high and competitiveness becomes low. To improve competitiveness, firms will reduce costs by pressing wages. Secondly, the tight competition among corporations can sacrifice the interests of workers because employers need huge profits to make their firms grow fast. Third, the wrong perception of the role of workers in the firm. Workers' welfare should be linked to productivity and efficiency that can support the competitiveness of firms. Fourth, the low skill of workers. Public schools have not been able to make workers as skilled workers. It takes skill training, but unfortunately no one is responsible for it. Fifth, the wage system that uses the minimum wage may be inappropriate because it is not dynamic and makes the worker a marginalized group of people.

If differentiated by occupation, production workers in the Indonesian apparel industry earn lower wages than non-production workers. The real wage difference between the two

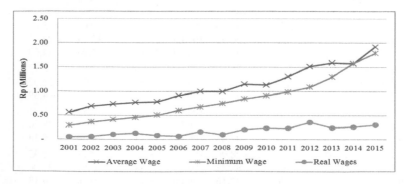

Figure 3. Average Wages, Real Wages, and Minimum Wages of Indonesian Apparel Industry, 2001–2015.
Source: IBS (2001–2015), author's calculation.

is twofold or even higher (Figure 4). This is due to differences in skills and share of female workers between the two. Production workers generally have lower skills and have higher share of female workers. Indonesia's apparel industry has not used high technology, so it does not require high skills for production workers, whereas women workers are preferred for production work due to their painstaking and nimble fingers.

The starting point for measuring employment quality is the level of earnings. High-quality employment should allow workers to support themselves and at least one other family member above the poverty threshold (Messier & Floro, 2008). The decent wages can also be seen through adequate earnings based on the concept of decent work. Workers must have adequate earnings to ensure the economic well-being of themselves and their households (ILO, 2012). This study calculated an adequate wages from adequate earnings with a minimum wage approach because the minimum wages are often formed through a democratic process and in fact almost half of them median wages (OECD Employment Outlook on Anker, 2002). The minimum wage rate in Indonesia was established taking into account the needs of workers and their families, living costs, economic development, employment levels, employers' capacity to pay, along with other factors (Van Klaveren, 2016). Therefore, adequate wages can be used to describe the ability of wages to meet the needs of decent living for workers and their families. If the average wage is lower than the national minimum wage or the ratio is below 100 then it can be categorized as inadequate wages.

Figure 5 shows that adequate wages in the Indonesian apparel industry differ between production and non-production workers, where the wages of non-production workers are more adequate. Even in recent years, the wages of production workers fall under inadequate wages. During the period 2001–2015, adequate wages for all workers in the Indonesian apparel industry experienced

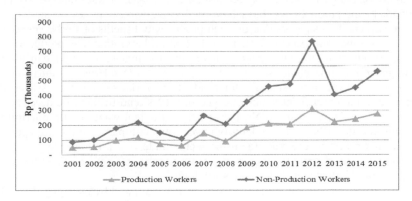

Figure 4.   Real wages of workers in Indonesian apparel industry by occupation, 2001–2015.
Source: IBS (2001–2015), author's calculation.

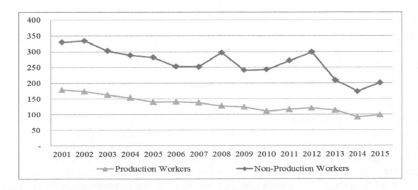

Figure 5.   Adequate wages of Indonesian apparel industry by occupation, 2001–2015.
Source: IBS (2001–2015), author's calculation.

downgrading. This condition illustrates the decline in the ability of workers to meet the needs of decent living for workers and their families. This condition can be explained by a simple supply-demand model of labor. The model states that with a higher minimum wage, fewer workers will be hired as costs increase or employers will replace workers with other inputs such as capital (Brown, Gilroy, & Kohen, 1982). If the firm chooses to keep the number of workers or even add to it, then the impact is wage suppression so that adequate wage value will fall.

The decent wages can also be determined by the characteristics of the firm either by firm size[4] (Appendix 5) and firm ownership[5] (Appendix 6). Based on firm size, wages of workers in large-firms are more decent than medium-firms. On the other hand, based on the structure of firm ownership, the wages of workers in foreign and joint investment firms are generally better than domestic firms. This is due to differences in the firm's capacity to pay workers. Large firms and foreign firms generally have larger capital, making it more likely to produce greater output and achieve higher profits. The profit can be used to provide compensation to workers so that workers benefit more. The benefits that workers receive differ by their type of work. The decent wages of non-production workers differed significantly by firm size and firm ownership, but not so for production workers. This shows that profits are more beneficial to non-production workers than production workers. However, the same pattern for all types of firms is that nominal and real wage trends tend to increase, while adequate wage trends tend to decline over the period 2001–2015.

Firms which engaged in GPNs were able to hire more workers, especially female workers, than firms which not engaged in GPNs (Table 2). The average productivity of workers in firms which engaged in GPNs is generally higher, and has an increasing tendency. This illustrates improving the quality of workers in the form of skills improvement. However, the gap between workers in firms engaged in GPNs is higher. This can be seen from the high value of wage ratio between non-production workers and production workers. Non-production workers' wages are more decent than production workers. This suggests that the high productivity benefits of GPNs engagement can be further transformed into benefits for non-production workers rather than production workers.

Increased productivity has an impact on increasing nominal wages, and it can also increase the real wages of the apparel industry workers engaged in GPNs. However, the real wages still tend to be low and stagnant especially for production workers (Figure 6). Wages of workers in Indonesian apparel firms which engaged in GPNs tend to be more decent compared to firms which not engaged in GPNs. Nevertheless, the real wages of production workers in firms engaged in GPN are still relatively lower when compared to the average real wage of production workers in the manufacturing sector under the foreman (Appendix 1). This condition means that despite the increasing trend, purchasing power owned by production workers in apparel industry in Indonesia is still low.

Figure 6 also shows that the wages of workers in the apparel firms engaged in GPNs are more adequate than those which not engaged in GPNs. Based on the type of work, it appears that non-production workers' wages are more adequate than production workers. This is due to differences in the level of education owned by both, the higher the level of education of workers then the standard wages received will be greater. Non-production workers in the manufacturing industry generally have higher levels of education than production workers (Amiti & Cameron, 2012). Unfortunately, adequate wages have a declining trend, both for production and non-production workers. This shows the decline in the ability of wages to meet the needs of decent living for workers and their families. In fact, in recent years the

---

4. The concept of firm size is based on the classification of the BPS manufacturing industry, which is solely based on the number of workers employed, regardless of whether the company uses electric machines or not, and without regard to the amount of capital the company has. The firm size classification used in this study is large firms (having 100 or more workers), and medium firms (with 20–99 workers).

5. The concept of firm ownership is based on the firm's capital with the source of capital. If the firm's capital is sourced from the central government, local government, and/or national private it is classified as a domestic firm. If the company's capital comes from a government or a foreign citizen, it belongs to a foreign firm. Whereas if the firm's capital is a mix between domestic and foreign, then the joint investment firm.

Table 2.   Wages characteristics of Indonesian apparel industry by status of GPNs in selected years.

Year	Firm category by status of GPNs	Share of production workers (%)	Share of female workers (%)	Average Number of Labor in Firms	Ratio of Wage between Non-Production and Production Workers	Average Productivity of Workers (000 Rp)
2001	Not-engaged in GPNs	89.44	81.33	253	1.80	53,059
	Engaged in GPNs	89.41	84.21	1,530	1.89	70,509
2005	Not-engaged in GPNs	89.42	80.75	238	1.90	71,770
	Engaged in GPNs	90.01	84.27	1,497	2.16	90,507
2009	Not-engaged in GPNs	89.47	81.13	292	1.87	86,297
	Engaged in GPNs	89.46	85.43	1,465	2.03	119,056
2015	Not-engaged in GPNs	90.73	75.14	179	1.37	218,115
	Engaged in GPNs	90.51	84.43	1,764	2.56	179,489

Source: IBS (2001–2015), author's calculation.

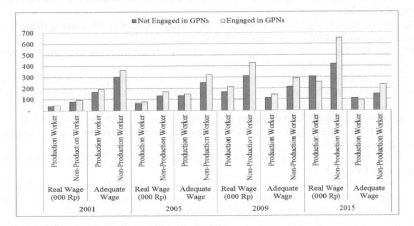

Figure 6.   Decent wages characteristics of Indonesian apparel industry by status of GPNs in selected years. Source: IBS (2001–2015), author's calculation.

wages of Indonesian apparel industry production workers engaged in GPNs fall under inadequate wages, which means that the wages are lower than the need for decent living.

5   CONCLUSION AND SOME REMAINING PROBLEMS

This study has examined whether the workers benefit greatly from GPNs in the Indonesian apparel industry, focusing on decent wages' two elements, namely, real wages (as an indicator of workers' productivity and skill), and adequate wage (as an indicator of a living wage). This study confirms that GPNs growth in the Indonesian apparel industry has created new jobs, especially for production workers. Nevertheless, this study offers specific evidence for GPNs growth implications on the welfare of workers.

The main findings are as follows. First, despite increasingly integrated in GPN, competitiveness of the Indonesian apparel industry has been stagnant and weakening. The increase in the value of vertical specialization accompanied by an increase in worker productivity has not been maximally implemented due to various constraints.

Second, real wages of Indonesian apparel industry experienced an upward trend along with their increased engagement in the GPNs, but the increase itself was not enough. The amount of real wages received by the local workers is still very low, even lower than the average wage received by a typical foreman in other manufacturing sector.

Third, the study also finds that adequate wage trends are declining. Increasing the minimum wage can actually depress workers' wages since many firms seek to maintain labor demand while keeping costs down. As a result, adequate wages for workers in the Indonesian apparel industry are experiencing a downward trend. Finally, it can be concluded that the engagement of the Indonesian apparel industry in GPNs during the period 2001–2015 has not been able to provide welfare for workers and their families, especially production workers.

This sector-level study only analyzes existing conditions based on aggregated-sectoral data in medium and large enterprises. Therefore, more-detail investigation using firm-level data is needed to examine the impact of an individual firm engagement in GPNs focusing on decent wages issue.

REFERENCES

Amiti, M., & Cameron, L. (2012). Trade liberalization and the wage skill premium: Evidence from Indonesia. *Journal of International Economics, 87*(2), 277–287.

Amiti, M., & Davis, D. R. (2011). Trade, firms, and wages: Theory and evidence. *The Review of economic studies, 79*(1), 1–36.

Anker, R., Chernyshev, I., Egger, P., Mehran, F., & Ritter, J.A. (2003). Measuring decent work with statistical indicators. *International Labour Review, 142*(2), 147–178.

Attanasio, O., Goldberg, P.K., & Pavcnik, N. (2004). Trade reforms and wage inequality in Colombia. *Journal of development Economics, 74*(2), 331–366.

Barrientos, S., Gereffi, G., & Rossi, A. (2010). Economic and social upgrading in global production networks: Developing a framework for analysis. *International Labor Review, 150*(3–4), 319–340.

Barrientos, S., Gereffi, G., & Rossi, A. (2011). Economic and social upgrading in global production networks: A new paradigm for a changing world. *International Labour Review, 150*(3–4), 319–340.

Bernhardt, T. (2013). Developing countries in the global apparel value chain: a tale of upgrading and downgrading experiences.

Bigsten, A., & Durevall, D. (2006). Openness and wage inequality in Kenya, 1964–2000. *World Development, 34*(3), 465–480.

Blanchard, Olivier. (2009). *Macroeconomics: Fifth Edition*. New Jersey: Pearson International Edition.

Boly, A., Coniglio, N.D., Prota, F., & Seric, A. (2014). Diaspora investments and firm export performance in selected sub-Saharan African countries. *World Development, 59*, 422–433.

BPS. (2017a). Large and Medium Industry: Number of Large and Medium Manufacturing Workers by Sub Sector, 2008–2015. Downloaded January 10, 2018, from Statistics Indonesia: https://www.bps.go.id/statictable/2011/02/14/1063/jumlah-tenaga-kerja-industri-besar-dan-sedang-menurut-subsektor-2000-2015.html.

BPS. (2017b). Apparel Exports by Major Destination Countries, 2000–2015. Downloaded January 10, 2018, from Statistics Indonesia: https://www.bps.go.id/statictable/2014/09/08/1024/ekspor-pakaian-jadi-menurut-negara-tujuan-utama-2000-2015.html

Cowgill, M., & Huynh, P. (2016). Weak minimum wage compliance in Asia's garment industry. *Asia-Pacific Garment and Footwear Sector Research Note, Issue 5.*

Currie, J. and A. Harrison (1997), 'Trade Reform and Labor Market Adjustment in Morocco', *Journal of Labor Economics*, 15, pp. 44–71.

Dicken, P., & Hassler, M. (2000). Organizing the Indonesian clothing industry in the global economy: the role of business networks. *Environment and Planning A, 32*(2), 263–280.

Ernst, D., & Kim, L. (2002). Global production networks, knowledge diffusion, and local capability formation. *Research policy, 31*(8–9), 1417–1429.

Feenstra, R., & Hanson, G. (2001). *Global production sharing and rising inequality: A survey of trade and wages* (No. w8372). National Bureau of Economic Research.

Fernandez-Stark, K., Frederick, S., & Gereffi, G. (2011). The Apparel Global Value Chain. *Duke Center on Globalization, Governance & Competitiveness.*

Galiani, S. and P. Sanguinetti (2003), 'The Impact of Trade Liberalization on Wage Inequality: Evidence from Argentina', *Journal of Development Economics*, 72, pp. 497–513.

Gereffi, G., & Memedovic, O. (2003). *The global apparel value chain: What prospects for upgrading by developing countries* (pp. 5–6). Vienna: United Nations Industrial Development Organization.

Goldberg, P.K., & Pavcnik, N. (2007). Distributional effects of globalization in developing countries. *Journal of economic Literature, 45*(1), 39–82.

Gotexshow. (2015). Market: Overview of the Textile and Clothing Sector. Downloaded January 10, 2018, from GOTEX SHOW: http://www.gotexshow.com.br/eng/mercado

Harrison, A., & Hanson, G. (1999). Who gains from trade reform? Some remaining puzzles1. *Journal of development Economics*, *59*(1), 125–154.

Hummels, D., Ishii, J., & Yi, K.M. (2001). The nature and growth of vertical specialization in world trade. *Journal of international Economics*, *54*(1), 75–96.

Kawakami, M., & Sturgeon, T.J. (Eds.). (2011). *The dynamics of local learning in global value chains: Experiences from East Asia*. Springer.

Kohpaiboon, A., & Jongwanich, J. (2013). Global Production Sharing and Wage Premium: Evidence from Thai Manufacturing. *Impact of Globalization on Labor Market*.

Kuroiwa, I., & Heng, T.M. (Eds.). (2008). *Production networks and industrial clusters: Integrating economies in Southeast Asia* (Vol. 334). Institute of Southeast Asian Studies.

Mankiw, N.G. (2003). *Introduction to Economics*. Second Edition, Erlangga Publishers, Jakarta.

Manoto, R.P., Wicaksono, P., & Bakhtiar, T. (2017). Indonesia in the global apparel value chain: Economic and social development experience.

Messier, J., & Floro, M. (2008). Measuring the quality of employment in the informal sector. *Department of Economics, Working Papers Series*, (2008Y10).

Mishra, P., & Kumar, U. (2005). *Trade Liberalization and Wage Inequality: Evidence from India (EPub)* (No. 5–20). International Monetary Fund.

Plank, L., Rossi, A., & Staritz, C. (2012). *Workers and social upgrading in "fast fashion": The case of the apparel industry in Morocco and Romania* (No. 33). Working Paper, Austrian Foundation for Development Research (ÖFSE).

Rossi, A. (2011). *Economic and social upgrading in global production networks: The case of the garment industry in Morocco* (Doctoral dissertation, University of Sussex).

Tager, S. (2016). *Women in the Global Clothing and Textile Industry*.

Van Klaveren, M. (2016) Wages in Context in the Garment Industry in Asia. Amsterdam: Wage Indicator Foundation, April. http://www.wageindicator.org/main/Wageindicatorfoundation/publications.

Wicaksono, P., & Priyadi, L. (2016). Decent Work in Global Production Network: Lessons Learnt from the Indonesian Automotive Sector. *Journal of Southeast Asian Economies*, 33(1), 95–110.

APPENDICES

Appendix 1.   Leading Manufacturer of World Garment in 2012.

Country	Share of world production
China	47.20%
India	7.10%
Pakistan	3.10%
Brazil	2.60%
Turkey	2.50%
South Korea	2.10%
Mexico	2.10%
Italy	1.90%
Malaysia	1.40%
Taiwan	1.40%
Poland	1.40%
Romania	1.20%
Indonesia	**1.10%**
Bangladesh	1.00%
Thailand	1.00%
Others	22.90%

Source: Gotexshow, 2015.

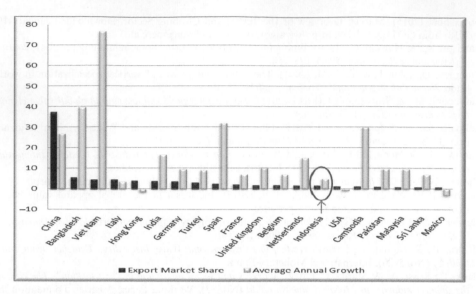

Appendix 2. Export Percentage of 20 World Largest Apparel Exporters by 2015, and Average Annual Growth during 2001–2015.
Source: UNComtrade Database (2001–2015), author's calculation.

Appendix 3. Classification of Indonesian apparel industry, 2001–2015.

ISIC Rev. 4	Description	ISIC Rev. 3	Description
(1)	(2)	(3)	(4)
13912	Manufacture of Embroidery Fabric	17293	Manufacture of Embroidery
13922	Manufacture of Embroidery Finished Goods	17293	Manufacture of Embroidery
14302	Manufacture of Embroidery Apparel	17293	Manufacture of Embroidery
14301	Manufacture of Knitwear	17302	Manufacture of Knitwear
14303	Manufacture of Knitted Socks and Similar Goods	17303	Manufacture of Knitted Socks
14111	Manufacture of Apparel (Convection) from Textile	18101	Manufacture of Apparel from Textiles and that Outfit
14131	Manufacture of Apparel Outfit from Textile	18101	Manufacture of Apparel from Textiles and that Outfit
14112	Manufacture of Apparel (Convection) from Leather	18102	Manufacture of Apparel (Convection) and That Outfit from Leather
14132	Manufacture of Apparel Outfit from Leather	18102	Manufacture of Apparel (Convection) and That Outfit from Leather
14200	Manufacture of Apparel and Goods from Peltry	18202	Manufacture of Apparel/Finished Goods from Peltry and Accessories
14120	Tailoring and Custom Made Clothes		

Source: BPS.

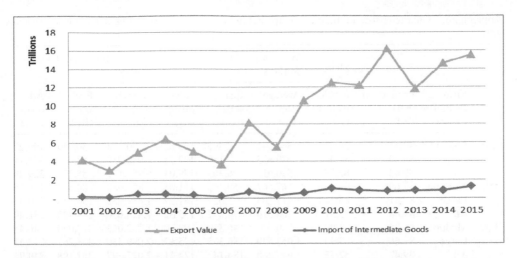

Appendix 4. Trend of intermediate imported and export value in Indonesian apparel industry, 2001–2015.
Source: IBS (2001–2015), Author's calculation.

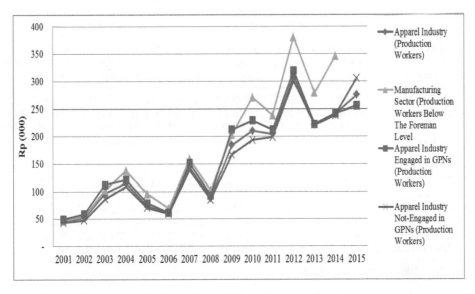

Appendix 5. Real wage of production worker in Indonesian apparel industry and manufacturing sector.
Source: IBS (2001–2015), Author's calculation.

Appendix 6. Characteristics of Indonesian apparel industry by firm size and occupation in selected years.

Year	Firm category by size	Share of production workers (%)	Share of female workers (%)	Wage of production workers			Wage of non-production workers		
				Average wage (Rp)	Real wage (Rp)	Ade-quate wage	Average wage (Rp)	Real wage (Rp)	Ade-quate wage
2001	Medium	88.41	58.77	364,941	32,495	125.62	570,174	50,769	196.27
	Large	89.49	84.07	529,623	47,158	182.31	983,950	87,612	338.71
	All	**89.43**	**82.58**	**520,012**	**46,303**	**179.01**	**957,172**	**85,228**	**329.49**
2005	Medium	87.69	60.27	567,867	59,405	111.85	807,988	84,524	159.15
	Large	89.81	83.74	717,205	75,027	141.27	1,477,250	154,536	290.97
	All	**89.68**	**82.31**	**708,316**	**74,097**	**139.52**	**1,428,656**	**149,453**	**281.40**
2009	Medium	86.25	61.90	830,623	146,329	98.70	1,232,062	217,050	146.41
	Large	89.66	83.99	1,057,539	186,305	125.67	2,090,430	368,267	248.41
	All	**89.47**	**82.75**	**1,045,268**	**184,143**	**124.21**	**2,027,597**	**357,198**	**240.94**
2015	Medium	86.99	63.82	1,654,018	259,938	92.39	2,405,989	378,115	134.39
	Large	90.81	81.93	1,757,267	276,164	98.15	3,697,858	581,139	206.54
	All	**90.59**	**80.82**	**1,750,979**	**275,176**	**97.80**	**3,589,637**	**564,132**	**200.50**

Source: IBS (2001–2015), author's calculation.

Appendix 7. Characteristics of Indonesian apparel industry by ownership and occupation in selected years.

Year	Firm category by ownership	Share of production workers (%)	Share of female workers (%)	Wage of production workers			Wage of non-production workers		
				Average wage (Rp)	Real wage (Rp)	Ade-quate wage	Average wage (Rp)	Real wage (Rp)	Ade-quate wage
2001	Domestic	88.59	80.12	490,820	43,703	168.96	863,502	76,888	297.25
	Foreign	93.10	91.90	526,242	46,857	181.15	975,281	86,840	335.72
	Joint Investment	90.84	87.91	665,900	59,293	229.23	1,565,552	139,399	538.92
2005	Domestic	88.44	79.29	666,192	69,691	131.22	1,183,097	123,764	233.03
	Foreign	93.88	89.63	780,488	81,647	153.73	2,784,249	291,262	548.41
	Joint Investment	88.95	84.87	781,943	81,800	154.02	1,453,925	152,096	286.38
2009	Domestic	86.30	79.12	977,438	172,193	116.15	1,939,590	341,694	230.48
	Foreign	94.51	86.06	1,235,247	217,611	146.79	2,148,645	378,523	255.33
	Joint Investment	91.07	87.14	980,734	172,774	116.54	2,240,193	394,650	266.20
2015	Domestic	89.95	77.05	1,739,037	273,299	97.13	2,843,710	446,905	158.84
	Foreign	94.58	85.97	2,040,146	320,620	113.95	2,450,555	385,118	136.88
	Joint Investment	86.93	83.33	1,368,188	215,018	76.42	5,619,609	883,153	313.88

Source: IBS (2001–2015), author's calculation.

Business Innovation and Development in Emerging Economies – Trinugroho & Lau (Eds)
© *2019 Taylor & Francis Group, London, ISBN 978-1-138-35996-3*

The implication of loan-to-value ratio on credit housing demand in Indonesia

L.R. Fauzia, S.A.T. Rahayu & A.A. Nugroho
Universitas Sebelas Maret, Surakarta, Indonesia

ABSTRACT: Housing is one of the basic needs in people's lives. Most people purchase a house or houses through a mortgage program. This encourages the Indonesian government to set the down payment based on the Loan-to-Value (LTV) ratio. This study aims to analyze the implication of the LTV ratio and other effects on housing demand. This study uses secondary data on a monthly basis between March 2012 and December 2017. Based on the results of estimation using the dynamic model Error-Correction Model–Engle Granger (ECM-EG), this study finds that in the short-term period, credit housing demand is influenced by income. The influence of mortgage rates and population also affect credit housing demand in the long term but are not statistically significant. However, housing price, income, and LTV ratio are found to have a statistically significant effect on credit housing demand in the long-term period. The theory of demand being influenced by high prices does not apply to housing demand. Therefore, the implications of the LTV ratio as a financial construct to control housing demand in the housing market is best observed in the long-term period. In addition, there are other factors that affect the demand for housing on credit which need to be considered.

Keywords: Housing demand, mortgage loan, ECM-EG, LTV

1 INTRODUCTION AND POLICY BACKGROUND

Housing is one of the main human needs. The large demand for housing along with population growth and people's financial capability make housing prices increase rapidly. However, home ownership has recently not only intended to meet the basic needs of housing, but has also become an attractive investment alternative. The rapid growth of housing prices is the reason for the purchase of housing, with the two motives that are consumption and investment (Arrondel et al., 2010).

Housing is one of the largest purchases for the community. This makes the purchase of housing credit sensitive. In line with Smith and Smith (2004), people purchase housing for both consumption and investment payoff. According to Figure 1, the purchase of housing credit is very common and widely done within the community.[1]

The purchase of housing credit certainly has considerable risks. As an example, the crisis that occurred in the United States in 2008 was caused by the distribution of sub-prime mortgages. Housing loans were given to all people, with either good or bad credit scores, and as a result the debtors have defaulted. The large number of housing units liquidated by banks for credit failure resulted in an oversupply in the housing market at that time. Lehman Brothers Holdings was one of the sub-prime lenders that had an enormous influence on the housing market in the United States. Lehman Brothers and other sub-prime lenders went bankrupt because the sale of liquidated houses had been unable to cover the losses on

---

1. Bank Indonesia as Central Bank of Indonesia conducted a residential property pricing survey, published quarterly beginning in 2002, which showed the majority of people use a mortgage program.

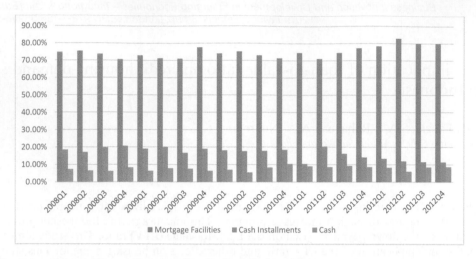

Figure 1. Sources of home financing. (Source: Bank Indonesia).

the debtor's default. This became one of the causes of the world financial crisis as the US economy deteriorated, and the US stock index fell drastically. This crisis demonstrates that the housing market has an influence on monetary policy. In addition, housing as a monetary transmission plays an important part in the economy (Demary, 2009). Moreover, the relationship between property prices and the amount of credit consumption of banking becomes an indicator of the health of a country's economy (Bunda & Ca'Zorzi, 2009).

The pattern of dualistic motive in housing purchase by the community encourages the government to regulate the housing market on monetary lines. Policies are made not only regarding the interest rate applied, but also about the credit advances submitted by communities in fulfilling these needs through Loan-to-Value (LTV) ratio. That this property characteristic is sensitive to credit is one of the reasons for a government to control the housing market through the LTV mechanism. Implementation of an LTV policy is expected to fit the property cycle of boom market, recession, and recovery.

Mishkin (2007a) and Wadud et al. (2012) mention that the direct monetary policy transmission comes from: (i) capital utilization cost; (ii) interest rate effect on expected housing price rises in the future; (iii) housing supply. Meanwhile the, indirect monetary policy transmission through; (i) the standar wealth effects on housing; (ii) the increase of credit balance affects consumption expenditures; (iii) the amount of mortgage loan on housing demand. On the basis of Figure 1, a low mortgage rate will cause the demand for home loans to increase, and then housing prices will increase as the price of land for housing and other housing facilities also increases. However, a credit channel can only work very well when there is financial stress (Hendricks & Kempa, 2009).

2 LITERATURE REVIEW

2.1 Housing demand

Demand occurs when there is a willingness and ability of buyers in the market (Turner, 1991). According to Turner, the obstacles that limit demand are low purchasing power and/or the expensive price of goods and services. Eckert (1990) mentioned that the factors that affect the housing market demand are economic, social, governmental, and environmental.

According to McKenzie and Betts (2006), housing demand should be studied from two points of view, the first being the total demand or the number of housing units that are clearly required in the market, while the second point of view is that of housing composition, such as unit size, location, condition, and whether the units are planned to be sold to consumers or are only for lease. This argument accords with Koutsoyiannis (1982), who stated that home consumption is a long-term consumption, and thus everyone will think about future benefits.

2.2 *Loan-to-value policy*

Loan-to-value is one of the government's policies that can fulfill two tasks simultaneously, first as an economic stimulus and second as the fulfillment of dwelling needs. As one of the property sectors, housing is a leading sector in economic recovery. Therefore, it is expected to have a spillover impact on other sectors such as construction, industry, and mining, as well as services. LTV easing includes two things: decreasing payouts of down payment, and tiering for first housing, second housing, and so on.

On March 15, 2012 Bank Indonesia issued Circular Letter of Bank Indonesia No. 14/10/ DPNP on loan-to-value ratio in order to increase caution by banks that provide payment services, mortgages or car loan services. Bank Indonesia regulates the limitation of credit granting. Based on Bank Indonesia's research results, and also considering that Indonesia still lacks available housing, the LTV policy therefore also aims to facilitate the government program in fulfilling the housing needs of the community.

In connection with this research, Rahal (2016) conducted a study using the Vector Auto-Regression (VAR) method of quarterly and monthly data of housing market datasets for eight Organisation for Economic Co-operation and Development (OECD) countries, showing that unconventional monetary policy can reduce real interest rates, reduce the cost of housing users, and therefore increase demand for housing and housing prices. Correspondingly, Xiao (2013) used the standard analysis of Evans and Honkapohja (2001) by using a log-linear proxy with the VAR method.

2.3 *Mortgage loan*

Credit is one of the sources of income for banks (Kasmir, 2010). Credit is classified into productive, consumptive, and trade credit. Credit that is intended to provide financing is called productive credit, whereas consumptive credit is used for personal consumption, financing or business entities, such as mortgages.

The prime lending interest rate is the basis for determining the interest rate conducted by each bank, and is divided into four types, namely (i) corporate credit; (ii) retail credit; (iii) micro credit; (iv) consumer credit. Consumer credit is divided into two, namely mortgage and non-mortgage. According to Utama (2012), the transmission of monetary policy can be seen from the influence of changes in interest rate that impact on real Gross Domestic Product (GDP) and inflation.

Muellbauer (2007) analyzed multi-country data and argues that credit liberalization contributes to the wealth effect that is caused by the appreciation in housing prices. Wadud et al. (2012) explained the influence of monetary policy on the housing market in Australia. The research method used was the Structural Vector Auto-Regression (SVAR) model. The study resulted in a contractionary monetary policy that significantly decreased housing activity. However, housing prices are not only affected by the monetary policy itself. The number of housing units and housing prices have a significant relationship to the housing offer, house demand, and other factors.

3 DATA

As indicated in Section 1, this research seeks to understand the impact of government policy on housing demand, especially from the perspective of mortgage programs. The effects on credit housing demand, besides that of the applied loan-to-value ratio, are prime lending rate on housing, income, residential property price index, and population. This study uses the dynamic model of Error-Correction Model–Engle Granger (ECM-EG) and uses secondary data on a monthly basis between March 2012 and December 2017. Despite the fact that not every variable can be found directly, due to both the definition of residential property and data collection methodologies, we attempt to harmonize data used in the estimation. One key insight to consider is that a series is available only in annual and quarterly frequency, and so we convert the data using linear interpolation to construct a monthly representation.

All of the data in this research is real data. Thus, if the data is not presented as real data, we divide it by the consumer price index. The monthly series for real GDP and the housing price index are interpolated with linear interpolation from quarterly representation. Real GDP data represents income, and the housing price index represents housing price. Meanwhile, the population's data is interpolated from annual data. For monthly mortgage rate series, we obtain the data from the average mortgage rate of every bank that has a mortgage program.

4 METHODOLOGY

Proposed as an alternative model, error-correction models–Engle Granger is chosen for this macroeconomic research. This study refers to the research of Lim and Nugraheni (2017), which had a primary research focus that used ECM-EG to explore the implications of loan-to-value ratio and other effects on credit housing demand.

4.1 Model specification

The analysis with statistical approach uses ECM-EG to capture the dynamics of credit housing demand, facing adjustment in LTV ratio, and other effects. The selection of independent variables in this research used constructs from Rahal (2016), Lim and Nugraheni (2017), and Eichholtz and Lindenthal (2014). Some adjustments were made to accommodate Indonesian data features. In order to understand better the dynamics of credit housing demand relations, the ECM will present short- and long-term analysis in this study.

In determining the impact of monetary policy on the housing market we adapt the model from Tsatsaronis and Zhu (2004), Andrews (2010), and Utama (2012). The short-term housing represents macroprudential policy as the loan-to-value ratio. The analysis of monthly data will be presented to provide a general impact of loan-to-value ratio on people's credit housing demand. The analysis continues with a statistical approach using ECM-EG to capture the dynamics of the housing cycle facing adjustments in the LTV ratio (three times adjustment after loan-to-value has been applied).

Firstly, we conduct an MWD test to determine the function of the empirical model with either log-linear or linear data. MWD test is one of method to choose the model function between a linear regression model or a log-linear regression model in empirical analysis. As a result, we determined to use log-linear data. Following the ECM-EG approach, unit root tests are used to indicate stationary data. The ECM provides an advantage in the form of extensive long-run and short-run time series analysis. The assumption made by the ECM model is that the market is not always in balance and alignment with the conditions that occur. For this reason, we estimate this model with lag lengths for the short run and long run.

4.2 Estimation and identification strategy

The ECM-EG method can be used after we consider the data stationarity using the MWD test. As our finding, the integration test is performed to find the data stationarity by looking at the degree of integration value. After the integration test is conducted we should have a co-integration test, which aims to determine what model can be used in the long run with the same stationary degree (Engle & Granger, 1987). Engle and Granger stated that if there's co-integration in the variables, then the dynamic model of ECM-EG can be used. As in

Section 2, we assume that the loan-to-value ratio has a time lag on housing demand; meanwhile, there are other factors that have an influence on credit housing demand.

Statistically, this study used integration tests known as Augmented Dickey–Fuller (ADF) tests for all of the variables. Thus, the formulations are:

$$DX_t = a_0 + a_1 BX_t + \sum_{i=1}^{k} b_i B^i DX_t \tag{1}$$

$$DX_t = c_0 + c_0 T + c_2 BX_t + \sum_{i=1}^{k} b_i B^i DX_t \tag{2}$$

The model used in this study can be expressed in the ECM model as follows. The first ECM-EG equation for the short term is:

$$D(LNCR)_t = \alpha_1 D(P)_t + \alpha_2 D(LNGDP)_t + \alpha_3 D(LNPOP)_t + \alpha_4 D(LTV)_t$$
$$+ \alpha_5 D(MR)_t + \alpha_6 (ECT) + \mu_t \tag{3}$$

The equation for ECM-EG in the long term is:

$$LNCR_t = \beta_0 + \beta_1 P + \beta_2 LNGDP_t + \beta_3 LNPOP_t + \beta_4 LTV_t + \beta_5 MR_t \tag{4}$$

5 DISCUSSION

The total of mortgage loan requested by the community in period 2012M03 to 2017M12 shows an increasing trend. This indicates that housing demand has tended to increase since the implementation of the loan-to-value ratio. In Figure 2, there are three circles that show that there is a change of LTV ratio. In the sample period it can be seen that the mortgages demanded by the community show a small fluctuation when there is a policy adjustment.

Based on the result of the analysis presented in Table 1, only the mortgage rate variable is stationary, with a significance level of 1%. Other variables besides mortgage rate are not significant at the level degree, and thus are tested in terms of the integration degree. After testing the integration degree, the result shows that all variables can be stationary at the first difference level. The probability value of the unit root test of all variables at the first difference is less than 1% significance, or all variables have an ADF value that is less than the

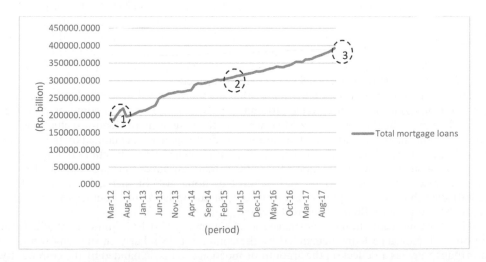

Figure 2.   Total mortgage loans. (Source: Bank Indonesia, 2017).

Table 1. Unit roots test.

| Variable | Level | | First difference | |
	ADF	Prob.	ADF	Prob.
LNCR	−1.55085	0.5021	−8.310759	0.0000***
P	−1.5807	0.4870	−8.072911	0.0000***
LNGDP	−1.27312	0.6376	−8.26619	0.0000***
LNPOP	−1.79667	0.3792	−9.613563	0.0000***
LTV	−1.31962	0.6160	−8.382172	0.0000***
MR	−4.28691	0.0010***	−8.281015	0.0000***

Notes: *** Significant in α: 5%, df = 64.

Table 2. Error-correction model–Engle Granger test results (Source: Author's calculation, 2018).

| Variable | Hypothesis | Short-term | | | Long-term | | |
		Coefficient	t-value	Prob.	Coefficient	t-value	Prob.
C	–				−1.41498	−0.51213	0.6103
P	–	0.178844	1.487000	0.1420	0.251702[a]	2.69816	0.0089[b]
LNGDP	+	0.728241[a]	4.282296	0.0001[b]	0.840699[a]	6.637562	0.0000[b]
LNPOP	+	2.251108[a]	0.927470	0.3572	0.001368[a]	0.00227	0.9982
LTV	+	0.001273[a]	0.729972	0.4681	0.005294[a]	2.944432	0.0045[b]
MR	–	0.00275[a]	−0.765955	0.4466	−0.00395	−0.56585	0.5735
E(−1)	–	−0.21261	−2.53587	0.0137			
R-squared		0.715601			0.980801		
Adjusted R-squared		0.693029			0.979301		
F-statistic					653.8874		
Prop (F-statistic)					0.000000		

Notes: a. The coefficient is the same as the hypothesis; b. Significant in α: 5%, df = 64.

critical ADF value at 5% significance. Table 2 shows the short-term and long-term ECM-EG test results.

The value of residual coefficient (ECT) as listed in Table 2 is –0.212608, and is significant at 5% confidence degree because the value of t-count is –2.535867 (< –1.66901). It is in accordance with the ECM-EG criteria that the result must be negative and significant. Thus, this research model is valid and has fulfilled the ECT requirement. Therefore, this research has passed the entire classical assumption test. The ECT value indicates that the dependent variable has adjusted the independent variable with one lag, or approximately 21% of the long-term and short-term mismatches that can be corrected.

The estimation result using ECM-EG for the short-term equation shows that in the short term, the government's loan-to-value ratio has a positive effect on the amount of mortgage loans demanded by the community. Meanwhile, in the long term, the loan-to-value ratio has a positive and significant effect on the amount of mortgage loans demanded by the community. The result of this research is in accordance with Rahal (2016), who states that the policy of loan-to-value is one of the monetary policies that can affect housing demand through housing prices, real interest rates, and the cost of home use. Similarly, Lewis and Mizen (2000) state that monetary policy affects the output and inflation in a country that responds to the banking sector, finance, and real sectors.

The short-term equation for mortgage rate effect using ECM-EG shows that it affects the amount of mortgage loans demanded by the community. Similarly, in the long term, the mortgage rate has an effect on the amount of mortgage loans demanded by the community. The hypothesis in this research predicts a negative relationship between the mortgage rate and the amount of mortgage loans demanded by the community; after the ECM-EG test is done, it turns out that the result obtained supports this hypothesis. This means that if the mortgage rate is increased by 1%, then the mortgage loans distributed to the community will decrease. The result of the analysis of the relationship between mortgage rate and the amount of credit distributed to the community supports the research conducted by Bunda and Ca'Zorzi (2009), which finds that the interest rate and credit channels are the influential channels in the real sector. A similar statement is also made by Koutsoyiannis (1982, p. 186), that the community will allocate the income consumed in a certain period with consideration of the prevailing interest rate. The mortgage rate assigned to the community using mortgage-backed housing facilities in Indonesia is enacted to a flat interest rate only for a few years at the beginning. In general, the community will pay interest on a flat interest rate system in the first two to five years, while in the following years they will use a floating interest rate.

The early hypothesis in this research predicts that there is a negative correlation between residential property price index and the number of mortgage loans demanded by the community. After the ECM-EG test is conducted it turns out that the results obtained do support this hypothesis. That means that if the value of residential property price index increases for one denomination then the amount of mortgage loans demanded by the community will increase, indicating a positive correlation.

Price is included in the category of economic factors in the housing market, as Eckert (1990) stated that factors affecting housing market demand include economic factors, social factors, governmental factors, and environmental factors. Eckert's research is also supported by Takatz's research (2012), which states that there is a significant effect on housing prices with housing demand. Meanwhile, Powell and Stringham (2008) stated that expensive housing prices or tend to increase steadily is not only because of a high demand, but also because of the intrinsic value of expensive land, regulations on the housing market, population density, attitude, home models, construction cost, building approval processes, and existing laws in the area. It also refers to determining the market value of housing that has various characteristics, one of them being scarcity.

Real GDP is a factor that greatly affects people's demand for housing credit consumption because real GDP has a positive and statistically significant effect in the long term and short term. A similar statement is expressed by Arrondel et al. (2010); that a person will allocate his income to buy a house as consumer goods but also as an investment asset. Meanwhile, Sabari (1994) states that housing has four dimensions, namely location, housing, life cycle, and income. The income dimension itself is related to the amount of a person's income multiplied by the duration of their stay in a city.

The estimation result using ECM-EG for the short-term equation shows that in the short term the rate of population growth has a positive effect on the amount of mortgage loans demanded by the community. This result is similar to the long-term estimation result; that the rate of population growth has a positive effect on the amount of mortgage loans demanded by the community. This supports Takatz's (2012) research, which shows that population growth has an influence on housing demand. McKenzie and Betts (2006) state that one of the most important considerations in housing demand is the number of housing units needed in the market, which are constrained by the development of the population, especially in terms of those who are already of working age.

6 CONCLUSION

Factors that influence people's motivation to make a home purchase on credit in the short term are a coefficient of real GDP, which represents a person's income to make a purchase, and it has a significant effect. An increased real GDP will effect an increasing pattern of home consumption. Meanwhile, the factors that significantly influence people's motivation to purchase housing on credit in the long term are the residential property price index, real gross domestic product, and loan-to-value ratio.

The government's effort is not only aimed at the fulfillment of there being dwellings for the community, especially for those who are impoverished citizens, but also at the country's stimulus. By providing the monetary policy of loan-to-value, the government can run two tasks simultaneously, which are as an economic stimulus and the fulfillment of housing needs. By relaxing macroprudential policy, it is expected to encourage the property demand. LTV easing includes two things: decreasing payouts of down payments and tiering for first housing, second housing, and so on.

7 RECOMMENDATION

In controlling the housing market from the housing demand side, most people use a mortgage that is not only based on the established LTV ratio, but where there are also other factors to be considered. Increased housing demand is also followed by rising housing prices, due to the increasing scarcity of vacant land or land that can be used to build houses. Thus, the government needs to re-examine the LTV, not only as one of the monetary policy transmissions but also as an economic stimulus.

Property market conditions, especially housing in various regions, have different characteristics. Some areas of Indonesia, such as Jabodetabek (Jakarta, Bogor, Depok, Tangerang, and Bekasi) as a megapolitan city, have different market behaviors compared to other regions such as Sumatera, Kalimantan, and other regions that have a lot of smaller cities. This obviously should be the government's consideration in establishing a general policy.

REFERENCES

Andrews, D. (2010). Real house prices in OECD countries: The role of demand shocks and structural and policy factors. *OECD Economics Department Working Papers, 831*, 1–34.

Arrondel, L., Badenes, N. & Spadaro, A. (2010). *Consumption and investment motives in housing wealth accumulation of Spanish households* (Working paper). Social Science Research Network (SSRN). doi:10.2139/ssrn.1597126.

Bank Indonesia. (2012). *Residential Property Survey Quarter IV-2012*. Jakarta, Indonesia: Bank Indonesia.

Bank Indonesia. (2013). *Residential Property Survey Quarter IV-2013*. Jakarta, Indonesia: Bank Indonesia.

Bank Indonesia. (2014). *Residential Property Survey Quarter IV-2014*. Jakarta, Indonesia: Bank Indonesia.

Bank Indonesia. (2015). *Residential Property Survey Quarter IV-2015*. Jakarta, Indonesia: Bank Indonesia.

Bank Indonesia. (2016). *Residential Property Survey Quarter IV-2016*. Jakarta, Indonesia: Bank Indonesia.

Bank Indonesia. (2017). *Residential Property Survey Quarter IV-2017*. Jakarta, Indonesia: Bank Indonesia.

Bunda, I. & Ca'Zorzi, M. (2009). *Signals from housing and lending booms*. Working Paper Series No. 1194. Frankfurt, Germany: European Central Bank.

Demary, M. (2009). The link between output, inflation, monetary policy and housing price dynamics. Munich Personal RePEc Archive (MPRA) paper. Retrieved from https://mpra.ub.uni-muenchen.de/15978/

Eckert, J.K., Gloudemans, R.J., & Almy, R.R. (1990). *Property appraisal and assessment administration*. Chicago, IL: International Association of Assessing Officers.

Eichholtz, P. & Lindenthal, T. (2014). Demographics, human capital, and the demand for housing. *Journal of Housing Economics, 26*, 19–32.

Engle, R.F. & Granger, C.W. (1987). Co-integration and error correction: Representation, estimation, and testing. *Econometrica, 55*(2), 251–276.

Evans, G.W. & Honkapohja, S. (2001). *Learning and expectations in macroeconomics*. Princeton, NJ: Princeton University Press.

Hendricks, W. & Kempa, B. (2009). The credit channel in US economic history. *Journal of Policy Modeling, 31*(1), 58–68.

Kasmir, S.E. (2010). *Pengantar Manajemen Keuangan* [Financial management introduction]. Jakarta, Indonesia: Kencana Prenada Media Group.

Koutsoyiannis, A. (1982). *Non-price decisions: The firm in a modern context.* Hong Kong: Styleset.

Lewis, K.M. & Mizen, D.P. (2000). *Monetary economics.* New York, NY: Oxford University Press.

Lim, C. & Nugraheni, S. (2017). Loan-to-value ratio and house price cycle: Empirical evidence from Indonesia. *Jurnal Ekonomi Pembangunan, 18*(2), 225–238.

McKenzie, D.J. & Betts, R. (2006). *Essentials of real estate economics* (5th ed.). Mason, OH: Thomson/South-Western.

Mishkin, F.S. (2007a). Housing and the monetary transmission mechanism. *Proceedings of the Federal Reserve Bank of Kansas City* (pp. 359–413).

Mishkin, F.S. (2007b). *Housing and the monetary transmission mechanism.* NBER Working Paper No. 13518. Cambridge, MA: National Bureau of Economic Research.

Muellbauer, J. (2007). Housing, credit and consumer expenditure. *Proceedings of the Federal Reserve Bank of Kansas City* (pp. 267–334).

Powell, B. & Stringham, E. (2008). Housing. In D.R. Henderson (Ed.), *The concise encyclopedia of economics.* Indianapolis, IN: Liberty Fund.

Rahal, C. (2016). Housing markets and unconventional monetary policy. *Journal of Housing Economics, 32,* 67–80.

Sabari, Y. (1994). *Teori and Model Struktur Keuangan Kota* [Theories and models of urban spatial structure]. Yogyakarta, Indonesia: Fakultas Geografi UGM.

Smets, F. & Peersman, G. (2001). *The monetary transmission mechanism in the Euro area: More evidence from VAR analysis.* Working Paper Series 091. Frankfurt, Germany: European Central Bank.

Smith, M.H. & Smith, G. (2004). Is a house a good investment? *Journal of Financial Planning, 17*(4), 68–75.

Takatz, E. (2012). Aging and house prices. *Journal of Housing Economics, 21*(2), 131–141.

Taylor, J. (2007). Housing and monetary policy. *Proceedings of the Federal Reserve Bank of Kansas City* (pp. 463–476).

Tsatsaronis, K. & Zhu, H. (2004). What drives housing price dynamics: Cross-country evidence. *BIS Quarterly Review, 2004* (March), 65–78.

Turner, J.F.C. (1991). *Housing by people. Towards autonomy in building environment.* London, UK: Marion Boyars.

Utama, Chandra. (2012). Transmisi kebijakan moneter melalui jalur perumahan [The monetary transmission on housing]. *Ekonomika-bisnis, 03*(1), 29–42.

Wadud, I.K.M., Bashar, O.H. & Ahmed, H.J.A. (2012). Monetary policy and the housing market in Australia. *Journal of Policy Modeling, 34*(6), 849–863.

Xiao, W. (2013). Learning about monetary policy rules when the housing market matters. *Journal of Economic Dynamics and Control, 37*(3), 500–515.

APPENDIX 1

Analysis Error-Correction Model

a. The first equation:

$$LNCR_t = \beta_0 + \beta_1\, LTV_t + \beta_2\, MR_t + \beta_3\, P_t + \beta_4\, LNPDB_t + \beta_5\, LNPENDUDUK_t + e_t \qquad (a)$$

b. Error-correction model formulation:

$$Ct^{de} = b_1\, (LNCR_t - LNCR_{t*})^2 + b_2\, ((1\text{-}B)\, LNCR_t - f_t\, (1\text{-}B)Zt)^2 \qquad (b)$$

c. Minimized quadratic cost single-period function:

$$LNCR_t = bLNCR_{t*} + (1\text{-}B)B\, LNCR + (1\text{-}B)f_t(1\text{-}B)Z_t \qquad (c)$$

d. Equation substitution:

$$LNCR_t = b(\alpha_0 + \alpha_1\, LTV_t + \alpha_2\, MR_t + \alpha_3\, P_t + \alpha_4\, LNGDP_t + \alpha_5\, LNPOP_t + (1\text{-}b)\, f_t\, (1\text{-}B)$$
$$(LTV, MR, P, LNGDP, LNPOP) \qquad (d)$$

341

as follows:

$$\text{LNCR}_t = \alpha_0\, b + \alpha_1\, b\, \text{LTV}_t + \alpha_2\, b\, \text{MR}_t + \alpha_3\, b\, P_t + \alpha_4\, b\, \text{LNGDP}_t + \alpha_5\, b\, \text{LNPOP}_t + (1-b)$$
$$\text{LNCR}_t - \text{LNCR}_{t-1} + (1-b)\, f_t\, (\text{LTV}_t - \text{LTV}_{t-1},\ \text{MR}_t - \text{MR}_{t-1},\ P_t - P_{t-1},$$
$$\text{LNGDP}_t - \text{LNGDP}_{t-1},\ \text{LNPOP}_t - \text{LNPOP}_{t-1}) \tag{e}$$

$$\text{LNCR}_t = \alpha_0\, b + \alpha_1\, b\, \text{LTV}_t + \alpha_2\, b\, \text{MR}_t + \alpha_3\, b\, P_t + \alpha_4\, b\, \text{LNGDP}_t + \alpha_5\, b\, \text{LNPOP}_t + (1-b)$$
$$\text{LNCR}_t - \text{LNCR}_{t-1} + (1-b)\, f_1\, (\text{LTV}_t - \text{LTV}_{t-1}) + (1-b)\, f_2\, (\text{MR}_t - \text{MR}_{t-1})$$
$$+ (1-b)\, f_3\, (P_t - P_{t-1}) + (1-b)\, f_4\, (\text{LNGDP}_t - \text{LNGDP}_{t-1}) + (1-b)\, f_5$$
$$(\text{LNPOP}_t - \text{LNPOP}_{t-1}) \tag{f}$$

$$\text{LNCR}_t = \alpha_0\, b + \alpha_1\, b\, \text{LTV}_t + \alpha_2\, b\, \text{MR}_t + \alpha_3\, b\, P_t + \alpha_4\, b\, \text{LNGDP}_t + \alpha_5\, b\, \text{LNPOP}_t + (1-b)$$
$$\text{LNCR}_t - \text{LNCR}_{t-1} + (1-b)\, f_1\, (\text{LTV}_t - (1-b)\, f_1\, \text{LTV}_{t-1}) + (1-b)\, f_2\, (\text{MR}_t$$
$$- (1-b)\, f_2\, \text{MR}_{t-1}) + (1-b)\, f_3\, (P_t - (1-b)\, f_3\, P_{t-1}) + (1-b)\, f_4\, (\text{LNGDP}_t - (1-b)\, f_4$$
$$\text{LNGDP}_{t-1}) + (1-b)\, f_5\, (\text{LNPOP}_t - (1-b)\, f_5\, \text{LNPOP}_{t-1}) \tag{g}$$

$$\text{LNCR}_t = \alpha_0\, b + (\,\alpha_1\, b + (1-b)\, f_1\,)\, \text{LTV}_t + (\,\alpha_2\, b + (1-b)\, f_2\,)\, \text{MR}_t + (\,\alpha_3\, b + (1-b)\, f_3\,)\, P_t +$$
$$(\,\alpha_4\, b + (1-b)\, f_4\,)\text{LNGDP}_t + (\,\alpha_5\, b + (1-b)\, f_5\,)\, \text{LNPOP}_t - (1-b)\, f_1\, \text{LTV}_{t-1}$$
$$- (1-b)\, f_2\, \text{MR}_{t-1} - (1-b)\, f_3\, P_{t-1} - (1-b)\, f_4\, \text{LNGDP}_{t-1} - (1-b)\, f_5\, (\text{LNPOP}_{t-1}) \tag{h}$$

or,

$$\text{LNCR}_t = c_0 + C_1\, \text{LTV}_t + C_2\, \text{MR}_t + C_3\, P_t + C_4\, \text{LNGDP}_t + C_5\, \text{LNPOP}_t + C_6\, \text{LTV}_{t-1} + C_7$$
$$\text{MR}_{t-1} + C_8\, P_{t-1} + C_9\, \text{LNGDP}_{t-1} + C_{10}\, \text{LNPOP}_{t-1} \tag{i}$$

with,
$$c_0 = \alpha_0\, b$$
$$c_1 = \alpha_1\, b + (1-b)\, f_1$$
$$c_2 = \alpha_2\, b + (1-b)\, f_2$$
$$c_3 = \alpha_3\, b + (1-b)\, f_3$$
$$c_4 = \alpha_4\, b + (1-b)\, f_4$$
$$c_5 = \alpha_5\, b + (1-b)\, f_5$$
$$c_6 = \alpha_6\, b + (1-b)\, f_6$$
$$c_7 = \alpha_7\, b + (1-b)\, f_7$$
$$c_8 = \alpha_8\, b + (1-b)\, f_8$$
$$c_9 = \alpha_9\, b + (1-b)\, f_9$$
$$c_{10} = \alpha_{10}\, b + (1-b)\, f_{10}$$

So, the function for mortgage loan becomes:

$$\text{LNCR}_{t-1} = C_1\, \text{LTV}_{t-1} + C_2\, \text{MR}_{t-1} + C_3\, P_{t-1} + C_4\, \text{LNGDP}_{t-1} + C_5\, \text{LNPOP}_{t-1} - C_1\, \text{LTV}_{t-1} - C_2$$
$$\text{MR}_{t-1} - C_3\, P_{t-1} - C_4\, \text{LNGDP}_{t-1} - C_5\, \text{LNPOP}_{t-1} + C_{11}\, \text{LTV}_{t-1} + C_{11}$$
$$\text{MR}_{t-1} + C_{11}\, P_{t-1} + C_{11}\, \text{LNGDP}_{t-1} + C_{11}\, \text{LNPOP}_{t-1} + C_{11}\, \text{LNCR}_{t-1} - C_{11}$$
$$\text{LTV}_{t-1} - C_{11}\, \text{MR}_{t-1} - C_{11}\, P_{t-1} - C_{11}\, \text{LNGDP}_{t-1} - C_{11}\, \text{LNPOP}_{t-1} - C_{11}\, \text{LTV}_{t-1}$$
$$- C_{11}\, \text{MR}_{t-1} - C_{11}\, P_{t-1} - C_{11}\, \text{LNGDP}_{t-1} - C_{11}\, \text{LNPOP}_{t-1} \tag{j}$$

$$\text{LNCR}_t - \text{LNCR}_{t-1} = C_0 + C_1\, \text{LTV}_t - C_1\, \text{LTV}_{t-1} + C_2\, \text{MR}_t - C_2\, \text{MR}_{t-1} + C_3\, P_t - C_3\, P_{t-1} + C_4$$
$$\text{LNGDP}_t - C_4\, \text{LNGDP}_{t-1} + C_5\, \text{LNPOP}_t - C_5\, \text{LNPOP}_{t-1} + C_1\, \text{LTV}_{t-1} + C_6$$
$$\text{LTV}_{t-1} + C_{11}\, \text{LTV}_{t-1} - \text{LTV}_{t-1} + C_2\, \text{MR}_{t-1} + C_7\, \text{MR}_{t-1} + C_{11}\, \text{MR}_{t-1} -$$
$$\text{MR}_{t-1} + C_3\, P_t + C_8\, P_{t-1} + C_{11}\, P_{t-1} - P_{t-1} + C_4\, \text{LNGDP}_t + C_9\, \text{LNGDP}_{t-1} + C_{11}$$
$$\text{LNGDP}_t - \text{LNGDP}_{t-1} + C_5\, \text{LNPOP}_t + C_{10}\, \text{LNPOP}_{t-1} + C_{11}\, \text{LNPOP}_t$$
$$- \text{LNPOP}_{t-1} + C_{11}\, \text{LTV}_{t-1} + C_{11}\, \text{MR}_{t-1} + C_{11}\, P_{t-1} + C_{11}\, \text{LNGDP}_{t-1} + C_{11}$$
$$\text{LNPOP}_{t-1} \tag{k}$$

Then, Equations j and k can be transformed:

$$
\begin{aligned}
LNCR_t - LNCR_{t-1} = &\ C_0 + C_1\,(LTV_t - LTV_{t-1}) + C_2\,(MR_t - MR_{t-1}) + C_3\,(P_t - P_{t-1}) \\
& + C_4\,(LNGDP_t - LNGDP_{t-1}) + C_5\,(LNPOP_t - LNPOP_{t-1}) + (C_1 + C_6 \\
& + C_{11} - 1)\,LTV_{t-1} + (C_2 + C_7 + C_{11} - 1)\,MR_{t-1} + (C_3 + C_8 + C_{11} - 1) \\
& P_{t-1} + (C_4 + C_9 + C_{11} - 1)\,LNGDP_{t-1} + (C_5 + C_{10} + C_{11} - 1)\,LNPOP_{t-1} \\
& + (1 - C_{11})\,(LTV_{t-1} + MR_{t-1} + P_{t-1} + LNGDP_{t-1} + LNPOP_{t-1} - LNCR_{t-1}) \quad \text{(l)}
\end{aligned}
$$

or,

$$
\begin{aligned}
D(LNCR_t) = &\ C_0 + C_1\,D(LTV_t) + C_2\,(MR_t) + C_3\,(P_t) + C_4\,(LNGDP_t) + C_5\,(LNPOP_t) + C_6 \\
& LTV_{t-1} + C_7\,MR_{t-1} + C_8\,P_{t-1} + C_9\,LNGDP_{t-1} + C_{10}\,LNPOP_{t-1} + C_{11}\,(LTV_{t-1} \\
& + MR_{t-1} + P_{t-1} + LNGDP_{t-1} + LNPOP_{t-1} - LNCR_{t-1}) \quad \text{(m)}
\end{aligned}
$$

The model used in this study can be expressed in the ECM model as follows. The ECM-EG equation in the short term is:

$$
\begin{aligned}
D(LNCR)_t = &\ \alpha_1 D(P)_t + \alpha_2 D(LNGDP)_t + \alpha_3 D(LNPOP)_t + \alpha_4 D(LTV)_t + \alpha_5 D(MR)_t \\
& + \alpha_6 (ECT) + \mu_t \quad \text{(n)}
\end{aligned}
$$

The equation for ECM-EG in the long term is:
$$
LNCR_t = \beta_0 + \beta_1\,P + \beta_2\,LNGDP_t + \beta_3\,LNPOP_t + \beta_4\,LTV_t + \beta_5\,MR_t \quad \text{(o)}
$$

APPENDIX 2

Variable definition.

Denotation	Variable	Note
CR	Total mortgage loans (Credit)	Total value of mortgage loans that people want. The data is the amount of total mortgage loans granted each month.
P	Housing Price	Measure from the averaged aggregated housing price from 16 major cities in Indonesia (for small, medium, and large houses); P is performed on log-linearization.
GDP	Real Gross Domestic Product	Reflects the housing consumption indicated by household income.
POP	Total Population	Population growth to measure the development of housing demand.
LTV	Loan-to-Value Ratio	A regulation set by the government to control down payments and tiering in the housing market.
MR	Short-term Mortgage Rate	Short-term rate for mortgage loans applied by every bank in Indonesia.

Business Innovation and Development in Emerging Economies – Trinugroho & Lau (Eds)
© *2019 Taylor & Francis Group, London, ISBN 978-1-138-35996-3*

Globalization in automotive industry: Can Indonesia catch-up with Thailand?

Agung Romy Hasiholan & Kiki Verico
University of Indonesia, Depok, Indonesia

ABSTRACT: Indonesia has a long story in participating the international automotive production network, but Indonesia is unable to reap gains from the outstanding growth in automotive and auto parts trade. In Southeast Asia region, Indonesia as the biggest economy is even surpassed by Thailand since Thailand has become major production base for many global OEMs. Therefore, this paper aims to answer a question why Indonesia could not become major player in ASEAN region as Thailand did. This paper argues that Indonesian industrial policies are the reason why Indonesian auto industry is lagging behind Thailand. Indonesia's national car project with its discriminatory policies adversely affects OEMs where they consider Indonesia not as a good place for investment. Empirical studies using fragmentation as logical framework shows the main determinants of Indonesian parts and components exports. The estimation using panel regression method shows improvement in service-link costs, improvement in road and port infrastructure, and availability of skilled labor become important prerequisite to boost parts and components exports, alongside with sound and export-oriented industrial policies.

Keywords: fragmentation, service-link costs, industrial policies, infrastructure

1 INTRODUCTION

Automotive industry plays a vital role in the context of international trade. Global sales of all types of vehicles hit 93.8 millions of units in 2016, increasing 42.4% from 65.9 million of units in 2005. The similar upward trend was also found in global automotive production where it approached 95 millions of units in 2016, increasing 42.35% from 66.7 millions of units in 2005. Another drastic change was found in the international trade of automotive parts and components (P/C). The global auto parts trade accounting for US$ 109 billion in 1988 increased up to US$ 1.6 trillion in 2016.

The drastic change in automotive trade especially in P/C are heavily influenced by the globalization. Under this phenomenon, countries show their effort to reduce trade barriers so that both trade and investment flows are not interrupted. Due to the globalization, MNCs from developed countries can fragment their production processes and send some stages of production primarily to developing countries whose cost of production is relatively low. This fragmentation allows countries to involve by specializing in one of the production processes which vary from production of high-valued components to assembly process.

ASEAN member states also have active participation in the international automotive production network, especially Thailand and Indonesia. Thailand is considered as key player of automotive trade in Southeast Asia region because it has been chosen by many OEMs as production bases. Now Thailand is known as 'Detroit of Asia' because of its success in developing the industry. Thailand had accounted 2% of the global auto parts trade in 2016.

* We gratefully acknowledge financial support from University of Indonesia.

The success story of Thai automotive industry does not happen in Indonesia. Although Indonesia has the biggest potential market indicated by its large population and rapid economic growth, the Indonesian automotive industry is still lagging behind Thailand. Many OEMs prefer Thailand as their regional hub more than Indonesia. Several companies which initially operated in Indonesia chose to relocate to Thailand. In 2016, share of Indonesia in the global auto parts trade only accounted 0.804%. It shows that Indonesia is outperformed by Thailand.

This paper examines Indonesian automotive industry to better understand why up until now Indonesian automotive sector, specifically parts and components, could not surpass Thailand. This paper contributes comparative analysis on industrial polices between Indonesia and Thailand. Another important contribution this paper provides is the empirical study concerning the key determinants of Indonesian P/C exports by using fragmentation theory as the logical framework.

The structure of this paper is as follows. In the following section, we provide brief explanation of the theoretical framework of globalization and fragmentation trade as well as its relevance in the automotive trade. Then, we explain the current situation of both Indonesian and Thai automotive industry by comparing selected indicators. Next, we give an overview of Indonesian and Thai industrial policies. We examine the key differences between Indonesian and Thai industrial policies which make them have different development path. Thereafter, we provide the empirical analysis on key determinants of Indonesian auto parts trade by using the regression analysis. In the end, we assess options Indonesia has in order to catch-up with Thailand.

2 THEORETICAL FRAMEWORK

2.1 Globalization and International trade

Globalization clearly had transformed the global trade. UNCTAD (2008) found that there was an increasing trend of national market integration and interdependence of countries for various goods and services. In the light of globalization, countries chose to liberalize their market in order to gain access to the global market. As a result, firms could enjoy lower trade and production costs.

Baldwin (2016) describes globalization as the effort to eradicate barriers separating the production and consumption activities. He describes the components of the unbundlings as the costs of moving goods, ideas, and people. Before the globalization took place, the production

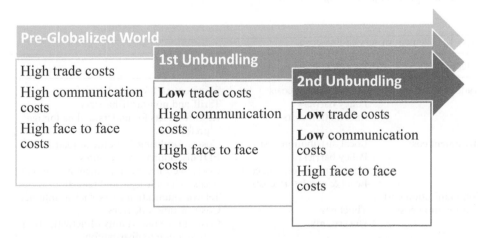

Pre-Globalized World

High trade costs

High communication costs

High face to face costs

1st Unbundling

Low trade costs

High communication costs

High face to face costs

2nd Unbundling

Low trade costs

Low communication costs

High face to face costs

Figure 1. Phase of globalization.
Source: Baldwin, 2016.

and consumption activities were limited by geographical proximity where goods produced in one location could not be consumed by others living at a significantly distant places. When the globalization started, there was improvement in trade costs resulting production and consumption activites were no longer bounded by distance.

Baldwin (2016) classified the phase of globalization into two stages. The initial stage, the *first unbundling*, was marked by the transport-technology revolution. In this stage, the transportation cost fell due to the improvement in transport technology. The lower transportation costs made the sales and distribution of goods to any locations possible. However, in this stage people was unable to move freely resulting production process could operate only in a certain place. The second stage, *the second unbundling*, was marked by the ICT revolution. The revolution made the production process able to be sliced and moved to other locations because ICT made the coordination from distant place possible.

2.2 *Fragmentation and International trade*

Globalization provides lots of benefits for MNCs. One major advantage that change MNCs' production pattern is that they have options to slice one of their stage of production to developing countries because. MNCs have interest to do so because developing countries offer lower production costs and access to the domestic market. As a result, developing countries now could join and be a part of international production networks.

The process of slicing production process as illustrated above is called fragmentation. Deardorff (2001) defined fragmentation as process of slicing stages of production into several locations. The production blocks then will generate different outputs, but lead to the same final output. Arndt & Kierzkowski (2001) defined fragmentation in similar way, but they emphasized the splitting up process from an integrated production system to the creation of fragments spread in various locations. In general, fragmentation is a phenomenon that change the global production pattern where MNCs from developed countries decide to move some stage of production to developing countries with their motive of seeking low production costs.

Service trade is important in the international fragmented trade. Firms that fragment their stage of productions will choose countries that provide efficient services used to coordinate one production block to another. Services that link and coordinate between production block is called service-link cost. Countries which are able to provide efficient service-link cost have active participation in international fragmented trade. Therefore, improvement in service-ink costs for countries to participate and become one of the fragments of the MNCs' production network. Table 1 below described the elements of service-link costs and Figure 2 explain the production process under fragmentation.

Table 1. Element of service-link costs.

Category	Subcategory	Details
Trade cost	Transportation costs	Shipment and freight charge
	Policy barriers	Tariff and non-tariff barriers
	Information costs	Research costs for understanding foreign preferences
Investment cost	Local distribution costs	Costs of utilizing domestic infrastructure
	Policy barriers	FDI discriminatory measures
	Contract enforcement costs	Costs to ensure the implementation of contract
	Legal & regulatory costs	Costs to handle legal and regulatory issues
Communication cost		Telecommunication costs, such as internet
Coordination cost	Timeliness	Costs of timely delivery
	Uncertainty	Costs due to uncertainty of activities from production to distribution

Source: Kimura & Takahashi (2004).

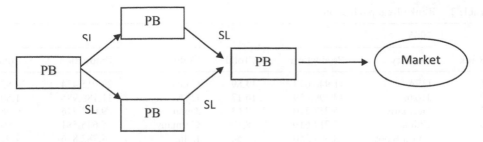

Figure 2. Fragmentation process.
Source: Jones & Kierzkowski (1990).

Service-link costs play important role in the production of automotive industry. Timeliness and transportation costs are very sensitive to the automotive production. Toyota as important OEMs for Indonesia and Thailand which use '*Just-in-Time*' inventory really relies on efficient infrastructure in order to minimize the production costs. Therefore, Japanese OEMs often consider quality of infrastructure as important prerequisite in deciding whether they want use a particular country to be part of its automotive production network.

3 INDONESIA AND THAILAND IN GLOBAL AUTOMOTIVE INDUSTRY

3.1 Car production

One of key indicators to see whether a country has an active participation in automotive production network is the automobile production. Table 2 provides the information regarding the global ranking of car production. In 2005, USA became the largest automobile production, followed by Japan, Germany, India, and South Korea. Thailand and Indonesia were the only Southeast Asia countries which become one of the top 25 automobile producer. Indonesia ranked 23rd with 500,710 cars (0.75% of world production), while Thailand ranked 14 with 1.2 million cars (1.83% of world production), more than double of Indonesia production.

In 2017, China became the largest automobile producer which accounted almost 30% of global automobile production. China had surpassed USA, Japan, and Germany where they accounted 11.5%, 9.96%, and 5.8% of global automobile production respectively. Indonesia and Thailand showed better performance. Indonesia had higher position in which it ranked 18th, accounting for 1.25% of global automobile production. Thailand also had higher position where it ranked 12th, accounting for 2.04% global automobile production. From the global ranking of automobile production, we could see that Thailand still outperforms Indonesia.

Another indicator used to evaluate both Indonesia and Thailand performance in automotive trade is the excess of car production. The indicator is calculated by differencing car production with car sales in a particular year. Negative excess car production means the automobile industry is net importer because domestic sales is higher than production, and vice versa. Figure 3 below describes the indicator for both Indonesia and Thailand. Indonesia's car production is smaller than the domestic car sales consistently since 2006 up to 2013. Indonesia starts to have an excess car production from 2014[1]. Unlike Indonesia, Thailand has already had excess of car production since 2006. Therefore, the indicator indicates that Thai automobile industry is more export-oriented than Indonesian automobile industry.

---

1. This trend made the domestic automotive players considered Indonesia as 'jago kandang' which means the output is domestic-oriented. This trend could exist because of Indonesia's large market that absorbs almost all the outputs. This is clearly different with Thailand where its domestic market is significantly smaller than Indonesia.

Table 2.   Rank of car production.

	2005			2017		
Rank	Country	Production	% Total	Country	Production	% Total
1	USA	11,946,653	17.89	China	29,015,434	29.82
2	Japan	10,799,659	16.17	USA	11,189,985	11.50
3	Germany	5,757,710	8.63	Japan	9,693,746	9.96
4	China	5,717,619	8.57	Germany	5,645,581	5.80
5	South Korea	3,699,350	5.54	India	4,782,896	4.92
6	France	3,549,008	5.32	South Korea	4,114,913	4.23
7	Spain	2,752,500	4.13	Mexico	4,068,415	4.18
8	Canada	2,687,892	4.03	Spain	2,848,335	2.92
9	Brazil	2,530,840	3.79	Brazil	2,699,672	2.77
10	UK	1,803,109	2.70	France	2,227,000	2.28
11	Mexico	1,684,238	2.52	Canada	2,199,789	2.26
12	India	1,638,674	2.45	**Thailand**	**1,988,823**	**2.04**
13	Russia	1,354,504	2.03	UK	1,749,385	1.79
14	**Thailand**	**1,222,712**	**1.83**	Turkey	1,695,731	1.74
15	Iran	1,077,190	1.61	Russia	1,551,293	1.59
16	Italy	1,038,352	1.56	Iran	1,515,396	1.56
17	Belgium	926,515	1.39	Czech Rep.	1,419,993	1.45
18	Turkey	879,452	1.32	**Indonesia**	**1,216,615**	**1.25**
19	Poland	613,200	0.92	Italy	1,142,210	1.17
20	Czech Rep.	602,237	0.90	Slovakia	1,000,520	1.03
21	Malaysia	563,408	0.84	Others	758,672	0.78
22	South Africa	525,227	0.79	Poland	689,729	0.71
23	**Indonesia**	**500,710**	**0.75**	South Africa	589,951	0.61
	World	66,719,519	100	World	97,302,534	100

Source: Author's calculation using OICA statistics.
Data source: http://www.oica.net/category/production-statistics/2017-statistics/.

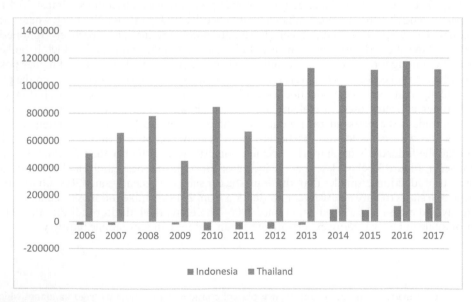

Figure 3.   Excess of car production for Indonesia & Thailand, 2006–2017.
Source: Author's calculation by using ASEAN Automotive Federation Statistics (Data: http://www.asean-autofed.com/statistics.html).

3.2　P/C exports of Indonesia and Thailand

P/C exports have played significant role in global automotive trade with its significant growth. Countries with high P/C exports arguably have intense participation in the automotive production networks. We could say higher P/C exports indicates that particular country has ability to provide more high-valued components, and vice versa. Hence, it is useful to see country's P/C exports for the purpose of evaluating its participation in automotive production network.

Figure 4 below depicts the automotive P/C exports for both Indonesia and Thailand. Since 1999, Thailand had higher P/C exports than Indonesia up to 2016. The difference between Indonesia and Thailand exports were relatively small. But Thailand had surpassed Indonesia at faster rate since 2001 where Thailand reached its highest P/C export growth in 2011. This clearly shows that Thailand is in the upper-hand than Indonesia in terms of P/C exports. But, it is important to note that Indonesia's P/C exports keep growing at moderate rate.

Indonesia and Thailand have relatively the same trading partners. In 2016, both countries exported to Japan, US, and Mexico. But Thailand was able to export more distant countries, such as China, Argentina, and South Africa. Thailand's ability to access more distant markets reflects the importance of Thailand in supplying automotive P/C in the international automotive production network.

3.3　RCA index and CMSA for Indonesian and Thai P/C

The calculation of revealed comparative advantage index (RCAI) is important to see at which goods or commodities that a country has advantage to trade. The calculation of RCAI follow the formula developed by (Balassa, 1965). The formula is shown below:

$$RCA_{ij} = \frac{\left[\dfrac{x_{ij}}{X_{it}} \right]}{\left[\dfrac{x_{wj}}{X_{wt}} \right]}$$

Country has comparative advantage at product i if the RCAI has value greater than one, and vice versa. We calculate the 57 P/C for Indonesia and Thailand and the results are shown

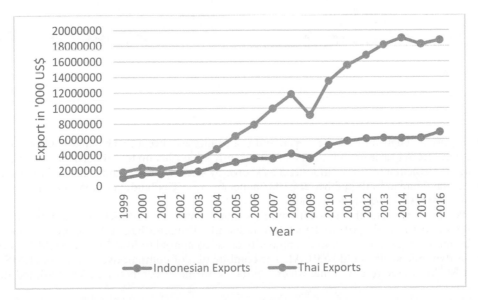

Figure 4.　Automotive P/C exports of Indonesia and Thailand, 1999–2016.
Source: Author's calculation using UN Comtrade Database.

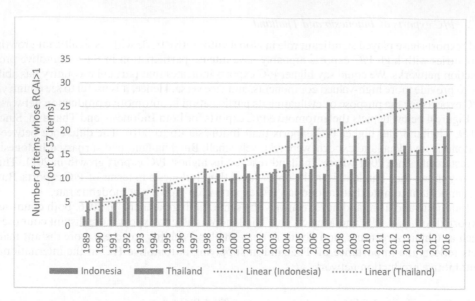

Figure 5. Number of P/C with RCA greater than one.
Source: Author's calculation by using UN Comtrade Database.

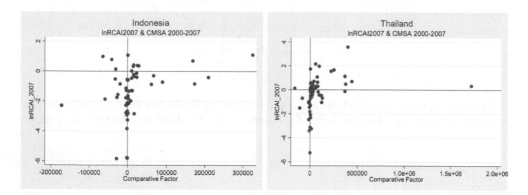

Figure 6. lnRCAI and CMSA for Indoesia and Thailand, 2007.
Source: Author's calculation using UN Comtrade Database.

by Figure 5. Thailand is in the upperhand than Indonesia because Thailand has more P/C that have RCAI value greater than one consistently since 1989 up to 2016. Thailand also had faster incremental rate than Indonesia as shown by steeper linear trend. This indicator suggests that Thailand still had better performance because Thailand had more trade potential with more P/Cs rather than Indonesia.

We also calculate the constant market share analysis (CMSA) in order to better understand the trade potential that Indonesia and Thailand had. CMSA provides more information regarding a country's competitiveness compared to RCAI because CMSA provides dynamic analysis unlike RCAI. The combination of comparative factor from CMSA and RCAI is used to determine which goods are the most competitive[2]. We modified the

---

2. The details on formula used to calculate the CMSA is shown in the appendix.

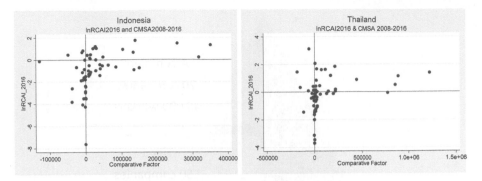

Figure 7. lnRCAI and CMSA of Indonesia and Thailand, 2016.
Source: Author's calculation using UN Comtrade Database.

model by constructing the RCAI in natural logarithm. Goods with positive comparative factor and positive natural logarithm of RCA are classified as *great*. Goods with negative comparative factor and positive natural logarithm of RCA are classified as *challenging*. Goods with positive comparative factor and negative natural logarithm of RCA are classified as *potential*. Goods with negative comparative factor and negative natural logarithm of RCA are classified as *none*. Therefore, goods classified as great are the most competitive product a country has. Figures 6 and 7 show the result of the calculation of CMSA and natural logarithm of RCA. List of products which are classified as great can be seen in the Tables A2 and A3. In short, Thailand has more products that are classified as great than Indonesia. This indicator shows once again that Thailand has better performance in P/C trade than Indonesia.

This section has explained the key indicators showing that Indonesia is lagging behind Thailand in automotive industry. In the following section, we are going to thoroughly overview the Indonesian and Thai industrial policies. It would help us to understand how significant past industrial policies affecting current automotive industry, especially its performance in P/C trade.

4   THE INDONESIAN AUTOMOTIVE INDUSTRY

4.1   *Structure of the Indonesian automotive industry*

The dominance of Japanese MNCs in Indonesian automotive industry heavily influences the structure of the industry. Within the business system called *keiretsu*[3], Japanese assemblers create vertical, three-level subcontracting system. The assemblers directly use the final components of first-tier suppliers, which will source the lower-valued components from the lower-tier suppliers. Both assemblers and first-tier suppliers are dominated by MNCs especially from Japan, while the second and third-tier suppliers are dominated by small medium enterprises (SMEs).

The following figure provides the latest information on Indonesian automotive industry. The industry consists of assemblers and component industries. There are 20 car assemblers whose shares are dominated by foreign firms (usually Japanese). These assemblers use output from first-tier suppliers which generate high-valued components, i.e. genuine parts. The second and third-tier suppliers generate lower-valued components whose output would be distributed either to the first-tier suppliers or directly to consumers. Most companies under the classification of second and third-tier suppliers are joint-venture between foreign and local firms where the local

---

3. Vertical keiretsu is relevant in explaining the production pattern of Japanese automotive industry (Branstetter, 2000). Within the organizational structure, the OEMs are acting as the large manufacturing company while the component companies are acting as supplier. This long-term and sustainable supplier-based relationship is due to the some share of ownership of one company in other companies.

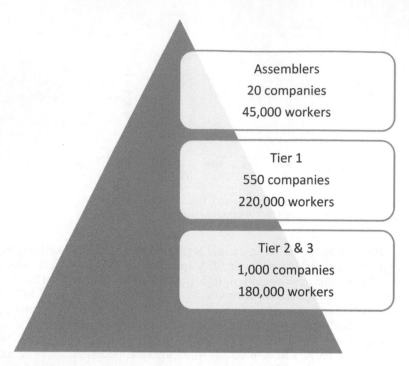

Figure 8. The structure of Indonesian automotive industry, 2015.
Source: Shephard & Soejachmoen, 2018.

firms often dominate. In 2015, the Indonesian automotive industries were able to employ 445 hundred thousand of people.

In January—April 2018, multi-purpose vehicles (MPVs) accounted for 56.7% of total retail sales, followed by pick-up trucks which accounted for 11.5%. MPVs are more popular among middle-income class due to its affordable price compared to sedans whose sales only accounted 0.8%. The share of each category for total production is also similar. MPVs dominated the total production whose share accounted 53.2%, followed by pick-up truck which accounted 12.7%. The production of sedans were the lowest among others which accounted only 2.9%. These sales and production pattern in Indonesia is quite different with the global trend where the global demands for sedans and SUVs are increasing.

4.2 *Stages of development of the Indonesian automotive industry*

4.2.1 *Early stages (1928–1969)*
The Indonesian automotive industry was started by the establishment of General Motor assembly plant in Tanjung Priok, Jakarta in 1928. This plant was able to produce 6000 units per year (Natsuda, Thoburn, & Otsuka, 2015). In 1950s, this assembly plant was nationalized under the Sukarno administration's program, *Program Benteng*, whose purpose was creating national car industry (Hale, 2001). The program failed to achieve its goal as the program remained stagnant in the early 1960s. Hale (2001) explained one of the reason is the limited foreign exchange. The end of this period was marked by low output where the industry could only produce 2000 vehicles annually and most of the components were imported. This shows that Indonesia could not reap gains from early industrialization in automotive industry[4].

---

4. Indonesia was one of the earliest ASEAN member state to have the assembly plant. Thailand's first assembly plant was established in 1975 and Malaysia's was established in 1960s.

4.2.2 Import-substitution through localization (1970–1992)

The development of automotive industry was different under the Suharto's administration. After Suharto came into power in 1976, the government implemented more liberal policies following the market principles. Suharto's liberal policies aimed to restore Indonesia's relationship with the world economy (Linbald, 2015). As a result, Indonesia was more open to foreign trade and investment under his administration.

In the 1970s, the Indonesian government implemented various localization policies in the automotive sector. In 1971, the CBUs imports were banned in Java and Sumatera. This policy was extended to a nation-wide ban in January 1974. It was designed to alter the CBUs imports and increase the production of the local assembly. Another localization policy was announced by the government in 1976 where the government targeted the auto parts and components to be produced locally. This policy was a response to the issue of high dependence on imported parts. In 1977, the government targeted paints, tires, and batteries to be locally produced. Thereafter, the government became more ambitious as the government targeted more parts and components, i.e. brakes, axles, engines and transmissions in 1984 (Aswicahyono, Basri, & Hill, 2000). This ambitious plan to localize some parts and components could not achieve its target due to the recession in the mid-1980s. However, these policies were maintained until 1993.

4.2.3 Inward-looking policies, national car projects, & liberalization (1993–2001)

In 1993, the Indonesian government ended the ban on CBU imports and implemented a new lower tariff and luxury tax on the imported components. The new tariff rate was based on the vehicle types and the local content level. This policy gives room for assemblers to choose which local components should be used.

Suharto administration implemented one of its controversial policy, the national car project, in February 1996. The scheme in conducting the national car project:

- The government gave three-year exemption from import duties and luxury taxes.
- The companies had to achieve LC of 60% within 3 years.

Suharto announced that the national car status was given to the local automobile companies, *Timor Putra National* (TPN), owned by his son. TPN unfortunately could not directly produce cars because the Indonesian automotive industry did not have technologies that match the international standard. Hence, TPN cooperated with Korean automobile manufactures by establishing the *PT Kia-Timor Motors*[5] (Aswicahyono, Basri, & Hill, 2000).

The protection to *PT Kia-Timor Motors* had created distortion for the automobile market. The product of *PT Kia-Timor Motors*, Timor S515, was half the price of its competitor such as Toyota Corolla. This adversely affected the automobile manufacturing companies, especially Toyota and Isuzu. These two companies were seriously affected because their products were relatively the same with PT Kia-Timor Motors. As a result, EU, Japan, and U.S filed a case to Dispute Settlement Body (DSB) of the World Trade Organization (WTO). The panel ruled out against Indonesia and ordered the Indonesian government to abolish the incentive system in 1999. *PT Kia-Timor Motors* declared bankruptcy in 2001.

The Asian Financial Crisis (AFC) in 1998 pushed the Indonesian government to liberalize its market. As the government received financial assistance from IMF, the government had to perform some programs of economic reforms. In automotive sector, the government had to reduce import tariffs, eliminate local content program, and commit to the WTO rules (Hale, 2001). Hence, the automotive industry could operate under the market principles with less government intervention than before.

---

5. Share of PT Kia-Timor Motors was owned by Kia Motors (30%), Timor Putra Nasional (35%), and Indauda Putra Nasional Motor (35%).

4.2.4 Low-cost green car and implementation of Euro-4 (2002–present)

The AFC in 1998 and the lost in DSB WTO had resulted structural reform in automotive industry where the incentive system was abolished. In post crisis, the Indonesian government used import duties based on weight, type, and engine of vehicles to influence the imports. In 2006, the government introduced a new policy, incomplete knocked-down (IKD) in order to increase LC ratio (Natsuda, Thoburn, & Otsuka, Dawn of Industrialization? The Indonesian Automotive Industry, 2015). The IKD system tried to encourage industries to import sub-components, so that the local industries could assemble CKD parts locally.

In 2009, the government announced *the low-cost green car* (LCGC) policy. This policy was similar with Thailand's *eco-car* project. The LCGC vehicles were designed to be economical and environmentally friendly vehicles. In order to encourage people to buy LCGC vehicles, the government exempted LCGCs from luxury tax so that the price would be affordable. As the middle-class grew, the demand for LCGCs increased sharply since they shifted from using motorcycles to cars. The LCGC policy was successful in attracting FDI because the demand for LCGC vehicles was very promising. LCGC policy was also a good response to Thailand's eco-car project (interview with GAIKINDO representative, 27 Apr. 2018). Thailand domestic market has already been saturated. If Indonesia did not have LCGC policies, Indonesia would import lots of vehicles from Thailand and domestic production would be less competitive.

LCGC policy showed that Indonesian government put the environmental issues as priority. Therefore, the policy was extended as the Minister of Environmental Affairs issued Regulation No. P.20/MENLHK/SETJEN/KUM.1/3/2017 concerning the implementation of Euro 4. The industry responded to this regulation positively because the industry could produce cars for both domestic and foreign market with the same standard (interview with GAIKINDO representative, 27 Apr. 2018). Although Indonesia had shown effort to increase the technology for its vehicles, this improvement was quiet late because other countries, such as Thailand, had used Euro 6.

Implementation of Euro 4 faced several challenges. One of the most important challenge is that Indonesia is lacking in the supporting infrastructure. Although the regulation has to be in effect in September 2018, PT Pertamina is still unable to provide the required fuel. If PT Pertamina is unable to provide the required fuel while the industry has produced cars with Euro-4 engine, the government has to import the fuels. This would worsen the national trade balance.

5 THE THAI AUTOMOTIVE INDUSTRY

5.1 Structure of the Thai automotive industry

The structure of Thai automotive industry is similar with the structure of Indonesian automotive industry. This is due to the fact that Japanese automotive MNCs have been so long dominating both domestic markets. The Thai automotive industry mainly consists of auto assemblers, 1st, 2nd and 3rd tier suppliers. The auto assemblers would directly source the final outputs from the 1st tier suppliers as their inputs for production. Thereafter, the 2nd and 3rd suppliers will deliver their output to either the 1st tier suppliers or consumers. The 1st tier suppliers consist of companies whose share are owned either by foreign majority, Thai majority, or pure Thai. The rest of 2nd and 3rd tier suppliers are locally owned.

The following figure provides information about Thai automotive industry structure in 2015. There are 18 companies operating as the auto assemblers. These companies provide jobs for 100,000 people. The 1st tier suppliers which have close relation with the auto assemblers consist of 462 companies and they provide 250,000 jobs. The 1st tier supplier creates high-valued components directly to the assemblers. The 2nd and 3rd tier suppliers comprise 1,137 companies which provide 175,000 jobs. In 2016, the Thai automotive industry are able to generate 525,000 employments.

The following information is about the sales and production of vehicles in Thailand. The recent data from *MarkLines* show that in 2017, 1-ton pickups are the main product in the

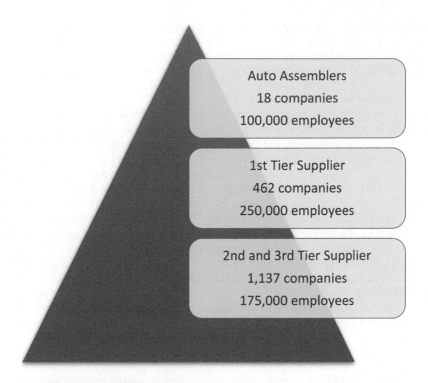

Figure 9.   Thai Automotive industry structure, 2015.
Source: Nitipathanapirak, 2017.

domestic market. The sales of 1-ton pickups reached 424,282 units. Passenger vehicles (PV) came the second as this vehicle type had sales of 346,244 units. The SUVs category had the least amount of sales which accounted for 53,431 units. In 2017, Thailand produced 1,988,823 units of new vehicles and exported 1,139,696 units.

5.2   Stages of development of the Thai automotive industry

5.2.1   Early period: Import substitution (1960–1970)
In 1960, the Thai government started to develop its industry by using import substitution strategy. The government planned to create protectionist trade-policies in order to encourage foreign investors to produce in Thailand. Therefore, the government through its Board of Investment (BoI) enacted the Investment Promotion Act in 1960 and its revision in 1961. The act covered some investment promotion measures, such as income tax break for the automotive sector (Kohpaiboon & Yamashita, 2010).

The Investment Promotion Act changed the course of history for Thai automotive industry which was only a repair business up to 1960 (Natsuda & Thoburn, 2012). In 1960s, some assembly plants for passenger and commercial vehicles, including Thai Motor Industry Co., Ltd., Karnasuta General Assembly Co., Ltd, and joint ventures between Nissan and Siam Motors) were set up (Fujita, 1998). In order to protect the domestic production, the government imposed high import tariff on CBU vehicles and CKD kits.

5.2.2   Local content requirement (1971–1977)
The government tried to extent the import substitution strategy on the automotive industry. It created localization policies which targeted the auto parts and components, such as tires, batteries, and radiators. This policy is necessary to be implemented because Thailand had a serious balance of payment deficit which was partially caused by the imported kits

(Techakanont, 2011). Thus, the government implemented the local content requirement (LCR) policy in 1975. Under this new regulation, the domestic auto assemblers had to use local parts at least 25% from total value of vehicle (Natsuda & Thoburn, 2012). However, the worsening automotive trade balance was still a serious problem (Kaosa-ard, 1993).

5.2.3 *Intensifying the localization policy (1978–1990)*
The deficits of automotive trade balance remained a problem and the government had a strong desire to reduce the deficits. As a consequence, the government decided to extent the localization policy. In order to reduce the deficits and also stimulate the domestic industry, the government imposed import ban on CBU passenger vehicles and increased the import duty on CKD kits from 50% to 80% (Abbot, 2003). In addition, the government also set 150% and 80% tariff for CBU PVs and CKD PVs respectively (Kohpaiboon & Yamashita, 2010).

Further localization was also conducted by setting higher LCR. In 1982, the government announced the LCR for all vehicles were set to 45%. In the next four years, the government imposed higher LCR for PVs which were set to 54%. Another policy that government imposed was the implementation of mandatory deletion scheme. This policy specifically targeted some parts, such as exhaust system and brake drums, to be produced locally (Natsuda & Thoburn, 2012). This policy resulted both positive and negative outcome. This policy was successful in encouraging the domestic assemblers to put more investment on component industries (Busser, 2008). Unfortunately, some small assemblers such as Simca and Hilman were eliminated from the market due to their inability to meet the LCR (Natsuda & Thoburn, 2012).

5.2.4 *Liberalization and Asian financial crisis (1991–2000)*
The liberalization especially in the automotive industry started after Anand Panyarachun came into power in 1991. Under his government, the industrial policies were designed to reduce the international trade barriers and attract more FDI especially from non-Japanese MNCs and to reduce the domestic automobile price (Natsuda & Thoburn, 2012). The high domestic automobile prices were resulted from high import tariff and inefficient local production. Initial liberal policies the new government implemented were to abolish the ban on CBU imports and to reduce tariff at substantial rate for all types of CBUs and CKD kits[6].

In 1993, the new government announced that the regulation concerning the local ownership requirement on the foreign-invested projects would be eliminated. The new regulation allowed the foreign investors to have 100% ownership of a particular project under the condition that they would export at least 60% from the total production. However, this regulation was finally abolished in 1997. Another important liberalization policy was the abolishment of LCR where the government announced that the LCR would be completely abolished in 1998. But the abolishment was realized in 2000.

The automotive industry was seriously hit by the Asian financial crisis. The substantial depreciation of Thai Baht and high inflation made the domestic market slumped. The industries could not rely on the domestic market anymore and they had to find the other alternative markets to offset the loss from downturned Thai market. The crisis then resulted the industries which were domestic-oriented to become an export-oriented industry. The Thai automotive export performance was unexpectedly helped by the crisis phenomenon. The large depreciation made the Thai products have more competitive prices.

5.2.5 *Post financial crisis (2000–present)*
Automotive sector had played important role for the Thai economy. Due to the Asian financial crisis that adversely affected the industry, the government implemented fiscal policies to foster the industry. The government implemented two significant fiscal policies, which were the increased tariff rate on CKDs and the reduced excise tax on pick-up trucks and passenger

---

6. According to Kohpaiboon & Yamashita (2010), tariff of CBUs over 2.3 liters was reduced from 300% to 100%, tariff of CBUs under 2.3 liters was reduced from 180% to 60%, and tariff of CKDs for both passenger and commercial vehicles was reduced from 112% to 20%.

vehicles (Natsuda & Thoburn, 2012). The first policy aimed to stimulate the local component industry while decreasing protection for the local assemblers. The second policy was designed to eliminate the consumer's burden resulted from the new high set import tariff on CKDs.

The industrial policies in developing the automotive industries were designed by Thaksin Shinawatra government which came into power in 2001. One of its most well-known policy was the development plan of Thai automotive industry, which was often referred as the 'Detroit of Asia' (Busser, 2008). The plan main target was to make Thailand known as the regional hub of Southeast Asia for automotive exports.

Another important industrial policy which was part of the automotive development plan was the selection of both pick-up production and its component industries as national priority. The chosen products is publicly known as a *product champion*. Here are the schemes for the development of product champion (Natsuda & Thoburn, 2012):

- Exemption of import tariffs on machinery.
- Three-year corporate tax exemption.
- Provided tax incentives for the establishment of R&D and regional operating headquarters.

The schemes were responded by big auto makers such as Toyota which decided to relocate its pick-up production and production development base from Japan to Thailand. In addition, Toyota also opened its R&D in Thailand. The main reason why Toyota chose Thailand to become its important production base was the fact that Thailand had large pick-up market that Toyota were interested to access. Today, Thailand plays a vital role in Toyota's Innovative International Motor Vehicle (IMV) project where Thailand has been chosen as Hilux pickup trucks and Fortuner SUV production base[7] (Kobayashi, 2014).

The other industrial policy which was significant in making Thailand as Detroit of Asia was the eco-car project. In order for the project to be successfully implemented, the government had to attract FDI from auto makers. The key feature of this project was that the government used the tax incentives which were linked to the localization of auto parts component (Natsuda & Thoburn, 2012). Companies which were interested and finally engaged with the project were all Japanese MNCs.

6 WHY INDONESIA IS LAGGING BEHIND THAILAND?

We have seen that Thailand had been successful in developing its automotive industry which it is known as 'Detroit of Asia'. Although Indonesia is now the biggest automotive market in ASEAN, Indonesia is still less preferable as main production base compared to Thailand. Therefore, this section is trying to seek the reasons why Thailand could have better performance than Indonesia in automotive trade.

One of the most important determinant that helps Thailand to become key automotive player in Southeast Asia is the industrial policies which were carefully designed by the government. Here are some comparison between Indonesia and Thailand industrial policies which make them have different path of automotive development.

- *Thailand liberalized its market before the Asian financial crisis*
Thailand had better performance than Indonesia is partly because Thailand decided to liberalize its market before the crisis in 1997–1998. The liberalization policies made many automotive MNCs decide to operate in Thailand. These MNCs then focused in exports due to the financial crisis which made the domestic market slumped. Fortunately, the Thai automotive exports could rise substantially because it was supported by substantial depreciation of

---

7. IMV project also involved some ASEAN member states. Kobayashi (2014) mentioned that Thailand and Indonesia had been chosen as strategic assembly bases for both American and Japanese OEMs. Thailand has more OEMs than Indonesia which clearly shows Thailand is more preferable than Indonesia

domestic currency. Unlike Thailand, Indonesia had not liberalized its market yet before the crisis. Inward-looking policies and *KKN*[8] under Suharto administration made uncertainty for investors resulting Indonesia could not enjoy the impact of substantial depreciation for its exports.

- *Thailand developed its product champion instead of firm champion*

Thailand had better position in attracting FDI because Thailand developed its *product champion*, unlike Indonesia which developed its *firm champion*, the controversial PT Kia-Timor Motor. As Thailand became more attractive for investment, automotive MNCs especially from Japan decided to use Thailand as their production bases. IMV project was a clear example that Thailand was used to produce and export its various pick-up trucks to over 90 countries (Kobayashi, 2014). The development of *firm champion* resulted different story for Indonesia. The national car project created distortion in the market where the assemblers were heavily disadvantaged. It resulted investors to have low confidence on Indonesia as a good place for investment. Hale (2001) reported that Chrysler canceled its plan to use assemble Neon in Indonesia due the implementation of national car project.

- *Thailand provided various fiscal incentives aiming to boost exports*

Thailand could have high automotive exports because Thailand provides lots of fiscal incentives aiming to boost exports. Thailand used various tax incentives, such as tax exemptions, and eliminated any business restriction to attract foreign investment. This policy was also assisted by the abolishment of foreign-ownership restriction where foreign-invested project had to export at least 60% from total production. Unlike Thailand, automotive players in Indonesia felt that Indonesian government did not provide fiscal incentives as the industries need. One of the Gaikindo member said that the new assembler required minimum three years to reach break-even point (Focus Group Discussion with Gaikindo, 12 April 2018).

Another important reason why Indonesia are lagging behind Thailand is because Indonesian automotive industries have small-export base. One of the reason to explain that phenomenon is that the pattern of automotive sales and production in Indonesia is not following the global trend (interview with Gaikindo representative, 27 April 2018). The domestic market is dominated by the MPVs where the component industries produced outputs mainly for the MPVs production, while the global trend is showing the increasing demand for sedan and SUVs. Back in the mid-1990s, Indonesia had an estimated 19 companies which produce sedans, and now Indonesia only has three companies. The other 16 companies decided to relocate to Thailand because Indonesian market is very small for sedan. The story of the component industry was quiet similar. There was once component-company which produced brake component. The company was able to supply the brake components to one of the top luxury-vehicle company. Later, the company could not supply the components anymore because the company could not fulfill the specifications which became more complex.

7 MODEL

Authors have to emphasize that performance in auto parts exports shows a level of participation in automotive production network. Higher value of auto parts exports and more distant trading partners show a particular country has high position in the network.

As we have already known that Indonesia is lagging behind Thailand, we would like seek what factors that could help Indonesia to have better auto parts exports and finally able to catch-up with Thailand. Therefore, this section is going to comprehend the determinants of Indonesia auto part exports through regression analysis.

The empirical model used in this research is based on the fragmentation theory (Jones & Kierzkowski, 1990). The theory postulates that the fragmented trade depends on the

---

8. KKN is and Indonesian term that stands for corruption, collusion, and nepotism.

service-link costs, i.e. trade costs, investment costs, communication costs, and coordination costs. In addition, the model would also employ the gravity equation as Kimura & Takahashi (2004) used in analyzing the machinery trade. The variable distance is important to represent some service-link costs which are not quantifiable yet. Therefore, the hypothesis of this research is that the service-link costs and labor costs as the important determinant of Indonesia auto part exports. The full model specification is shown below:

$$lnexports_{ijt} = \beta_0 + \beta_1 lndist_{ij} + \beta_2 lngdp\_reporter_{it} + \beta_3 lngdp\_partner_{jt}$$
$$+ \beta_4 endowment_{ijt} + \beta_5 lnlab\_cost_{it} + \beta_6 paved\_road_{it} + \beta_7 lnport_{it} + \varepsilon_{it}$$

where subscript i represents Indonesia, j represents Indonesia's 30 trading partners, and t represents the year, t = 2000, 2001, ..., 2014. Variable $lnexports$ is the natural logarithm of Indonesia's auto parts exports in real value to Indonesia's trading partners. Variable $lndist$ is the natural logarithm of weighted distance between Indonesia and its trading partners. Variable $endowment$ is the ratio of Indonesia's GDP per capita and its trading partners' GDP per capita. Variable $lnlab\_cost$ is the natural logarithm of unit labor cost. Variable $paved\_road$ is the ratio of the length of paved road to the total length of road in km. Variable $lnport$ is the total international cargo loading and unloading in Indonesia. The last two variables are used to proxy the quality of Indonesia's infrastructure.

The variable $lndist$ has a negative expected sign because it represents trade costs and other service-link costs. It is assumed that the further the distance, the higher the trade cost would be, and it will reduce the number of exports.

The variable $lnlab\_cost$ has a positive expected sign. The standard theory of microeconomics stated that real wage can represent the marginal productivity of labor (Pindyck & Rubinfeld, 2013). Industries which produce high valued-components require more productive labor. This results in a higher wage that the industries have to pay. In other words, exports of high-valued components are also resulted from the increasing real wage.

The variable $paved\_road$ has positive expected sign. The increasing ratio of paved road to the total road means that the road infrastructure is getting better. This would result in lowering the logistic costs as the distribution of goods from one production block to another would use land transportation. The *just-in-time inventory* (JIT), a well know Toyota management system, relies on timely delivery.

The variable $lnport$ also has a positive expected sign. The increasing amount of international loading and unloading throughout Indonesian seaport means the increasing capacity of Indonesian seaport to handle international transaction. In other words, the increasing capacity of Indonesian seaport would help to facilitate the auto part exports.

8 DATA

The model uses 30 Indonesia's trading partners with the time period starting from 2000 up to 2014. The list of parts and components used in this research are based on Soejachmoen (2016) and the lists can be found at the Appendix. The dataset for this research comes from various databases: CEPII Database, World Development Indicator, UN Comtrade, *Statistik Industri*, and Indonesian Statistical Agency (BPS).

Both variable $lnexports$ and $lnlab\_cost$ are reported in real value. The real value of auto part exports is estimated by dividing the nominal value of auto parts exports with Indonesia GDP deflator. The method to estimate the real wage is also similar. The nominal value of wages paid by the industries are divided by the wholesale price index. Table 3 provides summary of variables.

This study using panel regression method. Traditionally, there are three possible techniques in panel regression which are pooled ordinary least square, fixed effect, and random effects. We could not use fixed effect model because time-invariant variables such as distance will be omitted. By using the Breusch Pagan lagrangian multiplier tests, the result show the model should be estimated with random effect.

Table 3. Summary of variable.

Variable name	Data	Source	Expected Sign	Mean	Standard Deviation	Min	Max
Inexports (dep. var)	Real value of auto part exports	UN Comtrade		12.94306	1.887399	7.367195	16.60974
lndist	Weighted distance	CEPII	–	8.891448	0.7819354	6.920588	9.82425
gdp_reporter	Indonesia GDP	WDI	+	0.053362	0.0074888	0.0364347	0.0634502
lngdp_partner	GPD of Indonesia's trading partner	WDI	+	27.17821	1.359318	23.9202	30.41658
endowment	Ratio of GDP per capita of Indonesia & its trading partner	WDI	–	27.17278	1.111911	22.88743	30.3567
lnlab_cost	Unit labor cost	Statistik Industri	+	16.38858	0.2288217	15.94923	16.79359
paved_road	Ratio of paved road length to total road length	BPS	+	0.5757487	0.0169137	0.549336	0.6034635
lnport	International loading and unloading	BPS	+	19.45436	0.4104384	19.04431	20.21279

9 RESULTS

The regression result regarding the determinants of Indonesian P/C exports is presented above. Almost all of the coefficients of variables are consistent with the expected signs and statistically significant ranging from 10% to 1% level.

Variable distance has negative sign and it is statistically significant. This result is consistent with other studies employing the gravity equation. As Kimura, Takahashi, & Hayakawa (2007) stated that distance can be used as proxy of almost all service-link costs, we could use the coefficient estimate of distance variable above as the impact of service-link cost to Indonesia P/C exports. This study found that Indonesian P/C exports are inelastic to the change in service-link cost, as the coefficient can be interpreted as 1% increase in service link costs results in 0.9% in P/C exports, *ceteris paribus*. This finding is somehow different with Kimura, Takahashi, & Hayakawa (2007)'s where they report that P/C exports are elastic to the change in service-link costs. Nonetheless, it is important to note that improvement in service-link cost would significantly affect the Indonesian P/C exports. In addition, this variable is often used as a proxy for trade cost. The coefficient estimate tells us that trade cost is important hindrance for Indonesian P/C exports. Therefore, it is imperative to take a deeper look on factors affecting trade costs. They could be institutional problems and also the quality of infrastructure facilitating trade.

Variable GDP reporter has negative value and not statistically significant. There are several explanations for this result. First, ratio of Indonesia's P/C exports to Indonesia's GDP is extremely small. In period 2000–2014, Indonesia's P/C exports accounts, on average, only 0.77% of Indonesia's GDP and 3% of Indonesia's exports. This finding is consistent with Aswicahyono, Basri, & Hill (2000) study where they argue that Indonesian automotive industry had small-export base. This implies that large domestic market does not ensure the country to have high value of P/C exports. Second, Indonesia in only one of many countries that participate in global automotive production network. Countries in that network are chosen by the principal to supply various type of P/C. Once the principal has decided which P/C that Indonesian industry supply, the size and rapid growth of Indonesian economy do not affect its P/C exports. The principal would use a country to produce a particular P/C based on country's ability to meet the standard imposed by the principal. More advance and very technical P/C are often exported from countries with advance technological ability.

Table 4. Determinant of Indonesia P/C exports.

Variables	Generalized least squares
lndist	−0.903243***
	(0.2065485)
lngdp_reporter	−0.1962248
	(0.7139492)
lngdp_partner	0.742748***
	(0.1281091)
endowment	0.3902609*
	(0.2288004)
lnlab_cost	0.7982632***
	(0.1455317)
paved_road	11.74666**
	(5.215364)
lnport	0.1308781
	(0.2963789)
constant	−6.088571
	(17.09563)
R2	0.37
Observations	450
Number of partners	30

Robust standard errors in parentheses, ***p < 0.01, **p < 0.05, *p < 0.1.
Source: Author's calculation.

Indonesian P/C exports are quite sensitive to its partners' economy. The coefficient estimate shows that Indonesian P/C exports are inelastic to its partners' GDP where 1% increase in partners' GDP, Indonesian P/C exports will 0.74% increase, *ceteris paribus*. Hence, Indonesian P/C exports are really sensitive to the fluctuation of its parters' economy.

Interesting finding comes from the coefficient estimate of *endowment*. As the GDP per capita describes the level of economic development, the ratio tells us about the relative economic development between Indonesia and its trading partners. The coefficient estimate has positive value and statistically significant. This result shows that Indonesia has higher P/C export to developing countries which have similar or lower level of economic development. This finding is consistent with the current trend where Indonesia now is reaching African markets. Developing countries have relatively the same quality of vehicle, so that these countries could trade P/C that also have similar quality.

Variable labor cost has positive sign and it is statistically significant. This result is also consistent with the hypothesis that labor cost positively affects the P/C exports. Industries producing and exporting high-value components have to employ skilled labor that costs higher wage. It is important to note that the increasing labor cost has to be followed by the increasing labor productivity. The implication of this result is that the availability of skilled in P/C industries become important prerequisite for Indonesia to have higher position in international automotive production network. Unfortunately, some key automotive players in Indonesia complain that nominal wage set by politicians is too high and do not reflect the labor productivity (interview with GAIKINDO representative). These politicians often make populist policies by promising higher nominal wage as strategy maintain their constituents' vote.

Infrastructure also has significant effect to Indonesian P/C exports. The coefficient estimate for variable *paved road* is positive and statistically significant, but is statistically insignificant for variable *lnport*. It is interesting to find that road infrastructure has bigger impact to Indonesia P/C export than port infrastructure. The paved road is important for the industry since the distribution of P/C from one production block to another in Indonesia and also the access to the seaport is through the paved road. The problem of this road infrastructure in Indonesia is that there is limited space for the industry vehicles where they have to share the infrastructure with non-industry vehicle. As the companies often operate near the city, the industry vehicles have to suffer heavy traffic. This is actually a serious issue for automotive companies where they rely on efficient distribution in their production process[9]. The traffic creates cost for the companies. In other words, the provision of road infrastructure could increase the exports by lowering the cost of production, especially in terms of transportation costs. This results are consistent with Ismail & Mahyideen (2015) and Olarreaga (2016) that report infrastructure to have positive impact to countries' export performance.

10 CONCLUSION

This paper explains that Indonesia is clearly lagging behind Thailand in terms of automotive trade. There are several indicators to support it. First, Indonesia has lower position in global ranking of car production than Thailand. Second, Indonesian automobile industries started to be export-oriented in 2014 while Thai automobile industries had become export-oriented before 2006. Third, Thailand had export values which are substantially larger than Indonesia. Last but not least, Thailand also had various P/C that are classified as great, which shows that Thai P/C had more trade potential than Indonesia.

This paper argues that industrial policies are the reasons why Indonesia and Thailand have different development path in developing its automotive industries. Thailand promoted *product champion* rather than *firm champion* as what Indonesia did with its national car project. The implementation of national car project with various discriminatory policies resulted distortion for the industries and uncertainty for foreign investors. As a result, Indonesia was not considered as a good place to invest and the foreign investors chose Thailand for the main production base. In addition, Thailand are better than Indonesia in attracting FDI through providing fiscal incentives. Indonesian government is criticized because its inability to provide fiscal incentives the industries need.

Can Indonesia catch-up with Thailand, especially in P/C trade? The answer is yes. The comparative study on the difference between Indonesia and Thailand industrial policies show that sound and export-oriented industrial policies are essential in improving industry's exports. In addition, the empirical study on determinants of Indonesian P/C exports also provides the important factors that are significantly affecting the P/C exports. The estimation suggests that improvement in both service-link costs and the quality of infrastructure are important prerequisites to improve Indonesian P/C exports. In addition, the availability of high skilled labor is also necessary for the industries in order to be able to produce more high value components. The existing problem in labor market is not only about the limited supply of skilled labor, but also high nominal wage. The later issue resulted from the politicians' interest to gain political capital by promising higher minimum wage. Therefore it is important to ensure that the labor cost that industries paid reflect the real labor productivity.

REFERENCES

Abbot, J.P. (2003). *Developmentalism and Dependency in Southeast Asia: the Case of the Automotive Industry.* London: Routledge Curzon.

Arndt, S. W., & Kierzkowski, H. (2001). *Fragmentation: New Production Patterns in the World Economy.* Oxford: Oxford University Press.

Aswicahyono, H., Basri, M.C., & Hill, H. (2000). How not to Industrialize? Indonesia's Automotive Industry. *Bulletin of Indonesian Economic Studies*, 209–241.

Balassa, B. (1965). Trade Liberalization and Revealed Comparative Advantage. *Manchester School of Economic and Social Studies*, 99–123.

---

9. Kobayashi (2014) reports Toyota is affected by the heavy traffic near Tanjung Priok port where the traffic disturbs the implementation of JIT production system. The location of Tanjung Priok within Jakarta where non-industry vehicle also operates around Tanjung Priok makes the traffic issue is.

Baldwin, R.E. (2017). *The Great Convergence: Information Technology and the New Globalization*. Cambridge: Belknap Press.

Branstetter, L. (2000). Vertical Keiretsu and Knowledge Spillovers in Japanese Manufacturing: An Empirical Assessment. *Journal of the Japanese and International Economies*, 73–104.

Busser, R. (2008). 'Detroit of East Asia'? Industrial Upgrading, Japanese Car Producers and the Development of the Automotive Industry in Thailand. *Asia Pacific Business Review*, 29–45.

Deardorff, A.V. (2001). International Provision of Trade Services, Trade, and Fragmentation. *Review of International Economics*, 233–248.

Fujita, M. (1998). Industrial Policies and Trade Liberalization: The Automotive Industry in Thailand and Malaysia. In K. Omura, *The Deepening Economic Interdependence in the APEC Region* (pp. 149–187). Singapore: APEC Study Center, Institute of Developing Economies.

Hale, C.D. (2001). Indonesia's National Car Project Revisited. *Asian Survey*, 629–645.

Ismail, N.W., & Mahyideen, J.M. (2015). The Impact of Infrastructure on Trade and Economic Growth in Selected Economies in Asia. *ADBI Working Paper 553*, 1–28.

Jones, R.W., & Kierzkowski, H. (1990). The Role of Services in Production and International Trade: A Theoretical Framework. In R.W. Jones, & A. O. Krueger (Eds.), *The political economy of international trade: Essays in honor of R.E. Baldwin.* (pp. 31–48). Oxford: Basil Blackwell.

Kaosa-ard, M.S. (1993). TNC Involvement in the Thai Auto Industry. *TDRI Quarterly Review*, 9–16.

Kimura, F., & Takahashi, Y. (2004). International Trade and FDI with Fragmentation: The Gravity Model Approach.

Kimura, F., Takahashi, Y., & Hayakawa, K. (2007). Fragmentation and Parts and Components Trade: Comparison Between East Asia and Europe. *North American Journal of Economics and Finance*, 23–40.

Kobayashi, H. (2014). Current State and Issues of the Automobile and Auto Parts Industries in ASEAN. In *Automobile and Auto Component Industries in ASEAN: Current State and Issues* (pp. 1–24). ERIA and Waseda University.

Kohpaiboon, A., & Yamashita, N. (2010). FTAs and the Supply Chain in the Thai Automotive Industry. In C. Findlay, *FTAs and GLobal Value Chains in East Asia* (pp. 321–362). Jakarta: ERIA.

Linbald, J.T. (2015). Foreign Direct Investment in Indonesia: Fifty Years of Discourse. *Bulletin of Indonesian Economic Studies*, 217–237.

Marklines. (2018, January 23). *Automotive Industry Portal: Marklines.* Retrieved from Thailand, Flash Report, Sales Volume, 2017: https://www.marklines.com/en/statistics/flash_sales/salesfig_thailand_2017

Natsuda, K., & Thoburn, J. (2012). Industrial Policy and the Development of the Automotive Industry in Thailand. *Journal of the Asia Pacific Economy*, 413–437.

Natsuda, K., Thoburn, J., & Otsuka, K. (2015). Dawn of Industrialization? The Indonesian Automotive Industry. *Bulletin of Indonesian Economic Studies*, 47–68.

Nitipathanapirak, R. (2017). Retrieved from http://www.sti.or.th/uploads/files/files/20170427%20Automotive%20Industry%20Situation%2C%20Master%20Plan%20(K_Rachanida).pdf

Olarreaga, M. (2016). Trade, Infrastructure, and Development. *ADBI Working Paper 626*, 1–15.

Pindyck, R.S., & Rubinfeld, D.L. (2013). *Microeconomics, 8th ed.* Boston: Pearson.

Shephard, B., & Soejachmoen, M. (2018). Why is Indonesia Left Behind? In L.Y. Ing, G.H. Hanson, & S.M. Indrawati, *The Indonesian Economy: Trade and Industrial Policies* (pp. 114–135). New York: Routledge.

Soejachmoen, M. P. (2016). Globalization of the Automotive Industry: Is Indonesia Missing out? *Asian Economic Papers*, 1–19.

Techakanont, K. (2011). Thailand Automotive Parts Industry. In M. Kagami, *Intermediate Goods Trade in East Asia: Economic Deepening Through FTAs/EPAs* (pp. 193–229). Bangkok: Bangkok Research Center, IDE-JETRO.

UNCTAD. (2008). *Globalization for Development: The International Trade Perspective.* New York: United Nations.

Verico, K. (2017). Indonesia towards 2030 and beyond: A Long-Run International Trade Foresight. *MPRA Paper No. 79645*.

APPENDIX

Table A1.   List of P/C.

SITC3	Description
6251	Tires, pneumatic, new, of a kind used on motor cars (including station wagons and racing cars)
6252	Tires, pneumatic, new, of a kind used on buses or lorries
62541	Tires, pneumatic, new, of a kind used on motorcycles and bicycles of a kind used on motorcycles
62591	Inner tubes
62593	Used pneumatic tire
66471	Safety glass, consisting of toughened (tempered) or laminated glass of toughened (tempered) glass
66472	Safety glass, consisting of toughened (tempered) or laminated glass of laminated glass
66481	Rear-view mirrors for vehicles
69915	Other mountings, fittings and similar articles suitable for motor vehicles
69941	Springs and leaves for springs, of iron or steel
74315	Compressors of a kind used in refrigerating equipment
7438	Parts for the pumps, compressors, fans and hoods of subgroups 743.1 and 743.4
7481	Transmission shafts (including camshafts and crankshafts) and cranks
74821	Bearing housings, incorporating ball- or roller bearings
74822	Bearing housings, not incorporating ball- or roller bearings; plain shaft bearings
7485	Flywheels and pulleys (including pulley blocks)
7486	Clutches and shaft couplings (including universal joints)
7489	Parts, n.e.s., for the articles of group 748
74443	Other jacks and hoists, hydraulic
74363	Oil or petrol filters for internal combustion engines
74364	Intake air filters for internal combustion engines
71651	Electric generating sets with compression-ignition internal combustion piston engines (diesel or semi-diesel engines)
7169	Parts, n.e.s., suitable for use solely or principally with the machines falling within group 716
77812	Electric accumulators (storage batteries)
77821	Filament lamps (other than flash bulbs, infrared and ultraviolet lamps and sealed-beam lamp units)
77823	Sealed-beam lamp units
77833	Parts of the equipment of heading 778.31
77834	Electrical lighting or signaling equipment (excluding articles of subgroup 778.2), windscreen wipers, defrosters and demisters, of a kind used for cycles or motor vehicles
77835	Parts of the equipment of heading 778.34
77313	Ignition wiring sets and other wiring sets of a kind used in vehicles, aircraft or ships
76211	Radio-broadcast receivers not capable of operating without an external source of power, of a kind used in motor vehicles (including apparatus capable of receiving radio-telephony or radio-telegraphy) incorporating sound-recording or reproducing apparatus
76212	Radio-broadcast receivers not capable of operating without an external source of power, of a kind used in motor vehicles (including apparatus capable of receiving radio-telephony or radio-telegraphy) not incorporating sound-recording or reproducing apparatus
76422	Loudspeakers, mounted in their enclosures
76423	Loudspeakers, not mounted in their enclosures
76425	Audio-frequency electric amplifiers
88571	Instrument panel clocks and clocks of a similar type, for vehicles, aircraft, spacecraft or vessels
71391	Parts, n.e.s, for the internal combustion piston engines of subgroups 713.2, 713.3 and 713.8, suitable for use solely or principally with spark-ignition internal combustion piston engines
71392	Parts, n.e.s, for the internal combustion piston engines of subgroups 713.2, 713.3 and 713.8, suitable for use solely or principally with compression-ignition internal combustion piston engines
78431	Bumpers, and parts thereof
78432	Other parts and accessories of bodies (including cabs)
78433	Brakes and servo-brakes and parts thereof
78434	Gearboxes

(Continued)

Table A1.　(*Continued*).

SITC3	Description
78435	Drive-axles with differential, whether or not provided with other transmission components
78436	Non-driving axles, and parts thereof
78439	Other parts and accessories
78535	Parts and accessories of motorcycles (including mopeds)
78531	Invalid carriages, whether or not motorized or otherwise mechanically propelled
78536	Parts and accessories of invalid carriages
78537	Parts and accessories of other vehicles of group 785
82112	Seats of a kind used for motor vehicles
62593	Used pneumatic tire
7841	Chassis fitted with engines, for the motor vehicles of groups 722, 781, 782 and 783
71321	Reciprocating piston engines of a cylinder capacity not exceeding 1,000 cc
71322	Reciprocating piston engines of a cylinder capacity exceeding 1,000 cc
71323	Compression-ignition engines (diesel or semi-diesel engines)
78421	Bodies (including cabs), for the motor vehicles of groups 781
78425	Bodies (including cabs), for the motor vehicles of groups 722, 782 and 783

Source: Soejachmoen (2016).

Table A2.　P/C (Code) classified as great for Indonesia & Thailand based on lnRCAI 2007 and CMSA 2000–2007.

Country	Product code (SITC Rev. 3)
Indonesia	6251, 62591, 74363, 74364, 77313, 77812, 77821, 78535
Thailand	6251, 6252, 62541, 62591, 62593, 66471, 66481, 69915, 69941, 71321, 71323, 71391, 74315, 74364, 76211, 76212, 76423, 77823, 77833, 77834, 78431, 78439, 78535, 78537

Table A3.　P/C (Code) classified as great for Indonesia & Thailand based on lnRCAI 2008 and CMSA 2008–2016.

Country	Product code (SITC Rev. 3)
Indonesia	6251, 62541, 71321, 74315, 74363, 74364, 76211, 76212, 76422, 77313, 77833, 77835, 78434, 78353, 78536
Thailand	6251, 6252, 62541, 62591, 69915, 71321, 71323, 74315, 7438, 7485, 76211, 76423, 76425, 77833, 78432, 78433, 78435, 78439

Table A4.　List of Indonesia's trading partners.

ASEAN	Non-ASEAN
Malaysia, Philippines, Singapore, Thailand, Vietnam	Algeria, Argentina, Australia, Austria, Belgium, Brazil, Canada, Chile, China, Costa Rica, Denmark, Finland, France, Germany, Hong Kong, India, Italy, Japan, South Korea, Mexico, Netherlands, Saudi Arabia, UAE, UK, USA

Constant Market Share Analysis (CMSA)

$$X_{ijwt1} - X_{ijwt0} = \sum m_{iwj\Delta t}.X_{ijwt0} + \left(m_{iwj\Delta t} - \sum m_{iwj\Delta t}\right)X_{ijwt0} + \left(X_{ijwt1} - X_{ijwt0} - m_{iwj\Delta t}X_{ijwt0}\right)$$

$\sum m_{iwj\Delta t}.X_{ijwt0}$ 　　　　　　 : *general factor*

$\left(m_{iwj\Delta t} - \sum m_{iwj\Delta t}\right)X_{ijwt0}$ 　 : *composition factor*

$\left(X_{ijwt1} - X_{ijwt0} - m_{iwj\Delta t}X_{ijwt0}\right)$ 　 : *comparative factor*

X_{ijwt1} is export of commodity j of country i at time t1, X_{ijwt0} is nilai export of commodity j of country i at time t0, Σm_{iwjDt} is the change in world's export, and $m_{iwj\Delta t}$ is the change in world's export of commodity j.

Business Innovation and Development in Emerging Economies – Trinugroho & Lau (Eds)
© *2019 Taylor & Francis Group, London, ISBN 978-1-138-35996-3*

Panel data regression and support vector regression for Indonesian private external debt analysis

Janice Diani & Zuherman Rustam
Universitas Indonesia, Indonesia

ABSTRACT: Indonesian corporations have been borrowing large sums of money from foreign investors in the past decade, such that private debt ratio has reached 49% of Indonesia's total external debt by the end of 2017. This act of borrowing might improve the borrowing firms' performance which leads to increase in profit, but in other hand it might result on debt value expansion, due to the exchange rate depreciation trend in Indonesia. This paper employs Support Vector Regression, a machine-learning method, to study the relationship between factors that might affect corporate performance, and compares the results with that of the conventional panel data regression method. The study was done using data from annual financial statements of 189 firms in Indonesia during 2011–2017.

It is shown that the machine-learning approach discussed in this study gave better accuracy than the previously employed panel data regression method. Both methods generally showed that balance-sheet effect is more dominant in Indonesian corporations, and it is recommended for companies to minimize their foreign debts and imported purchases, and if possible, export more of their products.

Keywords: corporate debt, depreciation, exchange rate, external debt, machine learning, panel data regression, support vector regression

1 INTRODUCTION

Indonesia's external debt has been rising in the past decade, with 10.03% growth rate year on year. At first the debt was dominated by public debt, but the private debt ratio started increasing in 2012, and it has reached 49% of the total external debt by the end of 2017. (Indonesian Ministry of Finance, 2017).

The large amount of private debt, in one hand, might improve the borrowing companies' performance and increase their profit. This is caused by two main causes: firstly, it is expected from a company to produce more when it has more capital. Secondly, according to the Mundell-Fleming model, the depreciation trend increases the company's competitiveness towards foreign competitors, especially for exporter companies. This is known as competitiveness effect (Bleakley & Cowan, 2008).

On the other side, it also might expose said corporations to the latent risk of depreciation. The nominal value of external debt would inflate when depreciation happens (Cespedes, Chang, & Velasco, 2004). This is known as balance-sheet effect (Krugman, 1999). When depreciation occurs and the amount of debt gets too big, there is risk that the owing company might not be able to repay it. This phenomenon is known as currency mismatch, and it is very dangerous since it might cause bankruptcy to the company, and even economic contraction and major unemployment when it happens to a big number of companies at the same time.

The main objective of this study is to find out which between the balance-sheet and competitiveness effect is more dominant in Indonesian corporations, and calculate the threshold to which extent corporate external debt is allowed, so that the currency mismatch could be avoided and the best policy that maximizes the company's profit could be made.

The Central Bank of Indonesia have been analyzing this problem traditionally using multivariate statistics method, namely panel data regression (Central Bank of Indonesia, 2006, 2009, and 2011). The results didn't give satisfactory accuracy since the data was a combination of time series and cross-section data, which is why this study employs Support Vector Regression, a machine-learning method, to study the relationship between factors that might affect corporate performance, and compares the results with that of the conventional panel data regression method.

2 METHODOLOGY

The Central Bank of Indonesia (2011) derived the factors that affect a firm's performance from their balance sheet equation.

At time 0,

$$A = L + W_0 \tag{1}$$

where A represents corporate asset, L represents liabilities, and W represents net worth.

Firm's asset and liabilities might be in domestic or foreign currencies, so Equation (1) could be written as

$$A_d + e_0 A_f = L_d + e_0 L_f + W_0 \tag{2}$$

where

A_d = asset in domestic currency (Rupiah);
A_f = asset in foreign currencies;
L_d = liabilities in domestic currency (Rupiah);
L_f = liabilities in foreign currencies;
e_0 = exchange rate (to USD) in time 0.

Therefore, in time 1, Equation (2) could also be written as

$$(1+r_d^A)A_d + (1+r_f^A)\cdot e_1 A_f = (1+r_d^L)L_d + (1+r_f^L)\cdot e_1 L_f + W_1 \tag{3}$$

$$W_1 = (1+r_d^A)A_d + (1+r_f^A)\cdot e_1 A_f - (1+r_d^L)L_d - (1+r_f^L)\cdot e_1 L_f \tag{4}$$

where r_d^A and r_d^L are interest rates for domestic asset and liabilities, while r_f^A and r_f^L are interest rates for foreign asset and liabilities.

Suppose $\delta_d = r_d^A - r_d^L$ and $\delta_f = r_f^A - r_f^L$, then

$$
\begin{aligned}
W_1 &= (1+r_d^A)(A_d - L_d) + (1+r_f^L)\cdot e_1(A_f - L_f) + \delta_d A_d + \delta_f e_1 A_f \\
&= (1+r_d^A)(A_d - L_d) + (1+r_d^L)e_0(A_f - L_f) - (1+r_d^L)e_0(A-L) \\
&\quad + (1+r_f^L)\cdot e_1(A_f - L_f) + \delta_d A_d + \delta_f e_1 A_f \\
&= (1+r_d^L)W_0 - (1+r_d^L)e_0(A_f - L_f) + (1+r_f^L)e_f(A_f - L_f) + \delta_d L A_d + \delta_f e_1 A_f \\
&= (1+r_d^L)W_0 + \left[(1+r_f^L)e_1 - (1+r_d^L)e_0\right](A_f - L_f) - \delta_d A_d + \delta_f e_1 A_f
\end{aligned}
\tag{5}
$$

Based on Equation (5), we can build an empiric model:

$$EQ_{i,t} = \alpha_0 + \alpha_1\left[(A_f - L_f)_{i,t-1}\cdot \Delta e_t\right] + \alpha_2(A_f - L_f)_{i,t-1} + \alpha_3 \Delta e_t + \alpha_4 L_{i,t-1}^d + \alpha_5 rc_t \tag{6}$$

If net foreign asset or NFA is defined as $A_f - L_f$, then we can write Equation (6) as

$$EQ_{it} = \alpha_0 + \alpha_1\left[(NFA)_{i,t-1} \cdot \Delta e_t\right] + \alpha_2(NFA)_{i,t-1} + \alpha_3\Delta e_t + \alpha_4 L^d_{i,t-1} + \alpha_5 rc_t \qquad (7)$$

where

EQ_{it} = equity to asset ratio of company i in time t;
Δe_t = Rupiah exchange rate (to USD);
$NFA_{i,t-1}$ = NFA to asset ratio;
$L^d_{i,t-1}$ = domestic liabilities to asset ratio;
rc = credit interest rate.

The model in Equation (7) is expected to be able to measure corporate performance, which is represented in equity to asset ratio (EQ). In words, a firm's equity is determined by these variables: NFA, exchange rate fluctuation, amount of domestic debt, credit interest rate, and the interaction between NFA and exchange rate.

The main objective in this study is to find the value of regression coefficients (α_1, α_2, α_3, α_4, and α_5), which measure impacts of each factors towards the firm's equity, and later can be used to forecast the firm's equity in the following year, if the values of dependent variables are known. To find those coefficients, in this paper we used two methods, namely panel data regression and Support Vector Regression.

After obtaining those coefficients, we could also find to which extent a firm could have liabilities in foreign currencies without causing a decrease to its equity to asset ratio, by finding the first partial derivatives of Equation (7) with respect to the exchange rate difference (der):

$$\frac{\partial EQ}{\partial \Delta e} = \alpha_1\left[(NFA)_{i,t-1}\right] + \alpha_3 \qquad (8)$$

A firm's NFA threshold is the NFA value such that $\alpha_1\left[(NFA)_{i,t-1}\right] + \alpha_3 = 0$, or simply $-\dfrac{\alpha_3}{\alpha_1}$.

3   PANEL DATA REGRESSION

Panel data is defined as a dataset in which the behavior of entities is observed across time. It is also known as a combination between cross section and time series data. Panel data regression model is similar with the ordinary least-squares multiple linear regression model (Woolridge, 1999), which has the form

$$y = \beta_0 + \beta_1 x_1 + \beta_2 x_2 + \cdots + \beta_k x_k + \epsilon \qquad (9)$$

where $x_1, x_2, \ldots, x_k$ are regressor variables, y is the response variable, β_0 is known as intercept, β_j is known as slope, and ϵ is known as error, which is difference between observed value y and predicted value $\beta_0 + \beta_1 x_1 + \beta_2 x_2 + \cdots + \beta_k x_k$.

To estimate the value of unknown parameters $\beta_0, \beta_1, \ldots, \beta_k$, we need to minimize the least-squares function

$$S(\beta_0, \beta_1, \ldots, \beta_k) = \sum_{i=1}^{n} \epsilon^2 = \sum_{i=1}^{n}\left(y_i - \beta_0 - \sum_{j=1}^{k}\beta_j x_{ij}\right)^2 \qquad (10)$$

subject to

$$\frac{\partial S}{\partial \beta_0}\bigg|_{\hat{\beta}_0, \hat{\beta}_1, \ldots, \hat{\beta}_k} = -2\sum_{i=1}^{n}\left(y_i - \hat{\beta}_0 - \sum_{j=1}^{k}\hat{\beta}_j x_{ij}\right) = 0 \qquad (11.a)$$

and

$$\frac{\partial S}{\partial \beta_j}\bigg|_{\hat{\beta}_0, \hat{\beta}_1, \ldots, \hat{\beta}_k} = -2\sum_{i=1}^{n}\left(y_i - \hat{\beta}_0 - \sum_{j=1}^{k}\hat{\beta}_j x_{ij}\right)x_{ij} = 0, \quad j = 1, 2, \ldots, k \quad (11.b)$$

The least-squares regression is performed over some major assumptions:

1. The relationship between the response and the regressors is linear, at least approximately.
2. The error term ϵ has zero mean and constant variance σ^2.
3. The errors are uncorrelated and normally distributed.

Gross violations of the assumptions may yield an unstable model in the sense that a different sample could lead to a totally different model with opposite conclusions (Montgomery, 2012).

4 SUPPORT VECTOR REGRESSION

Support Vector Regression was found by Vapnik (1998) and is used to address target variables with real values. In contrast to the squared loss function in ordinary least squares regression, ϵ-Support Vector Regression (ϵ-SVR) uses ϵ-insensitive loss function, in which errors smaller than ϵ will be omitted. It has the form

$$\left|y - f(x)\right|_\epsilon := \max\left\{0, \left|y - f(x)\right| - \epsilon\right\} \quad (12)$$

The mathematical model to SVR is

$$f(x) = \mathbf{w} \cdot \mathbf{x} + \mathbf{b} \quad (13)$$

We need to find w (weight) and b (bias) that are solutions to

$$\min_{w,b}\frac{1}{2}\left\|\omega^2\right\| + C\sum_{i=1}^{n}\left(\xi_i^+, \xi_i^-\right)$$

$$\text{subject to } y_i - \omega \cdot x_i - b \le \epsilon + \xi_i^+$$

$$\omega \cdot x_i + b - y_i \le \epsilon + \xi_i^-$$
$$\xi_i^+, \xi_i^- \ge 0 \quad \forall i. \quad (14)$$

5 DATA

This study uses data from following sources:

Table 1. Sources of data.

Data	Source
Financial Statements from 189 nonfinancial firms that went public on IDX, from 2011 to 2017.	Indonesia Stock Exchange
Indonesia External Debt Statistics	Indonesian Ministry of Finance
Exchange Rates and Interest Rates	Central Bank of Indonesia

6 RESULTS

From estimating the parameters for the model in Equation (7) with both methods, we obtained

Table 2. Panel data and support vector regression coefficients.

Variable	Panel coefficient	SVR coefficient
Interaction between NFA and exchange rate difference (NFA × der)	0.0203029	0.652103159
NFA	0.6136893	0.793951493
Difference in exchange rates (der)	0.0701467	0.023527697
Domestic liabilities ratio to asset (Ld)	−0.6050588	−0.845872084
Credit interest rate	−0.2196315	−0.027195784
Intercept (panel data regression)/ Bias (SVR)	0.7845655	0.877221492

Therefore, we can formulate the equation to predict a firm's equity to asset ratio in time t as

$$EQ_t = 0.78456550 + 0.0203(NFA_{t-1} \times der) + 0.61369(NFA_{t-1})$$
$$+ 0.07015(der) - 0.60506(Ld_{t-1}) - 0.21963(cr_t) \tag{15}$$

for the model estimated with panel data regression, and

$$EQ_t = 0.877221 + 0.6521(NFA_{t-1} \times der) + 0.79395(NFA_{t-1})$$
$$+ 0.02353(der) - 0.84587(Ld_{t-1}) - 0.0272(cr_t) \tag{16}$$

for the model estimated with Support Vector Regression.

Firstly, we note from Table 2 that all the coefficients obtained from panel data regression have the same signs with their SVR counterpart. This means both methods generally gave the same information about the relationship between each regressor variables and the response variable. Positive coefficients indicate positive correlation between the regressor variable and the response variable, while negative coefficients indicate negative correlation between the regressor variable and the response variable.

From the results in Table 2, we conclude that a firm's NFA has a positive linear relationship with its equity, therefore the more assets in foreign currency they own in year t-1, the more equity they might have in year t. Conversely, more foreign liabilities they had in year t-1 might cause bigger decrease to their equity in year t.

Difference in exchange rates also has positive correlation with firm's equity. When the domestic currency in which the firm operates is appreciated, the firm's equity will increase, and when the currency is depreciated, its equity will decrease. This difference in exchange rates has bigger impacts to companies that perform foreign trade in large scale than to those that don't. When depreciation happens, exporters might gain more profit, since the value of their goods in the domestic currency will increase. On the other side, in depreciation times importers will spend more in domestic currency to buy the same amount of goods than they have to spend in normal times.

Interaction between NFA and difference in exchange rates (NFA × der) also has positive correlation with firm's equity. This proves that balance sheet effect is more dominant than competitiveness effect to the observed firms when depreciation occurs. In other words, when depreciation occurs, the effect of debt value expansion is bigger than the increase in profit it causes, such that the firm's equity will decrease.

A firm's domestic liability has negative correlation with its equity. A bigger amount of debt in year t-1 will cause bigger decrease in its equity in year t.

Credit interest rate also has negative correlation with firm's equity, especially for firms with domestic liabilities. This is because when credit rate increases, the amount of debt they need to repay expands. The increase of credit interest rate might also cause firms to opt for loans in foreign currencies, since foreign investors offer much lower interest rates than domestic credit rate.

To evaluate the accuracy of both methods, we predict each observed firms' equity to asset ratio in 2017 using Equation (15) for panel data regression and Equation (16) for SVR, and compare it to their actual equity to asset ratio in their 2017 financial statements. To compare those methods, we calculate their RMSE (root mean square error), with the formula

$$RMSE = \sqrt{\frac{1}{N} \sum_{i=1}^{n} (y_i - \hat{y}_i)^2}$$

where y_i is the actual equity to asset ratio of firm i, $\hat{y}_i$ is the predicted equity to asset ratio of firm i, and N is the number of observed firms.

The prediction obtained from panel data regression has the RMSE of 11.68%, while that of the SVR has the RMSE of 9.77%. Smaller RMSE value means the prediction obtained from the SVR model fits better than the panel regression model to the actual data.

From Equation (8) we also find the NFA threshold, which is −3.5% from the panel data regression model, and −3.61% from the SVR model. In other words, this threshold represents the maximum amount of debt a firm could have without causing loss in their equity. If a firm's foreign liability to asset ratio is more than 3.61%, their equity will decrease when depreciation happens. The observed firms in this study have the average NFA to asset ratio of −7.89%, which means they are exposed to the risk of equity loss when depreciation occurs.

We also separated the samples into groups according to their exporter and importer status, and then did the same estimation process with SVR as explained above. A firm is labeled as exporter if its export amount is more than 25% of its revenue, and labeled as non-exporter group otherwise, while a firm is labeled as an importer if it mentions import fee in its financial statement, and labeled as a non-importer otherwise.

We obtained

Table 3. Estimation on groups using support vector regression.

Status	NFA × der	NFA	der	Domestic liabilities	Credit rate	Bias	NFA Threshold
Exporter	−0.1137	0.7842	−0.0085	−0.8270	−0.0492	0.8163	−7.4%
Non-Exporter	0.6415	0.7867	0.0435	−0.8491	−0.015	0.8746	−6.7%
Importer	0.7405	0.7713	0.063	−0.8568	−0.0072	0.8891	−5.5%
Non-Importer	0.2224	0.8061	0.0124	−0.8441	−0.0078	0.8743	−5.6%

From Table 3, we can conclude that all groups have positive NFA coefficient, which indicates positive correlation between NFA and equity. It means more foreign debt they have in the previous year will cause further decrease in equity.

Difference in exchange rates has different effects on each group. When depreciation occurs, exporters might profit more, while other groups might experience loss. This is caused by the competitiveness effect. Exporters has more advantages than their competitors, since they become able to sell their product in lower prices, and the domestic value of their revenue also expands due to the depreciation. The other groups might experience decrease in equity when depreciation occurs. The importer group (which has the largest coefficient of all) might have the biggest loss out of all the groups, since their import is mainly done to purchase production factors. When depreciation happens, the domestic value of their purchase will expand, which leads to the increase on operational expenses and decrease on revenue.

It is also shown that interaction between NFA and exchange rate difference has positive correlation with equity on every observed group, except the exporters group. This proves the dominance of balance-sheet effect over competitiveness effect, which means when depreciation occurs, larger foreign debt to asset ratio will cause further decrease on equity to asset ratio. On the other side, exporters gain more when depreciation happens. We can say that competitiveness effect is more dominant than balance-sheet effect in this group. When depreciation occurs, the loss caused by debt value expansion could be balanced by the increase on revenue.

We can also note that exporters can have more debt in foreign currencies without causing decrease in equity to asset ratio compared to the non-exporters, while importers have the smallest amount of foreign debt allowance out of all groups. Importers should be very careful in managing their foreign debt level.

7 CONCLUSION

From this study, we can conclude that Support Vector Regression gives a better-fitting model than the conventionally used panel data regression, although both model generally still gives the same information, that the balance sheet effect is more dominant than competitiveness effect in Indonesian corporations when depreciation occurs. We also find that when depreciation occurs, exporter companies get more advantages than non-exporters, while importers experiences more loss than non-importers.

Therefore, the recommendation we offer is for the firms to carefully manage their foreign debt to asset ratio and minimize their import purchases so that the risk of equity loss and currency mismatch could be avoided, and to increase their exports if possible.

ACKNOWLEDGEMENT

The authors would like to thank the Economic and Monetary Policy Group of the Central Bank of Indonesia, for the guidance, information and references given to the authors while working on this study.

REFERENCES

Biro Riset Ekonomi Bank Indonesia (2006). *Balance-Sheet Effects of Exchange Rate Depreciation: Evidence from Non-Financial Firms in Indonesia*. Working Paper WP/06/2006, Bank Indonesia.
Biro Riset Ekonomi Bank Indonesia (2009). *Balance-Sheet Effects dari Depresiasi Nilai Tukar terhadap Perusahaan-Perusahaan Go Public di Indonesia*. Working Paper WP/04/2010, Bank Indonesia.
Biro Riset Ekonomi Bank Indonesia (2011). Analisa Perilaku Pembiayaan Asing dan Dampaknya Terhadap Ketahanan Perusahaan. Working Paper WP/12/2011, Bank Indonesia.
Bishop, C.M. (2006). Pattern Recognition and Machine Learning. UK: Springer.
Bleakley, H., Cowan, K. (2008). *Corporate Dollar Debt and Depreciations: Much Ado About Nothing?*. The Review of Economics and Statistics, MIT Press, vol. 90(4), pages 612–626, November.
Cespedes, L.F., Chang, R., Velasco, A. (2004): *Balance Sheets and Exchange Rate Policy*. American Economic Review vol. 94.
Cristianini, N., Shawe-Taylor, J. (2000): *An Introduction to Support Vector Machines and other kernel-based learning methods*. Cambridge University Press.
Kementerian Keuangan Republik Indonesia (2018): Statistik Utang Luar Negeri Indonesia. Edisi Maret 2018, v.02.
Krugman, P. (1999). *Balance Sheets, the Transfer Problem, and Financial Crises*. International Finance and Financial Crises. Springer, Dordrecht pp. 31–55.
Montgomery, D.C., Peck, E.A., Vining, G.G. (2012). Introduction to Linear Regression Analysis. Wiley.
Woolridge, J.M. (1999): Econometric Analysis of Cross Section and Panel Data. MIT Press.

Business Innovation and Development in Emerging Economies – Trinugroho & Lau (Eds)
© *2019 Taylor & Francis Group, London, ISBN 978-1-138-35996-3*

Determinant factors of external debt in ASEAN-8, 2005–2016

A. Ariani & M. Cahyadin
Faculty of Economics and Business UNS, Surakarta, Indonesia

ABSTRACT: In 2005–2016 the external debt of ASEAN-8 increased. It indicates that ASEAN-8 needs more money for the domestic economy. ASEAN-8 covers Indonesia, Lao PDR, Cambodia, Myanmar, Malaysia, Thailand, Vietnam, and Philippines. This research examines the impact of gross domestic product (GDP), export, reserved asset, corruption perception index (CPI), and budget deficit on external debt in ASEAN-8 in 2005–2016. It refers to Saleh (2008), Tarsilohadi (2005), Ouyang and Rajan (2013), Batubara and Saskara (2005), Cooray, Dzhumashev and Schneider (2017), and Qian and Steiner (2017). This research uses the *Fixed Effect Model* (FEM).

This research concludes that: GDP has a positive impact and significance on external debt; export has a negative impact and significance on external debt; CPI has a negative impact on external debt; while reserved asset and budget deficit have no significance on external debt. Based on this result governments of ASEAN-8 can use export and CPI to control external debt levels. In addition, the governments can also increase GDP to obtain the targeted value of external debt. Meanwhile, the government needs to manage the reserved asset and budget deficit for the domestic economy.

Keywords: External debt, ASEAN-8, FEM
JEL classification: F30, F34, H62, H63

1 INTRODUCTION

The external debt of ASEAN-8 increased in 2005–2016. In 2012–2016 the value of the external debt continued to increase becoming the highest level at the end of the research period. It requires deeper explanation. The development of the external debt in ASEAN-8 in 2005–2016 is displayed in Figure 1.

Figure 1 shows that the value of Indonesia's external debt in 2005 and 2016 was US$142.12 billion and US$ 316.43 billion, respectively. This means that Indonesia's external debt has grown about 122.65% over that time. Lao PDR's external debt in 2005 and 2016 was US$3.28 billion and US$14.16 billion respectively, with a growth rate of 331.71%.

In 2007 and 2016 the external debt in Cambodia was US$2.064 billion and US$10.23 billion, respectively (growth rate about 395.64%). In addition, Myanmar has the highest value of external debt in 2010 of about US$8.22 billion. Malaysia has a minimum and maximum external debt in 2005 and 2016 of about US$64.911 and US$200.364 billion, respectively (growth rate about 208.68%).

In 2005 and 2013 the external debt in Thailand was US$58.467 billion and US$137.353 billion respectively (growth rate about 134.924%). Vietnam has a minimum and maximum external debt in 2005 and 2016 of about US$18.53 billion and US$86.952 billion, respectively (growth rate about 369.25%). Finally, Philippines has a minimum and maximum external debt in 2009 and 2015 of about US$55.98 billion and US$80.62 billion respectively (growth rate about 44.016%).

Figure 2 describes the value of GDP, export, reserved asset, corruption Perception Index (CPI/CI), and budget deficit in ASEAN-8 in the year range of 2005–2016. Based on the

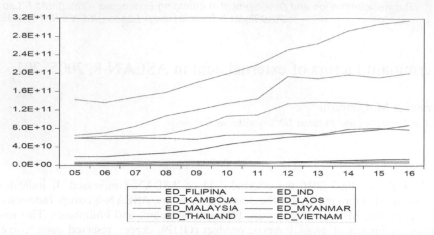

Figure 1. The development of external debt in ASEAN-8, 2005–2016 (US$).
Source: The World Bank.

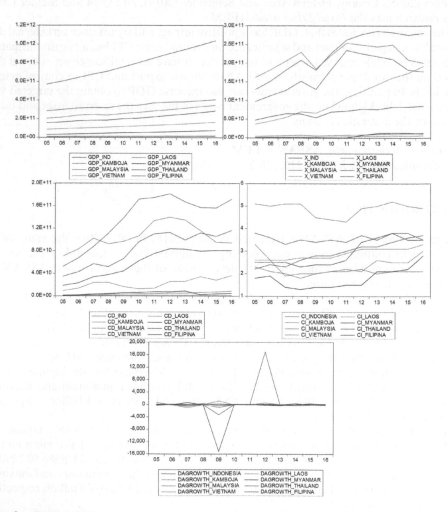

Figure 2. GDP, export, reserved asset (CD), corruption perception index (CPI/CI), and budget deficit (DA) in ASEAN-8, 2005–2016.
Source: The World Bank; Transparency International; www.countryeconomy.com

figure we can calculate the growth rate of variables in each of the countries. The growth rate of GDP, export, reserved asset, corruption perception index, and budget deficit in Indonesia between 2005 and 2016 were 81.67%, 82.65%, 235.06%, 68.182% and −102.82% respectively. Meanwhile, at the same time, the growth rate of variables in Lao PDR were 128.99%, 500.82%, 184.59%, −9.09% and 39.46% respectively.

The growth rate of GDP, export, reserved asset, corruption perception index, and budget deficit in Cambodia between 2005 and 2016 were 108.86%, 204.16%, 664.32%, −8.69% and −213.66% respectively. At the same time, the growth rate of variables in Myanmar were 156.61%, 51,462.26%, 449.37%, 55.56% and −102.92% respectively. Meanwhile, the growth rate of variables in Malaysia were 67.89%, 23.82%, 34.09%, −3.92% and 130.51% respectively. In addition, the growth rate of variables in Thailand were 43.21%, 116.56%, 229.85%, −7.89% and −401.18% respectively. Furthermore, the growth rate of variables in Vietnam were 92.26%, 423.49%, 303.59%, 26.92% and −101.23% respectively. Finally, the growth rate of variables in Philippines were 81.34%, 79.305%, 336.63%, 40% and 369.91% respectively.

Some studies identified that GDP or economic growth, export, reserved asset, corruption perception index and budget deficit have impacted on external debt (see Saleh, 2008; Tarsilohadi, 2005; Ouyang & Rajan, 2013; Batubara & Saskara, 2005; Cooray, Dzhumashev & Schneider, 2017; and Qian & Steiner, 2017). This research will focus on external debt in ASEAN-8 in the period of 2005–2016. Therefore, this research objectives are: a) to examine the impact of GDP on external debt; b) to examine the impact of export on external debt; c) to examine the impact of reserved asset on external debt; d) to examine the impact of corruption perception index on external debt; and e) to examine the impact of budget deficit on external debt.

This research is directed to support external debt management in ASEAN-8. The ability of ASEAN governments to control and allocate external debt in productive sectors is expected to reduce corruption and budget deficit. Furthermore, the ASEAN government should drive economic growth, export and reserved asset to control external debt.

2 LITERATURE REVIEW

Soeradi (2016) describes that the term foreign loan (or debt)—by government—can be defined as government revenue both in goods and services from foreign creditors that it must pay with specific requirements. Furthermore, Tribroto (2001) has identified two important aspects of foreign debt. The first is a material aspect that describes foreign debt as foreign capital for the domestic economy. The second is a formal aspect that explains foreign debt as investment for supporting economic growth. It means that foreign debt can be used as development financing.

Todaro (1994) explains two theories of government debt. The first theory is the Neoclassical Theorem that focuses on the role of debt to cover the saving gap. The second is a current theory that it focuses on the role of debt to cover the current account deficit (well known as gap filling). Furthermore, Saleh (2008), and Ouyang & Rajan (2013) suggest that government needs to manage budget properly (and discipline) and stabilize domestic currency. Government and private sectors decide to take foreign debt for stimulating the domestic economy and industry. The total of government and private debt from foreign countries can be called the total external debt.

The external debt can be determined by economic growth (or GDP), export, fiscal deficit, corruption and reserved asset. It refers to Amoateng & Amoako-Adu (2006); Ogunmuyiwa (2011); Lau & Kon (2014); Irwin (2015); Siddique, Selvanathan, & Selvanathan (2015); Cooray, Dzhumashev, & Schneider (2017); Kim, Ha & Kim (2017); Qian & Steiner (2017); Al Kharusi & Ada (2018). The researchers identify and analyze the relationship (impact) between economic growth (or GDP), export, fiscal deficit, corruption index and reserved asset on foreign (external) debt. Based on their empirical studies, governments need to manage and control external debt. In addition, the lowest level of corruption (the highest value of corruption index) is expected to create trust of creditors. Furthermore, the governments and central banks should use reserved asset properly to decide the level of external debt.

Anderson (2013) identified that foreign debt will create cost for fiscal. The governments need to prudent when using foreign debt. In addition CÂLEA argues that:

> It is very important for a country to determine and agree on the optimal level of indebtness which the economy can handle. This is precisely the reason why the external debt has to be contracted in close agreement with the specific needs of the economy and the loans to be proportional with the economy's ability to reimburse them. (2013, 32)

Meanwhile, Hur & Kondo (2013), Espinoza (2014), and Koc (2014) described the role of the reserved asset for domestic economy, foreign creditor and government (as well as the central bank).

3 METHOD

This research uses secondary data from the World Bank (data.worldbank.org), Transparency International (www.transparency.org) and www.countryeconomy.com. These data cover external debt, gross domestic product (GDP), export, reserved asset, corruption perception index (CPI) and the budget deficit of ASEAN-8 in the period 2005–2016. ASEAN-8 member countries are: Indonesia, Lao PDR, Cambodia, Myanmar, Malaysia, Thailand, Vietnam and Philippines. The number of research observation is about 92.

The research method uses Panel Data with Fixed Effect Model (FEM). It is based on the Chow and Hausman Tests (see Gujarati & Porter, 2009). In this research, the equation is as below:

$$ED = f (GDP, X, CD, CPI, GDA) \tag{1}$$

The equation 1 can be formulated in the logarithm model of data panel is as below:

$$LogED_{it} = \alpha + \beta_1 LogGDP_{it} + \beta_2 LogX_{it} + \beta_3 LogCD_{it} + \beta_4 CPI_{it} + \beta_5 GDA_{it} + \varepsilon_{it} \tag{2}$$

ED is external debt in US$, GDP is gross domestic product in US$, X is value of export in US$, CD is reserved asset in US$, CPI is corruption perception index, and GDA is growth of budget deficit in%. The α is a constant or an intercept, $\beta_{(1, 2, 3, 4, 5)}$ is a coefficient or slope, i is ASEAN-8 (Indonesia, Lao PDR, Cambodia, Myanmar, Malaysia, Thailand, Vietnam and Philippines), t is research period 2005–2016, Log is logarithm, and ε is *error term*.

4 RESULT AND DISCUSSION

Fixed Effect Model (FEM) was chosen as the best panel data model in this research. It is based on the Chow and Hausman Tests as displayed in Tables 1 and 2. The probability value from Table 1 and Table 2 are less than 1%. This means that we can employ FEM to examine the impact of GDP, export, reserved asset, corruption perception index and budget deficit on external debt in ASEAN-8.

Table 3 shows the result of fixed effect model on determinant factors of external debt in ASEAN-8. C has a negative impact and significance on external debt in ASEAN-8 and other independents are constant. The result indicates that the value of external debt tends to be negative while all independent variables are constant.

GDP has positive impact and significance on external debt in ASEAN-8. This means that a higher value of GDP will promote a higher value of external debt. This result becomes a driving factor for creditor countries to allocate debt. The allocation should be focused for developing countries with progressive economic growth. The empirical research from

Table 1.   Chow test result—Likelihood ratio.

Effect test	Statistic	Prob. value
Cross-section F	46.486746	0.0000
Cross-section Chi-square	150.233653	0.0000

Source: Secondary data (processed).

Table 2.   Hausman test result.

Correlayed random effects – Hausman test			
Test cross-section random effects			
Test summary	Chi-Sq. statistic	Chi-Sq. d.f.	Prob. value
Cross-section random	62.714553	5	0.0000

Source: Secondary data (processed).

Table 3.   Fixed effect method result.

Variable	Coeff.	Std. error	t-*statistic*	Prob. value
C	−26.04639	4.170140	−6.245928*	0.0000
LOGGDP?	2.224797	0.232494	9.569270*	0.0000
LOGX?	−0.172738	0.027027	−6.380617*	0.0000
LOGCD?	−0.055280	0.075466	−0.732523	0.4660
CPI?	−0.174102	0.070113	−2.483156**	0.0151
GDA?	1.16E-05	9.27E-06	1.247253	0.2160
Fixed Effects (Cross)				
_INDONESIA—C	−2.436888			
_LAOS—C	3.488266			
_KAMBOJA—C	2.156461			
_MYANMAR—C	−1.078329			
_MALAYSIA—C	−0.108421			
_THAILAND—C	−1.230799			
_VIETNAM—C	−0.092517			
_FILIPINA—C	−0.824793			
R-squared	0.983262			
Adjusted R-squared	0.980719			
F-statistic	386.7273*			0.000000

Source: Secondary data (processed).
Note: a. Dependent Variable = LogED,
b. *α = 1% and **α = 5%.

Saleh (2008) confirms that GDP has a significant impact on external debt. Meanwhile, Tarsilohadi (2005) describes that GDP does not have an impact on foreign debt.

Export has negative impact and significance on external debt in ASEAN-8. This means that a higher value of export will promote a lower value of external debt. The result can be used by ASEAN governments to support export orientation and competitiveness. Furthermore, the governments should focus on commodities competitiveness in the global market. This research was supported by Saleh (2008), Ouyang & Rajan (2013) and Batubara & Saskara (2005).

Reserved asset has no significant impact on external debt in ASEAN-8. This means that the value of external debt cannot be stimulated by reserved asset. This result indicates that governments of ASEAN tend to make an agreement on foreign debt while the reserved asset is available and adequate. Furthermore, the reserved asset should be directed to become buffer stock for price stabilization and import.

The corruption perception index (CPI) has a negative impact and significance on external debt in ASEAN-8. This means that a higher value of CPI will stimulate a lower value of external debt. Thus, the ASEAN governments can promote governance of the external debt. A clean and transparent external debt use and allocation should be conducted by the governments. The research result was in accordance with Cooray, Dzhumashev & Schneider (2017) and Henri (2018).

Budget deficit has no significant impact on external debt in ASEAN-8. This means that a decision on external debt does not always depend on the (growth) budget deficit. The result indicates that budgets do not become an urgent reason in external debt decision making. Meanwhile, the governments of ASEAN should keep the level of budget deficit prudently.

The probability value of the F test shows that the all independent variables have a significant impact on external debt. In addition, the value of adjusted R^2 is about 0.980719 or 98.0719%. This means that about 98.0719% of the variation of the dependent variable is determined by the variation of the independent variables. Conversely, about 1.9281% of the variation of the dependent variable is determined by other variables outside the FEM.

5 CONCLUSION AND POLICY RECOMMENDATION

External debt becomes a driving factor for the domestic economy of ASEAN-8. There are five results from this research that explain external debt. Firstly, GDP has a positive impact and significance on external debt in ASEAN-8. Secondly, export has a negative impact and significance on external debt. Thirdly, reserved asset has no significant impact on external debt. Fourthly, the corruption perception index has a negative and significance on external debt. Finally, growth of the budget deficit has no significant impact on external debt.

The governments of ASEAN-8 can manage and control external debt for the domestic economy. In addition, industry and the government should collaborate and cooperate to stimulate and increase export in the global market. Furthermore, the governments of ASEAN-8 should manage and use fiscal and debt in a good governance framework. The creditors will use CPI as an indicator of debt governance in ASEAN-8.

REFERENCES

Amoateng, K. & Amoako-Adu, B. (2006). Economic growth, export and external debt causality: the case of African countries. *Applied Economics, 28*(1), 21–27.

Anderson, J.E. (2013). Government debt and deficits. Association of Christian Economists. *Faith & Economics, 61/62*, 1–31.

Batubara, Dison M.H., & Saskara, I.A. Nyoman. (2005). The Relationship between export, import, GDP and Indonesian foreign debt in 1970–2013. *Jurnal Ekonomi Kuantitatif Terapan, 8*(1), 46–55.

CÂLEA, S.A. (2013). *Implications of External Debt on National Economy.* Rumania: Babeş-Bolyai University.

Cooray, A., Ratbek D. & Schneider, F. (2017). How does corruption affect public debt? An Empirical Analysis. *World Development, 90*, 115–127.

Espinoza, R. (2014). *A model of external debt and international reserves.* London: University College London.

Gujarati, D.N. & Porter, D.C. (2009). *Basic Econometrics.* Fifth Edition, USA: McGraw-Hill Companies.

Hur, S. & Kondo, I.O. (2013). A theory of rollover risk, sudden stops, and foreign reserves. *International Finance Discussion Papers* Number 1073 www.ssrn.com.

Irwin, T.C. (2015). Defining the government's debt and deficit. *IMF Working Paper* WP/15/238 pp. 1–35.

Kharusi, S.A. & Ada, M.S. (2018). External debt and economic growth: the case of emerging economy. *Journal of Economic Integration, 33*(1), 1141–1157.

Kim, E., Yoonhee, H. & Kim, S. (2017). Public debt, corruption and sustainable economic growth. *Sustainability, 9*(433), 1–30.

Koc, F. (2014). *Sovereign asset and liability management framework for DMOs: what do country experiences suggest?* Geneva: UNCTAD.

Lau, E. & Thian-Ling, K. (2014). External debt, export and growth in Asian countries: 1988–2006. *Journal of Applied Sciences, 14*, 2170–2176.

Ogunmuyiwa, M.S. (2011). Does fiscal deficit determine the size of external debt in Nigeria? *Journal of Economics and International Finance, 3*(10), 580–585.

Ouyang, A.Y., & Rajan, S.R. (2013). What determines external debt tipping points? *Journal of Macroeconomics, 39*, 215–225.

Qian, X. & Steiner, A. (2017). International reserves and the maturity of external debt. *Journal of International Money and Finance, 73*, 399–418.

Saleh, S. (2008). Factors affecting foreign debt and its impact on government budget). *UNISIA Journal of Social Sciences, 31*(70), 343–363.

Siddique, A., Selvanathan, E.A. & Selvanathan, S. (2015). The impact of external debt on economic growth: empirical evidence from highly indebted poor countries. *Discussion Paper 15.10.* The University of Western Australia.

Soeradi. (2016). *Implementation of foreign debt as development funding.* Yogyakarta: EKUILIBRA.

Tarsilohadi, E.R. (2005). Fund gap and domestic financing, must be covered by foreign debt?. *Jurnal Ekonomi Pembangunan, 6*(2), 206–226.

Todaro, M.P. (1994). *Economic Development,* (5th ed.). New York: Longman Publishing.

Tribroto. (2001). *Policy and management of foreign debt in Indonesian foreign debt profile and its problem.* Jakarta: Bank Indonesia.

Marketing

Business Innovation and Development in Emerging Economies – Trinugroho & Lau (Eds)
© *2019 Taylor & Francis Group, London, ISBN 978-1-138-35996-3*

A study of the role of altruism in the process of individual behavior in donating blood

B. Haryanto & P. Suryanadi
Fakultas Ekonomi dan Bisnis, Universitas Sebelas Maret, Indonesia

B. Setyanta
Universitas Janabadra, Indonesia

E. Cahyono
Fakultas Ekonomi dan Bisnis Universitas Sebelas Maret and Sekolah Tinggi Ilmu Ekonomi Atma Bhakti, Indonesia

ABSTRACT: Risk perceptions and incentives are two important predictors of individual intention to donate blood, but this requires further explanation due to the problem of inconsistencies that occur in relation to attitudes and behaviors. This study conceptualizes altruism as a moderator that will give a detailed explanation of the problem of inconsistency. It examines the effects of risk perceptions and incentives on positive attitudes and individual intentions in donating blood, moderated by altruism. A sample of 200 individuals was taken randomly at various public places in Surakarta, Indonesia, including malls, hospitals and during a car free day event. The multigroup structural equation model is a statistical tool used to describe the relationship between conceptualized variables. This study provides clarity that altruism is a moderating variable that can explain the intention of individuals in donating blood.

Keywords: altruism, risk perception, incentive, blood donors

1 INTRODUCTION

Palang Merah Indonesia (PMI) is a government institution that has the authority to manage blood banks in Indonesia. Statistical data on Indonesians in need of blood donations indicate 2.5% of the population in 2013, and this situation shows an increase in blood demand from year to year (PMI, 2016). Furthermore, it can be explained that the frequency of blood banks having a deficit of blood is due to the increasing number of requests from hospitals that deal with various cases related to the demand for blood that must be immediately cultivated (Ministry of Health RI, 2014). The situation, which is often a deficit in blood supply, is likely caused by high demand, which is not matched by the number of donors (Greinacher et al., 2007). Some literature explains that one of the causes is that not all individuals can donate blood due to relatively strict donor checks, as well as some other requirements to be met (Hinrichs et al., 2008; Volken et al., 2013).

This study does not reveal more about the problem of blood donation from the supply and demand aspect, but will focus on the problem of how to make people more willing to donate blood. Much research has been undertaken to uncover the problem, with very specific objectives and different research settings, some of which are interesting studies that offer an individualized model of behavior regarding blood donation, with various factors and variables included (Masser et al., 2008; Gader et al., 2011; Ngo et al., 2013). Based on these studies, what is interesting to note is that risk perceptions and incentives are the two important variables that individuals consider in relation to donating blood. There are several reasons, including: (1) blood donation is a humanitarian activity related to the process of blood transfusion with all the consequences of time, contracting disease, as well as other consequences; (2) as a humanitarian activity, the reward is a stimulus that is relatively attractive to individuals in donating blood. However, the effectiveness of these two stimuli still requires empirical clarity regarding the significance of their influence on positive attitudes and the intention to donate blood.

In this study, altruism is a variable that is conceptualized as a moderating variable in the process of individual behavior in donating blood. This is because altruism is the nature of an individual's basic character in relation to his/her concern for humanitarian issues. The logic offered in this study is related to different behaviors among individuals who have high and low altruism properties. At low altruism, individuals will usually consider incentives more than individuals with high altruism. Therefore, the response to risk is also different, in that individuals with low altruism are more concerned about risks than individuals with high altruism. This conceptual logic is expected to provide further explanation of the problem of the influence of risk perceptions and incentives on positive attitudes and intent to donate blood.

1.1 *The relationship between risk perceptions and positive attitudes toward blood donation*

Individual risk perception tends to be based on subjective factors. These include fear of loss emergence (Slovic et al., 2005; Ngo et al., 2013), individual's abilities to control risk (Ngo et al., 2013) and individual attention, that is not necessarily correct (Slovic et al., 2005), to the loss consequences (Sjöberg et al., 2004) that are influenced by knowledge, experience, values, attitudes and feelings (Wachinger & Renn, 2010).

Despite the perception of risk being different from the actual risk (Lowe & Ferguson, 2003; Ngo et al., 2013), but being influential in decision-making (Ngo et al., 2013), because people tend to avoid risk (Sjöberg et al., 2004; Menon et al., 2008), and pro-social behavior based on the calculation of benefits and costs received donor (Lyle et al., 2009). Slovic et al. (2005) identifies that the risks and benefits tend to be positively correlated, because high-risk activities tend to have greater benefits compared to low-risk activities, such as sweepstakes (Menon et al., 2008). However, the risks and benefits correlate negatively in the mind and individual assessment, because high risk is associated with low benefits, and vice versa (Slovic & Peters, 2006).

This shows that risk assessment involves evaluation aspects. When it is evaluated as profitable, it is considered low risk and high benefit, and vice versa. Therefore, in practice, considerations of the cost and benefit affect decisions of pro-social behavior (Lyle et al., 2009). If the perceived benefit is more valuable than the costs, someone will have a positive attitude to blood donation. However, if the benefits are perceived as being less than the cost, someone will have a negative attitude toward blood donation (Rodríguez & Hita, 2009).

Several previous studies have identified that risk perception has a significantly negative effect on attitude toward blood donation (Adam & Soutar, 1999; Shashahani, 2006; Abderrahman & Saleh, 2014), so that if the perception of risk increases, the attitude toward blood donation decreases. To explain the relationship between the variables, the following hypothesis was formulated:

Hypothesis 1: The perception of risk has a negative influence on positive attitude toward blood donation.

1.2 *The relationship between incentive and attitude toward blood donation*

Incentive is the individual's perception of certain rewards both material and non-material (Buchan et al., 2000). Previous research has identified that there is an inconsistent effect of incentives on attitude toward blood donation. Mellström and Johannesson (2008) state that incentives have the effect of decreasing the altruism of blood donors, because the incentives, in the form of cash, can decrease intentions to donate (Lacetera & Macis, 2009), as they can impact on the good name of the donor (Bénabou & Tirole, 2006. The study is contrary to research conducted by Goette and Stutzer (2008), which states that selective incentives have a positive influence on pro-social motives. Some previous studies identify that incentives have a positive effect on blood donation. For instance, if the incentives involve a lottery (Goette &

Stutzer, 2008; Lacetera & Macis, 2009), an extra day off work (Lacetera & Macis 2009), financial incentives (Lacetera & Macis 2009) or a free medical test (Kasraian & Maghsudlu, 2012), they can increase blood donation. Even incentive rated as important in influencing blood donor altruism (Yuan et al., 2011), if a monetary incentive is removed, the donor will reduce the frequency of blood donors (Buciuniene et al., 2006).

Various previous studies indicate that the effect of incentives on attitude toward blood donation depends on the types of incentives, blood donor objectives and characteristics of the donor (Errea & Cabasés, 2013). To explain the relationship between variables, the following hypotheses was formulated:

Hypothesis 2: Incentives have a positive influence on attitude toward blood donation.

1.3   *The relationship between attitude toward blood donation and intention to donate*

Attitude has a relationship with the social behavior (Wicker, 1969), because attitude has an important role in predicting behavior (Ajzen, 2005). Attitude can be a positive or negative disposition respond to evaluative things of an object (Fishbein, 1963; Ajzen, 2005). Attitude has a dynamic influence on behavior, because attitude is a reaction to the environment and the response to an object or situation (Jain, 2014). Attitude cannot be seen from physical characteristics, because it is hidden and can only be inferred from behavior (Ajzen, 2005).

Attitude is composed of cognitive and affective components (Fishbein, 1963; See & Petty, 2008). The cognitive component related to the perception and belief in the acquired object of knowledge and information, such as profit and loss, while the affective component are emotions and feelings to an object, such as fear and pain (See & Petty, 2008). The cognitive component is a precondition of the affective component (Fishbein, 1963; Lazarus, 1993), because the affective component is an evaluation of the cognitive component (Fishbein, 1963).

Attitude can have a direct effect on behavior, if the attitude and behavior there is a very close relationship and are at the same level specifications. A specific attitude can be a predictor of a specific behavior and is inconsistent when used to predict behavior in general (Wicker, 1969). This is evidenced by research by Fishbein and Ajzen (1974) who identified that general attitude toward religion is a predictor of religious behavior in general, but inconsistent in predicting the behavior toward religion in particular, because attitude influences behavior indirectly as it is mediated by intention (Ajzen, 2005). The study is consistent with the research of Bagozii (1989), Adam and Soutar (1999) and Giles et al. (2004). They identify that the attitude toward blood donation has a positive effect on the intention to donate and has an indirect effect on the behavior of blood donors because the intention to donate is mediated. If the attitude toward blood donation is more positive, then, the intention to donate is higher. To explain the relationship between variables in this study, the following hypothesis was formulated:

Hypothesis 3: The attitude toward blood donation has a positive influence on the intention to donate.

1.4   *The relationship between altruism with perception of risk and attitude toward blood donation*

Altruism is a deliberate act to help others and not oneself (Hoffman, 2011). Based on previous studies, it has been identified that altruism is the main reason for blood donation (Wells & Christenberry, 2002). However, this is contrary to research conducted by Andreoni (1990), which states that blood donation is more influenced by social pressure, guilt, sympathy and emotional motive gain compared with the motive of altruism. Ferguson et al. (2012) stated that the motive for blood donation was personal gain, combined with the motive of helping others, and blood donors are more influenced by personality than the motive of altruism (Ferguson et al., 2007), which shows that personality affects the level of individual altruism (Oda et al., 2014).

Other studies identify genetic and environmental factors as having an equally strong influence on pro-social behavior. Pro-social behavior by genetic factors have characteristics depending on the closeness of the relationship between donor and recipient, the benefit to the recipient, the risk to the donor, as well as accompanying environmental conditions. The relationship between personality variables and pro-social behavior is influenced by the attractiveness of the environmental situation. If the environmental situation exerts a strong influence, then the environment is more dominant in influencing behavior, but if the environmental situation is weak, then the more dominant personality influences pro-social behaviors (Carlo et al., 1991). Pro-social behavior is influenced by differences in time and situation (Bierhoff & Rohmann, 2004; Otto & Bolle, 2011) and personality traits (Penner et al., 2005), therefore, altruism can also occur within the family environment.

Parents who have the intention to reduce health risks to their children demonstrate a form of altruism and influence the allocation of household spending decisions. Parents will be increasing spending product that serves to reduce their health risks (Cai et al., 2008). This study consistent with research by Dickie and Gerking (2007) which identifies those parents make purchase protective lotion sunlight to reduce their risk of skin cancer. Research conducted by Cai et al. (2008), which examined the sensitivity of parents to the dangers faced by children, indicated that parents increase family expenditures by purchasing water treatment equipment to protect their children from toxins. Previous research indicates that altruistic behavior increases when the environment poses a perceived risk. To explain the relationship between variables in this study, the following hypotheses was formulated:

Hypothesis 4: Altruism moderates the influence of risk perception on attitude toward blood donation.

1.5 *The relationship between altruism with incentive and attitude toward blood donation*

A person with an altruistic personality has concern for the problems of others, so someone who has altruistic personality traits, has the ability to recognize the emotional state of others (Haas et al., 2015), have a social responsibility and high empathy (Bierhoff & Rohmann, 2004), so that pro-social behavior is inherent in people who have altruistic personality.

Personal traits of a donor are different from other donors because the closeness of the donor and recipient relationship influenced it (Rushton, 2004; Carlo et al., 1991; Záškodná 2010; Bierhoff & Rohmann, 2004; Guzmán et al., 2013; Oda et al., 2014). Personality is the foundation of thinking and feeling that influences preferences (Hill et al., 2014) and beliefs (Guzmán et al., 2013). According to Guzmán et al. (2013), preferences and beliefs describe the personality in response to the incentives of pro-social behavior.

Pro-social behavior is an economic concept according to Záškodná (2010), based on the calculation of profit and loss, so that social behavior will increase if the gain is greater than the cost. Previous research indicates that an altruistic personality increases the incentive effect on pro-social attitude. To explain the relationship between variables in this study, the following hypotheses was formulated:

Hypothesis 5: Altruism moderates the incentive effect on attitude toward blood donation.

1.6 *The relationship between altruism and attitude toward blood donation and intention to donate*

Altruism is an act to help others where personal norm affects. Altruism is also a moral obligation to do or not to do (Schwartz & Howard, 1981). Personal norms are influenced by four situations (Steg & de Groot, 2010). The first is the degree to which donors are aware of the positive or negative consequences of the pro-social actions undertaken, their sense of responsibility for the negative affect that occurs if the pro-social behavior is not done, the identification of action to relieve burdens of others and last, the ability to know oneself in pro-social action.

Ajzen and Fishbein (1970) identified altruism as playing a role as a moderating variable by manipulation of an intention to behave in a game with a different purpose. When the objective of the game is to obtain the highest individual scores, it will increase the individual's intention to obtain the highest value. However, if an assessment based group, the attitude to get the highest score had no significant effect on the intention to obtain the highest value, therapy increased the influence of subjective norms on attitude to obtain the highest score is rising hopes her partner to get the highest score.

Based on previous research which indicated that altruism can play a role as a moderating variable because it can increase or decrease the influence of attitude on the intention to behave, the following hypotheses was formulated:

Hypothesis 6: Altruism moderates the influence of attitude toward blood donation or the intention to donate.

2 RESEARCH METHODS

The population in this study were those who had the intention to donate blood, and met the requirements set by the PMI. The sample size was 200 people with the main reason for meeting the number of objects as required in Structural Equation Model (SEM) as a statistical tool selected (Loehlin, 1998). Data were collected through surveys using questionnaires given to respondents to complete. The questionnaire was immediately collected if all items were filled in completely. This technique is considered to be effective in maintaining a high response rate.

Data collection was done at a certain time so that the present study only describes the phenomenon of blood donor intent at that time. Sampling was conducted in four locations: a super market, a traditional market, a city square and a car free day. The fourth site was a gathering place in Klaten community, so that this sample can be representative of the population.

Perception of risk is a person's perception of the chances of loss (Sjöberg et al., 2004). In this study, the dimensions of the perception of risk consists of the risk of inconvenience, psychological risk, health risks and social risks (Adam & Soutar, 1999); a description of the dimensions of risk is outlined below. Perception of risk of inconvenience is the donor's perception of the service and process of blood donation. Indicators of the risk perception of inconvenience include tiring, boring, dull, convoluted, sucks. Psychological risk perception is the perception of the pain that is felt when donating. Indicators of the risk of psychological perception include anxiety, tension, restlessness, trembling, fearful. Health risk perception is the perception of the donation's negative influence on the health of blood donors. The indicators include, the body becomes weak, dizzy, faint, contracting disease, becoming sick. Social risk perception is the perception donors have of a negative reaction from the family and the environment as a result of blood donation. The indicators include, insults, scolding, reprimands, censure. Perception of incentives is the individual's perception of material and non-material rewards (Buchan et al., 2000; Errea & Cabasés, 2013), with each individual having a different perception of the incentive (Errea & Cabasés, 2013). In this study, the indicators of incentives include, suitable, reasonable, fair, decent, equitable. The attitude toward blood donation is a positive or negative disposition regarding evaluation of donation (Fishbein, 1963; Ajzen, 2005). Indicators of the attitude toward blood donation in this study were delighted, happy, excited, joy, positive thinking. The intention to donate is the extent to which individuals are motivated toward blood donation (Giles et al., 2004). Indicators of intention to donate in this study included, I want to donate blood, I will donate blood, I intend to donate blood, I am committed to donate blood, I promise to donate blood. Altruism is a deliberate act to help others and not oneself (Hoffman, 2011). Indicators of altruism in this study included feeling of sadness, empathy to disaster and famine, compassion toward an accident, and sympathy for suffering. In addition, observations of variables were measured using a Likert scale with five points, namely: 1: strongly disagree to 5: strongly agree.

3 RESULTS

The explanation of the test results begins with an explanation of the results of the goodness of fit model. Results of the test of goodness of fit indicate that AGFI which has marginal criteria while others have fit criteria. Research using maximum likelihood technique with a sample size < 250, research model indicated have fit the criteria, if the test results of CFI, GFI, TLI, RMSEA and IFI are fit (Hu & Bentler, 1999).

3.1   Behavioral process model of blood donors before altruism moderation

Results of the regression test between the variables of risk perception with the attitude toward blood donation indicate that the perception of risk has a negative influence on the attitude toward blood donation but is not significant ($\beta = -0.192$, Standart Error (SE) = 1.214 and Critical Ratio (CR) = -1.091); H1 is not supported. This study indicates that the influence of risk perception on the attitude toward blood donation tends to not have an effect.

Perception of risk has no effect on the attitude toward blood donation possibilities because the perception of risk in blood donors is low (Lowe & Ferguson, 2003; Marantidou et al., 2007), with a growing belief that blood donation has a positive effect on health, because the it may reduce the risk of heart disease (Desai & Satapara, 2014) and diabetes mellitus type II (Kumari & Raina, 2015). Additionally, it is also supposedly due to the increased confidence in the security system of blood donation due to the improvement of security procedures for blood donors (Amin et al., 2004).

Results of regression test, the variable incentive to the attitude toward blood donation indicates that incentives are positive and significant impact on the attitude toward blood donation ($\beta = 0.165$, SE = 0.057 and CR = 2.266), therefore, H2 is supported. Results of regression test, indicating that the attitude toward blood donation will be higher if the blood donor incentive is given even greater.

The results are consistent with previous research indicating that incentives have a positive effect on the attitude toward blood donation.

Results of regression test, among the variables the attitude toward blood donation with the intention of donating blood, indicating that the attitude toward blood donation has a positive and significant impact on the intention to blood donors ($\beta = 0.616$, SE = 0.068 and CR = 9.584), therefore, H3 is supported.

These findings support previous research phenomena, indicating that the attitude toward blood donation has a significantly positive effect on the intention to donate, so the more

Table 1.   Results of the test of goodness of fit.

Index	Cut-Off	Result	Conclusion
Chi-square	small	338.994	
Probability of chi-square (p)	≥ 0.05	1	Fit
CMIN/DF	≤ 2.00	0.752	Fit
Goodness of Fit Index (GFI)	≥ 0.90	0.917	Fit
Adjusted Goodness of Fit Index (AGFI)	≥ 0.90	0.884	Marginal
Comparative Fit Index (CFI)	≥ 0.95	1	Fit
Tucker-Lewis Index (TLI)	≥ 0.95	1.02	Fit
RMSEA	≤ 0.06	0	Fit
IFI	≥ 0.95	1.015	Fit

Source: Compiled by Authors, 2018.

Table 2.   Results of regression testing before moderation (*unconstrained model*).

			β	SE	CR
Attitude	<---	Perceived risk	−0.192	1.214	−1.091
Attitude	<---	Incentives	0.165	0.057	2.266
Intention	<---	Attitude	0.616	0.068	9.584

Source: Compiled by Authors, 2018.

positive the attitude toward blood donation, the higher the intention to donate (Bagozii et al., 1989; Adam & Soutar, 1999; Giles et al., 2004). Though it is tested in a different context, this study demonstrates that the attitude toward blood donation has a consistently positive influence on the intention to donate.

3.2 Behavioral process model of blood donors after altruism moderation

The moderator variable in this study was altruism, measured using a nominal scale based on the degree of altruism. Respondents who have an altruism value above average are classified as a group with high altruism, while respondents who have an altruism value below average are classified as a group with low altruism. Because of moderator variables are grouped based on the degree of altruism, therefore the test methods used in moderation is multigroup of SEM.

To test whether there were significant differences between the constrained models compared to the unconstrained models, a comparative test between the table values of chi-square ($\chi 2$) with the difference of the calculated value of chi-square ($\Delta \chi 2$) was used. If the value of the table of chi-square ($\chi 2$)> from the difference between the calculated value of chi-square ($\Delta \chi 2$), concluded that the constrained models have indicated significant differences with unconstrained model (Marsh & Scalas, 2010).

The results of the research model test after moderated altruism, it is known that research model is moderated (table value of chi-square ($\chi 2$)> the difference between the calculated value of chi-square ($\Delta \chi 2$)), therefore constrained models are significantly different from the unconstrained models.

These results are supported by the fact that, from the results of the multigroup regression test, in the group of high altruism, the perception of risk had a negative influence on the attitude toward blood donation but it was not significant ($\beta = -0.068$, SE = 0.034 and CR = -1.42). The findings indicate that in the group with high altruism, risk perception regarding the attitude toward blood donation tends to have no relation. At low altruism group, obtained by the fact that the perception of risk has a negative and significant impact on the attitude toward blood donation ($\beta = -0.258$, SE = 0.044 and CR = 2.996). The findings indicate that, in low altruism groups, the influence of perception of risk on the attitude toward blood donation tends to lead to a negative and consistent phenomenon. Therefore, if the perception of risk increases, the attitude toward blood donation decreases.

The results were consistent with previous studies indicating that blood donation is influenced by the personality of the donor (Penner et al., 2005). Individuals who have an altruistic personality, have the ability to recognize the emotional state of other people, have a social responsibility and high empathy (Bierhoff & Rohmann, 2004), therefore, pro-social behavior is inherent in people who have an altruistic personality. In the group with high altruism, the perception of risk does not have a negative effect on the attitude toward blood donation, because the perceived risk is low (Lowe & Ferguson, 2003; Marantidou et al., 2007), and previous studies sug-

Table 3.   Results of regression testing after moderation and chi-squared difference (*constrained model*).

			High altruism			Low altruism		
			β	SE	CR	β	SE	CR
Attitude	<--	Risk	−0.068	0.034	−1.42	−0.258	0.044	−2.996
Attitude	<--	Incentive	−0.172	0.121	−1.045	0.442	0.125	2.202
Intention	<--	Attitude	0.576	0.092	7.112	0.64	0.165	4.659

Chi-square calculation ($\Delta \chi 2$) = 834.798 − 826.166 = 8.632
df difference (Δdf) = 862 − 766 = 96
Chi-square table (96; 0,05) = 119,871
Chi-square table ($\chi 2$) > chi-square calculate ($\Delta \chi 2$)
Constrained model significantly differs from *unconstrained model*

Source: Compiled by Authors, 2018.

389

gest that the perception of risk can increase altruistic behavior of parents toward their children (Dickie & Gerking, 2007; Cai et al., 2008). In the group with low altruism, perception of risk has a negative and significant impact on the attitude toward blood donation, which includes the risk of inconvenience, psychological risk, health risks and social risks (Adam & Soutar, 1999).

The results of the multigroup regression test showed that in the group with high altruism, incentives have a negative effect on the attitude toward blood donation but it is not significant ($\beta = -0.172$, SE = 0.121 and CR = -1.045). The findings indicate that in the group with high altruism, the incentive effect on the attitude toward blood donation tends to lead to the phenomenon of no effect. In the group with low altruism, it was noted that incentives have a positive and significant impact on the attitude toward blood donation ($\beta = 0.442$, SE = 0.125 and CR = 2.202). The test results indicate that in the low altruism group, the incentive effect on the attitude toward blood donation tends to lead to a positive phenomenon, consistent with previous studies. Therefore, if a greater incentive is provided, then the attitude toward blood donation increases.

The results of the multigroup regression test indicate that in the group with high altruism, the attitude toward blood donation has a positive and significant impact on the intention to donate blood ($\beta = 0.576$, SE = 0.092, CR = 7.112). In the group with low altruism, the attitude toward blood donation has a positive and significant impact on the intention to donate blood ($\beta = 0.64$, SE = 0.165 and CR = 4.659).

The findings indicate that in both high and low altruism groups, the influence of the attitude toward blood donation on the intention to donate blood tends to lead to a consistent, positive phenomenon. Previous research indicates that attitude toward blood donation has a positive and significant impact on the intention to donate blood, so the more positive the attitude toward blood donation, the higher the intention to donate blood (Bagozii et al., 1989; Adam & Soutar, 1999; Giles et al., 2004).

4 CONCLUSION

It can be concluded here that blood donation behavior is distinguished on the basis of altruism levels and only incentives influence positive attitudes to donation, whereas risk perception is found not to affect a positive attitude to donation. However, when the behavior of donating blood is distinguished on the level of altruism, in low altruism groups, risk perceptions and incentives significantly influence a positive attitude toward donating blood. Whereas in high altruism groups, both variables were found not to affect a positive attitude to donate blood. Furthermore, the relationship between positive attitudes and intent to donate blood is found to be consistent, namely significant and positive, both for the behavior of donating blood before and after moderation by altruism.

5 IMPLICATIONS

Theoretically, this study contributes to the role of altruism as a moderating variable of individual behavior in donating blood. Through this role, the inconsistency of the relationship between risk perceptions and incentives with positive attitudes to donating blood can be well explained. It can be explained here that risk perceptions and incentives are stimuli that can significantly influence positive attitudes toward donating blood, only in groups of individuals who have low altruism traits, while in groups of individuals who have high altruism, those two variables are found not to have a significant effect on a positive attitude to donate blood.

In practice, this study can provide blood donor marketers with an account of the behavior of different individuals in donating blood, based on the nature of the altruism of each individual. These behavioral differences have implications for different marketing strategies, where for individuals with low altruism, information about the potential risk of blood donation needs to be elaborated in such a way that it does not deter people from donating blood. Other than that, incentives are also one effective stimulus to encourage individuals to be willing to donate blood. Whereas in individuals with high altruism traits, both risk perception and incentives are not effective stimuli to influence individuals to be willing to donate blood.

In the future, this research can be used as a basis for further research, because there are some concepts that still require further empirical explanation, that is in the phenomenon with high altruism, risk perception and incentive has been found not to significantly influence the positive attitude of the individual to donate blood. This phenomenon requires further explanation of its causative factors.

6 LIMITATIONS

The limitation of this study is the possibility of a lack of ability of the model to be applied to different contexts and settings. This is because the concepts built in this study rely on the behavior of Indonesian people in donating blood, especially Surakarta residents with all their characteristics. Therefore, to apply this research to different objects and settings requires caution, especially regarding the emergence of potential variables in forming behavioral models.

7 ORIGINALITY

In some references to consumer behavior, altruism is not a new and unfamiliar variable, but in the context of blood donations, the role of these variables becomes important as the relationship between risk perceptions and incentives with positive attitudes in donating blood needs further empirical explanation. Through the role of altruism, it is finally explained that risk perceptions and incentives only affect positive attitudes to donate blood if altruism is low, while at high altruism, those two variables do not affect positive attitudes.

REFERENCES

Abderrahman, H.B. & Saleh, M.Y. (2014). Investigating knowledge and attitudes of blood donors and barriers concerning blood donation in Jordan. *Procedia—Social and Behavioral Sciences, 116*, 2146–2154. doi: 10.1016/j.sbspro.2014.01.535.

Adam, D. & Soutar, G.N. (1999). *A proposed model of the blood donation process.* Retrieved from www.academia.edu/2721804/A_proposed_model_of_the_blood_donation_process.

Ajzen, I. & Fishbein, M. (1970). The prediction of behavior from attitudinal and normative variables. *Journal of Experimental Social Psychology, 6*, 466–487.

Ajzen I. (2005). *Attitudes, personality and behavior* (2nd ed.). Berkshire, England: Open University Press.

Amin, M., Wilson, K., Tinmouth, A. & Hébert, P. (2004). Does a perception of increased blood safety mean increased blood transfusion? An assessment of the risk compensation theory in Canada. *BMC Public Health, 4*(20), 1–4. http://biomedcentral.com/I47I-2458/4/20.

Andreoni, J. (1990). Impure altruism and donations to public goods: A theory of warm glow giving. *Economic Journal, 100*, 464–477.

Bagozii, R.P., Baumgartner, J. & Yi, Y. (1989). An investigation into the role of intentions as mediators of attitude–behavior relationship. *Journal of Economic Psychology, 10*, 35–62.

Bénabou, R. & Tirole, T. (2006). Incentives and prosocial behavior. *American Economic Review, 96*, 1652–1677.

Bierhoff, H.W. & Rohmann, E. (2004). Altruistic personality in the context of the empathy–altruism hypothesis. *European Journal of Personality, 18*, 351–365. doi: 10.1002/per.523.

Buchan, J., Thompson, M. & O'May, F. (2000). *Incentive and remuneration strategies: Health workforce incentive and remuneration.* A Research Review Evidence and Information for Policy Department of Organization of Health Services Delivery, World Health Organization, Geneva. Discussion Paper No. 4: 1–37. Retrieved from http://apps.who.int/iris/bitstream/10665/69777/1/WHO_EIP_OSD_00.14_eng.pdf.

Buciuniene, I., Stonienĕ, L., Blazeviciene, A., Kazlauskaite, R. & Skudiene, V. (2006). Blood donors' motivation and attitude to non-remunerated blood donation in Lithuania. *BMC Public Health, 6*. doi: 10.1186/1471-2458-6-166.

Cai, Y., Shaw, W.D. & Wu, X. (2008). *Risk perception and altruistic averting behavior: Removing arsenic in drinking water*. Selected paper prepared for presentation at the American Agricultural Economics Association Annual Meeting, Orlando, FL, July 27–29: 1–40. Retrieved from http://ageconsearch.umn.edu.

Carlo, G., Eisenberg, N., Troyer, D., Switzer, G. & Speer, A.L. (1991). The altruistic personality: In what contexts is it apparent? *Journal of Personality and Social Psychology*, *61*(3), 450–458.

Desai, K.N. & Satapara, V. (2014). A study on knowledge, attitude, and practice on blood donation among health professional students in Anand: Gujarat. *Journal of Applied Hematology*, *5*(2), 51–53. doi: 10.4103/1658–5127.137140.

Dickie, M. & Gerking, S. (2007). Altruism and environmental risks to health of parents and their children. *Journal of Environmental Economics and Management*, *53*, 323–341. doi: 10.1016/j.jeem.2006.09.005.

Errea, M. & Cabasés, J.M. (2013). Incentives when altruism is impure: The case of blood and living organ donations. DT, 1302, 1–23. Retrieved from ftp://ftp.econ.unavarra.es/pub/DocumentosTrab/DT1302.PDF.

Ferguson, E., Atsma, F., de Kort, W. & Veldhuizen, I. (2012). Exploring the pattern of blood donor beliefs in first-time, novice and experienced donors: Differentiating reluctant altruism, pure altruism, impure altruism and warm glow. *Transfusion*, *52*, 343–355. doi: 10.1111/j.1537-2995.2011.03279.x.

Ferguson, E., France, C.R., Abraham, C., Ditto, B. & Sheeran, P. (2007). Improving blood donor recruitment and retention: Integrating social and behavioral science agendas. *Transfusion*, *47*, 1999–2010. doi: 10.1111/j.1537-2995.2007.01423.x.

Fishbein, M. (1963). An investigation of the relationships between beliefs about an object and the attitude toward that object. *Human Relations*, *16*, 233–239.

Fishbein, M. & Ajzen, I. (1974). Attitudes toward objects as predictors of single and multiple behavioral criteria. *Psychological Review*, *81*, 59–74.

Gader, A.G., Osman, A.M., Al Gahtani, F.H., Farghali, M.N., Ramadan, A.H. & Al-Momen, A.K. (2011). Attitude to blood donation in Saudi Arabia. *Asian Journal of Transfusion Science*, *5*(2), 121–126. doi: 10.4103/0973-6247.83235.

Giles, M., Mc Clenahan, C., Cairns, E. & Mallet, J. (2004). An application of the theory of planned behaviour to blood donation: The importance of self-efficacy. *Health Education Research*, *19*, 380–391. doi: 10.1093/her/cyg063.

Goette, L. & Stutzer, A. (2008). *Blood donations and incentives: Evidence from a field experiment*. I.Z.A DP No. 3580, 1–31. Discussion Paper Series.

Greinacher, A., Fendrich, K., Alpen, U. & Hoffmann, W. (2007). Impact of demographic changes on the blood supply: Mecklenburg-West Pomerania as a model region for Europe. *Transfusion*, *47*, 395–401. doi: 10.1111/j.1537-2995.2007.01129.x.

Guzmán, R.A., Abarca, N., Harrison, R. & Villena, M.G. (2013). *Reciprocity and trust: Personality psychology meets behavioral economics*. Preliminary version: 1–28. Retrieved from http://economia.uc.cl/wp-content/uploads/2015/01/dt_439.pdf.

Haas, B.W., Brook, M., Remillard, L., Ishak, A., Anderson, I.W. & Megan, M. (2015). I know how you feel: The warm-altruistic personality profile and the empathic brain. *Plos One*, *10*(3), 1–15. doi: 10.1371/journal.pone.0120639.

Hill, N.L., Kolanowski, A.M., Fick, D., Chinchilli, V.M. & Jablonski, R.A. (2014.) Personality as a moderator of cognitive stimulation in older adults at high risk for cognitive decline. *Research in Gerontological Nursing*, *7*(4) 159–170. doi: 10.3928/19404921-20140311-01.

Hinrichs, A., Picker, S.M., Schneider, A., Lefering, R., Neugebauer, E.A.M. & Gathof, B.S. (2008). Effect of blood donation on well-being of blood donors. *Transfusion Medicine*, *18*, 40–48. doi: 10.1111/j.1365-3148.2007.00805.x.

Hoffman, M. (2011). Does higher income make you more altruistic? Evidence from The Holocaust. *The Review of Economics and Statistics*, *93*(3), 876–887. Retrieved from http://files.eric.ed.gov/fulltext/ED472935.pdfHu, L. & Bentler, P.M. (1998). Fit indices in covariance structure modeling: Sensitivity to underparameterized model misspecification. *Psychological Methods*, *3*(4), 525–453.

Hu, L.T., & Bentler, P.M. (1999). Cutoff criteria for fit indexes in covariance structure analysis: Conventional criteria versus new alternatives. *Structural Equation Modeling*, 6, 1–55.

Jain, V. (2014). 3D model of attitude. *International Journal of Advanced Research in Management and Social Sciences*, *3*(3), 1–12.

Kasraian, L. & Maghsudlu, M. (2012). Blood donors' attitudes towards incentives: Influence on motivation to donate. *Blood Transfusion*, *10*, 186–190. doi: 10.2450/2011.0039-11.

Kumari, S. & Raina, T.R. (2015). A comprehensive analysis of factors that motivate and hinder the blood donation decision among the younger population. *Journal of Behavioral Health*, *4*(4), 107–111. doi: 10.5455/jbh.187247.

Lacetera, N. & Macis, M. (2009). *Do all material incentives for prosocial activities backfire? The response to cash and non-cash incentives for blood donations.* I.Z.A DP No. 4458, 1–17. Discussion Paper Series.

Lazarus, R.S. (1993). From psychological stress to the emotions: A history of changing outlooks. *Annual Review of Psychology, 44*, 1–21.

Loehlin, J.C. (1998). *Latent variable models: An introduction to factor, path, and structural analysis.* Mahwah, NJ: Lawrence Erlbaum Associates.

Lowe, K.C. & Ferguson, E. (2003). Benefit and risk perceptions in transfusion medicine: Blood and blood substitutes. *Journal of Internal Medicine, 253*, 498–507.

Lyle, H.F., Smith, E.A. & Sullivan, R.J. (2009). Blood donations as costly signals of donor quality. *Journal of Evolutionary Psychology, 74*, 263–286. doi: 10.1556/JEP.7.2009.4.1.

Marantidou, O., Loukopoulou, L., Zervou, E., Martinis, G., Egglezou, A., Fountouli, P., Dimoxenous, P., Parara, M., Gavalaki, M. & Maniatis, A. (2007). Factors that motivate and hinder blood donation in Greece. *Transfusion Medicine, 17*(6), 443–450. doi: 10.1111/j.1365-3148.2007.00797.x.

Marsh, H.W. & Scalas, L.F. (2010). Longitudinal tests of competing factor structures for the Rosenberg Self-Esteem Scale: Traits, ephemeral artifacts, and stable response styles. *Psychological Assessment, 22*(2), 366–381. doi: 10.1037/a0019225.

Masser, B.M., White, K.M., Hyde, M.K. & Terry, D.J. (2008). The psychology of blood donation: Current research and future directions. *Transfusion Medicine Reviews, 2*, 215–233. doi: 10.1016/j.tmrv.2008.02.005.

Mellström, C. & Johannesson, M. (2008). Crowding out of blood donation: Was Titmuss right? *Journal of The European Economic Association, 6* (4), 845–863. doi: 10.1162/JEEA.2008.6.4.845.

Menon, G., Raghubir, P. & Agrawal, N. (2008). Health risk perception and consumer psychology. In C. Haugtvedt, P. Herr & F. Kardes (Eds.), *The handbook of consumer psychology.* Lawrence Erlbaum and Associates. New York. Retrieved from http://people.stern.nyu.edu/gmenon/Menon%20Raghubir%20Agrawal.pdf.

Ministry of Health RI. (2014). *Blood donors situation in Indonesia.* Data and Information Center in Ministry of Health Republic of Indonesia. Jakarta Selatan. Retrieved from http://www.depkes.go.id/download.php?file=download/pusdatin/infodatin/infodatin-donor-darah.pdf.

Ngo, L.T., Bruhn, R. & Custer, B. (2013). Risk perception and its role in attitudes toward blood transfusion: A qualitative systematic review. *Transfusion Medicine Reviews, 27*, 119–128. doi: 10.1016/j.tmrv.2013.02.003.

Oda, R., Machii, W., Takagi, S., Kato, Y., Takeda, M., Kiyonari, T., Fukukawa, Y. & Hiraishi, K. (2014). Personality and altruism in daily life. *Personality and Individual Differences, 56*, 206–209. doi: 10.1016/j.paid.2013.09.017.

Otto, P.E. & Bolle, F. (2011). Multiple facets of altruism and their influence on blood donation. *Journal od Socio-Economics, 40*, 558–563. doi: 10.1016/j.socec.2011.04.010.

Penner, L.A., Dovidio, J.F., Piliavin, J.A. & Schroeder, D.A. (2005). Prosocial behavior: Multilevel perspectives. *Annual Review of Psychology, 56*, 14.1–14.28. doi: 10.1146/annurev.psych.56.091103.070141.

PMI Jateng. (2016). Blood Stock of District/City. Retrived from www.udd.pmi-jateng.or.id.

Rodríguez, M.E. & Hita, J.M.C. (2009). *An economic model of behaviour: Attitudes towards altruistic blood and organ donations.* Documentos de Trabajo-Lan Gaiak Departamento de Economia – Universidad Publica de Navarra 0901. Departamento de Economia-Universidad Publica de Navarra. Retrieved from ftp://ftp.econ.unavarra.es/pub/DocumentosTrab/DT0901.PDF.

Rushton, J.R., (2004). Genetic and environmental contributions to pro-social attitudes: A twin study of social responsibility. *Proceedings of the Royal Society of London, B.271*, 2583–2585. doi: 10.1098/rspb.2004.2941.

Schwartz, S.H. & Howard, J.A. (1981). A normative decision-making model of altruism. In J.P. Rushton & R.M. Sorrentino (Eds.), *Altruism and helping behavior.* NJ: Hillsdale.

See, Y.H.M. & Petty, R.E. (2008). Affective and cognitive meta-bases of attitudes: Unique effects on information interest and persuasion. *Journal of Personality and Social Psychology, 94*(6), 938–955. doi: 10.1037/0022-3514.94.6.938.

Shashahani, H., Yavari, M.T., Attar, M. & Ahmadiyeh, M. (2006). Knowledge, attitude, and practice study about blood donation in the urban population of Yazd, Iran, 2004. *Transfusion Medicine, 16*, 403–409. doi: 10.1111/j.1365-3148.2006.00699.x.

Sjöberg, L., Elin Moen, B. & Rundmo, T. (2004). *Explaining risk perception. An evaluation of the psychometric paradigm in risk perception research.* Rotunde publikasjoner, 84, Department of Psychology, Norwegian University of Science and Technology. Retrieved from http://www.svt.ntnu.no/psy.

Slovic, P. & Peters, E. (2006). Risk perception and affect. *Current Directions in Psychological Science, 15*(6), 322–325. doi: 10.1111/j.1467-8721.2006.00461.x.

Slovic, P., Peters, E., Finucane, M.L. & MacGregor, D.G. (2005). Affect, risk, and decision making, *Health Psychology, 24*(4), S35–S40. doi: 10.1037/0278-6133.24.4.S35.

Steg, S. & de Groot, J. (2010). Explaining prosocial intentions: Testing causal relationships in the norm activation model. *British Journal of Social Psychology*, *49*, 725–743. doi: 10.1348/014466609X477745.

Volken, T., Weidmann, C., Bart, T., Fischer, Y., Klüter, H. & Rüesch, P. (2013). Individual characteristics associated with blood donation: A cross-national comparison of the German and Swiss population between 1994 and 2010. *Transfusion Medicine and Hemotherapy*, *40*, 133–138. doi: 10.1159/000349985.

Wachinger, G. & Renn, O. (2010). *Risk perception and natural hazards*, CapHaz-Net WP3 Report, DIALOGIK Non-Profit Institute for Communication and Cooperative Research, Stuttgart. Retrieved from http://caphaz-net.org/outcomes-results/CapHaz-Net_WP3_Risk-Perception.pdf.

Wells, A.G. & Christenberry, N. (2002). *Developing altruistic behavior and motivation to donate blood: A role for educators and service learning projects*. Presented November 8, at the Annual Meeting of the Mid-South Educational Research Association, at Chattanooga, TN.

Wicker, A.W. (1969). Attitudes versus actions: The relationship of verbal and overt behavioral responses to attitude objects. *Journal of Social Issues*, *25*(4), 41–78.

Yuan, S., Hoffman, M., Lu, Q., Goldfinger, D. & Ziman. A. (2011). Motivating factors and deterrents for blood donation among donors at a university campus-based collection center. *Transfusion*, *51*, 2438–2444. doi: 10.1111/j.1537-2995.2011.03174.x.

Záškodná, H. (2010). *Prosocial traits and tendencies of students of helping professions*. University of South Bohemia, Faculty of Health and Social Studies, Ceske Budejovice, Czech Republic. Retrieved from https://www.zsf.jcu.cz/Members/zaskodna/prosocial-traits-and-tendencies-of-students-of-helping-professions.

Business Innovation and Development in Emerging Economies – Trinugroho & Lau (Eds)
© *2019 Taylor & Francis Group, London, ISBN 978-1-138-35996-3*

Self-monitoring in impulse buying: Effect of religiosity

W. Maryati
Pesantren Tinggi Darul Ulum University, Airlangga University, Indonesia

S. Hartini & G.C. Premananto
Airlangga University, Indonesia

ABSTRACT: **Purpose:** An examination of the effect of religiosity on impulse buying by considering the mediating role of self-monitoring. **Research Design:** This study quantitatively investigated 270 undergraduate students in East Java using a path analysis approach. **Findings:** The results showed that religiosity had both direct and indirect effects on impulse buying. The extent of the direct effect of religiosity on impulse buying was 0.113, while that of the indirect one effect mediated by self-monitoring was $0.113 \times (-0.243) = -0.0275$. This indicated that although an individual was religious and had a good level of self-monitoring, this did not immediately lead to the alleviation of impulse buying. **Implications:** Business players who understand that religious individuals with good self-monitoring might still impulse buy may take these findings as providing opportunities to reach a specific targeted market by offering products which corresponded to the needs of their religiosity.

Keywords: religiosity, self-monitoring, impulse buying

1 INTRODUCTION

The buying behavior of consumers is subject to both internal and external stimul, including impulse buying. Which is defined as a sudden and spontaneous decision to buy specific products (Zeb et al., 2016). The mapping of theories on impulse buying has been mentioned in psychology, economic behavior, sociology, social psychology, and other consumer and marketing literatures (Xiao & Nicholson, 2011).

Some researchers have studied impulse buying by taking individual and personal aspects into account (Seinauskiene et al., 2015; Mathai & Haridas, 2014). Other researchers have examined impulse buying from the perspective of external stimuli (Akram et al., 2016; Dameyasani & Abraham, 2013) and the impulsivity of consumption (Podoshen & Andrzejewski, 2014). While some have combined the individual and external stimuli and impulsivity. perspectives (Khan et al., 2016; Moran & Kowak, 2015; Sharma et al., 2010a, 2010b, 2014). It is evidence from the existing research that until now, impulse buying had been analyzed based on the three perspectives, of the individual, external stimuli, and the impulsivity of consumption (Xiao & Nicholson, 2011).

Previous studies examining the relationship between religiosity and consumer's behavior have been restricted to purchasing behavior in general (Alam, et al, 2011; Fianto et al., 2014) related both to purchasing (Mokhlis, 2009) and consumption (Karaduman, 2016) orientations. The relevance of religiosity as an individual factor had been conceptually examined in relation to impulse buying, but not from an empirical point of view (Salwaa, et al, 2017). This article therefore attempts to build to build on existing research by examining impulse buying through a focus on religiosity from the perspective of the individual.

Religiosity in the behavior of consumers has been studied existing research using a variety of measurements. Karaduman (2016) studied religiosity in a specific manner by referring to one religion only, while Mokhlis (2009) studied religiosity in general asumes by considering the commitment to religiosity in various religious affiliations. Another researcher studied the connection between religiosity and consumer's behavior from the aspects of *muamalah* (Khraim, 2010), building a good relationship with God and others (Fianto et al., 2014; Mathras et al., 2016), and the aspects of relative and contextual factors (Alam, et al., 2011). In this present study, religiosity in general was examined by considering the commitment of each religious affiliation and involving individual's self-monitoring activity as a mediating variable for the connection between religiosity and impulse buying.

As previously described, the phenomenon of impulse buying is a common buying behavior which people may possess due to individual aspects such as emotion, personality, and impulsivity, and situational aspects such as store environment and promotion. From the individual aspect, the impact of individuals strong faith and commitment to religious tenets (that is, religiosity) on impulse buying can be questioned and investigated. This study assumes that religiosity might alleviate the drive toward impulse buying because it would lead to positive self-monitoring and thus help individual to bring their impulse buying under control. It can be suggested that the higher the religiosity of individuals, the stronger their self-monitoring will be and that is may alleviate the tendency toward impulse buying. In contrast therefore, the lower the religiosity of individuals, the lower their self-monitoring, thus increasing impulse buying. The research questions of this study are:

a. does religiosity affect individuals' self-monitoring?
b. does self-monitoring affect impulse buying? And
c. does religiosity with self-monitoring alleviate individuals' impulse buying?

2   RESEARCH MODEL

The research model and hypotheses of the research are presented in Figure

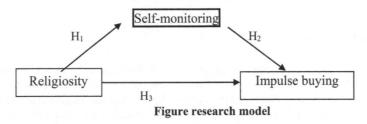

Figure research model

H_1: Religiosity has a positive effect on self-monitoring.
H_2: Self-monitoring has a negative effect on impulse buying.
H_3: Religiosity combined with self-monitoring has a negative effect on impulse buying.

3   RESEARCH METHODS

The research approach of this study is quantitative with causality. The population undergraduate students in East Java, particularly those studying management at any higher education institution classified as a university. In order to represent this very large population, particular universities in East Java were chosen and from these 270 students were enrolled as respondents.

Religiosity was designated as an exogenous variable coded by X, while two endogenous variables were designated self-monitoring, defined as an intervening variable coded by Z and impulse buying, defined as a dependent variable coded by Y. Religiosity (X) was measured using the indicators of RCI-10 (Religious Commitment Inventory-10) which relate to cognitive and behavioral aspects, as argued by Worthington, et al. (2003). Self-monitoring (Z) was measured using three factors as suggested by Carter, et al. (2012), while impulse buying (Y) was measured by indicators taken from Hausman (2000).

A Likert scale with interval options from "strongly disagree" to "strongly agree" was used to measure the variables. Data collection was through library and field researches and the analysis was conducted using procedures including (1) classifying the data into a table (tabulation) and (2) analyzing the data through calculations based on quantitative research with a path analysis technique using SPSS software.

4 FINDINGS AND DISCUSSION

4.1 *Respondents*

From among the 300 respondents who received questionnaires, 270 and from these the respondents was gathered as detailed below.

a. Respondents' university
As presented in Table 1, it can be seen that seven universities were considered as the research object, Universitas Negeri Surabaya being the largest of these (51 respondents or 18.9% of the sample). Table 1 shows the number and percentage of respondents from each university.
b. Respondents' study program and semester
Table 2 shows the respondents' study programs and semesters.

From Table 2 it can be seen that the largest group (120 students, [44.4%]) were in their sixth semester of study.

4.2 *Test of validity and reliability*

The test of validity showed that the instruments used to measure religiosity (X), self-monitoring (Z), and impulse buying (Y) were all valid, as the r count value at 0.138 was greater than

Table 1. Respondents university.

University	Number of respondents	%
Universitas Islam Negeri Malang	39	14.4%
Universitas Muhammadiyah Malang	35	13.0%
Universitas Negeri Surabaya	51	18.9%
Universitas Nahdatul Ulama Surabaya	36	13.3%
Universitas Hasyim Asy'ari	35	13.0%
Universitas Wahab Hasbullah	40	14.8%
Universitas Wiraraja	34	12.6%
Total	270	100%

Source: Questionnaire data, 2018.

Table 2. Respondents' study program and semester.

Study program and semester	Number of respondents	%
Management, semester 6	120	44.4%
Management, semester 2	15	5.6%
Management, semester 4	3	1.1%
Management, semester 8	21	7.8%
Management, semester 2	36	3.3%
Management, semester 2	35	13.0
Management, semester 4	40	14.8%
Total	270	100%

Source: Questionnaire data, 2018.

r table with a level of significance of 0.000 < α (0.05). Similarly, the value of Cronbach for all the items of statement in the instruments was > 0.6, making them all sufficiently reliable to be reused in different periods of time as they had strong consistency.

Table 3. Validity of test results.

Variable	Item	R count	R table	Result
Religiosity (X)	X1	0.586	0.138	Valid
	X2	0.652	0.138	Valid
	X3	0.650	0.138	Valid
	X4	0.724	0.138	Valid
	X5	0.708	0.138	Valid
	X6	0.669	0.138	Valid
	X7	0.653	0.138	Valid
	X8	0.619	0.138	Valid
	X9	0.631	0.138	Valid
	X10	0.608	0.138	Valid
Self-monitoring (Z)	Z1	0.868	0.138	Valid
	Z2	0.929	0.138	Valid
	Z3	0.882	0.138	Valid
Impulse buying (Y)	Y1	0.759	0.138	Valid
	Y2	0.365	0.138	Valid
	Y3	0.845	0.138	Valid
	Y4	0.788	0.138	Valid
	Y5	0.845	0.138	Valid
	Y6	0.757	0.138	Valid
	Y7	0.710	0.138	Valid

Source: Managed SPSS Data, 2018.

Table 4. Results of reliability test.

Variable	Alpha value	Result
Religiosity	0.846	Valid
Self-monitoring	0.872	Valid
Impulse buying	0.857	Valid

Source: Questionnaire data, 2018.

4.3 *Test of causality*

According to Table 5, religiosity (X) has a positive and significant effect on self-monitoring (Z). This is indicated by a value of t of 7.519 with a standard coefficient of 0.417 and significance value of 0.000. This value is smaller than the specified significance level [α] of 0.05. Thus the research hypothesis which states that religiosity has a positive effect on self-monitoring has been proven. The support a previous study by Carter, et al. (2012) which find that the more religious a person is, the more they will tend to monitor their existence according to their goals (self-monitoring), resulting in higher levels of self-control.

The results presented in Table 6 for hypothesis 2 and 3 explain the direct influence of self-monitoring and religiosity on impulse purchasing (H₂ and H₃, respectively). In proof of hypothesis 2, it can be seen that self-monitoring (Z) has a negative and significant effect on impulse purchasing (Y). This can be seen from the negative value of t of −3.695 with standard coefficient of −0.243 and obtained value of significance Of 0.000. This value is smaller than the specified significance level [α] of 0.05. Thus the research

Table 5. Test results for hypothesis 1.

Coefficients		Unstandardized coefficients		Standardized coefficients		
Model	B	Std. Error	B		T	Sig.
I (Constant)	4.180	1.080			3.870	0.000
Religiosity	0.200	0.027	0.417		7.519	0.000

a. Dependent variable: impulse buying.

Table 6. Test results for hypothesis 2 and 3 coefficients.

Coefficients		Unstandardized coefficients		Standardized coefficients		
Model	B	Std. Error	B		T	Sig.
I (Constant)	23.618	2.977			7.933	0.000
Religiosity	0.135	0.078	0.113		1.721	0.086
Self-monitoring	−0.605	0.064	−0.243		−3.695	0.000

a. Dependent variable: impulse buying.

hypothesis which states that self-monitoring has a negative effect on impulse purchasing has been proven. This result supports the theories of Sharma, et al. (2010b) which states that because of their desire to appear rational and cautious, a person with high levels of self-monitoring will be more motivated to self-control, in this case, to reducing their impulsive nature.

Similarly, for hypothesis 3, the path analysis results show that religiosity can directly affect impulse purchasing. In addition, it can also influence indirectly through self-monitoring as an intervening variable. Based on the direct influence with value of beta of 0.113, the magnitude of indirect influence is $0.113 \times (-0.243) = -0.0275$. These results show a negative effect on the indirect relationship. Thus the research hypothesis which states that there is a negative influence of religiosity on impulse purchase mediated by self-monitoring has been proven. Based on the results of testing, the direct influence of religiosity on the purchase of impulses that positive results and the indirect influence of religiosity on the purchase of impulse is proved that religiosity one will not be able to control the behavior of impulse purchases if self-monitoring of a person is still low. The results of this study support the research of Carter, et al. (2012) which states that the more religious a person is the more he or she will tend to monitor themselves and exercise self-control, including controlling their impulsive nature (Sharma, et al., 2010b).

5 IMPLICATIONS, LIMITATIONS, AND SUGGESTIONS FOR FUTURE RESEARCH

This present study provides both theoretical and practical contributions. In a theoretical context, it develops a model of impulse buying, particularly in terms of individual aspects. The results do not support previous studies, in that they show that the aspect of religiosity and self-monitoring did not definitely bring negative effects to impulse buying. In a practical context, if businesses know that religious individuals with good self-monitoring are also likely to impulse buy this might be a good opportunity for them to reach specific targeted markets by, for instance, offering particular products that fit the needs of individuals' religiosity.

The characteristics of the sample of this study are general and not focused on individuals in one specific religious group/organization in that it measure the general rather than specific religiosity of individuals. It is expected, therefore, that future studies will use more specific sampling. Furthermore, it is also suggested that experimental testing for data analysis to reach more specific measurement of the level of individuals' religiosity could be employed.

6 CONCLUSIONS

The findings and discussion reveal several conclusions. First, religiosity has a positive and significant effects on self-monitoring. This finding also shows that religiosity might make individuals control themselves as they believe that anything do is always being watched by God. The second conclusion is that self-monitoring has a negative and significant impact on impulse purchases, which supports the statement in hypothesis 2 that there is a negative influence of self-monitoring on impulse purchases. Thus the results show that self-monitoring can help a person control themselves in making impulse purchases. The third conclusion is that religiosity has a negative and significant effect on impulse purchases mediated by self-monitoring. The results of this study support hypothesis 3 which states that there is a negative influence of religiosity on the self-monitoring of impulse purchasing. But in testing the direct effect, religiosity has a positive and significant influence on impulse purchasing. This means that even if a person is religious they can still make impulse purchases if self-monitoring is not adequately developed or applied.

REFERENCES

Akram, U., Hui, P., Khan, M.K, Hashim, M. & Rasheed, S. (2016), "Impact of store atmosphere on impulse buying: Moderating effect on demographic variables", *International Journal of U-and e-Service, Science and Technology*, 9(7), 43–60.

Alam, S.S., Rohani, M. & Badrun, H. (2011), "Is religiosity an important determinant on Muslim consumer behaviour in Malaysia". *Journal of Islamic Marketing,* 2(1), 8396.

Carter, E.C., McCullough, M.E & Carver, C.S (2012), "The mediating role of monitoring in the association of religion with self–control", *Social Psychological and Personality Science*, 3(6), 691–697.

Dameyasani, A.W. & Abraham, J. (2013), "Impulse buying, cultural values dimensions and symbolic meaning of money: A study on college students in Indonesia's capital city and its surrounding", *International Journal of Research Studies in Psychology*, 2(4), 35–52.

Fianto, A.Y.A., Hadiwidjojo, D., Aisjah, S. & Solimun (2014), "Development and measurement of Islamic values in consumer behavior research", *International Journal of Business and Management Invention*, 3(9), 1–10.

Hausman, A. (2000), "A multi-method investigation of consumer motivations in impulse buying behavior", *Journal of Consumer Marketing,* 17(2), 403–419.

Karaduman, I. (2016), "The role of religious sensibilities on the relationship between religious rules and hedonic product consumption behavior in Turkey", *International Journal of Humanities And Social Science Invention,* 5(4), 12–20.

Khan, N., Hui, L.H., Chen, T.B. & Hoe, H.Y. (2016), "Impulse buying behavior of generation Y in fashion retail", *International Journal of Business and Management*, 11(1), 144–151.

Khraim, H. (2010), "Measuring religiosity in consumer research from Islamic perspective", *International Journal of Marketing Studies*, 2(2), 166–179.

Mathai, S.T.& Haridas, R. (2014), "Personality – its impact on impulse buying behaviour among the retail customer in Kochin City", *IOSR Journal of Business and Management*, 16(4), 48–55.

Mathras, D., Cohen, A.B., Mandel, N. & Mick, D.G. (2016), "The effects of religion on consumer behavior: A conceptual framework and research agenda", *Journal of Consumer Psychology*, 1–49.

Mokhlis, S. (2009), "Relevancy and measurement of religiosity in consumer behavior research", *International Business Research*, 2(3), 75–84.

Moran, B., Bryant, L. & Kwak, L.E. (2015), "Effect of stress, materialism and external stimuli on online impulse buying", *Journal of Research for Consumer, 27.*

Peter, J.P. & jery, C.O. (2013), *"Consumer Behavior And Marketing Strategy"*, Jakarta; Salemba Empat.

Podoshen, J.S. & Andrzejewski, S.A. (2014), "An examination of the relationships between materialism, conspicuous consumption, impulse buying and brand loyalty", *Journal of Marketing Theory and Practice*, 20(3), 319–333.

Salwa, S.H, Ahmad, M and Ilhaamie, A.G.A. (2017), "A Conceptual Paper: The Effect of Islamic Religiosity on Impulse Buying Behavior", *Journal of Global Business and Social Entrepreneurship (GBSE)*, 1(2), 137–147.

Seinauskiene, B., Jurate, M. & Indre, J. (2015), "The Relationship of Happiness, Impulse Buying and Brand Loyalty", *Procedia-Social and Behavioral Sciences*, 687–693.

Sharma, P., Bharadhwaj, S. & Marshall, R. (2010a), "Exploring Impulse buying and variety seeking by retail shoppers: Toward a common conceptual framework", *Journal of Marketing Management*, 16(5–6), 473–494.

——— (2010b), "Impulse buying and variety seeking: A trait correlates perspective", *Journal of Business Research*, 63, 276–283.

——— (2014), "Exploring impulse buying in service: Toward an integrative framework", *Journal of the Academic Marketing Science*, 42, 154–170.

Worthington, E.L., Jr., Wade, N.G., Hight, T.L., Ripley, J.S., McCullought, M.E. & Berry, J.W. …. O'Conner, L. (2003), "The religious commitment inventory-10: Development, refinement and validation of a brief scale for research and counseling", *Journal of Counseling Psychology*, 50, 84–96.

Xiao, S.H. & Nicholson, N. (2011), "Mapping impulse buying: A behaviour analysis framework for services marketing and consumer research", *The Service Industries Journal*, 31(15), 515–2528.

Zeb, A., Murad, A. & Khurshed, I. (2016), "Comparative study of traditional and online impulse buying in Pakistan", *City University Research Journal*, 6(1), 137–143.

Business Innovation and Development in Emerging Economies – Trinugroho & Lau (Eds)
© 2019 Taylor & Francis Group, London, ISBN 978-1-138-35996-3

Effect of religiosity and reference groups on intention to use Sharia bank products: Mediation role of cognitive and affective attitudes

Muthmainah & M. Cholil
Faculty of Economics and Business, Universitas Sebelas Maret, Indonesia

ABSTRACT: This research paper examines the effect of religiosity and reference groups on the intention to use Sharia bank products with a mediation role of cognitive and affective attitudes. A convenient sampling technique was used to study 300 micro, small, and medium-scale enterprise owners who were Moslems and who resided in Surakarta, Indonesia. Validity test of item instrument was based on loading factor > 0.50, reliability test was based on Cronbach's alpha of > 0.60 and the five variables fulfilled the specified criteria. Hierarchical regression analysis was used in this research. The results show that religiosity had a significant positive effect on affective attitude but did not affect cognitive attitude. The religiosity and the reference groups did not affect intention, while the cognitive and affective attitudes did. The mediation role was only shown by affective attitude, not by cognitive attitude. It means that only the affective attitude fully mediated in the relationship between religiosity and intention to use Sharia bank products. The result of this research provides an interesting probability for further discussions and research.

Keywords: religiosity, reference groups, cognitive and affective attitudes, intention

1 INTRODUCTION

To understand the factors affecting the intention to use Sharia bank products is an interesting subject to research (Sallam & Algammash, 2016). Among the factors affecting intention that is interesting to research according to Kahle et al. (2005) and Farhana (2014) is religiosity. Ajzen (1991) introduces planned behavior (The Theory of Planned Behavior (TPB)) where intention is influenced by attitudes, subjective norms and behavior control. Besides religiosity, the reference groups described by George (2004) is actually similar to subjective norms in the TPB. The attitudes of this study was based on the opinion of Kroenung and Eckhardt (2011), are differentiated into cognitive and affective attitudes. Crites et al. (1994) states that cognitive and affective attitudes affect different outcome consequences.

Wahid and Ahmed (2011), El-Ouafy and Chakir (2015), and Rohmatun and Dewi (2017) state that attitude mediates the relationship between religiosity and intention. Therefore, this research refer their studies, that examines attitude as a mediating variable in the relationship between religiosity and intention, and attitude in terms of cognitive and affective attitude.

According to Zamroni (2013), there is an intention for Micro, Small and Medium-Scale Enterprises' (MSMEs) to use Sharia bank products to expand their business. Bank Indonesia (2015) reported that most MSMEs do not have access to banks. Farhana (2014) explained that the rapid growth of Islamic finance institutions over the past 10 years is interesting enough to be researched.

This research is interesting enough to be conducted because of the reasons stated above. Besides that, there are many inconsistent results in previous research.

2. LITERATURE REVIEW

2.1 *Religiosity*

According to Weaver and Agle (2002), religiosity is defined as an individual faith to God and a commitment to act in accordance with the principles the said God defines. Kahle et al. (2005) explained that religiosity and value have implications for a customer's behavior and that this has not yet been researched widely. Metawa and Almossawi (1998) described that Sharia bank customers in Bahrain initially chose the banks for their religious factor, in which the customers stressed their obedience to Sharia principles.

According to Antonio (2001, p. 5), Islam is a comprehensive way of life. This means an Islamic finance institution must comply with Islamic law.

2.2 *Reference groups*

A reference group according to Bourn (1957) is a group whose perspective/opinion is taken by people, either an individual and/or groups of individuals, in developing openness in attitude, opinion, and behavior. It can be very small (close friends) or fairly big (sports team fan club or political party). As indicated by George (2004), an individual may be positively or negatively motivated by certain reference groups, such as close friends, family, and colleagues, so that they are encouraged to have certain intentions and to behave in a certain way for example, to buy something online. Fishbein and Ajzen (1975) explained that subjective norm reflects one's perception or assumption about expectation for others to behave in a certain way.

2.3 *Cognitive and affective attitudes*

Attitude, according to Fisbein Fishbein and Ajzen (1975) reflects on the evaluative thought or statement to an object, which has an either positive or negative nature, and attitude is an affection or a feeling of stimulation. Berg et al. (2006) explained that focus on aspects of cognitive and affective attitudes will have different consequences of effect.

Shook and Bratianu (2010) explicated that one of the forms of attitude is his/her faith to the possibilities of result achieved. Chaouch (2017) stated that society's diverse perception, attitude, and behavior to Sharia banks are caused by their lack of understanding which is primarily caused by the domination of conventional banks. Furthermore, it is also caused by the legal apparatus and prevailing legislation which have not yet fully accommodated Sharia banks' operation, the limited network of services, and the limited number of human resources and Sharia banking technology.

2.4 *Intention*

According to Ajzen (2015), intention to behave reflects on an individual's subjective probability that he/she will behave in a certain way. TPB predicts an individual's intentional behavior, because it can be considered and planned. This theory is based on the assumption that humans are rational beings and use information systematically. An individual considers the implication of his/her actions before deciding to conduct or not to conduct a certain behavior. Schiffman and Kanuk (2000, p. 208) defined intention as one's attitude in a conative component. Conner et al. (2015) and Ajzen (2015) stated that intention to buy is defined as a relative measurement intended to perform a certain behavior.

2.5 *Formulation of hypotheses*

2.5.1 *Religiosity, and cognitive and affective attitude*
Berg et al. (2006) explained that studies oriented toward cognitive and affective attitude will cause different consequences of attitude. There are a number of research results explaining

the existence of religiosity effect on attitude. One of them is brought forth by Gait and Worthington (2008) who in their study stated that in Australia, religious life becomes a factor affecting customers' behavior in performing transactions with Islamic financial institutions. Alam and Sayuti (2011) and Graafland (2017) concluded that religiosity positively affects attitude toward demands for social products, while Abou-Youssef et al. (2015) and Usman et al. (2017) also found links between religiosity and attitude in using banking services in Egypt.

2.5.2 *Religiosity and intention*
Metawa and Almossawi (1998), and Rehman and Shabbir (2010) concluded that religion plays a profound role in life, and becomes a factor which determines attitude toward choosing banks. Other researchers, such as Al-Khatib et al. (1999) and Metwally (2006) explained that religion and culture affirm the importance of the application of religious criteria in every aspect of a religious followers' behavior, including using banking services. As noted by Gerrard and Cunningham (1997) in their research in Singapore, it was agreed that Moslem customers' behavior was greatly affected by religiosity in each decision and behavior.

Existence of religiosity effect on intention was researched by El-Ouafy and Chakir (2015), and Khayruzzaman (2016). They concluded that there exists an effect of religiosity on intention and the behavior of saving in Islamic banks. Dekhil et al. (2017) found a positive effect of religious orientation on intention to buy branded luxury goods in Tunisia.

However, there also exists research which states otherwise. Alserhan et al. (2014) revealed that religiosity does not affect consumption of branded luxury goods, while Attia (2017) stated that religiosity does not affect intention.

2.5.3 *Reference groups and intention*
Research explaining the existence of the effect of reference groups on intention was conducted by Tse et al. (1997), Ramayah and Lo (2007), Taib et al. (2008), Teo and Pok (2003), and Emad et al. (2010). They found that reference groups had an effect on intention to use internet banking in Jordan. However, there is research that states otherwise, as conducted by Chau and Hu (2001). Omar et al. (2012) found that subjective norms do not have any effect on intention to buy food products.

2.5.4 *Cognitive and affective attitudes and intention*
Research explaining the effect of attitude on intention was conducted by Baumann et al. (2007), among others, who explained that the willingness to recommend intention is affected by attitude. Suki and Suki (2011) stated that attitude has direct, positive, and a strong effect on intention to use mobile devices or services. Mangin et al. (2012) stated that there existed a positive and significant relationship between attitude and intention to use online banking in North America.

Juwaheer et al. (2012) stated that there is a direct and significant effect between attitude and interest in using internet banking in Mauritania. Conner et al. (2015) explained that affective attitude becomes an important predictor to intention and behavior in the health field. Likewise, Hee-dong and Youngjin (2004) stated that cognitive attitude affects intention to buy, while Kroenung and Eckhardt (2011) showed that cognitive attitude with a beta value of –0.105, t-value of –1.173 and a p-value of 0.242 as independent variables, does not have any effect on the relationship of behavior and attitude; affective attitude has an effect on the relationship of behavior and attitude, with a beta value of 0.512, t-value of 5.726 and a p-value of 0.000. However, other researchers have found a different result. For instance, Ahmed et al. (2012), found a positive relationship between attitude and support of a celebrity and intention to buy, but this kind of relationship is significantly weak.

2.5.5 *Cognitive and affective attitude mediation toward effect of religiosity on intention*
Research results demonstrating the existence of the mediating role of attitude, which is then specified into cognitive and affective attitudes toward effect of religiosity on intention, are among others, explained by El-Ouafy and Chakir (2015). They explained a model that

allowed a mediating role of attitude to the relationship between religiosity and intention of using Sharia financial products by MSMEs in Morocco. Here, religiosity was distinguished into Religiosity Affiliation (RA) and Religiosity Commitment (RC). RA affects attitude, but does not affect intention, while RC affects intention, but does not affect attitude. Findings by Rohmatun and Dewi (2017) showed the existence of the effect of religiosity on intention to buy halal cosmetics through attitude.

Based on this theoretical review and supported from earlier research results, the following hypotheses were drawn up:

Hypothesis 1: Religiosity positively and significantly affects cognitive attitude.
Hypothesis 2: Religiosity positively and significantly affects affective attitude.
Hypothesis 3: Religiosity positively and significantly affects intention.
Hypothesis 4: Reference groups positively and significantly affects intention.
Hypothesis 5: Cognitive attitude positively and significantly affects intention.
Hypothesis 6: Affective attitude positively and significantly affects intention.
Hypothesis 7: Cognitive attitude mediates effect of religiosity on intention.
Hypothesis 8: Affective attitude mediates effect of religiosity on intention.

3 METHOD

The research population comprised of MSMEs owners who were Moslems and who resided in five districts of Surakarta city, Indonesia. Sampling was conducted by using the convenient sampling technique, which means that sampling was based on the willingness for people to fill out questionnaires voluntarily. 300 owners of MSMEs were expected to fill out the questionnaires.

Data collection was conducted by spreading questionnaires on each variable by using the Likert scale with five points. Religiosity was based on an opinion by Rehman and Shabbir (2010). The reference group was based on work by Witt (1969), while attitude was based on Taib et al. (2008). Intention to use Sharia bank products was adopted from research by Linan and Chen (2006).

Data analysis was conducted using hierarchical regression. Religiosity and reference groups were used as independent variables, and intention to use Sharia bank products as a dependent variable. Attitude (divided into cognitive and affective attitudes) was used as a mediating variable. In addition, validity and reliability tests were conducted in accordance with normative criteria points for research activities.

4 DATA ANALYSIS

4.1 *Validity, reliability, and classic assumption testing*

Testing of instrument validity was conducted by using the factor analysis technique with a loading factor criterion of > 0.50. The notion used is that the loading factor should be greater than 0.50 in absolute value. Testing of reliability was based on Cronbach's alpha criterion of > 0.60. From the estimation result assisted by IBM SPSS Statistics 22, two items on religiosity and one item of the reference groups were found to be not valid. In terms of reliability, the five variables fulfilled the specified criteria. The results also echoed in terms of the classic assumption test.

4.2 *Description of demography and research variables*

Demographically of all 300 respondents, 53% were women. 54% of total respondents were aged between 31 and 50. 77% of total respondents had MSMEs as their main occupation, and 40% of all respondents had ≤ 3 workers; 43% of them have 4–6 workers. Of all 300 respondents, 29% of them operated their MSMEs for ≤ 5 years, with 47% of them between

6 and 15 years. 76% of all respondents operated their MSMEs in stores, 69% of them operated in outside traditional markets, and 24% of them operated in outside stores. 73% of all respondents pioneered their MSMEs or did not continue it from their parents. The range of the mean research variables showed that most of them were in the medium range; between 3.231–3.659. For the reference group variable, it was considered low, amounting to 1.853 as shown in Table 1.

4.3 *Testing of hypotheses*

The testing of hypotheses was based on the estimation results of effect between variables and meditating effects, as shown in Table 2.

First, on the testing of religiosity effect on cognitive attitude, the coefficient of regression was 0.155, with a p-value of 0.102. Because $p > 0.05$ it was concluded that religiosity had no significant effect on cognitive attitude.

Second, on the testing of religiosity effect on affective attitude, the coefficient of regression was 0.297 with a p-value of 0.001. Due to $p < 0.05$ it was concluded that religiosity had a significant effect on affective attitude.

Third, on the testing of religiosity effect on intention, the coefficient of regression was 0.089, with a p-value of 0.287. Owing to $p > 0.05$ it was concluded that religiosity had no significant effect on intention.

Fourth, on the testing of the reference group effect on intention, the coefficient of regression was 0.020 with a p-value of 0.1711. On account of $p > 0.05$, it was concluded that the reference group had no significant effect on intention.

Table 1. Description of research variables.

Variable	Min.	Maxi.	Mean	SD
Religiosity	2.92	4.69	3.659	0.321
Reference groups	1.00	3.75	1.853	0.503
Cognitive attitude	1.80	4.40	3.351	0.513
Affective attitude	2.00	4.40	3.264	0.485
Intention	1.80	4.40	3.231	0.543

Source: treated primary data (2017).
Noted: Min. = Minimum
Maxi. = Maximum
SD = Standard of Deviation

Table 2. Results of Hypotheses testing.

Hyp.	Relationship path	Coefficient	p	95% CI Lower	Upper	Significance
H_1	Religiosity ➡ Cognitive attitude	0.155	0.102			Non-significant
H_2	Religiosity ➡ Affective attitude	0.297	0.001			Significant
H_3	Religiosity ➡ Intention	0.089	0.287			Non-significant
H_4	Reference groups ➡ Intention	0.020	0.711			Non-significant
H_5	Cognitive attitude ➡ Intention	0.236	0.000			Significant
H_6	Affective attitude ➡ Intention	0.476	0.000			Significant
H_7	Religiosity ➡ Cognitive attitude ➡ intention	0.037	95% CI	−0.007	0.092	Non-significant
H_8	Religiosity ➡ Affective attitude ➡ intention	0.141	95% CI	0.051	0.259	Significant

Source: treated primary data (2017).
Noted: Hyp. = Hypothesis
P = probability value
CI = Confidence Interval

Fifth, on the testing of cognitive attitude effect on intention, the coefficient of regression was 0.236 with a p-value of 0.000. In view of $p < 0.05$ it was concluded that cognitive attitude had a significant effect on intention.

Sixth, on the testing of affective attitude on intention, the coefficient of regression was 0.476 with a p-value of 0.287. As a consequence of $p > 0.05$, it was concluded that affective attitude had a significant effect on intention.

Seventh, on the testing of mediation to religiosity effect on intention, the indirect effect of religiosity on the intention through cognitive attitude mediation, was 0.037 (estimated). The indirect effect in population was situated between –0.007 and 0.092 in a 95% confidence interval. Since zero was contained in a 95% confidence interval, then it could be concluded that indirect effect was not significant at the level of 0.05. Therefore, cognitive attitude was not considered as a mediator of religiosity effect on intention.

Eighth, the indirect effect of religiosity on intention through the affective attitude mediator was 0.141 (estimated). The indirect effect on population was situated between 0.051 and 0.259 in a 95% confidence interval. Since zero was not contained in a 95% confidence interval, then it could be concluded that indirect effect was significant at the level of 0.05. Therefore, affective attitude was considered as a mediator of religiosity effect on intention. With the coefficient of direct effect of religiosity on intention being non-significant, it was concluded that complete mediation happened. This means that the effect of religiosity on intention is fully mediated by affective attitude.

5 DISCUSSION

Religiosity does not positively affect cognitive attitude, but affects affective attitude. For the affective attitude part, it corresponds to the research results of Worthington and Hoffman (2008) and Graafland (2017). For cognitive attitude, it indicates that religiosity values are less based on cognitive attitude (knowledge and perception), in which respondents still lack knowledge about Sharia banking. The cognitive component is built from belief and knowledge of physical product attributes and of their services. Sharia banks' network of services is considered less in meeting customers' aspirations. On the other hand, an affective component according to Lin et al. (2007) represents one's feeling toward a purpose since it is assumed that feelings are important components to hopes and purposes.

The reference groups and religiosity do not directly affect intention. The reference groups do not affect intention, which corresponds to the research results of Hui and Morrow (2001), and Sambamurthy et al. (2003). But the results do not correspond to those of Tse et al. (1997), and Ramayah and Lo (2007). This is because respondents are less oriented toward attitude and intention, which are quite important in performing transactions with Sharia banking institutions. This fact can also be seen from the low description of the perception of MSMEs from the reference group factor of 1.853 with a Standard of Deviation of 0.503. This shows that respondents did not consider it important to look for sources of reference for Islamic banking products and services in meeting funding needs for the development of their businesses. This means that the MSMEs studied are less concerned about the possibility of using Islamic bank products in funding their business. The reliability they do seem to use is still ritualistic and not contextual. The religiosity does not directly affect intention, which corresponds to the research results of Alam and Sayuti (2011), but does not correspond to those of Tse et al. (1997), Ramayah and Lo (2007), Alserhan et al. (2014), and Attia (2017). The religiosity may need to pass certain intervening variables, so that its effect is indirect.

Each of the cognitive and affective attitudes has a positive effect on intention. This corresponds to the research results of Baumann et al. (2007) and Mangin et al. (2012), but does not correspond to those by Ahmed et al. (2012). Such result shows that respondents' attitude, either cognitive or affective, can be directed by religiosity (especially if their religiosity level is built upon their serious and full obedience), and by employing reference groups to encourage the respondents to have a Sharia-conforming lifestyle, including funding their enterprises. Cognitive attitude does not mediate, while the affective attitude does mediate fully the relationship between religiosity and intention. This corresponds to the research results of El-Ouafi and Chakir (2015), Wahid and Ahmed (2011), Rohmatun and Dewi (2017). It indicates the importance of conducting socialization and education cognitively. This has an effect on affective attitude, which in turn affects the respondents' intention to use Sharia banking products in funding their enterprises. Cognitively, respondents did not have the right knowledge about the products and services of Islamic banks, and even carried the impression that Islamic banks were the same as conventional banks. But in preference, as an embodiment of affective attitudes, they intend to use Islamic bank products and services as an alternative source of business funding. From the point of view of knowledge as an aspect of cognitive attitudes of Moslem MSMEs, improvement is needed. Most Moslem MSMEs consider Islamic banks the same as conventional banks; their knowledge is still very limited and even Islamic scholars and community leaders also seem to lack attention, including for themselves and their families. From the affective aspect, their attitude is preferential, meaning that their conscience intends to fulfill Sharia provisions. Even MSMEs still subscribe to a *plecit* bank (money lender) and they are aware that Islam prohibits it because it contains elements of *riba*. From the mediating role of cognitive and affective attitudes toward the influence of religiosity on the intention to use Islamic banking products and services, the affective attitudes mediate, while cognitive attitudes do not mediate the relationship between religiosity and intention. In previous studies, the role of attitude as a mediating variable was not distinguished by cognitive and affective attitudes because according to Berg et al. (2006), the focus on cognitive and affective attitudes will have different consequences of effect.

6 CONCLUSION, IMPLICATION AND LIMITATIONS

6.1 *Conclusion*

Religiosity only positively affects affective attitude and does not have any effect on cognitive attitude. The cognitive and affective attitudes affect intention in a positive way. On the other hand, the reference group does not affect intention directly. The effect of religiosity on intention is only fully mediated by affective attitude, since religiosity does not directly affect intention.

6.2 *Implication*

The theoretical implication is that this research is interesting enough to be developed, since there are still some inconsistencies in the result of effect between variables of religiosity, reference groups, attitude, and intention. The practical implication is that there is a need to improve the socialization and education process, both from Sharia banking institutions and from respected individuals or groups, such as public figures, Islamic organizations, including business people, in order to give integrative, concrete, and active contributions according to their roles and functions.

6.3 Limitations

This research was conducted only in one period of time and therefore it did not fully reveal the result trends which can guarantee illustration in factual reality. The existence of three non-valid questionnaire items also indicated the lack of match between the researcher's expectations and research respondents' aspirations. Further research necessitates the need for a longer period of study time, and the use of better research instruments.

REFERENCES

Abou-Youssef, M.M.H. Abou-Aish, E. & El-Bassiouny, N. (2015) Effects of religiosity on consumer attitudes toward Islamic banking in Egypt. *International Journal of Bank Marketing, 33*(6), 786–807.

Ahmed, A., Mir, F.A. & Farooq, O. (2012). Effect of celebrity endorsement on customers' buying behaviour: A perspective from Pakistan. *Interdisciplinary Journal of Contemporary Research in Business*, *4*(2), 584–592.

Ajzen, I. (1991). The Theory of Planned Behavior. *Organizational Behavior and Human Decision Processes, 50*, 179–211.

Ajzen, I. (2015). Consumer attitudes and behavior: The theory of planned behavior applied to food consumption decisions. *Rivista di Economia Agraria, 70*(2), 121–138.

Alam, S.S. & Sayuti, N.M. (2011). Applying the theory of planned behavior (TPB) in halal food purchasing. *International Journal of Commerce and Management*, *21*(1), 8–20.

Al-Khatib, K., Naser, K. & Jamal, A. (1999). Islamic banking: A study of customer satisfaction and preferences in Jordan. *International Journal of Bank Marketing, 17*, 135–151.

Alserhan, B.A. Bataineh, M.K., Halkias, D. & Komodromos, M. (2014). Measuring luxury brand consumption and female consumers` religiosity in the UAE. *Journal of Developmental Entrepreneurship*, *19*(2), 1–16.

Antonio, M.S. (2001). *Islamic bank from theory to practice.* Jakarta, Indonesia: Gema Insani.

Attia, S.T. (2017). The impact of religiosity as a moderator on attitude towards celebrity endorsement-purchase intentions relationship. *American Journal of Management, 11*(1), 87–98.

Bank Indonesia. (2015). *Profil bisnis usaha mikro, kecil dan menengah, kerjasama lembaga pemgembangan perbankan Indonesia dan Bank Indonesia Business profile of micro, small and medium enterprises.* Jakarta, Indonesia: Bank Indonesia.

Baumann, C., Burton, S., Elliott,G. & Kehr, H.M. (2007). Prediction of attitude and behavioural intentions in retail banking. *International Journal of Bank Marketing*, *25*(2), 102–116.

Berg, H.V.D., Manstead, A.S.R., Pligt. J.V.D. & Wigboldus, D.H.J. (2006). The impact of affective and cognitive focus on attitude formation. *Journal of Experimental Social Psychology*, *42*, 373–379.

Bourne, F.S. (1957). Group influence in marketing and public relations. In R. Likert & S.P. Hayes (Eds.), *Some applications of behavioral research.* Paris, France: United Nation Educational Scientific and Cultural Organization.

Conner, M., McEachan, R., Taylor, N., O'Hara, J. & Lawton, R. (2015). Role of affective attitudes and anticipated affective reactions in predicting health behaviors. *Health Psychology, 34*(6), 642–652.

Chau, P.Y.K. & Hu, P.J. (2001). Information technology acceptance by individual professionals: A model comparison approach. *Decision Sciences*, *32*(4), 699–719.

Chaouch. N. (2017). An exploratory study of Tunisian customers' awareness and perception of Islamic banks. *International Journal of Islamic Economics and Finance Studies, 3*(2),7–32.

Crites, S.L., Fabrigar, L.R. & Petty, R.E. (1994). Measuring the affective and cognitive properties of attitudes: Conceptual and methodological issues. *Personality and Social Psychology Bulletin, 20*, 619–634.

Dekhil, F., Boulebech, H. & Bouslama, N. (2017). Effect of religiosity on luxury consumer behavior: The case of the Tunisian Muslim. *Journal of Islamic Marketing, 8*(1), 74–94.

El-Ouafy, S. & Chakir, A. (2015). The impact of religiosity in explanation of Moroccan very small businesses behaviour toward Islamic financial products. *Journal of Business and Management, 11*(7), 71–76.

Emad, A.S., Pearson, J. M. & Andrew J.S. (2010). Internet banking and customers' acceptance in Jordan: The unified model's perspective. *Communications of the Association for Information Systems*, *26*(23), 493–524.

Farhana, T. (2014). *Religiosity, generational cohort and buying behaviour of Islamic financial products in Bangladesh* (a submitted doctoral dissertation). Victoria University, Wellington, New Zealand.

Fishbein, M. & Ajzen, I. (1975). *Belief, attitude, intention, and behavior: An introduction to theory and research.* Reading, Massachusetts: Addison-Wesley.

Gait, A. & Worthington, A.C. (2008). An empirical survey of individual consumer, business firm and financial institution attitudes towards Islamic methods of finance. *International Journal of Social Economics*, *35*(11), 783–808.

George, J.F. (2004). The theory of planned behavior and internet purchasing. *Internet Research, 14*(3), 198–212.

Gerrard, J.P. & Cunningham, B. (1997). Islamic banking: A study in Singapore. *International Journal of Bank Marketing*, *15*(6), 204–216.

Graafland, J. (2017). Religiosity, attitude, and the demand for socially responsible products. *Journal of Business Ethics, 144*(1), 121–138.

Hee-dong, Y. & Youngjin, Y. (2004). It's all about attitude: Revisiting the technology acceptance model. *Decision Support Systems, 38*, 19–31.

Hui, S.S. & Morrow, J.R. (2001). Level of participation and knowledge of physical activity in Hong Kong Chinese adults and their association with age. *Journal of Aging and Physical Activity*, 9, 372–385.

Juwahir. T.D., Pudaruth, S. & Ramdin, P. (2012). Factors influencing the adoption of internet banking: A case study of commercial banks in Mauritius. *World Journal of Science, Technology and Sustainable Development*, 9(3), 204–234.

Kahle, L.R., Kau, A.K., Tambyah, S.K., Jun, S.J. & Jung, K. (2005). *Religion, religiosity, and values: Implications for consumer behavior*. Paper presented at the Fremantle, Australia and New Zealand Marketing Academy Conference.

Khayruzzaman, S.. (2016). Impact of religiosity on buying behavior of financial products: A literature review. *International Journal of Finance and Banking Research*, 2(1), 18–23.

Kroenung, J. & Eckhardt, A. (2011). Three classes of attitudes and their implications for IS research. *Proceedings of the 32nd International Conference on Information Systems*, Shanghai, China.

Lin, H.C., Morais, B.D., Kerstetter, L.D. & Hou, S.J. (2007). Examining the role of cognitive and affective image in predicting choice across natural, developed, and theme park destinations. *Journal of Travel Research*, 46(2), 183–194.

Linan, L.Y. & Chen, C.S. (2006). The influence of the country-of-origin image, product knowledge and product involvement on consumer purchase decisions: An empirical study of insurance and catering services in Taiwan. *Consumer Marketing*, 23, 1–13.

Mangin, J.L., Bourgault, N., Moriano León, J.A. & Mario Guerrero, M.M. (2012). Testing control, innovation and enjoy as external variables to the technology acceptance model in a North American French banking environment. *International Business Research* 5(2), 13–27.

Metawa, S. A. & Almossawi, M. (1998). Banking behaviour of Islamic bank customer: Perspectives and implication. *International Journal of Bank Marketing*, 16(7), 299–313.

Metwally, E. (2006). Kinetic studies for sorption of some metal ions from aqueous acid solutions onto TDA impregnated resin. *Journal of Radioanalytical and Nuclear Chemistry*, 270(3), 559–566.

Omar, K.M., Kamariah N.M., Imhemed, G.A. & Ahamed Ali, F.M. (2012). The direct effects of halal product actual purchase antecedents among the international Muslim consumers. *American Journal of Economics*, Special Issue, 6, 87–92.

Ramayah, T. & Lo, M.C. (2007). Impact of shared beliefs on perceived usefulness and ease of use in the implementation of an enterprise resource planning (ERP) system. *Management Research News*, 30(6), 420–431.

Rehman, A. & Shabbir, M.S. (2010). The relationship between religiosity and new product adoption. *Journal of Islamic Marketing*, 1(1), 63–69.

Rohmatun, K.I. & Dewi, C.K. (2017). Pengaruh pengetahuan dan religiusitas terhadap niat beli pada kosmetik halal melalui sikap The effect of knowledge and religiosity in buying intention on halal cosmetics through attitude. *Jurnal Ecodemica*, 1(1), 27–35.

Sallam, M.A. & Algammash, F.A. (2016). The effect of attitude toward advertisement on attitude toward brand and purchase intention. *International Journal of Economics, Commerce and Management*, 4(2), 509–520.

Sambamurthy, V., Bharadwaj, A. & Grover, V. (2003). Shaping agility through digital options: Re-conceptualizing the role of IT in contemporary firms. *Management Information System Quarterly*, 27, 237–263.

Schiffman, L.G. & Kanuk, L. (2000). *Consumer Behavior*. Prentice, Wisconsin: Hall International.

Shook C. & Bratianu, C. (2010). Entrepreneurial intent in a transitional economy: An application of the theory of planned behavior to Romanian students. *International Entrepreneurship and Management Journal*, 6, 231–247.

Suki, N.M. & Suki. N.M. (2011). Exploring the relationship between perceived usefulness, perceived ease of use, perceived enjoyment, attitude and subscribers' intention towards using 3G mobile services. *Journal of Information Technology Management*, 2(1), 1–7.

Taib, F.M., Ramayah, T. & Razak, D.A. (2008). Factor influencing intention to use diminishing partnership home financing. *International Journal of Islamic and Middle Eastern Finance and Management*, 1(3), 235–248.

Teo, T.S.H. & Pok, S.H. (2003). Adoption of WAP-enabled mobile phones among internet users. *The International Journal of Management Science*, 31, 483–498.

Tse, D.K., Pan, Y. & Y Au, K. (1997). How MNCs choose entry modes and form alliances: The China experience. *Journal of International Business Studies*, 28(4), 779–805.

Usman, H., Tjiptoherijanto, P., Ezni Balqiah, T. & Ngurah Agung, I.G. (2017). The role of religious norms, trust, importance of attributes and information sources in the relationship between religiosity and selection of the Islamic bank. *Journal of Islamic Marketing*, 8(2),158–186.

Wahid, N.A. & Ahmed, M. (2011). The effect of attitude toward advertisement on Yemeni female consumers' attitude toward brand and purchase intention. *Global Business and Management Research*, *3*(1), 21–29.

Weaver, G.R., & B.R. Agle. (2002). Religiosity and ethical behavior in organizations: A symbolic interactionist perspective. *Academy of management review, 27*(1), 77–97.

Witt, R.E. (1969). Informal social group influence on consumer behaviour choice. *Journal of Marketing Research*, *6*, 473–477.

Worthington, A.C. & Hoffman, M. (2008). An empirical survey of residential water demand modelling. *Journal of Economic Survey, 22*(5), 842–871.

Zamroni, Z. (2013). Peran bank Syariah dalam penyaluran dana bagi usaha mikro kecil dan menengah (Umkm) The role of Sharia bank in funding micro, small, and medium enterprises *iqtishadia.*, *6*(2), 225–240.

411

Business Innovation and Development in Emerging Economies – Trinugroho & Lau (Eds)
© 2019 Taylor & Francis Group, London, ISBN 978-1-138-35996-3

Identifying antecedents of loyalty to public transportation: A case study of the online taxi motorbike service GrabBike

Vita Briliana & Adelpho Billy Adkatin
Trisakti School of Management, Jakarta, Indonesia

ABSTRACT: The purpose of this study was to determine the impact of selected marketing constructs (perceived value, customer satisfaction, brand image, customer engagement, and trust) towards customer loyalty in the context of GrabBike, an online taxi transportation company in Jakarta. Data sampling for this study was purposive sampling via a questionnaire of 180 respondents. The methods for the data analysis was Structural Equation Modeling (SEM) using SmartPLS 2.0. The research has shown the impact of customer engagement, customer satisfaction, perceived value, and consumer trust toward customer loyalty are significant, and supported, while brand image have no significant effect.

1 INTRODUCTION

Transit demand increases as cities grow and government policies encourage modal shifts away from private automobiles. Hence the need for public transportation has grown. According to an online report by The Jakarta Post (JP, 2018), a survey from the Inrix 2017 Traffic Scorecard ranked Jakarta just below Bangkok and above Washington for traffic congestion. Jakarta's situation worsened in 2017, with the city moving into 12th position among the most congested cities in the world, compared to its 2016 position of 22nd. Meanwhile, the National Development Planning Agency (Bappenas) estimated that traffic congestion in Jakarta caused economic losses of about IDR67.5 trillion (US$4.73 billion) throughout 2017. In addition, as the capital of Indonesia, Jakarta has a large population, with over 10 million people across six regions in the central, east, west, south, north and Seribu Island (Jakarta in Figures, 2017). An online news report released by Suara Pembaharuan stated that this number can increase by up to 30% in the morning as the result of employees and students coming from outside the area (Beritasatu, 2017). Compounding this situation, the number of motor vehicles in Jakarta keeps growing. As of December 2016, Jakarta had over 18.6 million registered motor vehicles, and this number was growing by 100,000 per month (Jakarta in Figures, 2017).

As a result of the rising number of residents and increasing density, along with a high and growing number of vehicles, some people have begun to think up sophisticated new transport solutions. As a result, the creation of a unique service that never existed before was born. Recently, a public transportation service based on mobile apps began mass distribution. According to the Association of Internet Service Companies Indonesia (APJII, 2017), Indonesia was reported to have more than 143.25 million internet users in 2017. Of these, over 63.1 million (47.6 percent) used mobile devices to connect with the internet. Due to this high level of usage, there emerged a service, namely Grab, that offered one solution to the traffic woes in Jakarta through the use of advanced technology: a taxi motorbike accessed through a mobile device. Headquartered in Singapore, GrabTaxi Holdings Pte Ltd. is Southeast Asia's leading ride hailing platform, and has been available in Malaysia, Singapore, Thailand, Vietnam, Indonesia, and the Philippines. Indonesia has become Grab's largest market by offering rides across all platforms. Grab has primarily focused on the greater Jakarta market, which is home to more than 30 million people, where it offers bikes, private cars, and taxi-hailing, and expects to expand the multi-service platform.

To market this service, the innovators knew that they needed to expand the service thinking outside the narrow commercial orientation in order to meet the public need. Therefore, the service company must be able to make a breakthrough in which its services not only meet the need for usefulness and quality but also on the emotional side, because consumer attitudes and behavior has a strong impact on customer loyalty, especially in public transportation.

Customer loyalty may be defined as intention to re-purchase the good or service, as well as resistance to switching to competitors and a willingness to recommend the product to others. Factors affecting customer loyalty in Indonesia have been researched in areas such as the country's high-frills airline industry (Hapsari et al., 2017). However, there seems to be a knowledge gap in that customer loyalty has not been researched in the context of Indonesian public transportation. There are still very few empirical studies on trust of public transport. The importance of this present study can be seen from at least two viewpoints. First, discussion about the subject's interrelationship among customer trust, customer satisfaction, customer engagement, and customer loyalty, especially in the public transport context, is still limited, so this study can fill the gap in the literature. Second, this present study focuses on public transit in Indonesia, which saw its modal share (i.e., portion of travelers using the system) decline from 35.5% in 2002 to 12.9% in 2010 (JUTPI in Sumaedi et al., 2015), due to its underdevelopment, difficulty of access, inefficiency, and poor safety record. The aim of this study is to examine engagement on customer loyalty in a comprehensive modeling framework of Customer Engagement, Perceived Value, Customer Satisfaction, Brand Image, Consumer Trust, and Customer Loyalty in the context of an online taxi motorbike company.

2 LITERATURE REVIEW

Services companies need to focus on the criteria of customer loyalty as part of their effort to build a strong relationship with its customers. By applying customer relationship marketing, a business organization will obtain long-term customer profitability through customer loyalty (Kotler and Keller, 2016). Based on the service processes, public transportation services can be classified into type services in mass, for this organization that provide services with minimal customizations, little time interaction between customers and service personnel and involve many customers in the service delivery such as airlines, public transportation, restaurants etc.

2.1 Customer engagement

Kotler and Keller (2016, 33) define engagement as the extent of a customer's attention and active involvement with a communication. It reflects a much greater response than a mere impression and is more likely to create value for the firm. Schiffman and Wisenbilt (2015, 165) describe involvement as the degree of consumer information and communication towards products or services. A study by Hapsari et al., (2017) revealed that consumer involvement is the degree of personal relevance that the product or purchase holds for the consumer. The author gave a description that lists the five dimensions of the customer's touch points: identification, attention, enthusiasm, absorption, and interaction. Hence, in regards to the degree in which level of interaction relates to the loyalty of the service consumer, more interaction (attention, information search, etc.) means more loyalty.

H1: Customer Engagement has a significant impact on customer loyalty

2.2 Perceived value

According to Zeithaml et al., (2013, 453–455), perceived value is the difference between the cost (time, money, energy) and the benefit received by the customers. Lai and Chen (2011) found that a customer's perceived value has a positive effect on customer satisfaction: the higher the perception of value offered, the higher the satisfaction of the public transport user. The customers who are satisfied with the service will be more likely to have a long and positive interaction with the company (Van Doorn et al., 2010). The customer who is satisfied with the service tends to be loyal, reflected by favorable behavioral intention (Hu et al., 2009; Jen et al., 2011). From this, we can assume that a higher benefit means higher satisfaction, and the higher benefit will lead to a higher level of positive communication and loyalty.

H2: Perceived value has a significant impact on customer satisfaction
H3: Perceived value has a significant impact on customer loyalty
H4: Perceived value has a significant impact on customer engagement

2.3 Customer satisfaction

Kotler and Keller (2016, 153) noted that satisfaction is a person's feeling of pleasure or lack of disappointment that result from comparing a product or service's perceived performance (or outcome) to expectations. This means that we feel satisfied after we consume the product or service and that the result can become internal information to remind us of our experience to create new expectations in the future buying process. Rust and Oliver (1994, 2) defined customer satisfaction as a summary of cognitive and affective reactions to a service incident that results from the comparison of customers' perceptions of service quality with their expectations of service performance. In this research we can assume that satisfaction will lead to a positive image of brand performance, and also develop a positive interaction in service delivery which may lead to a commitment from the consumer.

H5: Customer satisfaction has a significant impact on brand image
H6: Customer satisfaction has a significant impact on customer engagement
H7: Customer satisfaction has a significant impact on customer loyalty

2.4 Brand image

According to Aaker's (1991, 109) definition, brand image is "a set of associations, usually organized in some meaningful way." Brand image is also how the brand is perceived. Basically, a consumer may compare the uniqueness of the brand if he already has a mental image of the brand itself. Furthermore, as described by Iversen and Hem (2008), a consumer could easily associate with the brand because the brand image acts as a personal symbol, which consists of all its descriptive and evaluative information. Brand image is an important factor in studying a buyer's behavior. The reason is that, when the consumer acquires his favorite brand, the researcher regards the brand as a measure of customer loyalty since it addresses brand performance. Brand image can also play a role in accessing touch points since it refers to psychological and social needs.

H8: Brand image has a significant impact on customer loyalty
H9: Brand image has a significant impact on customer engagement

2.5 Consumer trust

Izogo (2016) points out that it is important to acknowledge the existence of different approaches to the interactions among satisfaction, trust, and loyalty. Briliana (2017) explains that trust has been recognized as an important influence on customer commitment and hence on loyalty. Moreover, consumers will trust services if they feel are secure and reliable, they also believe that these services act in the consumers' best interests. Previous research from Jimenez et al., (2016) revealed that there is a relationship between satisfaction and trust on m-commerce patterns because previous positive experiences will allow for trade in the context of mobile purchases to be considered reliable. Consumer loyalty can be considered the final stage of a relational process in which trust has been built and customer satisfaction has been achieved (Jimenez et al., 2016).

H10: Consumer Satisfaction has a significant impact on consumer trust

2.6 Customer loyalty

Lovelock and Wirtz (2011, 338) defined loyalty as an old-fashioned word traditionally used to describe fidelity and enthusiastic devotion to a country, a cause, or an individual. More recently, it has been used in a business context to describe a customer's willingness to continue patronizing a firm over the long term, preferably on an exclusive basis, and recommending the firm's products to friends or associates. The loyal customer will keep informing the brand even in bad situations. Hapsari et al., (2017) define the measure of customer loyalty as a customer's intention to re-patronize and recommend a service to other people and remain loyal to the organization. Trust in recent years is one of many factors in marketing research to determine the important role of satisfaction, loyalty, and intention. The link between consumer trust and loyalty is also likely to endure in a service failure context.

H11: Consumer trust has a significant impact on customer loyalty

2.7 Mediation effect

Previous studies (e.g., Hapsari et al., 2017) have found that CS mediates the relationship between PV and CL in the airline sector. They also report that CE mediates the relationship between PV and CL. In addition CS partially mediates the relationship between CE and CL. Based on Hapsari and colleagues' (2017) findings, this study proposes the following hypotheses:

H12: Customer satisfaction mediates the relationship between perceived value and customer loyalty.
H13: Customer satisfaction mediates the relationship between perceived value and customer loyalty.
H14: Customer trust mediates the relationship between customer satisfaction and customer loyalty.
H15: Customer engagement mediates the relationship between customer satisfaction and customer loyalty.

3 THEORETICAL FRAMEWORK AND HYPOTHESIS FORMULATION

Based on the review of related literature (Hapsari et al., 2017; Jimenez et al., 2016), this study fills the gap by proposing a research objective to understand the relationships between constructs. The main objective is to investigate interrelationships among Customer Engagement, Perceived Value, Customer Satisfaction, Brand Image, Consumer Trust, and Customer Loyalty in the context of online taxi motorbike public transportation. This study has proposed a research model (Figure 1). The following hypotheses were put forward in this study:

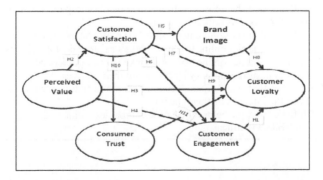

Figure 1. Conceptual framework.

Table 1. Measurement items of the constructs.

Constructs	Measurement items	Reference
Customer Loyalty	CL1: I intend to say positive thing about this online taxi motorbike CL2: I Intend to go with online taxi motorbike in the future CL3: I Intend to encourage relatives to go with online taxi motorbike. CL4: Overall, given the other choices of online taxi motorbike companies, I will remain riding with this online taxi motorbike.	Hapsari et al., (2017)
Customer Engagement	CE1: I am proud of this company success CE2: When someone praises this online taxi motorbike, it feels like a personal compliment CE3: I am passionate about this online taxi motorbike CE4: I Pay a lot of attention to any information about this online taxi motorbike CE5: When Interacting with this online taxi motorbike it is difficult to detach myself. CE6: I am immersed in my interaction with this online taxi motorbike CE7: I am someone who enjoys interacting with like-minded others that riding with this online taxi motorbike CE8: In general, I thoroughly enjoy exchanging ideas with other people that riding with this online taxi motorbike	Hapsari et al., (2017)
Perceived Value	PV1: Considering the voucher price I paid for riding the online taxi motorbike, I believe that the online taxi motorbike offers excellent services. PV2: Compared to what I have given up (including money, energy, time, and effort), the overall service of this online taxi motorbike is excellent. PV3: Overall, this online taxi motorbike offers good value for money. PV4: Overall, this online taxi motorbike's services and goods are valuable	Hapsari et al., (2017) and Brodie et al., (2009),
Brand Image	BI1: I believe that this online taxi motorbike has a better image than its competitors. BI2: This online taxi motorbike has a good reputation for safety. BI3: I have always had a good impression of this online taxi motorbike BI4: Overall, I believe that this online taxi motorbike has a positive image in the marketplace.	Hapsari et al., (2017)
Customer Satisfaction	CS1: I had a satisfying experience flying with this online taxi motorbike CS2: I did the right thing when I choose to riding with this online taxi motorbike CS3: I normally have a pleasant riding with this online taxi motorbike CS4: Overall, this online taxi motorbike provides a very satisfying experience.	Hapsari et al., (2017) Park et al.'s (2006), and Jimenez et al., (2016)
Consumer Trust	CT1: Belief that online taxi motorbike's will fulfill its promise CT2: Belief that the information provided by the companies online taxi motorbike's is honest CT3: Belief that the companies online taxi motorbike's are reliable CT4: the companies online taxi motorbike's never make false claims CT5: the companies online taxi motorbike's are characterized by sincerity and transparency by providing their services CT6: Belief that the companies online taxi motorbike's are concerned about the interests of their clients CT7: Belief that the behavior of t the companies online taxi motorbike's is ethical	Jimenez et al., (2016)

4 RESEARCH METHODOLOGY

Structural equation modeling (SEM) is a second generation multivariate data analysis method that simultaneously analyzes multiple variables representing measurements associated with individuals, companies, events, activities, situations, and so on. SEM is used to explore or confirm a theory (Hair et al., 2017, 31). As suggested by Hair et al., (2017), we conducted PLS-SEM to measure the characteristic of constructs in a model relationship (latent variables). PLS-SEM can better handle formative measurement models (validity assessment, significance and relevance of indicator weights, and indicator collinearity) and has advantages when sample sizes are relatively small.

4.1 *Sample and population*

Population is defined as the aggregate of all the elements that share some common set of characteristics and that comprise the universe for the purpose of the marketing research problem (Malhotra, 2015). According to Roscoe (1975) in Sekaran and Bougie (2016, 264), sample sizes larger than 30 and less than 500 are appropriate for most multivariate research projects. The sample size should be several times (preferably ten or more) larger than the number of variables in the study. Based on Hair et al., (2017, 24), the sample sizes should be equal to the larger of: 1) 10 times the largest number of formative indicators used to measure a single construct; or 2) 10 times the largest number of structural paths directed at a particular construct in the structural model. This rule of thumb is equivalent to saying that the minimum sample size should be 10 times the maximum number of arrowheads pointing at a latent variable anywhere in the PLS path model. Therefore, it can be concluded that the researcher will use a range of 100–200 samples based on the references above and by using Hair et al., (2017) as a basic indicator measurement (10 times 11 arrowheads) of GrabBike users. As suggested by Briliana et al., (2015), we conducted the justification of respondents' selection habits using application online is a respondent who has a motivation, ability, and opportunity for using application online of the online taxi motorbike service. Respondent criteria for this research are: minimum age of 17 years; living or working in Jakarta; occupational status of student or entrepreneur; and routinely using GrabBike and registered with GrabPay.

5 RESULT AND DISCUSSION

Based on data gathering process, the result consist of characteristics from respondents participation as Table 2 shown below:

Table 2. Characteristics sample.

Characteristics	Category	Frequency	%
Gender	Male	65	36
	Female	115	64
Age	17–21 years old	98	54.4
	22–26 years old	42	23.4
	27–31 years old	25	13.8
	>30 years old	15	8.4
Occupation	Student/College Student	126	70
	Student also employee	29	16.1
	Employee	18	10
	Entrepreneur	7	3.9
Admitted that they using	Less than one year	4	2.2
GrabBike	Since 1 until 2 years ago	75	41.7
	More than 2 years ago	101	56.1

Note: Figures in parentheses show the percentages to the total number of respondents.

Table 3 shows the results of the measurement model. The item reliability, composite reliability (CR), and average variance extracted (AVE) (Hair et al., 2017) support the convergent validity of the CFA results. The Composite Reliability, which shows the degree to which constructs indicators demonstrate the latent construct, ranged from 0.839 to 0.922, which was above the cutoff value of 0.7. The average variance extracted (AVE) ranged from 0.567 to 0.736, which was greater than 0.50, justifying the use of the construct. Overall, these results indicate that the measurement model has good convergent validity. The variance explained, R^2, are 0.348 (Brand Image), 0.354 (Consumer Trust), 0.257 (Customer Engagement), 0.385 (Customer Loyalty), and 0.111 (Customer Satisfaction) respectively.

The PLS analysis results shown in Table 4 illustrates all the hypothesized relationships in this study.

This research accepted the first hypothesis (H1) which stated that the customer engagement affects the customer loyalty of online taxi motorbike passengers positively and significantly ($t = 20.077$, $\beta = 0.565$, $p < 0.01$). These findings are consistent with previous studies about the relationship between customer engagement and customer loyalty (Hapsari et al., 2017). Furthermore, our finding showed that the CE has the strongest impact on customer loyalty. Given that the path coefficient indicates a positive sign, higher customer engagement results in higher customer loyalty this online taxi motorbike's passenger. Previous studies indicate that the higher the customer perception of the value of the service provided is, then the higher their possibility is to engage with a particular firm (Brodie et al., 2011).

Table 3.   Convergent validity and composite reliability.

Variable	Items	Loading	AVE	Composite reliability	R square	Cronbach' alpha
Brand	BI1	0.784	0.724	0.913		0.873
Image	BI2	0.897			0.348	
	BI3	0.889				
	BI4	0.827				
Customer	CE1	0.691	0.682	0.881		0.847
Engagement	CE2	0.637				
	CE3	0.653				
	CE4	0.733			0.257	
	CE5	0.660				
	CE6	0.746				
	CE7	0.707				
	CE8	0.719				
Customer	CL1	0.806	0.567	0.839		0.745
Loyalty	CL2	0.739			0.385	
	CL3	0.770				
	CL4	0.690				
Customer	CS1	0.815	0.736	0.917		0.880
Satisfaction	CS2	0.878			0.111	
	CS3	0.894				
	CS4	0.840				
Consumer	CT1	0.726	0.638	0.922		0.901
Trust	CT2	0.787				
	CT3	0.823				
	CT4	0.811			0.354	
	CT5	0.796				
	CT6	0.792				
	CT7	0.806				
Perceived	PV1	0.828	0.704	0.905		0.861
Value	PV2	0.875				
	PV3	0.824				
	PV4	0.829				

Table 4. Summary of the structural model.

Hypotheses	Path	Original sample (O)	Standard deviation (STDEV)	Standard error (STERR)	T statistics (|O/STDEV|)	Supported
H1	CE -> CL	0.565	0.028	0.028	20.077**	Yes
H2	PV -> CS	0.333	0.034	0.034	9.527**	Yes
H3	PV -> CL	0.113	0.040	0.040	2.822**	Yes
H4	PV -> CE	0.433	0.033	0.033	12.822**	Yes
H5	CS -> BI	0.590	0.029	0.029	20.249**	Yes
H6	CS -> CE	0.069	0.049	0.049	1.398	No
H7	CS -> CL	−0.067	0.040	0.040	1.676*	Yes
H8	BI -> CL	0.048	0.037	0.037	1.299	No
H9	BI -> CE	0.075	0.053	0.053	1.421	No
H10	CS -> CT	0.595	0.026	0.026	22.569**	Yes
H11	CT -> CL	−0.063	0.035	0.035	1.820*	Yes

Notes: If the t-value is greater than 1.65; *p < 0.05; 2.57; **p < 0.01

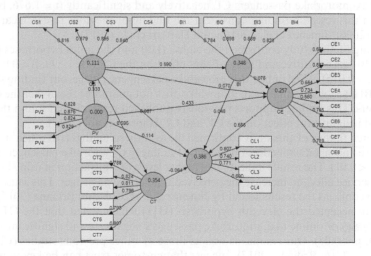

Figure 2. Results of the path analysis.

The second hypothesis (H2) of this study stated that perceived value has a positive and significant effect on the customer satisfaction of online taxi motorbike passengers. The results showed that the hypothesis were accepted (t = 9.527, β = 0.333, p < 0.01). The previous study found that perceived value is the strongest antecedent of customer satisfaction in restaurants, and heritage tourism (McDougall and Levesque, 2000). In the airline industry, perceived value is also noted as a significant antecedent of passenger loyalty (Hapsari et al., 2017; Park et al., 2006).

The research result accepted the third hypothesis (H3) which stated that PV affects the online taxi motorbike passengers CL positively and significantly (t = 2.822, β = 0.113, p < 0.01), our finding inconsistent the finding of previous research carried out by Chen (2008), and Hapsari et al., (2017). Based on a study by Park et al., (2006) found that perceived value had a direct and significant effect on customer loyalty. Hence, it is important to ensure that customers perceive a high value of service in order to create customer loyalty. The fourth hypothesis (H4) of this study stated that PV influences the CE of online taxi motorbike passengers positively and significantly (t = 12.822, β = 0.433, p < 0.01). Our finding is inconsistent with the finding of other researcher in airline transport services context (Hapsari et al., 2017).

The result of this study indicated that the fifth hypothesis (H5) which stated that CS affects the BI of the online taxi motorbike passengers positively and significantly was accepted

Table 5. Results of mediation test using PLS analysis.

	IV	M	DV	Point estimate	SE	t-value		Supported
H12	PV	CS	CL	0.051	0.014	3.567	Fully mediated	Yes
H13	PV	CE	CL	0.065	0.017	3.624	Fully mediated	Yes
H14	CS	CT	CL	0.105	0.022	4.474	Fully mediated	Yes
H15	CS	CE	CL	0.418	0.048	8.607	Fully mediated	Yes

Note: * Significance at p < 0.05, IV refers to independent variable; M refers to mediator; DV refers to dependent variable.

(t = 20.249, β = 0.590, p < 0.01). In literature of public transport services, our finding supports the finding of Park et al.'s (2006) and Hapsari et al., (2017). The finding of this study rejected the sixth hypothesis (H6) that states that CS affects the CE of the online taxi motorbike passengers positively and significantly (t = 1.398, β = 0.069). This finding is inconsistent with the finding of other researcher in airline transport services context (Hapsari et al., 2017). Furthermore the research result accepted the seventh hypothesis (H7) which stated that CS affects the online taxi motorbike passengers CL negatively and significantly (t = 1.676, β = − 0.067, p < 0.05). In this research finding consistent the finding of previous research carried out by Park et al., (2006), Jimenez et al., (2016), and Hapsari et al., (2017).

The finding of this study confirmed the eighth (H8) and the ninth hypotheses (H9) which stated that BI has a positive and insignificant effect on the CL (t = 1.299, β = 0.048) and CE (t = 1.421, β = 0.075) of the online taxi motorbike passengers. Hence, our finding showed that the higher BI level of an online taxi, motorbike does not guarantee the higher the CL and CE of passengers. This finding supports previously published research that found that BI has a positive and insignificant effect on the CL and CE Hapsari et al., (2017).

The research finding of this study confirmed the hypothesis tenth (H10) which stated that CS has a positive and significant effect on the CT of online taxi motorbike passengers (t = 22.569, β = 0.595, p < 0.01). The results indicate that higher levels of satisfaction in using of taxi online motorbike service affect an increase the perceived trust of the passenger. This satisfaction-trust relationship confirms the results found by Jimenez et al., (2016) and Briliana (2017).

This result accepted the eleventh hypothesis (H11) which stated that the CT affects the customer loyalty of online taxi, motorbike passengers negatively and significantly (t = 1.820, β = − 0.063, p < 0.05). Our results, as well as those of other previous studies (Chang, 2015; Jimenez et al., 2016; Briliana, 2017), suggest that customer trust can be key in maintaining relationships with the company.

The outcome of the mediation analysis (H12) shows that CS with the online taxi motorbike fully mediates PV and CL (t-value = 3.567). Similarly H13 reveals that CE fully mediates the relationship between PV and CL (t-value = 3.624). Lastly, the PLS analysis of H14 reveals that CT mediates the relationship between CS and CL (t-value = 4.474), and H15 is also supported (t-value = 8.607), concluding that CE mediates the relationship between CS and CL. This current study is supported by an earlier one that explains the importance of the effects of satisfaction and loyalty, which are mediated by CE and CT, respectively, especially in the transportation industry (Hapsari et al., 2017).

6 CONCLUSIONS

In the modern era of technology and global competition, product and services are becoming less-standardized and customers are demanding more personalized products and services. Hence, technological and communication urgently needed to make interactive dialogue between services provider and customer in identifying what the customer really needed for public transportation. When the customer is satisfied with their riding experience of the online taxi motorbike service and satisfied with the decision to ride with a particular public transportation services company, then they are more likely to say positive things about the public transportation services to other people. In addition, satisfied with the online taxi motorbike services, passengers tend to choose the same public transportation services over the other. Hence, it is essential for an online taxi, motorbike service company to satisfy its customers since its loyal customers benefit them in many aspects, especially in enhancing the firm's profitability.

REFERENCES

Aaker, David A. (1996). Building Strong Brands. United States: The Free Press.

Association of Internet Service Companies Indonesia – APJII (2017), Indonesia was reported to have more than 143.25 million internet users in 2017. Retrieved April 9, 2018 from https://apjii.or.id/survei2017/download/fqn7axhjkvARUVTmC2d8oi9EGgulz4.

BPS. (2017). Jakarta in Figures, 2017. available at https://jakarta.bps.go.id/publication/2017/08/16/3e3564fb6453d384983128b0/provinsi-dki-jakarta-dalam-angka-2017.html (accessed January 20, 2018).

Briliana Vita, NA, Wahid & Y, Fernando (2015). The Effect of Motivation, Opportunity, Ability and Social Identity Towards Customer-to-Customer Online Know-How Exchange. *Advanced Science Letters*, Volume 21, Number 4, April 2015, pp. 819–822(4). https://doi.org/10.1166/asl.2015.5887.

Briliana, Vita. (2017). Identifying Antecedents and Outcomes of Brand Loyalty: A Case of Apparel Brands in Social Media. Papers presented at *Global Conference on Business and Economics Research* (GCBER) 2017, held at Universiti Putra Malaysia, Malaysia 14–15 August 2017 (pp.319–326).

Brodie, R.J., Hollebeek, L.D., Juriæ, B., & Iliæ, A. (2011). Customer Engagement. Journal of Service Research, 14(3), 252–271.

Chang, C.C. (2015), "Exploring mobile application customer loyalty: the moderating effect of use contexts", Telecommunications Policy, Vol. 39 No. 8, pp. 678–690.

Chen, C.F. (2008). Investigating structural relationship between service quality, perceived value, satisfaction, and behavioral intentions for air passengers: Evidence from Taiwan. Transportation Research Part A: Policy and Practice, 42(4), 709–717.

Grab Indonesia. (2016). Gratis Naik GrabBike Khusus untuk Pengguna Pertama (14–20 Maret) https://www.grab.com/id/en/blog/gratis-naik-grabbike-khusus-untuk-pengguna pertama-14–20-maret-2016/.

Hair, Jr. Joseph F., G. Tomas M. Hult., Christian M. Ringle, and Marko Sarstedt. (2017). A Primer on Partial Least Squares Structural Equation Modeling (PLS-SEM). 2nd ed. Sage: United States.

Hapsari, R., Michael D. Clemes, and David Dean. (2017). The Impact of Service Quality and Customer Engagement and Selected Marketing Construct on Airline Passenger Loyalty. *International Journal of Quality and Service Sciences*. Vol. 9, No. 1:1–36.

Hayes, A.F. (2013). Introduction to mediation, moderation, and conditional process modeling: A regression-based approach. New York: The Guilford Press.

Hu, Hsin-Hui, Jay Kandampully and Thanika Devi Juwaheer. (2009). Relationships and Impacts of Service Quality, Perceived Value, Customer Satisfaction, and Image: An empirical study. *The Service Industries Journal*, Vol. 29, No. 2:111–125.

Iversen, N.M., & Hem, L.E. (2008). Provenance associations as core values of place umbrella brands. *European Journal of Marketing*, 42, 5/6,603–626.

Izogo, E.E. (2016), "Should relationship quality be measured as a disaggregated or a composite construct?" *Management Research Review*, Vol. 39 No. 1, pp. 115–131.

Jakarta post (2017), Jakarta's traffic worsens in 2017: Survey. Retrieved April 9, 2018 from www.thejakartapost.com/news/2018/02/26/jakartas-traffic-worse-in-2017-survey.html

Jen, William, Rungting Tu and Tim Lu. (2011). Managing passenger behavioral intention: An Integrated Framework for Service Quality, Satisfaction, Perceived Value, and Switching Barriers. *Transportation*, Vol. 38, No. 2:321–342.

Jimenez, Nadia, Sonia San-Martin. and Jose Ignacio Azuela. (2016). Trust and Satisfaction: The Keys to Client Loyalty in Mobile Commerce. *Academia Revista Latinoamericana de Administración*. Vol. 29, No. 4:486–510.

Kotler, Philip and Kevin L. Keller.(2016). Marketing Management. 15ed. Pearson: United States.

Lai, Wen-Tai, and Ching-Fu Chen. (2011). Behavioral intentions of public transit passengers: The roles of Service Quality, Perceived Value, Satisfaction and Involvement. *Transport Policy*. Vol. 18, No. 2:318–325.

Lovelock,Christopher and Jochen Wirtz. (2011). Services Marketing. 7th ed. Pearson: New Jersey.

Malhotra, Naresh K. 2015. Essentials of Marketing Research: *A Hands-On Orientation*. Pearson Global Edition: United States.

McDougall, G.H.G., & Levesque, T. (2000). Customer satisfaction with services: putting perceived value into the equation. *Journal of Services Marketing*, 14(5), 392–410.

Park, J.W., Robertson, R., & Wu, C.L. (2006). Modelling the impact of airline service quality and marketing variables on passengers' future behavioural intentions. *Transportation Planning and Technology, 29*(5), 359–381.

Rust, R.T. and Oliver, R.L. (1994). Service Quality: Insights and Managerial Implication from the Frontier. New Jersey.

Schiffman, Leon G. and Joseph L. Wisenbilt. (2015). Consumer Behavior. 11th ed. Pearson Global Edition: United States.

Sekaran,Uma and Roger Bougie. (2016). Research Methods for Business: A Skill-Building Approach. John Wiley & Sons: United Kingdom.

Tristia, L. Tambun. (2017). Pendatang Membanjir, Jakarta Mendekati Titik Kritis. BeritaSatu.com, 3 Juli, http://www.beritasatu.com/jakarta/439572-pendatang-membanjir-jakarta-%09mendekati-titik-kritis.html.

Tri Widianti Sik Sumaedi I Gede Mahatma Yuda Bakti Tri Rakhmawati Nidya Judhi Astrini Medi Yarmen, (2015), "Factors influencing the behavioral intention of public transport passengers", *International Journal of Quality & Reliability Management*, Vol. 32 Iss 7 pp. 666–692. http://dx.doi.org/10.1108/IJQRM-01-2013-0002.

Van Doorn, Jenny, Katherine N. Lemon, Vikas Mittal, Stephan Nass, Doreeen Pick, Peter Pirner, & Peter C. Verhoef. (2010). Customer Engagement Behavior: Theoretical foundations and research directions. *Journal of Service Research,* Vol. 13, No. 3: 253–266.

Zeithaml, Valerie, Mary Jo Bitner and Dwayne Gremler. (2013). Services marketing: Integrating customer focus across the firm. 6th ed. McGraw-Hill. New York.

Business Innovation and Development in Emerging Economies – Trinugroho & Lau (Eds)
© *2019 Taylor & Francis Group, London, ISBN 978-1-138-35996-3*

Celebgram endorsement: The influences of attractiveness, power, and credibility towards brand image and purchase intention

Christian Haposan Pangaribuan
Management Study Program, Faculty of Business, Sampoerna University, Jakarta, Indonesia

Alex Maulana
Management Department, BINUS Business School Undergraduate Program, Bina Nusantara University, Jakarta, Indonesia

ABSTRACT: The purpose of this study is to analyze the effect of the attractiveness, power, and credibility of a celebgram endorser on brand image and its impact on purchase intention. This research was conducted by distributing questionnaires to 401 respondents in Indonesia who had made a purchase through Instagram for fashion products. This research uses Structural Equalization Modeling (SEM) method which has been tested beforehand for its validity and reliability. The conclusion obtained from the results of this analysis is that there is a significant relationship and influence of attractiveness and credibility directly or through brand image, but the power factor did not play a significant role.

Keywords: celebgram, endorser, fashion, attractiveness, power, credibility, brand image, purchase intention

1 INTRODUCTION

Indonesia is one of the biggest countries in terms of the number of online population. In 2015, the Internet penetration in the country reached 29%, or about 73 million active Internet users, and most of them own at least one social media account (The Jakarta Post, 2015). Indonesians spend an average of 3 hours and 10 minutes to use mobile phones every day for various needs, such as watching videos or playing games, yet most of the time is to browse on social media, which amounted to 14% (We Are Social, 2015).

Social networks are now so well established and the top ranks tend to stay unchanged from year-to-year. Because of its quick growth, Indonesia's importance for marketers and localization of apps is quickly rising. Indonesia has become Facebook's third-largest country by monthly active users and is poised to pass the United Kingdom for second place (Morrison, 2010). The popularity of Instagram has been increasing steadily since its debut back in 2010. In the U.S., one-third of teenagers cite Instagram as their favorite social network, while 75% of its users are outside U.S. Many teenagers, including Indonesian teens, believe Facebook and Twitter are old-fashioned. The highest demographic to use Instagram was the 20 to 25 age group at 73.8% with one of the top pastimes on it include exploring online shopping accounts (Loras, 2016).

Celebrity endorsement is considered an effective promotional tool by marketers worldwide to represent a specific brand or product to attract potential customers. Lately, online shop owners are using celebgram (celebrity on Instagram) endorsers to promote their business lines (Maulidar and Irma, 2017). Other than being a powerful asset for marketers, testimonials about the benefits of using a product can be expected from a celebrity, who has the ability to endorse a product, or act as a spokesperson for a brand for an extended period (Blackwell, Miniard, & Engel, 2006). According to Solomon (2007), a spokesperson needs to possess attractiveness and credibility in delivering a message to the consumers.

A celebrity endorser is an individual who is known to the public for his or her achievement in areas others than of the product class endorsed (Friedman and Friedman, 1979). However, in a study conducted by Stephanie, Rumambi and Kunto (2013), the credibility of an endorser did not provide a significant effect to purchase intention. On the other hand, Hassan and Jamil's (2014) study found that there is a positive impact between the attractiveness of an endorser to purchase intention, but there is no positive impact between the credibility of an endorser to purchase intention.

To evaluate potential endorser, a celebrity needs to have the following:

1. Visibility, it refers to how recognizable the source is from the target audience perspective and helps to facilitate brand awareness, especially if a celebrity is used (Percy and Rosenbaum-Elliot, 2016).
2. Credibility, it denotes the tendency to believe or trust someone, expertise refers to the knowledge, experience or skill possessed by an endorser as they relate to the communication topic (Shimp, 2000a).
3. Attractiveness, it entails concepts such as intellectual skills, personality properties, way of living, athletic performances and skills of endorsers (Erdogan, 1999).
4. Power, it increases brand purchase intentions, though not attitude, by appearing to command the audience to act (Erdogan and Baker, 1999).

Research has shown that the use of celebrities in advertisements can have a positive influence on purchase intentions (Pornpitakpan, 2003; Pringle and Binet, 2005; Roy, 2006). Several studies have proven empirically related to the effectiveness and positive influence of endorsement made by celebrity to purchase intention (Sivesan, 2013). All of these arguments refer to a conclusion that celebrity endorsement has a positive relationship and influence on consumer buying behavior. Table 1 lists some of the studies together with the findings of research results to see the influence of celebrity endorsement on consumer buying intentions.

2 LITERATURE REVIEW

2.1 *Purchase intention*

Purchase intention represents to what consumers think they will buy (Blackwell et.al, 2006). According to Brown (2003), consumer with intentions to buy certain product will exhibit higher actual buying rates than those customers who demonstrate that they have no intention of buying. Purchase intention is defined as the willingness of consumers to plan the purchase of a particular product (Carrillat, Jaramillo, and Mulki, 2009).

According to Belch and Belch (2009), purchase intention is a tendency to buy a brand and, generally, it is based on the appropriateness between purchasing motives with attributes or brand characteristics that are considered. Thus, purchase intention can be summed up as the desire of consumers to buy a product based on their needs or characteristics of the brand.

2.2 *Brand Image*

Brand image represents the emotional aspects that identify the brand of a company or its products, and has a powerful impact on consumer buying behavior (Arora and Stoner, 2009). The American Marketing Association (AMA) defines brand as a "name, term, design, symbol, or any other feature that identifies one seller's good or service as distinct from those of other sellers," while brand image is defined as a "perception of a brand in the minds of persons. It is what people believe about a brand—their thoughts, feelings, expectations" (AMA). According to Keller (1993), positive brand image could be established by connecting the unique and strong brand association with consumers' memories about the brand through marketing campaigns. Based on those definitions, the authors conclude that brand image is the level of consumer's confidence in a brand.

Aaker (1997) stated the main dimensions that influence and shape the image of a brand are competence, sincerity, excitement, sophistication, and ruggedness.

Tabel 1. Previous Studies.

No.	Author(s), Title, Journal	Major findings
1.	Knoll and Matthes (2017), The effectiveness of celebrity endorsements: a meta-analysis, *Journal of the Academy of Marketing Science.*	Findings revealed strong positive and negative effects when theoretically relevant moderators were included in the analysis. The most positive attitudinal effect appeared for male actors who match well with an implicitly endorsed object. The most negative effect was found for female models not matching well with an explicitly endorsed object.
2.	Kaur and Garg (2016), Celebrity endorsement and buying behavior: A study of Panjab University students, *International Journal of Research.*	Celebrity endorsement enhances product information and creates awareness among consumers. The purchase attitude is more strongly influenced by product quality than endorsement factors, price, discounts. Respondents believe that the products advertised by celebrities are of good quality.
3.	Ziporah and Mberia (2014), The Effects of Celebrity Endorsement in Advertisements, *International Journal of Academic Research in Economics and Management Sciences.*	Celebrity appearance, knowledge, liking, and credibility of the celebrity are also highly correlated with advertising believability.
4.	Article I. Elberse and Verleun (2012), The Economic Value of Celebrity Endorsements, *Journal of Advertising Research.*	There is a positive payoff to a firm's decision to sign an endorser, and that endorsements are associated with increasing sales and relative to competing brands. Also, sales and stock returns jump noticeably with each major achievement by the athlete.
5.	Jain (2011), Celebrity Endorsement And Its Impact On Sales: A Research Analysis Carried Out In India, *Global Journal of Management and Business Research.*	Celebrity endorsement has an impact on sales to a little extent, celebrities should not always be used to endorse brands of various products, celebrities bring brand equity to the products.

1. Sincerity: a sincere brand personality is down to earth, honest and cheerful.
2. Excitement: brands which are daring, spirited, and imaginative are known to build excitement in consumers
3. Competence: brands which are competent are known to be reliable, intelligent, and successful.
4. Sophistication: the toughest personality to achieve in the five traits of brand personality that requires charm and a lot of patience to achieve.
5. Ruggedness: brand personality which is derived mainly from the nature of the products, which automatically builds the personality.

2.3 Elaboration Likelihood Model (ELM)

Persuasion occurs internally accessing your emotions. Someone you trust can persuade you by proposing an argument with facts. However, that persuasion does not always result in a successful way. At the end, the users have to persuade themselves with the validity of the message taken in.

The Elaboration Likelihood Model (ELM) is an information processing theory of persuasion and attempts to provide an integrative framework for understanding the antecedents and consequences of attitude change (Petty, Heesacker and Hughes, 1997). The ELM postulates that when a persuader presents information to an audience, a level of elaboration will be the outcome. Elaboration here means the amount of effort an audience member has to use in order to process and evaluate a message, remember it, and then accept or reject it.

Specifically, the ELM has determined that when facing a message, people react by using either of two channels (but sometimes a combination of both, too), reflecting the level of effort they need. As such, they either experience high or low elaboration, and whichever of these will determine whether they use central or peripheral route processing (Petty and Cacioppo, 1986). Celebrity endorsements can have a positive influence on the credibility of and preferences for advertising, and ultimately on purchase intentions (Wang, Cheng, and Chu, 2013). On the contrary, non-celebrity endorsers were more effective in creating a connection to the product as their characters could be developed and shaped to suit the brands and target audiences (Tom, Clark, Elmer, Grech, Masetti and Sandhar, 1992).

2.4 *Source attributes*

Based on extensive research, and as presented by the social psychologist Herbert Kelman, three basic source attributes contribute to a source's (e.g., an endorser's) effectiveness: attractiveness, power, and credibility (Kang and Herr, 2006).

Persuasion occurs through the identification process when receivers find something in the source that they like and consider attractive. This does not mean simply physical attractiveness, but includes any number of virtuous characteristics that consumers may perceive in an endorser: intellectual skills, personality properties, lifestyle characteristics, athletic prowess, and so on. When receivers perceive a source to be attractive, they are very likely to adopt the beliefs, attitudes, behaviors, interests, or preferences of the source (Shimp and Andrews, 2013).

In selecting celebrity spokespeople, advertising executives evaluate different aspects that can be taken together under the general label *attractiveness*, which is multifaceted, and includes more than physical attractiveness (Lee, 2016). It also is important to note that advertising executives generally regard attractiveness as less important than credibility and endorser matchup with the audience and the brand.

Source attractiveness consists of three related dimensions: similarity, familiarity, and liking. That is, a source (e.g., an endorser) is considered attractive to receivers if they share a sense of similarity or familiarity with the source or if they like the source regardless of whether the two are similar in any respect (Shimp and Andrews, 2013).

Kelman (1958) was among the first to address the underlying process responsible for the impact of power. Specifically, powerful sources produced influence by invoking *compliance*— going along with the source solely because of ability of the source to monitor one's actions (Briñol, Petty, Durso and Rucker, 2017). Compliance occurs when an individual is persuaded by an advertised source because they hope to achieve a favorable reaction or approval from this source. More specifically, this underlying process works through compliance with the perceived rewards and punishments from the source. Although this is not as likely to occur with mass media advertising, such processes may take place through personalized direct advertising materials, personal selling appeals, and/or social media messages (Shimp and Andrews, 2013).

Credibility refers to the tendency to believe or trust someone. When an information source, such as an endorser, is perceived as credible, audience attitudes are changed through a psychological process called internalization (Shimp and Andrews, 2013). *Internalization* occurs when the receiver accepts the source's position on an issue as his or her own. An internalized attitude tends to be maintained even if the source of the message is forgotten or if the source switches to a different position (Petty, Ostrom and Brock, 1981).

Two important dimensions of source credibility are expertise and trustworthiness. Expertise refers to the perceived knowledge, experience, or skills possessed by a source as they relate to the communications topic. Expertise is a perceived rather than an absolute phenomenon. Part of being an expert is learning what matters and what does not matter—including visual features (Rosen, 2006). Whether a source is indeed an expert is unimportant; all that matters is how the target audience perceives the source. For example, an endorser who is perceived as an expert on a given subject is more persuasive in changing audience opinions pertaining to his or her area of expertise than an endorser who is not perceived as an expert (Shimp and Andrews, 2013).

Trustworthiness refers to the perceived honesty, integrity, and believability of a source. Although expertise and trustworthiness are not mutually exclusive, often a particular source may be perceived as highly trustworthy, but not especially expert. The degree of honesty or trustworthiness of a source depends primarily on the audience's perception of the source's intent (Shimp, 2000b). If consumers believe that a source (e.g., an endorser) is motivated purely by self-interest, that source will be less persuasive than someone regarded as having nothing to gain by endorsing the brand (Shimp and Andrews, 2013).

A celebrity earns the audience's trust through the life he or she lives professionally (on the screen, on the sports field, in public office) and personally, as revealed to the general public through the mass media. Advertisers capitalize on the value of trustworthiness by selecting endorsers who are widely regarded as being honest, believable, and dependable people (Priester and Petty, 2003). In general, endorsers must establish that they are not attempting to manipulate the audience and that they are objective in their presentations. By doing so, they establish themselves as trustworthy and, therefore, credible. Also, although this represents more of a peripheral effect, an endorser has a greater likelihood of being perceived as trustworthy the more he or she matches the audience in terms of distinct demographic characteristics (Deshpande and Stayman, 1994).

3 MATERIALS AND METHODS

Based on the purpose of this quantitative study, the data collected is cross-sectional which is obtained from the questionnaires distributed online for one month. The 401 respondents selected are Indonesian people who have made a purchase of fashion products through Instagram. Sampling technique used is non-probability with purposive sampling with an error rate of 5% and the research uses Structural Equal Modeling (SEM) with Smart PLS. The variables in this research can be operationalized in Table 2 below:

Table 2. Variable Operationalization.

Variable	Dimension	Indicator
Attractiveness (X1)	Similarity	Similarity
	Likeability	Likeability
	Familiarity	Familiarity
Credibility (X2)	Expertise	Expert
		Knowledgeable
		Experienced
		Qualified
		Skilled
	Trustworthiness	Honest
		Trustworthy
		Sincere
Power (X3)	Expression	Expression
	Charisma	Charisma
	Physical Condition	Physical Condition
Brand Image (Y1)	Sincerity	Honesty
	Excitement	Courageous
		Imaginative
		Up-to-date
	Competence	Success
		Reliable
		Intelligent
	Sophistication	Prestige
Purchase Intention (Z1)	Transactional Intention	The tendency of someone to buy a product
	Referential Intention	The tendency of a person to refer the product to someone else
	Explorative Intention	The behavior of someone looking for information about the product he/she is interested in
	Preferential Intention	The behavior of a person who has a primary preference for a product

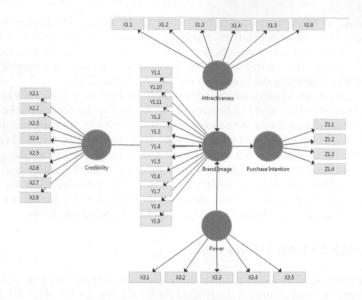

Table 3. Total Effect.

		Original sample (O)	Sample mean (M)	Standard deviation (STDEV)	T statistics (|O/STDEV|)	P values
H1	Attractiveness → Brand Image	0.319	0.322	0.050	6.431	0.000
H2	Credibility → Brand Image	0.486	0.482	0.045	10.878	0.000
H3	Power → Brand Image	0.016	0.026	0.044	0.366	0.714
H4	Brand Image → Purchase Intention	0.719	0.722	0.025	28.580	0.000
H5	Attractiveness → Purchase Intention	0.230	0.232	0.038	6.023	0.000
H6	Credibility → Purchase Intention	0.350	0.348	0.036	9.788	0.000
H7	Power → Purchase Intention	0.012	0.018	0.032	0.366	0.714

4 RESULTS AND DISCUSSION

After all statement of the questionnaire have been tested, and the final results show that all of the items included in the calculation are valid and reliable and all variables are normally distributed. Based on the result of the SEM analysis, the following results are obtained:

4.1 *Results*

Table 3 shows that the first hypothesis (H1) test result shows that the p-value of Attractiveness is 0.000 which is < α (0.005). Furthermore, the t value is higher than t table (6.431 > 1.960). The result indicates that the variable Attractiveness significantly affects Brand Image and null hypothesis (H0) should be rejected, while H1 is accepted.

The second hypothesis (H2) test result shows that the p-value of Credibility is 0.000 which is < α (0.005). Also, the t value is higher than t table (10.878 > 1.960). The result indicates that the variable Credibility significantly affects Brand Image and null hypothesis (H0) should be rejected, while H1 is accepted.

The third hypothesis (H3) test result shows that the p-value of Power is 0.714 which is > α (0.005). Moreover, the t value is higher than t table (0.366 < 1.960). The result indicates that the variable Power does not significantly affect Brand Image and null hypothesis (H0) should be accepted, while H1 is rejected.

The fourth hypothesis (H4) test result shows that the p-value of Brand Image is 0.000 which is $< \alpha$ (0.005). Furthermore, the t value is higher than t table (28.580 > 1.960). The result indicates that the variable Brand Image significantly affects Purchase Intention and null hypothesis (H0) should be rejected, while H1 is accepted.

The fifth hypothesis (H5) test result shows that the p-value of Attractiveness is 0.000 which is $< \alpha$ (0.005). Also, the t value is higher than t table (6.023 > 1.960). The result indicates that the variable Attractiveness significantly affects Purchase Intention and null hypothesis (H0) should be rejected, while H1 is accepted.

The sixth hypothesis (H6) test result shows that the p-value of Credibility is 0.000 which is $< \alpha$ (0.005). Moreover, the t value is higher than t table (9.788 > 1.960). The result indicates that the variable Credibility significantly affects Purchase Intention and null hypothesis (H0) should be rejected, while H1 is accepted.

The seventh hypothesis (H7) test result shows that the p-value of Power is 0.714 which is $> \alpha$ (0.005). Besides, the t value is higher than t table (0.366 < 1.960). The result indicates that the variable Power does not significantly affect Purchase Intention and null hypothesis (H0) should be accepted, while H1 is rejected.

4.2 Discussion

The result of this study shows that there is a positive influence between Attractiveness towards Brand Image as well as Credibility towards Brand Image. The result is consistent with Sivesan's (2013) study which found that there is a positive relationship between Celebrity Endorser to the Brand Equity. This concludes that there is a positive influence between endorsers and brand equity. On the other hand, the finding of this research on the relationship between Attractiveness and Purchase Intention is not consistent with what Hassan and Jamil (2014) studied. Their study of Celebrity Endorsement towards Purchase Intention used Attractiveness, Celebrity Expertise and Celebrity Congruence as the dimensions of the endorser. Their result concluded that there is a positive influence between Attractiveness and Purchase Intention, whereas Credibility has negative impact to Purchase Intention, while current study has shown that there is a positive and significant influence between the variables Credibility to the variable Purchase Intention.

Past research showed that the athlete's power relative to the brand is greater in an endorsement than in a sponsorship context (Carrillat and d'Astous, 2014). Hence, it was relevant for companies that considered associating their brands with athletes, both in sponsorship and endorsement, to gain commercial benefits. Even though sponsorships and endorsements may represent an effective brand utilizing marketing communications strategies, the result of present study is not consistent where it was found that Power does not have an impact towards the formation of Brand Image or Purchase Intention. This may partly be based on different business models between fashion and sports apparel.

Previous research found that there is a significant causal relationship between consumer-based Brand Equity and Purchase Intention in the fashion industry (Khan, Rahmani, Hoe and Chen, 2015). This result is in accordance with current research that Brand Image has a positive influence on Purchase Intention.

4.3 Suggestions

In choosing celebgram, it is recommended for the seller to pay more attention to the variable of Attractiveness with the dimension of *similarity* between the celebgram and the consumers. From this analysis, it is found that the similarity level is not too strong between the two, indicating that the number of online sellers in Indonesia paying attention to the level of *similarity* of celebgram and consumer is still very low. Yet, the factor is important in influencing the level of consumer buying interest. To improve the similarities between celebgram and consumer, online merchants can choose a celebgram that has character and lifestyle that match with the target consumers. The more fitting the character and lifestyle, the higher the level of similarity between celebgram and consumers that will encourage consumer buying interest.

In addition to the celebgram similarity level, the sellers need to pay attention to the brand image factor with sophistication dimension of the product. Sophistication level can be seen from how sellers upload photos of products on Instagram. Based on the data, it can be seen that the photographs uploaded do not provide strong influence on their buying interest, yet it is one of the key aspects for consumers to consider buying a product. In uploading photos, sellers can make the endorsed products look more exclusive which in turn would attractive potential buyers.

Celebrities on Instagrams, or celebgrams, need to carefully sort the product to be endorsed because if the product does not match the characteristics and lifestyle of the celebgram then it will reduce the level of similarity between the celebgram and the target market. In addition, it is important to create a more genuine 'partnership' so that the consumers are more likely to believe the celebrity is endorsing the product.

REFERENCES

Aaker, J.L. (1997). Dimensions of brand personality. *Journal of Marketing Research*, 34(3) 347–356.

Arora, R. and Stoner, C. (2009). A mixed method approach to understanding brand personality. *Journal of Product & Brand Management*, 18(4), 272–283.

Belch, G. and Belch, M. (2009). *Advertising and Promotion: An Integrated Marketing Communication Perspective*. New York: McGraw Hill.

Blackwell, R.D., Miniard, P.W. and Engel, J.F. (2006) *Consumer Behaviour* (10th Ed.), Thomson South-Western.

Briñol, P., Petty, R.E., Durso, G. and Rucker, D.D. (2017). Power and Persuasion: Processes by Which Perceived Power Can Influence Evaluative Judgments. *Review of General Psychology*, 21(3), 323–241.

Brown, M. (2003). Buying or browsing? An exploration of shopping orientations and online purchase intention. *European Journal of Marketing*, 37(11/12), 1666–1684.

Carrillat, F.A. & d'Astous, A. (2014). Power imbalance issues in athlete sponsorship versus endorsement in the context of a scandal. *European Journal of Marketing*, 48(5/6), 1070–1091.

Carrillat, F.A., Jaramillo, F., and Mulki, J.P. (2009). Examining the impact of service quality: A meta-analysis of empirical evidence. *Journal of Marketing Theory and Practice*, 17 (2), 95–110.

Deshpande, R. and Stayman, D. (1994). A Tale of Two Cities: Distinctiveness Theory and Advertising Effectiveness. *Journal of Marketing Research*, 31 (1994), 57–64.

Elberse, A. and Verleun, J. (2012). The Economic Value of Celebrity Endorsements, *Journal of Advertising Research*, 52(2), 149–165.

Erdogan, B.Z. (1999). Celebrity endorsement: A literature review. *Journal of Marketing Management*, 15(4), 291–314.

Erdogan, Z. and Baker, M. (1999). Celebrity Endorsement: Advertising Agency Managers Perspective. *The Cyber Journal of Sport Marketing*, Vol. 3, 356–372.

Friedman, H.H. and Friedman, L. (1979). Endorsers Effectiveness by Product Type. *Journal of Advertising Research*, 19(5).

Hassan, S.R. and Jamil, R.A. (2014). Influence of Celebrity Endorsement on Consumer Purchase Intention for Existing Products: A Comparative Study. *Journal of Management Info*, 4(1), 1–23.

Jain, V. (2011). Celebrity Endorsement And Its Impact On Sales: A Research Analysis Carried Out In India, *Global Journal of Management and Business Research*, 4(11).

Kang, Y-S. and Herr, P.M. (2006). Beauty and the Beholder: Toward an Integrative Model of Communication Source Effects. *Journal of Consumer Research*, 33 (2006), 123–30.

Kaur, S. and Garg, A. (2016). Celebrity endorsement and buying behavior: A study of Panjab University students. *International Journal of Research—Granthaalayah*, 4(11), 122–136.

Keller, K.L. (1993). Conceptualizing, Measuring, and Managing Customer-Based Brand Equity. *Journal of Marketing*, 57, 1–22.

Kelman, H.C. (1958). Compliance, identification, and internalization: Three processes of attitude change. *The Journal of Conflict Resolution*, 2, 51–60.

Khan, N., Rahmani, S.H.R., Hoe, H-Y and Chen, T-B. (2015). Causal Relationships among Dimensions of Consumer-Based Brand Equity and Purchase Intention: Fashion Industry. *International Journal of Business and Management*, 10(1), 172–181.

Knoll, J. and Matthes, J. (2017). The effectiveness of celebrity endorsements: a meta-analysis, *Journal of the Academy of Marketing Science*, 45, 55–75.

Lee, K-H (2016). The conceptualization of country attractiveness: a review of research. *International Review of administrative Sciences*, 82(4), 807–826.

Loras, S. (2016). Social media in Indonesia: big numbers with plenty of room to grow. Retrieved on 3 January 2017 from https://www.clickz.com/social-media-in-indonesia-big-numbers-with-plenty-of-room-to-grow/94062/

Maulidar and Irma, A. (2017). Role of Celebgram Endorser in the Process of Female Clothes Buying Decision through Instagram on the Female Students of Syiah Kuala University). *Jurnal Ilmiah Mahasiswa FISIP Unsyiah,* 1(1), 1–11.

Morrison, C. (2010). Inside the Numbers: Facebook's Third-Largest Country, Indonesia, Uses English Heavily. Retrieved on 10 December 2018 from http://www.adweek.com/digital/indonesia-facebook-english/

Percy, L. and Rosenbaum-Elliott, R. (2016). *Strategic Advertising Management,* 3rd Ed., Oxford University Press, U.K.

Petty, R.E. and Cacioppo, J.T. (1986). The Elaboration Likelihood Model of persuasion. In L. Berkowitz (Ed.), *Advances in experimental social psychology,* 19, 123–205.

Petty, R.E., Heesacker, M. and Hughes, J.N. (1997). The Elaboration Likelihood Model: Implications for the Practice of School Psychology. *Journal of School Psychology,* 35(2), 107–136.

Petty, R.E., Ostrom, T.M. and Brock, T.C. (1981, Eds.), *Cognitive Responses in Persuasion,* Hillsdale, N.J: Lawrence Erlbaum Associates, 143.

Pornpitakpan, C. (2003). The Effect of Celebrity Endorsers' Perceived Credibility on Product Purchase Intention: The Case of Singaporeans, *Journal of International Consumer Marketing,* 16(2), 55–74.

Priester, J.R. and Petty, R.E. (2003). The Influence of Spokesperson Trustworthiness on Message Elaboration, Attitude Strength, and Advertising Effectiveness. *Journal of Consumer Psychology,* 13(4), 408–21.

Pringle, H. and Binet, L. (2005). How Marketers Can Use Celebrities to Sell more Effectively, *Journal of Consumer Behaviour,* 4(3), 201–214.

Rosen, J. (2006). *Expertise and perception: How what we know can affect what we see.* Retrieved on 7 January 2017 from https://hub.jhu.edu/2016/03/02/recognizing-arabic-letters-expert-novice/

Roy, S. (2006). An Exploratory Study in Celebrity Endorsements. *Journal of Creative Communications,* 1, 2

Shimp, T.A. (2000a). Supplemental Aspects of Integrated Marketing Communications, 5th Ed., Fort Worth, TX: The Dryden Press.

Shimp, T.A. (2000b). *Advertising Promotion. Supplemental aspects of Integrated Marketing Communications* (5th Ed.), San Diego, CA: Harcourt College Publishers.

Shimp, T.A. and Andrews, C.J. (2013). *Advertising, promotion, and other aspects of integrated marketing communication,* 9th Ed., USA: South-Western Cengage Learning.

Sivesan, S. (2013). Impact of celebrity endorsement on brand equity in cosmetic product. *International Journal of Advanced Research in Management and Social Sciences,* 2(4), 1–11.

Solomon, M.R. (2007). *Consumer Behavior: Buying, Having, and Being.* Upper Saddle River, N.J.: Pearson Prentice Hall.

Stephanie, E., Rumambi, L.J. and Kunto, Y.S. (2013). Analisa Pengaruh Rio Dewanto dan Donita Sebagai Celebrity Endorser Terhadap Minat Beli Produk Axe Anarchy dengan Daya Tarik Iklan dan Efek Iklan Sebagai Variabel Intervening. *Jurnal Manajemen Pemasaran,* 1–9.

The Jakarta Post (2015). *Internet users in Indonesia reach 73 million.* Retrieved on 5 December 2017 from http://www.thejakartapost.com/news/2015/03/10/internet-users-indonesia-reach-73-million.html.

Tom, G., Clark, R., Elmer, L., Grech, E., Masetti, J. and Sandhar, H. (1992). The Use of Created versus Celebrity Spokesperson in Advertisement, *Journal of Consumer Research,* 20(4), 535–547.

Wang, J-S, Cheng, Y-F and Chu Y-L (2013). Effect of Celebrity Endorsements on Consumer Purchase Intentions: Advertising Effect and Advertising Appeal as Mediators. *Human Factors and ergonomics in Manufacturing & Service Industries,* 23(5), 357–367.

We Are Social (2015). *Digital, Social & Mobile in 2015.* Retrieved on 3 December 2017 from https://wearesocial.com/sg/special-reports/digital-social-mobile-2015.

Ziporah, M.M. and Mberia, H.K. (2014). The Effects of Celebrity Endorsement in Advertisements, *International Journal of Academic Research in Economics and Management Sciences,* 3(5), 178–188.

Business Innovation and Development in Emerging Economies – Trinugroho & Lau (Eds)
© *2019 Taylor & Francis Group, London, ISBN 978-1-138-35996-3*

The factors affecting customer satisfaction, loyalty, and word of mouth towards online shopping for millennial generation in Jakarta

Muhammad Gunawan Alif, Christian Haposan Pangaribuan & Novi Retno Wulandari
Sampoerna University, Indonesia

ABSTRACT: In this research, three studies examine the relations between customer satisfaction, loyalty, and word-of-mouth in the context of online shopping by evaluating several factors, i.e. perceived ease of use, design and feature, service security, time-savings, cost-savings, social influence, and performance expectancy. The data collected is from 230 millennials in Jakarta, Indonesia. Results from the "students" group indicate that performance expectancy influences customer satisfaction towards online shopping. While from the "non-students," cost savings, perceived ease of use, service security, and performance expectancy influence customer satisfaction. Further investigation of demographic factors shows how the differences can add to our understanding in the online shopping customer satisfaction.

Keywords: customer satisfaction, loyalty, word-of-mouth, online shopping, millennials

1 INTRODUCTION

The trend of shopping behavior has dramatically shifted into online shopping activities. Online shopping facilitates more efficient and convenient shopping experience in terms of eliminating the travel cost and time to go the physical store. According to a research by DBS (2017), although Indonesia had successfully attained US$ 1.1 billion from online transactions in 2014, the gain was only 0.7% from its total retail sales, way below Singapore and Thailand. Indonesia is enquired to increase its online retails sales as it can increase 56 million small and medium enterprise (SME) sales which may contribute to the 55% of the Indonesia's gross domestic product (DBS, 2017).

Millennial tends to have a more personal and emotional engagement towards a brand. This creates quite a challenge for online retailers to understand their needs and wants as well as develop strategies to attract this age group. Figure 1 shows the proportion of millennial both in the developed and emerging regions as well as the percentage of millennial's potential in the share of urban consumption growth.

Figure 1 shows that the millennials contribute to almost 2.30% out of 3.40% of urban consumption growth in Southeast Asia. Some people believe that tech-savvy characteristic-driven millennials use their gadget for shopping online rather than the older counterparts. However, KPMG's (2017) Global Online Consumer Report revealed that millennials serve as less active online shoppers. In 2016, KPMG conducted an online survey to 18,430 online transactions in the past 12 months in more than 50 countries. The result shows that generation X or people born in 1966–1981 becomes the most active online shoppers in the world while millennials (1982–2001) serve as the second, and baby boomers generation (1946–1965) came last. Although baby boomers are the least active online shoppers, they spent the highest average amount per transaction, reaching US$203. The millennials were the least, only US$173, while the generation X spent US$190. Nonetheless, the millennials were expected to surpass the older generation in terms of spending money for online shopping and becoming the most active online shoppers as they pursue their carriers and build a home and families in the upcoming years. Since the millennials will dominate the total population and consumption

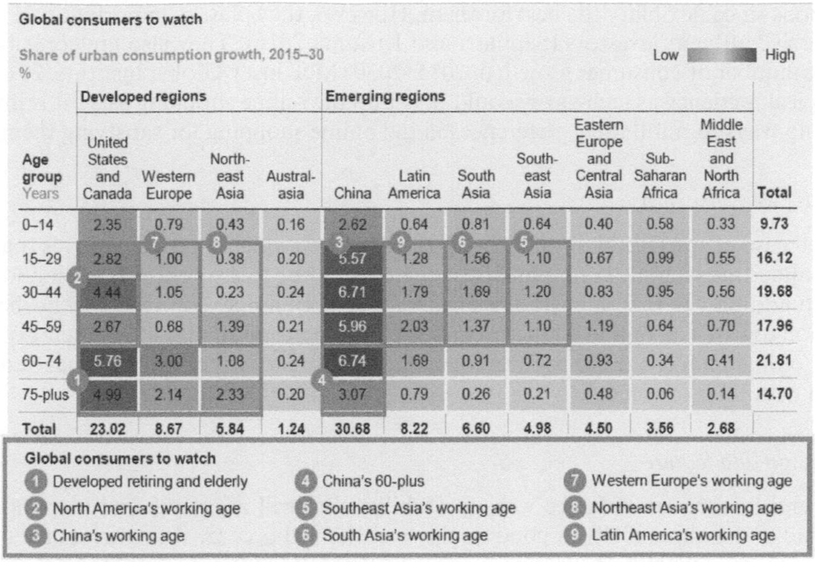

Global consumers to watch

Share of urban consumption growth, 2015–30
%

Low — High

	Developed regions				Emerging regions							
Age group Years	United States and Canada	Western Europe	North-east Asia	Austral-asia	China	Latin America	South Asia	South-east Asia	Eastern Europe and Central Asia	Sub-Saharan Africa	Middle East and North Africa	Total
0–14	2.35	0.79	0.43	0.16	2.62	0.64	0.81	0.64	0.40	0.58	0.33	9.73
15–29	2.82	1.00	0.38	0.20	5.57	1.28	1.56	1.10	0.67	0.99	0.55	16.12
30–44	4.44	1.05	0.23	0.24	6.71	1.79	1.69	1.20	0.83	0.95	0.56	19.68
45–59	2.67	0.68	1.39	0.21	5.96	2.03	1.37	1.10	1.19	0.64	0.70	17.96
60–74	5.76	3.00	1.08	0.24	6.74	1.69	0.91	0.72	0.93	0.34	0.41	21.81
75-plus	4.99	2.14	2.33	0.20	3.07	0.79	0.26	0.21	0.48	0.06	0.14	14.70
Total	23.02	8.67	5.84	1.24	30.68	8.22	6.60	4.98	4.50	3.56	2.68	

Figure 1. Global consumers to watch.

growth in Indonesia, understanding their characters and preferences will make them satisfy and increase their willingness to shop online more often (KPMG, 2017).

Accordingly, there is a problem in online shopping development in Indonesia. Despite supporting the growth of small and medium enterprises in Indonesia, online shopping total sales remains low, as it was only 0.7% from the total sales of online activities. Perhaps a study that investigates the factors affecting customer satisfaction as well as the post-purchase behavior; loyalty and word of mouth towards online shopping specifically for millennial generation by quantitative method through questionnaire could remedy this situation. Per initial investigation, a pre-research was distributed to 30 millennials around a local campus by asking the satisfying factors to shop online. Based on the result, the seven factors affecting customer satisfaction will be examined as the aim of this research whether the significance of perceived ease of use, design and feature, service security, time-savings, cost-savings, social influence, and performance expectancy have any influence towards customer satisfaction. In this research, the researchers will not only examine the aggregate sample but also identify how this sample from two different categories (students and non-students) might differ.

2 LITERATURE REVIEW

Online shopping as a form of electronic commerce allows the consumers to buy products directly from sellers over the Internet. People tend to correlate the number of Internet users with the number of potential Internet shoppers. This triggers them to acquire more online shoppers by attracting more active users to shop online. The involvement of smartphone in this entire path helps offline shoppers connected to the online platform mostly for comparing price, finding the product information, and reading the online review.

Durmus, Ulusu, Erdem, and Yalcin (2015) noted that customer satisfaction has strongly affected by the service quality. In the online shopping, the service quality are comprised of: e-service, which depends on the interaction between customers and service providers in exchanging information, and e-service quality which can enhance the strategic benefits delivery, operational efficiency, and profitability. Thus, maintaining and increasing customer satisfaction is important.

According to Septiatri and Kusuma (2016), the millennials is a group of people born in between 1980 and 2000. Despite many definitions, this generation shares the same characteristic: tech-savvy. Millennial's characteristics are harder to understand rather than other generations since flexibility fits best for them. However, they play an important role to be the next massive affluent investors (Septiatri and Kusuma 2016). They also are contributing to the large number of consumer growth in 2015–2030 (McKinsey Global Institute, 2016). Since their typical tendency as tech-savvy could boost up the online shopping sales, it is important to identify ways to fulfill their preferences in the online shopping for satisfying them.

2.1 Perceived ease of use

Perceived ease of use (PEOU) is defined as the degree to which a person believes that engaging in online transactions via mobile commerce would be free of effort (Gefen and Straub, 2000). In addition to that, a customer's general perception may contribute to his/her actual perception towards an online retailer website. A good perception may affect their expectation and it is online retailer's responsibility to manage customers' expectation to make customers satisfied with their services.

2.2 Design and feature

Design and feature can enhance website usability that will affect the end-user satisfaction (Hausman and Siekpe, 2009). A good quality website will become the main source of online retailer to interact with their customers, serve as their sales outlet, and represent their company's reputation. Nonetheless, enabling online communication with customers is not enough; online retailers have to ensure customers can find and access information they need through a single click (Hernandez, Jimenez, and Martin, 2009). Hsieh, Chiu, Tang, and Lee (2017) highlighted specific component of design and feature and examined the possibility of colors in changing the online shopping realities. Color can be a customer's main consideration after price and can influence their advertisement preference and perception of brand personality, which then affect their judgment and decision.

2.3 Service security

One of the key characteristics of business to consumers' website is security. Tontini (2016) defines security as trust and confidence conveyed by the site, including brand recognition. Since a website has to create trustworthiness, dependability, and reliability, the formal privacy policies in the online security system and the adoption of superior encryption technology have to be implemented by the online retailers (Clemes, Gan and Zhang, 2014). One of the examples is ensuring safe payment method to guarantee financial security and privacy of the customers. Phishing website, identity theft, and credit card theft are some customers' security concerns at the purchase stage (Pham and Ahammad, 2017). Beyond these security concerns, the encryption, authentication, and visual notification as the security system mechanisms positively influence customers' security perception (Vladlena, Saridakis, Tennakoon, and Ezingeard, 2015). As the security risk perception decreases, the customer satisfaction will increase (Pham and Ahammad, 2017).

2.4 Time savings

Online shopping may reduce travel time of the consumers rather than shopping in the brick-and-mortar stores (Miyatake, Nemoto, Nakaharai, and Hayashi, 2016). The consumers can also shop anytime they want by using their gadgets. The Internet utilization allows consumers access unlimited wide variety of products and services around the globe from various companies without visiting each physical store. This also reduces the consumers' effort in terms of time spending (Lim, Yap, and Lau, 2010). The search cost reduction is translated into a time saving (Kohli, Devaraj, and Mahmood, 2004).

2.5 Cost savings

Online shopping channels offer products with 9–16% prices lower than the conventional channels (Kohli, et al., 2004). Consumers may save cost by only buying products with the precise feature they need and compare prices with other online channels to get the most convenient price for them.

2.6 *Social influence*

Social influence is the degree to which an individual perceives that important others believe that he or she should use the system (Ashraf, Thongpapanl, and Spyropoulou, 2016). The form of reference groups is friendship, shopping, work, virtual or communities, and consumer-action group (Schiffman and Kanuk, 2009). An influencer is endorser such as celebrities, pop stars, fashion leaders, and familiar community members (Xu, Li, Peng, Hsia, Huang, and Wu, (2017). The endorser may strengthen consumers believe that they will make a better decision due to endorsers' expertise and credibility which accepted by consumers as their public image.

2.7 *Performance expectancy*

Performance expectancy is the degree to which an individual believes that using the system will help him or her attain gains in job performance (Singh, Verma, Kumar, Kaur, and Kaur, 2016). It is a summarized construct of perceived usefulness, relative advantage, outcome expectations, and extrinsic expectations.

2.8 *Customer satisfaction*

Kotler and Keller (2012) defined satisfaction as a person's feelings of pleasure or disappointment that result from comparing a product's perceived performance (or outcome) to expectations. The expectation may come from past buying experience, marketers' and competitors' information and promises as well as friends or associates. If customers set a high expectation but the performance is low, they will be disappointed with the products or services. If it is matching, they will be satisfied. When the performance is beyond customers' expectation, customers will be highly satisfied or delightful.

2.9 *Loyalty*

Loyalty is an attitude that refers to the positive feelings towards a brand in addition to repurchasing time after time and is achieved through providing customer satisfaction to retain the customers (Atulkar and Kesari, 2017). The special preference, attachment, and commitment will be shown by loyal customers by becoming the major driver of online shopping's success and major source of the online shopping growth and profit (Liao, Lin, Luo, and Chea, 2016).

2.10 *Word of mouth*

Word of mouth (WOM) is all informal communications directed at other consumers about ownership, usage, or characteristics of particular goods and services or their seller (Jung and Seock, 2017). Positive WOM will be shared by the satisfied customers while negative WOM will be shared by unsatisfied customers (Pham and Ahammad, 2017). Since attitudes and intentions are a part of consumer's behavior, satisfaction affects consumers' positive behavioral intentions including purchase repetition, loyalty, and positive WOM (Jung and Seock, 2017).

3 METHODOLOGY

The researchers used quantitative research method as well as both primary data (questionnaire) and secondary data to analyze the result of the research and support the background of the research. Operational variable test was performed to 30 respondents by distributing the questionnaire online (Google Form) and offline (printed questionnaire) to test the target respondent's understanding about the questions and statements in the questionnaire as well

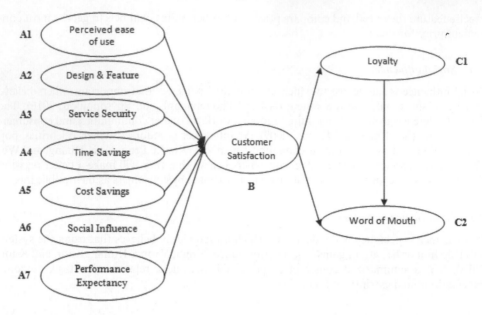

Figure 2. Framework model.

as to test the reliability and validity of the statement indicators. Afterwards, the survey was distributed to 275 millennials in Jakarta (230 valid responses). The framework model of this study can be seen in Figure 2.

3.1 *Population and samples*

According to Badan Pusat Statistik (2017), the latest data of Jakarta's population in 2017 is 10,177,924 (5,115,357 males). Specifically, there are 706,550 people in the age of 15–19 years old and 883,883 people in the age of 20–24 years old. Other age groups are 25–29, 30–34, and 35–39. The population of these age groups is 1,049,766; 1,044,047; and 927,120, respectively.

Total sample is determined by multiplying the total indicator item by 5 since the population of online shopping customers is difficult to predict (Hair, Black, Babin, and Anderson, 2010). In this research, the researchers prepared 46 statements as the indicator items, hence the 230 samples.

By using the above calculation, the minimum sample for this research is 230, consisting of 115 students and 115 non-students from Jakarta. Therefore, the researchers tried to identify how this sample from two different categories might differ.

At the end of January 2018, the researchers distributed online questionnaire by using Google form to the Whatsapp and LINE groups as well as personal message to the friends and relatives who have done online shopping. The relationship of customer satisfaction with loyalty and customer satisfaction with word of mouth will be analyzed using simple linear regression. Both multiple and simple linear regression are done by using **IBM SPSS Statistics 21**.

4 RESULTS AND DISCUSSIONS

4.1 *Operational variable test 1*

Operational variable test is aimed to examine the validity and reliability of each statement or question in the questionnaire as well as the understanding of the respondents about each words and sentences on the questionnaire.

4.2 *Validity test*

Table 1 shows the Kaiser-Meyer-Olkin (KMO) and anti-image correlation test result for each variable.

Table 1 shows two variables considered invalid (design & feature and time-savings) due to the values below 0.5 KMO. Also, 9 statement indicators considered invalid due to its anti-image correlation below 0.5. The variables having KMO and anti-image correlation above 0.5 are perceived ease of use, social influence, and performance expectancy. Since a variable requires at least two statement indicators as a representative, the researchers decided to take out statement indicator SS4 and CS4. Hence, the need to re-test service security and cost savings variables.

Although design and feature and time-savings variable were considered invalid where each of which had only a statement indicator with anti-image correlation value above 0.5, these variables

Table 1. Validity test result of all variables.

Variable	KMO	Statement indicator	Anti-image correlation	Note
A1: Perceived Ease of Use (PEOU)	0.875	PEOU1	0.840	Valid
		PEOU2	0.903	
		PEOU3	0.855	
		PEOU4	0.897	
		PEOU5	0.899	
A2: Design and Feature (DF)	0.451	DF1	0.422	Invalid
		DF2	0.415	
		DF3	0.758	Valid
		DF4	0.433	Invalid
		DF5	0.424	
A3: Service Security (SS)	0.804	SS1	0.768	Valid
		SS2	0.806	
		SS3	0.910	
		SS4	0.288	Invalid
		SS5	0.821	Valid
		SS6	0.895	
A4: Time Savings (TS)	0.360	TS1	0.392	Invalid
		TS2	0.406	
		TS3	0.121	
		TS4	0.841	Valid
A5: Cost Savings (CS)	0.652	CS1	0.636	
		CS2	0.748	
		CS3	0.713	
		CS4	0.152	Invalid
A6: Social Influence (SI)	0.524	SI1	0.561	Valid
		SI2	0.532	
		SI3	0.515	
		SI4	0.520	
A7: Performance Expectancy (PE)	0.844	PE1	0.813	Valid
		PE2	0.838	
		PE3	0.816	
		PE4	0.865	
		PE5	0.925	
		PE6	0.790	
		PE7	0.889	
B: Customer Satisfaction (Csat)	0.778	Csat1	0.817	Valid
		Csat2	0.827	
		Csat3	0.742	
		Csat4	0.876	
		Csat5	0.654	
		Csat6	0.742	
		Csat7	0.736	

were not taken out. Thus, the operational variable had to be re-tested only for design and feature as well as time-savings with changes on the statements and additional statement indicators to strengthen the representativeness of the statement indicators based on academic journals.

4.3 *Reliability test*

Table 2 shows that all variables are considered reliable; recalling the rule of reliability that the value of Cronbach's Alpha has to be more than 0.7 but if the value of Cronbach's Alpha is 0.6, it still can be accepted. In the accordance of the validity test result to delete statement indicators SS4 and CS4, the reliability test result supported this decision. If the statement indicator SS4 were to be deleted, the Cronbach's Alpha would increase to 0.901. Nevertheless, the Cronbach's

Table 2. Reliability test result of all variables.

Variable	Cronbach's alpha	Statement indicator	Cronbach's alpha if item deleted	Note
A1: Perceived Ease of Use (PEOU)	0.942	PEOU1	0.917	Reliable
		PEOU2	0.932	
		PEOU3	0.913	
		PEOU4	0.954	
		PEOU5	0.922	
A2: Design and Feature (DF)	0.751	DF1	0.687	Reliable
		DF2	0.668	
		DF3	0.731	
		DF4	0.752	
		DF5	0.676	
A3: Service Security (SS)	0.859	SS1	0.805	Reliable
		SS2	0.799	
		SS3	0.853	
		SS4	0.901	
		SS5	0.796	
		SS6	0.828	
A4: Time Savings (TS)	0.605	TS1	0.386	Reliable
		TS2	0.084	
		TS3	0.689	
		TS4	0.646	
A5: Cost Savings (CS)	0.657	CS1	0.532	Reliable
		CS2	0.372	
		CS3	0.355	
		CS4	0.905	
A6: Social Influence (SI)	0.610	SI1	0.582	Reliable
		SI2	0.666	
		SI3	0.367	
		SI4	0.464	
A7: Performance Expectancy (PE)	0.900	PE1	0.868	Reliable
		PE2	0.878	
		PE3	0.885	
		PE4	0.872	
		PE5	0.891	
		PE6	0.905	
		PE7	0.896	
B: Customer Satisfaction (Csat)	0.881	Csat1	0.833	Reliable
		Csat2	0.850	
		Csat3	0.872	
		Csat4	0.844	
		Csat5	0.873	
		Csat6	0.882	
		Csat7	0.882	

Alpha will also be increased to 0.905 if statement indicator CS4 to be deleted. Hence, the need to re-test service security and cost savings.

4.4 Operational variable test 2

The operational variable test 2 was aimed to re-test several invalid variables and repeat the test after deleting an operational variable for some variables. The re-tested variables are design and feature and time-savings while the repeated test variables are service security and cost savings. Before re-testing the variables to other 30 millennials, some items had to be changed as well as to add some statement indicators for design and feature and time-savings on the questionnaire while still keeping the valid and reliable statement of these variables based on the operational variable test 1.

Furthermore, after getting the result of operational variable test 1, the statement indicators on customer satisfaction were not representative enough for this variable. It is because the statements only identifying customer satisfaction based on the seven independent variables of this research; not their actual feeling after shopping online. Thus, the statement with a reference statement indicator for customer satisfaction had to be changed. Other changes also made on the statement indicator for loyalty and word of mouth. These two variables only have a statement indicator which did not represent the post-purchase behavior of online shoppers, thus it cannot be tested since the average value was unable to be counted.

Table 3. Validity Re-testing result.

Variable	KMO	Statement indicator	Anti-image correlation	Note
A2: Design and Feature (DF)	0.777	DF1	0.911	Valid
		DF2	0.751	
		DF3	0.812	
		DF4	0.761	
		DF5	0.773	
		DF6	0.724	
		DF7	0.833	
		DF8	0.713	
		DF9	0.408	Invalid
A3: Service Security (SS)	0.857	SS1	0.876	Valid
		SS2	0.810	
		SS3	0.940	
		SS4	0.821	
		SS5	0.907	
A4: Time Savings (TS)	0.572	TS1	0.815	Valid
		TS2	0.559	
		TS3	0.546	
		TS4	0.614	
A5: Cost Savings (CS)	0.752	CS1	0.770	Valid
		CS2	0.770	
		CS3	0.719	
B: Customer Satisfaction (Csat)	0.731	Csat1	0.768	Valid
		Csat2	0.692	
		Csat3	0.686	
		Csat4	0.827	
C1: Loyalty (LO)	0.761	LO1	0.737	Valid
		LO2	0.815	
		LO3	0.739	
C2: Word of Mouth (WOM)	0.541	WOM1	0.542	Valid
		WOM2	0.525	
		WOM3	0.757	
		WOM4	0.517	

4.5 *Validity test*

Table 3 shows that the value of **KMO** of all variables is above 0.5. However, there is only a statement indicator of DF9 considered as invalid since its anti-image correlation is 0.408 or below 0.5. Therefore, DF9 had to be deleted and re-tested. Nonetheless, after identifying the

Table 4.   Design and feature and time savings validity 2nd Re-testing result.

Variable	KMO	Statement indicator	Anti-image correlation	Note
A2: Design and	0.791	DF1	0.916	Valid
Feature (DF)		DF2	0.801	
		DF3	0.798	
		DF4	0.745	
		DF5	0.781	
		DF6	0.719	
		DF7	0.830	
		DF8	0.716	
A4: Time	0.547	TS2	0.546	Valid
Savings (TS)		TS3	0.530	
		TS4	0.595	

Table 5.   Reliability Re-testing result.

Variable	Cronbach's alpha	Statement indicator	Cronbach's alpha if item deleted	Note
A2: Design and	0.839	DF1	0.810	Reliable
Feature (DF)		DF2	0.814	
		DF3	0.804	
		DF4	0.822	
		DF5	0.824	
		DF6	0.810	
		DF7	0.804	
		DF8	0.825	
		DF9	0.881	
A3: Service	0.901	SS1	0.876	Reliable
Security (SS)		SS2	0.857	
		SS3	0.913	
		SS4	0.852	
		SS5	0.889	
A4: Time Savings (TS)	0.559	TS1	0.648	Unreliable
		TS2	0.459	
		TS3	0.325	
		TS4	0.500	
A5: Cost Savings (CS)	0.905	CS1	0.873	Reliable
		CS2	0.873	
		CS3	0.845	
B: Customer	0.865	CSat1	0.870	Reliable
Satisfaction (CSat)		CSat2	0.776	
		CSat3	0.813	
		CSat4	0.845	
C1: Loyalty (LO)	0.927	LO1	0.878	Reliable
		LO2	0.914	
		LO3	0.888	
C2: Word of	0.789	WOM1	0.749	Reliable
Mouth (WOM)		WOM2	0.607	
		WOM3	0.837	
		WOM4	0.692	

Table 6. Design and feature and time savings reliability 2nd Re-testing result.

Variable	Cronbach's alpha	Statement indicator	Cronbach's alpha if item deleted	Note
A2: Design and	0.881	DF1	0.860	Reliable
Feature (DF)		DF2	0.869	
		DF3	0.857	
		DF4	0.871	
		DF5	0.874	
		DF6	0.860	
		DF7	0.859	
		DF8	0.880	
A4: Time	0.648	TS2	0.580	Reliable
Savings (TS)		TS3	0.317	
		TS4	0.719	

value of reliability test, time-savings was not reliable since its Cronbach's Alpha was below 0.6. Hence, the deletion of TS1 and re-testing of time-savings.

The result revealed that design and feature is valid represented by 8 statement indicators while time-savings is valid represented by 3 indicators. Therefore, the real survey would have 46 statement indicators representing 10 variables in this research.

4.6 Reliability test

The result of operational variable test 2 for reliability of four variables (design and feature, service security, time-savings, and cost-savings) shows that all variables are reliable except for time savings, which Cronbach's Alpha is below 0.6. However, if statement indicator TS1 to be taken out, its Cronbach's Alpha will become 0.648. Referring to validity test result of the operational variable test 2, a statement indicator of design and feature had to be deleted. Hence, the retesting of the variables design & feature and time-savings (shown in Table 6). After conducting the operational variable test 2, the researchers decided to use 10 variables; perceived ease of use, design and feature, service security, time savings, cost savings, social influence, performance expectancy, customer satisfaction, loyalty, and word of mouth. The total statement indicator for the real survey is 46 representing each variable.

4.7 Study 1: The millennials in the greater Jakarta area

The first study reveals the factors affecting customer satisfaction, loyalty, and word of mouth towards online shopping for students and non-students in Jakarta. There are 230 valid responses analyzed, comprising of 115 students and 115 non-students. In this case, non-students are those who were born in between 1982 and 2016 and not currently enrolled as high school or university students, which professions can be workers, entrepreneurs, or housewives.

The aggregate sample of this research consists of 83% female and 17% male where in majority, they are 17–21 years old and 22–26 years old although there are also 14% of people who are in the age of 27–31 years old and 6% of them are 32–36 years old. These people are currently high school or university students (50%) while others are employees (49%) and entrepreneurs (1%). In addition, 17% of them are married while other 83% are single. In majority, 61% of them are living in Jakarta while other 39% respondents are living in Bogor, Depok, Tangerang, and Bekasi. Their last education attainment is mostly senior high school (41%), 26% both for diploma and bachelor degree, while 5% are junior high school and 2% are masters. Furthermore, the three biggest monthly spending of these respondents are IDR 1.000.001–3.500.000, < IDR 1.000.000, and IDR 3.500.000–5.500.000; 42%, 30%, and 16%, respectively. Instead of identifying the demographic data of the respondents, the researcher also identified their online shopping preference in terms of the product, access, device to access the online shopping, monthly shopping online frequency, and online shopping sites preference.

Figure 3 shows that people tend to buy clothes (24%), cosmetics (16%), and electronic devices (12%) through online shopping. The most favorite online shopping sites are Shopee (21%), Instagram (19%), and Lazada (18%). Instead of the listed online shopping sites, 2% respondents mentioned other shopping sites such as Gramedia, Make-up Indo, Althea, LINE Shopping, Youtube, Sociolla, HijUp, and KASKUS. The detail of the online shopping sites preference percentage is shown in Figure 4.

Seventy-two per cent customers access these online shopping sites through application while the other 28% through websites, while the devices to access these sites are smartphones (87%), laptops (7%), computers (5%), and only 1% accessing through a tablet. Although they can easily and mostly shop online through a smartphone, their monthly shopping online frequency is low. The 82% respondents shopped online in less than 3 times a month, 16% for 4–6 times a month, 1% for 7–10 times a month, and 3% for more than 10 times a month. This low online purchasing frequency contradicts with customers' needs and wants to buy products for daily needs. This highlights that customers' preference towards physical store is still big.

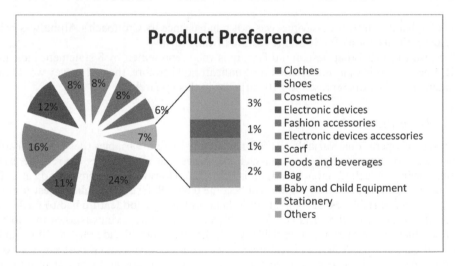

Figure 3.   Product preference.

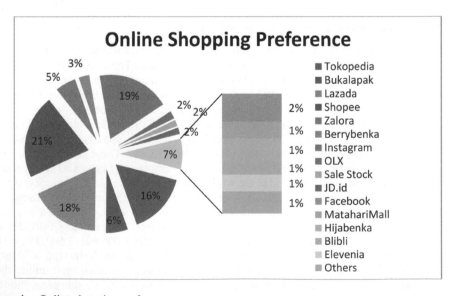

Figure 4.   Online shopping preference.

4.8 Linear regression of statistical model 1

The R-value in Table 7 (0.723) determined that independent variables A1-A7 have 72.3% influence towards dependent variable B (customer satisfaction) with $p < 0.001$. Meanwhile, the other 27.7% of customer satisfaction is influenced by other factors not mentioned in this research. However, not all independent variables A1-A7 have a significant influence towards customer satisfaction. According to the test, four out of seven variables that have significant influence towards customer satisfaction of online shopping (perceived ease of use, service security, cost savings, and performance expectancy).

From Table 8, the statistical model for factors affecting customer satisfaction towards online shopping for millennial generation in Jakarta is $Y_B = 0.443 + 0.153\ A1^* + 0.091\ A2 + 0.192\ A3^{**} + 0.046\ A4 + 0.191\ A5^{***} - 0.035\ A6 + 0.279\ A7^{***}$ (Note: * $p < 0.05$; ** $p < 0.01$; *** $p < 0.001$). Thus, it can be concluded that hypotheses 1, 3, 5, and 7 are accepted while 2, 4, and 6 are rejected.

4.9 Linear regression of statistical model 2

The R-value in Table 9 (0.635) determined that customer satisfaction has 63.5% influences towards loyalty with $p < 0.001$. Meanwhile, the 36.5% of loyalty was influenced by other factors other than customer satisfaction. The significance table of customer satisfaction towards loyalty is shown in Table 10.

Table 7. Model summary.

Model	R	R square	Adjusted R square	Std. error of the estimate
1	0.723	0.523	0.508	0.43056

Table 8. Multiple linear regression table.

Model	Unstandardized coefficients		Standardized coefficients		
	B	Std. Error	Beta	t	Sig.
1					
(Constant)	0.443	0.336		1.320	0.188
PEOUTot	0.153	0.073	0.132	2.082	0.038
DFTot	0.091	0.057	0.085	1.606	0.110
SSTot	0.192	0.068	0.179	2.805	0.005
TSTot	0.046	0.061	0.049	0.754	0.452
CSTot	0.191	0.052	0.229	3.654	0.000
SITot	−0.035	0.040	−0.043	−0.872	0.384
PETot	0.279	0.064	0.288	4.323	0.000

Table 9. Model summary.

Model	R	R square	Adjusted R square	Std. error of the estimate	Sig. F change
1	0.635	0.403	0.400	0.59005	0.000

Table 10. Simple linear regression table.

Model	Unstandardized coefficients		Standardized coefficients	
	B	Std. error	Beta	Sig.
1				
(Constant)	0.432	0.306		0.159
CSatTot	0.788	0.064	0.635	0.000

Table 11. Model summary.

Model	R	R square	Adjusted R square	Std. error of the estimate	Sig. F change
1	**0.694**	0.481	0.477	0.50448	0.000

Table 12. Multiple linear regression table.

Model	Unstandardized coefficients		Standardized coefficients	
	B	Std. Error	Beta	Sig.
(Constant)	0.940	0.263		0.000
CSatTot	0.445	0.070	0.391	0.000
LOTot	0.344	0.057	0.376	0.000

Based on Table 10, the statistical model for factors affecting loyalty towards online shopping for millennial generation in Jakarta is $Y_{C1} = 0.432 + 0.788\ B^{***}$ (p < 0.001). Thus, it can be concluded that hypothesis 8 is accepted.

4.10 *Linear regression of statistical model 3*

The R-value in Table 11, 0.694, determined that customer satisfaction has 69.4% influences towards loyalty with p < 0.001. Meanwhile, the 30.6% of loyalty was influenced by other factors other than customer satisfaction.

Based on Table 12, the statistical model for factors affecting loyalty towards online shopping for millennial generation in Jakarta is $Y_{C2} = 0.940 + 0.445\ B^{***} + 0.344\ C1^{***}$ (p < 0.001). Thus, it can be concluded that hypothesis 9 is accepted.

4.11 *Study 2: Students in the greater Jakarta Area*

The second study reveals the factors affecting customer satisfaction, loyalty, and word of mouth towards online shopping for students in Jakarta with 115 valid responses. The respondents are 82% female and 18% male. In majority, 82% of them are 17–21 year olds and 17% of them are 22–26, although there are also 1% of people in the age of 32–36. These students are currently enrolled in high schools or universities. Seventy per cent of them are living in Jakarta while the other 30% are living in Bogor, Depok, Tangerang, and Bekasi. Their last education attainment is mostly senior high school (74%), 10% holds diploma degree, 9% junior high school, while 6% bachelors and 1% masters. Furthermore, the three biggest monthly spending of these respondents are < IDR 1.000.000, IDR 1.000.001–3.500.000, and IDR 3.500.000–5.500.000; 56%, 38%, and 3%, respectively. Instead of identifying the demographic data of the respondents, the researchers also identified their online shopping preference in terms of the product, access, device to access the online shopping, monthly shopping online frequency, and online shopping sites preference.

According to Figure 5, people tend to buy clothes (24%), cosmetics (18%), and electronic devices (13%) through online shopping. The most favorite online shopping sites are Shopee (25%), Instagram (22%), and Lazada (20%). Instead of the listed online shopping sites, 3% respondents mentioned other shopping sites such as Gramedia, Make-up Indo, Althea, LINE Shopping, Blibli, and Elevenia. Figure 6 shows the detail of the online shopping sites preference percentage.

Seventy-eight per cent customers access online shopping sites through application while other 22% through websites using smartphones (92%), laptops (6%), and only 1% accessing both through tablets and laptops. Although they can easily and mostly shop online through a smartphone, their monthly shopping online frequency is low. There are 87% respondents

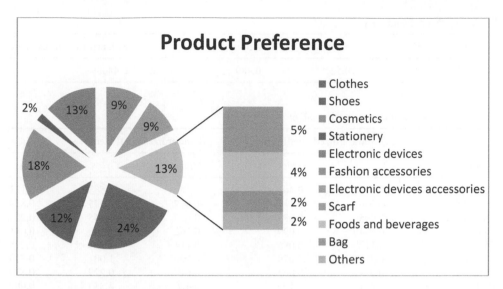

Figure 5.　Product preference.

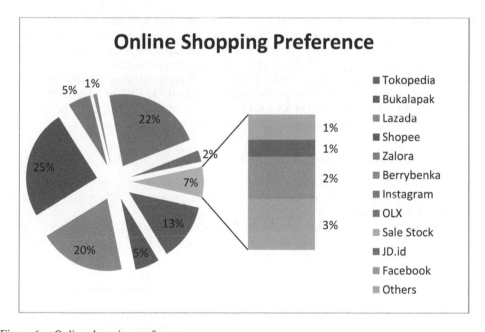

Figure 6.　Online shopping preference.

said shop online less than 3 times a month, 12% for 4–6 times a month, and 1% for more than 10 times a month. This low online purchasing frequency contradicts with customers' needs and wants to buy products for daily needs. Again, this highlights that customers' preference towards physical store is still big.

4.12　Linear regression of statistical model 1

The R-value in Table 13, 0.722, determined that independent variables A1–A7 have 72.2% influence towards dependent variable B or customer satisfaction with $p < 0.001$. Meanwhile, other 27.8% of customer satisfaction is influenced by other factors not mentioned in this

Table 13. Model summary.

Model	R	R square	Adjusted R square	Std. error of the estimate
1	0.722	0.521	0.489	0.44089

Table 14. Multiple linear regression table.

Model	Unstandardized coefficients		Standardized coefficients		
	B	Std. error	Beta	t	Sig.
(Constant)	0.720	0.461		1.563	0.121
PEOUTot	−0.020	0.114	−0.017	−0.177	0.859
DFTot	0.095	0.083	0.096	1.145	0.255
SSTot	0.174	0.099	0.162	1.760	0.081
TSTot	0.117	0.096	0.114	1.214	0.228
CSTot	0.079	0.079	0.096	1.001	0.319
SITot	0.014	0.063	0.017	0.229	0.819
PETot	0.431	0.100	0.426	4.333	0.000

research. However, this study found that not all independent variables A1-A7 have a significant influence towards customer satisfaction. According to the multiple linear regression result, there is only one of seven variables having significant influence towards customer satisfaction of online shopping. This variable is performance expectancy.

Based on Table 14, the statistical model for factors affecting customer satisfaction towards online shopping for millennial generation in Jakarta is $Y_B = 0.742 − 0.020 A1 + 0.095 A2 + 0.174 A3 + 0.117 A4 + 0.079 A5 + 0.014 X6 + 0.431 A7***$ ($p < 0.001$). Thus, it can be concluded that the hypothesis 1–6 are rejected while the hypothesis 7 is accepted.

4.13 *Linear regression of statistical model 2*

The R-value in Table 15, 0.646, determined that customer satisfaction has 64.6% influences towards loyalty with $p < 0.001$. Meanwhile, the 35.4% of loyalty was influenced by other factors other than customer satisfaction.

Based on Table 16, the statistical model for factors affecting loyalty towards online shopping for millennial generation in Jakarta is $Y_{C1} = 0.600 + 0.757 B***$ ($p < 0.001$. Thus, it can be concluded that the hypothesis 8 is accepted.

4.14 *Linear regression of statistical model 3*

The R-value in Table 17, 0.650, determined that customer satisfaction has 65% influences towards loyalty with $p < 0.001$. Meanwhile, the 35% of loyalty was influenced by other factors other than customer satisfaction.

Based on Table 18, statistical model for factors affecting loyalty towards online shopping for millennial generation in Jakarta is $Y_{C2} = 1.401 + 0.521 B*** + 0.156 C1$ ($p < 0.001$). Thus, it can be concluded that the hypothesis 9 is rejected.

4.15 *Study 3: Non-students in the greater Jakarta area*

The third study reveals the factors affecting customer satisfaction, loyalty, and word of mouth towards online shopping for non-students in Jakarta with 115 valid responses analyzed. In this case, the researchers determined non-students as people who were born in between 1982 and 2016, and not currently enrolled as either high school or university students. Their professions can be a worker, an entrepreneur, or a housewife.

Table 15. Model summary.

Model	R	R square	Adjusted R square	Std. error of the estimate	Sig. F change
1	0.646	0.417	0.412	0.55454	0.000

Table 16. Simple linear regression table.

Model	Unstandardized coefficients		Standardized coefficients		
	B	Std. error	Beta	t	Sig.
1					
(Constant)	0.600	0.405		1.482	0.141
CSatTot	0.757	0.084	0.646	8.995	0.000

Table 17. Model summary.

Model	R	R Square	Adjusted R Square	Std. Error of the Estimate	Sig. F Change
1	**0.650**[a]	0.423	0.413	0.47540368.5 pt	0.000

Table 18. Multiple linear regression table.

Model	Unstandardized coefficients		Standardized coefficients		
	B	Std. error	Beta	t	Sig.
1					
(Constant)	1.401	0.350		3.998	0.000
CSatTot	0.521	0.095	0.518	5.507	0.000
LOTot	0.156	0.081	0.182	1.935	0.056

The respondents are 83% female and 17% male. In majority, 58% of them are 22–26 years old and 27% of them are 27–31 although there is also 11% of people in the age of 32–36 and 4% of them are 17–21. There are 98% of them having a job as an employee and only 2% of them are entrepreneurs. In addition, 35% of them are married while other 65% single. According to the researchers' data, 53% of them are living in Jakarta while the other 47% in Bogor, Depok, Tangerang, and Bekasi. Their last education attainment is mostly bachelor degree (46%), 43% diplomas, 8% senior high schools, 2% masters, and 1% junior high schools. Furthermore, the three biggest monthly spending of these respondents are IDR 1.000.001–3.500.000, IDR 3.500.000–5.500.000, and IDR 5.500.001–8.000.000; 46%, 28%, and 14%, respectively. Instead of identifying the demographic data of the respondents, the researchers also identified their online shopping preference in terms of the product, access, device to access the online shopping, monthly shopping online frequency, and online shopping sites preference.

According to Figure 7, people tend to buy clothes (25%), cosmetics (15%), electronic devices as well as shoes both counted as 10% of product preferences through online shopping. The most favorite online shopping sites are Shopee (19%), Tokopedia (18%), and Lazada (17%). Instead of the listed online shopping sites, 1% respondents mentioned other shopping sites such as YouTube, Sociolla, HijUp, and KASKUS. The detail of the online shopping sites preference percentage can be seen in Figure 8.

447

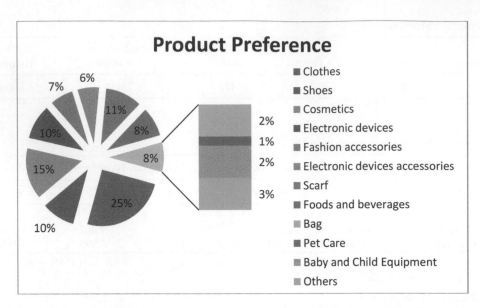

Figure 7. Product preference.

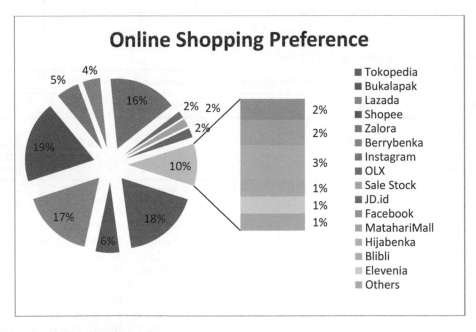

Figure 8. Online shopping preference.

Sixty-seven per cent customers access these online shopping sites through an application, while the other 33% through websites using their smartphones (82%), computers (9%), laptops (8%), and 1% tablets. Although they can easily and mostly shop online through smartphone, their monthly shopping online frequency is still low. There are 77% respondents shopped online in less than 3 times a month, 20% 4–6 times, and 1% 7–10 times, and 2% more than 10 times a month. This low online purchasing frequency contradicts with customers' needs and wants to buy products for daily needs. This highlights customers' preference towards physical store is still high.

Table 19.   Model summary.

Model	R	R square	Adjusted R square	Std. error of the estimate	Sig. F change
1	0.754	0.569	0.540	0.41542	0.000

Table 20.   Multiple linear regression table.

Model	Unstandardized coefficients		Standardized coefficients		
	B	Std. error	Beta	T	Sig.
1					
(Constant)	0.138	0.511		0.271	0.787
PEOUTot	0.243	0.097	0.217	2.499	0.014
DFTot	0.081	0.081	0.068	0.996	0.321
SSTot	0.215	0.097	0.200	2.215	0.029
TSTot	0.015	0.080	0.017	0.188	0.852
STot	0.271	0.072	0.315	3.767	0.000
SITot	−0.049	0.054	−0.064	−0.909	0.365
PETot	0.187	0.088	0.202	2.138	0.035

4.16   Linear regression of statistical model 1

The R-value in Table 19 (0.754) determined that independent variable A1-A7 has 75.4% influence towards dependent variable B or customer satisfaction with $p < 0.001$. Meanwhile, the 24.6% of customer satisfaction was influenced by other factors not mentioned in this research. However, this study found that not all independent variables A1-A7 have significant influence towards customer satisfaction. According to the regression test using 95% confidence level, four out of seven variables have significant influence towards customer satisfaction of online shopping (perceived ease of use, service security, cost savings, and performance expectancy).

Based on Table 20, the statistical model for factors affecting customer satisfaction towards online shopping for millennial generation in Jakarta is $Y_B = 0.138 + 0.243$ A1* $+ 0.081$ A2 $+ 0.215$ A3* $+ 0.015$ A4 $+ 0.271$ A5*** $- 0.049$ A6 $+ 0.187$ A7* (*$p < 0.05$; *** $p < 0.001$). Thus, it can be concluded that the hypothesis 1, 3, 5, and 7 are accepted while the hypothesis 2 and 6 are rejected.

4.17   Linear regression of statistical model 2

The R-value in Table 21 (0.627) determined that customer satisfaction has 62.7% influences towards loyalty with $p < 0.001$. Meanwhile, the 37.3% of loyalty was influenced by other factors other than customer satisfaction.

Based on Table 22, the statistical model for factors affecting loyalty towards online shopping for millennial generation in Jakarta is $Y_{C1} = 0.257 + 0.819$ B*** ($p < 0.001$). Thus, it can be concluded that hypothesis 8 is accepted.

4.18   Linear regression of statistical model 3

Before determining the significance relationship of customer satisfaction and loyalty as the independent variable towards word of mouth as the dependent variable, the percentage of customer satisfaction and loyalty affect word of mouth can be seen in Table 23.

The R-value in Table 23 (0.747) determined that customer satisfaction has 74.7% influences towards loyalty with $p < 0.001$. Meanwhile, the 25.3% of loyalty was influenced by other factors other than customer satisfaction.

Based on Table 24, the statistical model for factors affecting loyalty towards online shopping for millennial generation in Jakarta is $Y_{C2} = 0.547 + 0.399$ B*** $+ 0.484$ C1*** (***: $p < 0.001$). Thus, it can be concluded that hypothesis 9 is accepted.

Table 21. Model summary.

Model	R	R square	Adjusted R square	Std. error of the estimate	Sig. F change
1	0.627	0.393	0.388	0.62704	0.000

Table 22. Simple linear regression table.

Model	Unstandardized coefficients		Standardized coefficients		
	B	Std. error	Beta	t	Sig.
1					
(Constant)	0.257	0.462		0.557	0.578
CSatTot	0.819	0.096	0.627	8.552	0.000

Table 23. Model summary.

Model	R	R square	Adjusted R square	Std. error of the estimate	Sig. F change
1	0.747	0.558	0.550	0.51561	0.000

Table 24. Multiple linear regression table.

Model	Unstandardized coefficients		Standardized coefficients		
	B	Std. error	Beta	t	Sig.
1					
(Constant)	0.547	0.380		1.437	0.154
CSatTot	0.399	0.101	0.318	3.947	0.000
LOTot	0.484	0.077	0.505	6.259	0.000

4.19 *Discussion*

Based on the regression result, the outcome of regression model 1 for study 1 is the same with the outcome of regression model 1 for study 2. It reveals that there are only four variables significantly influence the customer satisfaction of online shoppers, i.e. perceived ease of use, service security, cost savings, and performance expectancy. Nonetheless, the result of regression model 2 for study 1–3 is the same. It reveals that customer satisfaction is significantly influenced the loyalty towards online shopping. While loyalty and customer satisfaction have significant influences towards word of mouth in studies 1 and 3, only satisfied students will spread positive word of mouth while loyal students do not.

Among seven variables identified in statistical model 1 in this research, design & feature, time-savings, and social influence do not affect customer satisfaction of online shoppers in study 1–3. According to KPMG (2017), the millennials expect a good service offers by the online retailers such as providing an excellent customer support and listening to customer feedback. Although most online shopping retailers equip their website and application with customer service and online chat feature, these features do not highly affect millennial's satisfaction towards online shopping since their expectation relies on the service not the feature nor the design of the website and application. Furthermore, millennial engages with brands "more extensively, personally, and emotionally" (Ordun, 2015). This personalized demand cannot be solved only by creating an excellent design and feature on the website and application. Six major reasons people switching from physical stores to online shopping: (1) convenience, (2) price related, (3) push from offline, (4) ease of selection, (5) free shipping, and (6) only option to buy (KPMG, 2017). Two items from the "ease of selection" and "price-related" categories reflect the design & feature variable, i.e. "to locate hard to find items" and "ability to compare prices". A good design of online shopping with complete feature will create an ease of choosing and finding products needed by customers. A complete feature in the website and application will also trigger customers to find and compare prices faster. However, it only contributes to 16.7% of the total percentage reasons people choosing to shop online. Thus, the existence of complete and interesting design and feature in the online shopping's website and application have no significant effect on customer satisfaction.

Another variable to be analyzed due to its insignificance is time-savings. Kohli, et al. (2004) justified time-savings as the "reduction in buyer search costs." They also related this justification to Achrol and Kotler's finding (1999) that the networked economy enables customers to get real-time information which decrease their time-savings. However, the digitalized and networked economy will depend on each country's capability in providing good internet service to their citizens. McKinsey Global Institute (2016) noted that the digital revolution including mobile internet, cloud technology, Internet of Things as well as big data and advanced analytics have arrived in Indonesia. According to World Robotics Report as cited by McKinsey Global Institute (2016), all of these digital revolutions kept improving from 2014 to 2015: the mobile Internet users increasing by 12 million people, the total cloud services vendor revenues increased by 1.4x, the total connected units increased by 7 million units, and the internet protocol traffic increased by 60% in 2015. These increasing numbers may cause by the affordable price of mobile broadband pricing in Indonesia. However, the quality of the Internet access in Indonesia is still lower than its neighboring countries such as Malaysia, Thailand, and Singapore (McKinsey Global Institute, 2016).

Indonesia's unstable Internet speed may influence the buyers' search cost and real-time information needed by online shoppers. However, their urgency towards a product may result in a negative shopping experience, which could switch their option from shopping online to shop in physical stores as it will save time. The low Internet bandwidth and average connection speed may cause longer time in browsing for a product and finding information. Nonetheless, online shoppers will also need to make a purchasing request, wait for the system and the seller to respond, and sometimes they also need to transfer the money and send the receipt to the seller (non-cash on delivery payment method). Not only that, product shipment regularly takes 1–3 days. Another issue to be considered is return privileges. Although a privilege to return goods is provided, the necessity to return goods for any reasons can be costly in terms of time and effort. Therefore, those with the greatest time pressure are precisely those who would suffer the most in time lost. Consequently, high levels of time pressure may inhibit them rather than encourage them to shop online. The greater the consumers' time pressure, the more impatient they become. Because of time required for shopping, delivery, online purchasing to meet a schedule or deadline requires planning in advance and willingness to wait for arrival of the goods, therefore, the most effective appeal is not time savings by purchasing online, but rather get more done by online shopping. Through these calculations, time-savings in the online shopping does not exist which cause its insignificance towards online shoppers' satisfaction. This proves Simon's (1955) Decision-Making Model for time-savings affecting customer satisfaction does not work in the Greater Jakarta area.

Despite design and feature as well as time-savings, social influence is also insignificance towards customer satisfaction. According to Kim, Kim, Choi, and Trivedi (2017), there are two types of social interactions studied by previous researchers: online and offline social interaction. Online social interaction increased the online sales as it increased the customer expenditure (Chen, Wang, and Xie, 2011; Manchanda, Packard, and Pattabhiramaiah, 2015). Meanwhile, offline social interaction may exert more reliable information as people meet physically and have face-to-face interaction (Ramirez and Wang, 2008). Offline social interaction may also better connect people since they live in the same region to better exchange information and experience in the locally based opinion. These two factors do not exist in the online social interaction. Thus, offline social interaction gets closer to individual and more associated with online demand (Choi, Bell, and Lodish, 2012). This justification had statistically proven Kim, et al.'s study (2017) with the p-value of 0.001. However, the offline social interaction becomes faded as people living in particular region have high online shopping preferences. Preferences are the accumulation of experience and information. As the online shopping preference in a region is high, consumers in this region tend to have greater knowledge and experience towards online product purchases (Herhausen, Binder, Schoegel, and Herrmann, 2015). This creates redundant information from other sources which makes shopping uncertainties will less effectively be reduced and shopping benefits will less effectively be increased (Kim, et al., 2017). Relating these past studies with the current social condition in Jakarta, it is concluded that the city has less offline social interaction and high online shopping preference. This triggers social influence insignificance towards customer satisfaction.

Comparing the result of studies 2 and 3, customer satisfaction is affected by more variables in study 3 rather than study 2 where only performance expectancy is statistically significant towards customer satisfaction. This finding can be related to the demographic information of the respondents in study 2 (students) and 3 (non-students). The biggest percentage of last education attainment for the students is senior high school while non-students are bachelor degree. People with higher educational background tend to have better and detailed logical thinking regarding the potential risks of activities and decisions they make. This may also influence their level of maturity. In addition, students mostly spend less than 1 million IDR a month while non-students mostly spend between 1 to 3.5 million IDR a month. The bigger their spending is, the bigger consideration towards perceived risk especially the service security in the online shopping. This makes service security only significant for non-students. In addition, according to the marital status, 35% of the non-student respondents are married, meaning that they have more complex needs rather than the students who are still single. This means non-students who are married need to take care of their children and spouse as well as other married couple's necessities. Thus, non-student respondents tend to consider cost savings as the important aspect affecting their satisfaction while students do not care with the cost savings. Another potential cause is the money the students get from. Students are mostly still supported by their parents or scholarships or loans to support their living cost and school fees. In majority, they do not need to think how to survive by themselves. Thus, whenever they need money they can possibly ask their parents or family members. This contradicts with non-students who are pushed to be independent in fulfilling their needs and wants.

5 CONCLUSIONS

Perceived ease of use, service security, cost savings, and performance expectancy are significant towards customer satisfaction for millennial generation in Jakarta. Customer satisfaction will generate loyalty of the millennial in Jakarta to repeat purchases in the same online shopping. Customer satisfaction and loyalty will generate positive word of mouth for millennial generation in Jakarta towards online shopping.

In addition, a comparison of study 2 (students) and study 3 (non-students), this research found some differences in these samples that the customer satisfaction of students towards online shopping is only influenced by performance expectancy. The customer satisfaction of non-students towards online shopping is influenced by cost savings, perceived ease of use, service security, and performance expectancy respectively. Customer satisfaction and loyalty will generate positive word of mouth for non-students towards online shopping. However, only customer satisfaction can generate positive word of mouth towards online shopping for students.

Based on the result of this research, if online retailers are targeting millennial in Jakarta, they have to maintain and improve cost savings, performance expectancy, service security, and perceived ease of use to satisfy the customers. The performance expectancy has the highest value among other factors. Another important point to boost online shopping's potential is being focused to the market segmentation. The findings of this research reveal students and non-students have different characteristics to be understood. Satisfying students is simpler rather than satisfying non-students due to more factors to be satisfied.

If the online retailers are targeting students, they need to focus only on meeting students' performance expectancy. On the other hand, if they want to target non-students, they need to focus on maintaining and improving cost savings, perceived ease of use, service security, and performance expectancy. Furthermore, if online retailers want to target a region similar to Jakarta that has a high online shopping preference and less offline social interaction, they do not need to focus on creating a strong social influence within their target market. Having a positive word of mouth is good to maintain their brand image, but it will not affect satisfaction.

Although this study did provide additional insight into the influence that perceived ease of use, design and feature, service security, time-savings, cost-savings, social influence, and performance expectancy have on customer satisfaction, a number of limitations do exist. First, this study examined the factors contributing to customer satisfaction towards online shopping for millennial generation or people born in between 1982 and 2001. Second, the use of primary data was from Greater Jakarta Area or Jabodetabek (Jakarta, Bogor, Depok, Tangerang, and Bekasi). Future research could target respondents from various generations, e.g. baby boomers, generation X, and generation Y.

REFERENCES

Achrol, R.S. and Kotler, P. (1999). marketing in the network economy. *Journal of marketing*, 63(1999), 146–163.

Ashraf, A., Thongpapanl, N. and Spyropoulou, S. (2016). The connection and disconnection between e-commerce business and their customers: exploring the role of engagement, perceived usefulness, and perceived ease-of-use. *Electronic Commerce Research and Applications*, 20(2016), 69–86.

Atulkar, S. and Kesari, B. (2017). Satisfaction, loyalty and repatronage intentions: role of hedonic shopping values. *Journal of Retailing and Consumer Services*, 39(2017), 23–34.

Badan Pusat Statistik (2017). *Jumlah penduduk menurut kelompok umur dan jenis kelamin di provinsi DKI Jakarta, 2015*. Retrieved on 14 February 2018 from https://jakarta.bps.go.id/statictable/2017/01/30/142/jumlah-penduduk-menurut-kelompok-umur-dan-jenis-kelamin-di-provinsi-dki-jakarta-2015.html

Chen, Y., Wang, Q., and Xie, J. (2011). Online social interactions: A natural experiment on word of mouth versus observational learning. *Journal of Marketing Research*, 48(2), 238–254.

Choi, J., Bell, D.R., and Lodish, L.M. (2012). Traditional and IS-enabled customer acquisition on the Internet. *Management Science*, 58(4), 754–769.

Clemes, M., Gan, C., and Zhang, J. (2014). An empirical analysis of online shopping adoption in Beijing, China. *Journal of Retailing and Consumer Services*, 21(3), 364–375.

DBS (2017). *Mendorong wirausaha baru di era digital*. Retrieved on 15 February 2018 from https://www.dbs.com/spark/index/id_id/dbs-yes-asset/files/(Riset%202)%20Mendorong%20Wirausaha%20Baru%20di%20Era%20Digital.pdf.

Durmus, B., Ulusu, Y., Erdem, S. and Yalcin, Y. (2015). Are private shopping sites really satisfied customers? *Social and Behavioral Sciences*, 175(2015), 84–89.

Gefen, D., & Straub, D. (2000). The relative importance of perceived ease of use in IS adoption: a study of e-commerce adoption. *Journal of the Association of Information Systems*, 1(8), 1–28.

Hair, J., Black, W., Babin, B., and Anderson, R. (2010). *Multivariate data analysis: a global perspective*, 7th Ed. New Jersey, USA: Pearson.

Hausman, A., and Siekpe, J. (2009). The effect of web interface features on consumer online purchase intentions. *Journal of Business Research*, 62(2009), 5–13.

Herhausen, D., Binder, J., Schoegel, M., and Herrmann, A. (2015). Integrating bricks with clicks: Retailer-level and channel-level outcomes of online–offline channel integration. *Journal of Retailing*, 91(2), 309–325.

Hernandez, B., Jimenez, J., and Martin, M. (2009). Key website factors in e-business strategy. *International Journal of Information Management*, 29(5), 362–371.

Hsieh, Y., Chiu, H., Tang, C., and Lee, M. (2018). Do colors change realities in online shopping? *Journal of Interactive Marketing*, 41(2018), 14–27.

Jung, N. and Seock, Y. (2017). Effect of service recovery on customers' perceived justice, satisfaction, and word-of-mouth, intentions on online shopping websites. *Journal of Retailing and Consumer Services*, 37(2017), 23–30.

Kim, J., Kim, M., Choi, J., and Trivedi, M. (2017). Offline social interactions and online shopping demand: Does the degree of social interactions matter? *Journal of Business Research,* 2017. doi: https://doi.org/10.1016/j.jbusres.2017.09.022.

Kohli, R., Devaraj, S. and Mahmood, M.A. (2004). Understanding determinants of online customer satisfaction: a decision process perspective. *Journal of Management Information Systems,* 20(1), 115–135.

Kotler, P. and Keller, K. (2012). *Marketing management, 14th Ed.* New Jersey, USA: Pearson.

KPMG (2017). *The truth about online consumers.* Retrieved on 15 February 2018 from https://assets. kpmg.com/content/dam/kpmg/xx/pdf/2017/01/the-truth-about-online-consumers.pdf.

Liao, C., Lin, H.-N., Luo, M.M. and Chea, S. (2017). Factors influencing online shoppers' repurchase intentions: the roles of satisfaction and regret. *Information and Management,* 54(5), 651–668.

Lim, Y.M., Yap, C.S. and Lau, T.C. (2010). Online search and Buying Behavior: Malaysian Experience. *Canadian Social Science,* 6(4), 154–166.

Manchanda, P., Packard, G., and Pattabhiramaiah, A. (2015). Social dollars: The economic impact of customer participation in a firm-sponsored online customer community. *Marketing Science,* 34(3), 367–387.

McKinsey Global Institute (2016). *Urban world: the global consumers to watch.* McKinsey & Company. Retrieved on 25 February 2018 from https://www.mckinsey.com/~/media/mckinsey/global%20 themes/urbanization/urban%20world%20the%20global%20consumers%20to%20watch/urban-world-global-consumers-executive-summary.ashx.

Miyatake, K., Nemoto, T., Nakaharai, S. and Hayashi, K. (2016). Reduction in consumers' purchasing cost by online shopping. *Transportation Research Procedia,* 12 (2016), 656–666.

Ordun, G. (2015). Millennial (Gen Y) Consumer Behavior Their Shopping Preferences and Perceptual Maps Associated With Brand Loyalty. *Canadian Social Science,* 11(4), 1–16.

Pham, T., and Ahammad, M. (2017). Antecedents and consequences of online customer satisfaction: a holistic process perspective. *Technological Forecasting and Social Change,* 124(2017), 332–342.

Ramirez, A., and Wang, Z. (2008). When online meets offline: An expectancy violations theory perspective on modality switching. *Journal of Communication,* 58(1), 20–39.

Septiatri, E., and Kusuma, G. (2016). Understanding the perception of millennial generation toward traditional market (a study in Yogyakarta). *Review of Integrative Business and Economics Research,* 5(1), 30–43.

Schiffman L.G. and Kanuk, L.L. (2009). *Consumer Behavior,* 10th Edition, London: Prentice Hall.

Simon, H.A. (1955). A behavioral model of rational choice. *Quarterly Journal of Economics,* 59(1955), 99–118.

Singh, J., Verma, R., Kumar, J., Kaur, R., and Kaur, P. (2016). Consumer Behavior and Perception with Respect to M-Commerce in Indian B2C Retail. *International Journal of Applied Business and Economic Research,* 14(7) 5071–5083.

Tontini, G. (2016). Identifying opportunities for improvement in online shopping sites. *Journal of Retailing and Consumer Services,* 31(2016), 228–238.

Vladlena, B., Saridakis, G., Tennakoon, H., and Ezingeard, J. (2015). The role of security notices and online consumer behavior: an empirical study of social networking users. *International Journal of Human-Computer Studies,* 80(2015), 36–44.

Xu, X., Li, Q., Peng, L., Hsia, T., Huang, C., and Wu, J. (2017). The impact of informational incentives and social influence on consumer behavior during alibaba's online shopping carnival. *Computers in Human Behavior,* 76(2017), 245–254.

Business Innovation and Development in Emerging Economies – Trinugroho & Lau (Eds)
© 2019 Taylor & Francis Group, London, ISBN 978-1-138-35996-3

How value co-creation works in agri-food context? A future thinking for horticulture product marketing

Hesty Nurul Utami & Eleftherios Alamanos
Newcastle University Business School, Newcastle University, Newcastle Upon Tyne, UK

Sharron Kuznesof
Applied Social Sciences—School of Natural and Environmental Sciences, Newcastle University,
Newcastle upon Tyne, UK

ABSTRACT: Value co-creation has gained increased attention among scholars of business management, marketing and supply chain management since the early 2000s. It represents a movement away from organisations defining value to their customers to a more participative process in which people and companies generate and develop meaning. Within the field of agribusiness, the dynamic market situation makes the concept of value co-creation worthy of investigation. The aim of this paper is to review the value co-creation literature in relation to the agri-food sector with a particular focus on the horticulture market. The study identified and analysed peer-reviewed journal articles published between 2000 and 2018, which were retrieved from two online subscription databases. The papers were classified according to the value co-creation issue, focus, orientation, business setting, methodology, and the directions for future research. The analysis showed that value co-creation research in the agribusiness sector is limited but emerging, with the majority of papers published in the past five years. There are a limited number of studies in the horticultural industry. Agri-food along with horticulture has shifted from conventional to a modern marketing perspective, twinned with new marketing providers and channels. Behavioural change, innovation, and environmental issues were also key foci of these value co-creation studies. Recommendations for future research include an empirical focus on the horticulture industry especially in developing countries.

Keywords: value co-creation, innovation, marketing, agribusiness, behaviour

1   INTRODUCTION

Value co-creation has gained increased attention among scholars of business management, marketing and supply chain management since the early 2000s. The concept represents a movement away from organisations defining value to their customers, to a more participative process in which people and companies generate and develop meaning. Despite the increasing scholarship in the area of value co-creation, there is no precise definition of the processes and of the concept of co-creation itself (Alves et al., 2016). Key to the definition is the examination of the different perspectives, processes and roles of both business actors and customers (Alves et al., 2016). However, as an emerging concept, it has continued to develop and broaden (Ind and Coates 2013) in relation to the disciplines that have adopted it and the and the empirical focus. For example, value co-creation has been prominent in the manufacturing, services, and information technology industries. However, studies which focus on the agri-food sector remain scarce and ill-defined despite the importance that agricultural products play in the global economy, particularly as raw ingredients in food processing and manufacturing, and biofuels. The shift in what drives value in relation to food for consumers highlights the need for further understanding of health, safety, experience, social change, and transparency related issues (Ringquist et al. 2016). In the agri-food context, studies on innovation (Yu and Nagurney 2013) and new ways to provide and deliver foods through various activities within marketing and supply chains has drawn researchers' attention (Koutsimanis et al., 2012; Cagnon et al., 2013; Utami et al., 2016).

However, the recent research lacks focus on the relationships within the food business ecosystem, possibility due to complex characteristics of food such as seasonality, perishability, authenticity and safety, which distinguishes food from the manufacturing or the services industries. The study aims to investigate the implementation of the value co-creation concept and process in marketing within the agri-food sector. Thus, the objective of this literature review study is to identify the research streams and the perspective of value co-creation for horticultural products; and to see the emerging trends and gaps of the recent publications during the past decade in this area.

2 LITERATURE REVIEW

2.1 *The development of value co-creation in marketing*

The notion of value co-creation has experienced a long journey prior to its recognition as a central element within marketing strategies (Anderson et al., 1992). The idea of value started with the concept of the exchange of value between a company and its customer (Zeithaml 1988). Thus, under the Good-Dominant (G-D) logic, a physical good which has a value for customers is typically exchanged for money, which is of value to the seller. Therefore, the exchange of value facilitates the role of marketing by creating and delivering customer value that is superior to competitors and facilitating the development of long-term relationships with customers (Zeithaml 1988; Woodall 2003; Lindgreen and Wynstra 2005). However, the G-D logic does not account for exchange in a service context where the production and consumption of a service typically happens simultaneously and by implication. Instead, such situations require the co-creation of value rather between company and consumer instead of the exchange of value (Grönroos and Ravald 2011). This services perspective has led to the evolution of a new logic in marketing is known as Service-Dominant (S-D) logic which has a focus on value co-creation, intangible resources, and the relationships among business actors and customers (Vargo and Lusch 2004).

Under the S-D logic, the concept of value has moved away from being an overall assessment made by customers on a good's 'utilities' (Porter 1985; Zeithaml 1988) in which a company would attempt to increase market share or sales by offering increased perceived benefits (Payne et al., 2008), towards a more interactive and realistic approach (Holbrook 2005) involving customers creating value through their experiences (Prahalad and Ramaswamy 2004; Grönroos 2011); referred to a 'value-in-use'. The service-dominant logic has become the central proposition in the co-creation concept that considers customers as co-creators of value (Payne et al., 2008). The logic gives further insight on the value co-creation concept by suggesting that the process of creating value is not solely created during the production process, but mainly, continues in the consumption process (Gummesson, 2008). The changes in how value is perceived from a marketing perspective is the effect of the transformation of the marketing orientation from being company-driven to being customer-driven (Gummerus, 2013). A customer focus foreshadows the emergence of value co-creation to maximise customer lifetime value by managing the involvement of customer into a company's activities (A. Payne and Frow 2005; Martinez 2014).

The concept continues to develop by considering not only the creation of value for the customer but also for the business actors and the marketers. The value proposition occurs through the development of interactions and relationships (Lambert and Enz 2012). The shift from creating and delivering **value for the customer**, to creating **value with the customer** is the result of the interaction between marketers and customers in an on-going relationship (Grönroos 2011). Integrating customers in co-creation activities requires an effort to bound the company and customer in order to address the challenges and propose mutual benefits and value (Kumar Agrawal and Rahman 2015). The value co-creation is still a developing concept, because value is determined and perceived by customers in unique ways both contextually and experientially (Grönroos 2011).

As a process, value co-creation can be defined in three primary stages: 1) value proposition through joint crafting activities; 2) actualisation of the value; and 3) determination of value (Lambert and Enz 2012). The processes of defining value co-creation shed the light on co-creation activities as a lively and creative action in which companies have improved communication with their customers through their participation and involvement.

2.2 *Value co-creation within agribusiness*

As the value co-creation concept continues to find its shape and consensus among scholars, the practical implementation of this concept in the various industries also continues to develop. Understanding the concept of value creation in the agriculture context facilitates the presentation of the relationships between the producers or farmers and the agribusiness actor knowledge that is considered as the innovation target (Bitzer and Bijman 2015). The current understanding in defining value creation in the agribusiness context shows an abstract and broad definition that requires greater coordination within the value chain which sometimes still considers value creation similar to the financial value (Cucagna and Goldsmith 2018). However, the issues which are experienced in the agribusiness sector are not only related to food safety and food security, but also other more significant problems such as environmental sustainability, biodiversity, health, and even poverty (Dentoni et al., 2012). The value co-creation concept should also offer solutions to customer problems by utilising consumer information (Cui and Wu 2016). In the agribusiness context, the delivery of such solutions might also involve the chain actors solving the business obstacles by developing closer collaboration within the chain providers (Kottila and Rönni 2008).

Modern consumer food preferences such as demand for better food quality and safety, has triggered the industry to drive differentiation by adding value to raw or fresh products (Cucagna and Goldsmith 2018). This customer-centric approach is a strategic way to retain or increase market share, and focusing on the customer is the key to thrive in a competitive market (Martinez 2014). To obtain a competitive advantage the food industry needs to continuously innovate to satisfy individual consumers by providing a personalised offer (Cacciolatti, Garcia, and Kalantzakis 2015). There are different ways of utilising and implementing co-creation. Company resources through physical resources and capabilities will determine the type and outcome of the co-creation (Agrawal et al., 2015). Such implementation can be in the form of modern technology adoption (Agrawal, Kaushik, and Rahman 2015; Lal et al. 2016); innovation and open innovation (Matthyssens, Vandenbempt, and Berghman 2008; Martinez 2014); developing cooperation with business partners (Lawson et al. 2008); and service and relationships (Lusch et al., 2007; Battaglia et al., 2014).

3 METHODOLOGY

The selected literature was analysed to give further explanation, and insight on the concept of value co-creation in the agribusiness context by exploring the concept of adaptation applied to horticultural products. The reviewed papers were selected from two online subscription-based databases, namely, ProQuest Natural Sciences and Web of Science. ProQuest on the available agribusiness and food marketing scholarly journals collection that offer a comprehensive body of literature. Web of Science offers a scientific database with a comprehensive citation search. The selection of the two online scientific databases led to a comprehensive list of publications in marketing in general and agribusiness and food marketing in particular.

The review included only peer-reviewed journal articles published between 2000 and 2018. The selection process followed the suggested steps of conducting a literature review that focus on paper identification, selection, assessment and synthesis (Petticrew and Roberts 2006). The two databases were accessed by subscription provided by Newcastle University.

The selection of the journal articles was made by using keywords 'value co-creation', 'marketing' and 'horticulture' for ProQuest Natural Sciences, and 'value co-creation', 'marketing', and 'food' for Web of Science. The different keywords 'horticulture' and 'food' were used to expand the range of potential articles. The search was also refined by limiting the subject areas to 'business', 'management', 'food', 'food technology', and 'agriculture multidisciplinary'.

The search strategy produced twenty-six journal articles from the two online databases; twelve journal articles in ProQuest and fourteen journal articles in Web of Science. Thirteen articles were excluded due to subject irrelevance, leaving thirteen articles for analysis. These articles were synthesised into a table based upon the issues, focus, orientation, methodology, business setting, key findings and future research suggestion. The articles were then analysed according to the research stream within value co-creation in the agribusiness sector to identify potential research gaps and future research avenues.

4 FINDINGS

The low number of papers addressing marketing value co-creation in the context of food and horticulture suggests that the topic has not been investigated in great depth and there is scope for future research in this area. Although the first paper to be published in this area was a decade ago in 2008, the remaining research has been published in the past five to six years, indicating this is a relatively new and developing area of study (see Table 1). The empirical research also has a western or global north focus, with most empirical work conducted in the USA (see Hunt et al., 2012; Weiler, Moore and Moyle, 2013; Erickson et al., 2015; Delate et al., 2016), EU (see Xie et al., 2008; Bush and Oosterveer, 2015; Makri and Koutsouris, 2015; Delate et al., 2016; Wikström and L'Espoir Decosta, 2018) and Australia (Alonso and Northcote 2013). Only two studies were conducted in the global south: Macau (Ji et al., 2018); and Ethiopia, Benin and South Africa (Bitzer and Bijman, 2015). Only one study (Ji et al., 2018) was conducted in Asia, despite many countries in this area relying heavily on the contribution of their agricultural sector to gross domestic product. Thus, value co-creation which is fundamental in business management (Corsaro and Snehota 2010) is still unexplored in the context of agribusiness in South East Asia.

Based on the identification and analysis of previous studies which were classified according to issues, value co-creation focus, orientation, business setting, and methodology (see Figure 1), the most preferred methodologies were statistical modelling and case studies, followed by simulation modelling, which reflects a growing interest among scholars to integrate such results through network analysis (Makri and Koutsouris 2015; Wikström and L'Espoir Decosta 2018) and innovation modelling (De Koning et al. 2016). Recent studies on marketing concepts and theories on utilising value co-creation as a marketing strategy is considered as way to solve various business and marketing issues such as economic change, environment, innovation, health and lifestyle, changes in consumer behaviour, the role of the consumer in co-creation, and the research approach for co-creation within the agribusiness actor. It should be noted that the results from value co-creation research reveal other agriculture business settings besides agri-food, such as research on urban forestry (see Weiler, Moore and Moyle, 2013; Campbell et al., 2016) and fisheries (see Bush and Oosterveer, 2015). The papers were considered in the analysis due to the insights on the value co-creation research in the agribusiness context.

The orientation of the research on value co-creation has not only extended the concerns on end-users, but also on business markets and other stakeholders. Most previous studies focused on business markets, such as farmers as producers, the distribution channels, and the community. Unit analysis on business markets led to further insights on the span of value co-creation activities such as innovation (Makri and Koutsouris 2015; Delate et al. 2016; Wikström and L'Espoir Decosta 2018) and product quality improvement (Bush and Oosterveer 2015) which can be disseminated within the agribusiness actors. Previous research with end-consumer orientation examined changes in consumer behaviour such as attitudes, consumer experience, customer interaction, prosumption (production and consumption)

Table 1. Summary of paper extraction for value co-creation in marketing and agri-food scope.

No	Year	Authors	Issues	Value co-creation focus	Orientation	Methodology	Business setting	Key findings	Future research	Research location
1	2008	Xie et al.	Customer prosumption propensity (behavioural-based) to verify customer as co-creators of value	Food prosumption	End-consumers	Statistical modelling	Household meal or food preparation	• Prosumption intention (domain specific values) influenced by self-efficacy, attitudes, and past behaviour • Self-efficacy, attitudes, and past behaviour have the propensity to shape future consumer engagement in prosumption • Domain-specific values affected by global values in food prosumption	• Investigation on the interaction impact of global values and behaviour and attitudes on intention • Investigation of the relationship between intention and actual decision making in prosumption behaviour • Investigation of prosumption behaviour to different empirical cases	Norway
2	2012	Hunt et al.	Customer behaviour and value co-creation in different cross-channel	Customer co-production	End-consumers	Statistical modelling	Community-supported agriculture (CSA) for fresh produce	• Co-production provides customer value • Service convenience provision necessity refer to customer non-monetary value	Incorporation of customer satisfaction on value creation study that draws upon customer attitude toward co-production activities	The USA
3	2013	Weiler et al.	National park survival in the global rapid economic change	Benefits and threats of customer experience	End-consumer (national park visitor)	Conceptual model	National park	Threats reduction to enhance customer experiences and visiting benefits	• Visitor experiences threats clarification • Synchronise benefits accrual and management setting • Intervention design analysis to enhance and manage visiting benefits	The USA

(Continued)

Table 1. (Continued).

No	Year	Authors	Issues	Value co-creation focus	Orientation	Methodology	Business setting	Key findings	Future research	Research location
4	2013	Alonso & Northcote	• Multifunctional agriculture • Food crisis • Food prices	Value-added agriculture	Business market (farming operators)	Statistical modelling and content analysis	Small-scale orchard operators	Cost–benefits rationalisation effected less interest in adding value to food production due to cost expenses, lack of knowledge, time, and markets	Clarification the needs of value adding in agriculture production in other agricultural region and sectors	Australia
5	2015	Bitzer & Bijman	• Innovation initiatives benefits • Agri-food chain co-innovation	Co-innovation in the value chain	Business market (agriculture value chain)	Exploratory study through case studies	Potato, pineapple, and citrus	Co-innovation promoting agriculture require three components of collaboration, coordination, and complementary	• Explore success cases of co-innovation • An in-depth study on processes on the knowledge exchange in co-innovation • Analysis of actor's interaction thru the power dimensions	Africa (Ethiopia, Benin, and South Africa)
6	2015	Makri & Koutsouris	Transfer innovation among farmer producers from a network perspective	Innovation networking	Business market (farmer group)	Simulation modelling through social network analysis	Horticulture PGs (Producer Groups)	Innovations dissemination in the network require coordination and support instead rely only on organisation leadership	• Exploring the needs of know-how skills • New roles for innovation dissemination for farmer-producers necessity	The EU and Greece
7	2015	Bush & Oosterveer	Environmental certification standards	Vertical differentiation in the value chain	Business market (fisheries and NGOs)	Exploratory study through case studies	Fisheries industry	Standardisation gives effect on market, trade, and internal organisation governance to response global value chain threats	Dynamic behaviour and role of employing standardisation on the value chain	The Netherland

460

No.	Year	Author	Topic	Concept	Stakeholders	Method	Context	Findings	Future research	Country
8	2015	Erickson et al.	• Food literacy • Children healthy eating choices • Local food	Value creation for social marketing	Stakeholders	Exploratory study through case studies	Elementary school garden	• Verify the distinction between institution centred and institution target-centred • Consumer involvement in value-in-use relevance on social marketing • The needs of community support enhanced social networks • Consideration of barriers to exchange the desired and competitive behaviours on value creation in social marketing	Research extension to verify the analysis on other settings	The USA
9	2016	Campbell et al.	Green infrastructure through urban tree planting	Knowledge co-production	Academics and practitioners	Comparative case studies	Urban forestry	• Main keys of knowledge co-production are embeddedness and trust • Co-production new approaches require the blending form of expertise, capability and how to know	Exploration of knowledge co-production from a different perspective of beliefs, knowledge, values and management such as profit organisations	The USA
10	2016	Delate et al.	Participatory organic research	Co-innovation	Academics	Statistical modelling	Organic agriculture	• A similar result on participatory organic farming research in the US and Italy based on farmer knowledge and participation enhancement with farming system focus • Differences appear in term of pest management in the US and farming equipment in Italy	• Inspiration to create a useful participatory research methodology • Enhance the research focus on innovation partnership (co-innovation) for organic farming	The USA and the EU (Italy)

(*Continued*)

461

Table 1. (Continued).

No	Year	Authors	Issues	Value co-creation focus	Orientation	Methodology	Business setting	Key findings	Future research	Research location
11	2016	De Koning et al.	Mental innovation space of agri-food with small-scale industry	Open innovation	Business markets (agri-food with SMEs)	Simulation modelling through Mental Innovation Space	Fresh produce (fruits & vegetables), fish & meat, processed product (tea & coffee)	• The unfamiliarity of value co-creation • Inexperienced with new product innovation among agri-food with SMEs • Recognition of innovation necessary to maintain customer • SMEs enthusiasm on value co-creation	Further exploration thru qualitative research on innovation and co-creation among agri-food with SMEs to strengthen the results	Vietnam
12	2018	Wikström & Decosta	Consumer role in value creation	Value creation process	Business markets (suppliers, retailers) and end-consumer	Simulation modelling through the network-integrated hierarchy	Food industry	• Hierarchical goals of consumer value creation • Consumers resources and capacity identification • Firm-to-frim interaction necessity as value creation support	Validation relevant solution for consumer value creation process-based value-in-use on other food industries or countries	Sweden
13	2018	Ji et al.	Economic change effect on personal experience (customer escape experience)	• Customer-to-customer interaction (C2CI) • Food attachment	End-consumer	Statistical simulation through multi-methods	Food at fine dining restaurant	• Customer escape experience mediated food attachment and C2CI • Economic regional condition mediated escape experience and C2CI • Customer experience on food attachment also become part of the regional economic situation	• Investigation of other social (cultural values, preferences, lifestyle) and economic factors on escape experience and C2CI	Macau

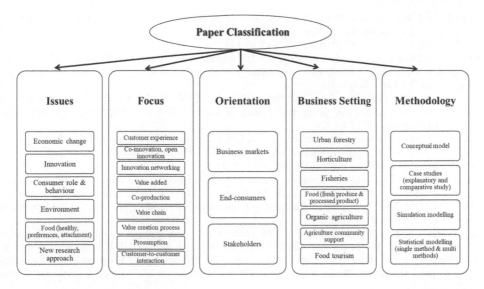

Figure 1. Paper classification summary.

and co-production during the co-creation process. The studies explored the causal effect or relationships alongside changes in economic situation, customer value in different channels, and customers as co-creators. Consumer experience can be used to align customer benefits and company setting (Weiler et al., 2013); and to establish customers as co-creator of value by facilitating behavioural change (Xie et al., 2008; Hunt, Geiger-Oneto and Varca, 2012). The stakeholder perspective was also of interested to some scholars who investigated the term co-innovation and co-production as the new approach to blending the involved stakeholders in co-creation by utilising resources (Campbell et al., 2016; Delate et al., 2016), and industry dynamic behaviour on employing co-creation on the value chain (Bush and Oosterveer 2015).

5 CONCLUSION

The paper highlights the research gaps in the existing literature on marketing value co-creation for agri-food sector. The classification of the previous literatures (see Figure 1) highlights the gaps in the topic related to the dynamic change of both business and end-consumer behaviour within complex agribusiness channels. This gap still remains unexplored especially the process of value co-creation in an agribusiness context. Value is always determined differently by each beneficiary in the specific business context of the value usage (Grönroos 2011). The analysis showed that value co-creation research in the agribusiness sector is limited do-date, with particular gaps in the global south including Asia, which are predominantly agrarian countries. Moreover, analysis on the business setting taken from previous research also identified limited research on horticulture products, both as fresh produce or processed products. Further research related to the horticulture industry especially in developing countries can contribute to further conceptual development of value co-creation in the agri-food chain (Bitzer and Bijman 2015).

REFERENCES

Aarikka-Stenroos, Leena, and Elina Jaakkola. 2012. "Value Co-Creation in Knowledge Intensive Business Services: A Dyadic Perspective on the Joint Problem Solving Process." *Industrial Marketing Management* 41: 15–26. https://doi.org/10.1016/j.indmarman.2011.11.008.

Agrawal, Amit Kumar, Arun Kumar Kaushik, and Zillur Rahman. 2015. "ScienceDirect Co-Creation of Social Value through Integration of Stakeholders." *Procedia -Social and Behavioral Sciences* 189: 442–48. https://doi.org/10.1016/j.sbspro.2015.03.198.

Alonso, Abel Duarte, and Jeremy Northcote. 2013. "Investigating Farmers' Involvement in Value-Added Activities A Preliminary Study from Australia." *British Food Journal* 115 (10): 1407–27. https://doi.org/10.1108/BFJ-04-2011-0104.

Alves, Helena, Cristina Fernandes, and Mário Raposo. 2016. "Value Co-Creation: Concept and Contexts of Application and Study ☆." *Journal of Business Research* 69 (5): 1626–33. https://doi.org/10.1016/j.jbusres.2015.10.029.

Anderson, James C, Dipak C Jam, and Pradeep K Chintagunta. 1992. "Customer Value Assessment in Business Markets: A State-of-Practice Study." *Journal of Business-to-Business Marketin* 1 (1): 1–29. https://doi.org/https://doi.org/10.1300/J033v01n01_02.

Battaglia, Daniel, Cristiano D. Schimith, Marcelo A. Marciano, Sandro A.M. Bittencourt, Leticia Diesel, Miriam Borchardt, and Giancarlo M. Pereira. 2014. "Creating Value through Services and Relationships: The Perception of Purchasing Companies." *Procedia CIRP* 16: 26–31. https://doi.org/10.1016/j.procir.2014.01.004.

Bitzer, Verena, and Jos Bijman. 2015. "From Innovation to Co-Innovation? An Exploration of African Agrifood Chains." *British Food Journal* 117 (8): 2182–99. https://doi.org/10.1108/BFJ-12-2014-0403.

Bush, Simon R, and Peter Oosterveer. 2015. "Vertically Differentiating Environmental Standards: The Case of the Marine Stewardship Council." *Sustainability* 7: 1861–83. https://doi.org/10.3390/su7021861.

Cacciolatti, Luca A., Claire C. Garcia, and Marios Kalantzakis. 2015. "Traditional Food Products: The Effect of Consumers' Characteristics, Product Knowledge, and Perceived Value on Actual Purchase." *Journal of International Food and Agribusiness Marketing* 27 (3): 155–76. https://doi.org/10.1080/08974438.2013.807416.

Cagnon, Thibaut, Aurore Méry, Pascale Chalier, Carole Guillaume, and Nathalie Gontard. 2013. "Fresh Food Packaging Design: A Requirement Driven Approach Applied to Strawberries and Agro-Based Materials." *Innovative Food Science & Emerging Technologies* 20 (October): 288–98. https://doi.org/10.1016/j.ifset.2013.05.009.

Campbell, Lindsay K, Erika S Svendsen, and Lara A Roman. 2016. "Knowledge Co-Production at the Research—Practice Interface: Embedded Case Studies from Urban Forestry." *Environmental Management* 57 (6): 1262–80. https://doi.org/10.1007/s00267-016-0680-8.

Corsaro, Daniela, and Ivan Snehota. 2010. "Searching for Relationship Value in Business Markets: Are We Missing Something?" *Industrial Marketing Management* 39: 986–95. https://doi.org/10.1016/j.indmarman.2010.06.018.

Cucagna, Maria Emilia, and Peter D Goldsmith. 2018. "Value Adding in the Agri-Food Value Chain International Food and Agribusiness Management Review." *International Food and Agribusiness Management Review Cucagna and Goldsmith* 21 (3): 293–316. https://doi.org/10.22434/IFAMR2017.0051.

Cui, Anna S, and Fang Wu. 2016. "Utilizing Customer Knowledge in Innovation: Antecedents and Impact of Customer Involvement on New Product Performance." *Journal of the Academy of Marketing Science* 44: 516–538. https://doi.org/10.1007/s11747-015-0433-x.

Delate, Kathleen, Stefano Canali, Robert Turnbull, Rachel Tan, and Luca Colombo. 2016. "Participatory Organic Research in the USA and Italy: Across a Continuum of Farmer—Researcher Partnerships." *Renewable Agriculture and Food Systems* 32 (4): 331–348. https://doi.org/10.1017/S1742170516000247.

Dentoni, Domenico, Otto Hospes, and R Brent Ross. 2012. "Managing Wicked Problems in Agribusiness: The Role of Multi-Stakeholder Engagements in Value Creation EDITOR'S INTRODUCTION." *International Food and Agribusiness Management Review Special Issue B* 15 (B): 1–12. http://ageconsearch.umn.edu/record/142273/files/introR.pdf.

Erickson, G. Scott, Marlene Barken, and David Barken. 2015. "Caroline Elementary School's Hybrid Garden: A Case Study in Social Marketing." *Journal of Social Marketing* 5 (4): 324–37. https://doi.org/10.1108/EL-01-2014-0022.

Grönroos, Christian. 2011. "A Service Perspective on Business Relationships: The Value Creation, Interaction and Marketing Interface." *Industrial Marketing Management* 40: 240–47. https://doi.org/10.1016/j.indmarman.2010.06.036.

Grönroos, Christian, and Annika Ravald. 2011. "Service as Business Logic: Implications for Value Creation and Marketing." *Journal of Service Management* 22 (1): 5–22. https://doi.org/http://dx.doi.org/10.1108/MRR-09-2015-0216.

Gummerus, Johanna. 2013. "Value Creation Processes and Value Outcomes in Marketing Theory: Strangers or Siblings?" *Marketing Theory* 13 (1): 19–46. https://doi.org/10.1177/1470593112467267.

Gummesson, Evert. 2008. "Customer Centricity: Reality or a Wild Goose Chase?" *European Business Review* 20 (4): 315–30. https://doi.org/10.1108/09555340810886594.

Holbrook, Morris B. 2005. "Customer Value and Autoethnography: Subjective Personal Introspection and the Meanings of a Photograph Collection." *Journal of Business Research* 58: 45–61. https://doi.org/10.1016/S0148-2963(03)00079-1.

Hunt, David M., Stephanie Geiger-Oneto, and Philip E. Varca. 2012. "Satisfaction in the Context of Customer Co-Production: A Behavioral Involvement Perspective." *Journal of Consumer Behaviour* 11: 347–56. https://doi.org/10.1002/cb.

Ind, Nicholas, and Nick Coates. 2013. "The Meanings of Co-Creation." *European Business Review* 25 (1): 86–95. https://doi.org/https://doi.org/10.1108/09555341311287754 Permanent.

Ji, Mingjie, Ipkin Anthony Wong, Anita Eves, and Aliana Man Wai Leong. 2018. "International Journal of Contemporary Hospitality Management A Multilevel Investigation of China's Regional Economic Conditions on Co-Creation of Dining Experience and Outcomes." *International Journal of Contemporary Hospitality Management International Journal of Contemporary Hospitality Management Iss International Journal of Contemporary Hospitality Management* 30 (4): 2132–52. https://doi.org/https://doi.org/10.1108/IJCHM-08-2016-0474.

Koning, Jotte De, Marcel Crul, Jo Van Engelen, Renee Wever, and Johannes Brezet. 2016. "Mental Innovation Space of Vietnamese Agro-Food Firms." *British Food Journal* 118 (6): 1516–32. https://doi.org/10.1108/BFJ-10-2015-0400.

Kottila, Marja - Riitta, and Päivi Rönni. 2008. "Collaboration and Trust in Two Organic Food Chains." *British Food Journal* 110 (4/5): 376–94. https://doi.org/10.1108/00070700810868915.

Koutsimanis, Georgios, Kristin Getter, Bridget Behe, Janice Harte, and Eva Almenar. 2012. "Influences of Packaging Attributes on Consumer Purchase Decisions for Fresh Produce." *Appetite* 59 (2): 270–80. https://doi.org/10.1016/j.appet.2012.05.012.

Kumar Agrawal, Amit, and Zillur Rahman. 2015. "Roles and Resource Contributions of Customers in Value Co-Creation." *International Strategic Management Review* 3: 144–60. https://doi.org/10.1016/j.ism.2015.03.001.

Lal, Bidit, Ameet Pandit, Mike Saren, Sanjay Bhowmick, and Helen Woodruffe-burton. 2016. "Journal of Retailing and Consumer Services Co-Creation of Value at the Bottom of the Pyramid: Analysing Bangladeshi Farmers' Use of Mobile Telephony." *Journal of Retailing and Consumer Services* 29: 40–48. https://doi.org/10.1016/j.jretconser.2015.10.009.

Lambert, Douglas M, and Matias G Enz. 2012. "Managing and Measuring Value Co-Creation in Business-to-Business Relationships." *Journal of Marketing Management* 28: 1588–1625. https://doi.org/10.1080/0267257X.2012.736877.

Lawson, Rob, John Guthrie, Alan Cameron, and Wolfgang Chr Fischer. 2008. "Creating Value through Cooperation An Investigation of Farmers' Markets in New Zealand." *British Food Journal* 110 (1): 11–25. https://doi.org/10.1108/00070700810844768.

Lindgreen, Adam, and Finn Wynstra. 2005. "Value in Business Markets: What Do We Know? Where Are We Going?" *Industrial Marketing Management* 34: 732–48. https://doi.org/10.1016/j.indmarman.2005.01.001.

Lusch, Robert F, Stephen L Vargo, and Matthew O'brien. 2007. "Competing through Service: Insights from Service-Dominant Logic." *Journal of Retailing* 83 (1): 5–18. https://doi.org/10.1016/j.jretai.2006.10.002.

Makri, Anastasia, and Alex Koutsouris. 2015. "Innovation Networking within Producer Groups (PGs): The Case of Two PGs in Ierapetra, Crete." *Agricultural Economics Review* 16 (1): 88–97.

Martinez, Marian Garcia. 2014. "Co-Creation of Value by Open Innovation: Unlocking New Sources of Competitive Advantage." *Agribusiness An International Journal* 30 (2): 132–47. https://doi.org/10.1002/agr.21347.

Matthyssens, Paul, Koen Vandenbempt, and Liselore Berghman. 2008. "Value Innovation in the Functional Foods Industry." *British Food Journal* 110 (1): 144–55. https://doi.org/10.1108/00070700810844830.

Payne, Adrian F, Kaj Storbacka, and Pennie Frow. 2008. "Managing the Co-Creation of Value." *Journal of the Academy Marketing Science* 36: 83–96. https://doi.org/10.1007/s11747-007-0070-0.

Payne, Adrian, and Pennie Frow. 2005. "A Strategic Framework for Customer Relationship Management." *Source Journal of Marketing* 69 (4): 167–76. http://www.jstor.org/stable/30166559.

Petticrew, Mark, and Helen Roberts. 2006. *Systematic Reviews in the Social Sciences: A Practical Guide.* Oxford: Blackwell Publishing Ltd.

Porter, Michael E. 1985. Competitive Advantage: Creating and Sustaining Superior Performance-Free Press. New York: The Free Press.

Prahalad, C K, and Venkat Ramaswamy. 2004. "Co-Creating Unique Value with Customers" 32 (3): 4–9.

Ringquist, Jack, Tom Phillips, Barb Renner, Rod Sides, Kristen Stuart, Mark Baum, and Jim Flannery. 2016. "Capitalizing on the Shifting Consumer Food Value Equation." United Kingdom. https://www2.deloitte.com/content/dam/Deloitte/us/Documents/consumer-business/us-fmi-gma-report.pdf.

Utami, H.N., A.H. Sadeli, and T. Perdana. 2016. "Customer Value Creation of Fresh Tomatoes through Branding and Packaging as Customer Perceived Quality." *Journal of the International Society for Southeast Asian Agricultural Sciences* 22 (1).

Vargo, Stephen L, and Robert F Lusch. 2004. "Evolving to a New Dominant Logic for Marketing." *Source Journal of Marketing* 68 (1): 1–17. http://www.jstor.org/stable/30161971.

Weiler, Betty, Susan A Moore, and Brent D Moyle. 2013. "Building and Sustaining Support for National Parks in the 21st Century: Why and How to Save the National Park Experience from Extinction." *Journal of Park and Recreation Administration* 31 (2): 115–31.

Wikström, Solveig, and Patrick L'Espoir Decosta. 2018. "How Is Value Created? — Extending the Value Concept in the Swedish Context." *Journal of Retailing and Consumer Services* 40 (May 2017): 249–60. https://doi.org/10.1016/j.jretconser.2017.10.010.

Woodall, Tony. 2003. "Conceptualising 'Value for the Customer': An Attributional, Structural and Dispositional Analysis." *Academy of Marketing Science Review* 12: 1–45.

Xie, Chunyan, Richard P. Bagozzi, and Sigurd V. Troye. 2008. "Trying to Prosume: Toward a Theory of Consumers as Co-Creators of Value." *Journal of the Academy of Marketing Science* 36 (1): 109–22. https://doi.org/10.1007/s11747-007-0060-2.

Yu, Min, and Anna Nagurney. 2013. "Competitive Food Supply Chain Networks with Application to Fresh Produce." *European Journal of Operational Research* 224: 273–82. https://doi.org/10.1016/j.ejor.2012.07.033.

Zeithaml, Valarie A. 1988. "Consumer Perceptions of Price, Quality, and Value: A Means-End Model and Synthesis of Evidence." *Journal of Marketing* 52 (3): 2–22. http://www.jstor.org/stable/1251446.

Business Innovation and Development in Emerging Economies – Trinugroho & Lau (Eds)
© *2019 Taylor & Francis Group, London, ISBN 978-1-138-35996-3*

The impact of service innovativeness on self-congruity and functional congruity with motivation to innovate as moderation variable

Kristiningsih
Wijaya Kusuma Surabaya University, Indonesia

S. Hartini & U. Indrianawati
Airlangga University, Indonesia

ABSTRACT: The aim of this research was to examine the role of service innovativeness in building image congruity consisting of self-congruity and functional congruity. This research also tested the moderation effect of motivation to innovate on the relationship between service innovativeness to self-congruity and functional congruity. Based on the basic theory of Elaboration Likeli-hood Model (ELM) from Petty and Cacioppo (1979) this research tried to understand the role of consumer motivation that would influence consumer perception on service innovativeness to image congruity (both self-congruity and functional congruity) that had not been studied previously. The population of this research was consumer of skin care and beauty services in several big cities in East Java. To test the hypothesis, this study used moderating regression analysis.

Keywords: service innovativeness, self-congruity, functional congruity, motivation to innovate

1 INTRODUCTION

An understanding of the service innovativeness felt by consumers associated with their behavior is important for several reasons. First, services innovativeness allow companies to continue to create and improve their offerings, and these offerings can be executed in the service business only if the company's customers regard them as innovative (O'Cass & Carlson, 2012). Second, the service innovativeness approach of consumer perception is more important than the perception within the company itself because the innovation is needed to improve service delivery for consumers (Raju & Lonial, 2001). Third, understanding service innovativeness from the customer's point of view is critical to the company's successful (Keller, 1993). Fourth, seeing service innovativeness from the customer point of view is appropriate for the company (Magnusson et al., 2003) and reflects the adoption of the concept of market orientation (Narver & Slater, 1990).

Although some research on service innovativeness that focuses on customer-centric perspectives is beginning to develop, the numbers are still lacking (Zolfagharian & Paswan, 2009).

In line with customer-centric orientation, several studies emphasize the role of image congruity as important. The mechanism by which customers process information to create behavioral outcomes, such as attitudes, satisfaction and loyalty (Ericksen, 1997; Sirgy, 1986, Sreejesh, 2016).

As customers experience innovation in services, derived from various aspects of services innovation, consumers will develop a sense of fun and excitement when thinking of them as potential of functional and symbolic value to the consumer (Fu & Elliott, 2013). This functional and symbolic element arises from the service innovativeness signals that are being offered which are the main reasons why consumers adopt products/services (Castaño, 2008). If the service is capable of generating excitement and enjoyment to the consumer, then their thought processes will analyze the information cognitively, affectively or both. When consumers describe or are motivated by a symbolic aspect, they will interpret it as their stereotypes and self-concept (also referred to as self-congruity). However, if consumers process information through motivation and their ability to research information relevant to the service in the form of functional conformity, which is referred to as functional congruity (Kang et al., 2012).

In today's challenging business environment, it is not enough for service providers to deliver quality service for customers at the right time. But companies must also find ways to innovate delivering new services that will be made to customers. Service innovation is defined as additional services from existing services (incremental improvements to existing services). Service innovation is not easy to do but increasingly considered. Because it is an important element of the company's competitive strategy (Mac Donough, Zack, Lin, & Berdrow, 2008). Therefore service innovativeness becomes important in creating an image that ultimately encourages consumer behavior.

This research tried to understand the role of consumer motivation that would influence consumer perception on service innovativeness to image congruity (both self-congruity and functional congruity). That had not been studied in previous studies.

In this context, service innovativeness was a potential determinant of an image of skin care and beauty services, so it could be stated that investigating service innovativeness comprehensively would feel the effect on the image of a skin care and beauty services. Therefore, the purposed of this research was to test empirically the role of service innovativeness in forming image congruity consisting of self-congruity and functional congruity in skin care and beauty services.

2 TEORITICAL BACKGROUND

2.1 *Service innovativeness*

The concept of service innovativeness had a diversity in definition. This diversity arose from two different perspectives in viewing it. Currently, there are two ideological perspectives with regard to service innovation, which were explored in several relevant studies.

First, the perspective of innovativeness from the service provider's point of view. Studies in this stream define innovativeness as a strategic tool for service providers to understand and determine the extent to which services are new (Garcia & Calantone, 2002).

Second, the perspective of innovativeness from the consumer's point of view is often referred to as customer perceived services innovativeness (Lundkvist and Yakhlef; 2004; Zolfagharian & Paswan, 2009). The Customer Perceived Service Innovativeness (CPSI) reflects the extent to which customers evaluate service offer dimensions meaningfully, different from existing ones (Zolfagharian & Paswan, 2009).

2.2 *Self-congruity*

Kang, Tang & Lee (2012) stated that image congruence is defined at the level of conformity between the consumer description with a product or brand image. Consumers make the decision to buy through two concepts of measuring instruments about image congruence, that is self-congruity and functional congruity (Sirgy, 1986; Kang, Tang & Lee, 2012).

Self-congruity explains the comparison between self image and image of a product owned by the consumer. Consumers focus on attributes symbolic of the product, which can be explained by various adjectives such as friendly, modern, or traditional that reflect the image of the user of the product (Sirgy, 1999; Kang, Tang & Lee, 2012). explains also that self-congruity includes the cognitive and affective judgments of the brand name/service/product with characteristics that match the consumer.

2.3 *Functional congruity*

The functional congruity explains the comparison between consumer perceptions of a product attributes and consumer evaluation of the product. An example of a functional attribute is a cauldron bag products, price, shop atmosphere, and performance (Koolivandi

& Lotfizadeh, 2015). According Kang, Tang & Lee (2012) the more positive image congruence shows the greater the image of conformity so that the more positive consumer behavior (Sirgy 1999; Kang, Tang & Lee, 2012) explained that functional congruity refers to conformity and discrepancy between consumer perceptions of product attributes before actual purchase and post purchase evaluation.

Kressman (2006) described functional congruity with other points of view as a fit between the consumer's ideal expectation and the consumer's perception of features or matters according to the consumer's judgment of a brand and product. In contrast to Sirgy (1999) that functional congruity involved an evaluation process after a consumer makes a purchase. Judging from the equation both explanations above stated that consumer perceptions are used when functional congruity is done.

2.4 *Motivation to innovate*

First, many consumer innovativeness scales include a hedonic dimension. One example often used to measure innovativeness (Steenkamp, 1996; Baumgartner, H., & Steenkamp, 1996) is the Exploratory Consumer Buying Behavior scale. One of their subscales was the Exploratory Acquisition of Products, which refers to buying innovations intended to stimulate the senses. (Venkatraman & Price, 1990) also include a Sensory Innovativeness dimension in their concept of innovativeness, and defined his hedonic innovativeness dimension as the drive to adopt innovations for hedonic reasons, such as to enjoy the newness of the product.

Working from a different perspective, Hirschman Venkatraman (1990) point to innovative consumers who were attracted to functional or useful new products. Babin, Darden, and Griffin (1994) and Voss, Spangenberg, and Grohmann (2003) proposed a similar distinction in emphasizing utilitarian reasons for buying products (as opposed to hedonic or affective reasons).

Of course, products were not always purchased for their hedonic or functional value alone. Consumers also want to impress others and raise their social status (Brown & Venkatesh, 2005; Foxall et al., 1998). Thus, innovativeness researchers stressed the importance of the social or symbolic component of consumer innovativeness (Venkatraman, 1990). Arnould (1989) observed that social rewards and social differentiation may both stimulate new product adoption. Simonson and Nowlis (2000) stated that the possession of innovations was a socially accepted way of making a unique impression. Consumers built a certain identity through the possession of these visible new products.

Finally, Cognitive Innovativeness was a distinct dimension of innovativeness in the scale by Venkatraman & Price (1990) defined it as "the desire for new experiences with the objective of stimulating the mind". Baumgartner and Steenkamp's (1996) exploratory information-seeking was also defined as providing mental stimulation, although it is focused on information-seeking rather than on measuring consumer innovativeness.

2.5 *Hypothesis building*

As customers experience innovation in services, derived from various aspects of services innovation, consumers would develop a sense of fun and excitement when thinking of them as potential for functional and symbolic value to the consumer (Fu & Elliott, 2013). This functional and symbolic element arose from the service innovativeness signals being offered which are the main reasons consumers adopt products/services (Castaño, 2008).

If the service was capable of generating excitement and enjoyment to the consumer, then their thought processes would analyze the information cognitively, affectively or both. When consumers described or are motivated by a symbolic aspect, they would interpret it as their stereotypes and self-concept (also referred to as self-congruity).

H1: Service innovativeness had significant effect on self-congruity

However, if consumers processed information through motivation and ability to research Information relevant to the service, in the form of functional conformity, which was referred to as functional congruity (Kang, 2012).

H2: Service innovativeness had significant effect on functional congruity

If motivation to innovativeness was high, then the level of consumer perception on the acceptance of innovation activities conducted by the company will be high as well, so these would move the central consumer route towards more complex considerations that lead to functional congruity. Conversely, if customer motivation to innovate was low, consumers on the acceptance of innovation activities conducted by the company would be low as well, which would move the central consumer route toward the consideration on the basis of conformity to his self-concept that leads to self-congruity. (Petty and Caccioppo, 1986).

H3: Motivation to innovate moderated the relationship between service innovativeness perceived by the consumer towards self—congruity

H4: Motivation to innovate moderated the relationship between service innovativeness perceived by the consumer towards self—congruity

Conceptual framework

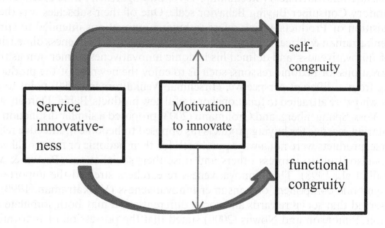

3 RESEARCH METHOD

3.1 *Population and sample*

The population of this study was consumer of skin care and beauty services in several big city cities in East Java because businesses were qualified for this service innovativeness based on the dimensions used in the measurement of this study. Samples were taken using purposive sampling. This research took 100 respondents to be sampled research.

3.2 *Variable and measurement*

1. Service innovativeness was defined as consumer perception about additional services from existing services (incremental improvements to existing services). Service innovativeness dimensions consist of: (a) administration, (b) interior facilitation, (c) exterior facilities, (d) service score, (e) employee, (f) technology and (g) responsiveness are used to examine the variables that make up service innovativeness. (Zolfagharian, 2009).
2. Self-congruity, was a suitability between an object description with a self-image formed by consumers when choosing a beauty skincare product so that consumers feel fit or fit the object. Wang & Hsu (2015) measured self-congruity with the following indicators: (a). products/services in accordance with the characters (b). products/services according to show the best according to the consumer (c). products/services show the identity of the consumer.
3. Functional congruity indicated compatibility between functional perceptions or characteristics of appearance and consumer desires or important functional characteristics

(Sirgy & Johar, 1999). The indicators that measure the functional congruity were: (a). socio-phsychological value (b) purposive value, and (c) entertainment value.
4. Motivation to innovate could be considered as antecedent condition that forces humans to behave in an innovative way. Motivation explained why people behave in a certain way, what energized their behavior and what drove them voluntarily to do so

The four dimensions reflected in the innovativeness literature also correspond to more general theories of values, goals, and motivation (Vandecasteele, 2010): (a) functional motives (b) hedonic motives (c) social motives and (d) cognitive motives.

3.3 Technique analysis

In accordance with the purpose of the research, which was finding out the impact of the independent variable, service innovativeness on self congruity and functional congruity by looking at the moderation effects of the motivation to innovative, then the statistical test used was Moderating Regression Analysis (MRA).

4 RESULT OF RESEARCH

Of the 100 respondents who met the criteria specified in the study, only 95 respondents filled in the questionnaire completely. The remaining five respondents did not fill it in completely, so that the respondent's answer was cancelled for further processing in this study

Based on the data in Table 1 it can be seen that service innovativeness has a significant effect on self-congruity because the value of p value $0.015 < 0.05$ (X to Y1). This means that when consumers perceive service innovation of skincare services were high, it would make consumers feel that it was in accordance with their self-concept (self-congruity).

Motivation to innovate (Z) in the table did not moderate the relationship between service innovativeness (X) to self-congruity (Y1), because the value of p value $0.054 > 0.05$. This means that the motivation of consumers to be innovative did not strengthen or weaken the relationship of service innovativeness on self-congruity.

Based on the data in Table 2 it can be seen that service innovativeness had significant effect on functional congruity because p value $0.000 < 0.05$ (X to Y2). This means that when

Table 1. Relationship between service innovativeness, motivation and self-congruity.

Model		Unstandardized coefficients		Standardized coefficients		
		B	Std. Error	Beta	t	Sig.
1	C	2.673	0.463		5.774	0.000
	X	0.368	0.148	0.312	2.491	0.015
	Xz	054	0.028	0.244	1.953	0.054

a. Dependent Variable: Self-congruity (Y1).

Table 2 Relationship between service innovativeness, motivation and functional congruity.

Model		Unstandardized coefficients		Standardized coefficients		
		B	Std. Error	Beta	t	Sig.
1	C	2.504	0.458		5.472	0.000
	X	0.538	0.146	0.443	3.677	0.000
	xz	0.081	0.028	0.353	2.934	0.004

a. Dependent Variable: Functional congruity (Y2).

consumers perceived service innovation from skincare services was high, it would make the consumer feel that it was in accordance with the ideal function they wanted (functional congruity).

Motivation to innovate (Z) in the table moderated the relationship between service innovativeness (X) to functional congruity (Y), because p value $0.004 < 0.05$. This means that consumer motivation to innovative strengthened the relationship of service innovativeness on functional congruity.

5 IMPLICATION, CONCLUSION AND LIMITATION

5.1 *Implication*

Based on the results of research that has been stated in the previous chapter it can be seen that service innovativeness had a significant effect on self-congruity. This supports the opinion of Sreejesh (2015). When consumers perceived the innovations made by the skincare object well then the information would be captured by consumers as something in accordance with her self-concept, such as a modern consumer self, sophisticated and like something better than before.

The results also showed that service innovativeness had a significant effect on functional congruity. This supports the opinion of Sreejesh (2015). When consumers perceived the innovations made by the skincare object well then the information would be captured by consumers as something in accordance with the function they wanted to look for in the object, such as wanting healthier skin, to be more beautiful and overcome the skin problem of the consumer.

The moderation effects shown in this study suggest that motivation to innovate did not moderate the relationship between service innovativeness and self-congruity, but merely moderated the relationship between service innovativeness and functional congruity. This supported the opinion of Petty and Caccioppo (1986), which stated that motivation, ability and opportunity will strengthen the perception of the consumer receipt of the information. This also supported the opinion of Vandercasteele (2010) that one's motivation to be innovative was determined by four determinants, such as hedonic value, functional value, social value and cognitive value. In this research, cognitive value and functional value were more dominant, thus moderating the relationship among service innovativeness to functional congruity rather than on self-congruity.

5.2 *Conclusion*

From the research results and discussion described in the previous section, several conclusions can be drawn from the study as follows:

1. It was found that service innovativeness had significant effect on self-congruity.
2. It supported the hypothesis that stated that service innovativeness had a significant effect on functional congruity
3. It rejected the hypothesis that motivation to innovate moderated the relationship between perceived service innovativeness by the consumer towards self-congruity
4. It supported the hypothesis that motivation to innovate moderated the relationship between perceived service innovativeness by the consumer towards self-congruity

5.3 *Lim itation*

From the study results, discussion and conclusions that have been described in the previous section, some suggestions can be given for further research. They were:

1. This research followed up the theories of Petty & Cacciopo (1986) about the Elaboration Likelihood Model which stated that with high motivation, ability and opportunity (MAO) consumers would more strongly perceive the information it captures. In this study only one moderation variable that was used and that was motivation, so further research can add two other variables, ability and opportunity.

2. The subsequent research can also expand the objects of the research into two groups of consumers, such as woman and man
3. Further studies can explore the consequence variables of self-congruity and functional congruity

REFERENCES

Ahmad, J. & Goode, M. (2001). Consumers and brands: a study of the impact of self-image congruence on brand preference and satisfaction. *Marketing Intelligence & Planning, 19*(7), 482–492.

Back, K.J. (2005). The effects of image congruence on customers brand loyalty in the upper middle-class hotel industry. *Journal of Hospitality Tourism Research, 29* (4), 448–467.

Baumgartner, H., & Steenkamp, J.B.E. (1996). Exploratory consumer buying behavior: Conceptualization and measurement. *International journal of Research in marketing, 13*(2), 121–137.

Castaño, R., Sujan, M., Kacker, M. & Sujan, H. (2008). Managing consumer uncertainty in the adoption of new products: temporal distance and mental simulation, *Journal of Marketing Research*, Vol. 45 No. 3, pp. 320–336.

Chan, R.Y. & Lau, L. (1998). A test of the Fishbein-Ajzen behavioral intentions model under Chinese cultural settings: are there any differences between PRC and Hong Kong consumers? *Journal of Marketing Practice: Applied Marketing Science, 4*(3), 85–101.

Ericksen, M.K. (1997). Using self-congruity and ideal congruity to predict purchase intention: an European perspective, *Journal of Euromarketing*, Vol. 6 No. 1, pp. 41–56.

Foxall, G.R. & Yani-de-Soriano (2005). Situational influence on consumers' attitude and behaviour, *Journal of Business Research, 58*, 518–525.

Fu, F.Q. & Elliott, M.T. (2013). The moderating effect of perceived product innovativeness and product knowledge on new product adoption: an integrated model, *The Journal of Marketing Theory and Practice*, Vol. 21 No. 3, pp. 257–272.

Garcia, R., & Calantone, R. (2002). A critical look at technological innovation typology and innovativeness terminology: a literature review. *Journal of Product Innovation Management: An International Publication of the Product Development & Management Association, 19*(2), 110–132.

Hair, J.F. Jr., Anderson, R.E., Tatham, R.L., & Black, W.C. (1998), Multivariate data analysis 5th Ed., Englewood Cliffs, New Jersey: Prentice Hall.

Hu, J. (2012). The role of brand image congruity in Chinese consumers' brand preference. *Journal of product & brand management, 21*(1), 26–34.

Hung, K. & Petrick, J.F. (2011). The role of self and functional congruity in cruising intentions, *Journal of Travel Research, 50*(1), 100–112.

Jacoby, J., Olson, J.C., & Haddock, R.A. (1971). Price, brand name, and product composition characteristics as determinants of perceived quality, *Journal of Applied Psychology, 6*(2), pp. 111–124.

Jianyao, Liu, Li Mizerski & Huangting S. (2012). Self-congruity, brand attitude, and brand loyalty: a study on luxury brands, *European Journal of Marketing*, Vol. 46 Iss 7/8 pp. 922–937.

Johar, J.S., & Sirgy, M.J. (1991). Value-expressive versus utilitarian advertising appeals: When and why to use which appeal, *Journal of advertising, 20*(3), 23.

Johnson, S.P., Menor, L.J., Roth, A.V. & Chase, R.B. (2000). A critical evaluation of the new services development process: integrating service innovation and service design. In: Fitzsimmons, J.A., Fitzsimmons, M.J. (Eds.), *New Service Development, Creating Memorable Experiences*. Sage, Thousand Oak, CA, pp. 1–32.

Kang, J., Tang, L., Lee, J.Y. and Bosselman, R.H. (2012). Understanding customer behavior in name-brand Korean coffee shops: The role of self-congruity and functional congruity, *International Journal of Hospitality Management, 31* (3), 809–818.

Keller, K & Lane (1998). Building, Measuring, and Managing Brand Equity. New Jersey: Prentice Hall.

Koolivandi, S., & Lotfizadeh, F. (2015). Effects of Actual Self and Ideal Self Image on Consumer Responses: The Moderating Effect of Store Image. *British journal of marketing studies, 3*(8), 1–16.

Kressmann, F., Sirgy, M.J., Herrmann, A., Huber, F., Huber, S. & Lee, D.J. (2006). Direct and indirect effects of self-image congruence on brand loyalty, *Journal of Business Research, 59* (9), 955–964.

Levy, J, Louie, & Curren (1994). How Does the Congruity of Brand Names Affect Evaluations of Brand Name Extensions? *Journal of Applied Psychology, 79*(1), 46–53.

Liljander, V., Gillberg, F., Gummerus, J., & Riel, A. (2006). Technology readiness and the evaluation and adoption of self-service technologies. *Journal of Retailing and Consumer Services, 13* (3), 177–191.

Lundkvist, A., & Yakhlef, A. (2004). Customer involvement in new service development: a conversational approach. *Managing Service Quality: An International Journal, 14*(2/3), 249–257.

473

Magnusson, P., Matthing, J. and Kristensson, P. (2003), Managing user involvement in service innovation, *Journal of Service Research,* Vol. 6 No. 2, pp. 111–124.

Malhotra (2004), Marketing Research: An Applied Orientation, 4th ed., Prentice Hall, Inc.

McDonough, E.F., Zack, M.H., Lin, H.E., & Berdrow, I. (2008). Integrating innovation style and knowledge into strategy. *MIT Sloan Management Review, 50*(1), 53.

Narver, J. and Slater, S. (1990). The effect of a market orientation on business profitability, *Journal of Marketing,* 54(4), pp. 20–35.

Osgood, E., Percy H.T. (1955). The Principle of congruity in the prediction of attitude change. *Psychological Review,* 62(1), 42–55.

O'Cass, A. & Carlson, J. (2012). An e-retailing assessment of perceived website-service innovativeness: implications for website quality evaluations, trust, loyalty and word of mouth, *Australasian Marketing Journal (AMJ),* 20 (1), pp. 28–36.

Petty, R.E. and Cacioppo, J.T. (1979), Issue involvement can increase or decrease persuasion by enhancing message-relevant cognitive responses, *Journal of Personality and Social Psychology,* 37(10), pp. 1915–1926.

Petty, R.E., & Cacioppo, J.T. (1986). The elaboration likelihood model of persuasion. *Advances in Experimental Social Psychology,* 19, 123–205.

Prahalad, C.K. & Ramaswamy, V. (2004), The Future of Competition: Co creating Unique Value with Customers, *Harvard Business School Press,* Boston, MA.

Raju, P.S. & Lonial, S.C. (2001), The impact of environmental uncertainty on the market orientation–performance relationship: a study of the hospital industry, *Journal of Economic and Social Research,* 3(1), pp. 5–27.

Roberts, D & Hughes, M (2014). Exploring consumers motivations to engage in innovation through co-creation activities. *European Journal of Marketing, 48(12)* pp. 147–169.

Rossiter, J.R., & Percy, L. (1998), Advertising communication and promotion management, Singapore; McGraw-Hill.

Simonson, I., & Nowlis, S.M. (2000). The role of explanations and need for uniqueness in consumer decision making: Unconventional choices based on reasons. *Journal of Consumer Research, 27*(1), 49–68.

Sirgy, M.J. (1986). Self-congruity: toward a theory of personality and cybernetics. Praeger Publishers/Greenwood Publishing Group.

Sirgy, M.J., & Johar, J.S. (1999). Toward an integrated model of self-congruity and functional congruity. *ACR European Advances.*

Sreejesh, S. Mitra, A & Sahoo, D. (2015), The impact of customer's perceived service innovativeness on image congruence, satisfaction and behavioral outcomes, *Journal of Hospitality and Tourism Technology,* 6 (3) pp. 288–310.

Sreejesh S, Sahoo, & Mitra, A. (2016), Can healthcare servicescape affect customer's attitude? a study of the mediating role of image congruence and moderating role of customer's prior experience, *Asia-Pacific Journal of Business Administration,* 8 (2) pp. 106–126.

Steenkamp, J.B.E., Baumgartner, H., & Van der Wulp, E. (1996). The relationships among arousal potential, arousal and stimulus evaluation, and the moderating role of need for stimulation. *International Journal of Research in Marketing, 13*(4), 319–329.

Vandecasteele, Bert & Geuens, M. (2010). Motivated consumer innovativeness: concept, measurement, and validation. *Intern. Journal. of Research in Marketing.* Vol 27 (2010) 308–318.

Venkatraman, M.P., & Price, L.L. (1990). Differentiating between cognitive and sensory innovativeness: concepts, measurement, and implications. *Journal of Business research, 20*(4), 293–315.

Wang, S.J, Hsu, C.P. & Chen, H.C. (2015), How readers' perceived self-congruity and functional congruity affect bloggers' informational influence perceived interactivity as a moderator, *Online Information Review,* 39 (4), pp. 537–555.

Webster, J. & Trevino, L. (1995), Rational and social theories as complementary explanations of communication media choices: two policy-capturing studies, *Academy of Management Journal,* 38 (6), pp. 1544–73.

Zeithaml VA, (2009). Services marketing, McGraw-Hill, 5th Edition.

Zolfagharian, M. & Audhesh, P. (2009). Perceived service innovativeness, consumer trait innovativeness and patronage intention. *Journal of Retailing and Consumer Services, 16,* 155–16.

Micro, small and medium enterprises

Business Innovation and Development in Emerging Economies – Trinugroho & Lau (Eds)
© *2019 Taylor & Francis Group, London, ISBN 978-1-138-35996-3*

Training in bookkeeping and financial management for revitalization of traditional market (study case: Sukatani Market Depok, West Java)

M. Safitry
IISMI STIAMI, Indonesia

R.F. Kaban
Perbanas Institute, Jakarta, Indonesia

ABSTRACT: This research conducted at Sukatani Market Depok, West Java. The purpose of this study is to give the traders training for simple bookkeeping and financial. In this study the authors conducted research on consumers and traders to measure the potential Sukatani Market Depok whose purpose is to know consumer perceptions (buyer) in market revitalization at Sukatani Market Depok. after that serials of training were given regards Simple Bookkeeping, Concept and Financial Report and Socialization of Financial Bank Product. At the end of the program there was a survey about training participant satisfactory. From the results of the research, the authors take the conclusion that this training gives positive impacts to increase motivation, knowledge, and skill of the traders in traditional market especially at Sukatani Market Depok. Finally, the authors hope one day similar improved trainings may be applied in the other traditional markets in Depok area.

Keywords: Training, Bookkeeping, finance, traditional market

1 INTRODUCTION

The traditional market is gaining increasing competition from the modern market due to the development of times and lifestyle changes. This has the potential to threaten the existence of the traditional market itself. Soliha (2008) in her research about Retail Industry Analysis in Indonesia shows data of AC Nielsen's research in 2003 which informs that Indonesia's total retail sales per year above Rp 600 Trillion. In Indonesia in 2003 recorded 267 department stores, 683 supermarkets, 972 mini markets, and 43 hypermarkets. The AC Nielsen survey noted in between some modern retail forms such as supermarkets, mini markets, wholesale centers, and hypermarkets, the fastest growth experienced by hypermarkets that increased in number in 2005 to 83 units.

Table 1. The ratio of people's interest to shop in traditional and modern market.

Year	Traditional market	Modern market
1999	65%	35%
2000	63%	37%
2001	60%	40%
2002	52%	48%
2003	56%	44%
2004	53%	47%

Source: AC Nielsen 2005 in Soliha, 2008.

Based on AC Nielsen data, the contribution of traditional market sales continues decreased. In 2002 the dominance of sales in the traditional market segment reached 75% then in the next year fell by only 70%. In contrast, modern retail hypermarket in 2002 was 3% sales share up by 5% in 2003, and 7% in 2004. According to AC Nielsen Asia Pacific research report Retail and Shopper Trend 2005 mentioned that in Asia Pacific countries (except Japan) the ratio of people desire to shop in traditional markets and modern markets as follows (Soliha, 2008).

Based on the table above the writer and team choose Sukatani Market Depok as Community Engagement (CEG) activity to ensure the sustainability of traditional markets. It is necessary to improve traditional market management systems that allow them to compete and exist despite of the presence of supermarkets. A good traditional market management system can be started from the availability of reliable human resources from market managers, especially business players in the market (retail merchants). Training on bookkeeping and financial management that is periodic and appropriate and in cooperation with financial institution such as bank BJB can provide solution for capital market problems in Sukatani Market Depok, so they can become "agent of change" in progress of Depok city. In line with this research conducted by Pradipta and Wirawan (2016) entitled *"The Influence of Traditional Market Revitalization And Trader's Resource To The Performance Of Market Traders In Denpasar,"* affirms that the development of merchant resources is basically a continuous series with training, and the training development needs to be done to increase the merchant's income. Good resources have the quality of physical health and intellectual quality that can develop the knowledge and skills possessed in order to obtain maximum results in accordance with the expectations of the traders. Furthermore, Aryoko et al. (2016) explains the importance of training on the management of traditional markets with environmentally friendly needs to be held continuously for traditional market traders in their research on traditional market traders in Tanjungpandan, Belitung.

Traditional markets have a very strategic role in the framework of improving income and employment, for which efforts are needed in order to increase the competitiveness of traditional markets that have been synonymous with location a slum, chaotic, dirty trade, and a source of traffic congestion (Dewi & Winarni, 2007). The image of the unfavorable traditional market should have received quite large attention because it related to the livelihood of many people. Settling the traditional market into a positive-looking shopping spot is a challenge which is quite heavy and should be pursued by all stakeholders to synergize for deleting the negative impression so that traditional markets still exist in the midst of the tighter competition (Rozaki, 2012).

The rivalry between the modern market and the traditional market is getting out of control, include in Depok City. As a buffer city of Jakarta, Depok has experienced growth economy and population at a rapid pace. Until 2007 there were 62 modern retail business in Depok, 46 of which are mini markets. While four big supermarkets such as: Carrefour, Hypermart, Giant, and Superindo have also existed in Depok. Number of modern retailers in Depok is expected to increase in the future especially mini-markets, as a result of the relative easy to get permit and relatively small amount of capital needed for opening mini market.

Human resources (HR) in this case the market managers both the office/market office, market leader and retribution officer and traders in the market are continuously given training and knowledge management (management know how) and management skills that educate and train the bookkeeping and took the bank BJB for the practice in the future. The simple bookkeeping and financial management training explains the basic concept of bookkeeping because there are some traders close their stalls due to capital run out while the customers added a lot and they do not understand whether in their trade gain or loss. In addition, in good bookkeeping administration, one of the indicators is the procedure of recording financial transactions both in and out with evidence of receipts or sales invoices from suppliers, so that traders can more easily do financial analysis with precise and accurate. Including the need for traders to set up a Reserve fund to pay for the kiosk/stall at the right time with the right amount.

Assistance in bookkeeping and finance management are in the form of: 1) Basic simple bookkeeping concepts, that is by providing knowledge about the concept Assets, Liabilities (debt), Equity (capital), Revenue, and Expense. 2) Training to make cash flow (cash inflows and outflows), Income Statement and Balance Sheet (financial reporting statement). Traders need to be given knowledge about this so that their business is more profitable and not mixed between business and personal finance and have the confidence to apply for funding to investors or bank in the capital issue. 3) CEGs team in collaboration with BJB bank provides socialization about bank features and procedures for applying for microcredit and Micro and Small Medium Enterprises (MSMEs) with low interest rate and a program to save money.

1.1 *Expected changes can happen*

Expected changes can occur from giving bookkeeping and financial management training is by increasing the capacity and potential bookkeeping of traders and managers of Sukatani traditional markets. A modern management but still maintain the peculiarities of traditional market and create a traditional market that has a clean image, honest, beautiful and profitable business. The other expected changes can happen from this training are the traders survive and buyers still love the traditional market especially in the era of market openness which is very competitive ie ASEAN Economic Community.

2 RESEARCH METHOD

Community engagement program in the form of training Bookkeeping and financial management for Sukatani market traders in Depok city is the result of Community Engagement Grants (Research Based) University of Indonesia, where the authors are alumni of the campus. Before the community service program is conducted, firstly conducted research on consumers and traders to measure the market potential Sukatani Depok whose purpose is to know consumer perceptions (buyers) in revitalization of Sukatani Market Depok. The information collected through this research helps the Sukatani Market Depok i.e., *Unit Pelaksana Teknis* (*UPT*) Market as a market manager and traders as business actors in the market revitalization.

Bookkeeping and financial management training provided to traders in form of:

1. Consultation and discussion
2. Education knowledge and skills
3. Survey and evaluation

At the end of the program again conducted research to measure or evaluate the satisfaction of the trainees. The main objective of the questionnaire given to trainee traders is to gain marketers' perception of training implementation. The information obtained becomes the material of evaluation and recommendation of decision making in quality improvement of Sukatani Market Depok.

3 RESULT AND DISCUSSION

3.1 *Market profile*

Sukatani Market Depok was established in 1990 which is the result of self-help community, this market stands on an area of 2,892 M² and building area of 1.039 M². The market location is at the Department of Sukatani Complex, Cimanggis, Depok.

Based on preliminary survey conducted to buyers and traders regarding market revitalization of Sukatani, in general (68%) expressed satisfaction while 32% still not satisfied. Results of the research are divided into two circumstances, pre-training and post training circumstances.

3.2 *Pre-training circumstances*

1. Not all trainees have understood whether the business they are working on gets a profit or loss?
2. Mixed funds of merchant business and personal.

Table 2. Data of Sukatani market.

Status	City's property
Number of Stalls	42
Number of Kiosks size 0–5 m	204

Source: Division of Cooperative, MSMEs and Market Depok, 2014.

Table 3. Data of Sukatani market traders.

Status	Number	Open	Close
Kiosks size 0–5 m	204	147	57
Stalls	42	30	12
Total	246	177	69

Source: Division of Cooperative, MSMEs and Market Depok, 2014.

3. Traders need additional capital for business
4. There are several requirements that are predicted to prevent merchants from obtaining products KCR (Credit of Love People) from BJB banks include:

- Kiosk that rent system must be obtained to the cooperative service of SMEs and the market
- Many traders whose Identity cards are not residents of West Java
- Complaint of merchant if must use guarantee system in process of loan or additional capital

5. Merchants want, daily billing. Because this can ease the merchant in paying the loan installments

The traders were given two trainings:

Table 4. Types of training.

Date	Type of training	Number of participant
21-9-2015	Simple Bookkeeping Concept and Financial Reporting	8
12-10-2015	Socialization of Financial Bank Product	11

3.3 Post-training circumstances

1. The training participants understand how to make good bookkeeping in order to plan it in their business
2. Participants understand the training can sort and do not mix between personal funds and business funds
3. The training participants intend to continue to improve the bookkeeping in order to apply for additional capital in their business to a trustworthy financial institution so as not to become entangled by moneylenders
4. With the socialization of bank BJB trainees know access to micro credit and SMEs with low interest rates
5. Participants understand the importance of saving from some of the profits earned in order to make auto debit payment for loan installment to the bank

3.4 Implementation constraints

1. Self-awareness of merchants to participate in community engagement program is still lacking the reason that no one keeps kiosks or stalls

2. The time duration of activities that cannot be too long because the traders have to keep their kiosks/stalls

3.5 *Program implementation constraints*

1. There is no complete coordination or involvement of the market managers in training activities. In each training activity, there is no member of the market manager is involved, so the ideas and results of the training are difficult to communicate between the materials giver and the trainee to the market manager.
2. The interest of market traders is not too big in training, this is seen from the number of participants who participated in the training with the total of all traders in the Sukatani Market Depok. Traders have difficulty in dividing their time between selling and training. Nevertheless, the merchant who is serious to get knowledge and skills in bookkeeping and finance management, strives to remain present in every activity.

Based on the final survey conducted to Sukatani market traders, bookkeeping training participants in general (80.28%) expressed satisfaction. The Things measured related to this level of satisfaction are as follows:

No	Assessment component	Satisfaction level
1	Time	80%
2	Material Completeness	80%
3	Material Comprehension	80%
4	Delivery	80%
5	Usefulness	80%
6	Friendliness	83%
7	Numbers of Training	80%

4 CONCLUSIONS

4.1 *Summary*

Based on the results of the implementation of the activities that have been conducted, then there are some things that can be concluded from the activities of bookkeeping and financial management training, namely:

1. Knowledge and understanding of small business actors or market traders are not qualified related to the spirit of making good simple bookkeeping and seeking additional venture capital.
2. Training conducted to improve the business spirit and motivation of the traders in running business well and optimal.

4.2 *Suggestion*

The suggestions for improving the quality of training activities of Community Engagement include:

1. It should be strived to arouse awareness of market traders about the urgency of bookkeeping and finance training, this should be exemplified also by the market managers (UPT)
2. Periodic evaluations and controls may continue to monitor whether Sukatani market traders have been applying the knowledge and skills gained from the training provided
3. Community service activities can be developed by adding skills training that can be applied in the business, and extending the reach of the target to other traditional market in Depok city

REFERENCES

Aryoko Hagoes, Tartini, and Djunaedi Djafar. 2016. Training and assistance multipurpose business cooperation (savings and loans, traditional market management, and mini market). *EKSIS* Volume XI N0 II, October 2016. Jakarta, Indonesia.

Dewi, Utami & F. Winarni. *"Traditional market development facing the attack of modern market in Yogyakarta City"*. 2007. Yogyakarta, Indonesia.

Dewi, Utami & F. Winarni. *"Results summary of the traditional market versus modern market: A case study of market management policy in Yogyakarta City"*. 2012. Yogyakarta, Indonesia.

Division of Cooperative, MSMEs and Market in Depok City. *"Market Revitalization in Depok City"*. 2014. Depok, Indonesia.

Division of Cooperative, MSMEs and Market in Depok City. *"Compilation of traditional market profile of Depok City"* 2014. Depok, Indonesia.

Pradipta AAGP & Irawan IGPN. 2016. The influence of traditional market revitalization and trader's resource to the performance of market traders in Denpasar. *E-Journal of Economic Development Udayana University* Vol.5 No.4 Page 460–479. April 2016. Denpasar, Indonesia.

Rozaki, Abdur. *"Traditional market: Under the shadow of the dominance of the role of modern market"*. 2012. IRE, Yogyakarta, Indonesia.

Semeru. *"Supermarket impact on traditional markets and retailers in urban areas in Indonesia, Research Reports"*. 2007. Jakarta, Indonesia.

Soliha, Euis. 2008. Retail industry analysis in Indonesia, *Business and Economy Journal (JBE)*, September 2008, Page 128–142, Vol.15 No 2. Jakarta, Indonesia.

Discussion & Interview (dated May 28, 2014, May 31, 2014, and June 14, 2014)

Mr. Ahmad Kafrawi Head Division of Cooperative, MSMEs and Market in Depok City
Mr. Supomo, Market Section Head Division of Cooperative, MSMEs and Market in Depok City
Traders of Sukatani Market Depok.
Research Address:
Sukatani Market
Jl. Bunga 1
Sukatani, Depok, West Java

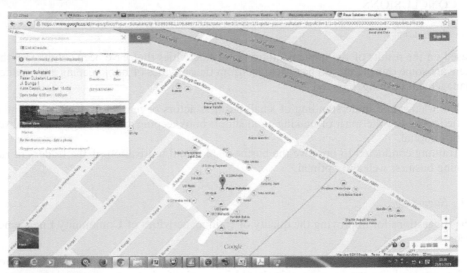

Picture 1. Location Map.
Source: Google Map.

Operation management

Business Innovation and Development in Emerging Economies – Trinugroho & Lau (Eds)
© *2019 Taylor & Francis Group, London, ISBN 978-1-138-35996-3*

Freedom of parties to determine the form and content of the agreement in a contract of construction services

Dewi Anggraeni
Pamulang University, Indonesia

ABSTRACT: There are many assumptions regarding an imbalance in the position of a construction service contract between the service user and the service provider due to the non-reflection of the principle of freedom of contract and the absence of clarity regarding the rights and obligations of the parties concerned at the time of the contract's implementation. The objective of this study is to obtain evidence of the absence of the principle of freedom of contract as in article 1338 Civil Code in Law No. 2, year 2017 on construction services. Whereas article 47 of Law No. 2, year 2017 only explains the clauses that must exist in a construction service contract, the application of the principle of freedom of contract, which is the basis for the making of a contract, must also be reflected, not only through the content of the contract but also in the execution of the contract to determine the clauses will be agreed upon in the contract so that the position between service users and service providers is well balanced in the construction of the contracts and also at the time of execution of such contracts. The results of the study show that: (1) article 47 of Law No. 2, year 2017 concerning construction services has explained the clauses that must exist in a construction service contract. However, there needs to be an explanation of the rights and obligations of the parties at the time of contracting and the stages of how the agreement between the parties concerned in the contract can formulate the contents of these provisions in the Construction Services Act, so that article 1338 Civil Code on freedom of contract is also reflected in a contract construction service; (2) that negotiations in a contract, especially a construction service contract, not only to determine the amount of rupiah or the price and value of the contract, but there should also be negotiation on the content of the contract clause.

Keywords: contract, construction services, freedom of contract

1 INTRODUCTION

Development in Indonesia which is one of the development process of Indonesia follows the development of the era, which runs more rapidly. The demands of the development of a country in the current era of globalization support the pace of development in Indonesia. Along with the rapid development in Indonesia, it is also followed by many local construction services that become pillars in supporting development in Indonesia, whether it is in the form of small, medium or large scale. Moreover the entry of foreign construction services will create the competition in the construction market of our country.

In fact, commercial contracts are becoming increasingly sophisticated, indeed irrefutable facts, such as those concerning franchising, join ventures, contract construction and engineering, license, agency and distribution, power plants, hire purchase, trustee agreements, merger agreements, loan syndication and many more. Generally, in Indonesia, contractors, especially those who offer the contract only accept the draft contract from the party who gave the job and signed it without conducting a through and jurisdictional analysis of the law to be used. Whereas in conducting a business relationship, the parties have equal opportunity and rights in determining what will be agreed. In fact, according to Imam Syaukani and A.

AhsinThohari, the law should not be taken for granted without considering the non-legal background, which is then very determinant in influencing the form and content of a particular legal product.[1]

Development in Indonesia also creates complex problems and problems in terms of implementation of such construction services. So, in this case, by using laws and regulations that are formalized to protect all Indonesians, while supporting the development of the country following the rapid growth of the world economy, the government issued Law No. 18 of 1999 on construction services, which was later revised into Law No. 2, year 2017, which contains definitions of construction services, construction service contracts and construction work procedures.

One of the definitions of a contract is that a contract is a promissory agreement between two or more parties that may cause, modify or remove legal relationships.[2] Further, there is also an understanding of the contract as an agreement, or set of agreements in which the law provides redress against the breach of such contract, or against the execution of such contract by law regarded as a duty.[3] From here, the importance of making a contract for all parties involved can be seen.

In other words, agreement is a legal relationship between legal subjects with one another in the field of property, in which the subject of law which one is entitled to achievement and so are other legal subjects are obliged to carry out its achievements in accordance with what has been agreed.[4] In such terms, there are several elements, namely: the existence of legal relationships, about wealth, between two or people, giving rights, putting the obligation on the other side, the achievement. The validity of the agreement can be seen in article 1320 Civil Code which states that: 'In order for a valid agreement, it is necessary to meet four conditions: their binding agreement; the ability to make an engagement; a certain subject matter; an unlawful cause'.

Meanwhile, in construction services law, there is a construction work contract, which is defined as the entire document governing the legal relationship between service users and service providers in the implementation of construction works (article 1 number 8 of Law No. 2, year 2017). While in article 1 number 17, Regulation of the President of the Republic of Indonesia no. 8 of 2006 concerning the fourth amendment to Presidential Decree no. 80 of 2003 concerning Guidelines on Procurement of Government Goods/Services, a contract is a commitment between a committing officer and a provider of goods/services in the procurement of goods/services. Thus, it can be seen that the definition of a construction work contract is a legal act between the user of the service with the service provider in a legal relationship that is regulated in relation to the rights and obligations of the parties.

Construction work contracts performed by service providers and service users occur due to an agreement between the two parties. The agreement itself is the conformity of the will statement between the parties. Any agreement made by the parties shall have legal consequences. The consequences of law are the emergence of rights and obligations, the right is a pleasure and a duty is a burden.[5]

The bargaining position in a construction service contract is expressed only in the value area of the contract itself, not the contract clauses. Therefore, the weak position of service providers in considering or refuting the contents of a construction service contract clause became a problem later in the implementation of the contract when the service provider was unable to implement clauses that had been made earlier. In fact, in Law No. 2, year 2017 on construction services, article 3 stated that:

The arrangement of construction services aims to:

---

1. Imam Syaukani dan A. Ahsin Thohari, *Dasar – Dasar Politik Hukum,* Jakarta; Raja Grafindo Persada, 2004, hal 54.
2. Henry Campbell, *Black's Law Dictionary, Seventh Edition: Black*, USA: West Publishing, 1968, hal. 394.
3. Gifis, Seteven H., *Law Dictionary,* New York: Barron's Educational Series Inc, 1968, hal. 94.
4. H. Salim HS, H. Abdullah, danWiwiekWahyuningsih, *PerancanganKontrakdan Memorandum of Understanding (MoU). Cet. 2.* Jakarta; SinarGrafika, 2007.hal. 9.
5. Salim, *HukumKontrak (Teori & TeknikPenyusunanKontrak)*, Jakarta: SinarGrafika, 2009, hal. 5.

a. Provide the direction of growth and development of construction services to realize a robust business structure, reliable, highly competitive and which results in quality construction work.
b. Realizing the order of construction services that ensures equality of service users and service providers in rights and obligations, and improves compliance with applicable laws and regulations.
c. Realizing the increasing role of society in the field of construction services.
d. Arranging a construction service system capable of realizing public safety and creating environmentally sound comfort.
e. Ensure good governance of construction services.
f. Creating value-added integration of all stages of construction services.

The existing form of a construction service agreement is a standard contract form, with the aim of keeping the contract and its implementation in compliance with the laws and regulations. The parties, especially the service providers, have no freedom in determining the contract of construction work. Since all proceedings from the initial stages of registration to the designation of the tender winner are all regulated by law, its implementing rules included in the construction work contract agreement are set in the form of contractual standards. The service user in this case, the government and or state institutions, are more dominant in determining the contents of the agreement.

The loss of the terms of the validity of the agreement in article 1320 Civil Code and article 1338 Civil Code, regarding the freedom to determine the form and content of the agreement seems to have a major impact on the implementation of construction services work, and also on all parties concerned. Keep in mind in the opinion of John Adriaanse (2010):

> A variety of factors makes a construction contract different from most other types of contracts. These include the length of the project, its complexity, its size and the fact that the price agreed and the amount of work done may change it as it proceeds.[6]

Generally, the service user has prepared the substance of the contract unilaterally, while the service provider just learns about the substance of the contract. This is coupled with the many models of construction contracts that exist in the world, which provide a free choice on the part of service users to choose a contract model, once it has not deviated from the Construction Services Act mentioned above.

2 DISCUSSION

2.1 Overview of contracts

Contract is a term that refers to the concept of agreement in general. The word contract itself comes from the English contract or *overennkomst* in Dutch, which means agreement. The definition of a treaty itself is set forth in article 1313 Civil Code which reads 'a Covenant is an act by which one or more parties commit themselves to one or more persons'.

Charles L. Knapp and Nathan M. Crystal, claim that a contract is an agreement between two or more persons not only to give confidence but mutual understanding to do something in the future by someone or both of them.[7]

In a legal dictionary, a contract is mentioned as a written agreement between two parties in a trade, lease and so on. In the same definition, the contract is a legal–legal agreement between two or more to engage or not in particular activities.[8] While the meaning of agreement in the same dictionary is an agreement made by two or more parties, written

6. John Adriaanse, *Construction Contract Law: The Essential s*, Wales: Palgrave Macmillan, 2010, hal.1.
7. Salim H.S. Abdullah, dkk, *PerancanganKontrak & Memorandum Of Understanding (MOU)*, Jakarta; SinarGrafika, 2011, hal. 8.
8. Sudarsono, *Kamus Hukum*, Jakarta; PT.Rineka Cipta, 1992, hal. 228.

or oral, each agreeing to comply with the content of the agreement that has been made together.[9]

This is based on the theory that in contract law there are five principles known according to the science of civil law. The five principles are: the principle of freedom of contract, the principle of consensualism (*concsensualism*), the principle of legal certainty (*pacta sunt servanda*), the principle of good faith and the principle of personality.

The principle of freedom of contract can be analyzed from the provisions of article 1338 paragraph 1 of the Civil Code, which reads: 'All legally-made agreements act as laws for those who make them'. This principle is one that gives freedom to the parties to:

a. make or not make agreements
b. enter into an agreement with anyone
c. determine the terms of the agreement and its implementation
d. determine whether the agreement is written or oral

2.2 *Forms of construction services contracts*

In general, users of construction services are divided into two types, namely service users who are not experienced, where users of this service used experts who are experienced in preparing and making contracts, which then became a consultant for service users. Another type of service user are experienced service users. Usually included in this type is the government and developers who have long been in the world of construction. Experienced service users and consultants for inexperienced service users have their own contractual standards, usually called tailor-made contracts, which can be tailored to the interests of contracting parties and do not need to follow existing contractual standards. Users of this type of contract draft are common in Indonesia, and the problem is that this type of contract often raises ambiguities caused by mistakes in translating and spelling, in order to prioritize one-sided interests so that the basic principle of contracting justice is not achieved.

In choosing the form of construction contract to be applied, service users need to consider the following:

1. The complexity and uniqueness of the project.
2. The ability of service users to manage design and construction.
3. Service user tolerance of risk.
4. Availability of resources.
5. The ability of service users to control the project.
6. The ability of service users to select service providers.
7. Possible changes and delays in employment.
8. The total duration of work required.
9. The financial condition of service users.

The forms of construction service contracts continue to develop along with the trend of existing construction methods. In fact, how the project will be implemented affects the form of construction contract to be applied.

The Government of Indonesia has issued Regulation of the President of the Republic of Indonesia no. 70 of 2012 on the Second Amendment of Presidential Regulation of the Republic of Indonesia No. 54 of 2010 on the Presidential Regulation on the Procurement of Government Goods/Services, where in this rule, in article 50 paragraph 2 concerns the distribution of types of contracts and paragraphs 3 to 6 describe the form of construction contracts according to the type of contact contained in paragraph 2. The contract scheme of Presidential Regulation no. 70 is:

---

9. *Ibid*. Hal. 355.

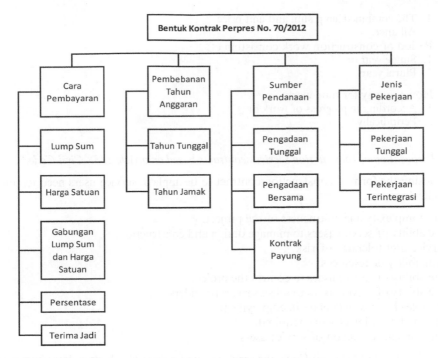

Figure 1.   Contract form scheme in Presidential Regulation no. 70.

2.3   *Contents of construction services contracts*

The main principles in construction contracts must be grounded in equality and clarity. In Law No. 2, year 2017, construction services are stipulated on construction work contract, as a basis for the relationship between legal subject perpetrators of construction services or procurement of goods/services. The location of the relationship lies in the concept of agreement between legal subjects in construction, implementation and supervision projects. The construction work contract under article 1, paragraph 8, shall be construed as the entire document governing the legal relationship between the service user and the service provider in the operation of the construction work.

In the case of the contents of further construction contracts, Law No. 2, year 2017, section 51 states that further provisions on construction work contracts as referred to in article 46 to article 50 shall be regulated by government regulation. In addition, the previous government regulation in the case of the determination of the working contract of construction service, namely Government Regulation no. 29 of 2000 regarding the implementation of construction services, in article 20, states several provisions, as the following:

1. Construction work contracts are basically made separately according to stages in construction work consisting of construction work contracts for planning work, construction work contracts for execution work and construction work contracts for supervisory work.
2. In the case of integrated work, construction work contracts as referred to in paragraph 1 may be set forth in one construction work contract.
3. The construction work contract as referred to in paragraph 1 shall be differentiated based on:

   a. Forms of rewards consisting of:
      1. Lump sum
      2. Unit price
      3. Fee added service fee

489

4. The combined lump sum and unit price
5. Alliance.
  b. Period of construction work consisting of:
    1. Single year or
    2. Plural year.

  c. How you pay for your work:
    1. According to progress of work or
    2. Periodically

2.4 *Preparation of construction services contracts based on article 1338 Civil Code*

In choosing the form of construction contract to be applied, service users need to consider the following:

1. The complexity and uniqueness of the project.
2. The ability of service users to manage design and construction.
3. Service user tolerance of risk.
4. Availability of resources.
5. The ability of service users to control the project.
6. The ability of service users to select service providers.
7. Possible changes and delays in employment.
8. The total duration of work required.
9. The financial condition of service users.

By using article 1338 Civil Code, the service provider must also have a share in approving the contract form itself by identifying the above matters to anticipate, prepare things to avoid the worst thing happen, as the following:

1. The complexity and uniqueness of the project that can lead to changes in specifications, prices, work time increases.
2. The inability of service users in managing the design and construction that leads to increased work time and irregular construction work that ends as a cost to be incurred by service providers.
3. The extent to which service users tolerate the risks that will occur and must be borne by service providers.
4. The ability of service users to control the project should be known to the service provider because a project that cannot be controlled properly can result in a cost to the service provider as a direct implementer of a project.
5. The possibility of change and delay in work, because any changes and delays in the work will certainly provide changes to the cost of work.
6. The total duration of work required.
7. The financial condition of the service user, which the service provider must know in order to predict and prepare the job billing plan, so that there is no inability of the service user to make payment for all expenses incurred by the service provider, so the financial rotation of the service provider remains stable.

2.5 *Good faith and compassion in construction services contracts*

The principle of good faith can be inferred from the provisions of article 1338 paragraph 3 of the Civil Code, which reads: 'An agreement shall be executed in good faith'. The goodwill, according to that chapter, is that the execution of the treaty must proceed with regard to the norms of propriety and decency.[10]

Regarding the meaning of propriety and morality, the law does not provide a formula. Therefore, there is no precise definition of the term. However, based on the meaning of the word, propriety means propriety; appropriateness; conformity; compatibility. While decency means

10. Abdulkadir Muhammad, 1990, *HukumPerikatan*, Bandung: PT Citra AdityaBakti, hal. 99.

decency; civilization. Based on the meaning of these two words, presumably it can be described that propriety and decency are worthy value; worth it; feasible; corresponding; suitable; polite and civilized, as equally desired by each of the parties who involve in the covenant.[11]

In general, good faith (article 1338 paragraph 3 of the Civil Code) and propriety (article 1339 of the Civil Code) are mentioned simultaneously and HogeRaad (HR) in the decision of January 11, 1924 has agreed that if testing with the appropriateness of an agreement given by the judge cannot be conducted by the parties, it means that the covenant is against the public order and morality. Thus, in the execution of the agreement there is a close relationship between justice, propriety and decency in good faith.

Based on article 1338 paragraph 3 of the Civil Code, if there is a disagreement about the implementation of the agreement in good faith (decency and morality), then the judge is authorized by law to supervise and assess the implementation of the agreement, whether there is a violation of the norms of propriety and decency. This means that the judge is authorized to deviate from the contents of the covenant according to his words, if the execution of the covenant according to his words will be contrary to good faith (if the execution according to the norms of decency and morality is considered fair). This is understandable because the legal purpose is to guarantee certainty (order) and to create justice.[12]

Justice in law requires certainty that what is promised must be fulfilled because the promise is binding as law (article 1338 paragraph 1 Civil Code). While the things that must be fulfilled is in accordance with propriety and decency (article 1338 paragraph 3 Civil Code, the principle of justice). The judge has the authority to prevent an unfair execution of the agreement, which is not in accordance with propriety and decency.[13]

Goodwill and decency are principles that provide protection to the weak in an agreement from the actions of the strong. This is understandable because the legal purpose is to guarantee certainty (order) and to create justice. The conclusion drawn from the foregoing is that the strong party on the grounds of freedom of contract may determine the articles in a treaty whose contents are more profitable to his party, but the judge has the right to assess and annul the articles in a biased agreement, which can be expected to bring about greater benefits on the one hand by harming the other.

Furthermore, in conjunction with the determination of the form of the contract, the tender will also commence where the tender type which has been determined from the beginning shall clarify the instrument of agreement, the conditions of the contract in detail, the correspondence between the contracting party, the work item, Bill of Quantity, the specification, contract documents.

3 CONCLUSION

The form of freedom of the parties to determine the form and content of the agreement in a construction service contract, pursuant to article 1338 Civil Code on the principle of freedom of contract, is that the parties are given the maximum freedom to determine the form and content of the construction service contract as clearly as possible and given the freedom to process bargain, not only related to the price or value of the construction service contract, but also to the contents of the construction services contract so that both parties understand their rights and obligations without avoiding the clauses required under the Construction Services Act.

Preparation of construction service contracts in accordance with article 47, Law No. 2, year 2017 concerning construction services shall only clarify the clauses which shall be contained in the construction service contract. In relation to the rights and obligations of the drafting contractor, the contract bidder in the article is not explained. While the contents of Law No. 2, year 2017 explains the authority of each party in the construction activities, the authority of the government in the construction activities, structures and conditions, the form of construction

11. *Ibid*, hal. 99.
12. Subekti, 1987, *HukumPerjanjian*, Jakarta: PT. Intermasa, hal. 40.
13. Abdulkadir Muhammad, *Op. Cit*, hal. 100.

services and foreign construction services, the holding, construction services, construction and financing of construction services, there is no mention of the rights and obligations of bidders and contractors, so the principle of freedom of contract in which the freedom of the parties to enter into a contract, including the freedom to decide what they agree on will be very hard to realize due to the absence of the above explanation.

4 SUGGESTIONS

In support of Law No.2 of 2017, the authors advise the government to enact government regulations that support the enforcement of the law, in particular with respect to the process of contracting for construction services (rights and obligations of construction service contracts and contract auction participants construction services until construction service contracts), auction of construction service contracts and their execution.

Construction service contracts related to government agencies or sourced from government funds have been widely accommodated in Law No. 2, year 2017 on construction services as well as other supporting government regulations. However, a binding contract between private parties or private sources of funds should also take into account the rules, as well as other international rules that may serve as a source of guidelines for contracting of construction services, including the use of foreign terms and international rules.

REFERENCES

Abdulkadir, Muhammad. (1990). *Law of alliance* (p. 99). Bandung: PT Citra Aditya Bakti.
Adriaanse, J. (2010). *Construction contract law: The essentials* (p. 1). Wales: Palgrave Macmillan.
Campbell. (1968). *Black's law dictionary* (7th ed., p. 394). Black, USA: West Publishing.
Gifis, Seteven. (1984). *Law dictionary* (p. 94). New York: Barron's Educational Series Inc.
Law no. 2 year 2017 on Construction Services.
Salim. (2009). *Contract law (theory & contract preparation techniques)* (p. 5). Jakarta: Sinar Grafika.
Salim H.S, Abdullah, et al. (2011). *Contract design and memorandum of understanding (MOU)* (p. 8). Jakarta: Sinar Grafika.
Salim HS, Abdullah, & WiwiekWahyuningsih. (2007). *Contract design and memorandum of understanding (MoU)* (p. 9). Cet. 2. Jakarta: Sinar Grafika.
Subekti. (1987). *Legal agreement* (p. 40). Jakarta: PT. Intermasa.
Sudarsono. (1992). *Legal dictionary* (p. 228). Jakarta: PT. Rineka Cipta.
Syaukani, Imam and A.AhsinThohari. (2004). *Political legal basics* (p. 54). Jakarta: King Grafindo Persada.

Business Innovation and Development in Emerging Economies – Trinugroho & Lau (Eds)
© *2019 Taylor & Francis Group, London, ISBN 978-1-138-35996-3*

Engineering–Procurement–Construction (EPC) mega project: Analysis of the dominant cost factors and viable solutions

Muhammad Andry Rezky & Arviansyah
University of Indonesia, Depok, Indonesia

ABSTRACT: The Mega Construction Project (MCP) is vital to help a nation meets its economic and social needs, to raise the nation's social status, and to lead its society to follow international developments. MCP has significant contract value; hence, the contractors have a chance to get a significant profit. However, MCP has several challenges compared with ordinary projects, such as having more complex work with larger capacity, involving multiple parties, bearing high risk, and raising public concern. The purpose of this study is to determine the dominant cost factors in the Engineering–Procurement–Construction (EPC) mega project and future viable practice for companies to deal with these factors. This study applies a qualitative approach by utilizing the Delphi method followed by in-depth interviews. The study was conducted in Company X which is one of the largest state-owned construction companies engaged in EPC projects. The results of this study are five dominant cost factors: (1) miscalculation in the tender process, (2) subsystem integration, (3) corruption in the tender process, (4) corruption in ongoing projects, and (5) project delay. The consensus degree is 0.78 showing a high agreement among experts. In-depth interviews were held with six experts from two different EPC mega projects. The interviews revealed miscalculation in the tender process and project delay was the main cost-wasting factors in the two EPC mega projects.

Keywords: Mega construction project, Engineering-Procurement-Construction (EPC) Project, Cost overrun factor, Delphi method

1 INTRODUCTION

The Mega Construction Project (MCP) is necessary because it has three roles: meeting economic and social needs of a country, raising the country's social status, and leading the country to follow international developments (Mok et al., 2015). MCP usually has several challenges such as including complex work with large capacity, involving multiple parties, high risk, and public concern (Mok et al., 2015). MCP is a massive infrastructure investment initiated by the government; it has a long and rigorous work schedule with high-complexity work (Mok et al., 2015).

Harris and McCaffer (2013) showed that the productivity of construction can be improved by applying the method from manufacturing industries associated with the machine. Koskela (1992) revealed that an important factor of lean production is the presence of two aspects in the production system: conversions and flows. Conversions are value-added activities or activities required by customers, while flows are activities that do not add value or that require cost, time, and resources but do not add value to the final result. Activities that do not add value are often referred to as waste. Construction projects implementing lean construction should reduce waste and non-value-added activities.

MCP has several traits that involve multiple stakeholders from various occupations and professional backgrounds, socially influential, and having an economic impact (Mok et al., 2015). The Engineering–Procurement–Construction (EPC) project contractors perform work from design to procurement to construction.

Table 1. Methods utilized in the research.

Systematic literature review	Delphi method	In-depth interview
Objective: Creating a list of cost-wasting factors	Objective: Analyzing the dominant cost-wasting factors	Objective: Analyzing the dominant cost-wasting factors in real-life cases and finding solutions
Description in the literature review section	Description of research methodology	Description of research methodology

Westin and Päivärinta (2011) found the problems in the engineering and construction project using the Delphi method. However, the engineering and construction project has a smaller scope than the EPC project. The scope of the EPC project ranges from design to procurement to construction, while the scope of the engineering and construction project is only design and construction.

Research related to waste and cost in a construction project was carried out by Lind and Brunes (2015). The results of this research revealed that cost-wasting factors in construction projects include design changes, price changes, and unforeseen technical problems. The research was undertaken on construction projects that did not include mega projects, thus having a lower level of work complexity and fewer related parties.

The waste of costs on EPC mega projects needs to be prevented and solutions sought so that contractors can get planned benefits. The high complexity of the EPC mega project makes it difficult for contractors to identify cost-wasting factors quickly. The benefit of this research is to help EPC mega project contractors identify cost-wasting factors that are frequent and have a significant dominant impact. Cost-wasting factors that have been identified can be prevented and overcome so that the project does not require a higher cost than the initial plan and contractor profit can be maintained and increased.

Referring to the explanation of previous research and the characteristics of the EPC mega project, the problems of this research are as follows: (1) What is the most dominant cost-wasting factor in the EPC mega project? (2) What solutions can be found to prevent and overcome the most dominant cost-wasting factors that occur? The methods utilized to achieve the research objectives are shown in Table 1.

2 LITERATURE REVIEW

The EPC project is the design and construction work arranged and done by one organization (Harris & McCaffer, 2013). This project model is usually used for industrial plant projects that require a more complete engineering design than an architectural design. The EPC project usually includes commissioning and improvements to the specification of facilities that have been built according to the needs of the project owner.

Contracts of the EPC project often use FIDIC (2005) guidelines. FIDIC is an international federation that provides contract content guidance depending on the type of construction project. Based on the type of contract, the construction project is divided into four categories: (1) construction, (2) small construction, (3) plant and design-build, and (4) EPC/ Turnkey projects. In the construction project, the contractor performs construction work based on the design from the project owner. In the small construction project, the contractor carries out construction or design work that has small costs, is simple, has a short duration, and/or is repetitive. In a plant and design-build project, the contractor does the design and construction work based on the needs of the project owner. Finally, EPC projects are those that have a higher cost and time frame than other projects; the contractor is entirely responsible for the design and implementation of the project development.

The Mega Construction Project (MCP) is a project with more complex construction works and the contractor is connected to multiple stakeholders (Mok et al., 2015). The rapid development of MCP causes the contractor to need advanced technologies and work with global companies; for instance, the contractor needs to work with technology companies from abroad. The MCP has a large capacity or land area with high levels of complex work.

The MCP has three challenges in project management: (1) the involvement of many stakeholders resulting in the complex relationships and conflict of interest, (2) the dynamics of work and capacity growth due to the high uncertainty of the project owner's desire, and (3) the government applying many administrative rules (Mok et al., 2015). The MCP can be divided into two infrastructure functions: (1) one or a group of projects built for one infrastructure function; and (2) group of projects that have different infrastructure functions but are integrated into a single development plan.

The cost-wasting factors are synthesized from 22 articles related to the MCP. The articles are selected from five prominent construction and project management journals (Aarseth

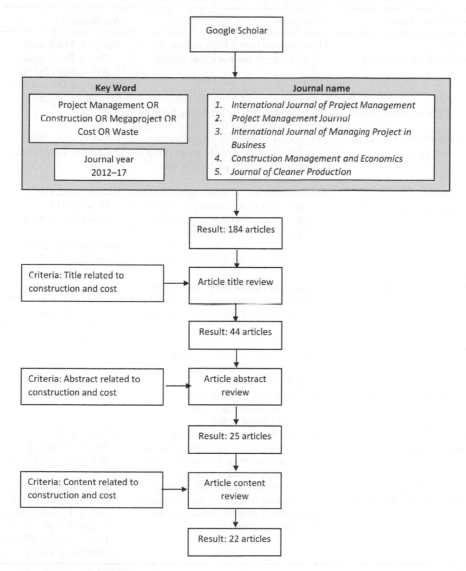

Figure 1. Systematic literature review.

et al., 2017). These five journals are: (1) *International Journal of Project Management*, (2) *Project Management Journal*, (3) *International Journal of Managing Projects in Business*, (4) *Construction Management & Economics*, and (5) *Journal of Cleaner Production.*

This research used a systematic literature review flow as shown in Figure 1 (Arviansyah et al., 2011). The systematic literature review reveals 20 cost-wasting factors synthesized from the selected 22 articles. The cost-wasting factors as revealed in the systematic literature review is described in Table 2.

An example of the first cost-wasting factor is the construction project of the London Olympic facility. The project needed to be divided into several small systems to make analysis and coordination relationships between systems easier (Davies & Mackenzie, 2014). Another example is the need for other projects to integrate between subsystems, with the coordination of all parties involved for achieving good project performance (Demirkesen & Ozorhon, 2017).

Table 2. Description of cost-wasting factors revealed in the literature review.

Cost-wasting factor	Description	Article source
Factor 1: Subsystem integration	Mega projects have a high degree of complexity in the work and system.	Davies & Mackenzie (2014), Demirkesen & Ozorhon (2017), Brady & Davies) (2014)
Factor 2: Corruption in the project	Corruption occurs when the project runs, after a kick-off meeting between the contractor and the project owner.	Locatelli et al. (2017)
Factor 3: Corruption in the bidding process	The process of uncooperative contractor tender, which causes the project cost to be higher than the price offer that should be provided.	Locatelli et al. (2017), Chou & Yang (2012)
Factor 4: Dispute between parties	The occurrence of disputes or differences of opinion between contractors with other parties, resulting in legal costs, administration costs, and others.	Boateng et al. (2015), Winch (2013), Wang et al. (2016)
Factor 5: Work result not a match to the owner's request	The contractor's work does not meet the standards used by the project owner.	Boateng et al. (2015), Lei et al. (2017)
Factor 6: Project delay	Project delays cause the contractor to pay fines for delays and other penalties, including additional costs to pursue project delays.	Boateng et al. (2015), Brady & Davies (2014), Wauters & Vanhoucke (2016), Wang et al. (2016)
Factor 7: Bad weather	Unfavorable climatic conditions cause contractors to use additional equipment, materials, or facilities to keep doing the work.	Boateng et al. (2015)
Factor 8: Political issue	The existence of political problems in local and national governments that cause additional costs.	Boateng et al. (2015)
Factor 9: Project manager ability	The skills of project managers in understanding the dynamics that occur in the project can help the contractor solve the problem immediately.	Winch (2013), Banihashemi et al. (2017), Rolstadås et al. (2014)
Factor 10: Miscalculation in the bidding document	Errors in the estimated cost and risk calculations that cause the contractor to underestimate the actual cost to finish the project.	Demirkesen & Ozorhon (2017), Chih & Zwikael (2015), Flyvbjerg (2014), Locatelli et al. (2017), Lind & Brunes (2015), Wang et al. (2016)

(Continued)

Table 2. (*Continued*).

Cost-wasting factor	Description	Article source
Factor 11: Communication	Problematic communication and exchange of knowledge/information among contractors, project owners, and vendors/subcontractors lead to data dispute among parties.	Demirkesen & Ozorhon (2017), Banihashemi et al. (2017), Liu et al. (2014), Eduardo Yamasaki Sato & de Freitas Chagas Jr. (2014), Rolstadås et al. (2014), Zidane & Olsson (2017), Pasquire (2012)
Factor 12: Health, Safety & Environment (HSE) incident	The occurrence of work accidents can cause excess costs incurred by the contractor.	Banihashemi et al. (2017)
Factor 13: Uniqueness of local employees	Unique characteristics of local employees who are different from employees in the central contractor company.	Lei et al. (2017)
Factor 14: Transportation issue	Project locations that are difficult to reach by regular modes of transportation cause contractors to spend more on transportation.	Lei et al. (2017)
Factor 15: Additional policies from government	The government makes additional regulations on the materials used in the project once the materials are already purchased.	Mok et al. (2017)
Factor 16: Site location issue	There is a risk of uncertainty in local culture and profile of local workers.	Saunders et al. (2016), Wang et al. (2016)
Factor 17: Project management team	Project team activity that affects team performance, such as senior management assistance, management abilities to follow up on changes, clear and realistic team goals, and detailed team work plans.	Banihashemi et al. (2017), Rolstadås et al. (2014), Zidane & Olsson (2017), Lind & Brunes (2015), Pasquire (2012)
Factor 18: Similar project benchmarking	Similar projects that the company does can be examples to imitate or prevent unfortunate situations. In the absence of similar projects, companies should try with problem-solving approaches that waste money and time.	Rolstadås et al. (2014)
Factor 19: Project scope	There is a change in the scope of work that makes the contractor cost spend more than planned.	Lind & Brunes (2015)
Factor 20: Preliminary work	The work undertaken by the project owner as a project preliminary has not been done, thus hampering project implementation and increasing project costs.	Wang et al. (2016)

An example of the second cost-wasting factor is shown by Locatelli et al. (2017). The EPC project records the increased costs that exceed realization in the field. Another example is the decline in the quality of technical specifications during the construction phase.

An example of the third cost-wasting factor presented by Locatelli et al. (2017) is the occurrence of kickback (return of excess prices) from vendors or subcontractors directly to private individuals without going through the company. Chou and Yang (2012) show that improper procurement procedures result in legal issues and tender procurement tenders. The consequences are costs and time are doubled compared with the initial plan.

Boateng et al. (2015) give examples of the fourth cost-wasting factor that occurred with Infrastructural Contract (INFRACO) project owners and contractors claiming from each other as a result of disputes in work. In Winch (2013), another example is the difference of opinion between the Transmanche-Link (TML) contractor and the project owner, causing the project to be late; also, the contractor must face excess costs of EUR 1,000 million to accelerate the completion of the project.

Furthermore, Boateng et al. (2015) provide an example of the fifth cost-wasting factor, that is, the ETN project not meeting the standards of tram-tracks work. Lei et al. (2017) provide an example of Chinese contractors experiencing differences between Chinese standards and international standards; the result is an increase of time and costs needed in the implementation of these international standards.

Boateng et al. (2015) give an example of the sixth cost-wasting factor in the ETN project. The project was delayed due to land, environmental, political, and social conditions so that ETN had to pay the penalty for delays. Brady and Davies (2014) provide an example of project delays due to unexpected events in the project leading to a delay penalty and considerable project costs for catching-up delays.

Boateng et al. (2015) also report an example of the seventh cost-wasting factor. Difficulties in road repair works at ETN projects resulted from extreme weather, causing project overheads and overdue projects.

Boateng et al. (2015) provide an example of the eighth cost-wasting factor. The ETN project experienced political problems with the central government causing workers to demonstrate and stop working on projects.

An example of the ninth cost-wasting factor is presented by Winch (2013). The TML contractor and project manager are inexperienced, and he cannot act quickly on the occurrence of a difference in opinion between TML and the project owner. This causes TML excess costs of EUR 1,000 million. In another example, Rolstadås et al. (2014) show that the leadership of project managers determines the application of the Manage by results (MBR) and Manage by means (MBM) approaches to solving internal and external project issues.

An example of the tenth cost-wasting factor found by Chih and Zwikael (2015) is an UK government project contractor underestimating the cost and time to finish the project. Another example is found by Flyvbjerg (2014), where the contractor manipulated the calculations resulting in very optimistic estimations of time, cost, and profit. Errors are also due to underestimating future costs and increasing profit estimates to please the project promoter.

Banihashemi et al. (2017) provide an example of the eleventh cost-wasting factor. Integration of information and knowledge between the parties in the project affects project performance with regard to cost, time, and quality. Liu et al. (2014) also provide another example of inadequate communication between INVAP (contractor company from Argentina) and the Australian Nuclear Science and Technology Organization (ANSTO) as the project owner. This interferes with understanding the scope of work between the contractor and the project owner, thus causing the project to stop for one year. The contractor goes over the budget to accelerate the completion of the project.

An example of the twelfth cost-wasting factor taken from Banihashemi et al. (2017) is inadequate health care and safety protocols by contractors, which results in accidents and additional project costs.

Local workers used by Chinese contractors cannot meet the job standards specified by the project owner. This is an example of the thirteenth cost-wasting from Lei et al. (2017).

The fourteenth cost-wasting factor is the difficulty of transportation at the project site. Project locations that are difficult to reach by regular modes of transportation make contractors spend more on transportation. Lei et al. (2017) provide an example of Chinese contractors experiencing poor cost and time performance due to the difficulty of transportation from the city to the project site.

An example of the fifteenth cost-wasting factor from Mok et al. (2017) is that contractors bought imported vessels as there are no local vessels suitable for the project specifications. After the vessels are imported, the vessels must pass a license test conducted by the local government.

An example of the sixteenth cost-wasting factor described by Saunders et al. (2016) is the large number of requests by (local) residents around the project that led to cost-wasting. Another example given by Wang et al. (2016) is the lack of local workers available in the project, causing the company to take time to recruit and mobilize workers from outside the project area.

Rolstadås et al. (2014) provide an example of the seventeenth cost-wasting factor. Infrastructure projects in the United Kingdom are improving mutual trust among project team members, resulting in improvement of cooperation and project performance. Another example in Lind and Brunes (2015) is the speed of following up on issues in projects; with appropriate change, management can bring the project to run more smoothly.

An example of the eighteenth cost-wasting factor by Rolstadås et al. (2014) is the selection of work methods and procedures for performing a job more efficiently by using similar successful projects. Other examples include a company creating a project manual for project management from similar successful projects.

Lind and Brunes (2015) provide an example of the nineteenth cost-wasting factor. The London 2012 Olympic project experienced cost overruns due to changes in the scope of work. Another example is the changes in specifications and new designs that lead to additional costs.

Finally, an example of the twentieth cost-wasting factor is taken from Wang et al. (2016). The project owner had not completed clearing the land nor the initial design of the project; temporary facilities had not been made so the daily construction works could not be done and added extra costs to deal with the delay.

3 RESEARCH METHODOLOGY

This study uses a sample from an EPC mega project in Indonesia. Company X is a state-owned construction company engaged in EPC projects. Company X's vision for 2020 is to be one of the best Engineering–Procurement–Construction (EPC) and investment companies in Southeast Asia. Based on the regulation of Company X, mega projects have the following characteristics: (1) project contracts valued above IDR 1 trillion, (2) production per month above IDR 50 million, (3) labor density managed by more than 100,000 workers monthly per year, and (4) the number of service providers managed by 50 or more service providers. For the last five years Company X has received many government, state-owned, and private enterprise development projects. Company X projects include a sugar factory, cooking oil factory, smelter factory, and oil and gas facilities from gas station to pipes to refineries.

In five years of working on these projects, Company X often suffered cost and time wastage. Wastage of project costs can lead to a reduced project or loss of project profit. The percentage of Company X's EPC project cost realization will always be higher than the plan from 2012 to 2016. The average realization of the cost of the last five years' project is 102% of the initial project cost plan.

The method used in this research is the Delphi method. The Delphi method is a structured group decision-making process for finding solutions to problems that do not have a definitive answer. This method is chosen because there are experts who can make consensus, the required data is not available and the results can be used for future application. The Delphi method is used to order the cost-wasting factor based on frequency and impact.

The EPC cost-wasting factor used in the Delphi method is the result of the literature review conducted using the Google Scholar tool. Article searches are limited to those from the *International Journal of Project Management, Project Management Journal, International Journal of Managing Projects in Business, Construction Management and Economics*, and *Journal of Cleaner Production*. Articles were sought for the last five years (i.e., 2012–2017). Selected articles passed the selection of titles, abstracts, and contents.

The experts for the survey were selected from among company employees. The experts have at least ten years of work experience and participate in the development of at least three EPC mega projects. The expert will be a resource person in the Delphi method. The number of experts used depends on the number of topics or questions and the number of experts that are available (Cuhls, 2004). This research employed 15 experts for the Delphi method.

The Delphi method uses questionnaires that are distributed via a survey website. The Delphi method is done in three stages. The first-stage Delphi method of each expert shows 20 cost-wasting factors resulting from the literature review. Experts may add cost-wasting factors other than those lists. In the second stage, an expert assesses the frequency and impact of the list of cost-wasting factors of the first-stage results. The cost-wasting factor that has a low mean frequency and a low mean impact are omitted from the list. When the expert is shown a second-stage Delphi recap, the name of the expert is anonymized to keep objectivity in the expert's answer. Next, the expert is shown a recap of the answers to the third-stage Delphi method. The expert may replace or remain with the answer.

The fourth stage is used to get a level of consensus among experts on the order of cost-wasting factors from the Delphi method third-stage result. The fourth-stage questionnaire uses a Likert scale interpolated to the Kendall W coefficient. The consensus rate is also calculated using Schmidt (1997) calculations from the answer to the sequence of cost-wasting factors per expert. Skinner et al. (2015) explain that the Kendall W coefficient is suitable for determining the level of Delphi consensus. A reasonable level of consensus is that there are 0.500 for the middle level and 0.700 for the high level. The Delphi method will stop once the level of consensus is reached.

The flow of the Delphi method uses three stages adapted from the research (Keil et al. 2013). The difference in the research is found in the first-stage Delphi method. The first stage in this research shows a list of cost-effectiveness factor results from the literature review. Experts can add cost-wasting factors in addition to the 20 factors of the wasteful cost of literature review results.

An in-depth interview was used to find out cost-wasting solutions and real-life examples that occur in the project. Interviews were conducted with six experts from two different EPC mega projects. Experts were carefully selected from commercial, financial, and procurement divisions. The experts chose to share the example and solution of the cost-wasting factor which includes the five most dominant factors based on the Delphi method. The experts presented three solutions: (1) a solution used by the project management team, (2) a solution recommended by experts, (3) future best-practice solution based on Gerbert (2016).

4 RESULTS AND DISCUSSION

Results of this research are from the Delphi method and in-depth interviews. The Delphi method was used to order cost-wasting factors based on frequency and impact. The in-depth interview was used to provide real-life examples and solutions to overcome cost-wasting factors.

The first method discussed here is the Delphi method. There are three stages of the Delphi method. In the first stage, experts share their biography, such as the amount of work experience, education obtained, and the number of mega projects undertaken. The average expert work experience is 12.8 years and includes three postgraduate experts, ten bachelor degree experts, one third-diploma expert, and one senior high school educator. The average value of mega projects done by experts is IDR 2.3 trillion. Experts come from different divisions such as commercial, finance, engineering, project control, construction, and procurement.

Experts in the first stage provided 11 additional cost-wasting factors beyond the 20 cost-wasting factors resulting from the literature review. Once analyzed, the 11 factors were found to be real-life cases whereas the 20 cost-wasting factors were from the literature review. So, the 11 factors from the experts were not added to the cost-wasting factor used in second-stage of the Delphi method.

In the second stage, experts were asked to sequence 20 cost-wasting. The sequencing result of the 20 cost-wasting factors based on the frequency of occurrence and the resulting impact can be seen in Table 3.

In the third stage, the 11–20 cost-wasting factors ordered in the second stage were eliminated. Experts made a new order of the top 10 dominant cost-wasting factors with regard to frequency and impact, the results of which are described in Table 4.

The fourth stage of the Delphi method is the determination of the level of consensus from the experts. This was calculated using the Likert scale questionnaire and Kendall W coefficient (Schmidt 1997). The average consensus rate obtained from the Likert questionnaire is 4.4. The consensus level is interpolated by the Kendall W coefficient value. The resulting consensus level is 0.78, with strong agreement on interpretation and a high level of trust. The Kendall W coefficient calculation based on Schmidt (1997) is 0.75, with a high confidence level interpretation.

The second method discussed here is the in-depth interview. Interviews were conducted into two of the most substantial EPC mega projects in the last five years. The first project was a ferronickel plant construction project funded by the government, with a duration of 30 months. The project had a value of IDR 3 trillion, with monthly progress payments.

Table 3. The order of 20 cost-wasting factors based on the second-stage Delphi method.

Order	Dominant cost-wasting factor
1	Factor 10: Miscalculation in the bidding document
2	Factor 6: Project delay
3	Factor 2: Corruption in the project
4	Factor 1: Subsystem integration
5	Factor 3: Corruption in the bidding process
6	Factor 11: Communication
7	Factor 9: Project manager ability
8	Factor 17: Project management team
9	Factor 5: Work result not a match to the owner's request
10	Factor 19: Project scope
11	Factor 8: Political issue
12	Factor 16: Site location issue
13	Factor 14: Transportation issue
14	Factor 4: Dispute between parties
15	Factor 12: HSE incident
16	Factor 18: Similar project benchmarking
17	Factor 20: Preliminary work
18	Factor 15: Additional policies from government
19	Factor 7: Bad weather
20	Factor 13: Uniqueness of local employees

Table 4. The order of 10 cost-wasting factors based on the third-stage Delphi method.

Order	Dominant cost-wasting factor
1	Factor 10: Miscalculation in the bidding document
2	Factor 1: Subsystem integration
3	Factor 3: Corruption in the bidding process
4	Factor 2: Corruption in the project
5	Factor 6: Project delay
6	Factor 11: Communication
7	Factor 5: Work result not a match to owner's request
8	Factor 9: Project manager ability
9	Factor 17: Project management team
10	Factor 19: Project scope

Company X was responsible for engineering, procurement, and construction work. Company X cooperated with a Japanese construction company. The second project, started in 2014, was a Chemical Grade Alumina (CGA) plant construction project financed by private companies, with a duration of 32 months. The project had a contract value of USD 450 million, with monthly progress payments. Company X worked with Indonesian private construction companies and Japanese construction companies.

The experts interviewed were positioned as financial managers, commercial managers, and procurement managers. The interviews were conducted with the experts in these positions because the divisions held by the experts were directly involved with the costs of the project. So, the experts really knew if there was a waste of costs. Experts chosen had to present real-life examples and solutions of the cost-wasting factors, including the five most dominant cost-wasting factors based on the results of the Delphi method. Experts examined the sequencing of cost-wasting factors from the Delphi method. They had to explain three solutions: (1) a solution used by the project management team to overcome the problem, (2) a solution recommended by the experts, and (3) a future best-practice solution based on Gerbert's (2016) recommendation from the experts.

Company X in April 2018 has been working on this project for 15 months. Company X's actual cost exceeded the initial cost planned, but profit was still obtained. Company X used Primavera software support for work schedules. By the 15th month, the Primavera simulation identified the project to be 1.5 months late. Project delays beyond the duration of the contract will cause Company X to pay a delay penalty.

The three experts from project 1 explained the waste of costs due to cost-wasting factor 10: errors in the tender calculation. These errors included cost estimation errors and engineering design errors when bidding to get the project. The three experts experienced different events.

A project manager who previously acted as a commercial manager knows that several buildings and equipment were not accounted for by the team that made the bidding documents. This fact caused the existence of buildings and equipment that were not included in the cost plan but must be realized in the actual project costs. The actual cost of these works and equipment caused project profits to decrease. Equipment not taken into account are a cover blow, slag pot, and weight bridge. Buildings that have not been taken into account are a cable pit for a utility system, drainage system, seawater intake, and weight bridge. The solution used by the project management team was to keep working on the building and purchase the equipment, but this has an impact on the inefficiency to the order of IDR 34 billion. The solution the expert proposes is to request a change in order to the project owner for the building and equipment.

The financial manager found a miscalculation in the bidding documents with regard to the estimated rupiah exchange rate against the dollar. The rupiah exchange rate used is 13,250, while until the 15th month the current project reached IDR 14,000 per dollar. This fact causes cost-wasting when paying vendors from an international company. Vendors paid with dollars are off-gas equipment, refractory, copper cooler, rotary kiln, rotary dryer, and electric smelting furnaces. The total additional cost is IDR 15 billion. The solution of the project team is to use the reserve cost to pay the foreign exchange incurred. Reserve cost is a money reserve, and if not used, will increase the project's profit. The expert recommends that the project make a contract addendum with the vendor to change contract value to rupiah.

The procurement manager found that the electrical and instrument equipment cost planned was 20% lower than the original vendor's price. The bidding team did this to win the tender by lowering the bid. Inefficiency caused by the purchase of this equipment is IDR 13 billion. The solution of the project team is to keep buying equipment because it is included in the scope of the project. The expert's recommended solution is to request a changed vendor listed in the tender document and then replace it with a cheaper vendor.

Future best practice from Gerbert (2016) could be used by the project team to overcome miscalculation of the bidding document. Project plans with an engineering design with costs that could overcome several buildings and equipment was not calculated by a team that made the bidding documents. Industry-to-industry collaboration in the supply chain to share project risks such as exchange rate risk could solve miscalculation on the estimated rupiah exchange rate against the dollar. Improving data sharing, examples of similar projects, and the best solution to solve project problems across projects and subsidiaries could prevent miscalculation in estimating costs for electrical and instrument equipment.

Three experts explained that the central problem of project 1 was cost-wasting factor 10: miscalculation of the tender with the order of the most dominant cost-effectiveness factor resulting from the Delphi method. Several studies have also found errors in the tender calculation as a major factor in cost wastage. Flyvbjerg's (2014) study provides solutions by cost estimation according to procedures, correct assumptions, and non-manipulation. Projects can work with third parties to help estimate costs so as not to bias optimism and with additional views from experts (Lind & Brunes, 2015). Appropriate tools and techniques, such as analogous, parametric, and bottom-up estimation, can be used to improve the accuracy of estimated costs (PMI, 2017).

Three experts interviewed in project 2 were also in the same position as those in project 1: financial managers, commercial managers, and procurement managers. All three experts provided different cost-effectiveness samples with different cost-effectiveness factors as well. However, there is one factor that all three experts explained, that cost-wasting factor is factor 6: project delay.

The CGA project's financial manager was aware of costs incurred by cost-wasting factor 6: project delays. The delayed project caused Company X to pay a delay penalty and made additional labor pursue work late to finish the project. The project team's solution is to continue to pay the delay penalty and recruit a special task force to finish Company X's project quickly. This solution will result in additional costs of IDR 28 billion. The solution recommended by an expert is to recruit workers with high productivity at the beginning of the project.

The CGA project's commercial managers found cost-wasting due to factor 2: corruption during project work. The existence of corruption in the estimation of subcontractor progress is higher than the actual work. XYZ internal auditor also knew this and immediately asked the subcontractor to refund the overpayment. The expert recommends that subcontractors also have to pay a delay penalty due to 3 months delay.

Commercial managers also found cost-wasting due to factor 6: project delays. Project delay was caused by the addition of barracks from four to ten barracks.

The CGA project's procurement managers found cost-wasting due to factor 1: integration errors between project subsystems. Integration errors caused hydro test failure and resulted in damaged pipes and gaskets needing to be replaced. This replacement costs IDR 5 billion. The expert recommended solution is for the engineering team to hold meetings frequently to integrate the systems into the project.

Procurement managers also experienced cost-wasting due to factor 6: project delays. Installation of a pressure gauge and limit switch failure caused both devices to be repurchased from Germany. It takes five months to complete the procurement process.

Three experts identified that project delay was the main cost-wasting factor in project 2. All three explained that different factors caused the delay in project 2. Delay was caused by (1) rework in the installation of piping due to failure in the hydro test in the utility system, (2) the addition of six barracks, and (3) repurchasing of the limit switch and pressure gauge. Cost-wasting due to project delays can be prevented by identifying possible risks in the project, such as (1) imported materials required, (2) necessary tests or licenses, and (3) potential events that could cause the project to be delayed (Wauters & Vanhoucke, 2016). Another solution is to improve communication and coordination among stakeholders in the project so that the project team stays informed of potential job delays and can follow up earlier (Brady & Davies, 2014). The appropriate business process PMBOK to solve project delay is to control schedule. This business process aims to (1) monitor the status of project activities, (2) update the progress of the project, and (3) adjust changes to the baseline schedule.

Future best practice from Gerbert (2016) could be used by the project team to prevent and overcome the problem in project 2. Optimal workforce and recruitment planning could be used to prevent project delay because of low productivity by workers. The company needs to record employee performance data throughout the project to find out the right worker profile for a particular job. Implementation of strict and anti-corruption standards could prevent corruption during project work. Procedures for surveillance in the workplace should be accompanied by strict punishment. Closely monitoring the time and cost of projects with the help of technologies like closed-circuit television and drones could assist the project team in evaluating project schedule and preventing project delay. Digital technology and big data to control the supply chain in the project could prevent failure of subsystem integration. This suggestion led to the sharing of data between the subsystem engineering team more efficiently. Creating a component that is standard, modular, and prefabricated to minimize high-risk field work could overcome project delay caused by installation failure at the project site.

Cost-wasting due to project corruption can be solved by making official procedures from company headquarters and closely monitoring the process. Headquarters needs to monitor the qualifications of bidders and analyze document engineering (Locatelli et al., 2017). Business process PMBOK is appropriate to solve the problem of corruption in quality control. This business process can be used to monitor the work results correctly and make a match to project owner requirements (PMI, 2017). If the work results are checked correctly to match the request of the project owner, then cheating in the estimation of progress can be avoided.

Cost-wasting due to incorrect subsystem integration can be prevented by coordinating all subsystem teams and related parties effectively. Related parties are internal project management, project owners, subcontractors, and vendors (Brady & Davies, 2014). The appropriate business process in Project Management Body of Knowledge (PMBOK) to solve the problem of subsystem integration in the project is managing communications. This business process creates, collects, distributes, and stores project information from related parties. Well-distributed project information to relevant parties such as engineering teams, vendors, and subcontractors makes integration between systems easier (PMI, 2017).

Projects 1 and 2 have similar conditions, such as (1) scope of work, (2) project owner, (3) location profile, and (4) terms of payment. Both projects have different causes of cost-wasting. Cost-wasting in project 1 is because of miscalculation in the bidding document. Cost-wasting in project 2 is mostly because of project delay.

5 CONCLUSION

Previous research indicates the MCP had several challenges such as complex work with large capacity, involving multiple parties, high risk, and public concern. MCP is a massive infrastructure investment initiated by the government; it has a long and rigorous work schedule with high complexity work (Mok et al., 2015). The EPC mega project contractor usually experienced cost-wasting in their project. Research related to waste and cost in a construction project has also been done by Lind and Brunes (2015). Based on the previous research and the EPC mega project contractor experience, the objectives of this research were (1) to make an order of the most dominant cost-wasting factor in the EPC mega project; and (2) to provide solutions to prevent and overcome the most dominant cost-wasting factor that occurs.

This research used three methods to fulfill the research objectives. The first method is a systematic literature review, used to identify cost-wasting factors in the EPC mega project. The second is the Delphi method, used to make a cost-wasting factor order based on frequency and impact. The third method is in-depth interviews, used to know real-life examples and solutions to overcome cost-wasting factors.

The most dominant cost-wasting factor is the miscalculation of the tender. These errors include cost estimation errors and fundamental engineering design errors. A consensus level is achieved based on coefficient Kendall W. The Likert scale questionnaire score is 0.78 and Schmidt (1997) calculation is 0.75, both with high consensus interpretation. This cost-wasting factor is chosen based on the order of frequency and impact levels.

Bidding calculation errors can be solved by estimating costs by procedures, by correct assumptions, and by not manipulating (Flyvbjerg 2014). Projects can work with third parties to help estimate costs so as not to bias optimism and with additional views from experts

(Lind & Brunes, 2015). The business process of **PMBOK** estimate costs is suitable to solve the problem of tender calculation error. Appropriate tools and techniques, such as analogous, parametric, and bottom-up estimation, can be used to improve the accuracy of cost estimates.

Recommendations were made based on this conclusion. The EPC mega project contractor company had to know about all the project management business process. Contractor companies should train for Project Management Professional certification. Contractor companies could use business processes that suit their projects.

REFERENCES

Aarseth, W., Ahola, T., Aaltonen, K., Økland, A., & Andersen, B. (2017). Project sustainability strategies: A systematic literature review. *International Journal of Project Management*, *35*(6), 1071–1083. https://doi.org/10.1016/j.ijproman.2016.11.006.

Arviansyah, A., Berghout, E., & Tan, C.-W. (2011). Evaluation of ICT investment in healthcare: Insights and agenda for future research. *Proceedings of the 5th European Conference on Information Management and Evaluation*, (2005), 53–64.

Banihashemi, S., Hosseini, M. R., Golizadeh, H., & Sankaran, S. (2017). Critical success factors (CSFs) for integration of sustainability into construction project management practices in developing countries. *International Journal of Project Management*, *35*(6), 1103–1119. https://doi.org/10.1016/j.ijproman.2017.01.014.

Boateng, P., Chen, Z., & Ogunlana, S. O. (2015). An analytical network process model for risks prioritisation in megaprojects. *International Journal of Project Management*, *33*(8), 1795–1811. https://doi.org/10.1016/j.ijproman.2015.08.007.

Brady, T., & Davies, A. (2014). Managing structural and dynamic complexity: A tale of two projects. *Project Management Journal*, *45*(4), 21–38. https://doi.org/10.1002/pmj.

Chih, Y., & Zwikael, O. (2015). Project benefit management: A conceptual framework of target benefit formulation. *International Journal of Project Management*, *33*(1), 352–362.

Chou, J.-S., & Yang, J.-G. (2012). Project management knowledge and effects on construction project outcomes. *Project Management Journal*, *43*(5), 47–67.

Cuhls, K. (2004). Delphi method. *Fraunhofer Institute for Systems and Innovation Research, Germany*, 93–113. https://doi.org/10.1111/j.1539-6924.2009.01325.x.

Davies, A., & Mackenzie, I. (2014). Project complexity and systems integration: Constructing the London 2012 olympics and paralympics games. *International Journal of Project Management*, *32*(5), 773–790.

Demirkesen, S., & Ozorhon, B. (2017). Impact of integration management on construction project management performance. *International Journal of Project Management*, *35*(8), 1639–1654. https://doi.org/10.1016/j.ijproman.2017.09.008.

Eduardo Yamasaki Sato, C., & de Freitas Chagas Jr, M. (2014). When do megaprojects start and finish? Redefining project lead time for megaproject success. *International Journal of Managing Projects in Business*, *7*(4), 624–637. https://doi.org/10.1108/IJMPB-07-2012-0040.

Flyvbjerg, B. (2014). What you should know about megaprojects and why: An overview. *Project Management Institute*, *45*(2), 6–19. https://doi.org/10.1002/pmj.

Harris, F., & McCaffer, R. (2013). *Modern construction management (Google eBook)*. https://doi.org/10.1515/9783990434550.

International Federation of Consulting Engineers. (2005). Conditions of contract for construction for building and engineering works designed by the employer. FIDIC.

Keil, M., Lee, H. K., & Deng, T. (2013). Understanding the most critical skills for managing IT projects: A Delphi study of IT project managers. *Information and Management*, *50*(7), 398–414. https://doi.org/10.1016/j.im.2013.05.005.

Koskela, L. (1992). Application of the new production philosophy to construction. *Stanford University, Center for Integrated Facility Engineering, Technical Report*, 1–81.

Lei, Z., Tang, W., Duffield, C., Zhang, L., & Hui, F. K. P. (2017). The impact of technical standards on international project performance: Chinese contractors' experience. *International Journal of Project Management*, *35*(8), 1597–1607. https://doi.org/10.1016/j.ijproman.2017.09.002.

Lind, H., & Brunes, F. (2015). Explaining cost overruns in infrastructure projects: A new framework with applications to Sweden. *Construction Management and Economics*, *33*(7), 554–568. https://doi.org/10.1080/01446193.2015.1064983.

Liu, L., Borman, M., & Gao, J. (2014). Delivering complex engineering projects: Reexamining organizational control theory. *International Journal of Project Management, 32*(5), 791–802. https://doi.org/10.1016/j.ijproman.2013.10.006.

Locatelli, G., Mikic, M., Kovacevic, M., Brookes, N., & Ivanisevic, N. (2017). The successful delivery of megaprojects: A novel research method. *Project Management Journal, 48*(5), 78–94.

Mok, K. Y., Shen, G. Q., & Yang, J. (2015). Stakeholder management studies in mega construction projects: A review and future directions. *International Journal of Project Management, 33*(2), 446–457. https://doi.org/10.1016/j.ijproman.2014.08.007.

Mok, K. Y., Shen, G. Q., Yang, R. J., & Li, C. Z. (2017). Investigating key challenges in major public engineering projects by a network-theory based analysis of stakeholder concerns: A case study. *International Journal of Project Management, 35*(1), 78–94. https://doi.org/10.1016/j.ijproman.2016.10.017.

Pasquire, C. (2012). Positioning lean within an exploration of engineering construction. *Construction Management and Economics, 30*(8), 673–685. https://doi.org/10.1080/01446193.2012.689431.

PMI. (2017). PMBOK 6: *The 10 knowledge areas & 49 processes*.

Rolstadås, A., Tommelein, I., Morten Schiefloe, P., & Ballard, G. (2014). Understanding project success through analysis of project management approach. *International Journal of Managing Projects in Business, 7*(4), 638–660. https://doi.org/10.1108/IJMPB-09-2013-0048.

Saunders, F. C., Gale, A. W., & Sherry, A. H. (2016). Mapping the multi-faceted: Determinants of uncertainty in safety-critical projects. *International Journal of Project Management, 34*(6), 1057–1070. https://doi.org/10.1016/j.ijproman.2016.02.003.

Schmidt, R. C. (1997). Managing Delphi surveys using nonparametric statistical techniques. *Decision Sciences, 28*(3), 763–774. https://doi.org/10.1111/j.1540-5915.1997.tb01330.x.

Skinner, R.., Nelson, R. R.., Chin, W. W.., & Land, L.. (2015). The Delphi method research strategy in studies of information systems. *Communications of the Association for Information Systems, 37*(July), 31–63.

Wang, J., Shou, W., Wang, X., & Wu, P. (2016). Developing and evaluating a framework of total constraint management for improving workflow in liquefied natural gas construction. *Construction Management and Economics, 34*(12), 859–874. https://doi.org/10.1080/01446193.2016.1227460.

Wauters, M., & Vanhoucke, M. (2016). A study on complexity and uncertainty perception and solution strategies for the time/cost trade-off problem. *Project Management Journal, 47*(4), 29–50.

Westin, S., & Päivärinta, T. (2011). Information quality in large engineering and construction projects : a Delphi case study. *ECIS Proceedings*.

Winch, G. M. (2013). Escalation in major projects: Lessons from the Channel Fixed Link. *International Journal of Project Management, 31*(5), 724–734. https://doi.org/10.1016/j.ijproman.2013.01.012.

Zidane, Y. J.-T., & Olsson, N. O. E. (2017). Defining project efficiency, effectiveness and efficacy. *International Journal of Managing Projects in Business, 10*(3), 621–641. https://doi.org/10.1108/IJMPB-10-2016-0085.

Cost inefficiency of large and medium industries in Indonesia

A. Riyardi
Economics Study Program, Universitas Muhammadiyah Surakarta, Surakarta, Indonesia

Triyono
Accounting Study Program, Universitas Muhammadiyah Surakarta, Surakarta, Indonesia

Triyono
Economics Study Program, Universitas Muhammadiyah Surakarta, Surakarta, Indonesia

ABSTRACT: Since 2009, the number of large and medium industries in Surakarta has decreased by 27.4% because of cost inefficiency problems. The purpose of this research is to estimate and analyze the cost inefficiency. The method is specifying a 2010–2015 panel data stochastic cost frontier model. The results show that the model is characterized by a linear homogeneity in factor price, a truncated half-normal distribution of cost inefficiency error term data and time-varying panel data. The results also show that in 2010, the industry experienced very high cost inefficiency. Since 2011, cost inefficiency had decreased sharply. The cause was governmental policies such as monetary and fiscal policies, and efforts of the industry such as cost and production adjustments. Another result is that textile (International Standard Industrial Classification (ISIC 13), textile product (ISIC 14), printing (ISIC 18), chemistry product, (ISIC 20), rubber product (ISIC 22) and other (ISIC 32) industries have a higher cost inefficiency level than food (ISIC 10) and furniture (ISIC 31) industries. This indicates that the first group finds it more difficult to adjust to a crisis, costs and production than the second group.

Keywords: stochastic cost frontier, linear homogeneity in the factor price function, time-varying panel data model, cost inefficiency, industry

1 INTRODUCTION

The economic activity of an industry is processing raw materials. The processing can be mechanical, chemical or manual. This includes changing from a low value product to a higher value product. The result of the process can be intermediate or end-user products. Therefore, an industry is specified by inputs, production processes and product characteristics (Wikipedia Indonesia, 2016).

The large and medium industries (henceforth is called the industry) of Surakarta are classified according to a global classification called the International Standard of Industrial Classification (ISIC). The industry consists of eight classes, including food (ISIC 10), textile, (ISIC 13), textile product (ISIC 14), paper and printing (ISIC 18), chemistry product (ISIC 20), rubber product (ISIC 22), furniture (ISIC 31) and other (ISIC 32) industries (Badan Pusat Statistik Kota Surakarta, 2015, p. 15).

The industry contributes to Surakarta city's economy. It contributes over 19% to Surakarta's gross domestic product. The contribution is higher than other sectors such as the electricity, gas and water sectors, the building sector, the transportation and communication sectors or the finance and service sectors that contribute 10% to 15% to Surakarta's gross domestic product. Moreover, industry contribution is far higher than the agriculture and mining sectors that contribute less than 1% to Surakarta's gross domestic product. The contribution percentage of the industry sector is lower than the trade, hotel and restaurant sectors that contributes more than 26% to Surakarta's gross domestic product (Badan Pusat Statistik Kota Surakarta, 2012).

The industry has decreased since 2009. At that time, the number of firms was 182, but declined to 132 in 2014 (Badan Pusat Statistik Kota Surakarta, 2015, p. 13). The industry fell by about 27.47%. This means that the industry experienced an economic problem.

The economic problem faced by the industry should be noticed. The reason is, as shown by Table 1, other municipalities around Surakarta city such as Sukoharjo, Boyolali, Klaten and Sragen municipalities experienced an industrial increase. Since 2009, the Klaten municipalities' industrial increase was 4%, Sukoharjo municipality's industrial increase was 76.7%, while the Boyolali municipality's industrial increase was more than 100%. Even, Sragen municipality that does not focus on large and medium industries, experienced an industrial increase.

Table 1 also shows industry in Karanganyar and Wonogiri municipalities. They seem to have a similar problem to the industry sector. However, the problem is different, indicating that attention should be paid to the industry problem.

The decrease of the Wonogiri municipality industry was more than 45%. However, the number of large and medium industrial firms is less than 30. This indicates that Wonogiri municipality does not focus on industry. It is different to the industry number of Surakarta city where the number is greater than 100 firms, indicating that Surakarta city focuses on industry. Therefore, the decrease in the number of large and medium industries in Wonogiri is different to the decrease in number of the industry sector.

The decreasing number of firms faced by the large and medium industries of Karanganyar municipality is also different to the decreasing number faced by the industry. As Table 1 shows, the number of large and medium industries in the two regions is relatively similar. In addition, the decrease rate can be seen as a similar decreasing. However, the industry is in a city, while industry in Karanganyar municipality is in a rural area. Therefore, the problem faced by the industry is different to its counterpart in the Karanganyar municipality.

The problem faced by the industry is a unique problem because it only occurs in Surakarta city. The problem does not occur in other regions around Surakarta city. The problem should be noticed and solved because the industry contributes to Surakarta city's economy.

The problem experienced by the industry includes increased costs caused by the global and Indonesian 2008 financial crisis. Some industrial firms felt that the increasing costs caused their average costs to be higher than their prices. They chose to close down or relocate their firms outside Surakarta city. Other firms tried to operate as usual by adjusting costs and products, although they experienced a cost inefficiency problem. Focusing on the industrial firms that chose to operate as usual, the objective of this research is analyzing the cost inefficiency of the industry after the 2008 financial crisis.

A cost efficiency framework is expected to explain and analyze the problem faced by the industry. It is derived from cost theory and function that assumes a firm always minimizes its costs. At a given production, the firm buys the cheapest priced inputs so that its cots are

Table 1.   The change of Surakarta's large and medium industries.

City/Municipality	Number of large and medium industrial firms		Change
	2009	2014	
Surakarta	182	132	−27.5%
Sukoharjo	235	413	76.7%
Klaten	126	131	4%
Sragen	13	19	46.15%
Boyolali	65	144	121.54%
Wonogiri	29*	16	−44.83%
Karanganyar	155**	122	−21.3%

Note: The data are sourced from the statistics for each city and municipality in 2009 and 2014, published by Badan Pusat Statistik. * = 2007. ** = 2011.

minimized. The theory can be written mathematically as Equation 1, which is modified from Silberberg (1978, p. 176):

$$C = C^*\left(pi_1, pi_2, \ldots pi_n, q_0 \right) \tag{1}$$

where C is the cost paid by the firm, C (.) is the cost function, * means the cost is minimized, pi_1 is the price of input 1, pi_2 is the price of input 2, pi_n is the price of input n and q_0 is a given product produced by the firm.

Imposing a linear homogeneity function assumption, Equation 1, except q_0, should be divided by one of the input prices. Supposing that Equation 1, except q_0, is divided by pi_1, the resulting equation is:

$$c = c^*(p'_{i_2}, \ldots p'_{in}, y_0) \tag{2}$$

where c is C/Pi_1, c^*(.) is a linear homogeneous cost function, p'_{i_2} is p'_{i_2}/p_{i_1}, p'_{i_n}/p_{i_1} is p_{i_n}/p_{i_1} and q_0 is similar to the previous definition.

The framework can be divided into a cost efficiency method and model. The cost efficiency method focuses on the method to measure the cost efficiency, while the cost efficiency model tries to model the cost efficiency equation.

Farrell (1957) was the first to construct and measure the efficiency method. Using an iso-quant–isocost approach he argued a productive efficiency measurement as an alternative to the old and incomplete measurement. He interpreted efficiency as price and technical efficiencies. After that, he measured the technical efficiency of US agriculture based on the Cobb–Douglas production function assumption.

Aigner, Lovell and Schmidt (1977) developed a stochastic frontier model. They found that efficiency can be modeled econometrically called a Stochastic Production Frontier (SPF) model. An efficiency parameter can be involved in the parametric production function. The econometric equation can be run by Maximum Likelihood Estimation (MLE) and the technical efficiency can be measured.

Schmidt and Lovell (1979) developed a Stochastic Cost Frontier (SCF) model. The development was based on the SPF model. The efficiency error term data were distributed by a half-normal distribution, $u_i \sim N^+(0, \sigma_u^2)$. The assumption behind the SCF model is the indirect cost function as required by the cost theory. The efficiency measured is the allocative efficiency that a firm will allocate its cost to the cheapest price of inputs.

Stevenson (1980) generalized the distribution of the efficiency error term data. He estimated a truncated half-normal distribution, $u_i \sim N^+(\mu_i, \sigma_u^2)$. He found that the truncated half-normal distribution of the efficiency error term data can be used to form the SPF model as well as the half-normal distribution. However, the truncated half-normal distribution is appropriate for the SCF model.

Liu and Zhuang (1998) analyzed Chinese state owned manufacturing enterprises' cost efficiency after the new regulation of ownership and privatization was launched. The SCF model was derived from a linier homogeneous translog cost function. The efficiency error term data distribution was assumed as truncated half-normal data distribution. The cost efficiency was measured based on the assumption of time-invariant panel data. The observed data were 769 panel data from 1980 to 1989. The result was that cost efficiency increased by 1.18% annually.

Mac Mullen and Lee (1999) compared US trucking industry cost efficiency before and after the 1980 deregulation. The SCF model was a cross section SCF model. The SCF model was formed for each year. The model was a translog cost function. The efficiency error term data distribution was assumed as a half-normal data distribution. The data included all motor carrier firms data from 1976 to 1987. The data were divided into 1976–1981 data and 1982–1987 data. The result showed that the deregulation could not improve the industrial cost efficiency.

Khanna, Mundru and Ullah (1999) estimated the cost inefficiency of Indian electricity firms. The estimation was approached by parametric/frontier and semi parametric/Generalized

Least Square (GLS) approaches. The SCF model was derived from a linear homogeneous translog cost function. The efficiency error term data distribution was assumed as a truncated half-normal data distribution. The cost efficiency was measured based on the assumption of time-invariant panel data assumption. The data observed included 198 panel data, consisting of 66 electricity firms' data for three years. The result was the high cost inefficiency of Indian electricity firms. Based on the frontier approach, the cost was 43% to 48% higher than the minimum cost.

Anderson and Kabir (2000) analyzed cost efficiency of public schools in Nebraska. The data observed included 274 panel data, consisting of 90 schools' data for three years. The SCF model was derived from a cost function. Two models were developed. The first was the panel data SCF model, and the second was the cross section of a yearly observation SCF model. The efficiency error term data distribution was assumed as a truncated half-normal data distribution. The cost efficiency was measured based on the assumption of a time-varying panel data assumption. The result showed that the cost efficiency of Nebraska public schools can be increased from 33.59% to 24.83% above the minimum cost.

Li and Rosenman (2001) analyzed the cost efficiency of Washington state hospitals. The SCF model was derived from a linear homogeneous translog cost function. The efficiency error term data distribution was assumed as a truncated half-normal data distribution. A Feasible Generalized Least Square (FGLS) approach was used to estimate the cost efficiency. The data observed included 90 panel data consisting of 15 hospitals for six years. The result showed that the cost inefficiency was 33% more expensive than the minimum cost.

Farsi and Filippini (2003) analyzed Swiss electricity firms' cost efficiency. The SCF model was derived from a linear homogeneous translog cost function. The efficiency error term data distribution was assumed as a truncated half-normal data distribution. The Deterministic Cost Frontier (DCF) function was also determined. The cost efficiency was measured based on the assumption of a time-invariant panel data assumption. The data observed included 380 panel data panel consisting of 59 firms for nine years. The result showed that the SCF model was better than the DCF model. Based on the SCF model, the average cost of Swiss electricity firms was 1.15% more expensive than the minimum cost.

Hormazábal and Van de Wyngard (2007) analyzed Chilean electricity firms' cost efficiency after efficiency regulation launching. The SCF model was derived from a cost function. The efficiency error term data distribution was assumed as a truncated half-normal data distribution. The data observed included 35 electricity firms. The result showed that the cost inefficiency was 15.9% more expensive than the minimum cost.

Ogundari (2010) analyzed Nigerian sawmill firms' cost efficiency. The SCF model was derived from a linear homogeneous translog cost function. The efficiency error term data distribution was assumed as a truncated half-normal data distribution. The data observed included 160 sawmill firms' data. The result showed that the cost inefficiency was 26% more expensive than the minimum cost.

Hidayah, Hanani, Aninditya and Budi (2013) analyzed the integral management of the Buru municipality of Maluku province of Indonesia paddy farming. The SCF model was derived from a linear homogeneous translog cost function. The efficiency error term data distribution was assumed as a truncated half-normal data distribution. The efficiencies analyzed were cost and production. The data observed included 120 paddy farms' data. The result indicated the cost inefficiency was 20% more expensive than the minimum cost.

Heimeshoff and Schreyögg (2014) analyzed technical and cost efficiencies of physician practices. The SCF model was derived from a linear homogeneous translog cost function. The efficiency error term data distribution was assumed as a truncated half-normal data distribution. The cost efficiency was measured based on the assumption of a time-varying panel data assumption. The data observed included 4,003 physician practices for three years from 2006 to 2008. The result showed that the cost efficiency was related to technical efficiency. If the technical efficiency increased, then the cost efficiency was stable or decreased. Cost efficiency was influenced by capital goods buying.

Based on the above research, the usage of the SCF model based on the MLE regression equation should recognize five important things. The first is recognizing the linear homogeneous cost function as shown above. Recognition is needed in order to fulfill the return to scale assumption stating that a change in an input price causes a change in cost proportionally. The second is recognizing the large number of data. Almost of all the research employs large data numbers. This is to avoid inefficient use of the data processing tool. As mentioned by Gujarati (2003, pp. 616–619, p. 484), the MLE regression equation is appropriate to process a large number of data. The third is recognizing the type of SCF regression equation. The system equation as suggested by Bauer (1987) is an alternative, however, the practice one is the standard simultaneous equation used by all the above research. The fourth is recognizing the cost efficiency error term data distribution. Almost all the research employed truncated half-normal distribution as suggested by Stevenson (1980). They did not employ a half-normal distribution that is appropriate to estimate technical efficiency as suggested by Aigner, Lovell and Schmidt (1977) and Jondrow, Lovell, Materov and Schmidt (1982). The fifth is recognizing the technique to estimate the cost efficiency level for panel data. If panel data are assumed as time-varying data, then the cost efficiency is measured by $U_{it} = (U_i exp(-\eta(t - T)))$ as mentioned by Coelli (2007, p. 5). However, if panel data are assumed as time-invariant data, then the level of the cost efficiency is measured by assuming $\eta = 0$.

This research follows previous SCF research. The SCF model is built based on the linear homogeneous cost function to fulfill the assumption of the return to scale, the truncated half-normal data distribution of cost efficiency error term to fulfill the assumption that industry cannot completely minimize the cost, and the time-varying panel data cost efficiency estimation. Imposing all of these assumptions ensures that the industry is well described.

To the best of our knowledge, this research is the first in Indonesia to analyze the cost efficiency of the industry based on the SCF approach. Other research analyzed the cost efficiency of the industry based on a Corrected Ordinary Least Square (COLS) approach, analyzed the technical efficiency of the industry based on SPF approach or analyzed the cost efficiency in the agriculture sector. Research analyzing the cost efficiency of the industry based on COLS was conducted by Riyardi, Triyono and Triyono (2017), research that analyzed the technical efficiency of the industry based on the SPF approach was conducted by Pitt and Lee (1981), Wajdi (2012) or Riyardi, Wardhono, Wahyuddin and Romdhoni (2015), while research analyzing the cost efficiency of the agriculture sector was conducted by Hidayah, Hanani, Aninditya and Budi (2013).

The aim of this paper is to construct a SCF model of Surakarta city industry and to measure the industry cost inefficiency. The assumption of the SCF model is a Cobb–Douglas, homogeneous in factor price, truncated half-normal and time-varying frontier model. Another assumption is that cost inefficiency is measured as the rest of the cost efficiency.

The results show that the cost inefficiency of the industry was affected by the 2008 financial crisis. A treatment and adjustment to the 2008 financial crisis causes the cost inefficiency to decrease. Another result indicates that the industry can be classified based on the difficulty of the cost inefficiency to adjust to the financial crisis. All the results show the importance of understanding the industry cost inefficiency and its source, such as a financial crisis.

Section 2 introduces the method to develop the SCF model and to measure the cost inefficiency level. Section 3 introduces the results that are continued by the analysis. Section 4 concludes the analysis of the cost inefficiency of the industry after the 2008 financial crisis.

2 RESEARCH METHOD

The SCF model of the industry, characterized by a panel of cross section and time series data, is developed. In addition, the model is characterized by seven input prices and output variables relying on the Cobb–Douglas function. Furthermore, the composite error term consists of statistical noise and cost efficiency error terms. The cost efficiency error term has a truncated half-normal data distribution. The model is:

$$C_{it}(Q, Pi) = A_{it} q_{it}^{\beta_1} P1_{it}^{\beta_2} P2_{it}^{\beta_3} P3_{it}^{\beta_4} P4_{it}^{\beta_5} P5_{it}^{\beta_6} P6_{it}^{\beta_7} P7_{it}^{\beta_8} e_{it}^{v+u} \qquad (3)$$

where C is the industry real cost, C(.) is a Cobb–Douglas cost function, A is a constant, Q is the industrial real value of the output, P1 is the industrial real wage, P2 is the industrial gasoline real price, P3 is the diesel fuel real price, P4 is the industrial oil real price, P5 is the industrial electricity real price, P6 is the industrial raw materials real price, P7 is the industrial capital goods real price, βi is the coefficient of variables, v is the two-sided statistical noise *error term* that is assumed as independent and identically distributed at $\mu_{it} \_ N(0, \sigma_u^2)$, u is the one-sided positive *error term* showing the cost efficiency that is assumed as having a truncated half-normal distribution, $\mu_{it} \_ |N^+(\mu > 0, \sigma_u^2)|$, i is the sub industry and t is time from $2010 = 1$ to 2015.

Input price data are collected from a division of the industrial expenditure of inputs to the amount of industrial inputs. The division is done for each industrial input. Equation 4 shows the division formula:

$$Pi_{it} = \frac{Er_{it}}{N_{it}} \tag{4}$$

where Pi is the real price of an industrial input, Er is the industrial expenditure of an input and N is the amount of the industrial input.

The industrial raw materials and capital goods real prices are approached from their real expenditures. That is because their units are varied. Some of the raw material inputs are in meter, kilogram or number units, while the capital goods inputs are in Indonesian Rupiah (IDR) currency.

The real value is better than the current value. The industrial cost, output, input prices and expenditures are better described in the real value measurement. The inflation effect can be minimized. The real value is obtained by deflating the current value. The deflation is done by using the 2010 inflation as the base for other years. Equations 5 to 7 show the formulas for the real cost, output and expenditure of inputs:

$$C = \frac{If_{2010}}{If_t} X \, C_{nm} \tag{5}$$

$$Q = \frac{If_{2010}}{If_t} X \, Q_{nm} \tag{6}$$

$$Pi = \frac{If_{2010}}{If_t} X \, Pi_{nm} \tag{7}$$

where C, Q and Pi are as mentioned above, If_{2010} is the 2010 inflation, If_t is the year t inflation and nm is the current value.

Table 2 shows the variables and their specific information. Every variable has a symbol and unit. In addition, they are separated based on the variable type. The variable explanation clarifies the variable definition.

The equation should be imposed by $P1_{it}$ dividing for both sides of the equation, except the Q variable. This is in order to fulfill the linear homogeneity assumption. Equation 8 expresses the function:

$$c_{it} = a_{it} q_{it}^{\beta_1} p2_{it}^{\beta_3} \, p3_{it}^{\beta_4} \, p4_{it}^{\beta_5} \, p5_{it}^{\beta_6} \, p6_{it}^{\beta_7} \, p7_{it}^{\beta_8} \, e_{it}^{v+u} \tag{8}$$

where $c_{it} = C_{it}/P1_{it}$, $a_{it} = A_{it}/P1_{it}$, $p2_{it} = P2_{it}/P1_{it}$, $p3_{it} = P3_{it}/P1_{it}$, $p4_{it} = P4_{it}/P1_{it}$, $p5_{it} = P5_{it}/P1_{it}$, $p6_{it} = P6_{it}/P1_{it}$ and $p7_{it} = P7_{it}/P1_{it}$.

Equation 8 can be transformed into a linear equation. Transforming into a natural logarithm, Equation 8 becomes a linear equation. The linear equation is:

$$\begin{aligned} Lnc_{it} = {} & Lna_{it} + \beta_1 \, Lnq_{it} + \beta_3 \, Lnp2_{it} + \beta_4 \, Lnp3_{it} + \beta_5 \, Lnp4_{it} + \beta_6 \, Lnp5_{it} \\ & + \beta_7 \, Lnp6_{it} + \beta_8 \, Lnp7_{it} + v_{it} + u_{it} \end{aligned} \tag{9}$$

Equation 9 will be applied to the industrial data. The result is the SCF model for Surakarta's large and medium industries. After that, the cost inefficiency can be estimated and analyzed.

Four types of statistical test will be conducted on the model. The first is the SCF model specification test. The Likelihood Ratio (LR) or LR one-sided error values will be compared to the Kodde and Palm statistic or $\chi 2$ statistic to reject the null hypothesis that the

Table 2. Variables and their specific information.

Name	Symbol	Variable type	Unit	Explanation
Real cost	C	Dependent	IDR	Real cost is the deflated current total cost
Real output value	Q	Independent	IDR	Real output value is the deflated current output value
Real wage	P1	Independent	IDR	Real wage is the deflated current wage
Gasoline real price	P2	Independent	IDR/liter	Gasoline real price is the deflated gasoline current price
Diesel fuel real price	P3	Independent	IDR/liter	Diesel fuel real price is the deflated diesel fuel current price
Oil real price	P4	Independent	IDR/liter	Oil real price is the deflated oil current price
Electricity real price	P5	Independent	IDR/KwH	Electricity real price is the deflated electricity current price
Raw material real price	P6	Independent	IDR/unit	Raw material real price is the deflated raw material current price
Capital goods real price	P7	Independent	IDR/unit	Capital goods real price is the deflated capital goods current price

model is not the SCF model. The second is the cost efficiency error term test. The t-value of the composite error term variance (σ^2) will be compared to its t-statistic to reject the null hypothesis that the composite error term variance is formed by statistical noise error term variance (σ_v^2). In addition, the cost efficiency error term variance (γ) should be tested. The t-value of the cost efficiency error term variance will be compared to its t-statistic to reject the null hypothesis that the cost efficiency error term variance does not exist. Furthermore, the data distribution of the cost efficiency error term (μ) should be checked. The t-value of the data distribution of the cost efficiency error term will be compared to its t-statistic to reject the null hypothesis that the cost efficiency error term data distribution is a half-normal distribution. The third is the panel data (η) test. The t-value of the panel data parameter will be compared to its t-statistic to reject the null hypothesis that the panel data are not time-varying panel data. The fourth is the independent variable signification test. The t-value of each independent variable will be compared to its t-statistic to reject the null hypothesis that an independent variable does not influence the dependent variable.

After checking, the model is used to estimate the cost efficiency and inefficiency. Two stages will be conducted. The first stage is the cost efficiency estimation. The annual sub industry cost efficiency is measured as well the average cost efficiency. Equation 10 shows the formula, as explained by Coelli (2007, p. 9). The second stage is the cost inefficiency estimation. The estimation is done by assuming that it is the residual of the estimated cost efficiency (Battese & Collie, 1995). Equation 11 is the formula to estimate the cost inefficiency:

$$Eff_{it} = \frac{E\left(C_{it} \mid v_{it}, u_{it}, Q_{it}, Pi_{it}\right)}{E\left(C_{it} \mid v_{it}, Q_{it}, Pi_{it}\right)} = u_{it}(\eta(t-T)) \tag{10}$$

$$\text{Ineff}_{it} = \text{Eff}_{it} - 1 \tag{11}$$

where Eff is the cost efficiency, Ineff is the cost inefficiency, and i,t, Q, Pi and v are similar to the previous definitions.

3 RESULT AND ANALYSIS

3.1 Results

The industry SCF linear model is specified. The SCF model is developed from the industry Ordinary Least Square (OLS) cost function model. Equation 12 shows the OLS cost function model, while Equation 13 shows the SCF model:

$Lnc_{it} = -6.41 - 0.009\ LnQ_{it} + 0.842\ Lnp2_{it} + 0.006\ Lnp3_{it} + 0.450\ Lnp4_{it} - 0.001\ Lnp5_{it}$

$+ SD:\quad (1.41)\ (0.004)\qquad (0.076)\qquad\quad (0.005)\qquad (0.453)\qquad (0.007)$

t-value: $(-4.56)^*\ (-1.79)^{**}\qquad (11.08)^*\qquad\quad (1.36)\qquad (0.993)\qquad (-0.227)$

$\qquad\qquad 0.326\ Lnp6_{it} - 0.011\ Lnp7_{it} + u_{it}$ (12)

SD: (0.005)   (0.006)

t-value: (0.708)   (−1.823)**

* = Significant at α level = 5%

** = Significant at α level = 10%

$Lnc_{it} = -6.55 - 0.006\ LnQ_{it} + 0.83\ Lnp2_{it} + 0.005\ Lnp3_{it} + 0.005\ Lnp4_{it} - 0.002\ Lnp5_{it} +$

$SD:\quad (0.03)\ (0.003)\qquad (0.006)\qquad (0.004)\qquad\quad (0.004)\qquad (0.006)$

t-value: $(-.18)^*\ (-1.620)^{**}\qquad (13.55)^*\quad (1.27)\qquad\quad (1.28)\qquad (-0.418)$

$\qquad\qquad 0.415\ Lnp6_{it} - 0.008\ Lnp7_{it} + v_{it} + u_{it}$ (13)

SD: (0.414) (0.005)

t-value: (1.001) (−1.591)

Sigma Squared (σ^2): 0.42   Gamma (γ): 0.002   Mu (μ): 0.052   Eta (η): 0.466

SD: (0.108) SD: (0.007) SD: (0.058) SD: (0.178)

t-value: (3.886)* t-value: (0.243) t-value: (0.898) t-value: (2.611)**

LRrs: 49.332   LRur: 47.733   LR one-sided error: 3.20

* = Significant at α level = 5%

** = Significant at α level = 10%

Statistical tests were conducted on the industrial SCF linear model as specified by Equation 13. The first test was the specification test. The statistical test to test that the SCF model informs that the null hypothesis was accepted. The model was not the SCF model. This can be known from the LR test. As shown by Table 3, the $-2(-LR_2+LR_{ur})$ value was smaller than the Kodde and Palm statistic. In addition, the LR *one-sided error* value was smaller than χ^2 statistic. The LR *one-sided error* was 3.2, which is smaller than 7.045 as in the Kodde and Palm statistic (Kodde & Palm, 1986). It was also smaller than 7.185 as in the χ^2 statistic (Gujarati, 2003, p. 969).

The second test was the cost efficiency error term test. The test consists of the overall error term variance, the cost efficiency error term variance and the cost efficiency error term data distribution tests. The test of the overall error term variance (σ^2) shows that the null hypothesis that the error term is formed by only statistical noise error term was rejected. It means that the error term is a composite error. As shown by Table 3, the t-value of σ^2 is 3.886, while its t-table is 1.96 at a 5% α level. The test of the cost efficiency error term variance (γ) shows that the null hypothesis was accepted. The cost efficiency error term variance (γ) does not exist. As shown by Table 3, the t-value of γ is 0.243, while the t-table is 1.96 at a 5% α level. The test of the cost efficiency error term data distribution (μ_v) shows that the null hypothesis

Table 3.   Statistical tests of Surakarta's large and medium industries' SCF model.

Test	Coefficient	Error standard	t-value	
			Statistic	Table
Sigma Squared (σ^2)	0.42	0.108	3.886	1.96*
Gamma (γ)	0.002	0.007	0.243	1.96
Mu (μ)	0.052	0.058	0.898	1.96
Eta (η)	0.466	0.178	2.611	1.645**
$-2(LR_{rs}-LR_{ur})$	–	–	$-2(-49.332 + 47.733) = 3.198$	7.045[]
LR *one-sided error*	–	–	3.20	or 7.815 [][]

Note: * = Significant at 5%, ** = Significant at 10%, [] = Table of Kodde and Palm (1986), 3 restriction, 5% [][] = Table of χ^2, 3 restriction, 5% (Gujarati, 2003, p. 969).

of the greater than zero mean was rejected. It means that the data distribution is not a truncated half-normal data distribution. As shown by Table 3, the t-value of μ_v is 0.898, while its t-table is 1.96 at a 5% α level.

The third was the time-varying panel data test. The test shows that the null hypothesis of the time-invariant panel data was rejected. It means that that the panel data were time-varying panel data. As shown by Table 3, the t-value of η is higher than its t-table at a 10% α level.

The fourth was the test of the signification of the intercept and independent variables. The individual test of the intercept, real output (q) and gasoline real price (p3), shows that the null hypothesis of no relationship to the real cost (c) was rejected. The intercept, real output and gasoline real price influence the industry real cost. However, the sign of the intercept and real output was negative. As Equation 13 shows, the t-values of them were higher than their t-table at a 5% and 10% α level.

The individual test of the diesel fuel real price (p3), the oil real price (p4), the electricity real price (p5), the raw material real price (p6) and the capital goods real price (p7) shows that null hypothesis of no relationship to the real cost was accepted. This means that the diesel fuel, the oil, the electricity, the raw material (p6) and the capital goods real prices do not influence the industrial real cost. As shown by Table 3, their t-values are smaller than the t-table at a 5% α level. Equation 13 also shows that the signs of electricity and capital goods real prices are negative.

3.2 *Efficiency and inefficiency levels of Surakarta's large and medium industries*

The industry was not efficient in 2010. The average cost efficiency was 1.795. It was 0.795 higher than the minimum cost. Even, the cost efficiency of the rubber product industry (ISIC 22) was 2.07, while the cost efficiency of chemistry product industry (ISIC 20) was two. The average cost efficiency in 2011 was 1.44, while in 2015 it was 1.056. The industry cost efficiency from 2010 to 2015 increased. Table 4 shows the annual industry cost efficiency.

The increasing industry cost efficiency means that the cost inefficiency for those years had decreased. The decrease informs two things. The first is informing the industry group based on their cost inefficiency. The textile (ISIC 13), textile product (ISIC 14), paper and printing (ISIC 18), chemistry product (ISIC 20), rubber product (ISIC 22) and other (ISIC 32) indus-

Table 4. Efficiency level of Surakarta's large and medium industries.

ISIC	2010	2011	2012	2013	2014	2015
10	1.54	1.31	1.18	1.11	1.07	1.04
13	1.9	1.49	1.28	1.17	1.1	1.06
14	2	1.54	1.31	1.18	1.11	1.07
18	1.82	1.45	1.26	1.15	1.09	1.06
20	1.76	1.42	1.24	1.14	1.09	1.05
22	2.07	1.57	1.32	1.19	1.12	1.07
31	1.57	1.32	1.19	1.11	1.07	1.04
32	1.7	1.39	1.22	1.13	1.55	1.05

Table 5. Inefficiency level of Surakarta's large and medium industries.

ISIC	2010	2011	2012	2013	2014	2015
10	0.54	0.31	0.18	0.11	0.07	0.04
13	0.9	0.49	0.28	0.17	0.1	0.06
14	1	0.54	0.31	0.18	0.11	0.07
18	0.82	0.45	0.26	0.15	0.09	0.06
20	0.76	0.42	0.24	0.14	0.09	0.05
22	1.07	0.57	0.32	0.19	0.12	0.07
31	0.57	0.32	0.19	0.11	0.07	0.04
32	0.7	0.39	0.22	0.13	0.08	0.05

tries are a higher cost inefficiency group. The lowest level of cost inefficiency in this group is 0.7 for the chemistry product industry. The food (ISIC 10) and furniture (ISIC 31) industries are in the lower cost inefficiency group. Since 2011, the cost inefficiency has decreased. However, the group was similar to 2010.

The second is the level of the cost inefficiency decrease. The decrease level can be divided into three stages. The first is a very sharp cost inefficiency decrease from 2010 to 2011, the second is a sharp cost inefficiency decrease from 2011 to 2012 and the third is the slowing of the cost inefficiency decrease from 2012 to 2015. Figure 1 shows the level of the cost inefficiency decrease.

3.3 *Analysis*

The industry SCF model suffers from the model specification, the existence of cost efficiency error term and the significance of independent variables. The SCF model cannot be differentiated from the OLS model. The cost efficiency error term contributes to the composite error term, but its existence is not significant and its data are not distributed as a truncated half-normal distribution. Moreover, some independent variables do not influence the real cost, or their signs are negative.

The cause of the problems is the data number. The data number is 48. It is too small to be used in the SCF modeling that is based on the MLE. As explained by Gujarati (2003, p. 117), MLE needs a large number of data; 55 may be insufficient, 48 perhaps is also insufficient. The previous research outlined also employ a large number of data. Only Hormazábal and Van de Wyngard (2007) used a small number of data. However, their data were greater than this study.

Other effects of the small number of data lay on the slope coefficients of the model and the significance of the independent variables. As seen from Equations 12 and 13, the slope coefficients of the SCF model are different to the OLS model, whereas according to Coelli, Rao, Christhopher and Batesse (2005, pp. 223–224) the slopes should be similar. In addition, some variables are not significant, whereas all research, such as by Farsi and Filippini (2003), Khanna, Mundru and Ullah (1999), Liu and Zhuang (1998), Mac Mullen and Lee (1999) or Ogundari (2010) that observed the industrial sector with a large data number confirmed the significance of the variables.

The very high industry cost inefficiency since 2010 was caused by the industry cost increasing. The production cost increased due to the increasing input prices such as wages (p1), energy (p2, p3, p4 and p5), raw materials (p6) and capital goods (p7) price increases. Referring to Murtiningsih (2015), the input price increase was caused by the international and Indonesian 2008 crisis. When the crisis occurred in 2008, the exchange rate was more expensive, and the input prices also were more expensive.

An effort to handle the crisis was attempted by government and industry. The government launched monetary and fiscal policies. The government effort caused the industry costs to decline sharply. In 2015, the cost inefficiency was only 5% above minimum cost.

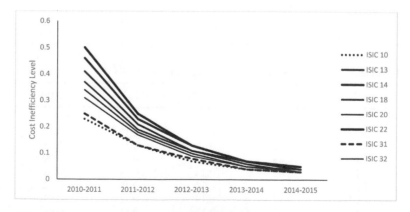

Figure 1.   The level of the industry cost inefficiency decrease.

516

The industry also handled the crisis. Some of them closed their industrial firms, while others relocated their firms outside Surakarta city. Consequently, as mentioned above, the number of industry firms declined by 27.4%.

Firms that still operated in Surakarta city handled the crisis by adjusting costs. They adjusted costs that were not related to their production, such as transportation and distribution costs. This is indicated by the significance of the intercept (A), but with a negative sign. The costs that were not related to production decreased, while in the same cases the total costs increased. This means that the industry tried to cut down costs that were not related to production.

Another cost adjustment was the reduction of production costs, especially the cost of electricity and capital goods usage. Their usage can be adjusted because they were not a stock. The electricity cost usage was adjusted based on the production need and its regulated price. The capital goods cost usage was adjusted in the form of adjusting the loan or rent costs of the capital goods. Consequently, a negative significant relationship exists between industry electricity real price – that was estimated from the division of its real spending to its usage and the capital goods real price that involved the loan and rent of the capital goods – to the industry real cost.

Another adjustment was the industry production adjustment. The industry output produced (q) decreased. Consequently, the real value of industry output influenced the real cost, but the sign was negative. The real value of output decreased, while the real cost increased.

Based on the decreasing cost inefficiency, the industry can be divided into two groups. The first group includes textile (ISIC 13), textile product, (ISIC 14), paper and printing (ISIC 18), chemistry product (ISIC 20), rubber product (ISIC 22) and other (ISIC 32) industries. The second group includes the food (ISIC 10) and furniture (ISIC 31) industries. The cost inefficiency of the first group is higher than the second group. The first group had more difficulty than the second group in terms of handling the crisis, cost adjusting and product adjustment.

4 CONCLUSION

This research analyzed Surakarta's large and medium industries' cost inefficiency. The industry SCF model was developed and cost inefficiency was estimated. The SCF model is characterized by a linear homogeneous Cobb–Douglas cost function, time-varying panel data and a truncated half-normal distribution.

The 2008 global and Indonesian crisis caused the industry to experience a very high cost inefficiency. The cost inefficiency decreased sharply because of government and industry efforts. The government implemented monetary and fiscal policies, while the industry adjusted costs and production.

The textile (ISIC 13), textile product, (ISIC 14), paper and printing (ISIC 18), chemistry product (ISIC 20), rubber product (ISIC 22) and other (ISIC 32) industries were different to the food (ISIC 10) and furniture (ISIC 31) industries. The first had higher cost inefficiency than the second. The first had difficultly handling the crisis, cost adjustment and production adjustment

The weakness of this research lies in the industry SCF model that is developed based on a small data number. The data number is only 48. Consequently, the SCF model cannot be differentiated from the OLS model, the cost efficiency error term is not significant to form the composite error and the variables relationship cannot be verified exactly.

REFERENCES

Aigner, D., Lovell, C.K. & Schmidt, P. (1977). Formulation and estimation of stochastic frontier production function models. *Journal of Econometrics*, 6, 21–37.
Anderson, J.E. & Kabir, M. (2000). *Public education cost frontier model: Theory and an application*. Nebraska, NE: Economics Department University of Nebraska-Lincoln.
Surakarta Statistic. (2012). *Surakarta in Figures 2012*. Surakarta, Indonesia: Surakarta Statistic.
Surakarta Statistic. (2015). *Statistic of Large and Medium Industries in Surakarta 2015. Surakarta*, Indonesia: Surakarta Statistic.

Battese, G.E. & Collie, T. (1995). A model for technical inefficiency effects in a stochastic frontier production function for data panel. *Empirical Economics*, 20, 325–332.

Bauer, P.W. (1987). *A technique for estimating a cost system that allows inefficiency*. Cleveland Ohio, OH: Federal Reserve Bank of Cleveland.

Coelli, T. (2007). *A guide to frontier 4.1: A computer program for stochastic frontier production and cost function estimation*. Armidale, Australia: Centre for Productivity Analysis (CEPA).

Coelli, T.J., Rao, D.P., Christhopher, O. & Batesse, G.E. (2005). *An introduction to efficiency and productivity analysis*. New York, NY: Springer.

Farrell, M. (1957). The measurement of productive efficiency. *Journal of the Royal Statistical Society*, 120(3), 253–290.

Farsi, M. & Filippini, M. (2003). *Regulation and measuring cost efficiency with panel data models: Application to electricity distribution utilities*. Zurich, Swiss: Centre of Energy Policy and Economics Swiss Federal Institute of Technology.

Greene, W.H. (2007, August Friday). *Stochastic frontier models*. Diambil kembali dari stern.nyu.edu: www.stern.nyu.edu

Gujarati, D.N. (2003). *Basic econometrics*. Boston, MA: McGraw-Hill.

Heimeshoff, M. & Schreyögg, J. (2014). Cost and technical efficiency of physician practices: A stochastic frontier approach using panel data. *Health Care Management Science*, 17(2), 150–161.

Hidayah, I., Hanani, N., Aninditya, R. & Budi, S. (2013). Production and cost efficiency analysis using frontier stochastic approach, a case study on paddy farming system with integrated plant and resource management (IPRM) approach in Buru District Maluku Province Indonesia. *Journal of Economics and Sustainable Development*, 4(1), 78–84.

Hormazábal, R.S. & Van de Wyngard, H.R. (2007). Frontier methodologies for the determination of efficiencies in distribution costs. *Ingeniare. Revista Chelina de Ingenieria*, 15(3), 220–226.

Jondrow, J., Lovell, C.K., Materov, I.S. & Schmidt, P. (1982). On the estimation of technical inefficiency in the stochastic frontier production function model. *Journal of Econometrics*, 19, 233–238.

Khanna, M., Mundru, K. & Ullah, A. (1999). Parametric and semi-parametric estimation of the effect of firm attributes on efficiency: The electricity generating industry in India. *Journal International Trade and Economic Development*, 8(4), 419–436.

Kodde, D.A. & Palm, F.C. (1986). Wald criteria for jointly testing for equality and inequality restrictions. *Econometrica*, 54(5), 1243–1248.

Li, T. & Rosenman, R. (2001). Cost inefficiency in Washington hospitals: A stochastic cost frontier with panel data. *Health Care Management Science*, 4, 73–81.

Liu, Z. & Zhuang, J. (1998). Evaluating partial reforms in the Chinese state industrial sector: A stochastic frontier cost function approach. *International Review of Applied Economics*, 12(1), 9–24.

Mac Mullen, B.S. & Lee, M.-K. (1999, September). Cost efficiency in the US motor carrier industry before and after deregulation: A stochastic frontier approach. *Journal of Transport Economics and Policy*, 33(3), 303–318.

Murtiningsih. (2015). Globalization and Its Correlation to Manufacturing Industry as a Leading Sector. *Jurnal Ilmiah Bisnis, Ekonomi dan Akuntansi*, 9(2), 33–39.

Ogundari, K. (2010). Estimating and analysing cost efficiency of sawmill industry in Nigeria. *China Agricultural Economic Review*, 2(4), 420–432. doi:10.1.108/17156371011097731.

Pitt, M.M. & Lee, L.-F. (1981). The measurement and sources of technical inefficiency in the Indonesian weaving industry. *Journal of Development Economics*, 9, 43–64.

Riyardi, A., Triyono & Triyono. (2017). Cost inefficiency estimation of large and medium industry in Surakarta based on corrected ordinary least square estimation. Samarinda, Indonesia: Fakultas Ekonomi dan Bisnis Universitas Mulawarman Kalimantan Timur.

Riyardi, A., Wardhono, A., Wahyuddin, M. & Romdhoni, A.H. (2015). Analysis of technical inefficiency of food and textile industries in Central Java Province. *Kongres ISEI XIX*. Surabaya, Indonesia: ISEI.

Schmidt, P. & Lovell, C.K. (1979). Estimating technical and allocative inefficiency relative to stochastic production and cost frontiers. *Journal of Econometrics*, 9, 343–366.

Silberberg, E. (1978). *The structure of economics*. New York, NY: McGraw-Hill Book Company.

Stevenson, R.E. (1980). Likelihood functions for generalized stochastic frontier estimation. *Journal of Econometrics*, 13(1), 57–66.

Wajdi, F. (2012). Efficiency analysis of small industry based on Stochastic Frontier Stochastic frontier analysis. *Benefit*, 16(1), 10–22.

Wikipedia Indonesia. (2016, Desember). *Wikipedia: about*. Retrieved from wiki/Final_good: https://id.wikipedia.org

Performance analysis of tinplate's main raw material procurement process using the SCOR (Supply Chain Operations Reference) model

R.D. Wulandari & R.D. Kusumastuti

Department of Management, Faculty of Economics and Business, Universitas Indonesia, West Java, Indonesia

ABSTRACT: This article analyzes the performance of tinplate's main raw material procurement process using the SCOR 11.0 framework. This analysis is conducted at PT X Tbk, the only tinplate manufacturer in Indonesia. As competition becomes tighter due to the global steel overcapacity which affects Indonesia's tinplate market (weakened by the dumping price issue from three countries: China, South Korea, and Taiwan), PT X Tbk keeps suffering financial injury despite Anti Dumping Import Duty (BMAD) having been applied since 2014. With tighter competition, PT X Tbk needs to excel in its supply chain performance, specifically in raw material procurement as the company fully relies on imports with a limited supply and a long lead time. Analysis is conducted using SCOR (Supply Chain Operations Reference) strategic metrics (reliability, responsiveness, and asset management efficiency), which all gave results below the stated targets. The results indicate that improper inventory management is the main factor causing poor performance.

Keywords: Performance analysis, procurement, tinplate, SCOR model, inventory management

1 INTRODUCTION

The global sector of the steel industry is still facing overcapacity problems that have lasted since China's massive expansion in 2000, which is producing over 2.300 m MT when global demand is only 1.500 m MT (Brun, 2016). This overcapacity greatly influences the profitability of the steel industry, due to the inclusion of steel at very cheap prices, and inflicts financial losses on many companies, as well as the notion of fairness in the steel industry.

In capital-intensive industries, such as the steel industry, high maintenance utilization is the key to success (Hill, 1994). Excess capacity in some countries far outstrips their domestic needs, inducing the need for expansion in new markets to absorb their production capacity in an effort to increase economic utilization, which is around 80% for steel mill (Brun, 2016). Indonesia, in particular, is a very attractive market for the world due to its very large population. This research will focus on tinplate, a derivative product of steel that is commonly used in the canned packaging industry because of its durability and resistance to rust. The main raw material of tinplate is Tin Mill Black Plate (TMBP).

It is undeniable that overcapacity in the global steel industry affects Indonesia's tinplate domestic market, which is also flooded with imported products from several countries. Especially with national consumption of around 250,000 tons/year and production capacity of only about 160,000 tons/year, the inclusion of imported products in the domestic market cannot be avoided. The huge amount of imported tinplate is the reason that PT X Tbk, as the only tinplate manufacturer in Indonesia, requested an investigation of tinplate imports from China, South Korea and Taiwan which were allegedly conducting the practice of dumping in 2012.

In February 2014, the government of Indonesia issued the regulation regarding the imposition of an anti-dumping import duty (BMAD) for the above three countries. Enforcement of the regulation was expected to protect the domestic industry from continuous losses. However, the implementation of this policy has not been effective as indicated by the financial losses which still happen, as well as by the high volumes of imports which remain dominated by these three countries in the last four years.

The above condition causes very tight competition for the price of tinplate. In addition, global business competition gives consumers many choices with their own competitive advantage. Thus, companies need to improve their operations to reduce costs and become profitable. One aspect that is very critical to the operations of tinplate producers is supply chain management. In the modern world, competition is no longer between organizations, but between supply chains (Trkman et al., 2010). Supply chain management serves as the foundation of overall organizational strategy to achieve and maintain competitive advantage (Kristal et al., 2010). The goal of supply chain management is to structure the supply chain to maximize its competitiveness and benefits for consumers (Heizer et al., 2017). Supply chain surplus is the difference that arises between the value generated for consumers and overall costs arising at all stages of the supply chain (Chopra & Meindl, 2013). Effective supply chain management is fundamental for companies in maintaining a sustainable competitive advantage, and to achieve it, overall supply chain performance measurement is indispensable (Hernandez et al., 2014).

Inventory management is also one of the most important operating management decisions and associated with supply chain management. In general, efficient or inefficient inventory management is one of the factors affecting the company's performance (Koumanakos, 2008). The purpose of inventory management is to balance between inventory investment and service to consumers (Heizer et al., 2017). Procurement of key raw materials is a critical component of the supply chain. Active management of upstream sourcing can create more value and produce more efficient supply chains (Agrawal, 2014).

In running its business, PT X Tbk is highly dependent on tinplate raw materials, namely TMBP, whose procurement can only be done through an import mechanism with very limited suppliers. Performance evaluation of the supply chain, especially for its main raw material's procurement process, is one of the strategic steps that can be done to ensure that the company's supply capability is sustainable. The performance evaluation also needs to be conducted to increase cost efficiency since the main raw material is the largest component of the production cost, even more than 50% of the total cost.

The objective of this research is to analyze the performance of the procurement process of TMBP for PT X Tbk, and to identify ways to improve the process. The analysis is conducted by using the Supply Chain Operational Reference (SCOR) model. It is expected that this paper will provide insights into the complexities faced by the tinplate manufacturer for survival in a very competitive market and offer ways to improve its supply chain performance.

The remainder of this paper is structured as follows: relevant literature is explained next, followed by discussions on research methodology, results and discussions, and conclusions.

2 LITERATURE REVIEW

Operations management is a series of activities to create value in the form of goods or services by converting inputs into outputs (Heizer et al., 2017). The entire operating process that transforms the input resource into an output can be analyzed at three levels; as a business itself, as part of a larger operating network, or at the level of individual processes in its operations (Slack & Lewis, 2015). The operations manager will apply the management process to any decisions made on the operations management function. One of the decision areas in operations management is supply chain management.

The supply chain is the last important frontier of business process engineering. The modern manufacturing process method has optimized its production process to reach the point where a competitive advantage will be obtained by companies that are able to optimize their distribution process (Taylor, 2004). In this highly competitive business condition, the company has no choice but to run the supply chain process well. The modern manufacturing process has reduced the time and cost of the production process and has made the supply chain the final frontier for cost reduction and competitive advantage (Taylor, 2004). The supply chain is a process that transfers information and materials from and to the process of manufacturing/service in a company (Jacobs & Chase, 2014). Therefore, actors in the supply chain include suppliers, manufacturing companies or service providers, distributors, wholesalers, and retailers distributing products to customers (Heizer et al., 2014). Supply chain management can be viewed as a philosophy which believes that every actor in the supply chain directly and indirectly affects the performance of other supply chain members, and eventually the overall supply chain performance (Cooper et al., 1997).

Supply chain management is a set of approaches that are used to efficiently integrate suppliers, manufacturing companies, warehouses and stores, until the goods are produced and distributed, at the right quantity, time and place with the goal of minimizing costs while meeting the service level (Simchi-Levi et al., 2004). Meanwhile, according to Vokurka et al. (2002), supply chain management is defined as all activities that include the delivery of goods from raw materials to consumers, including the procurement of raw materials and parts, manufacturing and assembly, inventory tracking and warehousing, order management and order entry, distribution through channels, delivery to customers, and the information systems needed to monitor the entire activity. According to Beamon (1998), the supply chain is divided into two main processes: production planning and inventory control processes. Nowadays, procurement teams face an increasing number of complex challenges, as reported by CASME members (the global membership network for corporate procurement) around the world during a series of round table meetings, such as risk management, reputation and brand image, CSR (corporate social responsibility), becoming a customer of choice and centers of excellence (CIPS, 2017).

Organizational performance measurement systems have an important role in business regulation, such as "No Measure, No Improvement" (Kaplan & Norton, 1992). Therefore, to evolve into an efficient and effective supply chain, a supply chain needs to be measured for its performance (Gunasekaran et al., 2001). Referring to Balfaqih et al. (2016), classification of a supply chain performance system is carried out on three main things: approach, technique, and performance criteria. The approach classification is divided into three main areas, namely the perspective-based approach, the process-based approach, and the hierarchical-based approach. The perspective-based approach is the most popular, and Balance Scorecard (BSC) and SCOR are examples of this approach. The process-based approach considers the performance of the core operation process in the supply chain, while the hierarchical-based approach evaluates the supply chain's performance through various levels of hierarchy. The most popular performance criteria are cost/finance, followed by internal process, customers, flexibility, innovation, reliability, response, time, asset management, quality, efficiency, resources, information and output. These criteria tend to be part of the BSC and SCOR models. The performance criteria which are associated with the SCOR and BSC models are generally the most appropriate for several supply chain contexts.

SCOR is a product of Supply Chain Council Inc. (SCC) that can be applied to review and contrast the tasks of the supply chain and its performance. The SCOR method serves a structure that connects business processes, metrics, best practice and technologies to an incorporated structure to promote the relationship among the supply chain actors, and to enhance the efficacy of supply chain management refinement (SCC, 2012). The SCOR method was evolved to portray related business tasks throughout the phases to meet customer needs. Improvement of the model are created as a resolution by SCC members, those alteration should be created to simplify the application of the model in practice. The SCOR model version 11.0 is the 13th revision since the model's launch in 1996. Revision 11.0 makes changes to three parts, namely metrics, processes and practices (SCC, 2012)

Researchers and practitioners found the SCOR model as a reference that integrates most of the business processes of an organization within a cross-functional framework (Kocaoglu et al., 2013). Measurement of the supply chain performance by the SCOR model is the most comprehensive, is well recognized in various industries and has been used by many companies to improve supply chain management (Kocaoglu et al., 2013). The SCOR model provides a systematic approach toe explaining, classifying, and evaluating complex supply chain processes (Erkan & Baç, 2011). Process modeling and performance measurement are crucial subjects in the SCOR model (Esin & Kocaoglu, 2016). Various industry verticals have implemented the SCOR model, such as aerospace & defense, automotive & industrial, computers, consumer goods, electronic equipment, energy/chemical & applied materials, medical devices & equipment, semiconductors and telecoms & network equipment (APICS, 2017).

3 METHODS

This study uses a combination of qualitative and quantitative approaches. The data used in this study are primary and secondary data. The primary data is obtained through observation in the area in the company which directly handles the procurement process for raw materials, and by interviewing informants that directly manage all the supply chain activities, namely sales and marketing, production planning control, logistics, and accounting divisions. The secondary data, on the other hand, are the marketing data to capture the atmosphere of tight business competition that the company needs to deal with, the sales data to get an idea of overall demand, the logistics data to capture the historical activities for raw material procurement, the warehouse data to understand the inventory policy, as well as the financial data to capture the loss that the company has experienced. All secondary data is recorded from 2015 until 2017.

The research stages start with the literature review and initial data collection to identify the main problems, then continue with an assessment of the supply chain performance in the procurement area by using SCOR 11.0 which consists of three levels as follows:

3.1 SCOR level 1

In this level, the objective is to describe the main scope and configuration of the supply chain, as well as evaluate strategic metrics against the company's targets or benchmark data. The performance target of the supply chain is determined as the basis for competing in the market. The scope of discussion of SCOR Level 1 includes six primary management processes, namely plan, source, make, deliver, return, and enable.

At this level, the performance attribute along with the strategic metric to be evaluated are determined. Concerning the customer-focused performance attribute, reliability is used as one of the attributes, with Perfect Order Fulfillment (POF) as its strategic metric. Another attribute is customer-focused responsiveness, with the Order Fulfillment Cycle Time (OFCT) as its strategic metric. Regarding the internal-focused performance, asset management efficiency is used as the last attribute, with Cash-to-Cash cycle time (C2C) and Return on Working Capital (RWC) specifically for inventory, as its strategic metrics. Table 1 shows the calculation formula that are used for the predefined strategic metrics, both referring to the SCOR guidelines as well as their adaptations for PT X Tbk.

3.2 SCOR level 2

The objective at this level is to describe all of the company's supply chain activities and identify potentially problematic activities for overall supply chain management. In SCOR Level 2, a more detailed configuration of SCOR Level 1 is made, and a capability assessment of the supply chain is made in the Make-to-Stock, Make-to-Order, or Engineer-to-Order environment. SCOR Level 2 mapping is conducted using a geographic map and a thread diagram.

3.3 SCOR level 3

The objective at this level is to describe the details of the potentially problematic activity identified at SCOR Level 2, focusing on plan and source activity, as well as to analyze the factors that cause the problem by using a fishbone diagram.

Table 1. SCOR strategic metric level 1 – PT X Tbk.

SCOR 11.0			PT X Tbk		
Performance attribute	Strategic metric—Lv 1	Formula	Strategic metric— Lv 1	Formula	Definition
Reliability	Perfect Order Fulfillment [RL. 1.1]	([Total Perfect Orders]/ [Total Number of Orders]) 100%	POF (%)	([Total quantity of TMBP received on-time]/ [Total quantity shipment plan]) 100%	Percentage of actual TMBP received on-time as planned without split shipments
Responsiveness	Order Fulfillment Cycle Time [RS. 1.1]	[Sum of Actual Cycle Times for All Orders Delivered] / [Total Number of Orders Delivered] in Days	OFCT (days)	[Actual Received Date TMBP] – [PO TMBP Issue Date]	Total time needed from Purchase Order TMBP (PO) issued until its arrival
Asset management efficiency	Cash-to-Cash Cycle Time [AM. 1. 1]	[Inventory Days of Supply] + [Days Sales Outstanding] – [Days Payable Outstanding] in days	C2C (days)	[Actual Sales Date] - [Actual Received Date TMBP]	Total time needed (aging period) from TMBP arrival until sold as finished goods to customer
	Return on Working Capital [AM. 1.3]	([Supply Chain Revenue] – [Total Cost to Serve])/([Inventory] + [Accounts Receivable] – [Accounts Payable])	Inventory (MT)	[Actual TMBP Stock]/ [Production Target]	Inventory stock level of TMBP as tonnage compared to production target

After calculating the score for each metric, benchmark analysis is carried out to determine the company's achievements. As the detailed problem is illustrated, the required improvements are identified.

4 RESULTS AND DISCUSSION

The object of this research, PT X Tbk, is the first and the only company in Indonesia to produce high quality tinplate at an international standard. In 2012, the company was modernized and increased its production capacity from 130,000 tons per year to 160,000 tons per year. The company is supported by high technology production facilities, with three main production lines: the electrolytic tinning line with 160,000 tons capacity/year, the shearing line with 70,000 tons capacity/year, and the scroll cutting line with 37,000 tons capacity/year. The company runs its business activities in a make-to-order manner, where orders from consumers trigger the production process.

In carrying out its business activities, the company engages many actors in its supply chain management from suppliers to final customers as can be seen in Figure 1. Steel works are the factories of steel mills, which directly undertake the production process of TMBP order from the company. Each steel works has its own production range as a guideline for the procurement of raw materials. All steel works which becomes the raw material suppliers of the company are currently located outside of Indonesia. Steel mills is the holding company that allocates TMBP raw material production to each steel works. The trading house works as the mediator between companies and steel mills and performs the related commercial activities. The shipping line which is designated by the trading house in coordination with the steel mills, is responsible for delivering raw materials by sea in vessels/containers to ports in Indonesia in accordance with the delivery schedule, agreed both by the steel mills and the company. The forwarder is an agency designated by the company to manage the process of receiving the imported raw materials in Indonesian ports, including customs services. 3PL is a

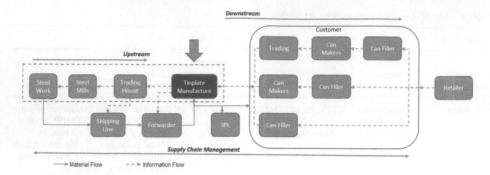

Figure 1. Supply chain PT X, Tbk.

logistics company which is responsible for delivering tinplate to the consumer warehouse. As for the customer, they are classified into three types, namely the trading one, the can maker and the can filler, from which finished goods will be distributed to the retailers.

TMBP, which is the main raw material of tinplate, can only be procured from foreign countries as there is currently no steel manufacturer in Indonesia able to produce it. TMBP is classified as special steel with a limited number of suppliers in the world, because it requires special technology in its manufacturing process. The limited number suppliers and high level of dependence on outside parties, make the process of procurement of this main raw material very crucial. Due to the characteristics of the supply which are quite different, especially in terms of lead time, in this study, the steel mill is divided into two main types, namely the integrated mill and the non-integrated mill. Broadly speaking, the integrated mill is an integrated production line, from steel-making to blast furnace to hot rolling, cold rolling, and finally TMBP. An integrated mill tends to have a number of steel works scattered in several locations, so that the production capacity tends to be larger. While a non-integrated mill is a mill which is not integrated with steel-making, so for the TMBP manufacturing process, it needs to procure raw material from other mills, such as Hot Rolled Coil (HRC) or Cold Rolled Coil (CRC). Table 2 presents the main raw material supplier data, along with the trading houses.

4.1 SCOR level 1

For SCOR Level 1, gap analysis is performed for each selected strategic metric. For the analysis of tinplate's main raw material procurement performance, four strategic metrics are selected. Gap analysis is performed to compare the actual historical conditions for the past three years, and the company's targets or the industry's benchmarks (if available) obtained from APQC (American Productivity & Quality Center), which is one of the most frequently used benchmark data sources that focuses on productivity and quality (Bolstorff & Rosenbaum, 2007). The metrics and their benchmarks are as follows:

1. Perfect Order Fulfillment (POF)
This strategic metric measures the company's ability to procure TMBP raw materials, which can be met in a timely manner and an appropriate amount, according to the original delivery plan. In the calculation process, split shipment for the balance quantity of each lot is not considered. Gap analysis is done by benchmarking with the industry, where for the top performer, the POF reaches 95%, the median performer is 90%, while the bottom performer is only about 82%.

2. Order Fulfillment Cycle Time (OFCT)
This strategic metric measures the time it takes from ordering the TMBP until it is received at the Indonesian port. Due to the very different mill characteristics of the lead time between integrated mills and non-integrated mills, the OFCT is calculated with separate data between the mills in order not to cause bias, and misinterpretation.

524

Table 2. Upstream players.

Type	Steel works	Steel mill	Country of origin	Trading house
Integrated	IT.A.1	IT.A	Japan	AA
	IT.A.2			AB
	IT.A.3			AC
	IT.B.1	IT.B		B
	IT.B.2			
	IT.C.1	IT.C	South Korea	C
	IT.C.2			
	IT.F.1	IT.F	China	F
	IT.F.2			
Non-integrated	NIT.E.1	NIT.E	Japan	AB
	NIT.D.1	NIT.D	Taiwan	D
	NIT.G.1	NIT.G	Thailand	AC

Table 3. Gap analysis of strategic metrics.

Performance attribute	Level 1 metric	Actual performance	Benchmark	Company's target	Gap
Reliability	Perfect Order Fulfillment (POF)	90%	95%	–	5%
Responsiveness	Order Fulfillment Cycle Time (OFCT)	81 days	–	75 days	6 days
Asset management efficiency	Cash-to-Cash Cycle Time (C2C)	64 days	30 days	–	34 days
	Return on Working Capital (Inventory)	Avg 1.5, St Dev 0.3	–	1.5	Std. Dev 0.3

3. Cash-to-Cash cycle time (C2C)

This strategic metric is used to measure the time required from when the TMBP raw materials arrived at the company warehouse until they are finally sold as finished goods in the form of tinplate to consumers. This metric can show the stock period/aging of TMBP raw material before it can be processed and sold as finished goods, and thus, becomes a cash inflow for the company. Gap analysis is performed with the industry, where it is 30-days for top performers, 50-days for median performers and the bottom performers are over 85 days.

4. Return on Working Capital (Inventory)

This strategic metric shows the amount of raw material inventory owned by the company. To analyze it, the amount of raw material inventory is compared with the company's production requirement. Inventories that are too high are certainly not very efficient, because working capital is stuck in inventory, but low inventory can greatly affect the production process, where the company runs 24 hours a day, 7 days a week, apart from during overhaul, maintenance or public holidays.

Based on the four strategic metrics, the summary of the gap analysis which shows the need for future improvement for all metrics, is presented in Table 3.

4.2 SCOR level 2

In SCOR Level 2, the mapping is conducted in the form of geographical mapping and a thread diagram. The geographical mapping for the main raw material procurement process is presented in Figure 2.

Referring to Figure 2, it can be clearly seen that the main raw material suppliers are from outside Indonesia, namely South Korea, Japan, China, Taiwan and Thailand. Reliance

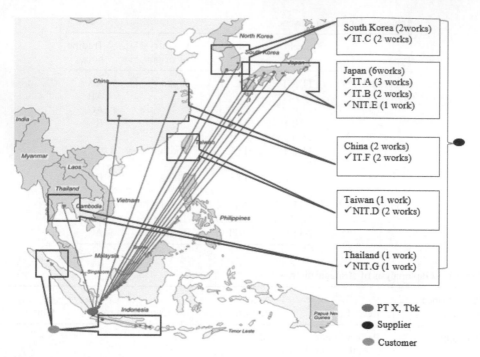

Figure 2. Geographical map.

on foreign companies is a major problem in the procurement of the company's main raw materials. Japan, South Korea and China are major steel players in the Asian market. Having a very limited number of suppliers makes suppliers and companies more dependent on each other and increases the risk for both parties. Risks to the supply chain can arise in various ways. Here are possible supply chain risks and mitigation strategies for the company:

- Failure of suppliers to ship raw material
- Supplier quality failures
- Delayed shipment
- Transportation damage
- Information loss (distortion)
- Competitive price

Figure 3 shows the overall activities carried out in the procurement process of raw materials as well as the relationship of each activity, the actual conditions and the potential problems that may happened in the future.

Based on Figure 3, the most likely problem is the Source Make-to-Order (sS2) activity in the SCM (Supply Chain Management) division that is directly related to the Deliver Make-to-Order (sD2) activity in the supplier area. Issues that occur in this activity are deviation of the strategic metrics perfect order fulfillment and order fulfillment cycle time. While the deviations in cash-to-cash cycle time and return on working capital (inventory) are most likely to arise in the internal coordination of Plan Source activities (SP2) between the SCM division and internal operations. To determine the root cause of the problem, the next mapping will focus on the activities of Plan Source (sP2) and Source Make-to-Order (sS2).

4.3 SCOR level 3

The SCOR Level 3 mapping is performed to identify all the Plan Source activities carried out by the supply chain. In this level, the causes of problems that occur can be identified so that

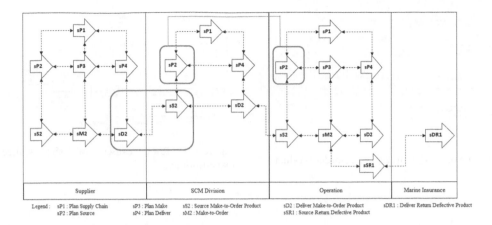

Supplier | SCM Division | Operation | Marine Insurance

Legend : sP1 : Plan Supply Chain sP3 : Plan Make sS2 : Source Make-to-Order Product sD2 : Deliver Make-to-Order Product sDR1 : Deliver Return Defective Product
sP2 : Plan Source sP4 : Plan Deliver sM2 : Make-to-Order sSR1 : Source Return Defective Product

Figure 3. Thread diagram.

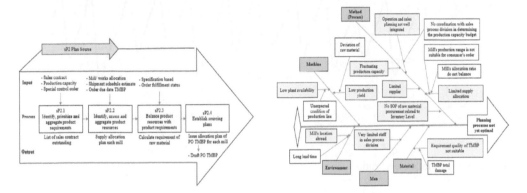

Figure 4. sP2 plan source.

Figure 5. Root cause sP2 plan source.

remedial measures can be determined to overcome these from both within the company and externally.

Based on the evaluated strategic metrics, it appears that the company needs to make improvements in those four metrics. The four metrics are perfect order fulfillment, order fulfillment cycle time cash-to-cash cycle time, and return on working capital (inventory), all caused by the inventory problem. The company needs to set an optimal inventory policy to improve all four metrics.

One way to overcome the inventory problem at the company is to use a probabilistic inventory model, in which demand and lead time are not always known and constant. Based on the demand data and lead time in the last three years (2015 to 2017), both the demand and the lead time are normally distributed. However, a conventional probabilistic inventory model cannot be applied due to the following aspects:

– The procurement lead time of raw material is very long, about 81 days.
– The company has limited warehouse capacity for raw materials, a maximum of 25,000 MT, which is not sufficient to bear the inventory load on such long lead times.
– Also, from the financial perspective, having a high level of inventory to accommodate probabilistic demand and lead time will not be beneficial for the company.

Hence, the appropriate inventory level is determined by simulating the production capacity budget in 2018 using historical lead times, and by comparing the company's current inventory policy (which is in the ratio 1,5) with a gradual decrease, to see if there will be

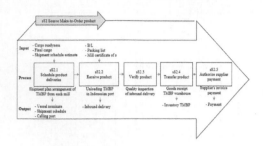

Figure 6. sS2 source make-to-order product.

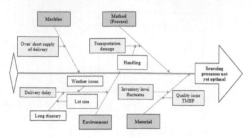

Figure 7. Root causes of sS2 source make-to-order.

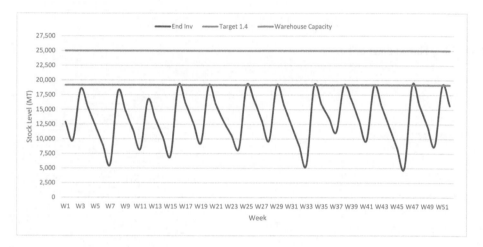

Figure 8. Inventory level ratio 1,4.

Table 4. Comparison of costs.

	Stock level		Material cost	Carrying cost	Total cost	Cost difference	
Ratio	Min	Max	(a)	(b)	(a + b)	(to current)	Saving
1,5	5,768	20,597	$ 13,743,297	$ 3,573,257	$ 17,316,554	–	
1,4	5,083	19,097	$ 12,742,403	$ 3,313,025	$ 16,055,428	$ (1,261,126)	–7%
1,3	3,883	18,441	$ 12,305,203	$ 3,199,353	$ 15,504,556	$ (1,811,997)	–10%
1,2	2,583	18,441	$ 12,305,203	$ 3,199,353	$ 15,504,556	$ (1,811,997)	–10%

shortage of raw material if the inventory level is lowered. With a ratio of 1,5, the company can pursue inventory coverage for one month and half the production target capacity. Simulation is conducted for 0,1 gradual decrease in the ratio from the current ratio of 1,5 down to 1,2. The results (see Figure 8 and Table 4) show that reducing the current inventory level to a ratio of 1,4 would be suitable for the company. The minimum stock level (around 5,000 tons) is the acceptable level for company, to cover production needs for around two weeks, equal to one production cycle. As for the cost, with the carrying cost around 26% (Heizer et al, 2014), by reducing the stock ratio to 1,4, it will reduce its material and carrying costs by 7%, and thus, be quite beneficial for the company. Though reducing the ratio to 1,3 or 1,2 will bring much more benefit, this can't be applied, as the stock level will be very low, hence the possibility of shortages will be higher if the lead time much deviates much from the average norm.

5 CONCLUSION

This study's main purpose is to evaluate the raw material procurement process of tinplate using SCOR 11.0. As the company's fully reliant for its sourcing from abroad which has a very long lead time, the risk to its supply chain can arise in various way. Its dependence on imported raw material and its very limited supply, greatly affects the company's sustainability.

As the one and only tinplate manufacture in Indonesia, its sustainability should be maintained and protected in order to fulfill Indonesia's national demand. Especially in a very tight business competition as result of the dumping price issue which affects Indonesia's tinplate market, this study will be helpful for the business.

This research provides an insight that the main problem of PT X Tbk's procurement process is the limited number of suppliers which are in foreign countries and, thus, causing a very long lead time and high level of inventory. The simulation results show that by reducing the inventory ratio to 1:4, it will lead to better financial performance for the company.

Further research is still needed to focus on the downstream players, as this business has a probabilistic lead time and demand, the supply ability for finished goods for the end customer is very critical and requires proper study for a more comprehensive view.

REFERENCES

Agrawal, A. (2014). Managing raw material in supply chains. *European Journal of Operational Research*, *239*(3), 685–698.

APICS. (2017). SCORmark Process—Industry Vertical. Retrieved from: https://www.apics.org/apics-for-business/benchmarking/scormark-process.

Balfaqih, H., Saibani, N. & Al-Nory, M.T. (2016). Review of supply chain performance measurement systems: 1998–2015. *Computers in Industry*, *82*, 135–150.

Beamon, B.M. (1998). Supply Chain Design and Analysis: Models and Methods. *International Journal of Production Economics*, *55*(3), 281–294.

Bolstorff, P. & Rosenbaum, R. (2007). *Supply Chain Excellence: A Handbook for Dramatic Improvement Using the SCOR Model* (2nd ed.). AMACOM.

Brun, L. (2016). Overcapacity in Steel, China's Role in a Global Problem. *Alliance for American Manufacturing*, (August).

CIPS. (2017). The top six challenges facing procurement. Retrieved from: https://www.cips.org/supply-management/opinion/2017/september/the-top-six-challenges-facing-procurement-/.

Chopra, S. & Meindl, P. (2013). *Supply Chain Management Strategy, Planning, and Operation* (5th ed.). New Jersey: Pearson Education Inc.

Cooper, M.C., Lambert, D.M. & Pagh, J.D. (1997). Supply chain management: more than a new name for logistics. *The International Journal of Logistics Management*, *8*(1), 1–14.

Erkan, T.E. & Baç, U. (2011). Supply chain performance measurement: A case study about applicability of SCOR model in a manufacturing industry firm. *International Journal of Business and Management Studies*, *3*(1).

Esin, G. & Kocaoglu, B. (2016). Using SCOR model to gain competitive advantage: A literature review. *Procedia—Social and Behavioral Sciences*, *229*, 398–406.

Gunasekaran, A., Patel, C. & Tirtiroglu, E. (2001). Performance measures and metrics in a supply chain environment. *International Journal of Operations & Production Management*, *21*(1/2), 71–87.

Heizer, J., Render, B. & Munson, C. (2017). *Operations Management Sustainability and Supply Chain Management* (12th ed.) P. E. Inc., Ed.

Hill, T. (1994). *Manufacturing Strategy—Text and Cases*. Palgrave, Basingstoke.

Hernandez, J.E., Lyons, A.C., Zarate, P. & Dargam, F. (2014). Collaborative decision-making and decision support systems for enhancing operations management in industrial environments. *Production Planning & Control*, *25*(8), 636–638.

Jacobs, F.R. & Chase, R. (2014). *Operations and Supply Chain Management* (14th ed.). New York: McGraw Hill Education.

Kaplan, R. & Norton, D. (1992). The balanced scorecard measures that drive performance. *Harvard Business Review*, January–February.

Kocaoglu, B., Gülsün, B. & Mehmet, T. (2013). A SCOR based approach for measuring a benchmarkable supply chain performance. *Journal of Intelligent Manufacturing*, *24*, 113–132.

Koumanakos, D.P. (2008). The effect of inventory management on firm performance. *International Journal of Productivity and Performance Management, 57*(5), 355–369.

Kristal, M.M., Huang, X. & Roth, A.V. (2010). The effect of an ambidextrous supply chain strategy on combinative competitive capabilities and business performance. *Journal of Operations Management, 28*(5), 415–429.

SCC. (2012). *Supply Chain Operations Reference Model Revision 11.0.*

Simchi-Levi, D., Kaminsky, P. & Simchi-Levi, E. (2004). *Managing Supply Chain—The Definitive Guide for The Business Professional.* New York: McGraw-Hill.

Slack, N. & Lewis, M. (2015). *Operations Strategy* (4th ed.). London: Pearson Education Limited.

Taylor, D.A. (2004). *Supply chains: A manager's guide.* London: Pearson Education Limited.

Trkman, P., McCormack, K., Oliveira, M.P.V. de & Ladeira, M.B. (2010). The impact of business analytics on supply chain performance. *Decision Support Systems, 49*(3), 318–327.

Vokurka, R.J., Zank, G.M. & Lund III, C.M. (2002). Improving competitiveness through supply chain management: A cumulative improvement approach. *Competitiveness Review, 12*(1), 14–25.

Business Innovation and Development in Emerging Economies – Trinugroho & Lau (Eds)
© *2019 Taylor & Francis Group, London, ISBN 978-1-138-35996-3*

Offshore workover rig barge optimization in support of oil and gas well operation activity in CNOOC SES Ltd using mixed-integer programming method

M. Abdillah
Program Studi Magister Manajemen FEB UI, Jakarta, Indonesia

M.E. Harahap
Program Studi Magister Manajemen FEB UI, Jakarta, Indonesia
Pusat Teknologi Material, Deputi Bidang TIEM, BPPT, Banten, Indonesia

ABSTRACT: The production activity of offshore oil and gas cannot be sustained without the vital role of the offshore workover rig barge that is used as a main tool in oil and gas well intervention operation activity. Rental cost of this equipment is considered to be one of the biggest cost components in every well intervention activity, which has an objective to maintain the company's oil and gas production rate. Every effort to minimize the rental cost of workover rig barge operation will have a significant effect on the cost in every well intervention activity. The objective of this research is to optimize offshore workover rig barge operations in CNOOC SES Ltd. A mathematical model will be formulated prior to using the mixed-integer programming as the usage methodology. The decision variables' values of the mathematical model are computed by Lingo 17.0 optimization software. The optimization process delivers a result of cost reduction for every well intervention, with an average cost reduction of 3.59%, or about US$8,518.

Keywords: Mixed-integer programming, assignment problem, optimization, workover rig barge

1   INTRODUCTION

The oil and gas industry is an important industry for other industries. It is also important for civilization sustainability, because oil and gas are still important energy sources that are being used for industries and humankind for living sustainability, such as for daily operational fuel. The oil and gas industry consists of three main sectors: upstream, middle, and downstream. The middle operation is also sometimes included in the downstream sector. The upstream oil and gas industry includes the oil and gas exploration process on the earth's surface, and oil and gas exploitation by moving oil and gas from underground to the earth's surface. As for the downstream oil and gas sector, this includes the oil and gas processing and refining phase, which transforms the oil and gas into various ready-to-use products such as fuels, solvents, medicines, fertilizers, pesticides, and plastics. The downstream sector also includes the transportation and marketing process of oil and gas.

CNOOC SES Ltd (China National Offshore Oil Corporation, South East Sumatera) is an oil and gas company which operates in Indonesia. Oil and gas production from wells is an important task in the daily operation of an oil and gas company. To sustain the operation, the main offshore operation activity includes drilling, well intervention, production, and facility maintenance. This research will be focused on the well intervention operation service activity. Well intervention operation activity is done continuously to compensate for the well mortality rate, which is caused by interference from both factors internal to the well itself and also the mechanical factor of the well production supporting equipment.

Furthermore, the well intervention activity consists of the workover and well service activities. The workover operation relates to the effort to change the oil and gas well configuration into a more suitable design. The well service operation relates to the repairing or reactivation effort on an oil and gas well, which had already been shut previously without changing the well's configuration and design.

Offshore oil production activity is an activity requiring intensive capital and high technology. One of the most significant and expensive operations in the oil and gas field development and production phases relates to rig usage (Bassi et al., 2012). As an example, the average cost for every well intervention activity in CNOOC SES Ltd for 2016 was US $237,349. Of that cost, the biggest component was offshore workover rig barge usage. The workover barge is a flat-bottomed vessel or ship without a locomotive engine. It is used as the main tool to conduct well intervention operations. Almost all types of working activity are possible with the presence of a workover barge, which functions as working tools storage, office quarters, living quarters, and as an accommodation place for all workover crew. Regularly, it moves around from one platform to another to carry out operations for wells that need to be serviced.

In CNOOC SES Ltd, well intervention operation activity is supported by four or five workover barges, depending on the company's daily production target rate and the well mortality rate. All barges are distributed and allocated to each operation area based on daily operational needs. The workover barge has a very high daily usage rental price. This price varies between barges, mainly due to differences in valid contract clauses, which are based on each barge's specification. CNOOC SES Ltd's working zone is divided into three operational areas, which are the Northern Business Unit (NBU), Central Business Unit (CBU), and Southern Business Unit (SBU). Each operational area has different characteristics in its underground geological and geophysical conditions, which impact on different well characteristics in each operational area. Well differences include well depth, the complexity of problems inside the well, and several other factors. These factors cause the duration of each well intervention job to be different.

The workover barge requirement for each operational area is different. It depends on each operational area's production target, which is cascaded from the company's oil and gas production target. Each operational area also has a different well natural mortality rate, which is caused by the geological and geophysical differences of each operational area.

With all the differences mentioned—workover barge daily rental cost, well intervention duration, and workover barge requirement in each operational area—the appropriate workover barge placement in the appropriate operational area becomes very important. The barge arrangement and placement will give savings on the average intervention well cost if done correctly. Despite the importance of the well intervention activity, which is routinely done to maintain the company's oil and gas production sustainability, the workover barge placement and arrangement are not currently carried out or formulated well: it has been shown that no quantitative optimization process has been conducted in the company's workover barge operation. Such quantitative optimization is needed to determine the allocation and assignment of workover barges to each available operational area, so minimizing average well intervention cost as a contribution to maintaining the company's oil and gas production rate.

Offshore workover rig barge optimization is rarely done. Previous research has involved only a land workover rig (Ribeiro et al., 2012b); however, this was using a different method. While integer programming optimization has been done for the offshore supply vessel armada (Prasetyo, 2013), this research used a slightly different method and is applied to different objects, parameters, and variables.

2 LITERATURE REVIEW

2.1 *Logistics and transportation in supply chain management*

The supply chain is defined as a network of partners which collectively converts basic commodities (upstream) into finished products (downstream) that are valued by the end-customer, and who manages returns at each stages (Harrison & van Hoek, 2014). Logistics concerns the tasks associated with coordination of materials and information flows in a system of supply chains to fulfill the needs of an end customer (Harrison & van Hoek, 2014).

2.2 *Logistics network configuration*

Configuring a logistics network in a supply chain requires data, analytical techniques, computer software, and an information system (Li, 2007). Logistics network design requires a lot of business data. Once the related data and information has been collected, the process of network design can begin. The process used to configure a logistics network involves determining material and service flows. The process is complicated and is usually helped by using a mathematical model and computer simulation. The technique that is commonly used in industry includes an optimization model, computer simulation, a heuristic model, plus an expert and decision support system. From those logistic network configuration techniques, an optimization model method is identified as being the most suitable to be used in the research. From the model complexity point of view, a problem-solving method using linear programming is considered able to give the best solution mathematically for a workover barge problem.

2.3 *Linear programming*

Linear programming can be defined as a mathematical technique created to assist managers of operations in planning, and to assist in making necessary decisions in allocating resources (Heizer & Render, 2014). Computational results on real problems present better solutions than all the other available approaches that are usually used in research (Ribeiro et al., 2011). Many decisions of operations management are related to the effort to create the most effective usage of an organization's resources. Such resources typically consist of machinery (for example, a workover rig, in the case of an oil and gas company), labor (for example, drillers), time, money, and raw materials (for example, diesel fuel). These resources can be utilized in producing products (for example, oil, gas, or other petroleum products) or services (for example, workover rig schedules, or maintenance policy). Linear programming has been utilized routinely in the worlds of military, finance, accountancy, agricultural and industrial problems. Actual situational experimentation shows the simulation–optimization approach to be effective (Bassi et al., 2012). Linear programming has multiple usages, but still involves characteristic similarities and assumptions (Render et al., 2012), including:

- Being a function with one objective, which means that linear programming is used in minimizing or maximizing a parameter, commonly cost or profit.
- Having one or more constraints, which will limit achievement rate of an object.
- Alternatives available on decision direction. For example, if a company has three products, management will decide resource allocation between those three product types.
- Mathematical relationship is linear.
- There is certainty. This means that as the study goes along, the parameters of the objective and constraints are known for certain and do not change.
- Divisibility: the given solution can be fractional.
- The variables are non-negative.

2.4 *Integer programming*

The integer programming model is a model which has identical constraints and objective function to that of linear programming. A distinguished factor only exists in one or two decision variables, which must be integers (Render et al., 2012). Integer programming consists of three types:

- Pure integer programming, in which all variables must be integers.
- Mixed-integer programming, in which not all decision variables are integers.
- One-zero integer programming, which is a special case where all decision variables are integers, valued 1 or 0.

Integer programming problem-solving requires more time compared to common integer programming, due to its more complicated level (Render et al., 2012). Experimentation using computational methods demonstrates the exact algorithm able to solve a practical-sized problem, while still having a computation time that is reasonable (Ribeiro et al., 2012a).

2.5 *Integer programming solution with the branch and bound method*

The branch and bound method is an integer programming solution approach based on the principle that the total solution compilation may be able to be partitioned into a subset of smaller solutions (Bradley et al., 1977). Its general procedure can be seen in Figure 1.

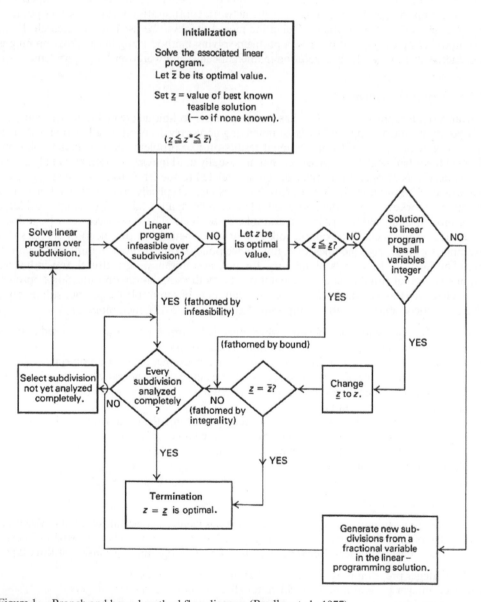

Figure 1.   Branch and bound method flow diagram (Bradley et al., 1977).

2.6 Assignment problem

The assignment problem can appear in various conditions of decision-making in which the common problem might happen, including allocating a machine to do a certain job, marketing personnel allocation in a certain location, or the decision to allocate a contract to a bidder. Specifically, the goal objective in an assignment problem is to minimize cost or time, or to maximize profit from the assignment arrangement of an existing activity group.

The assignment problem can be formulated mathematically as follows (Markland, 1983):

Objective function:

$$Minimize \ \ Z = \sum_{i=1}^{n} \sum_{j=1}^{n} c_{ij} x_{ij} \tag{1}$$

with constraints:

$$\sum_{i=1}^{n} x_{ij} = 1 \quad for \ j = 1,2,...,n \tag{2}$$

$$\sum_{i=1}^{n} x_{ij} = 1 \quad for \ i = 1,2,...,n \tag{3}$$

in which: $x_{ij} = 0 \ or \ 1 \ for \ all \ i \ and \ j$

where:
x_{ij} = resource assignment i to activity j
c_{ij} = related cost with resource allocation i to activity j

In an assignment problem, as explained by Equations 1, 2, and 3, it can be seen that the problem is a balance assignment problem, due to the activity amount being similar to the resource amount.

The assignment problem is a special case of the transportation problem, where the number of source points and demand points is the same ($m = n$). If $a_1 = 1$ for $i = 1, 2, ..., n$ is source availability, and $b_j = 1$ for $j = 1, 2, ..., m$ is demand requirement, then the statement can be mathematically stated as follows:

$$\sum_{i=1}^{n} a_i = \sum_{j=1}^{n} b_j = n \tag{4}$$

Assignment problem-solving using an integer programming method is expected to be able to optimize the allocation cost of a workover rig. Lingo 17.0 software is an optimization tool that will be used for the problem-solving.

3 RESEARCH METHODOLOGY

3.1 Research object and research methodology flow diagram

The research object includes all workover rig barge units and all field locations in CNOOC SES Ltd. The research process is done according to the flow diagram in Figure 2.

3.2 Data collection method

Secondary data is acquired from daily reports of workover and well service operation activity in CNOOC SES Ltd. Data is collected from the Drilling and Completion department, as the section that conducts the activity. The focus of the data collection includes the following:

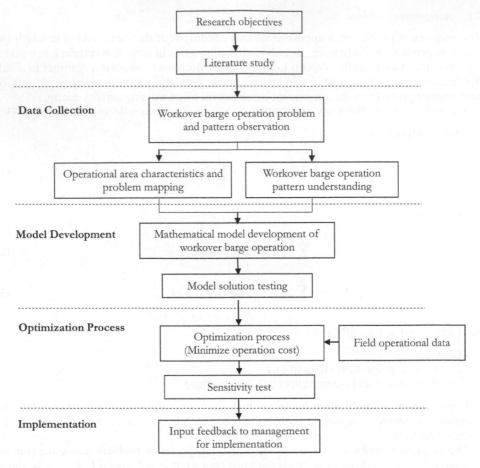

Figure 2.   Research methodology flow diagram.

- Yearly workover barge usage requirement for each operation area.
- Duration of well intervention activity in each operation area.
- Total yearly usage allocation for each workover barge in CNOOC SES Ltd offshore area.
- Daily workover barge rental cost value.
- Total well intervention job cost value for each well.

The collection of well intervention historical data was done in the period between January 1st, 2016 and December 31st, 2017. The source was the daily report database of the Drilling and Completion department. It should be possible to get an understanding of the workover barge operation pattern in each operational area from the data collection process, and hence the existing problem can be charted.

3.3   *Model formulation*

The model is a representation of an object or actual situation which can be served in various forms, such as in a mathematical equation (Anderson et al., 2012). Model formulation is done after all necessary data and operational patterns have been understood. Operational patterns are formulated into model constraints. Filtering is done by identifying operational data parameters to find the ones that have a significant effect on the objective. Accuracy is guaranteed by testing the model using actual operational parameters, which are then compared to the output generated by real conditions. If the model is not accurate enough, then the model formulation process will be repeated until a minimum accuracy level is achieved.

3.4 *Optimization process*

The optimization process is done after the workover barge system model is acquired, which in this case is minimization. It uses the integer programming method, where the searching process of decision variables is done by using optimization software Lingo 17.0. Prior to the optimization process using Lingo 17.0 software, the transformation process from mathematical model into programming language must be done. The validation of the process is done during the syntax compilation process by the software and also through a manual check for mathematical logic conformity.

The feasibility of decision variables resulting from the optimization process is then tested. It is a requirement that the variables meet with the model constraints. If the optimization solution has already passed the feasibility test, then the next step is to conduct the sensitivity test.

4 RESULTS AND DISCUSSION

4.1 *Operational area description*

The working area of CNOOC SES Ltd in South East Sumatera block consisted of three main operational areas, which are:

1. Northern Business Unit;
2. Central Business Unit;
3. Southern Business Unit.

Each area consisted of platforms, which are a cluster of oil and gas wells.

4.2 *Workover barge operation problem and pattern observation*

Historically, there are four or five workover barges which are used for daily operations in a year. The barge number is dependent on the company's daily production target and the well mortality rate. All barges will be placed and allocated to each operational area based on daily operational needs. A workover barge has a very high daily rental cost, as shown in Table 1, which forms one of the main cost components for well intervention activities.

Of all workover barges, only four are used for the entire full-year period, namely COSL 221, COSL 222, COSL 223, and COSL 225. A fifth barge, HWO, is a kind of workover rig of a snubbing unit type and is only used to fulfill the remaining requirement. This allocation arrangement, as shown in Table 2, is as per company policy.

As mentioned before, the duration of each well intervention in each operational area is different, as shown in Table 3. In addition, the workover barge usage requirement is different for each operational area, as shown in Table 4. This yearly workover barge requirement for each operational area is not an integer, but a fraction: each number represents the requirement of one barge to be used for a whole full year. Thus, a fractional number represents a barge requirement duration of less than a year. An example for the NBU area with average workover barge requirement is 1.62 units per year, which means that for each annual period one unit workover barge is required to be used for the whole year and one other barge to be used for only 0.62 of the year.

4.3 *Important variables mapping*

Important variables for workover barge operation activity are as follows:

B_k Workover barge requirement per year in assignment k
R_m Daily rental cost of workover barge number m (US$/day)

Table 1. Daily rental cost rate of workover barge in CNOOC SES Ltd for 2016–2017 (US$).

Line	Description	COSL 221	COSL 222	COSL 223	COSL 225	HWO
Controllable sosts						
21	Contract Rig *(Barge Cost + Catering)*	16,630	16,630	16,630	13,385	11,694
	Barge Cost	15,037	15,037	15,037	11,738	*10,419*
	Catering	1,593	1,593	1,593	1,647	*1,275*
22	Drilling Crew (Workover Rig/HWO Unit + Crew)	7,850	7,850	7,850	7,850	8,500
23	Mud, Chemical, Engineering and Services	318	318	318	318	318
26	Equipment Rent	2,944	2,944	2,944	2,944	
	Slickline Services	1,178	1,178	1,178	1,178	1,000
	Logging Unit	925	925	925	925	390
	Venting	–		–	–	
	Stimulation Package	385	385	385	385	*308*
	Fishing Tools	187	187	187	187	*150*
	Drill Pipe rental	–	–	–	–	–
	Wellhead Maintenance	237	237	237	237	*372*
	Explosive	33	33	33	33	*26*
57	Fuel and Lubricants	3,093	3,347	3,252	3,192	2,139
50	Supervision	748	732	734	730	
	Safety Officer	279	279	279	279	*176*
	WO Consultant	243	243	243	243	*194*
	Doctor	226	210	212	208	*107*
	Total	31,583	31,820	31,727	28,419	25,169
Uncontrollable costs						
24	Water	113	113	113	113	113
53	Marine Rental and Charters	13,913	13,913	13,913	13,913	13,913
60	Allocation Overheads— Jakarta Office	4,544	4,544	4,544	4,544	4,544
	Total	18,570	18,570	18,570	18,570	18,570
Total		50,153	50,390	50,298	46,989	43,740

Table 2. Workover barge usage allocation in CNOOC SES Ltd for 2016–2017 (unit).

Barge	Barge allocation per year
COSL-221	1
COSL-222	1
COSL-223	1
COSL-225	1
HWO	Remainder

D_k Duration of well intervention in assignment k (day)
L_{km} Binary logic for workover barge m usage in assignment k ($0 = no/1 = yes$)
M Number of used workover barges
K Number of assignments

Table 3. Summary of well intervention activity duration in each operation area, 2016–2017.

Area	Duration for each well intervention (Day)		
	2016	2017	Average
NBU	4.64	5.66	5.15
CBU	5.38	6.10	5.74
SBU	5.04	6.31	5.68

Table 4. Summary of workover barge usage requirement in each operation area, 2016–2017.

Area	Yearly barge requirement (Unit)		
	2016	2017	Average
NBU	1.49	1.75	1.62
CBU	1.23	1.36	1.30
SBU	0.95	1.56	1.26

Table 5. Workover barge numbering list.

m	Workover barge usage
1	COSL 221
2	COSL 222
3	COSL 223
4	COSL 225
5	HWO

where variable L_{km} is the decision variable of the optimization process. Table 5 shows the workover barge numbering in the modeling, and Table 6 shows the assignment grouping.

4.4 Mixed-Integer model formulation of workover barge operation

Mixed-integer programming model formulation of workover barge operation is directed to minimize the average cost of well intervention jobs, which are due to workover barge usage. Briefly, the cost function is the average total cost expended to pay workover barge rental, based on its allocation and assignment in each operational area, and based on well intervention job duration in each assignment type and operational area.

Based on operational data analysis, the objective function can be formulated as follows:

$$Well\ intervention\ cost = \frac{1}{\sum_{k=1}^{K} B_k} \times \sum_{k=1}^{K} \sum_{m=1}^{M} \left(B_k \times R_m \times D_k \times L_{km} \right) \tag{5}$$

Formulation of Equation 5 is an assignment problem in which the decision is the workover barge placement and assignment type that will give the lowest intervention well cost.

Constraints are based on the actual operational condition of the workover barge in CNOOC SES Ltd, which are fulfilled by the following equations:

$$\sum_{k=1}^{K} \sum_{m=1}^{M} L_{km} \geq M \tag{6}$$

Table 6. Workover barge assignment grouping list.

K	Assignment	Barge requirement per year	Duration per well intervention (Day)
1	NBU1	1	5.15
2	NBU2	0.62	5.15
3	CBU1	1	5.74
4	CBU2	0.30	5.74
5	SBU1	1	5.68
6	SBU2	0.26	5.68

$$L_{1m}, L_{3m}, L_{5m}, \in [1,0] \tag{7}$$

$$\sum_{m=1}^{M} L_{1m} = 1 \tag{8}$$

$$\sum_{m=1}^{M} L_{2m} = 0.62 \tag{9}$$

$$\sum_{m=1}^{M} L_{3m} = 1 \tag{10}$$

$$\sum_{m=1}^{M} L_{4m} = 0.3 \tag{11}$$

$$\sum_{m=1}^{M} L_{5m} = 1 \tag{12}$$

$$\sum_{m=1}^{M} L_{6m} = 0.26 \tag{13}$$

$$\sum_{k=1}^{K} L_{k1} = 1 \tag{14}$$

$$\sum_{k=1}^{K} L_{k2} = 1 \tag{15}$$

$$\sum_{k=1}^{K} L_{k3} = 1 \tag{16}$$

$$\sum_{k=1}^{K} L_{k4} = 1 \tag{17}$$

$$\sum_{k=1}^{K} L_{k5} < 1 \tag{18}$$

Every assignment is served by M number of available workover barges (Equation 6). Decision variables L_{1m}, L_{3m}, and L_{5m} are binary variables (0 or 1) (Equation 7). The constraints in Equations 8 to 13 show the workover barge requirement for each assignment. The constraints in Equations 14 to 17 represent the condition that workover barge numbers 1 to 4 must be allocated for one whole year's assignment, and the constraint of Equation 18 shows the condition that workover barge number 5 can only be allocated for less than a year in each annual period.

4.5 Variable constant calculation

Tables 7 to 10 show the data calculation for barge requirement per year (B_k), daily rental cost of workover barge (R_m), and duration of well intervention (D_k). The end result is variable

540

Table 7. Workover barge requirement (B_k) number m per year on assignment k (unit).

B_k		m				
		1	2	3	4	5
k	1	1.00	1.00	1.00	1.00	1.00
	2	0.62	0.62	0.62	0.62	0.62
	3	1.00	1.00	1.00	1.00	1.00
	4	0.30	0.30	0.30	0.30	0.30
	5	1.00	1.00	1.00	1.00	1.00
	6	0.26	0.26	0.26	0.26	0.26

Table 8. Daily rental cost (R_m) of workover barge number m on assignment k (US$/day).

R_m		m				
		1	2	3	4	5
k	1	50,153	50,390	50,298	46,989	43,740
	2	50,153	50,390	50,298	46,989	43,740
	3	50,153	50,390	50,298	46,989	43,740
	4	50,153	50,390	50,298	46,989	43,740
	5	50,153	50,390	50,298	46,989	43,740
	6	50,153	50,390	50,298	46,989	43,740

Table 9. Duration of well intervention (D_k) with workover barge usage number m on assignment k (day).

D_k		m				
		1	2	3	4	5
k	1	5.15	5.15	5.15	5.15	5.15
	2	5.15	5.15	5.15	5.15	5.15
	3	5.74	5.74	5.74	5.74	5.74
	4	5.74	5.74	5.74	5.74	5.74
	5	5.68	5.68	5.68	5.68	5.68
	6	5.68	5.68	5.68	5.68	5.68

Table 10. Variable constant L_{km}, binary logic for workover barge usage number m on assignment k.

L_{km}		m				
		1	2	3	4	5
k	1	258,366	259,589	259,113	242,069	225,328
	2	159,855	160,612	160,317	149,772	139,414
	3	287,943	289,306	288,775	269,780	251,122
	4	85,305	85,709	85,551	79,924	74,397
	5	284,735	286,082	285,558	266,774	248,325
	6	73,191	73,537	73,402	68,574	63,832

constants for binary logic (L_{km}) for workover barge m on assignment k. These variable constants will be used as the main input in the programming algorithm that is used in the Lingo 17.0 optimization software.

4.6 *Modeling assumptions*

In mathematical model formulation for a workover barge operation, there are several assumptions made in order to simplify the formulation process, while also considering its relevance to the fundamental problem. Some of these assumptions are as follows:

- Calculation is using yearly period basis, which refers to the cost calculation method that is used in the company and which is calculated once a year.
- The daily rental cost value of a workover barge uses one average rental price, which is assumed to be constant during the data research period.
- All of the five workover barges are assumed to be available and able to be operated.
- The yearly workover barge requirement for each operational area is assumed to be constant throughout the yearly period, which refers to data observation results from January 1st, 2016 to December 31st, 2017.
- The job duration for every well intervention type is assumed to be the same, which is using the average duration value for every intervention job in each of the operational areas.
- The main objective of the optimization process is workover barge allocation and placement (assignment problem) to minimize the average cost of the well intervention job, which is mainly attributed to workover barge usage.

4.7 *Optimization through computer simulation*

The optimum solution for the formulated problems is solved with the help of Lingo 17.0 optimization software. The solution result report of the optimization process consists of two main outputs, which are the variable values (Table 11) and the reduced cost (Table 12). The variable values show the best combination within the constraints producing the optimal

Table 11. Decision variable values of computer simulation results (Objective value: 228,831).

L_{km}		m				
		1	2	3	4	5
k	1	0.00	0.00	1.00	0.00	0.00
	2	0.00	0.44	0.00	0.00	0.18
	3	0.00	0.00	0.00	1.00	0.00
	4	0.00	0.30	0.00	0.00	0.00
	5	1.00	0.00	0.00	0.00	0.00
	6	0.00	0.26	0.00	0.00	0.00

Table 12. Reduced cost variables of computer simulation results.

Reduced cost		m				
		1	2	3	4	5
k	1	61,958	57,168	62,137	24,618	54,035
	2	4,902	0	5,013	35,917	0
	3	69,051	64,294	69,251	64,695	60,221
	4	4,987	0	5,046	3,696	2,371
	5	68,282	63,521	68,479	63,975	59,550
	6	5,000	0	5,051	3,893	2,756

solution. The reduced cost for any variable shows how much the value of the objective function would increase if one unit of that variable were to be included in the solution.

4.8 Sensitivity test

The sensitivity test for a linear programming problem often becomes more critical, compared to the linear programming problem itself (Anderson et al., 2012). This can occur just as a result of a small change to the coefficient in the constraint, which can lead to a relatively big change in the optimal solution value. Due to the extreme sensitivity level in the linear programming problem for the constraint coefficient, many expert practitioners usually recommend performing problem re-solving on the linear program several times, to ensure minimal variation in the coefficient prior to choosing the optimum solution (Anderson et al., 2012).

For the workover barge operation case in CNOOC SES Ltd, variation will be done on constraints (Equations) 14 to 18, which show the maximum allocation number for each workover barge. Variation is done by changing the assignment limit to be less than one unit a year for workover barge numbers 1 to 4, which are COSL 221, COSL 222, COSL 223, and COSL 225, respectively, and also by changing the assignment limit to one unit a year for workover barge number 5, which is HWO.

This constraints modification is entered into the Lingo 17.0 program. Then the computer simulation is rerun for each constraint coefficient variation. For each simulation result, each workover barge assignment will be assessed and compared to the actual operational condition. The total allocation of the workover barge requirement from simulation results will also be assessed. If it is more than the actual barge requirement for all operational areas, then the solution can be considered as being *not feasible*.

Table 13 shows simulation results of the coefficient variation. It is shown that there is no significant variable sensitivity. Variation of the constraint coefficient does not cause relatively big changes in the optimum solution value. With this low sensitivity level, there is no need to do problem re-solving on the existing mixed-integer linear program and the chosen optimum solution from the computer simulation can be considered as *feasible*.

4.9 Optimization result analysis

The computer simulation results (Tables 11 and 12) show that the optimum solution gives the minimum average cost of a well intervention job to be US$228,831. This can be achieved through using the workover barge assignment scenario in Table 14. By following this scenario, a reduction in the average well intervention cost can be expected. This can happen because the workover barge assignment configuration in the scenario is the configuration in which the workover barges with the lowest rental cost are assigned to the longest assignments in terms of yearly usage. It can be seen that workover barges COSL 223 and HWO

Table 13. Sensitivity test through constraints variable variation.

Constraint variable	Maximum assignment	Total workover barge requirement	Objective value (US$)	Feasible (Yes/No)
Workover Barge 1	2	4.18	220,932	Yes if workover barge 1 not fully used and workover barge 5 fully used
Workover Barge 2	2	4.18	220,727	Yes if workover barge 2 not fully used and workover barge 5 fully used
Workover Barge 3	2	4.18	220,772	Yes if workover barge 3 not fully used and workover barge 5 fully used
Workover Barge 4	2	4.18	225,022	Yes if workover barge 4 not fully used and workover barge 5 fully used
Workover Barge 5	1	5	254,473	No

Table 14. Optimum assignment scenario of workover barges for CNOOC SES Ltd.

k	Assignment	Workover rig usage
1	NBU1	COSL 223
2	NBU2	COSL 222 & HWO
3	CBU1	COSL 225
4	CBU2	COSL 222
5	SBU1	COSL 221
6	SBU2	COSL 222

are assigned solely in operational area NBU, that COSL 225 is assigned for a whole year in operational area CBU, and that COSL 221 is assigned for a whole year in operational area SBU. In contrast, workover barge COSL 222 is assigned as a floating barge, which moves around the operational areas depending on well intervention operational need.

Referring to the current ongoing operational scheme in the company, the workover barge placement in a certain operational area has not yet been considering usage rental cost and barge requirement in each operational area. Workover barge assignment to an operational area has been done only by referring to the suitable name and number order, and is even sometimes done randomly. It is all conducted without a proper study of the optimum operational scenario.

As a comparison, the average cost of a well intervention job in CNOOC SES Ltd in 2016 is US$237,349. With ideal conditions as per the modeling assumption, application of the optimum operational scheme shown in Table 13 will result in a reduction of average well intervention cost by 3.59%, or about US$8,518 per well.

4.10 Optimization result implementation strategy

In order to make sure that the optimization result of the workover barge operation can be implemented effectively, it requires management involvement and commitment in supporting the optimization program implementation. This is at least present at the Drilling and Completion department manager level, which manages the well intervention job execution. Optimization program socialization for all involved parties in the well intervention operation should also be done, especially for those that handle the daily operation of workover barges.

Observation of the average well intervention operation activity cost during the optimization program implementation should also be done in order to acquire information as to whether the program does or does not create a positive impact on the company. Ideally, this observation should be done for at least a one-year period. An adjustment to the model formulation can be done if it is found that the implementation has not had a positive impact on reducing operational cost.

The workover barge operation model formulation done in this research is based on historical data. However, the formulation can also be applied to forecast future operations. One thing to be considered is the determination of every important variable value. The workover barge assignment scenario can then be derived from simulation, which is done based on the new variable constants data. Observation also has to be done in the scenario implementation, for instance, regarding a suitable adjustment that needs to be done in order to make sure implementation gives the expected results.

5 CONCLUSION

Offshore oil and gas production activity cannot be done without the important role of workover rig barge usage as the main tool used in well intervention operation activity. The cost of using workover barges is one of the biggest cost components in every well intervention activity. Every effort to minimize the rental cost of workover rig barge operation can have a significant impact in cost reduction for every well intervention job.

Current operating workover barge numbers using a configuration of four barges that are rented for a whole full year, and one barge that is rented for part of a year, is still sufficient for well intervention operation activity requirements, which consist of three operational areas with differing characteristics in barge requirement and well intervention job duration.

The optimization process for the offshore workover rig barges in CNOOC SES Ltd can be done using mixed-integer programming method, and by formulating its mathematical model beforehand. One tool that can be used to simulate the model and acquire decision variables values is Lingo 17.0 optimization software. Through the optimization of workover barges on given assignments, the calculation result can be obtained that the average well intervention operation activity cost can be reduced at 3.59% per well, or about US$8,518. This reduction can be achieved by using a workover barge assignment configuration which considers placing barges with lower rental cost on the longer assignments on a yearly usage basis.

REFERENCES

Aloise, D.J., Aloise, D., Rocha, C.T., Ribeiro, C.C., Ribeiro Filho, J.C. & Moura, L.S. (2006). Scheduling workover rigs for onshore oil production. *Discrete Applied Mathematics, 154*(5), 695–702.

Anderson, D., Sweeney, D., Williams, T., Camm, J. & Martin, K. (2012). *An introduction to management science: Quantitative approaches to decision making* (Revised 13th ed.). Mason, OH: South-Western Cengage Learning.

Atamtürk, A. & Savelsbergh, M. (2005). Integer-programming software systems. *Annals of Operations Research, 140*(1), 67–124.

Bassi, H.V., Ferreira Filho, V.J.M. & Bahiense, L. (2012). Planning and scheduling a fleet of rigs using simulation-optimization. *Computers and Industrial Engineering, 63*(4), 1074–1088.

Bosch, R. & Trick, M. (1989). Integer programming. *Handbooks in operation research and management science* (pp. 69–95). Springer, Boston, MA.

Bradley, S.P., Hax, A.C. & Magnanti, T.L. (1977). *Applied mathematical programming*. Boston, MA: Addison-Wesley.

Chopra, S. & Meindl, P. (2016). *Supply chain management: Strategy, planning & operations* (6th ed.). New Jersey, NJ: Pearson Prentice Hall.

Harrison, A. & van Hoek, R. (2014). *Logistics management & strategy: Competing through the supply chain* (5th ed.). Harlow, UK: Pearson Education.

Heizer, J. & Render, B. (2014). *Operations management: Sustainability and supply chain management* (11th global ed.). Harlow, UK: Pearson Education.

Kersting, K., Mladenov, M. & Tokmakov, P. (2017). Relational linear programming. *Artificial Intelligence, 244*, 188–216.

Laporte, G. (2016). Scheduling issues in vehicle routing. *Annals of Operations Research 236*(2), 463–474.

Li, L. (2007). *Supply chain management: Concepts, techniques and practices*. London, UK: World Scientific.

Markland, R.E. (1983). *Topics in management science* (2nd ed.). New York, NY: John Wiley & Sons.

Prasetyo, N. & Harahap, M. (2013). *Optimization of offshore supply vessel fleet in supporting the logistic operation of oil and gas production at CNOOC SES company by utilizing integer programming method* (Master's thesis, Universitas Indonesia, Jakarta, Indonesia).

Render, B., Stair, R. & Hanna, M. (2012). *Quantitative analysis for management* (11th ed.). Upper Saddle River, New Jersey: Pearson Education.

Ribeiro, G.M., Desaulniers, G. & Desrosiers, J. (2012a). A branch-price-and-cut algorithm for the workover rig routing problem. *Computers & Operations Research, 39*(12), 3305–3315.

Ribeiro, G.M., Desaulniers, G., Desrosiers, J., Vidal, T. & Vieira, B.S. (2014). Efficient heuristics for the workover rig routing problem with a heterogeneous fleet and a finite horizon. *Journal of Heuristics, 20*(6), 677–708.

Ribeiro, G.M., Laporte, G. & Mauri, G.R. (2012b). A comparison of three metaheuristics for the workover rig routing problem. *European Journal of Operational Research, 220*(1), 28–36.

Ribeiro, G.M., Mauri, G.R. & Lorena, L.A.N. (2011). A simple and robust simulated annealing algorithm for scheduling workover rigs on onshore oil fields. *Computers & Industrial Engineering, 60*(4), 519–526.

Schrijver, A. (1998). *Theory of linear and integer programming*. Chichester, UK: John Wiley & Sons.

Sebastian, K.R. (2011). *Future characteristic of offshore support vessels* (Master's thesis, Massachusetts Institute of Technology, Cambridge, MA).

Taha, H.A. (2007). *Operations research: An introduction* (8th ed.). New Jersey, NJ: Pearson Prentice Hall.

Thornburg, K. & Hummel, A. (2010). *Lingo 8.0 tutorial*. Chicago, IL: Lindo Systems.

Wang, T., Meskens, N. & Duvivier, D. (2015). Scheduling operating theatres: Mixed integer programming vs. constraint programming. *European Journal of Operational Research*, 247(2), 401–413.

A. Programming Algorithm Lingo 17.0

```
!Objective Function;
MIN = 1/4.17* (

    258366*L11 + 259589*L12 + 259113*L13 + 242069*L14 + 225328*L15 +
    159855*L21 + 160612*L22 + 160317*L23 + 149772*L24 + 139414*L25 +
    287943*L31 + 289306*L32 + 288775*L33 + 269780*L34 + 251122*L35 +
    85305*L41 + 85709*L42 + 85551*L43 + 79924*L44 + 74397*L45 +
    284735*L51 + 286082*L52 + 285558*L53 + 266774*L54 + 248325*L55 +
    73191*L61 + 73537*L62 + 73402*L63 + 68574*L64 + 63832*L65 );

!Integer Variable Declaration;
@bin(L11); @bin(L12); @bin(L13); @bin(L14); @bin(L15);
@bin(L31); @bin(L32); @bin(L33); @bin(L34); @bin(L35);
@bin(L51); @bin(L52); @bin(L53); @bin(L54); @bin(L55);

!Constraints;
L11 + L12 + L13 + L14 + L15 +
L21 + L22 + L23 + L24 + L25 +
L31 + L32 + L33 + L34 + L35 +
L41 + L42 + L43 + L44 + L45 +
L51 + L52 + L53 + L54 + L55 +
L61 + L62 + L63 + L64 + L65 > = 4.17; !At least there are workover
barge available;

L11 + L12 + L13 + L14 + L15 = 1; !Task-1 serviced by one workover barge;
L21 + L22 + L23 + L14 + L25 = 0.62; !Task-2 serviced by less than one
workover barge;
L31 + L32 + L33 + L34 + L35 = 1; !Task-3 serviced by one workover barge;
L41 + L42 + L43 + L44 + L45 = 0.3; !Task-4 serviced by less than one
workover barge;
L51 + L52 + L53 + L54 + L55 = 1; !Task-5 serviced by one workover barge;
L61 + L62 + L63 + L64 + L65 = 0.26; !Task-6 serviced by less than one
workover barge;

L11 + L21 + L31 + L41 + L51 + L61 = 1; !workover barge number 1 assigned
to one task;
L12 + L22 + L32 + L42 + L52 + L62 = 1; !workover barge number 2 assigned
to one task;
L13 + L23 + L33 + L43 + L53 + L63 = 1; !workover barge number 3 assigned
to one task;
L14 + L24 + L34 + L44 + L54 + L64 = 1; !workover barge number 4 assigned
to one task;
L15 + L25 + L35 + L45 + L55 + L65 <1; !workover barge number 5 assigned
to less than one task;
```

B. Simulation Solution Results Report Program Lingo 17.0

```
Global optimal solution found.
Objective value:                    228830.7
Objective bound:                    228830.7
Infeasibilities:                    0.000000
Extended solver steps:                     2
```

```
Total solver iterations:                    73
Elapsed runtime seconds:                     0.11

Model Class:                                MILP

Total variables:               30
Nonlinear variables:            0
Integer variables:             15

Total constraints:             13
Nonlinear constraints:          0

Total nonzeros:               120
Nonlinear nonzeros:             0
```

Variable	Value	Reduced cost
L11	0.000000	61958.27
L12	0.000000	57168.11
L13	1.000000	62137.41
L14	0.000000	24617.51
L15	0.000000	54035.49
L21	0.000000	4901.918
L22	0.4400000	0.000000
L23	0.000000	5012.710
L24	0.000000	35916.55
L25	0.1800000	0.000000
L31	0.000000	69051.08
L32	0.000000	64294.48
L33	0.000000	69250.60
L34	1.000000	64695.44
L35	0.000000	60221.10
L41	0.000000	4986.571
L42	0.3000000	0.000000
L43	0.000000	5045.564
L44	0.000000	3696.163
L45	0.000000	2370.743
L51	1.000000	68281.77
L52	0.000000	63521.34
L53	0.000000	68479.14
L54	0.000000	63974.58
L55	0.000000	59550.36
L61	0.000000	5000.480
L62	0.2600000	0.000000
L63	0.000000	5051.079
L64	0.000000	3893.285
L65	0.000000	2756.115

Row	Slack or surplus	Dual price
1	228830.7	-1.000000
2	0.1000000E-01	0.000000
3	0.000000	0.000000
4	0.000000	-33432.61
5	0.000000	0.000000
6	0.000000	-15470.26
7	0.000000	0.000000
8	0.000000	-12551.32
9	0.000000	0.000000
10	0.000000	-5083.453
11	0.000000	0.000000
12	0.000000	0.000000
13	0.8200000	0.000000

Business Innovation and Development in Emerging Economies – Trinugroho & Lau (Eds)
© *2019 Taylor & Francis Group, London, ISBN 978-1-138-35996-3*

Development strategies for the sharia banking industry in Indonesia using an analytic network process method

A.S. Rusydiana
SMART Indonesia, Indonesia

F.F. Hasib & Lina N. Rani
Faculty of Economics and Business, Airlangga University, Surabaya, Indonesia

ABSTRACT: The number of Islamic banks in Indonesia is growing. However, the industry has experienced a decrease in performance compared to conventional banking. This research aimed to address the problems faced by Islamic banking institutions in Indonesia, using an Analytical Network Process (ANP) framework. The results show that the problems that arise in the development of Islamic banks in Indonesia consist of four important aspects, namely: human, technical, legal/structural and market/communal. Decomposition of the problem highlighted the following: 1) there is not yet enough Islamic bank capital; 2) there is a lack of understanding of Islamic bank practitioners; 3) there is a lack of government support; 4) trust and public interest in Islamic banks tend to be low. Meanwhile, the priorities of strategic policies, which are considered qualified to solve the problems of the sharia banking industry in Indonesia, are as follows: 1) to strengthen funding and business scale and increase efficiency; 2) to improve the quantity and quality of sharia banks' human resources, as well as their information systems and technology; 3) to improve sharia banks' funding structure and to align ruling and supervision.

1   INTRODUCTION

The Islamic economy is widespread globally and has made progress to date. According to the latest data published in the Global Islamic Finance Report (2018), the Islamic banking industry currently exists in at least 45 countries of the world including the United States, the United Kingdom and China. In Indonesia, the development of the Islamic financial economy began in around 1992 when the first sharia bank, Bank Muamalat Indonesia, was established.

In its recent development, sharia banking encompasses, according to data from the Financial Services Authority (OJK) up to December 2017, 13 sharia public banks, 23 sharia business units and 163 sharia public financing banks, with a total of 2,189 offices networked nationwide.

However, the industry has experienced a decrease in performance compared to conventional banking. This is reflected by, among others, relatively high non-performing financing or lower efficiency assessment compared to the conventional banking industry. The existence of sharia banks in Indonesia has not yet been reinforced by supporting factors which enable sharia banking to keep developing and running well.

There are several factors which keep the sharia banking industry in Indonesia from developing. Among these are: an insufficient number of educated and professional human resources; poor human resource management; underdeveloped entrepreneurship spirit and culture of our people; relatively small and limited financing; ambivalence between Islamic management concepts of sharia banks and operation in the field; trust issues by Muslims; and lack of a perfect academic formula to develop sharia financial institutions systematically and proportionally. These complex problems impact public confidence in the existence of sharia banks among conventional financial institutions.

On the contrary, considering the background of its establishment, sharia banking is the answer to Muslims' demands and needs. Sharia banks arose when Muslims required an Islamic-based financial institution which is free from the considered haram, *riba*. Considering the data, sharia banks in Indonesia have been thriving so far. Sharia banking keeps making progress every year in terms of assets, which have grown above 30% on average (2007–2013 data). Their financing to deposit ratio (fund disbursed) also remains around 100%, with the most part of it financing Small and Medium Enterprises (SMEs). This proves that the sharia banking industry is acceptable to the public as an institution capable of empowering small societies.

The existence of sharia banking is obviously essential to developing a sharia-based economy, especially in providing solutions to empower small and medium-sized enterprises, in being the core of populist-based economic power as well as in becoming the main support of a national economic system. This proves that the role of sharia banks is considerable for the people, since they serve as an intermediary capable of solving fundamental problems faced by small and medium-sized entrepreneurs, particularly in terms of financing. Not only can sharia banks disburse capital, they also deal with social activities.

From the concept side, sharia banks are an institution whose existence is highly needed by the people. However, on the other hand, there are still substantial weaknesses in their operation. Hence, the problematic aspects have to be overcome satisfactorily in order to build a positive image of sharia banks as clean and reliable sharia financial institutions.

Therefore, according to the aforementioned rationale, the problem formulations discussed in this study are: What are the problems faced by sharia banking institutions in Indonesia specifically in the last one to two years? What are the effective solutions? What are the strategies best applied in the long term strategic framework? Using an Analytic Network Process (ANP) approach, some of these questions will be examined and solved. This aim of this paper is to help sharia banks in Indonesia have a better performance by: 1) identifying the issues/problems faced by the sharia bank industry; 2) proposing alternative strategies for the Islamic banking industry.

2 THEORETICAL FRAMEWORK

Sharia financial institutions are those that operate based on the sharia concept under the principle of profit and loss sharing as their main method. The structure of sharia financial institutions includes sharia public banks, sharia rural banks, sharia insurance and Baitul Maal Wat Tamwil (BMT). Each of these institutions has different products and market share. However, from the side of principle and instruments, there is no significant difference among the said sharia financial institutions, only in terms of their operational working scope.

The principle of sharia finance can be widely applied in an economic system which not only focuses on the profit sharing system, but also impeccably instills ethical codes (moral, social and religious) in promoting fairness and welfare to the wider community. There is no difference in principle among the sharia financial institutions (insurance, banks and BMT) as these institutions emphasize the partnership relation (mutual investor relationship) with a profit sharing scheme as its fundamental basis.

In short, sharia banks are financial institutions which operate under sharia principles. The difference with conventional banks lies in the interest system, where sharia banks do not charge interest since to them interest system means *riba* (Sholahuddin & Hakim, 2008).

Regarding their basic concept, sharia banks are financial institutions which serve to support the economic mechanism in the real sector through business activities (investments, trades, etc.) based on the sharia principle that involves agreement rules based on sharia law between the banks and other parties in terms of fund saving and/or business activity financing, or other activities which meet the micro or macro sharia values (Ascarya & Yumanita, 2005).

Sharia banks are one of the sharia-based financial institutions which include banks, finance and investment institutions, and insurance. These three kinds of financial institutions have been operating in Indonesia. The purpose of developing them is to improve social welfare, materially and spiritually, according to Al-Qur'an and *hadiths*. The three fundamental principles of sharia banking are: productivity oriented; fairness; and halal investments.

Among other aspects included in the productivity oriented principle are: (a) capital and human resources are deployed in production and distribution which generate welfare; (b) there cannot be idle capital or resources; and (c) profit taking is allowed merely to stimulate business. From those aspects, it can be said that even sharia financial institutions are welfare oriented, but do not take account of the motivation to yield a financial return.

To uphold the fairness principle, the interest system is considered haram; whereas investments are made under the principle of risk sharing. The sanctity of the contract/covenant must be maintained where transparency and openness between both parties are highly important to reduce the risks posed from unequal information and moral hazard. In terms of halal investments, sharia banks are prohibited from investments in haram sectors such as liquor, gambling, prostitution and so on. They are also not allowed to make investments in speculative activities (Manurung & Raharja, 2004).

Sharia banking in Indonesia has made good progress. However, there are numerous obstacles to its further development, namely the *fiqh* obstacle such as the opinion from the *ulama* on whether interests are halal, haram or *syubhat* (Muhammad, 2004); inadequate dissemination on sharia banking; insufficient human resources and expertise; limited sharia banks' office network; complicated liquidity; and asymmetric information possibility (Karim, 2003).

Siswanto studies had the purpose of identifying and analyzing Sharia Financial Institutions (SFIs) models which were able to empower small enterprises, as well as finding strategies and means for SFIs to empower small and medium-sized enterprises (Siswanto, 2009). The study was undertaken under the descriptive method with the analysis technique of examining the theme content of the previous literature and studies' data concerning his study. This study analyzed the weakness and potential strength development of Baitul Maal Wat Tamwil (BMT) institutions using the SWOT (Strength, Weakness, Opportunity, Threat) technique, which was then continued by proposing solutions and strategies in developing the SFIs. The weaknesses of the SFIs, among others, were; a) external factors (competition level with competitors, collaboration or cooperation with financial institutions, government policies and other external factors such as non-governmental organizations); and b) internal factors (products of saving and financing programs, managerial competency and financial management). The suggested solution regarding the issues is that the SFIs should focus themselves on their vision and positive image building to the community, business prospects, managerial capacity and the system of technology, operation and risk.

Conversely, Susilo (2008) in his study tried to formulate applicable strategies for Sharia Public Financing Banks (BPRS) to develop business credit for SMEs. The purpose of the study was to discover the internal factors (strength and weakness) and external factors (opportunity and threat), to formulate development strategies based on the internal and external factors, as well as to determine priorities in the development strategies for PT BPRS Amanah Ummah. The result of the study suggested that the main strength factors of PT BPRS Amanah Ummah were their position and strategies, regarding nearby to customers; their weakness was the insufficient quality of human resources; their opportunity was the potential market share of Muslims around the Islamic boarding school area; while the threat for the BPRS was the numerous competitors of small and medium-sized enterprises. The study explained that the strategic location, market share, human resources quality and number of competitors became the factors to develop BPRS. This information can also be associated with sharia banking institutions in general which are part of sharia financial institutions. Therefore, in developing sharia banks, the four factors should be addressed and dealt with.

3   METHOD

This paper uses a qualitative method including problem identification from a previous study, interviews with experts and other relevant reports, with three phases in the ANP:

1. Construct model (issues and strategies)
2. Quantify the model
3. Result analysis

4 ANALYSIS AND DISCUSSION

4.1 *Decomposition*

4.1.1 *Problem identification*
The problems in developing sharia banks in Indonesia could be categorized into four aspects namely Human Resources (HR) aspects, technical aspects, legal/structural aspects and market/communal aspects. The categorization of the aspect relevant with the study conducted by Rusydiana and Devi (2013). In general, the aspects were divided into a problem criteria and a strategy criteria. Based on problem identification and policy direction in developing Indonesian sharia banks, the following ANP structural network was then formulated.

4.2 *Overall result of geometric mean*

The results obtained show statistically the consensus among experts and practitioners regarding the problems and alternatives to developing sharia banks in Indonesia. In Figure 2, show the priority problems. The result shows that the limited understanding from sharia bank practitioners becomes the main problem in the HR aspect. In the structural aspects, the lack of government support or help for the sharia banks' development becomes the main problem. From the market aspect, public trust issues and relatively low interest are the main problems. Meanwhile, from the technical aspect, the problem of inadequate sharia banks' funding becomes the most important subject of concern. Each of the rater agreement values from the four clusters varies between $W = 0.135$–0.592. The complete calculation result regarding problem clusters and their priorities can be seen as follows.

Regarding strategies, as shown by Figure 3 which depicts the overall result of geometric mean, it can be seen that to experts and practitioners there are three top strategic priorities. Firstly, to strengthen funding and business scale and increase efficiency. This strategy is the

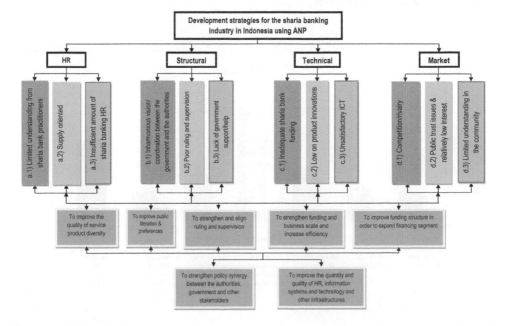

Figure 1.   ANP network.

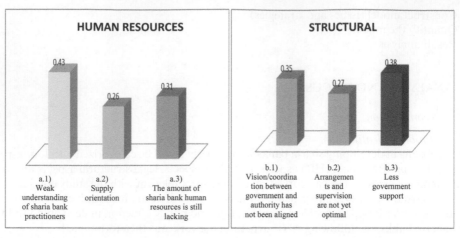

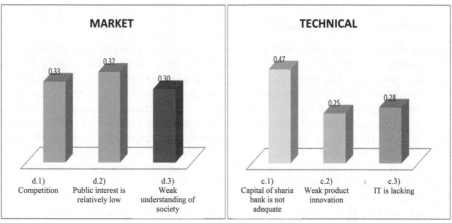

Figure 2.   Problem clusters' priorities.

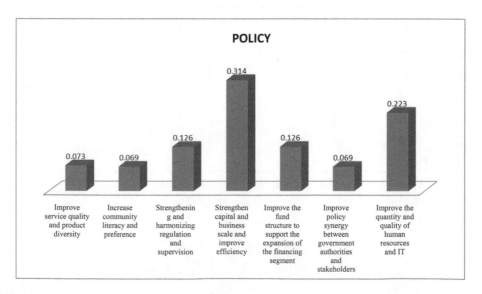

Figure 3.   Strategic policy priorities.

answer to a few problems currently faced by sharia banks namely inadequate funding, the high cost of production and other related issues.

The second strategic priority is to improve the quantity and quality of sharia banks' HR, as well as their information systems and technology. It is well known that the HR factor is a typical problem faced by sharia banks which up to now has been a major concern for sharia financial industry stakeholders. It is essential to address this priority as well as making general improvements in sharia banks' ICT.

The third and fourth strategies are to improve sharia banks' funding structure and to align ruling and supervision. With a better funding structure, sharia banks are expected to be able to achieve competitive rivalry to conventional banks. Moreover, ruling and supervision from the authorities are of high importance. Indeed, support and alignment from the government are key to the development of sharia financial institutions in Indonesia.

Overall, in terms of strategic priority order with a low rater agreement value ($W = 0.017$), the result shows that the respondents' answers regarding the priority of the strategies are highly varied.

5 RESULTS

The results of this study show that the problems arising in developing sharia banks in Indonesia consist of four important aspects, namely: HR aspects, technical aspects, legal/structural aspects and market/communal aspects. The elaboration of the problems on the whole generated the following priority order: 1) inadequate sharia banks' funding; 2) limited understanding from sharia bank practitioners; 3) lack of government support; and 4) public trust issues and relatively low interest in sharia banks.

Meanwhile, the priorities of strategic policies, which are considered qualified to solve the problems of the sharia banking industry in Indonesia, are as follows: 1) to strengthen funding and business scale and increase efficiency; 2) to improve the quantity and quality of sharia banks' HR, as well as their information systems and technology; 3) to improve sharia banks' funding structure and align ruling and supervision.

As for the level of conformity and consensus among the respondents, a relatively moderate Kendall's coefficient value among the respondents who were practitioners and experts of sharia banks is demonstrated. This means that the practitioners and experts have relatively similar opinions on identifying the problems and solutions in developing sharia banks in Indonesia.

In the meantime, several suggestions and recommendations, in accordance with the results of this study among others, are:

1. A mutual commitment between policy makers is necessary in supporting and encouraging the efforts to develop the sharia financial industry, particularly sharia banks in this case.
2. This study is expected to broaden the academic research on sharia banks especially using an ANP method. The prioritization of problems and solutions in developing sharia banks elaborated in this study is intended to be effective input for all stakeholders, so that it is clear which problem needs to be dealt with first and which strategy can be competently applied for that.
3. A further study under the same approach (ANP) will be needed to add to the number of respondents from all stakeholders who are considered to have a good understanding of the problems of sharia banks in Indonesia. Similar expectations apply regarding the use of a BOCR model.

REFERENCES

Ascarya, (2005). Analytic Network Process (ANP) New Approach to Qualitative Studies. Paper presented at the Intern Seminar of the Accounting Master's Program in the Faculty of Economics at Trisakti University, Jakarta.

Ascarya, (2011). The persistence of low profit and loss sharing financing in Islamic banking: The case of Indonesia. *Review of Indonesian Economic and Business Studies, 1.* LIPI economic research center.

Ascarya and Yumanita, D. (2005). *Islamic Banking: General Overview*. Bank Indonesia, Center of Central Bank Education.

Ascarya and Yumanita, D. (2010). "Determinants and Persistence of Conventional and Sharia Banking Margin in Indonesia" *working paper series* No.WP/10/04. Bank Indonesia, Center of Central Bank Education.

Islamic Banker Association. (2018). *Global Islamic Finance Report 2018*.

Karim, A.A. (2003), *Islamic Bank Jurisprudence and Financial Analysis* (Jakarta:IIIT. Indonesia).

Manurung, M. and Rahardja, P. (2004). "*Money, Banking and Economics. Monetary (Indonesian Contextual Study)*". Institute of Publishing of the Faculty of Economics, University of Indonesia. Jakarta

Muhammad. (2004). *Islamic Bank Fund Management*. Yogyakarta: Ekonisia.

Mu'allim, A. (2003). "Community Perception of Islamic Financial Institutions". Journal Al-Mawarid Ed X, year 2003.

OJK, (2015). *Indonesian Islamic Banking Roadmap 2015–2019*. OJK (The Financial Services Authority) Islamic Banking Department.

Rusydiana, A.S. and Abrista, D. (2013). Challenges in developing BaitulMaalwatTamwiil in Indonesia using analytic network process. *Business Management Quarterly Review, 4*(2).

Saaty, R.W. (2003). *The Analytic Hierarchy Process (AHP) for decision making and the Analytic Network Process (ANP) for decision making with dependence and feedback*. Pittsburgh: Springer, Creative Decision Foundation.

Saaty, T.L. and Vargas, L.G. (2006). *Decision making with the analytic network process. Economic, political, social and technological applications with benefits, opportunities, costs and risks*. Pittsburgh: Springer, RWS Publication.

Sholahuddin, M. dan Hakim, L. (2008). *Contemporary Islamic Economics and Finance Institute*. Surakarta: Muhammadiyah University Press.

Siswanto. (2009). *Development Strategies for Sharia Financial Institutions in Empowering Small and Medium-Sized Enterprises*. Thesis on Postgraduate Program at Diponegoro University.

Susilo, J. (2008). *The Development Strategies Formula for PT BPRS Amanah Ummah using the Approach of Analytic Network Process*. Thesis on Postgraduate Program at Bogor Agricultural University.

Wibowo, H. (2006). "The Role of Sharia Banking in Moving the Real Sector." *Paper*, presented at National Seminar and Colloquium; "Perkembangan Sistem Keuangan Syariah di Indonesia Kini dan Tantangan Hari Esok", Bandung Institute of Technology, September 30 (2006).

Business Innovation and Development in Emerging Economies – Trinugroho & Lau (Eds)
© 2019 Taylor & Francis Group, London, ISBN 978-1-138-35996-3

Production allocation and scheduling at a pharmaceutical plant

Citra Permata Sari & Ratih Dyah Kusumastuti
Department of Management, Faculty of Economics and Business, Universitas Indonesia, West Java, Indonesia

ABSTRACT: Production allocation and scheduling is the phase in production management that creates a detailed description of operations to be executed in a given period of time. The complexity of the production process and on-time delivery issue makes pharmaceutical companies had difficulties to have scheduling plans which are well balanced with actual production capacity. The objective of the research is to propose a production allocation model that captures the key aspects of the pharmaceutical industry that maximizes the profit and can be used for scheduling activities. The model is implemented to determine product allocation and schedule at PT. X*, a pharmaceutical company in Indonesia. The secondary data is collected from 1-month production report and via in-depth interview with planner of PT.X. The model is formulated using Integer Linear Programming (IP), and then, Gantt chart is used to develop the scheduling plan. The numerical example shows that the model solutions have higher profit than the current production allocation and scheduling practiced by the company.

Keywords: Production Allocation, Scheduling, Optimization, Integer Linear Programming, Pharmaceutical

1   INTRODUCTION

Global health care spending is expected to increase at an annual rate of 4.1% in 2017–2021, up from just 1.3% in 2012–2016 with aging and increasing populations, developing market expansion, advances in medical treatments, and rising labour costs as the driving factors (Deloitte, 2018). Furthermore, health care spending in the world's major regions increases from 2.4 percent to 7.5 percent between 2015 and 2020 (Deloitte, 2018). As one of the emerging countries in the world, Indonesia is expected to become the major drive of this rise (Deloitte, 2018).

The Indonesian population, which is expected to grow at a substantial rate of 1.48%, and also the implementation of National Health Service with its main goal to provide health insurance to all Indonesians (which means covering more than 260 million people) by January 2019, have increased the demand for pharmaceuticals, medical devices, and other healthcare products (Frost & Sullivian, 2015). These facts make Indonesia become a huge market for pharmaceuticals which is valued at $6.5 billion with an annual growth rate of 12.5% and is expected to continue through 2018 (Pacific Bridge Medical, 2014). It becomes a motivation for pharmaceutical firms, both domestic and foreign, to grow as large as they can. However, pharmaceutical manufacturing is not a simple straight- forward production. The manufacturing processes are regulated by the standards of Good Manufacturing Practice (Gallus, Sugiyama, & Schmidt, 2014). Moreover, due to the economical and legal implications of late deliveries and stock-outs at the final customers, the pharmaceutical industry is focusing more and more on on-time delivery issue (Venditti, 2010). The complexity of the production process and on-time delivery issues make pharmaceutical industry had

---

*Real name of the pharmaceutical company is not disclosed in terms of confidentiality reason.

difficulties to have scheduling plans which are well balanced with actual production (Venditti, 2010).

Many methods have already been applied for production and scheduling optimization such as Monte Carlo Simulation for instance by Gallus et al., 2014), Mixed-Integer Programming such as by Castro & Grossmann (2006), Integer Programming (such as by Roushdy, 2016). However, only a small proportion of these works directly addresses the scheduling issues faced by the pharmaceutical companies. Hence, the purpose of this research is to develop a production allocation model that captures the specific characteristics of the pharmaceutical industry, that can be used to develop a schedule for batch processes in the industry. The model is formulated using Integer Programming (IP) and Gannt Chart is then used to develop the schedule. The approach is implemented at PT.X to maximize its profit.

The remainder of the paper is structured as follows. Literature review is explained next, followed by discussions of research methodology, numerical example, and conclusions and recommendations.

2 LITERATURE REVIEW

2.1 Scheduling in batch process

As one of the operations management aspect, scheduling is related to how an organization arranging the timing of the use of specific resources such as facilities, equipment, human activities, etc. (Nigel, Alistair, & Robert, 2016). The constraints (i.e. capacity of the system, equipment selection, design of products and services, etc) need to be established before scheduling decisions to be made, to ensure that the scheduling decision is narrow in scope and latitude (Nigel et al., 2016).

In manufacturing system, demands or orders are interpreted as jobs along with the associated due dates (Pinedo, 2004). Then, all of the jobs are processed on the machines in a certain order or sequence (Pinedo, 2004). Pinedo (2004) also mentions that a comprehensive scheduling process of the jobs needs to be done in order to uphold efficiency and operations control.

Several manufacturing system have an integrated site for its industrial processes such as sites for chemical, food, or pulp and paper processing (Lindholm & Giselsson, 2013). This means that the products of some production areas may become the raw materials for other areas and changing the production rate in one area may therefore affect several other areas at the site (Lindholm & Giselsson, 2013). This industry commonly operating in batch mode (Susarla & Karimi, 2011). Batch processing mode consists of a set of operations which are carried out over a period of time on a separate, identifiable item or parcel of material (Fernández, et al., 2012). This processing mode enables a process to be modified without any significant equipment changes occurred (Fernández et al., 2012). Batch processing mode offers an operational flexibility and suitability, which makes this kind of process attractive when product demands change quickly, or when a small production is needed (Fernández, et al., 2012).

A production schedule in batch processing mode specifies the sequence in which products have to be produced and also the times when the processing operations should be carried out (Fernández et al., 2012). Production scheduling becomes a critical point in batch processing mode in order to achieve overall productivity and economic effectiveness because all of the operation processes are time dependent (Majozi, 2005). Finding an optimal production scheduling in batch processing mode with respect to a certain objective can be trivial or very complex process depending on the amount of available resources and time (Castro & Grossmann, 2006).

Many methods have already been applied for production and scheduling optimization especially in pharmaceutical plant as batch processing manufacturer. Investigation which relates the effects of total lead time variability with the output of tablet manufacturing plant at Eli Lilly has already been performed by Blocher, Garrett, & Schmenner (1999), by using top level information to successfully reduce the throughput time by one third. Then, a simple mathematical approach to integrated campaign planning and resource allocation in multi-product batch plants is introduced by Susarla & Karimi in 2011. Moreover, a recourse-based stochastic model is introduced by Hansen, Grunow, & Gani in 2011 as an approach that can be used in scheduling drug launching activities. Resource-Task Network is also introduced by Moniz, Barbosa-Póvoa, & de Sousa in 2014 for scheduling of campaign and short-term products in multipurpose batch plants such as in chemical-pharmaceutical company. Lastly, Brown & Vondráček (2013) introduce the time-based manufacturing (TBM) practice that successfully improves the productivity performance of two pharmaceutical preparation manufacturers in the Netherland.

2.2 *Campaign process*

Operational planning in a multi-product, multi-stage batch plants is intrinsically complex (Susarla & Karimi, 2011). The frequent product changeovers and also the long cleaning times involved are two main characteristics that must be considered in such plants (Susarla & Karimi, 2011). To minimize cost and time consumption, some plants are operated in campaign mode (Susarla & Karimi, 2011). In a campaign mode, several batches of a particular product are run sequentially to produce appropriate quantity in order to reduce set-up costs incurred due to switching between different products (Rajaram & Karmarkar, 2004). Reducing the variance in conformance quality and associated rework costs also become the advantages that a plant can get by implementing the campaign mode (Rajaram & Karmarkar, 2004). However, sufficient inventory of raw materials needs to be build up during this campaign mode which resulted in holding costs (Rajaram & Karmarkar, 2004). Thus, an optimal trade-off is required between campaign lengths and inventories (Susarla & Karimi, 2011).

The importance of campaign process is to solve scheduling problem in production plant operating in batch mode is firstly pointed out by Mauderli & Rippin (1979). Similarly, Lázaro, Espuña, & Puigjaner (1989) also implement campaign process to consider the effect of limited utilities in multipurpose batch plants. In 2002, campaign planning for multi-stage batch processes in the chemical industry is introduced by Grunow, Günther, & Lehmann. And, recently, Susarla & Karimi (2011) implement campaign process to improve the scheduling process in a pharmaceutical manufacturing plant.

3 RESEARCH METHOD

This research proposes a mathematical model (formulated using IP) for the problem of resource allocation in a pharmaceutical plant as a multi-stage batch plant. The model is developed in order to meet the special condition of PT. X as the object of this research. This research presents a general framework based on the variation of productivity with different resource allocation profile by allowing several real-life scenarios such as maintenance, safety stock limits, and minimum campaign lengths.

3.1 *IP formulations*

The problem can be defined as follows. Given the following parameters:

1. Production units, raw materials, and products.
2. Planning horizon (H), batch sizes, processing times, and cleaning times.
3. Demand forecasts.
4. Planned or anticipated maintenance schedule.
5. Product prices and manufacturing costs.
6. Electricity resources and their availability profiles.
7. Warehouse capacity.

Determine the optimal production plan (number of batches for each product and number of batches to be campaigned) that will maximize the net revenue (or profit).

The notations for the formulation are as follows:

Set & Indices:
i : product; $i \in I$
j : raw material; $j \in J$
I : set of products
J : set of raw material

Decision Variable:
B_i : number of batch units of product i that will be produced.
G_{ik} : number of batch unit of product i that will be campaigned in sequence k.

Parameters
T_i : time needed to produce batch i (hours)
T^m : maintenance time (hours)
TC : time needed for cleaning process of each batch (hours)
T_k^c : time needed for cleaning in sequence k (hours)
E_i : electricity consumption to produce batch i (hours)
E^m : electricity consumption for maintenance process (kVA)
EC : electricity consumption for cleaning process (kVA)
E_k^c : electricity consumption for cleaning process in sequence k (kVA)
M_{ji} : amount of material j needed to produce batch i (kg)
P_i : price of each batch unit of product i (Rupiah)
D_i : demand for batch units of product i per month in batch
Z_{min} : minimum number of batch units that can be campaigned (batch)
Z_{max} : maximum number of batch units that can be campaigned (batch)
T^{max} : maximum available working hours per month (hours)
E^{max} : maximum electricity consumption per month (Rupiah/kVA)
W^{max}: maximum warehouse capacity (kg)
C_i^M : production cost for batch i (Rupiah)
C^E : cost of electricity per hour (Rupiah)

Objective Function

Maximize profit:

$$\Sigma_i(B_iP_i) - \Sigma_i(C_i^M B_i) - \left((\Sigma_i)(\Sigma_k G_{ik}E_i + E_k^c) + E^m \right)C^E \tag{1}$$

Subject to:

$$B_i \le D_i \tag{2}$$

$$\sum_k G_{ik} \le D_i \tag{3}$$

$$\Sigma_i (T^C + \Sigma_i B_i T_i) + \Sigma_k (T_k^c + \Sigma_i \Sigma_k G_{ik} T_i) + T^m \le T^{max} \tag{4}$$

$$\Sigma_i(EC + B_i E_i) + \Sigma_k(E_k^c + \Sigma_i G_{ik}E_i) + E^m \le E^{max} \tag{5}$$

$$\sum_j \sum_i M_{ji}B_i \le W^{max} \tag{6}$$

$$Z_{min} \le G_{ik} \le Z_{max} \tag{7}$$
$$B_i \ge 0 \tag{8}$$

B_i and G_{ik} are integers   (9)

Two objectives have been used widely in the planning literature which are revenue and profit maximization. In this research, maximization of profit is preferred.

Maximum profit is calculated as total profit, which is the sum of the quantity of each product multiplied by the price of each batch unit of the respective product, minus the costs which consist of production cost and electricity cost (see Formulation 1).

At the beginning of the month, the forecast demand data of 10 brands of hormonal products which should be fulfilled by PT.X is available. Formulation (2) and formulation (3) ensure that the batch quantity of each product which is produced on that month cannot exceed the forecast demand.

Formulation (4) ensures that all of manufacturing processes (including: cleaning and production processes) and also maintenance process is not exceeding the maximum available working hours (T^{max}).

In addition to working hours, electricity consumption is also a matter that needs to be considered. The electricity consumption of manufacturing process and maintenance process cannot be more than the maximum electricity capacity. This requirement is shown in Formulation (5).

For inventory, the quantity of raw material that being stored in warehouse must not exceed the maximum capacity of the warehouse. This requirement is shown in Formulation (6).

Generally, manufacturing process being executed by performing a cleaning and a production process. However, the manufacturing process can also be executed by a campaign with certain minimum and maximum number of batch units that can be campaigned. This requirement is shown in Formulation (7).

Lastly, number of batch unit of product i that will be produced both by single manufacturing process or by campaigned process must be non-negative variables and integers (see Formulation 8 and 9).

The model is solved using Open Solver based in MS. Excel environment. After the number of batches for all products to be produced in the scheduling period have already been determined, the scheduling plan is developed by using Gantt Chart. All products which already named with numbers will be scheduled based on the optimization result.

Product which has smaller number will be produced first. For example, Product 1 will be produced first then followed by Product 2, Product 3, etc.

3.2 Numerical example

To analyse the performance of the mathematical model, it is applied to find the optimal production of 10 brands of hormonal products in three consecutive months. Sensitivity analysis is then performed by changing demand for each product by increasing and decreasing by 10%, 20% and 30% from the base case to see the impact of the changing of the demand to the profit generated and composition of the product that being allocated to produced.

3.3 Model application

We consider a facility of PT. X with the following information:

- Forecast demand of 10 brands of hormonal product (i_1–i_{10}) on three consecutive months which are available in Table 1.
- The batch quantity of each product which being produced on that month cannot exceed the forecast demand except the demand of one product which is in Health National Services program must be fulfilled entirely ($i4$). This matter caused the Formulation (10) is divided into two parts.

$$B_i \leq D_i \quad \forall \ i \in \{1,2.3,5,6,7,8,9,10\} \tag{10a}$$

$$B_4 = D_4 \tag{10b}$$

Table 1. Data for the optimization models.

Product	Demand (Batch) Month 1	Month 2	Month 3	Completion Time (Hours) Type of Process	Process Time	Cleaning Time	Electricity Consumption (kVA) Production Process	Cleaning Process	Manufacturing Cost (Rupiahs) Production Process	Cleaning Process	Electricity Cost (Rupiahs)	Sell Price (Rupiahs)	Profit (Rupiahs)
Product 1	12	10	10	Single	18	2	17	8	8,718,159	2,340,000	25,900	11,757,588	673,529
				Campaign (1 C + 2 P)	36	2	34	8	17,436,318	2,340,000	43,512	23,515,175	3,695,345
				Campaign (1 C + 3 P)	54	2	51	8	26,154,477	2,340,000	61,124	35,272,763	6,717,162
				Campaign (1 C + 4 P)	72	2	68	8	34,872,636	2,340,000	78,736	47,030,350	9,738,978
Product 2	14	11		Single	17	2	17	8	179,969,748	2,340,000	25,900	252,766,500	70,430,852
				Campaign (1 C + 2 P)	34	2	34	8	359,939,496	2,340,000	43,512	505,533,000	143,209,992
				Campaign (1 C + 3 P)	51	2	51	8	539,909,244	2,340,000	61,124	758,299,500	215,989,132
				Campaign (1 C + 4 P)	68	2	68	8	719,878,992	2,340,000	78,736	1,011,066,000	288,768,272
Product 3	13	10		Single	18	2	17	8	8,718,159	2,340,000	25,900	222,208,182	211,124,123
				Campaign (1 C + 2 P)	36	2	34	8	17,436,318	2,340,000	43,512	444,416,364	424,596,534
				Campaign (1 C + 3 P)	54	2	51	8	26,154,477	2,340,000	61,124	666,624,546	638,068,945
				Campaign (1 C + 4 P)	72	2	68	8	34,872,636	2,340,000	78,736	888,832,728	851,541,356
Product 4	5	5		Single	13	2	12	8	282,592,800	2,340,000	20,720	396,900,000	111,946,480
				Campaign (1 C + 2 P)	26	2	24	8	565,185,600	2,340,000	33,152	793,800,000	226,241,248
				Campaign (1 C + 3 P)	39	2	36	8	847,778,400	2,340,000	45,584	1,190,700,000	340,536,016
				Campaign (1 C + 4 P)	52	2	48	8	1,130,371,200	2,340,000	58,016	1,587,600,000	454,830,784

560

			Single	17	2	17	8	78,881,768	2,340,000	25,900	110,789,000	29,541,332
			Campaign (1 C + 2 P)	34	2	34	8	157,763,536	2,340,000	43,512	221,578,000	61,430,952
Product 5	5	4	Campaign (1 C + 3 P)	51	2	51	8	236,645,304	2,340,000	61,124	332,367,000	93,320,572
			Campaign (1 C + 4 P)	68	2	68	8	315,527,072	2,340,000	78,736	443,156,000	125,210,192
			Single	21	2	23	8	269,136,000	2,340,000	32,116	378,000,000	106,491,884
			Campaign (1 C + 2 P)	42	2	46	8	538,272,000	2,340,000	55,944	756,000,000	215,332,056
Product 6	6	5	Campaign (1 C + 3 P)	63	2	69	8	807,408,000	2,340,000	79,772	1,134,000,000	324,172,228
			Campaign (1 C + 4 P)	84	2	92	8	1,076,544,000	2,340,000	103,600	1,512,000,000	433,012,400
			Single	14	2	15	8	89,934,500	2,340,000	23,828	126,312,500	34,014,172
			Campaign (1 C + 2 P)	28	2	30	8	179,869,000	2,340,000	39,368	252,625,000	70,376,632
Product 7	6	8	Campaign (1 C + 3 P)	42	2	45	8	269,803,500	2,340,000	54,908	378,937,500	106,739,092
			Campaign (1 C + 4 P)	56	2	60	8	359,738,000	2,340,000	70,448	505,250,000	143,101,552
Product 8	2		Single	12	2	12	8	59,612,200	2,340,000	20,720	83,725,000	21,752,080
Product 9	2	2	Single	13	2	12	8	127,935,720	2,340,000	20,720	179,685,000	49,388,560
			Single	10	2	10	8	33,092,743	2,340,000	18,648	45,274,555	9,823,164
			Campaign (1 C + 2 P)	20	2	20	8	66,185,486	2,340,000	29,008	90,549,111	21,994,617
Product 10	2	4	Campaign (1 C + 3 P)	30	2	30	8	99,278,229	2,340,000	39,368	135,823,666	34,166,069
			Campaign (1 C + 4 P)	40	2	40	8	132,370,971	2,340,000	49,728	181,098,221	46,337,522

- From 10 brands, there are two products which cannot be manufactured by campaign process (i_8 and i_9) because there still no further validation study conducted in order to allowed these two products being campaigned. So, the formulation (11) become:

$$\Sigma_k G_{ik} \leq D_i \ \forall i \in \{1,2,3,4,5,6,7,10\} \tag{11}$$

- Available working hours for production process is 400 hours which include 6 working hours dedicated for maintenance process which needs to be done at the end of the month.
- Resource consumption for production process and maintenance cannot exceed 4000kVa per month with a tariff per kVa is Rp. 1,036.
- For a campaign process, a minimum number of batches that can be campaigned is 2 batches, with the maximum number of batches that can be campaigned is 4 batches. The minimum and maximum number of batches that can be campaigned resulted from a proc-

Table 2. Comparison between historical data and the optimization model solution.

	Historical Data		Optimization Model Solution			
	Quantity of Batch Unit of Product to Produce	Total Profit in Rupiah (P1)	Quantity of Batch Unit of Product to Produce	Total Profit in Rupiah (P2)	Profit Differences in Rupiah (P2–P1)	Increase Profit (%)
Month 1	9 batches of Product 1 5 batches of Product 4 2 batches of Product 8 2 batches of Product 9 2 batches of Product 10	727,721,766	13 batches of Product 3 5 batches of Product 4 3 batches of Product 6 1 batch of Product 9	3,706,086,243	2,978,364,477	509.27
Month 2	10 batches of Product 2 2 batches of Product 3 5 batches of Product 4 2 batches of Product 8 1 batches of Product 9 1 batches of Product 10	1,789,005,050	10 batches of Product 3 5 batches of Product 4 5 batches of Product 6 2 batches of Product 9	3,332,737,914	1,543,732,864	186.29
Month 3	7 batches of Product 2 2 batches of Product 3 5 batches of Product 4 3 batches of Product 6 2 batches of Product 9 2 batches of Product 10	1,912,895,711	3 batches of Product 2 10 batches of Product 3 5 batches of Product 4 4 batches of Product 6	3,343,458,042	1,430,562,331	174.79

ess validation study which has already been conducted to define the procedure of campaign process, which still need to comply with the regulation and suitable with the actual capacity of PT.X. So, the formulation (12) become:

$$2 \leq G_{ik} \leq 4 \qquad (12)$$

- Several raw materials being needed to produce each product (j_1–j_{32}). The quantity being stored for the raw materials cannot exceed the warehouse capacity (880 tons ~ 880.000 kg).

Details data related to forecast demand of each product and other supporting data are given in Table 1.

The problem consists of 16,595 numeric, 80 constraints, 69 integers, and 885 coefficients. Results of the optimization model and its comparison with historical data of manufacturing process at PT. X can be seen in Table 2. The safety stock limits for each raw material based on the optimization model can be seen in Table 3, while the scheduling plan can be seen Figure 1.

From the Table 2, it can be seen that there are significant differences between the profit generated from the optimization model with the profit from the historical data. Profit

Table 3. Safety stock limit of the optimization model results.

Material	Quantity (kg)			Material	Quantity (kg)		
	Month 1	Month 2	Month 3		Month 1	Month 2	Month 3
Material 1	0.00	0.00	0.00	Material 17	43.20	36.00	28.80
Material 2	0.00	0.00	0.00	Material 18	0.00	0.00	0.00
Material 3	2135.94	1727.97	1718.17	Material 19	48.00	40.00	32.00
Material 4	18.45	15.08	15.94	Material 20	9.60	8.00	6.40
Material 5	446.24	349.60	368.58	Material 21	1.87	1.56	1.24
Material 6	5.46	4.20	4.20	Material 22	9.60	8.00	6.40
Material 7	0.27	0.20	0.20	Material 23	29.66	24.72	19.78
Material 8	29.87	22.98	26.36	Material 24	0.00	0.00	0.00
Material 9	62.40	48.00	55.20	Material 25	16.33	32.67	0.00
Material 10	1.65	1.27	1.27	Material 26	20.49	40.98	0.00
Material 11	0.00	0.00	9216.87	Material 27	1.56	3.12	0.00
Material 12	0.97	0.81	1.06	Material 28	1.95	3.90	0.00
Material 13	3.89	3.89	3.89	Material 29	0.39	0.78	0.00
Material 14	0.30	0.30	0.30	Material 30	1.58	3.15	0.00
Material 15	15.75	15.75	15.75	Material 31	0.06	0.12	0.00
Material 16	0.36	0.30	0.24	Material 32	2.25	4.50	0.00

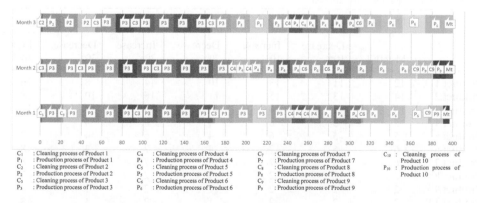

Figure 1. Scheduling plan for the optimization model solution.

generated from the historical data is lower, mainly due to miscalculation of safety stock limit of raw materials needed which resulted in the allocation of production process in low profit products. Moreover, the miscalculation of safety stock limit of raw materials needed also causes the failure of the implementation of campaign process which has already been scheduled.

From the scheduling plan, as depicted in Figure 1, scheduling plan for the three consecutive months is established. For example, Month 1 starts by producing 1 batch unit of Product 3 then followed by 3 consecutive campaign processes which producing 4 batch units of Product 3 in each process. Then, manufacturing process is continued by producing 5 batch units of Product 4 which start by producing 1 batch unit of Product 4 then followed with a campaign process of 4 batch units of Product 4. Furthermore, 3 batch units of Product 6 is produced by performing a campaign process. Then, followed by manufacturing process of 1 batch unit of Product 9. The scheduling plan of Month 1 ends with maintenance process for 6 hours.

The amount of each raw material needed is provided in Table 3. By precisely knowing the quantity of each raw material, the inventory shortage can be prevented to ensure that the manufacturing process goes according to plan.

3.4 *Sensitivity analysis*

The changing of the demand for each product is used as the reference in sensitivity analysis process because in the actual condition, the demand for each product can change significantly and more fluctuated than the changing of the selling price of each product.

Sensitivity analysis is performed in two steps as follow: 1) Decrease the demand proportionally as 10%, 20% and 30% from the base case; 2) Increase the demand proportionally as 10%, 20% and 30% from the base case (See Table 4).

The profit comparison and the number of batches manufactured for each product can be seen in Table 5 and Figure 2.

From Table 5, it can be seen that profit is directly proportional with the changing of the demand. It also can be seen that as the amount of demand increased, the number of product variety which can be produced will be decreased. This is occurred because by increasing the demand of Product 4 which needs to be fulfilled entirely, the available working hours will be spent to meet the demand as the first priority. Otherwise, when the demand of product 4 is decreased, the number of product variety which can be produced increases as seen in Table 5.

It also can be seen in Table 5, that in addition to Product 4, Product 3 and Product 6 are also commonly allocated to be produced. This is because the profit generated by these

Table 4. Data for sensitivity analysis.

	Demand (Batch)						
	Month 1	Decrease 10%	Increase 10%	Decrease 20%	Increase 20%	Decrease 30%	Increase 30%
Product 1	12	11	13	10	14	8	16
Product 2	14	13	15	11	16	10	18
Product 3	13	12	14	10	15	9	17
Product 4	5	5	6	4	6	4	7
Product 5	5	5	6	4	6	4	7
Product 6	6	5	7	4	7	4	8
Product 7	6	5	7	4	7	4	8
Product 8	2	2	2	1	2	1	3
Product 9	2	2	2	1	2	1	3
Product 10	2	2	2	1	2	1	3

Table 5. Profit comparison and the number of batches generated for each product related sensitivity analysis results.

	Profit (Rupiah)	Quantity of Batch Unit of Product to Produce
Base Case	3,706,086,243	13 batches of Product 3 5 batches of Product 4 3 batches of Product 6 1 batch of Product 9
Decreased by 10%	3,601,454,004	12 batches of Product 3 5 batches of Product 4 4 batches of Product 6 1 batch of Product 9
Increased by 10%	3,873,276,402	14 batches of Product 3 6 batches of Product 4 2 batches of Product 6
Decreased by 20%	3,280,900,122	3 batches of Product 2 10 batches of Product 3 4 batches of Product 4 4 batches of Product 6 1 batch of Product 9
Increased by 20%	3,993,584,457	1 batch of Product 2 15 batches of Product 3 6 batches of Product 4 1 batch of Product 9
Decreased by 30%	3,124,832,463.00	4 batches of Product 2 9 batches of Product 3 4 batches of Product 4 4 batch of Product 6 1 batch of Product 7
Increased by 30%	4,201,532,224.00	16 batches of Product 3 7 batches of Product 4

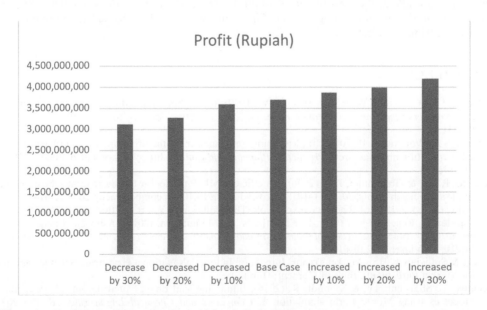

Figure 2. Profit comparison related sensitivity analysis results.

products are the largest. As also can be seen in Table 5 that in case where the demand being increased by 30%, the products that are allocated to be produced only Product 4 (which the demand must be fulfilled entirely) and Product 3 which has higher profit.

4 CONCLUSIONS AND RECOMMENDATIONS

Production allocation and scheduling are very important for pharmaceutical companies as the production process are very complex and the companies must have an on-time delivery of the finished goods at the customer sites. We propose an integer programming model that accommodates and presents novel treatment for key aspects of a pharmaceutical plant, such as sequence-dependent changeovers, maintenance, resource allocations, safety stock, etc. Compared to the historical data, the optimization model generates higher profit and also calculate the amount of raw data which prevents the inventory shortage to ensure the manufacturing process goes according to plan. From the sensitivity analysis, it can be concluded that the model will always choose product with the highest priority and those with higher profit. Thus, the model implementation shows that it successfully demonstrates the usefulness of the model as it is able to quickly optimize the production allocation for any given scenario, and thus has a potential to serve as a decision support tool for planners and other stake holders in practice.

The Gantt Chart model which being used to develop the schedule plan also helps to represent the sequence of production processes that must be done. It helps planners and other stake holders in terms of maintaining and monitoring the manufacturing process during 1-month period to ensure everything goes according to the optimization result given by the IP model.

Although the optimization model successfully gives significant results in terms of generated profit and also appropriately illustrates the relationship between the parameters, this model still has several limitations. First the model only considers a single line production process, and only takes into account certain resources. Water, for instance, is not consider in the model due to the difficulty to clearly differentiate the water consumption for manufacturing process and non-manufacturing process. The delivery dates are also not included as parameter in this model because there are no pre- specified delivery dates for each product. Lastly, the model still assumes that there is no initial inventory available and the unit of machines are idle at the start of the planning horizon, which may not be the actual representation of the manufacturing system. Thus, further research is needed to overcome the above-mentioned limitations.

REFERENCES

Blocher, J. D., Garrett, R. W., & Schmenner, R. W. (1999). Throughput time reduction: Taking one's medicine. *Production and Operations Management, 8*(4), 357–373. https://doi.org/10.1111/j.1937–5956.1999.tb00313.x.

Brown, S., & Vondráček, P. (2013). Implementing time-based manufacturing practices in pharmaceutical preparation manufacturers. *Production Planning and Control.* https://doi.org/10.1080/09537287.2011.598267.

Castro, P. M., & Grossmann, I. E. (2006). An efficient MILP model for the short-term scheduling of single stage batch plants, *30*, 1003–1018. https://doi.org/10.1016/j.compchemeng.2005.12.014.

Deloitte. (2018). 2018 Global health care outlook: The evolution of smart health care, 1–31.

Fernández, I., Renedo, C. J., Pérez, S. F., Ortiz, A., & Mañana, M. (2012). A review: Energy recovery in batch processes. *Renewable and Sustainable Energy Reviews, 16*(4), 2260–2277. https://doi.org/10.1016/j.rser.2012.01.017.

Frost & Sullivan. (2015). Market Trends: Impact of Indonesia's National Healthcare Scheme Insights for Market Participants Table of Content, (July).

Gallus, L., Sugiyama, H., & Schmidt, R. (2014). Improving lead time of pharmaceutical production processes using Monte Carlo simulation &. *Computers and Chemical Engineering, 68*, 255–263. https://doi.org/10.1016/j.compchemeng.2014.05.017.

Grunow, M., Günther, H. O., & Lehmann, M. (2002). Campaign planning for multi- stage batch processes in the chemical industry. *OR Spectrum, 24*(3), 281–314. https://doi.org/10.1007/s00291-002-0098-y.

Hansen, K. R. N., Grunow, M., & Gani, R. (2011). *Robust Market Launch Planning for a Multi-Echelon Pharmaceutical Supply Chain. Computer Aided Chemical Engineering* (Vol. 29). https://doi.org/10.1016/B978-0-444-53711-9.50187-5.

Indonesian Pharmaceuticals 2014 Update. (2014). Retrieved March 17, 2018, from https://www.pacific-bridgemedical.com/publication/indonesian-pharmaceuticals-2014-update/.

Lázaro, M., Espuña, A., & Puigjaner, L. (1989). A comprehensive approach to production planning in multipurpose batch plants. *Computers and Chemical Engineering, 13*(9), 1031–1047. https://doi.org/10.1016/0098–1354(89)87045-0.

Lindholm, A., & Giselsson, P. (2013). Production scheduling in the process industry. *22nd International Conference on Production Research*. Retrieved from http://lup.lub.lu.se/record/3736888.

Majozi, T. (2005). Wastewater minimisation using central reusable water storage in batch plants. *Computers and Chemical Engineering, 29*(7), 1631–1646. https://doi.org/10.1016/j.compchemeng.2005.01.003.

Mauderli, A., & Rippin, D. W. T. (1979). Production planning and scheduling for multi- purpose batch chemical plants. *Computers and Chemical Engineering, 3*(1–4), 199–206. https://doi.org/10.1016/0098–1354(79)80033-2.

Moniz, S., Barbosa-Póvoa, A. P., & de Sousa, J. P. (2014). Simultaneous regular and non-regular production scheduling of multipurpose batch plants: A real chemical- pharmaceutical case study. *Computers and Chemical Engineering, 67*, 83–102. https://doi.org/10.1016/j.compchemeng.2014.03.017.

Nigel, S., Alistair, B.-J., & Robert, J. (2016). *Operations Management*. (S. Wall, Ed.) (12 ed). Pearson Education, Inc.

Pinedo, M. (2004). *Planning and Scheduling in Manufacturing and Services* (2nd ed.). New York: Springer.

Rajaram, K., & Karmarkar, U. S. (2004). Campaign Planning and Scheduling for Multiproduct Batch Operations with Applications to the Food-Processing Industry. *Manufacturing & Service Operations Management, 6*(3), 253–269. https://doi.org/10.1287/msom.1040.0045.

Roushdy, B. H. (2016). Integer Programming Model for Inventory Optimization for a Multi Echelon System, *4*(1), 47–52. https://doi.org/10.12720/joams.4.1.47–52.

Susarla, N., & Karimi, I. A. (2011). Integrated campaign planning and resource allocation in batch plants. *Computers and Chemical Engineering, 35*(12), 2990–3001. https://doi.org/10.1016/j.compchemeng.2011.03.025.

Venditti, L. (2010). *Production scheduling in pharmaceutical industry*. Roma Tre University.

Business Innovation and Development in Emerging Economies – Trinugroho & Lau (Eds)
© *2019 Taylor & Francis Group, London, ISBN 978-1-138-35996-3*

Application of fuzzy kernel robust clustering for evaluating the internationalization success of companies

Z. Rustam & F. Yaurita
Department of Mathematics, Universitas Indonesia, Depok, Indonesia

M.J. Segovia-Vergas
Department of Financial Economy and Accounting I, Facultad de Ciencias Económicas y Empresariales, Universidad Complutense de Madrid, Spain

ABSTRACT: Internationalization started to be seen as an opportunity for many companies. This is one of the most crucial growth strategies for companies. Internationalization can be defined as a corporative strategy for growing through foreign markets. It can enhance the product lifetime, and improve productivity and business efficiency. However, there is no general model for a successful international company. Therefore, the success of an internationalization procedure must be estimated based on different variables, such as the status, strategy, and market characteristics of the company. In this paper, we try to build a model for evaluating the internationalization success of a company based on existing past data by using Fuzzy Kernel Robust C-Means. The results are very encouraging and show that Fuzzy Kernel Robust C-Means can be a useful tool in this sector. We found that Fuzzy Kernel Robust C-Means achieved 85.40% accuracy rate with RBF kernel, 70% data training, and $\sigma = 0.05$.

1 INTRODUCTION

Internationalization started to be seen as an opportunity for many companies. It is an important strategic choice of the companies to face the economic globalization challenge (Behyan, 2016). Internationalization is defined as the process of increasing involvement in international operations (Welch & Luostarinen, 1988; Jiang & Xu, 2011). It refers to the degree of sale incomes or operations which companies are directing in the foreign market (Holmlund et al., 2007). It can enhance the product lifetime, and improve productivity and business efficiency. Thomson & Strickland in 2015 said there are five reasons why a company may opt to enter the foreign market. There are: to gain access to new customers, to gain access to resources in foreign market, to gain access to low-cost inputs of production, to spread the business risk, and the last is to achieve lower costs through the increased of purchasing power However, internationalization is a high-risk business strategy. The companies should prepare for political and economic risks: they will face international market change, technology, market and financial risks.

There is no general model for a successful international company. The success of an internationalization procedure must be estimated based on different variables such as the status, strategy, and market characteristics of the company (Elango & Pattnaik, 2007). In this study, we will build a model by using existing past data of companies that have faced an internationalization process before. This model will be able to classify internationalization success or failure of a company. It will help a company to predict the internationalization success of the company based on the company's current condition.

We used machine learning techniques to build the model. Machine learning is an application of artificial intelligence that provides systems with the ability to automatically learn and improve from experience without being explicitly programmed. Fuzzy C-Means (FCM) is known as being a powerful machine learning tool for classification. Application of FCM has been used by Rustam et al. (2017), Fanita and Rustam (2017), Penawar and Rustam (2016), Rustam and Talita (2015), and Rustam and Yaurita (2017) for predicting the direction of Indonesian stock price movement, predicting the composite index price, forecasting stock market momentum, intrusion detection systems, and insolvency prediction in insurance companies, respectively.

In this study, we try to improve the FCM algorithm by using several theories. As mentioned above, the algorithm used past data of previous companies which have faced an internationalization process. We found that Fuzzy Kernel Robust C-Means can be a useful tool to evaluate the internationalization success of companies.

The rest of the paper is structured as follows. In the next section, there is a brief introduction to Fuzzy Kernel Robust C-Means. The third section describes the variables that we used and the results of our experiment, and finally, the last section contains some concluding remarks.

2 EXPERIMENTAL DETAILS

2.1 *Fuzzy C-Means*

The Fuzzy C-Means (FCM) is one of a fuzzy clustering method (Bishop, 2006) first introduced by Dunn (Dunn, 1973), later extended by Bezdek (Bezdek, 1981). FCM is a technique of data grouping where each data point in a cluster is determined by the membership values (Murisdah & Murfi, 2017). Given data set $X = \{(x_i, y_i) : i = 1, 2, \ldots, n\}$, where x_i is vectors data from sample i, and $y_i \in (-1, +1)$ is the associated label of x_i. Each data x_i has a variable indicator $U = [u_{ij}], u_{ij} \in [0,1], 1 \le i \le n, 1 \le j \le c$ that denotes the membership values of data x_i to cluster v_j. Each object in $V = \{v_1, v_2, \ldots, v_c\}$ is an element of d-dimensional Euclidean space (Bezdek, 1981). The mathematical model of FCM can be expressed as (Bezdek, 1981):

$$J(x, v) = min \sum_{i=1}^{n} \sum_{j=1}^{c} \left(u_{ij} \right)^m d^2 \left(x_i, v_j \right) \tag{1}$$

with constraints:

$$\sum_{j=1}^{c} u_{ij} = 1, i = 1, 2, \ldots, n$$

$$\sum_{i=1}^{n} u_{ij} > 0, j = 1, 2, \ldots, c \tag{2}$$

$$u_{ij} \in \left[0,1 \right], j = 1, 2, \ldots, c$$

where c is the number of cluster, $2 < c \le n$, n is the number of data, d is distance or dissimilarity function, and $m \in [1, \infty]$ is the fuzziness degree for the cluster partition. By applying the technique of Lagrange multiplier in Equation 1 and Equation 2, cluster center and membership values are updated by using:

$$v_j = \frac{\sum_{i=1}^{n} \left(u_{ij} \right)^m x_i}{\sum_{i=1}^{n} (u_{ij})^m}, j = 1, 2, \ldots, c \tag{3}$$

and

$$u_{ij} = \left(\sum_{l=1}^{c} \left(\frac{d\left(x_i, v_j' \right)}{d\left(x_i, v_l' \right)} \right)^{\frac{2}{m-1}} \right)^{-1}, 1 \le i \le n \; j = 1, 2, \ldots, c \tag{4}$$

In the first condition, it is obvious that the location of cluster center still not accurate. By repairing the cluster center and membership value of each data repeatedly, the cluster center will move to the right location. Figure 1 shows the algorithm of FCM.

2.2 *Fuzzy robust clustering*

Fuzzy Robust Clustering is the reformulation of Fuzzy C-Means. The classical FCM is not robust against noise and outliers (Cimino et al., 2004). Thus, in this study we used the robust version of FCM, this method can reduce the noise and also the outliers, it is called Fuzzy Robust Clustering. The algorithm works almost the same as FCM. Given the data set $X = \{(x_i, y_i) : i = 1, 2, \ldots, n\}$, where x_i is vectors data from sample i and $y_i \in (-1, +1)$ is the associated label of x_i. Similarly to FCM, there will be a membership matrix $U = [u_{ij}], 1 \leq i \leq n, 1 \leq j \leq c$, and the set of prototype $V = \{v_1, v_2, \ldots, v_c\}$. The general form of the proposed objective function is:

$$J(x, v) = \sum_{i=1}^{n} \sum_{j=1}^{c} (u_{ij})^m d^2(x_i, v_j) + \eta_j f^m(u_{ij}) \tag{5}$$

with constraints:

$$\sum_{j=1}^{c} u_{ij} = 1, i = 1, 2, \ldots, n$$

$$\sum_{i=1}^{n} u_{ij} > 0, j = 1, 2, \ldots, c \tag{6}$$

$$u_{ij} \in [0, 1], j = 1, 2, \ldots, c$$

where c is the number of cluster, n is the number of data, $d(x_i, v_j)$ is the distance measure from the data point x_i to $v_j \in V$. Assume that there is a noise cluster outside each data cluster. $f(u_{ij})$ stands for the fuzzy complement of u_{ij}, it may be interpreted as the degree to which x_i does not belong to the j-th data cluster (Yang et al., 2002). Its formula defined as (Krishnapuram & Keller, 1993):

$$f(u_{ij}) = (1 - u_{ij}) \tag{7}$$

```
Input    : X, c, m, ε, T
Output   : U and V
   1. Initial condition:
      U¹ = [u_ij], 1 ≤ i ≤ n, 1 ≤ j ≤ c
   2. For i = 1 to T
   3. Calculate Vᵗ = [v_j], 1 ≤ j ≤ c
```

$$v_j = \frac{\sum_{i=1}^{n} (u_{ij})^m x_i}{\sum_{i=1}^{n} (u_{ij})^m}, j = 1, 2, \ldots, c$$

```
   4. Calculate membership
```

$$u_{ij} = \left(\sum_{l=1}^{c} \left(\frac{d(x_i, v_j{}^t)}{d(x_i, v_l{}^t)} \right)^{\frac{2}{m-1}} \right)^{-1}$$

```
   5. If ‖U^{t+1} − Uᵗ‖ ≤ ε stop
```

Figure 1.   Fuzzy C-Means algorithm.

or (Yang et al., 2002):

$$f\left(u_{ij}\right) = \left(1 + u_{ij}\ log\ u_{ij} - u_{ij}\right) \tag{8}$$

For the fuzzy complement $f(u_{ij})$ we select the formula of Yang et al. (Equation 8) because of the simplicity of its derivative. It is easy to prove that $f(u_{ij})$ satisfies the axioms of fuzzy complements, boundary conditions, and monotonicity (Penawar & Rustam, 2016). For the membership matrix $U = [u_{ij}]$, , we differentiate the objective function $J(x,v)$ with respect to u_{ij} and then we get:

$$u_{ij} = exp\left(-\frac{d^2\left(x_i, v_j\right)}{\eta_j}\right) \tag{9}$$

and the prototype will be:

$$v_j = v_j + \alpha_t\left(x_i - v_j\right) exp\left(-\frac{d^2\left(x_i, v_j\right)}{\eta_j}\right) \tag{10}$$

where $\alpha_t = \alpha_0(1 - \frac{t}{T}), T$ is the maximum iteration and t denotes the t-th iteration, and the value of η_j is calculated using:

$$\eta_j = {}^{min}_k d^2\left(v_k, v_j\right), k \neq j \tag{11}$$

2.3 Fuzzy kernel robust clustering

Most of the data set is not linearly separable. Therefore the data set should be transformed into multidimensional form. We need another tool as a 'connector' between the data space and the feature space, so we can obtain better accuracy without directly working in feature space (Bezdek, 1981). This concept is called the kernel method, which was defined by Vapnik (2000) and elaborated further by Platt (1999), Christianini and Taylor (2000), and Schölkopf et al. (1998).

Set a nonlinear mapping φ from input data space $\mathbb{R}^d$ into feature space F. Let x and y are objects in data space. We need to find a way to measure the distance between transformed data $\varphi(x)$ and $\varphi(y)$ without knowing the explicit form of φ. To solve this problem we use kernel function K, as in Vapnik (2000). By using the kernel function, the distance between $\varphi(x)$ and $\varphi(y)$ can be measured by:

$$\begin{aligned} d^2(\varphi(x), \varphi(y)) &= \|\varphi(x) - \varphi(y)\|^2 = \varphi(x)^t \varphi(x) - 2\varphi(x)^t \varphi(y) + \varphi(y)^t \varphi(y) \\ &= k(x,x) - 2k(x,y) + k(y,y) \end{aligned} \tag{12}$$

The success of the kernel method on the classification problem (Christianini & Taylor, 2000; Schölkopf et al., 1998) inspired other researchers to apply the kernel method to the classic classification method, as in Krishnapuram and Keller (1993), that combines the Fuzzy K-Medoids algorithm and kernel method to solve a multi-class, multidimensional data classification problem. Therefore, in this paper we used Fuzzy Kernel Robust C-Means, which is obtained by applying the kernel method to the Fuzzy Robust C-Means method.

Given the data set $X = \{(x_i, y_i) : i = 1,2,...,n\}$, let a set of companies be represented by vectors data $X' = \{x_1, x_2,...,x_n\}$ where n is the number of samples and every sample is described by 30 variables. There is an associated label $y_i \in \{-1,1\}$ for each of the samples, and this represents the success or failure of the company in its internationalization process. Similarly to

Fuzzy Robust C-Means, there will be membership matrix $U = [u_{ij}]$, and the set of prototypes $V = \{v_1, v_2, \ldots, v_c\}$. The general form of the proposed objective function is:

$$J(x,v) = \sum_{i=1}^{c} \sum_{j=1}^{n} \left(u_{ij} \right)^m d^2 \left(x_i, v_j \right) + \eta_j f^m \left(u_{ij} \right) \tag{13}$$

with constraints:

$$\sum_{j=1}^{c} u_{ij} = 1, i = 1, 2, \ldots, n$$

$$\sum_{i=1}^{n} u_{ij} > 0, j = 1, 2, \ldots, c \tag{14}$$

$$u_{ij} \in [0,1], j = 1, 2, \ldots, c$$

In Karayiannis and Bezdek's research about Fuzzy Learning Vector Quantization (LVQ) (Yang et al., 2002), for each iteration a different fuzziness degree m is used:

$$m = m_i + \frac{t}{T} \left(m_f - m_i \right) \tag{15}$$

where T is the maximum iteration, t is denoting the t-th iteration, m_i and m_f are initial value and end value of m, respectively. When the value of m_f is small and m_i is quite big then it is expected that m will be decreasing and vice versa (Bezdek, 1981).

Let $w = [w_1, w_2, \ldots, w_c]$ where $w_j \in [0,1], w_1 \geq w_2 \geq \ldots \geq w_c$ and $\sum_{j=1}^{c} w_j = 1$. For every iteration, the distance between x_i and prototype $v_j(v_1, v_2, \ldots, v_c,) \in V \subset X$ is sorted so that $d^2(x_i, v_1) \leq d^2(x_i, v_2) \leq \ldots \leq d^2(x_i, v_c)$ [Karayiannis & Bezdek, 1997]. This approach will improve the accuracy of LVQ in the (MRI) segmentation.

Based on the Karayiannis experiment (Karayiannis & Bezdek, 1997; Karayiannis, 2000), in this paper, we placed the weight on updating the prototype process. Thus,

$$v_j = v_j + w_j u_{ij} \left(x_i - v_j \right) \tag{16}$$

```
Input   : X, c, mᵢ, m_f, ε, T
Output  : U and V
   1. Initial condition:
      V⁰ = [v₁, v₂, …, v_c]
   2. For t = 1 to T
   3. m = mᵢ + t(m_f−mᵢ)/T
   4. Calculate membership

      u_ij = (r²(xᵢ,vⱼ)/ηⱼ)^(1/(m−1)) , 1 ≤ i ≤ n, 1 ≤ j ≤ c

      where ηⱼ = min_k r²(vₖ,vⱼ), k ≠ j
   5. Update prototype
      Vᵗ = [v₁, v₂, …, v_c], where vⱼ = vⱼ + wⱼu_ij(xᵢ − vⱼ)
   6. If E = Σ_{j=1}^{c} r²(v_jt, v_jt−1) ≤ ε stop
```

Figure 2. Fuzzy Kernel Robust C-Means algorithm.

and for updating the membership matrix, we used:

$$u_{ij} = \left[\frac{r^2\left(x_i, v_j \right)}{\eta_j} \right]^{\frac{1}{m-1}}, 1 \leq i \leq n, 1 \leq j \leq n \qquad (17)$$

where

$$\eta_j = {}^{min}_{k} d^2\left(v_k, v_j \right), k \neq j \qquad (18)$$

By combining the Fuzziness Degree concept, Ordered Weight, Possibilitic C-Means, Fuzzy K-Medoids, kernel function, and Fuzzy Robust Clustering, the algorithm of Fuzzy Kernel Robust C-Means can be seen in Figure 2.

3 METHODOLOGY AND RESULTS

3.1 *Data and variables*

In this study, we built the model by using the data set from Segovia (Landa-Torres et al., 2012). In prior research, she used the Hybrid Grouping Harmony Search method for evaluating the internationalization success of companies, as we can see in Landa-Torres et al. (2012). This data set is elaborated annually by the Spanish Foundation of State Industrial Participation (SEPI), from several enquiries to manufacturing companies with more than ten employees. This database is elaborated for the whole country and all variables involved are annual. In this experiment, there is data from 595 Spanish companies, including the success or failure of the company in its internationalization process. Each company is described by 30 variables. All the respective variables describe the companies themselves and their internationalization strategy. Table 1 shows a list that briefly describes every factor taken into account in the proposed algorithm for assessing the success or failure of internationalization companies.

In machine learning, before we use the algorithm, we should split the data set into a training set and a testing set. The training set is implemented to build up a model while the testing set is to validate the model built (check the accuracy). Based on the prior experiments, most

Table 1. List of variables for assessment of success of corporate internationalization.

Variable	Definition
General features of the company	
V1	Categorical variable that indicates if the company is integrated in any group of societies. State of the variable: No—Yes.
V2	Percentage of share capital which corresponds to the company with equity interest in share capital.
V3	Percentage of direct or indirect foreign capital share in the share capital of the company.
V4	Categorical variable that indicates if the company has equity interest in the capital share of other companies located in foreign countries. State of the variable: No—Yes.
V5	Owners and family subsidies, employees up to December 31st, 2008.
V6	Owners and family subsidies occupying management positions up to December 31st, 2008.
Business abilities and skills	
V7	Average percentages for the year in which the standard production capacity of the company is used.
V8	Added value, in miles of Euros, divided by the average total personnel. This variable is only calculated for companies showing non-negative added value.
V9	Percentages that represent the total Research and Develompment (R&D) expenses over sales volume.

(*Continued*)

Table 1. (*Continued*).

Variable	Definition
V10	Categorical variable that indicates if the company introduced innovations during the production process, consisting of the incorporation of new machinery and equipment.
V11	Categorical variable that indicates if the company introduced innovations for the production process, consisting of the use of new software for industrial processing.
V12	Number of markets defined by the company, which geographical scope is exterior or interior and exterior.
V13	Percentage of the company's importations over total sales figure.
V14	Categorical variables that indicates if the company had some technological collaboration with its customers. States of the variable: No—Yes.
V15	Categorical variables that indicates if the company had some technological collaboration with its suppliers. States of the variable: No—Yes.
V16	Categorical variable that indicates if the company uses the head company located in foreign countries as a means of access to the international markets.
V17	Investment intensity of equipment.
V18	Added value over sales.
V19	Increase of the company's total sales.
V20	Percentages of production and other incomes less total cost. States of the variable: Less than 5%—From 5 to 15%—From 15 to 25%—Over 25%.

Human capital

V21	Percentage of non-skilled personnel over total personnel up to December 31st, 2008.

Strategic decisions or export policy

V22	Percentage of exportations to the European Economic Community (EEC) over total exportations.

Competitive environment or structure

V23	Code for representing the main activity of the company, according to the addition of the three digit codes CNAE-93 to 20 manufacturing industrial sectors.
V24	Categorial variable that includes the geographical scope of main selling market of the company. States of the variable: Local—Provincial—Regional—National—Exterior—Interior and exterior.
V25	Categorial variable that includes the geographical scope of number two selling market of the company. States of the variable: Local—Provincial—Regional—National—Exterior—Interior and exterior.
V26	Categorical variable that includes the geographical scope of number three selling market of the company. States of the variable: Local—Provincial—Regional—National—Exterior—Interior and exterior.
V27	Categorical variable that includes the geographical scope of number four selling market of the company. States of the variable: Local—Provincial—Regional—National—Exterior—Interior and exterior.
V28	Categorical variable that classifies the company according to the value of the market share evolution index for the fiscal year. States of the variable: Increased (index rated 65–100)—Constant (index rated 35–65)—Decreased (index rated 0–35).
V29	Sum of the market share evolution indexes corresponding to all the markets served by the company during the fiscal year.
V30	Categorical variable that indicates the number of competitors in the main selling market for the company's products. The market is considered to be atomized when the company reports no presence of other companies with significant market share and the company itself has a market share less or equal to 10% of its main market. States of the variable: 10 or less—From 11 to 25—More than 25—Atomized market.

of the training set is not linearly separable. To solve this problem, the data set should be transformed into another space (feature space), with a dimension that is much higher than the data space. It is expected that the transformed data behavior can approach the linearly separable data, so that classification accuracy can be improved. In this experiment, we used the RBF kernel, which is defined in Equation 19:

$$k\left(\mathbf{x}_i, \mathbf{x}_j\right) = \exp\left(-\frac{\left\|x_i - x_j^2\right\|}{2\sigma^2}\right) \tag{19}$$

3.2   Laplacian score as feature selection

A lot of classical feature selection methods have been proposed. Feature selection is used to eliminate irrelevant and redundant variables of every company and regularly improves the model performance. The remaining variables are used by the Fuzzy Kernel Robust Clustering for the classification process. In this study we used the Laplacian score. The objective function of the Laplacian score is defined as follows:

$$L_r^{LS} = \frac{\sum_{ij}\left(f_{ri} - f_{rj}\right)^2 S_{ij}}{Var\left(f_r\right)} \tag{20}$$

where L_r^{LS} is the Laplacian score of the r-th feature, f_{ri} and f_{rj} denote the r-th feature of the samples x_i and x_j, $var(f_r)$ is the estimated variance of the r-th feature, and S is the unsupervised weight matrix of the k nearest neighbor graph, which is defined as:

$$S_{ij} = exp\left(-\frac{\|x_i - x_j\|^2}{t}\right) \tag{21}$$

if x_i and x_j are k nearest neighbors, and $S_{ij} = 0$ otherwise. Take a look at He et al. (2005) for more information about this feature selection.

3.3   Results

As previously described, the aim of this paper is to build a model that can classify the internationalization success or failure of a company. We solved this problem by using the Fuzzy Kernel Robust C-Means method. We have two clusters here: the first cluster is successful companies and the second cluster is unsuccessful companies. The first step is to split the data set into training data and test data. Let $p\%$ be the percent data training that we took from both classes of data. From the training data we build the model by using the Fuzzy Kernel Robust C-Means algorithm, as shown in Figure 2. The last step is to calculate the accuracy. We used a $(100-p)\%$ data set from the first and second clusters. We can calculate the accuracy by using this form:

Table 2.   Accuracy and running time results of internationalization success evaluation of Spanish companies using Fuzzy Kernel Robust C-Means, kernel: RBF, and $\sigma = 0.05$.

% Data training	% Accuracy	Running time (seconds)	Parameter (σ)
10	58.69	0.09	0.05
20	63.45	0.12	0.05
30	66.35	0.1	0.05
40	71.99	0.20	0.05
50	73.74	10.13	0.05
60	83.19	12.09	0.05
70	82.58	0.33	0.05
80	83.19	0.39	0.05
90	76.27	18.72	0.05

Table 3. Accuracy and runningtime results of internationalization success evaluation of Spanish companies using Fuzzy Kernel Robust C-Means, kernel: RBF, feature selection: Laplacian
score, and $\sigma = 0.05$.

% Data training	% Accuracy	Running time (seconds)	Parameter (σ)
10	63.74	0.09	0.05
20	66.81	0.13	0.05
30	68.99	0.13	0.05
40	73.67	0.20	0.05
50	80.13	10.14	0.05
60	85.29	12.75	0.05
70	85.39	0.31	0.05
80	81.51	16.28	0.05
90	79.66	0.6	0.05

$$Accuracy = \frac{Number\ of\ true\ prediction}{Number\ of\ data\ testing} \times 100\% \qquad (22)$$

Tables 2 and 3 show the summary of our experiments, respectively, for the experiment without using feature selection (using all features/variables), and the experiment using feature selection. As previously described, feature selection is used to eliminate irrelevant and redundant variables of every company, and regularly improves the model performance. In this study we used Laplacian Score as the feature selection. Note that the best solution achieved by Laplacian Score involved 20 variables: V1, V2, V3, V4, V6, V8, V9, V10, V11, V12, V13, V14, V15, V18, V21, V24, V27, V28, V29, V30.

Both tables list the accuracy result of evaluating the internationalization success of companies. The first column is % data training, and it represents the percentage of companies that were included to build up a model. The second column is accuracy. In the third column, we can see the running time in seconds, which shows how long the program took to provide the result. The last column is the parameter (σ) of RBF kernel that we used.

Take a look at the first row in both tables: this means that we took 10% data training from successful and unsuccessful companies to build up a model. Then we used the rest of the data set (100–10)% for the data testing to validate the model built. In Table 2, we used RBF kernel, a constant parameter of 0.005, and chose percentages for data training ranging from 10% to 90%. From Table 2, we see that the best accuracy achieved was when we used 60% and 80% data training, and the worst accuracy achieved was when we used 10% data training.

In Table 3, we used RBF kernel, a constant parameter of 0.005, percentages for data training ranging from 10% to 90%, and Laplacian Score feature selection. The best accuracy achieved when we used 70% data training, and the worst accuracy achieved when we used 10% data training. The average accuracy in Table 2 is 73.27%, while the average accuracy in Table 3 is 76.1%. This means that the better model for Fuzzy Kernel Robust C-Means is obtained by using feature selection.

4 CONCLUSION

In this research, we constructed a model for evaluating the internationalization success of companies by using data from 595 companies that have faced an internationalization process. This new model can help a company to predict its internationalization success. The highest accuracy is obtained by the experiment that uses 70% data training and feature selection, it is 85.39%. The average accuracy of experiment that uses feature selection is 71.93%. In future studies, we are interested in trying other machine learning algorithms, such as Support Vector Machines, AdaBoost, kNN, Random Forest and Naïve Bayes.

REFERENCES

Behyan, M. (2016). Investigating the influence of external networks on internationalization success of firms. *IEEE 978-1-5090-2172-7/16*.

Bezdek, J. (1981). *Pattern recognition with fuzzy objective function algorithms*. New York, NY: Plenum Press.

Bishop, C.M. (2006). *Pattern recognition and machine learning*. London, UK: Springer.

Christianini, N. & Taylor, J.S. (2000). *An introduction to support vector machines and other kernel-based learning methods*. Cambridge, UK: Cambridge University Press.

Cimino, Maria G.C.A., Lazzerini, B., & Marcelloni, F. (2004). *A novel approach to robust fuzzy clustering of relational data. IEEE 0-7830-8376-1/04.*

Elango, B. & Pattnaik, C. (2007). Building capabilities for international operations through networks: A study of Indian firms. *Journal of International Business Studies, 38*(4), 541–555.

Fanita, F. & Rustam, Z. (2017). Predicting the Jakarta composite index price using ANFIS and classifying prediction result based on relative error by fuzzy kernel C-means. *AIP Conference proceedings of 3rd International Symposium on Current Progress in Mathematics and Sciences, 2023(1), 020206(2018).*

He, X.F., Cai, D. & Niyogi, P. (2005). Laplacian score for feature selection. In *Proceedings of the 18th International Conference on Neural Information Processing Systems* (pp. 507–514). Cambridge, MA: MIT Press.

Holmlund, M., Kock, S. & Vanyushyn, V. (2007). Small and medium-sized enterprises' internationalization and the influence of importing on exporting. *International Small Business Journal, 25*(5), 459–477.

J.C. Dunn (1973). A Fuzzy Relative of the ISODATA Process and Its Use in Detecting Compact Well-Separated Clusters. *Journal of Cybernetics, 3(1)*, 32–57.

Jiang, M. & Xu, M. (2011). Chinese firms. Internationalization and performance on the basis of learning capability. *International Conference of Information Technology, Computer Engineering and Management Sciences. IEEE 978-0-7695-4522-6/11.*

Karayiannis, N.B. (2000). Soft learning vector quantization and clustering algorithms based on ordered aggregation operators. *IEEE Transactions on Neural Networks, 11*(5), 1093–1105.

Karayiannis, N.B. & Bezdek, J.C. (1997). An integrated approach to fuzzy learning vector quantization and fuzzy c-means clustering. *IEEE Transactions on Fuzzy Systems, 5*(4), 622–628.

Krishnapuram, R. & Keller, J, (1993). A possibilistic approach to clustering. *IEEE Transactions on Fuzzy Systems, 1*(2), 98–110.

Landa-Torres, I., Ortiz-García, E.G., Salcedo-Sanz, S., Segovia-Vargas, M.J., Gil-López, S., Miranda, M., … Scr, J.D. (2012). Evaluating the internationalization success of companies through a hybrid grouping harmony search. *IEEE Journal of Selected Topics in Signal Processing, 6*(4), 388–398.

Mursidah, I. & Murfi, H. (2017). Analysis of Initialization Method of Fuzzy C-Means Algorithm Based on Singular Value Decomposition for Topic Detection. *International Conference on Informatics and Computational Sciences, IEEE 978-1-5386-0903-3/17.*

Penawar, H.K. & Rustam, Z. (2016). A fuzzy logic model to forecast stock market momentum in Indonesia's property and real estate sector. *AIP Conference proceedings of International Symposium on Current Progress in Mathematics and Sciences, 1862(1), 030125(2017).*

Platt, J.C. (1999). Fast training of support vector machines using sequential minimal optimization. In B. Schölkopf, C.J. Burges & A.J. Smola (Eds.), *Advances in kernel methods: Support vector learning*. Cambridge, MA: MIT Press.

Rustam, Z. & Talita, A.S. (2015). Fuzzy kernel C-means algorithm for intrusion detection systems. *Journal of Theoretical and Applied Information Technology, 81*(1), 161.

Rustam, Z., Vibranti, D.F. & Widya, D. (2017). Predicting the direction of Indonesian stock price movement using support vector machines and fuzzy kernel C-means. *AIP Conference proceedings of 3rd International Symposium on Current Progress in Mathematics and Sciences, 2023(1), 020208(2018).*

Rustam, Z. & Yaurita, F. (2017). Insolvency prediction in insurance companies using support vector machines and fuzzy kernel C-means. *Journal of physics: conf. series, 1028(1).*

Schölkopf, B., Smola, A. & Müller, K.R. (1998). Nonlinear component analysis as a kernel eigenvalue problem. *Neural Computation, 10*(5), 1299–1319.

Thompson, et al. & Strickland, A.J. (2015). *Crafting and Executing Strategy: The quest for competitive advantage*. New York: McGraw-Hill Education.

Vapnik, V.N. (2000). *The nature of statistical learning theory*. New York, NY: Springer-Verlag.

Welch, L.S. & Luostarinen, R. (1988). Internationalization: Evolution of a concept. *Journal of General Management, 14*(2), 34–55.

Yang, T.N., Wang, S.H. & Yen, S.J. (2002). Fuzzy algorithms for robust clustering. Dept. of Computer Science, Chinese Culture University.

Public sector accounting and governance

Business Innovation and Development in Emerging Economies – Trinugroho & Lau (Eds)
© 2019 Taylor & Francis Group, London, ISBN 978-1-138-35996-3

Open government: Does local wisdom matter? (Case from East Java, Indonesia)

I. Supheni
Universitas Sebelas Maret Surakarta, Surakarta, Indonesia
Sekolah Tinggi Ilmu Ekonomi Nganjuk, Ploso, Indonesia

A.N. Probohudono & A.K. Widagdo
Universitas Sebelas Maret Surakarta, Surakarta, Indonesia

ABSTRACT: The purpose of this study is to study the existing policy in Bojonegoro District of incorporating local wisdom in Bojonegoro Regency. This can be done effectively in strong developing countries. Previously, Bojonegoro was faced with low public prosperity and high public doubt about the legislature. However, this has now changed. Since 2008, Bojonegoro has upgraded its organization rules to restore public confidence through Open Government execution.

This study of uniqueness will improve the thinking around the domestic principles of Open Government. The qualitative research here is also explorative. Data analysis is done by stages of data collection, categorization, searching for the relationship between the factors that have been identified, and then lastly drawing a conclusion. Data are collected in two categories: the primary data and the secondary data. The primary data is collected using in-depth interviews with the head of the region (*bupati*), authorized officials and the public. The secondary data is collected from related documents such as articles, books, websites, government reports, and Non-Governmental Organizations (NGOs). The results of this study are one of the first improvements to connect communities with administrative reforms by increasing local wisdom. Through public communication called 'Public Dialog' (which is local wisdom), Bojonegoro people can now communicate directly with the government at a '*Sobo Pendopo*' hall. The dialog is broadcast live through public and private radio stations. Public dialog is considered to be the most effective platform for addressing issues in a participatory way. This allows the public to voice their complaints and aspirations, so that the government can directly appreciate them. The district government of Bojonegoro has decided to be more open as a way to allow the government to regulate policies.

Keywords: Open Government, local wisdom, governance

1 INTRODUCTION

As one of the largest democracies in the world, Indonesia is often faced with various challenges, regarding issues ranging from poverty, education, and public services, to disaster management issues such as floods. These challenges must not only be overcome by the government but also community participation. For example, during floods, the government's response will be more effective with community participation. One factor that can support such collaboration is the government itself. The existence of public disorder and public criticism are useful indicators to aid the government in overcoming the challenges, in order to make Indonesia better.

The movement to build openness is not something new. In 2011, Indonesia launched the Open Government Indonesia (OGI) initiative, aimed at promoting transparency and public participation in the country's development. Open Government Indonesia is a joint government movement, working with the community to achieve openness in the Indonesian government and to accelerate the improvement of public services in Indonesia, as mandated by Law No. 14 of 2008 on Public Information Disclosure (KIP) and Law No. 25 of 2009 on Public Service. The OGI movement was launched by Vice President Boediono in the Vice-Presidential Palace in January 2012. Through OGI, governmental and non-governmental institutions can sit together to determine the appropriate measures to encourage broader access to information on the activities of publicly funded bodies, and to make public services economical, easier to access and effective. The agreed-upon steps are then outlined in the OGI Action Plan. OGI is committed to implement the programs based on three pillars: transparency, participation, and innovation.

Bojonegoro District was selected to represent Indonesia as a pilot area in the Open Government Partnership (OGP) Subnational Government Pilot Program, also referred to as the Open Government Pilot Program. Bojonegoro District, along with Seoul (South Korea) and Tbilisi (Georgia), is the first regional government pilot in Asia, along with 13 of the world's major cities out of 45 cities enrolled in the event. This confirms that the district government of Bojonegoro is an example for other districts in Indonesia in implementing Open Government.

With justice and openness comes prosperity. Bojonegoro District, Indonesia, was formerly known as one of the poorest districts in East Java Province. Previously, it faced low levels of public welfare and high public distrust of the government. However, that is no longer the case. Since 2008, Bojonegoro has reformed the district administration to regain public confidence through the implementation of Open Government.

Research into Open Government is very important because it has brought changes for Indonesia. In less than a decade of endorsement, public information disclosure laws have begun to bring about change in Indonesia. Communities can provide input on public services and improve governance. To initiate Open Government, the first step taken by Indonesia is to open up data—that is, to provide access to government data in a format that is easy to use by the public. The government's perceived open data benefits are not limited to increased transparency and accountability, but also improve public services with community participation, as well as increasing social and economic innovation.

The World Bank supports the initiative by providing technical assistance, including establishing its online One Data Portal, which acts as a data center for various government agencies. The World Bank also supports efforts to encourage central and local government agencies to participate in the open data movement. Various competitions and events were created to raise public awareness about the availability of government data and to stimulate its usage.

The main question of our contribution is: what can we learn from the literature on local wisdom for the realization of Open Government? The purpose of this research is to study the local wisdom of the Open Government policy applicable in Bojonegoro Regency, Indonesia. Open Government can apply appropriately in strong developing countries. This research is an explorative case study highlighting Bojonegoro District. Data are collected in two categories: primary and secondary data. The primary data is collected using in-depth interviews. The secondary data is collected from related documents such as articles, books, websites, government reports, and Non-Governmental Organizations (NGOs). What can we learn from the literature on transparency and participation for the realization of the Open Government?

2 LITERATURE REVIEW

2.1 Concept of open government

The Organization for Economic Cooperation and Development (OECD) defines Open Government as being transparency of government action, access to services and information from government, and government responsiveness to new ideas, requests, and needs. Open Government has a major focus on transparency and public participation at four points: the basic community services, corruption-prone areas, other strategic issues of public concern, and areas of Open Government institutional and governance infrastructure in government public agencies. The key to successful implementation of the Open Government agenda is the realization of the innovative breakthroughs created by everybody. So far there have been several innovations that have been generated to support transparency and collaboration: one service portal, one government portal, Report, and an open data portal. In a spirit of transparency, Open Government provides various information related to government agencies. The information provided is the work program, budget and its absorption, and other information. In addition to this information, we can also easily access information about public services through a single service. We can easily find information on all public services available in Indonesia. In addition, we can also provide input if there is an incomplete public service. If the available information is different from the facts in the field we can use the People's Online Aspiration and Complaint Service (LAPOR) application to notify the government directly, for processing and action by the institution concerned. The open data portal allows free access to public data, which can be reprocessed by everybody to make breakthroughs that can help society. With the opening of access to information by the government, we can all now collaborate effortlessly to support government performance, starting from channeling opinions, watching, reporting, and even contributing work. Starting from an idea that is done individually, the spirit of Open Government is a communication door to connect every individual to each other, in order to refine their ideas and collaborate to advance Indonesia.

Open Government is a term used internationally to encourage efforts to improve the quality of government openness and public services. According to Harrison et al. (2012), Open Government is based on several principles, such as transparency, participation, and collaboration. By opening and expanding access to information about government (including reducing bureaucratic processes in accessing information), the government will automatically be required to further improve the quality of public services as well as the quality of the information itself. Poor performance and bad information from government will increasingly come under political pressure from the public. Conversely, if the government is transparent and continues to improve public services as well as the quality of information, it will increasingly gain high trust from the public.

This will stimulate and create a more open participation environment. Ultimately, widespread participation has the potential to encourage and strengthen the role of communities in their development, and allow researchers, the private sector, and community organizations to collaborate with the government. Some scientists argue that Open Government is also very open to technological developments (Harrison et al., 2012; Rajshree & Srivastava, 2012). Information and Communication Technology (ICT) is considered crucial in reducing not only the bureaucratic structure of government bureaucracy but also in improving efficiency and effectiveness, as well as the ease of use of public services themselves. Efficient services will also automatically further reduce the procedural costs that often waste government budgets (Janssen et al., 2012). A case in point are public services that have been made available in electronic format. The electronic or 'e' format allows governments and communities to more easily analyze the progress of existing or ongoing developments, including the detection of possible corruption. Even so, the implementation of Open Government in a region does not mean that the people must first have high-level ICT skills (Yu & Robinson, 2012, p. 181). In other words, in starting an Open Government initiative, ICT control is not an absolute requirement.

To start an Open Government initiative, the first necessary step is a high-level commitment to the transparency of information generated by an institution, and then to publishing it publicly. Such publication does not always have to be done through sophisticated or up-to-date ICT. Research conducted by Pawelke and Canares (2016, p. 15) shows that the openness of fiscal data in conventional format, issued by the local government in the interior of Yogyakarta, can stimulate the process of raising awareness and public participation in relation to budget control.

The research also shows that the role of NGOs also plays an important part in shaping the ecosystem of government openness, by facilitating increased data literacy as well as local community information. Given empowered and 'information-literate' communities, demands

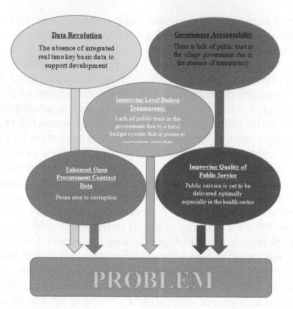

Figure 1.   Framework of thinking.

increase for local budgets to be utilized as efficiently and effectively as possible. The focus of Open Government is not limited to how the government can improve its public information disclosure governance. More recently, the OGP's agenda has focused on government collaboration with non-governmental bodies through democratic, accountable and transparent processes.

2.2   Local wisdom

Local wisdom is part of a community that cannot be separated from the language of society itself. Local wisdom (local wisdom) is usually inherited literally from story to story by word of mouth. Local wisdom is in folklore, proverbs, songs, and people's games. Local wisdom, called knowledge, is acquired by local people who collect information from home and abroad (Baedowi, 2015).

According to the Kamus Besar Bahasa Indonesia dictionary, wisdom is a series of concepts and principles that form the outline and basis of the plan for implementing a job, leadership, and how to act. The local word, which means a place or somewhere in a place to grow, is present, the life of something that may be different from another place or contained in a place of a value that may be local or universally applicable (Fahmal, 2006: 30–31).

This research took place in one region in Indonesia. (Efferin & Hartono, 2015) explained that Indonesia consists of many ethnic groups, distributed across the country. More than 1,000 ethnic groups are scattered throughout Indonesia. These ethnic groups all have different local pearls of wisdom and cultures. Among these 1,000 different ethnic groups in Indonesia, the country is dominated by the Javanese, who comprise 41.71% of the population.

2.3   Methodology

The chosen type of research is case study research. Patton (2002) describes case studies as an effort to collect and then organize and analyze data on specific cases with regard to issues of concern to the researcher. The data is then compared or linked with each other (in more cases than one) by adhering to a holistic and contextual principle. This study uses standard methods of observation, interviews, and the incorporation of several methods.

This research takes as its object Bojonegoro District, and is conducted by interviewing the head of the region (bupati) and officials in the district and in the community in Bojonegoro

Table 1. Summary of resources.

Working group	Origin	Duration of interview
Government	7	2 hours
Entrepreneur	5	2 hours
Labor	5	2 hours
Contractor	3	2 hours
Household	7	2 hours

Regency. Data retrieval is also done by observation, that is, by doing direct observation in the research location, and recording systematically observed phenomena. Interview is data collection through direct question and answer, where the interviewer meets face to face physically with the interviewee. In the execution of the interview, in-depth interviews using interview guides are undertaken with parties directly related to the implementation of Open Government.

3 RESULTS AND DISCUSSION

3.1 Action plan

These commitments are from the most recently processed action plan. The action plan submitted by the government is set out below (RPJMD, 2016). Because these commitments have not been reviewed by the Independent Reporting Mechanism (IRM), the data available may be limited. IRM-reviewed commitments appear in the next section.

1. Data Cycle
Issues to be addressed: The absence of integrated real-time key basic data to support the development program's primary objective; the development of integrated real-time data collected by the Dasa Wisma (group of ten households at the village and subdistrict/urban ward level) data application. Description of commitment: The availability of integrated, real-time, verified data in the Dasa Wisma application, incorporated into the data.go.id portal will facilitate better access and utilization of data by all parties in the decision-making process. OGP challenge: Promoting data openness, based on real-time integrated data relevance; the availability of integrated data is critical for the decision-making process, and the integration of information available at Dasa Wisma into the data.go.id portal will improve data accessibility for all parties. Ambition: To strengthen village administration data governance toward 'One Data Bojonegoro', integrated into data.go.id national portal, which will strengthen Open Government practices in Bojonegoro.

2. Enhancing Village Government Accountability
Issues to be addressed: There is a lack of public trust in the village government due to the absence of transparency. Primary objective: Open, accountable and transparent village administration and village community building capacity. Commitment description: To enhance village administration accountability and community capacity, by opening up planning and budgeting processes, village assets data and active public participation in every cycle of the decision-making process, which will lead to open village government. OGP challenge: To increase village administration accountability and transparency. Relevance: Village level accountability and transparency strengthen open village government. Ambition: A village government that is open, transparent and accountable will support corruption-prevention efforts and increase public trust in the village government.

3. Improving Local Budget Transparency
Issues to be addressed: Lack of public trust in the government due to a local budget system that is prone to corrupt activities. Primary objective: Increased public trust in the local budget system, and strengthening corruption-prevention efforts through fiscal policies that are transparent in accordance with Indonesia's Freedom Of Information (FOI) Law (Law No. 14/2008). Commitment description: We are committed to engaging the four stakeholder

groups in society (academia, private sector, government, and community) in every cycle of policymaking, from planning and implementation, and then, finally, reporting. We are also committed to publishing the key output of each stage of this cycle in accordance with public information disclosure law (Law No. 14/2008). OGP challenge: Promotion of greater transparency, participation and innovation. Relevance: A government that is more open and accountable. Ambition: Fiscal transparency policies will strengthen corruption-prevention efforts and increase public trust in the local government.

4. Enhanced Open Procurement Contract Data

Issues to be addressed: Areas prone to corruption, lack of competency of goods and services procurement providers, and lack of public oversight throughout procurement activities. Primary objective: To develop an application that would allow procurement contract data to be opened to the public in accordance with Law No. 14/2008; improve the competency of goods and services procurement providers; and increase public participation in procurement activities. Commitment description: We are committed to developing an application and appropriate business process that would allow for a more transparent procurement system. Existing procurement activities are very prone to corruption; there is lack of transparency as to how and why certain goods and services are being procured; whether or not things being procured are based on actual needs of the government and their work performance. It is worth noting that direct appointment activities are the activities most prone to corruption. OGP challenge: Strengthening transparency, participation and innovation. Relevance: Stronger regulation to promote open procurement policies. The latter will provide more space for the public to monitor, provide inputs, and oversee. Ambition: Innovation in open contract/procurement policies helps to increase transparency and accountability of the overall procurement activities and thus also serves as corruption-prevention efforts. Through this application, we wish to increase public oversight and public participation throughout the cycle.

5. Improving Quality of Public Service

Issues to be addressed: Public service is yet to be delivered optimally, especially in the health sector. Primary objective: The commitment seeks to improve public-service standards through an effective periodical evaluation, and seeks to increase public participation in the public-service delivery schemes/public-service policymaking process. Commitment description: We are committed to improve the quality of public service through the implementation of Service Standard Evaluation and a public-service standard that is jointly developed and agreed, both by the government and local society in two community health centers. OGP challenge: Strengthening public-service delivery and government accountability. Relevance: An improved public-service standards formulation process through a collaborative approach aims to improve public trust and satisfaction in the government. Ambition: Stronger collaboration between the four stakeholders with the transparent and accountable manner that can foster public participation and better public-service delivery.

3.2 *Open government based on local wisdom in Bojonegoro*

Bojonegoro Regency is located in East Java Province, consisting of 28 sub-districts with 11 urban villages and 419 villages. Bojonegoro is aiming to provide the best service to the public so as to apply the principle of public services properly. Therefore, in practice, the Bojonegoro government applies the concept of the Open Government Partnership Pilot Project. Broadly speaking, the concept of the Pilot Project of the Open Government Partnership disclosure includes financial management, human resource management, and asset management. The latter can be seen from the planning stage, the later stage of implementation, and the accountability stage. In addition, the fact is that the data is owned by the district government for all types of data that are not state secrets. The information can be viewed by society, and this is called passive openness. On the other hand, there is also active transparency, where the Bojonegoro Regency explains things to the public. Then, third, openness to all complaints and aspirations of the people. The public can make a complaint, report, and see what can be found out—publicly, either online or manually, such as through dialog or radio. Thus, the report can be idea, can be a matter of the complaint, and it can be integrated with an online system. Thus, we have the innovation management based on public participation or public complaints.

Applying the Open Government project can further solidify a collaboration between the government and community action. The task of government is as a facilitator. So, if the government comes, then when the residents say there is a problem, it is discussed together to find solutions. The government has what potential, for example, the government makes a budget plan and communicates together starting from a lower level.

Mechanisms are implemented in dialog, not through debate. Local government should present to the public, but not only present but feel, and not only feel but show how it will jointly formulate understanding of complexity of the problems that are likely to occur. Then, the government tries to brainstorm and find solutions together with the community, and the basic purpose of all this is that a better life is achieved. Programs are held every Friday, named '*Dialog Jumat*', after prayers from 1 pm until 3 pm, and open for people to share their aspirations. Various things discussed at the meeting, for example, discussed budget issues. This is a concrete form, since people can know the targets and the budget constraints of the government itself. Then, on Friday morning, the regional government also has a program management review. There, local governments jointly evaluate the form of the public response. So the government has a mechanism to check everything every Friday morning at 8 am.

The Open Government Partnership is launching an exciting new pilot program designed to more proactively involve subnational governments in the initiative. OGP is a 69-country partnership aiming to secure concrete commitments from governments to promote transparency, empower citizens, fight corruption and harness new technologies to strengthen governance. OGP is looking for subnational governments with committed political and working-level reformers, and for engaged and energetic partners in civil society, to take part in a pilot program designed to advance Open Government reform.

Bojonegoro, Indonesia, is one of the subnational governments that successfully applied to engage directly with OGP in a pilot 'pioneer' program. Participants will receive dedicated assistance and advice from the OGP Support Unit and OGP Steering Committee to develop and fulfill independent Open Government commitments in action plans, in partnership with civil society organizations. They will actively contribute to peer learning and networking activities with other subnational governments. The commitments and short action plans developed by the pioneers will be assessed by OGP's independent reporting mechanism. The pilot will give OGP the opportunity to test and assess the IRM's capacity to act as the accountability mechanism for subnational government participation.

Bojonegoro local government joined the Open Government Partnership 2016 on the Pilot Project Transparency of Government at the local government level. In general, Bojonegoro has a population of 1,450,889 inhabitants, an area of 230,706 hectares, of which 40.15% is forest area, while 32.58% are agricultural areas. It also has oil- and gas-producing regions that constitute around 20% of Indonesian oil and gas reserves. Most of the residents are farmers and farm workers. Bojonegoro also has disaster areas that flood during the rainy season and suffer drought in the dry season. It is administratively divided into 28 districts, 419 villages, and 11 urban villages.

In the discussion of the Open Government Partnership, based on the idea that openness is the main principle in public welfare. This problem will develop smoothly if all elements of society are involved at every stage. Openness will manifest in dialogue on governance, distribution and change in democratic governance. The Bojonegoro Government is committed to sustainable Open Government implementation.

Bojonegoro Regency government has implemented governmental transparency by conducting '*Sobo Pendopo*', packaged in 'Public Dialog', since March 14th, 2008. The community is given the flexibility to raise with the Regent and the entire staff a variety of problems that have occurred in order to find a solution. Public dialog is held every Friday, starting at 13:00, except on national holidays and in the month of Ramadhan. As well as the on-air broadcast on Radio Malowopati Madani 95.8 FM and 102.5 FM, there is attendance in a marquee, averaging 100–150 people.

The implementation of public dialogue, until January 2016, was recorded to have entered episode II/126, which relates to most of the problems raised by the community are infrastructure, bureaucratic reform, and good governance. Various problems were expressed by the community through public dialogue, most of which were followed up to get a solution.

To facilitate access to open government, the Bojonegoro government has also implemented the LAPOR application Open Data and Work Unit Monitoring System and Presidential Work Unit Development Control System (UKP-4), now changed to the Office of Staff of the President (KSP). Thereby the Bojonegoro people, the majority of whom are subsistence farmers and farm workers, can access open government via information technology, integrated with access to information disclosure by non-information technology existing in Bojonegoro. Examples are: *'ngetril'* (ride the bike trail), conducted by the Regent and its board, primarily to see first-hand the condition of society in areas that are difficult to reach; SMS direct access to the Regent/Deputy Regent, the regional Secretary and the heads of regional apparatus, SMS and complaints through Malowopati Radio 95.8 FM; and the use of social media, Twitter, Facebook and WhatsApp. All information and complaints are entered into the system menu of the LAPOR app. Development of integration with information technology-based information access is called SIAP LAPOR (System Integration Aspiration-Service Complaints Online People), according to the implementation of Bojonegoro Regent Regulation No. 30 of 2013 on Innovation Management Development based on Public Participation.

Follow-up of community complaints through SIAP LAPOR: Work Units (SKPD) should provide an answer within five working days and upload it into the system as part of the follow-up process. Activities addressing the many problems in government openness: evaluation is conducted by the Regent regularly every Friday from 08.00 until 12.00 under Performance Evaluation activities. Several groups can take advantage of WhatsApp to share issues, to be followed up by the Head SKPD and the parties who are members of the WhatsApp group.

The implementation of government openness has changed the culture of Bojonegoro. It is not just 'complain and complaints', but has developed into learning from each other to provide solutions and inspiration/ideas. The activities program of the government comes from the inspiration/ideas of society that have been implemented by the Bojonegoro government. Examples of these are the construction of roads by paving (more scalable and minimizes corruption), the construction of reservoirs, the management of oil and gas resources, allocating grants for two million high school students to increase long-term learning, and mentoring Civil Society Organizations (CSOs) in rural community development activities. In addition, the openness of government has transformed the delivery mechanism of the aspirations of the people through mass mobilization (demonstration), and has been shifted by means of public dialog and access to existing information disclosure in Bojonegoro. Therefore, in one year the average aspirations of the people are relatively small, which is less than five times.

The needs of the public for information and data organization in government has been uploaded onto the Bojonegoro Regency government website (http://bojonegorokab.go.id/), as well as the website of the Documentation Information Management Officer (PPID) (http://PPID.bojonegorokab.go.id). The practice of government openness in Bojonegoro has become a study for various regions. The study became part of the democratic implementation of research conducted by Otto Scharmer and Katrin Kaufer in the book *Leading from the Emerging Future: From EgoSystem to EcoSystem Economies* (Scharmer & Kaufer, 2013). In addition, the practice of openness in the Bojonegoro Regency administration has been published through other various media.

Various civil society organizations in Bojonegoro have participated in joint ventures to engage in the development and openness of government. Among these are Bojonegoro Institute (BI), and CSO Institute of Development of Society (IDF). CSO Bojonegoro Institute has, among others, been involved in the formulation of Bojonegoro Regent Regulation No. 40 of 2014 on Guidelines for Information and Documentation Services in the Environment Government of Bojonegoro. In addition, it is also involved in the formation of a documentation and information management officer website in each SKPD, has been involved in the preparation of the list of Public Information (DIP), in the increased service capacity of public information, and in the formulation of the endowment. The IDF has, among others, been involved in the formulation of Bojonegoro Regional Regulation No. 5 of 2015 concerning Corporate Social Responsibility (CSR).

4 CONCLUSION

Bojonegoro Regency, Indonesia, is a great example of this case. Previously known as one of the poorest districts in the Province of East Java, Bojonegoro has steadily climbed its way out of poverty. Previously, Bojonegoro faced having low public welfare and high public distrust toward the government, but that is no longer the case. Since 2008, Bojonegoro has reformed regency administration to regain public trust through Open Government implementation. Low public welfare in Bojonegoro was caused by trust issues. Thus, one of the first refinements was aimed at connecting the public with the administration, through a public communication initiative named 'Public Dialog', which became the local wisdom. Bojonegoro people are now able to have face-to-face communication with their government in the *pendopos* (common halls), which are a type of town hall; these are better known as '*Dialog Jumat*' and '*Sobo Pendopo*'. The dialog is broadcast live on government and private radio stations. Public dialog is considered the most effective platform for participatory problem-solving. It allows the public to voice their complaints and aspirations, and the government to respond directly to their concerns. The district government of Bojonegoro chooses to be open as a way to allow the government to find solutions to the current challenges faced.

REFERENCES

Ayat, R., 1986. *Kepribadian Budaya Bangsa (Local genius)*. Jakarta, Indonesia: Pustaka Jaya.
Baedowi, A., 2015. *Calak Edu 4: Esai-esai Pendidikan 2012–2014*. Pustaka Alvabet. 61. ISBN 978-602-9193-65-7.
Bowie, N. (1990). Equity and access to information technology. *Annual Review of the Institute for Information Studies*, 131–167.
Chadwick, A. & May, C. (2003). Interaction between states and citizens in the age of the Internet: 'e-government' in the United States, Britain, and the European Union. *Governance, 16*(2), 271–300.
Cholisin & Nasiwan, 2012. *Dasar Ilmu Politik*. Yogyakarta, Indonesia: Ombak.
Clarke, M. & Stewart, J. (2003). Handling the wicked issues. In *The managing care reader* (pp. 273–280), London.
Comfort, L.K. & Kapucu, N. (2006). Inter-organizational coordination in extreme events: The World Trade Center attack, September 11, 2001. *Natural Hazards, 39*(2), 309–327.
Comfort, L.K. (2002). Rethinking security: Organizational fragility in extreme events. *Public Administration Review, 62*, 98–107.
Coppola, D.P. (2007). *Introduction to international disaster management*. Burlington, MA: Elsevier.
Cramton, R.C. (1971). The why, where, and how of broadened public participation in the administrative process, *The Georgetown Law Journal, 60*(3), 525.
Creighton, J.L. (2005). *The public participation handbook: Making better decisions through citizen involvement*. San Francisco, CA: Jossey-Bass.
Cresswell, A.M. (2010). *Public value, and government ICT investment*. Antalya, Turkey.
Cresswell, A.M., Burke, G.B. & Pardo, T. (2006). *Advancing return on investment, analysis for government IT: A public value framework*. Albany, NY: Center for Technology in Government, University at Albany.
Cullen, R. (2010). Defining the transformation of government: Government or e-governance paradigm. In H.J. Scholl (Ed.), *E-Government: Information, technology, and transformation* (pp. 57–71). M.E. Sharpe, IOS Press Amsterdam, The Netherlands, The Netherlands.
Curtin, D. & Meijer, A.J. (2006). Does transparency strengthen legitimacy? *Information Polity, 11*(2), 109–122.
Drabek, T.E. (1986). *Human system responses to disaster: An inventory of sociological findings*. New York, NY: Springer-Verlag.
Efferin S., Monika S. Hartono, (2015) *"Management control and leadership styles in family business: An Indonesian case study"*, Journal of Accounting & Organizational Change, Vol. 11 Issue: 1, pp.130-159
Fahmal, M. (2006). Peran Asas-asas Umum Pemerintahan yang Layak Dalam Mewujudkan Pemerintahan yang Bersih. Yogyakarta, Indonesia: UII Press.

Field, J. (2008). *Social capital*. London, UK: Routledge.

Godschalk, D.R., Kaiser, E.J. & Berke, P.R. (1998). Hazard assessment: The factual basis for planning and mitigation. In R. Burby (Ed.), *Cooperating with nature: Confronting natural hazards with land-use planning for sustainable communities* (pp. 85–118). Washington, DC: Joseph Henry.

Gopalakrishnan, C. & Okada, N. (2007). Designing new institutions for implementing integrated disaster risk management: Key elements and future directions. *Disasters, 31*(4), 353–372.

Guha-Sapir, D., Hargitt, D. & Hoyois, P. (2004). *Thirty years of natural disasters 1974–2003: The numbers*. Centre for Research on the Epidemiology of Disasters, UCL Presses Universitaires De Louvain.

Harrison, T.M., Guerrero, S., Burke, G.B., Cook, M., Cresswell, A., Helbig, N., ... Pardo, T. (2012). Open government and e-government: Democratic challenges from a public value perspective. *Information Polity, 17*(2), 83–97.

Janssen, M., Charalabidis, Y. & Zuiderwijk, A. (2012). Benefits, adoption barriers and myths of open data and open government. *Information Systems Management, 29*(4), 258–268.

KBBI, 2017. Kamus Besar Bahasa Indonesia (KBBI). [Online] Available at: http://kbbi.web.id/stroke [Accessed 21 Juni 2017].

Kurniawan, T. (2012). Peranan akuntabilitas publik dan partisipasi masyarakat dalam pemberantasan korupsi di pemerintahan. *Bisnis & Birokrasi Jurnal, Mei–Agustus 2009*, 116–121.

Lane, M.R. (2014). *Decentralization & its discontents: An essay on class, political agency and national perspective in Indonesian politics*. Pasir Panjang, Singapore: ISEAS Publishing.

Layne, K. & Lee, J. (2011). Developing fully functional e-government: A four-stage model. *Government Information Quarterly, 18*(2), 122–136.

McKinsey & Company. (2013). *Open data: Unlocking innovation and performance with liquid information*. Retrieved from http://www.mckinsey.com/~/media/McKinsey/Business%20Functions/McKinsey%20Digital/Our%20Insights/Open%20data%20Unlocking%20innovation%20and%20performance%20with%20liquid%20information/MGI_Open_data_Executive_summary_Oct_2013.ashx.

Mietzner, M. (2015). Jokowi's challenge: The structural problems of governance in democratic Indonesia. *Governance: An International Journal of Policy, Administration, and Institutions, 28*(1), 1–3.

Moleong, L.J. (2004). *Metodologi Penelitian Kualitatif*. Bandung, Indonesia: PT Remaja Rosdakarya.

Noveck, B. (2012). *Demand a more open-source government*. Retrieved from https://www.ted.com/talks/beth_noveck_demand_a_more_open_source_government

Open Data Barometer (n.d.). Retrieved from http://opendatabarometer.org.

Open Government Indonesia. (2016). *Rencana Aksi Nasional Keterbukaan Pemerintah 2016–2017*. Retrieved from http://www.opengovpartnership.org/sites/default/files/31102016_Renaksi%20OGI%20 2016-2017.pdf.

Padmanugraha, A.S. (2010). *Common sense outlook on local wisdom and identity: A contemporary Javanese native's experience*. Paper presented in international conference on 'Local Wisdom for Character Building', Yogyakarta, Indonesia.

Patton, M & Cochran M., 2002, *A Guide to Using Qualitative Research Methodology*, https://evaluation.msf.org/sites/evaluation/files/a_guide_to_using_qualitative_research_methodology.pdf. (Accessed 21 Juni 2017).

Permana, C.E. (2010). *Kearifan Lokal Masyarakat Baduy dalam Mengatasi Bencana*. Jakarta, Indonesia: Wedatama Widia Sastra.

Probohudono, N.A., Muqofah, M., Wardojo, W.W. & Wibowo, A. (2017). Dimensions of Javanese culture as social control in water conflict (story from Indonesia). *Taiwan Water Conservancy, 65*(1), 80–94.

Rajshree, N. & Srivastava, B. (2012). Open government data for tackling corruption – A perspective. Semantic Cities AAAI Technical Report WS-12-13, USA.

Rosidi, A. (2011). *Kearifan Lokal Dalam Perspektif Budaya Sunda*. Bandung, Indonesia: Kiblat Buku Utama.

Scharmer and Katrin Kaufer, 2013, *Leading from the Emerging Future: From EgoSystem to EcoSystem Economies*, Berrett – Kohler Publisher, Inc, San Fransisco.

Sedyawati, E. (2006). *Budaya Indonesia, Kajian Arkeologi, Seni, dan Sejarah*. Jakarta, Indonesia: Raja Grafindo Persada.

Stagars, M. (2016). *Open data in Southeast Asia: Towards economic prosperity, government transparency, and citizen participation in the ASEAN*. Palgrave Pivot, Southeast Asia.

Transparency International France. (2017). Open data against corruption in France. Retrieved from http://webfoundation.org/docs/2017/04/2017_OpenDataFrance_EN-3.pdf.

Wihantoro, Y., Lowe, A., Cooper, S. & Manochin, M. (2015). Bureaucratic reform in post-Asian crisis Indonesia: The directorate general of tax. *Critical Perspectives on Accounting, 31*, 44–63.

Yu, H. & Robinson, D.G. (2012). The new ambiguity of 'open government'. *UCLA Law Review Discourse, 59*, 178–208. doi:10.2139/ssrn.2012489.

Risk management

Business Innovation and Development in Emerging Economies – Trinugroho & Lau (Eds)
© 2019 Taylor & Francis Group, London, ISBN 978-1-138-35996-3

Corporate governance, risk, firm size, financial performance and social performance: Granger causality and path analysis

Lidia Desiana, Fernando Africano & Aryanti
UIN Raden Fatah Palembang, Indonesia

ABSTRACT: This research aims to analyze the cause–effect relationship between financial performance and corporate social performance, which refers to research conducted by Makni, Francoeur and Bellavance (2009). In addition, this study examines the influence of corporate governance, risk and firm size on financial performance and corporate social performance. The population of this research are all companies listed on the Indonesian Stock Exchange. The sample is chosen using a purposive sampling method. The total sample of the research is 19 companies registered in the Jakarta Islamic Index. The data analysis method used is a Granger causality and path analysis technique. Prior to hypotheses testing, a classic assumption test is performed. The results of the research found that corporate governance (board of commissioners and audit committee) and firm size influence social performance. Corporate governance (board of commissioners, independent board of commissioners and audit committee) and risk (debt to total assets ratio and debt to equity ratio) affect financial performance (return on assets). Corporate governance (board of commissioners and independent board of commissioners) affects financial performance (return on equity). Only social performance variables mediate the influence of corporate governance (board of commissioners) on financial performance (return on assets and return on equity).

Keywords: corporate governance, risk, firm size, financial performance, social performance

1 INTRODUCTION

The phenomenon of global warming is increasingly widespread both in Indonesia and around the world. Both natural and environmental damage are the drivers of global warming. Damage to nature or the environment can be caused by the acts of man. Lack of human awareness of maintaining and preserving the environment accelerate global warming. Therefore, it takes high awareness of and attention to the environment to maintain and preserve it.

Discourse on social responsibility in Indonesia is growing. According to Ulfah (2008) since the 1980s, It has been discussed on Corporate Social Responsibility (CSR) and social accounting in Indonesia. This is supported by the existing provisions in Indonesia, namely Law No. 40 of 2007 on Limited Liability Companies in article 66 paragraph 2, article 74 and Law No. 25 of 2007 on Capital Investment in article 15 part b, article 16 parts d and e, and article 17. The provision affirms the obligation of companies to undertake social responsibility and to report it.

CSR is an idea that makes the company no longer faced with responsibility based on the single bottom line, the value of the company (corporate value) is reflected in the financial condition only. However, the responsibility of a company should be based on triple bottom lines that also pay attention to social and environmental issues (Daniri, 2008). While corporations are no longer entities that are only self-centered so they must alienate themselves from the community where they work. Business entities that are obliged to carry out cultural adaptation within their social environment. Corporate Social Responsibility (CSR) Asia as quoted by Darwin (2008) provides a definition of CSR as follows: "CSR is a company's commitment to operating in an economically, socially and environmentally sustainable manner while balancing the interests of diverse stakeholders". Utama (2007) states that the development of CSR is related to the increasingly severe environmental damage occurring in Indonesia and the world, ranging from deforestation, air and water pollution, to climate change.

The realization of corporate social-environmental responsibility is reflected through CSR. CSR is the responsibility of the organization regarding the impact of decisions and activities on society and the environment, embodied in the form of transparent and ethical behavior in line with sustainable development and welfare of the community, considering the expectations of stakeholders in line with established law and international behavior norms, and integrated with the organization as a whole (ISO, 2010). Currently, companies in Indonesian Stock Exchange reveal their CSR through their annual reports. Social and environmental disclosure is a way for companies to show good performance to the public and investors.

Nelling and Webb (2006) examined the causal relationship between social performance and financial performance by introducing a new econometrics technique, the Granger causality approach. Using the Ordinary Least Square (OLS) regression model, social performance is related to firm performance. They found a low relationship between social performance and Corporate Financial Performance (CFP) when using the time series effect approach. They found the same results when introducing the Granger causality model. In addition, with a focus on each of the social performance measures, the causality that occurred from the stock market performance on the social performance assessment of employee relations (Makni et al., 2009).

The study by Makni et al. (2009) examined the causality relationship between social performance and financial performance of 179 companies in Canada using the Granger causality approach. They found no association between the combined measurements of social performance and financial performance of firms, except for market returns. According to researchers, so far in Indonesia there is no any causality research on social performance and CFP using the Granger causality approach.

There are number of guidelines, criteria, indicate or assessment aspects in disclosing social responsibility such as the Global Reporting Initiative (GRI), an indicator developed by Gray et al. (1995), items developed by Michael Jantzi Research Associates, Inc. and others. However, in the absence of a single guideline for a company in disclosing its social responsibility, the researcher tried to use the GRI indicator. The GRI indicator was selected because it is an international rule that has been recognized by companies around the world.

Various results are shown by studies that examine the relationship of corporate governance, risk, firm size and financial performance to CSR disclosure. Barnea and Rubin (2005) and Machmud and Djakman (2008) stated that corporate governance does not affect CSR disclosure. On the other hand, Farook and Lanis (2007) found a link between corporate governance and social performance. Anggraini (2006) and Sari (2012) stated that there is no relationship between risk and corporate social performance. However, Meek et al. (1995) stated that leverage has a positive effect on social performance. Sembiring (2006) found that firm size had a positive effect on a company's social performance. This study is in contrast to the results of research from Anggraini (2006) which states that firm size does not affect the social performance of the company. Yusoff et al. (2013), in Malaysia, found a positive influence between financial performance and corporate social performance. On the other hand, Luethge and Guohong (2012) did not find any influence between a company's financial performance level and its social performance in China. The same was also found by Lanis and Richardson (2012) in Australia, who found that financial performance had no effect on corporate social performance.

Diverse results are shown by studies that examine the relationship between corporate governance, risk, firm size and social performance on financial performance. Research by Yaparto et al. (2013) states that social performance has a significant effect on financial performance proxied with return on assets and return on equity. Meanwhile, in research by Wijayanti and Prabowo (2011), social performance only had a significant effect on financial performance return on equity and no significant effect on financial performance return on assets and earnings per share. Dalton et al. (1998) found a positive relationship between corporate governance and company performance. While, Eisenberg et al. (1998) found a negative relationship between corporate and company performance. Research on risk variables conducted by Yahya (2011) found that risk positively influenced financial performance but the review put forward by Sunarto and Budi (2009) found that risk does not have an effect on financial performance. Sunarto and Budi (2009) proposed that firm size has a positive effect on financial performance, but research from Oktariani and Mimba (2014) found that the size of the company had no effect on financial performance.

2 LITERATURE REVIEW AND HYPOTHESIS FORMULATION

In this research, several theories are used as the foundation underlying the field of CSR, namely stakeholder theory and legitimacy theory. According to Deegan (2013), stakeholder theory is closely related to legitimacy theory.

2.1 *Stakeholder*

The term stakeholder, based on the definition by Gray et al. (1995), states that "stakeholders are parties interested in a company that can affect or can be influenced by the activities of the company, stakeholders such as society, employees, government, suppliers, capital markets and others". According to Ghozali and Chariri (2007), stakeholder theory states that a company is not an entity that only operates for its own interests but must provide benefits to its stakeholders (shareholders, creditors, consumers, suppliers, government, community, analysts and others). Thus, the existence of a company is strongly influenced by the support provided by stakeholders in the company.

Gray et al. (1995) in Ghozali and Chariri (2007) state that the survival of a company depends on stakeholder support and that support should be sought so that the company's activity is to seek that support. The more powerful the stakeholders, the greater the company's efforts to adapt. Social disclosure is considered part of a dialog between the company and its stakeholders (Ghozali & Chariri, 2007). According to the stakeholder approach, the organization chooses to respond to the many demands made by stakeholders in any group from outside the organization's environment that is affected by the actions and decisions of the organization. According to this approach, organizations will seek to meet the environmental demands of groups such as employees, suppliers, investors and society (Robbins & Coulter, 1999).

There are several reasons why companies need to pay attention to the interests of stakeholders, namely: (1) environmental issues involve the interests of various groups in society that can disrupt their quality of life; (2) the era of globalization has encouraged traded products to be friendly to the environment; 3) investors in investing tend to choose companies that own and develop environmental policies and programs; (4) civil society organizations and environmentalists are increasingly vocal in criticizing companies that are less concerned about the environment.

1. Furthermore, Gray et al. (1995) note that, in order for a company to know what its stakeholders want, the company should be able to assess the substantive environment consisting of: (1) the primary level describing the company's social interaction in the use of infrastructure, the influence of aesthetics, employee health and status satisfaction, consumer and welfare options, advertisement, residual waste, new technology and resources and social opportunity costs; (2) the tertiary level describes interactions in more complex organizational systems that are related to the quality of freedom (moral, educational, cultural and aesthetic), level of information such as news, cultural heritage, third world, system, individual choice, health, environment, legal system.
2. Lepineux (2005) divides stakeholders into two main types: stakeholders and business stakeholders. Societal stakeholders are divided into three types: the global community; the civil society of the countries where the company operates; the local communities around the company area, international institutions, governments, activists, NGOs, civil association and the media. While business stakeholders consist of three players: shareholders, internal stakeholders and external stakeholders. In detail, these stakeholders consist of shareholders, managers and executives, employees and workers, trade

unions, customers, suppliers, subcontractors, banks, investors, competitors and business organizations.

Broadly speaking, the parties included in the above stakeholders can be seen in Figure 1, which illustrates the complexity of the relationship between an organization and its stakeholders. Arrows pointing to each other shows that both types are interconnected and there is an attachment between the organization and its stakeholders.

2.2 *Legitimacy*

O'Donovan (2002) states that: Legitimacy theory as the idea that acceptable for societies deems socially. Further Suchman (1995) in Barkemeyer (2007) defines organizational legitimacy as follows: "Legitimacy is a generalized perception or assumption that some actions construct systems of norms, values, beliefs and definitions".

Lindblom (1994) in Deegan et al. (2002) defines legitimacy theory as follows is a condition or status that exists when a congruent entity value system with a broader community value system in which society as a part. When a difference, whether real or potential, exists between the two value systems, there will be a threat to the legitimacy of the company.

The postulate of the theory of legitimacy is that organizations should not only be concerned about investor rights but in general should also pay attention to the rights of the public (Deegan & Rankin, 1996).

Based on these definitions, the objectives, methods of operation and the output of an organization must be in accordance with social norms and values. More importantly, organizations must conform to the rules of society to ensure social approval so they can continue to exist. Accordingly, accountability and social accounting systems are essential for the acceptance of an organization's ongoing operations (continued approval of the organization's operations).

The theory of legitimacy asserts that companies continue to work to ensure that they operate within the frames and norms that exist in the society or environment in which the company is located, where they strive to ensure that their activities are accepted by outsiders as 'legitimate' (Deegan, 2013). The same opinion is expressed by Tilt (1994) in Haniffa and Cooke (2005), which stated that the company has a contract with the community to conduct its activities based on justice values, and to respond various interest groups regarding its actions. The theory of legitimacy, in relation to social performance and financial performance, is that if there is an inconsistency between the corporate value system and the community value system, then the company may lose its legitimacy, which will further threaten its survival (Lindblom, 1994 in Haniffa & Cooke, 2005).

Ghozali and Chariri (2007) argue that the underlying theory of legitimacy is the 'social contract' that occurs between companies and communities in which firms operate and use

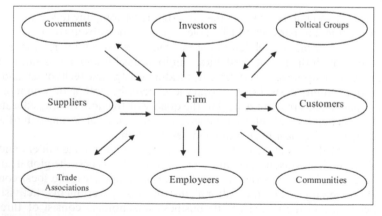

Figure 1. Chart/schematic of theory.
Source: Preston and O'Bannon (1997).

economic resources. Shocker and Sethi (1973) in Ghozali and Chariri (2007) provide an explanation of the concept of social contracts, all social institutions does not have exception to companies operating in society through social contracts, both explicit and implicit, where survival and growth based on outcomes social can be given to the wider community and the distribution of economic, social or political benefits to the group in accordance with the power held.

2.3 *Corporate social responsibility*

There are various definitions of CSR, including that of the World Business Council for Sustainability Development (WBSCD), which is as follows: 'Corporate social responsibility is the continuing commitment by the business to behave ethically and contribute to economic development while improving the quality of life of the workforce and their families as well as the local community and society at large'.

Based on this understanding, social responsibility is a sustainable business commitment to contribute to economic development, through collaboration with employees and their representatives, their families, local communities and the general public to improve the quality of life in a way that is beneficial to both their own businesses and development. In agreement with that, Rosmasita (2007) defined CSR as an attempt to balance its commitments to groups and individuals within the enterprise, including customers, other companies, employees and investors.

CSR strives to give environmental and social attention to its operations. As explained by Anggraini (2006), social responsibility is a mechanism for an organization to voluntarily integrate environmental and social concerns into its operations and its interactions with interested parties, which exceeds its legal responsibilities. Thus, the company's business operations are not only committed to the size of the financial gain alone, but also to overall socio-economic development and sustainability.

According to Wood's model of the concept of CSR, companies and society are interwoven; not only society has certain expectations about the company but also the company has a responsibility toward society (Pierick et al., 2004). According to Frederick, the basic notion of CSR is that the business enterprise has a duty to work for the advancement of society (Pierick et al., 2004). The description by Davis (Pierick et al., 2004) is more instructive: 'The firm's obligation evaluate in its decision-making process the effects of its decisions on the external social system in a manner that accompany social benefits along with the traditional economic gains which the firm seeks. It means that social responsibility begins where the law ends. Social responsibility goes one step further. It is a firm's acceptance of a social obligation beyond the requirements of the law'.

Hill (Uddin et al., 2008) explains that CSR is a set of practices that are in part a form of good management or business practice that are mostly about transparency and disclosure. According to Mahoney and Thorne (Uddin et al., 2008), social responsibility consider about attention to social and environmental factors. Uddin et al. (2008) stated that the goal of CSR is to create a business enterprise activity and a sustainable corporate culture in three aspects: economic, social, and environmental and ecological.

The Ministry of Social Affairs states that CSR is the company's commitment to implement social obligations to its environment, to pay attention to the ethics and regulations that apply in order to improve the welfare of the community and maintain a sustainable balance of ecosystem life. Based on the above notions, CSR can be summed up as a concept of corporate commitment to good management practices with social and environmental concerns aimed at meeting people's expectations, advancing community life, maintaining environmental ecosystems and maximizing the company's long-term finances.

2.4 *Corporate social performance*

According to Orlitzky (2000), Corporate Social Performance (CSP) defined as 'a configuration of business organizational principles of social responsibility, social response processes, and observable policies, programs and outcomes as to those relationships in the community'. The 'corporate social performance' model developed by Wood (Meehan et al., 2006), offers a conceptual blend of existing developments in an effort to give academics the concept of CSR in a broader sense. However, the main purpose of the CSP model is to replace Ackerman and Bauer's attention with a focus on results: 'performance terms talk about actions and outcomes, not interactions or integration'.

Carroll in Fauzi (2009) defines the CSP as meet a certain moment in time from three dimensions: the principles of CSR, which will be held at four different levels (economic, legal, ethical and discretionary); the total number of social problems facing the company (e.g. racial discrimination, etc.); and the philosophy underlying its response, which can range as long as the continuum goes from anticipating company problems to direct rejection of responsibility in all companies. Most understanding, environmental aspect is rarely mention because CSP perspective has included environmental aspect in it. Based on the above, CSP can be interpreted as an action and result when a company has undertaken CSR (social and environmental aspects contained).

Orlitztky in Fauzi (2009) classifies the CSP measurement approach into four types of measurement strategies:

1. Disclosure
2. Level of reputation
3. Social audit, CSP process and observable results
4. CSP managerial principles and values.

Similarly, Cochran and Wood (Fauzi, 2009) argue that there are two generally accepted methods for measuring the CSP of content analysis and reputation index.

This research will use a type of measurement approach of annual report content with aspects of social responsibility assessment issued by the GRI and obtained from the website www.globalreporting.org. The GRI standard is chosen because it focuses more on the disclosure standards of various economic, social and environmental performances of the company in order to improve the quality, rigor and utilization of sustainability reporting.

In GRI standards, performance indicators are divided into three main components, namely economic, environmental and social covering human rights, labor and work practices, product responsibility and society. The total indicator performance reaches 79 indicators, consisting of nine economic indicators, 30 environmental indicators, 14 labor practice indicators, nine human rights indicators, eight community indicators and nine indicators of product responsibility.

2.5 *Corporate financial performance*

Bird et al. (2006) a proposition has good management will invest a wider range of CSR activities to seek satisfaction from the interests of large stakeholder groups that prerequisite for creating environmental needs that enable the company to produce strong financial performance. Fauzi (2009) the responsibility of management to improve higher financial performance toward the direction of increasing wealth from the stakeholders. Therefore, in terms of economic aspects, the company will certainly continue to strive financial performance.

According to Fauzi (2009), there are many measurements used to represent financial performance. These are divided into three categories: Return On Assets (ROA) and Return On Equity (ROE) (Waddock et al. in Fauzi, 2009); profitability (Stanwick in Fauzi, 2009); and multiplication of accounting based on measurement with an overall index using a score of 0–10 (Moore in Fauzi, 2009). Griffin, Mahon, Orlitzky et al. (Fauzi, 2009) in terms of enterprise performance measurement, there is a high degree of consensus on the basis of variables reflecting financial performance: profit-related indicators with the most fundamental are asset investments, growth, liquidity and profitability risks.

Financial performance, in this research, will be measured from ROA and ROE. ROA shows the capability of invested capital in the overall assets held to generate profit. ROE measures how well a company uses revenue reinvested to generate additional revenue, giving a general indication of company efficiency.

2.6   Factors affecting CSP and CFP

2.6.1   Corporate governance

Kaen (2003) defines corporate governance concern on who controls the company. The Cadburry committee, defined corporate governance as the principle that directs and controls a company to achieve a balance between the strength and authority of the company in providing accountability to its shareholders in particular, and stakeholders in general. Meanwhile, the Forum for Corporate Governance in Indonesia defined corporate governance as the system that directs and controls a company. Shleifer and Vishny (1997) defined corporate governance as ways to give confidence to the company's suppliers of funds for returns on their investments (Darmawati et al., 2005).

Good Corporate Governance (GCG), in this research, is a corporate governance mechanism. The factors of the corporate governance mechanism are also correlated with the level of CSR disclosure in the company's annual report. The size of the board of commissioners, the size of the audit committee, the quality of the external auditor and the ownership structure are positively correlated with CSR disclosure (Haniffa & Cooke, 2005; Sembiring, 2006; Anggraini, 2006). Machmud and Djakman (2008) linked foreign ownership and institutional ownership to CSR disclosure. Farook and Lanis (2007) correlated corporate governance and CSR disclosure in Islamic banks with the size of the Islamic Governance Score. The results indicate a positive correlation. The current study will use institutional ownership and composition of the independent board of commissioners as a proxy for corporate governance mechanisms. This is to re-examine the results of previous research.

2.6.2   Firm size

The firm size indicates in terms of total assets, sales level, and stock of market value. In legitimacy theory, large corporations are more visible in activity than small firms so that the demands and pressures of stakeholders and the public are increased. Luo et al. (2012) stated that large companies will be under greater pressure from the public and stakeholders who have high expectations about management practices. To respond these pressures, the company conducts social disclosure regarding the environment in order to gain support from stakeholders and legitimacy from the community. Brammer et al. (2006) found that companies in the United Kingdom were encouraged to provide voluntary disclosure in order to gain legitimacy.

According to Waddock, Graves and Itkonen (Fauzi, 2009), firm size has a relationship with CSP in that larger companies tend to behave more socially responsibly than smaller companies. According to Orlitzky and Itkonen (Fauzi, 2009), CSP related to firm size from the beginning, with entrepreneurial strategies focus on basic economic survival and philanthropic responsibility. Based on that argument, it is expected that firm size can be attributed to CFPs resulting from, for example, economies of scale (Orlitzky & Itkonen, in Fauzi, 2009).

2.6.3   Risk

According to Moore and Itkonen (Fauzi, 2009), in order for companies to have low risk, they should consider and manage social responsibility. Consequently, a company with a low CSP will have an adverse impact in terms of risk. Financial risk can be seen from financial leverage. Financial leverage shows the proportion of debt usage to finance investment. Companies do not have leverage, use 100% of their own capital (Husnan & Enny, 2004). Some industries have different values of Debt to Equity Ratio (DER) and can be larger than others. This is because these industries have a smaller business risk, thus daring to use a larger proportion of debt (Husnan & Enny, 2004). According to Moore and Itkonen, in order for companies to have low risk, they should consider and manage social responsibility; consequently, a company with low CSP will have an adverse impact in terms of risk (Fauzi, 2009).

2.7   Research accomplished

This research is a development of previous work conducted by:

1. Fauzi (2009) who compared social and environmental performance between Indonesian companies and multinational companies operating in Indonesia. The results indicated that relationship between social variables, and the performance of Indonesian firms and multinationals is exactly the same. Meanwhile, the relationship from an environmental perspective, showed that performance of multinational is better than Indonesian companies. In Indonesian companies, corporate performance (financial) and social and environmental performance were unrelated. Meanwhile, multinational corporations related to corporate finances performance partly supported by social perspectives and fully supported by an environmental perspective.
2. Preston and O'Bannon (1997) developed five hypotheses concerning the relationship between social and financial performance. The five hypotheses are named as the social impact hypothesis, trade-off hypothesis, available fund hypothesis, managerial opportunism hypothesis and positive or negative synergies. The social impact hypotheses and available funds hypotheses explained that indicate a positive relationship between social and financial performance. While the trade-off hypotheses and managerial opportunities hypotheses are explained that indicate a negative relationship with social and financial performance. The results of these studies indicate that the relationship between social and financial performance is best explained by positive synergies or with available funds. Meanwhile positive synergies show the relationship of social performance and financial performance will produce a positive synergy together. This means that social performance and financial performance will improve with each other. Availability of funds emphasize companies may want to follow the normative rules of good corporate citizenship, but their actual behavior depends on available resources.
3. Waddock and Graves (1997) found a significant positive relationship between the CSP index and performance benchmarks, such as ROA in the following year (cited by Tsoutsoura, 2004). Tsoutsoura found a significant and positive relationship between CSP and CFP. A research company in France, led by Charles & Stephane. (2002) found that no relationship between CSP and CFP with industrial control variables, risk, firm size, R & D.
4. Research by Nelling et al. (2006) examined the causal relationship between CSP and CFP by introducing a new econometric technique, the Granger causality approach. Their findings suggest that, using the OLS regression model, CSP and CFP are interrelated. Disagreeing with previous research, they found a low relationship between CSP and CFP when using the time series effect approach. Similar results were also found when introducing the Granger causality model. In addition, by focus on each size of the CSP, found causality that occurs in the stock market performance to the CSP rating in employee relations (Makni et al., 2009).
5. Research by Makni et al. (2009) examined the causality between social performance and financial performance of Canadian firms using the Granger causality approach, with control variables including size, corporate and industry risk. They found no relationship between a number of CSP and CFP measurements, except for market returns.

The differences with previous research is a sample company in Jakarta Islamic Index, this study examines the influence of corporate governance, risk and firm size on financial performance and corporate social performance and this study will test the good relationship of social performance to financial performance and performance financial to social performance using the Granger causality approach. Most previous studies only tested the relationship of social performance to financial performance alone using control variables. Research conducted by Preston and O'Bannon (1997) also examined the relationship of financial performance to social performance. As mentioned above, the results obtained from Preston's research, positive synergies with available funds show the best relationship between financial performance and social performance. This is in accordance with the slack resources theory, which states that the company will contribute to social performance if it has a good financial position. It can be said that financial performance has a positive relationship to social performance. In addition, this study also uses aspects of the CSR assessment issued by the GRI as a measurement or indicator of CSP.

2.8 Conceptual framework

This research was undertaken to provide an overview of the social performance of Indonesian and multinational companies operating in Indonesia. In addition, to determine the good relationship between social performance and CFP and the relationship between financial performance and CSP. CSP is derived from aspects of the CSR assessment issued by the GRI. These aspects affect CSP so that social performance can be determined from these aspects, while CFP is proxied from ROA and ROE. The independent variables used in this study are corporate governance proxied by board of commissioners, independent board of commissioners, audit committee and board meetings, risk proxied with Debt to Total Assets Ratio (DAR) and DER and firm size proxied from total asset log (SIZE).

2.9 Hypotheses

Based on previous research and the facts in the field as described above, this research will examine causality between social performance and financial performance in companies operating in Indonesia using Granger causality, with the following hypotheses:

2.9.1 The effect of corporate governance on CFP and CSP

The Indonesian Institute for Corporate Governance (IICG) states that a company's financial performance is determined by implementing GCG. Companies listed in the corporate governance rating scores conducted by the IICG have implemented GCG by indirectly raising share value. The higher implementation of GCG as measured by the Corporate Governance Perception Index (CGPI) is also higher in producing good corporate performance. Implementation of GCG will give a good impact on the company, so it can indirectly improve financial performance, and can raise the image of a company from investors opinion and parties who lend money to the company because of the factor of trust so that the company can be more easy to get loans for operational processes and reduce the risk for shareholders and be able to improve the ability to compete in the global market. Darmawati et al. (2005) discloses that the better implementation of GCG in a company will affect its financial performance. Because the results of the analysis show that corporate governance affect significantly to financial performance of the company. Hastuti (2005) conducted a study on the relationship between GCG and ownership structure with financial performance, stating that there is a significant relationship between the disclosure of financial statements with company performance. The concept of GCG support to produce good corporate performance in the management that must be apply one of the principles of GCG which is transparency.

The practice and disclosure of CSR is a logical consequence of the implementation of the concept of corporate governance, which states that companies need to pay attention to the interests of their stakeholders, in accordance with existing rules and establish active

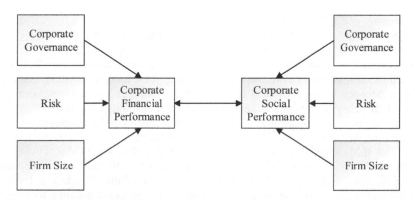

Figure 2. Conceptual framework.
Source: Self-made for this research.

cooperation with them for the long-term viability of the company (Ratnasari & Prastiwi, 2010). Corporate governance is very effective in ensuring that stakeholder interests are protected. Therefore, companies should disclose a company's economic, social and environmental performance to its stakeholders (Said et al., 2009 in Ratnasari & Prastiwi, 2010). Implementation of the concept of GCG is expected to improve the implementation and disclosure of CSR. The results of research conducted by Murwaningsari (2010) indicate that corporate governance has a significant influence on social responsibility.

H1: Corporate governance affects to CSP.
H2: Corporate governance affects to CFP.

2.9.2 The effect of risk on CFP and CSP

If the industry has a low leverage position will fund assets through own capital. According to agency theory that leverage management is similar to shrink CSR implementation that was created to avoid the attention of debtholders. Research conducted by Subroto (in Triyanto, 2010) found that firm leverage level positively affects the width of disclosure social responsibility.

The amount of debt used and the short repayment time will arise fixed expenses of a company. Additionally, note the benefits of loyalty so that the use of debt can advance the assets of the company and will increase its profitability. Research conducted by Sunarto and Prasetyo found that leverage had a positive effect on company profitability and Yahya (2011) found that leverage had a positive effect on profitability.

H3: Risk affects to CSP.
H4: Risk affects to CFP.

2.9.3 The effect of firm size on CFP and CSP

Firm size will be used as the estimator variable to describe how various companies report disclosures, because according to agency theory, a growing company has a substantial agency budget in unpacking the information necessary to reduce its agency budget. Research conducted by Kamil and Herusetya (2012) and Rahajeng and Marsono (2011) found that firm size had a significant positive effect on CSRD.

The size of the company if it is linked in the agency theory so that the growing company has a large enough cost can be explain the important information to reduce agency costs. Size can also be used as a proxy in describing various disclosures of annual reports on information from firms. Research conducted by Utama (2007) found that firm size had a significant positive effect on profitability, while research conducted by Sunarto and Budi (2009) found that firm size had a significant positive effect on profitability.

H5: Firm size affects to CSP.
H6: Firm size affects to CFP.

2.9.4 Causality of CFP and CSP

Based on a study conducted by Nelling and Webb (2006), which examined the causal relationship between CSP and CFP using the Granger causality approach, it was found that by using OLS regression models, CSP and CFP are interrelated. Mahoney and Roberts (2007) investigated the relationship between CSP and CFP in the Canadian context, examining the relationship between these constructs by measuring Canadian Social Investment Database (CSID) on CSP. They found that there is no significant relationship between CSP and CFP firms. However, using a one-year lag, their findings indicate a significant positive relationship between each of the company's CSP measurements on the environment, international activity and CFP. This study tested only one-way causality: from CSP to CFP (Makni et al., 2009).

Legitimacy theory explains that a company should strive to maintain its legitimacy in the eyes of all stakeholders. In order to maintain its sustainability (Guthrie & Parker, 1989; Deegan et al., 2002; Gamerschlag et al., 2011). Generally, the way companies to achieve commonality with the values held by the public is disclose the activities of its operations that have a social impact on the environment through annual reports or advertising in other media

(Gray et al., 1995; Deegan, 2013; Gamerschlag et al., 2011). Companies generally disclose more social responsibility in order to achieve social expectations of communities and governments around their business operations to increase stakeholder confidence (Bewley & Li, 2000; Islam & Deegan, 2010; Gamerschlag et al., 2011).

Some researchers, such as Davis (1973), suggest to practicing voluntary responsibility to make the company gain the advantage of its competitors in the competitive side in the short term such as increases in productivity to develop the ability to attract large amounts of human resources, buyers may be very sensitive to social issues and reduce the expected costs can affect the relationship with potential creditors and suppliers. Therefore, the better a company performs its disclosure of social responsibility, it will build a good corporate image in the eyes of consumers. Consumers will have a good view because the company has shown the public interest, thus consumers do not mind using the product. If there are so many consumers using the products, it will increase the company's sales.

H7: Financial performance affects CSP.
H8: Social performance affects CFP.

Based on the theory and previous research of H1–H7, so the hypotheses as follows:

H9: CFP mediates the effect of corporate governance to CSP.
H10: CFP mediates the effect of risk to CSP.
H11: CFP mediates the effect of firm size to CSP.
H12: CSP mediates the effect of corporate governance to CFP.
H13: CSP mediates the effect of risk to CFP.
H14: CSP mediates the effect of firm size to CFP.

3 RESEARCH METHODS

3.1 *Research variables and definition of the operational variable*

The Granger causality analysis of social performance and financial performance company uses dependent variable, independent and control variables. Because this research examines the relationship of causality as the dependent variable is CSP and the independent variable is CFP,. The CSP is measured using the GRI indicator, while the CFP is proxied with ROA and ROE. The independent variables for this research are corporate governance, risk and firm size.

3.2 *Corporate social performance*

CSP variables are measured by using the items in the aspects of the CSR assessment issued by the GRI. The company's social performance will be calculated by comparing how many items are disclosed from a total of 79 disclosure items, including: economic, environmental, labor practices, human rights, community and product responsibility. If the specified item of information is disclosed in the annual report it is given a score of one, and if it is not disclosed in the annual report it is given a score of zero. The calculation of CSR Disclosure Index (CSRI) is formulated as follows:

$$CSRI\ t = \frac{\text{Total disclosure items}}{79}$$

3.3 *Corporate financial performance*

The CFP variables used are ROA, ROE and Earnings Per Share (EPS). The measurement of a firm's financial performance with ROA shows the capability of the invested capital in its total assets to generate profits. ROA is the ratio of net profit after tax to assess the return of assets owned by the company. Calculating ROA can be formulated as follows:

$$ROA = \frac{\text{Earning After Tax}}{\text{Total Assets}} \times 100\%$$

ROE is a ratio that measures how much profit the owner owns. It is considered that own capital is common stock, share premium, retained earnings, preferred stock and other reserves. ROE measures how well a company uses reinvested revenue to generate additional revenue, giving a general indication of company efficiency. ROE can be calculated as follows:

$$ROE = \frac{\text{Earning After Tax}}{\text{Total Equity}} \times 100\%$$

3.4 Independent variables

The independent variables used in this research are corporate governance, risk and firm size.

3.4.1 Corporate governance

The corporate governance variables referred to in this research are the corporate governance mechanism measured by board of commissioners, independent board of commissioners, audit committee and board of commissioners meeting.

3.4.2 Risk

Financial risk can be seen from financial leverage. Financial leverage shows the proportion of debt usage to finance investment. The financial risks used in this research are: DER to measure loan rates from corporate finance and calculated based on the ratio of total liabilities compared to total equity:

$$Leverage = \frac{\text{Total Liabilities}}{\text{Total Equity}} \times 100\%$$

DAR is one of the ratios used to measure the level of corporate solvency. The company's solvency level is its ability to pay its long-term liabilities.

$$Leverage = \frac{\text{Total Liabilities}}{\text{Total Assets}} \times 100\%$$

3.4.3 Firm size

Firm size, according to Waddock, Graves and Itkonen (Fauzi, 2009), has a relationship with CSP, with larger companies tending to behave more socially responsibly than smaller companies. The firm size used in this research is calculated as follows:

$$Size = \log Total\ Assets$$

3.5 Population and sample research

The population of this research were all companies listed on the Indonesian Stock Exchange. The sample was chosen using a purposive sampling method based on the following criteria:

a. Companies listed in the Jakarta Islamic Index from 2013–2015.
b. The company discloses its CSR report in its annual report for the 2003–2015 accounting period, accessible through the Jakarta Islamic Index website.

Based on these criteria, the sample that meets the requirements and is used in this research includes 19 companies registered in the Jakarta Islamic Index.

3.6 Types of data and sources

The data used in this research are documentary data in the form of annual reports from 2013–2015 for companies listed in the Jakarta Islamic Index. While the data source used is secondary data that data obtained indirectly. The data used in this research comes from:

a. The official website of the Indonesian Stock Exchange (www.idx.co.id).
b. The companies' official websites.
c. Indonesia Capital Market Directory.

3.7 Method of data collection

The data used in this research are secondary data in the form of documentation and direct quotations. The method used in measuring social performance in the annual report is content analysis. The development of hypotheses and frameworks is qualitative data which obtained with documentation and direct quotes from several books, journals and internet media.

The content of analysis method is a research technique to generate conclusions based on data that can be repeated and valid. The content of analysis method in this research is comes from the content from annual report which is adjusted to aspects of social responsibility assessment of business world issued by GRI. Annual reports reviewed by indicate a checklist based on the assessment aspect. Each item reported will be assigned a value of one and the unreported item will be assigned a value of zero. Then, the item disclosed is summed and divided by the total items of disclosure available, yielding the company's social performance index.

3.8 Data analysis method

The data analysis used in this research involves Granger causality and a path analysis technique. Here is some of granger causality and a path analysis equals:

$$CSP = \alpha + \beta1\ CG\ (DK) + \beta2\ CG\ (DKI) + \beta3\ CG\ (KA) + \beta4\ CG\ (RDK) + \beta5\ Risk$$
$$(DAR) + \beta6\ Risk\ (DER) + \beta7\ Firm\ Size + e_1 \qquad \text{(Equation 1)}$$

$$CFP\ (ROA) = \alpha + \beta1\ CG\ (DK) + \beta2\ CG\ (DKI) + \beta3\ CG\ (KA) + \beta4\ CG\ (RDK) + \beta5\ Risk$$
$$(DAR) + \beta6\ Risk\ (DER) + \beta7\ Firm\ Size + e_1 \qquad \text{Equation 2}$$

$$CFP\ (ROE) = \alpha + \beta1\ CG\ (DK) + \beta2\ CG\ (DKI) + \beta3\ CG\ (KA) + \beta4\ CG\ (RDK) + \beta5\ Risk$$
$$(DAR) + \beta6\ Risk\ (DER) + \beta7\ Firm\ Size + e_1 \qquad \text{(Equation 3)}$$

$$CSP = \alpha + \beta1\ CG\ (DK) + \beta2\ CG\ (DKI) + \beta3\ CG\ (KA) + \beta4\ CG\ (RDK) + \beta5\ Risk$$
$$(DAR) + \beta6\ Risk\ (DER) + \beta7\ Firm\ Size + \beta8\ CFP\ (ROA) + \beta9\ CFP\ (ROE) + e_1$$
$$\text{(Equation 4)}$$

$$CFP\ (ROA) = \alpha + \beta1\ CG\ (DK) + \beta2\ CG\ (DKI) + \beta3\ CG\ (KA) + \beta4\ CG\ (RDK) + \beta5\ Risk$$
$$(DAR) + \beta6\ Risk\ (DER) + \beta7\ Firm\ Size + \beta8\ CSP + e_1 \qquad \text{(Equation 5)}$$

$$CFP\ (ROE) = \alpha + \beta1\ CG\ (DK) + \beta2\ CG\ (DKI) + \beta3\ CG\ (KA) + \beta4\ CG\ (RDK) + \beta5\ Risk$$
$$(DAR) + \beta6\ Risk\ (DER) + \beta7\ Firm\ Size + \beta8\ CSP + e_1 \qquad \text{(Equation 6)}$$

Note:
CSP: Social performance
CFP: Financial performance
ROA: Return on assets
ROE: Return on equity
CG: Financial performance of firm
DK: Board of commissioners
DKI: Independent board of commissioners
KA: Audit committee
RDK: Board of commissioners meting

Risk: Financial pressures
DAR: Debt to total assets ratio
DER: Debt to equity ratio
Firm Size: Firm size

4 DATA ANALYSIS AND DISCUSSION OF RESEARCH

This research conducted several tests, namely the classical assumption tests (normality test, multicollinearity, autocorrelation, heteroscedasticity and linearity) and hypotheses testing (coefficient of determination test, F-test and t-test), which are discussed and presented in Tables 1 and 2.

4.1 Assumption of classical linear regression

Based on Table 1 that obtained Asymp. Sig. greater than 0.05 can be inferred from Equation 1 to Equation 6 of the normal distributed data.

Table 1. Classical assumption test.

Kolmogorov–Smirnov (Normality test)

	Equation 1	Equation 2	Equation 3	Equation 4	Equation 5	Equation 6
Asymp.Sig	0.991	0.85	0.144	0.995	0.859	0.209

Multiplier Lagrange test (Linearity test)

	Equation 1	Equation 2	Equation 3	Equation 4	Equation 5	Equation 6
R-square	0.001	0.051	0.033	0.002	0.053	0.034

Tolerance and Value Inflation Factor (VIF) (Multicollinearity test)

	Equation 1		Equation 2		Equation 3		Equation 4		Equation 5		Equation 6	
Model	Tol	VIF	Tol	VIF	Tol	VIF	Tol	VIF	Tol	VIF	Tol	VIF
CG(DK)	0.42	2.38	0.42	2.38	0.42	2.38	0.377	2.651	0.372	2.687	0.372	2.687
CG(DKI)	0.412	2.429	0.412	2.429	0.412	2.429	0.372	2.689	0.404	2.475	0.404	2.475
CG(KA)	0.715	1.399	0.715	1.399	0.715	1.399	0.643	1.556	0.647	1.544	0.647	1.544
CG(RDK)	0.796	1.256	0.796	1.256	0.796	1.256	0.76	1.316	0.796	1.257	0.796	1.257
Risk(DAR)	0.134	7.462	0.134	7.462	0.134	7.462	0.111	8.977	0.131	7.651	0.131	7.651
Risk(DER)	0.156	6.415	0.156	6.415	0.156	6.415	0.118	8.477	0.155	6.462	0.155	6.462
Firm size	0.864	1.158	0.864	1.158	0.864	1.158	0.795	1.258	0.741	1.349	0.741	1.349
FP(ROA)	–	–	–	–	–	–	0.197	9.337	–	–	–	–
FP(ROE)	–	–	–	–	–	–	0.106	9.408	–	–	–	–
SP	–	–	–	–	–	–	–	–	0.51	1.961	0.51	1.961

Durbin–Watson (Autocorrelation test)

	Equation 1	Equation 2	Equation 3	Equation 4	Equation 5	Equation 6
DW	1.81	0.907	0.623	1.903	1.002	0.7

Uji White (Heteroscedasticity test)

	Equation 1	Equation 2	Equation 3	Equation 4	Equation 5	Equation 6
R-square	0.236	0.647	0.518	0.277	0.71	0.555

Source: Data processed.

The value of R^2 in Equation 1 = 0.001, Equation 2 = 0.051, Equation 3 = 0.033, Equation 4 = 0.002, Equation 5 = 0.053 and Equation 6 = 0.034, with n observation number of 57; then, the value of $c^2 = n \times R^2$. This value is compared with c^2 table (n–k) Equations 1, 2 and 3 = 66.339, Equation 4 = 64.001 and Equations 5 and 6 = 65.171. Since the value of c2 is smaller than the c^2 table, the correct model is a linear model.

The tolerance values of all independent variables > 0.10 and Value Inflation Factor of all independent variables <10.00 can be concluded that there is no multicollinearity.

Durbin–Watson's equation of Equation 1 to Equation 6 lies between –2 to +2 so it can be concluded there is no autocorrelation.

The value of R^2 in Equation 1 = 0.236, Equation 2 = 0.647, Equation 3 = 0.518, Equation 4 = 0.277, Equation 5 = 0.710 and Equation 6 = 0.555, with n observation number of 57; then, the value of $c^2 = nx\ R^2$. This value is compared with c^2 table (n–k) Equations 1, 2 and 3 = 56.942, Equation 4 = 52.192 and Equations 5 and 6 = 54.572. Since the value of c2 is smaller than the c2 table, the heteroskedasticity in the model is rejected.

4.2 Hypothesis test

Based on Table 2, the effect of corporate governance (DK, DKI, KA and RDK), risk and firm size to CSP is 41.7%. The magnitude of corporate governance (DK, DKI, KA and RDK), risk and firm size influence to CFP (ROA) is 33.0%. The amount of corporate governance (DK, DKI, KA and RDK), risk and firm size influence to CFP (ROE) is 26.4%. The magnitude of corporate governance (DK, DKI, KA and RDK), risk, firm size and CFP (ROA and ROE) influence to CSP is 42.4%. The amount of corporate governance (DK, DKI, KA and RDK), risk, firm size and CSP influence to CFP (ROA) is 35.2%. The amount of corporate governance (DK, DKI, KA and RDK), risk, firm size and CSP influence to CFP (ROE) is 28.5%.

Table 2. Hypothesis test.

Adjusted R-square	Equation 1			0.417	
	Equation 2			0.330	
	Equation 3			0.264	
	Equation 4			0.424	
	Equation 5			0.352	
	Equation 6			0.285	
F-test	Equation 1	F		6.724	
		Sig		0.000	
	Equation 2	F		4.948	
		Sig		0.000	
	Equation 3	F		3.874	
		Sig		0.002	
	Equation 4	F		5.590	
		Sig		0.000	
	Equation 5	F		4.809	
		Sig		0.000	
	Equation 6	F		3.789	
		Sig		0.002	
	Equation 1	beta	0.008	Sig. CG (DK)	0.015
			0.005	Sig. CG (DKI)	0.340
			0.013	Sig. CG (KA)	0.029
			0.000	Sig. CG (RDK)	0.860
			0.062	Sig. Risk (DAR)	0.271
			–0.008	Sig. Risk (DER)	0.552
			–0.008	Sig. Firm Size	0.006
	Equation 2	beta	–2.221	Sig. CG (DK)	0.023
			3.418	Sig. CG (DKI)	0.030

(Continued)

607

Table 2. (*Continued*).

			3.974	Sig. CG (KA)	0.028
			−0.421	Sig. CG (RDK)	0.244
			43.527	Sig. Risk (DAR)	0.017
			−8.631	Sig. Risk (DER)	0.048
			−1.698	Sig. Firm Size	0.064
t-test	Equation 3	beta	−5.091	Sig. CG (DK)	0.030
			8.277	Sig. CG (DKI)	0.029
			7.734	Sig. CG (KA)	0.075
			−1.271	Sig. CG (RDK)	0.147
			62.460	Sig. Risk (DAR)	0.148
			−4.922	Sig. Risk (DER)	0.635
			−3.061	Sig. Firm Size	0.164
	Equation 4	beta	0.009	Sig. CG (DK)	0.005
			0.002	Sig. CG (DKI)	0.677
			0.010	Sig. CG (KA)	0.101
			0.001	Sig. CG (RDK)	0.646
			0.033	Sig. Risk (DAR)	0.590
			−0.003	Sig. Risk (DER)	0.853
			−0.007	Sig. Firm Size	0.023
			0.001	Sig. CFP (ROA)	0.614
			7.494	Sig. CFP (ROE)	0.872
	Equation 5	beta	−2.765	Sig. CG (DK)	0.007
			3.081	Sig. CG (DKI)	0.047
			3.067	Sig. CG (KA)	0.097
			−0.436	Sig. CG (RDK)	0.221
			39.042	Sig. Risk (DAR)	0.030
			−8.046	Sig. Risk (DER)	0.062
			−1.114	Sig. Firm Size	0.247
			71.770	Sig. CSP	0.109
	Equation 6	beta	−6.347	Sig. CG (DK)	0.011
			7.500	Sig. CG (DKI)	0.047
			5.638	Sig. CG (KA)	0.206
			−1.305	Sig. CG (RDK)	0.132
			52.101	Sig. Risk (DAR)	0.226
			−3.572	Sig. Risk (DER)	0.728
			−1.713	Sig. Firm Size	0.461
			165.756	Sig. CSP	0.127

Source: Data processed.

The result of F-test in Table 2 can be F-value of 6.724 with Sig. 0.000 (Equation 1); 4.948 with Sig. 0.000 (Equation 2); 3.874 with Sig. 0.002 (Equation 3); 5.590 with Sig. 0.000 (Equation 4); 4.809 with Sig. 0.000 (Equation 5) and 3.798 with Sig. 0.002 (Equation 6). because the probability value of significance is smaller than 0.05 indicating that the regression model is feasible and correct.

The result of statistic test t in Table 2, for Equation 1, showed that corporate governance (DK and KA) and firm size, is significant because it has significance values below 0.05 (0.015, 0.029 and 0.006). Therefore, it can be concluded that from the seven independent variables in the regression model in Equation 1, there are three independent variables, namely corporate governance (DK and KA) and firm size, which affect the CSP.

In Equation 2, corporate governance (DK, DKI and KA) and risk (DAR and DER) are significant because they have significance values that are below 0.05 (0.023, 0.030, 0.028, 0.017 and 0.048). Therefore, it can be concluded that from the seven independent variables in the regression model in Equation 2, there are five independent variables, namely corporate governance (DK, DKI and KA) and risk (DAR and DER), which affect the dependent variable CFP (ROA).

In Equation 3, corporate governance (DK and DKI) is significant because it has significance values below 0.05 (0.030 and 0.029). Therefore, it can be concluded that from the seven independent variables in the regression model in Equation 3, there are two independent variables, namely corporate governance (DK and DKI), which affect the dependent variable CFP (ROE).

In Equation 4, corporate governance (DK) and firm size are significant because they have significance values below 0.05 (0.005 and 0.023). Therefore, it can be concluded that from the nine independent variables in the regression model in Equation 4, there are two independent variables, corporate governance (DK) and firm size, which affect the dependent variable CSP.

In Equation 5, corporate governance (DK and DKI) and risk (DAR) are significant because they have significance values below 0.05 (0.007, 0.047 and 0.030). Therefore, it can be concluded that from the eight independent variables in the regression model in Equation 5, there are three independent variables, namely corporate governance (DK and DKI) and risk (DAR), which affect the dependent variable CFP (ROA).

In Equation 6, corporate governance (DK and DKI) is significant because it has significance values below 0.05 (0.011 and 0.047). Therefore, it can be concluded that from the eight independent variables in the regression model in Equation 6, there are two independent variables, namely corporate governance (DK and DKI), which affect the dependent variable CFP (ROE).

4.3 *Mediated variable test causal step strategy*

Coefficients a and b are Significant to indicate the existence of mediation. If c is significant then there is partial mediation, but if c is not significant then there is full mediation (Preacher & Hayes, 2004).

Based on Table 3 it can be concluded that only Corporate Social Performance variables is mediating Corporate Governance (Commissioners) to Corporate Financial Performance (Return on Assets and Return on Equity).

4.4 *Discussion*

4.4.1 *The effect of corporate governance on CSP*

Corporate governance (DK and KA) affects CSP. The practice and disclosure of CSP is a logical consequence of the implementation of the concept of corporate governance, which states that a company needs to pay attention to its stakeholder interests, in accordance with existing rules and establish active cooperation with its stakeholders for long-term corporate survival (Ratnasari & Prastiwi, 2010). Corporate governance is very effective in ensuring that stakeholder interests are protected. Therefore, companies should disclose their economic, social and environmental performance to their stakeholders (Said et al., 2009 in Ratnasari & Prastiwi, 2010). Implementation of the GCG concept is expected to improve the implementation and disclosure of CSP. The results of this research states that corporate governance has a significant influence to CSP is conducted by Murwaningsari (2010).

Corporate governance (DKI and RDK) has no effect on CSP. Che Ahmad et al. in Hashim and Devi (2004) explain that the reasons for this insignificant outcome is probably because corporate governance (DKI and RDK) does not seem to affect decision-making, as it not

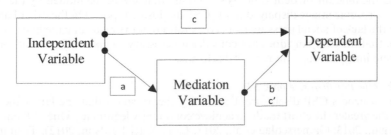

Figure 3. Causal step strategy.

Table 3. Causal step test.

Variable	A	B	c	c'
Sig. CG (DK) effect on CSP mediated by CFP (ROA)	0.2358	0.0029	0.0000	0.0000
Sig. CG (DKI) effect on CSP mediated by CFP (ROA)	0.0231	0.4555	0.0004	0.0016
Sig. CG (KA) effect on CSP mediated by CFP (ROA)	0.4089	0.1452	0.0104	0.0155
Sig. CG (RDK) effect on CSP mediated by CFP (ROA)	0.5005	0.0647	0.1070	0.0711
Sig. Risk (DAR) effect on CSP mediated by CFP (ROA)	0.0234	0.0734	0.8462	0.4639
Sig. Risk (DER) effect on CSP mediated by CFP (ROA)	0.6670	0.0923	0.6009	0.5286
Sig. Firm size effect on CSP mediated by CFP (ROA)	0.9750	0.0863	0.0243	0.0223
Sig. CG (DK) effect on CSP mediated by CFP (ROE)	0.1369	0.0069	0.0000	0.0000
Sig. CG (DKI) effect on CSP mediated by CFP (ROE)	0.0542	0.6437	0.0004	0.0010
Sig. CG (KA) effect on CSP mediated by CFP (ROE)	0.9325	0.1931	0.0104	0.0104
Sig. CG (RDK) effect on CSP mediated by CFP (ROE)	0.4245	0.1411	0.1070	0.0761
Sig. Risk (DAR) effect on CSP mediated by CFP (ROE)	0.0015	0.1426	0.8462	0.4323
Sig. Risk (DER) effect on CSP mediated by CFP (ROE)	0.0677	0.1527	0.6009	0.3887
Sig. Firm size effect on CSP mediated by CFP (ROE)	0.8803	0.1732	0.0243	0.0217
Sig. CG (DK) effect on CFP (ROA) mediated by CSP	0.0000	0.0029	0.2358	0.0061
Sig. CG (DKI) effect on CFP (ROA) mediated by CSP	0.0004	0.4555	0.0231	0.0878
Sig. CG (KA) effect on CFP (ROA) mediated by CSP	0.0104	0.1452	0.4089	0.7698
Sig. CG (RDK) effect on CFP (ROA) mediated by CSP	0.1070	0.0647	0.5005	0.2830
Sig. Risk (DAR) effect on CFP (ROA) mediated by CSP	0.8462	0.0734	0.0234	0.0185
Sig. Risk (DER) effect on CFP (ROA) mediated by CSP	0.6009	0.0923	0.6670	0.5576
Sig. Firm size effect on CFP (ROA) mediated by CSP	0.0243	0.0863	0.9750	0.6258
Sig. CG (DK) effect on CFP (ROE) mediated by CSP	0.0000	0.0069	0.1369	0.0049
Sig. CG (DKI) effect on CFP (ROE) mediated by CSP	0.0004	0.6437	0.0542	0.1309
Sig. CG (KA) effect on CFP (ROE) mediated by CSP	0.0104	0.1913	0.9325	0.7179
Sig. CG (RDK) effect on CFP (ROE) mediated by CSP	0.1070	0.1411	0.4245	0.2691
Sig. Risk (DAR) effect on CFP (ROE) mediated by CSP	0.8462	0.1426	0.0015	0.0012
Sig. Risk (DER) effect on CFP (ROE) mediated by CSP	0.6009	0.1527	0.0677	0.0527
Sig. Firm size effect on CFP (ROE) mediated by CSP	0.0243	0.1732	0.8803	0.5759

Source: Data processed.

involved in routine company operations. The capability of corporate governance (DKI and RDK) for monitoring will also be limited if affiliated parties dominate and control (Abdullah et al., in Hashim & Devi, 2004). This can also be attributed to the fact that corporate governance (DKI and RDK) is not very influential in making decisions about CSP, as many companies have established other committees that deal directly with CSR, such as CSR committees and GCG committees.

4.4.2 The effect of risk on CSP

Risk (DAR and DER) has no effect on CSP. The results of this research differ from some previous research that found that a high risk (DAR and DER) tends to make the company reveal more CSPs, in order to keep its image before lenders and other stakeholders (Clarkson et al., 2008; Lanis & Richardson, 2012). This indicates that companies in Indonesia have not considered the amount of debt ratio as something that should be hidden by the disclosure of a good corporate image displayed in the CSP disclosure report. On the other hand, it also illustrates the lack of social pressure from corporate creditors regarding the issue of CSR disclosure, in that the company does not consider it necessary to make a focus shift or improve its corporate image through more CSP disclosure.

4.4.3 The effect of firm size on CSP

Firm size influences CSP disclosure. Legitimacy theory states that the larger the size of a company the greater the effort made to achieve community legitimacy. One of them expresses CSP (Deegan, 2013; Gamerschlag et al., 2011; Lanis & Richardson, 2012). Then implies an increase CSP disclosure. The results of this research are consistent with the research findings of Gamerschlag et al. (2011), Luethge and Han (2012) and Lanis and Richardson (2012). The researchers also add that the greater the size of a company necessitates it having the ability to finance the disclosure of CSP for the better.

4.4.4 *The effect of corporate governance on CFPs (ROA and ROE)*

Corporate governance (DK) negatively affects CFP (ROA and ROE). The results of this research are consistent which research conducted by Allen and Gale in Beiner et al. (2006), which asserts that the board of directors is an important corporate governance mechanism, as the board of directors can ensure that managers pursue the board's interests. They also suggest that large boards of directors are less effective than small boards of directors. This is because the large number of boards of directors will increase the agency problem. Loderer and Peyer in Beiner et al. (2006) also found evidence that a large number of boards of directors will result in low firm performance.

Corporate governance (DKI) positively affects CFP (ROA and ROE). This indicates that a company has an independent board of commissioners only to meet the minimum requirements of the independent board of commissioners set by the Indonesia Stock Exchange, thus making the company's performance ineffective. It will be difficult to get a good response from investors to assess the company's value higher than company book value. This is in line with research conducted by Juwitasari (2008).

Corporate governance (KA) positively affects CFP (ROA). This shows that having an audit committee that is expert in finance, can automatically realize more effective supervision so that easily to achieved the goals of a company and to improve the CFP (ROA). While corporate governance (KA) has no effect on CFP (ROE). This is because the number of audit committees cannot guarantee the effectiveness of the audit committee's performance in monitoring CFP (ROE). The establishment of an audit committee within an enterprise is solely on the grounds of compliance with regulations requiring that an enterprise should establish one. The results of this research are in line with Diandono (2012), which shows that an audit committee has no effect on ROE. There is no influence from the number of audit committees in a company due to the role of the audit committee is not optimal in carrying out supervisory and control functions in the company management. In addition, maintained the quality of financial statements and assisted the board of commissioners has not been fully achieved by the audit committee so it has not been able to improve the profitability of the company. The selection of audit committee members is still based on kinship so that the monitoring of the board of directors is not maximal.

Corporate governance (RDK) has no effect on CFP (ROA and ROE). The reason for this is that in conducting its duties within the company, the board of commissioners will hold a meeting at least once a month and at any time if deemed necessary to discuss various problems and business. A large number of board meetings indicates the possibility that meetings conducted by the board of commissioners are less effective, due to the dominance of votes from commissioners who prioritize their personal or group interests to the exclusion of corporate interests (Muntoro, 2006). In addition, a board of commissioners meeting is not always intensive in discussing the improvement of corporate performance but also discusses other matters outside the company's performance. The meeting held too frequent or may suddenly can made the members of the meeting cannot learn the material of the meeting in depth, so that it can lead to less effective meetings due to the limited knowledge of commissioners to the material to be discussed in the meeting.

4.4.5 *The effect of risk on CFP (ROA and ROE)*

Risk (DAR) positively affects CFP (ROA). This is in contrast with the results of research conducted by Puspitasari and Ernawati (2010) but is similar to the results of research conducted by Reddy et al. (2010) and Ehikioya (2009). Risk (DAR) acts as a corporate governance mechanism used to transfer the function of monitoring and evaluation of managerial performance to the lender (Reddy et al., 2010). This is because lenders have a tendency to protect their investments by monitoring company performance on a regular basis. Thus, the higher the risk (DAR), the higher the demand for performance improvements made by the lender. The higher the risk (DAR), the greater the increase in CFP (ROA) (Ehikioya, 2009; Reddy et al., 2010). Different results were obtained in the test using risk (DER). Risk (DER) negatively affects CFP (ROA). The results of this research are in line with the results of research conducted by Puspitasari and Ernawati (2010), especially in terms of profitability. Sanda et al. in Puspitasari and Ernawati (2010) stated that this negative effect is caused by the possibility of a conflict of interest between the debt holder party (which usually has representation on the board of commissioners) with shareholders. The debt holder wants the business entity to have stability in order to repay the debt, so that the debtor does not want a business enterprise strategy that threatens the ability of the business entity to repay the debt. While the shareholder aims to maximize prosperity, the shareholders want the business entities to implement a strategy that can raise stock prices.

Risk (DAR and DER) has no effect on CFP (ROE). The results of this research indicate that the risk (DAR and DER) that can be obtained from the financial statements does not affect CFP (ROE). This is because the higher the level of debt in the capital structure, the interest expense will increase so that the discretionary expense increases, but does not increase the CFP (ROE) net income compared to shareholder equity. Discretionary expenses are incurred under management policies, including operating expenses, non-operating expenses, interest expenses, and salaries and wages. In this case, the discretionary expense is the total operating expense charged with interest. These findings are in line with the research by Kusumasari et al. (2010) which states that risk (DAR and DER) has no effect on CFP (ROE).

4.4.6 *Firm size effect on CFP (ROA and ROE)*

Firm size has no effect on CFP (ROA and ROE). This means that the size of the company has no significant effect on the high level of CFP (ROA and ROE). This is because during the research period, the general condition of Indonesia's macro economy was still not very well marked by a high inflation rate, high interest rate and rupiah exchange rate against the US dollar, which was still high and unstable. The economic condition is still not too good that many large companies cannot produce in economies of scale, because the demand or consumer consumption of products which produced by manufacturing companies. Large companies also dare to expand business through new investment projects because, in general, large companies are still concentrating on solving the debt problem and also because of low purchasing power and low levels of public consumption of the products which produced by the company. Therefore, the size of the firm size does not significantly affect the profitability of the company. The results of the authors studies were inconsistent with those of Ville (1984), Eljelly (2004) and Abor (2005) who found firm size has a positively significant effect to firm profitability.

4.4.7 *Causality of CFP and CSP*

The analysis of the causality of financial performance on the company's social performance found no significant results on CFP proxy (ROA and ROE). Waddock and Graves (1997) found a significant positive relationship between the CSP and CFP indices (Tsoutsoura, 2004). Tsoutsoura found a significant and positive relationship between CSP and CFP. Opinions that support this view are that companies which have good financial performance will invest in the availability of resources on social performance, such as employee relations, environmental concerns or community relations. Companies with good financial performance will invest in generating long-term strategies, such as providing services to communities or their employees. This theory is called the slack resource theory, in other words, Tsoutsoura's research results contradict the results of research analysis conducted by authors who are more supportive of the social impact hypothesis.

While research conducted by Preston and O'Bannon (1997) also examines the relationship of CFP to CSP. The results obtained from Preston's research, positive synergies with available funds show the best relationship between CFP and CSP. This is in accordance with the slack resources theory, which states that the company will contribute to social performance if it has a good financial position. It can be said that CFP has a positive relationship with CSP. The research results contradict those of research conducted by the author, that CSP does not affect CFP.

Research by Makni et al. (2009) tested the causality between CSP and CFP of Canadian companies using the Granger causality approach, with control variables including size, corporate and industry risk. They found no relationship between a number of CSP and CFP measurements, except for market returns. This is in line with the results of research that found no causal relationship between CSP and CFP.

Research by Nelling and Webb (2006) examined the causal relationship between CSP and CFP by introducing a new econometric technique, the Granger causality approach. Their findings suggest that, using the OLS regression model, CSP and CFP are interrelated. Disagreeing with previous research, they found a low relationship between CSP and CFP when using the time series effect approach. Similar results were also found when introducing the Granger causality model. In addition, by focusing on each size of the CSP, they found that causality occurs in the stock market performance against the CSP rating in employee relations (in Makni et al., 2009). This is contrary to the results of research conducted by the author, that CFP is not significant to CSP.

The results of this research are not in accordance with the theory of slack resources which states that companies must have a good financial position to contribute to their social performance. Better financial performance leads to the management of the available investment opportunities, as well as the allocation of resources for socially responsible activities (Bird et al., 2006). This can be due to a lack of firm commitment and a paradigm of social responsibility. Therefore, even though a company has a good financial position, the funds spent on social activities and the environment are a fraction of those owned. This means that the percentage of funds spent on social and environmental activities does not affect the company's overall finances (such as donations or charities). It can be said that the company does not have a strategic approach to social responsibility. This can be seen from one of the aspects of assessment items such as vision, mission and company policy. Companies that do not have a strong commitment to CSP, can be seen from the vision, mission and policy that is not listed strategic approach in strong vision, mission or policy to the CSP. Therefore, even if the company has a good financial position, when performance of social responsibility only in order to meet the requirements of regulation. Therefore, the funds spent are only slightly or a fraction of the funds contained in the company as a whole. It can be said that there is still a low pressure both from stakeholders and the community on social performance by the company. This is consistent with Sudibyo (quoted from Ulfah, 2008) who concludes that there are two things that become difficult obstacles to the application of social accounting in Indonesia: the weak of social pressure requiring CSR and low awareness of companies in Indonesia about the importance of CSR. In addition, there is no regulations that require the company to publish its annual report. In accordance with Law No. 40 of 2007, there is no provision for a company to publish its social responsibility report as it only requires companies to engage in social responsibility and report it but not publish (publish to the public). Based on the data obtained, it is said that a low level of companies publish annual reports. Therefore, researchers cannot determine the overall social responsibility that has been undertaken by the companies.

4.4.8 *Mediated variable test causal step strategy*

The conclusion that CSP mediating the influence of CG (DK)to CFP (ROA and ROE) and models include full mediation. Accordance with stakeholder theory, states that the parties who have an interest in the company can influence or can be influenced by the activities of the company, the stakeholders among others community, employees, government, suppliers, capital markets and others. This shows that the role of CSP is quite important on the influence of CG (DK) in determining CFP (ROA and ROE).

5 CONCLUSIONS AND IMPLICATIONS

5.1 *Conclusions*

Based on the results of the analysis, hypotheses testing, research discussion, the conclusion of research as follows: corporate governance (DK and KA) and firm size influence CSP.

Corporate governance (DK, DKI and KA) and risk (DAR and DER) affect CFP (ROA). Corporate governance (DK and DKI) affects CFP (ROE). Only CSP variables mediate the influence of CG (DK) on CFP (ROA and ROE).

5.2 Implications

The implications of this research are considered by companies registered in the Jakarta Islamic Index to conduct periodic substantive and revision of CSP which has become the operational basis of the company in Jakarta Islamic Index in order to support the achievement of Good Corporate Governance

5.3 Limitations

The limitations of this research that should be considered for further research are:

1. The sample in this study focuses only on companies registered in the Jakarta Islamic Index.
2. The number of samples used in this research is relatively small.
3. This research only use data published in annual reports and annual financial statements which may not fully reflect the actual conditions.

REFERENCES

Abor, J. (2005). The effect of capital structure on profitability: An empirical analysis of listed firms in Ghana. *The Journal of Risk Finance*, 6(5), 438–445.

Anggraini, F. R. R. (2006). Social Information Disclosure and The Factors Influencing Social Information Disclosure in Annual Report (Empirical Study on JSX). In *9th National Symposium on Accounting.. Simposium Nasional Akuntansi*, 9, 23–26.

Barkemeyer, R. (2007). Legitimacy as a key driver and determinant of CSR in developing countries. Paper for the.

Beiner, S., Drobetz, W., Schmid, M.M. & Zimmermann, H. (2006). An integrated framework of corporate governance and firm valuation. *European Financial Management*, 12(2), 249–283.

Bewley, K., & Li, Y. (2000). Disclosure of environmental information by Canadian manufacturing companies: a voluntary disclosure perspective. In Advances in environmental accounting & management (pp. 201–226). Emerald Group Publishing Limited.

Bird, R., Casavecchia, L., & Reggiani, F. (2006). Corporate social responsibility and corporate performance: where to begin. University of Technology, Sydney Bocconi University, Milan.

Brammer, S., Brooks, C. & Pavelin, S. (2006). Corporate social performance and stock returns: UK evidence from disaggregate measures. *Financial Management*, 35(3), 97–116.

Clarkson, P.M., Li, Y., Richardson, G.D. & Vasvari, F.P. (2008). Revisiting the relation between environmental performance and environmental disclosure: An empirical analysis. *Accounting, Organizations and Society*, 33(4–5), 303–327.

ContitutionNo.25 tahun 2007 about Capital Investment.

Dalton, D.R., Daily, C.M., Ellstrand, A.E. & Johnson, J.L. (1998). Meta-analytic reviews of board composition, leadership structure, and financial performance. *Strategic Management Journal*, 19(3), 269–290.

Daniri, M. A. (2008). The standardization of corporate social responsibility (part 1). www. madani-ri. com/2008/01/17/ The standardization of corporate social responsibility (part 1).

Darmawati, D., Khomsiyah, K. & Rahayu, R.G. (2005). The Relationship Between Corporate Governance and Corporate Performance. *The Indonesian Journal of Accounting Research*, 8(1), 101–120.

Darwin, A. (2008, June). CSR, Standards & Reporting. In *Seminar Nasional Universitas Katolik Soegijapranata*.

Davis, K. (1973). The case for and against business assumption of social responsibilities. *Academy of Management Journal*, 16(2), 312–322.

Deegan, C. & Rankin, M. (1996). Do Australian companies report environmental news objectively? An analysis of environmental disclosures by firms prosecuted successfully by the Environmental Protection Authority. *Accounting, Auditing & Accountability journal*, 9(2), 50–67.

Deegan, C. (2013). *Financial accounting theory*. Australia: McGraw-Hill Education.

Deegan, C., Rankin, M. & Tobin, J. (2002). An examination of the corporate social and environmental disclosures of BHP from 1983–1997: A test of legitimacy theory. *Accounting, Auditing & Accountability Journal*, *15*(3), 312–343.

Diandono, H. (2012). The Influence of Mechanism of Good Corporate Governance To Financial Performance On Jakarta Islamic Index Company Listed In Period 2006–2011. *Sunan Kalijaga State University of Yogyakarta*

Ehikioya, B.I. (2009). Corporate governance structure and firm performance in developing economies: Evidence from Nigeria. *Corporate Governance: The International Journal of Business in Society*, *9*(3), 231–243.

Eisenberg, T., Sundgren, S. & Wells, M.T. (1998). Larger board size and decreasing firm value in small firms. *Journal of Financial Economics*, *48*(1), 35–54.

Eljelly, A.M. (2004). Liquidity–profitability tradeoff: An empirical investigation in an emerging market. *International Journal of Commerce and Management, 14*(2), 48–61.

Farook, S., & Lanis, R. (2007). Banking on Islam? Determinants of corporate social responsibility disclosure. Islamic Economics and Finance, 217.

Fauzi, H., & Idris, K. (2009). The relationship of CSR and financial performance: New evidence from Indonesian companies. Issues in Social and Environmental Accounting, 3(1).

Gamerschlag, R., Möller, K. & Verbeeten, F. (2011). Determinants of voluntary CSR disclosure: empirical evidence from Germany. *Review of Managerial Science*, *5*(2–3), 233–262.

Ghozali, I., & Chariri, A. (2007). Accounting theory. *Semarang: BP*.

Gray, R., Kouhy, R. & Lavers, S. (1995). Corporate social and environmental reporting: A review of the literature and a longitudinal study of UK disclosure. *Accounting, Auditing & Accountability Journal*, *8*(2), 47–77.

Guthrie, J. & Parker, L.D. (1989). Corporate social reporting: A rebuttal of legitimacy theory. *Accounting and Business Research*, *19*(76), 343–352.

Haniffa, R.M. & Cooke, T.E. (2005). The impact of culture and governance on corporate social reporting. *Journal of Accounting and Public Policy*, *24*(5), 391–430.

Hashim, H., & Devi, S. Corporate Governance, Ownership Structure and Earnings Quality: Malaysian Evidence, (2004).

Hastuti, T.D. (2005). The Relationship Between Good Corporate Governance and Ownership Structure On Financial Performance (Case Study On Jakarta Stock Exchange Company Listed). National Accounting Symposium *VIII*, 238–247.

Husnan, S. & Enny, P. (2004). Basic of Financial Management. Yogyakarta: UPP AMP YKPN.

Indonesia, R. (2007). Contitution No. 40 Tahun 2007 about PT.

Islam, M.A. & Deegan, C. (2010). Media pressures and corporate disclosure of social responsibility performance information: A study of two global clothing and sports retail companies. *Accounting and Business Research*, *40*(2), 131–148.

ISO, I. (2010). 26000 *Guidance on social responsibility*. Ginebra: ISO.

Juwitasari, R. (2008). The Influence of Independent, Frequency of Meeting and Commissioner Remuneration To Company Value On BEI Listed On Period 2007. *Doctoral dissertation, Indonesia University. Faculty of Economics).*

Kaen, F. (2003). A blueprint for corporate governance: Strategy, accountability, and the preservation of shareholder value. *Amacom.*

Kamil, A. & Herusetya, A. (2012). The Influence of Company Characteristic To Extensive Disclosure of Corporate Social Responsibility Activity. *Media Riset Akuntansi*, *2*(1), 1–17.

Kusumasari, B., Alam, Q. & Siddiqui, K. (2010). Resource capability for local government in managing disaster. *Disaster Prevention and Management: An International Journal*, *19*(4), 438–451.

Lanis, R. & Richardson, G. (2012). Corporate social responsibility and tax aggressiveness: A test of legitimacy theory. *Accounting, Auditing & Accountability Journal*, *26*(1), 75–100.

Lepineux, F. (2005). Stakeholder theory, society and social cohesion. *Corporate Governance: The International Journal of Business in Society*, *5*(2), 99–110.

Lindblom, C.K. (1994). The implications of organizational legitimacy for corporate social performance and disclosure. In *Critical Perspectives on Accounting Conference, New York, 1994*.

Luethge, D. & Guohong Han, H. (2012). Assessing corporate social and financial performance in China. *Social Responsibility Journal*, *8*(3), 389–403.

Luo, L., Lan, Y.C. & Tang, Q. (2012). Corporate incentives to disclose carbon information: Evidence from the CDP Global 500 report. *Journal of International Financial Management & Accounting*, *23*(2), 93–120.

Machmud, N. & Djakman, C.D. (2008). The Influence of Ownership Structure To CSR Disclosure On Annual Report. Empirical Study On Public Company On Indonesia Stock Exchange Company Listed In 2006. *Simposium Nasional Akuntansi, 11*, 50–63.

Mahoney, L. & Roberts, R.W. (2007). Corporate social performance, financial performance and institutional ownership in Canadian firms. *Accounting Forum, 31*(3), 233–253. Elsevier.

Makni, R., Francoeur, C. & Bellavance, F. (2009). Causality between corporate social performance and financial performance: Evidence from Canadian firms. *Journal of Business Ethics, 89*(3), 409.

Meehan, J., Meehan, K. & Richards, A. (2006). Corporate social responsibility: The 3C-SR model. *International Journal of Social Economics, 33*(5/6), 386–398.

Meek, G.K., Roberts, C.B. & Gray, S.J., (1995). Factors influencing voluntary annual report disclosures by US, UK and continental European multinational corporations. *Journal of International Business Studies, 26*(3), 555–572.

Muntoro, R.K. (2006). Develop Effective Commissioners. Management Institution Articles. Faculty of Economics University of Indonesia.

Murwaningsari, E. (2016). The Relationship of Corporate Governance, Corporate Social Responsibilities and Corporate Financial Performance in One Continuum. *Indonesian Management and Accounting Research (IMAR), 9*(1), 78–98.

Nelling, E., & Webb, E. (2009). Corporate social responsibility and financial performance: the "virtuous circle" revisited. *Review of Quantitative Finance and Accounting, 32*(2), 197–209.

Oktariani, N.W. & Mimba, N.P.S.H. (2014). The Influence of Company Characteristics and Environment Responsibility On Company Social Responsibility. *E-Jurnal Akuntansi*, 402–418.

Orlitzky, M. (2000). Corporate social performance: developing effective strategies. *Centre for Corporate Change*, Australian Graduate School of Management.

O'Donovan, G. (2002). Environmental disclosures in the annual report: Extending the applicability and predictive power of legitimacy theory. *Accounting, Auditing & Accountability Journal, 15*(3), 344–371.

Pierick, E.T., Beekman, V., van der Weele, C.N., Meeusen, M.J.G. & de Graaff, R.P.M. (2004). *A framework for analysing corporate social performance Beyond the Wood model*. The Hague: Agricultural Economics Research Institute (LEI).

Prabowo, A. S. P., Titisari, K. H., & Wijayanti, A. (2018, August). The Effect of Good Corporate Governance on Financial Performance of The Company (Empirical Study on Manufacturing Company of Consumer Goods Sector Industry Listed On Indonesia Stock Exchange Year 2015–2016). In *PROCEEDING ICTESS (Internasional Conference on Technology, Education and Social Sciences)*.

Preacher, K. J and Hayes, A. F., 2004. SPSS and SAS Procedures for Estimating Indirect Effects in Simple Mediation Models. *Behavior Research Methods*, Instruments, & Computers, 36 (4): 717–731. Psychonomic Society, Inc.

Preston, L.E. & O'Bannon, D.P. (1997). The corporate social-financial performance relationship: A typology and analysis. *Business & Society, 36*(4), 419–429.

Puspitasari, F. & Ernawati, E. (2010). The Influence of Corporate Governance Mechanism To Financial Performance of Bussiness Entity. *Jurnal Manajemen Teori dan Terapan| Journal of Theory and Applied Management*. 3(2), 189–215.

Rahajeng, R.G. & Marsono, M. (2011). The Influence Factors of Social Disclosure In Annual Report. (Empirical Study On Manufacture Company In Indonesia Stock Exchange). *Doctoral dissertation, DiponegoroUniversity*.

Ratnasari, Y. & Prastiwi, A. (2010). The Influence of Corporate Governance To CSR Disclosure In Sustainability Report. (*Doctoral dissertation, Universitas Diponegoro*).

Reddy, K., Locke, S. & Scrimgeour, F. (2010). The efficacy of principle-based corporate governance practices and firm financial performance: An empirical investigation. *International Journal of Managerial Finance, 6*(3), 190–219.

Robbins, S. P. coulter, M. (1999) Management.

Rosmasita, H. (2007). The Influence Factors of Social Disclosure On Annual Report of Manufacture Company In Indonesia Stock Exchange. National Accounting Symposium *X*. Makasar.

Rubin, A. (2005). Corporate social responsibility as a conflict between owners.

Said, R., Hj Zainuddin, Y. & Haron, H. (2009). The relationship between corporate social responsibility disclosure and corporate governance characteristics in Malaysian public listed companies. *Social Responsibility Journal, 5*(2), 212–226.

Sari, R. A. (2012). The Influence of Company Characteristic To CSR On Manufacture Company in BEI Listed. *Jurnal Nominal*, 1(1), 28.46.

Sembiring, E. R. (2005). Characteristics and Corporate Social Responsibility Disclosure: Empirical Study on Companies Listed in Jakarta Stock Exchange. *Accounting VIII National Symposium*. Solo.

Shleifer, A. & Vishny, R.W. (1997). A survey of corporate governance. *The Journal of Finance, 52*(2), 737–783.

Shocker, A.D. & Sethi, S.P. (1973). An approach to incorporating societal preferences in developing corporate action strategies. *California Management Review, 15*(4), 97–105.

State Secretariat. Jakarta.

Suchman, M.C. (1995). Managing legitimacy: Strategic and institutional approaches. *Academy of Management Review, 20*(3), 571–610.

Sunarto, S., & Budi, A. P. (2009). The Influence of Leverage, Firm Size and Growth Company To Profitability. Management Review Scientific Journal, 6(1), 86–103.

Tilt, C.A. (1994). The influence of external pressure groups on corporate social disclosure: Some empirical evidence. *Accounting, Auditing & Accountability Journal, 7*(4), 47–72.

Trebucq, S., & d'Arcimoles, C. H. (2002). The corporate social performance-financial performance link: evidence from France. Univ. of Bordeaux Dept. of Int'l Acc'tg Working Paper, (02-01).

Triyanto, E. (2010). The Influence Factors of CSR Disclosure. (Empirical Study On BEI Company Listed In 2005–2008. (*Doctoral dissertation, Universitas Sebelas Maret*).

Tsoutsoura, M. (2004). Corporate social responsibility and financial performance.

Uddin, M. B., Tarique, K. M., & Hassan, M. (2008). Three dimensional aspects of corporate social responsibility.

Ulfah, M. (2008). The Analysis of CSR and Social Accounting. Case Study On PT Jamsostek. (*Doctoral dissertation, Universitas Muhammadiyah Surakarta*).

Utama, S. (2011). An evaluation of support infrastructures for corporate responsibility reporting in Indonesia. *Asian Business & Management, 10*(3), 405–424.

Ville, S. (1984). Note: Size and profitability of English colliers in the eighteenth century—A reappraisal. *Business History Review, 58*(1), 103–120.

Waddock, S. A., & Graves, S. B. (1997). The corporate social performance–financial performance link. *Strategic management journal, 18*(4), 303–319.

Yahya, S. D. (2011). The Analysis of Financial Leverage Effect To Profitability On Telecommunication Company BEI Listed. *Skripsi. Faculty of Economics. Hasanuddin University. Makassar.*

Yaparto, M., Frisko, D., & Eriandani, R. (2013). The Influence of CSR To Financial Performance On Manufacture Sector In BEI Listed In period 2010–2011. *Calyptra,* 2(1), 1–19.

Yusoff, H., Mohamad, S.S. & Darus, F. (2013). The influence of CSR disclosure structure on corporate financial performance: Evidence from stakeholders' perspectives. *Procedia Economics and Finance, 7,* 213–220.

Business Innovation and Development in Emerging Economies – Trinugroho & Lau (Eds)
© 2019 Taylor & Francis Group, London, ISBN 978-1-138-35996-3

Default risk modeling of working capital loans to palm oil farmers in Nagan Raya Regency, Aceh Province, Indonesia

Y. Nugroho & K.S. Maifianti
Department of Agribusiness, Faculty of Agriculture, Universitas Teuku Umar, West Aceh, Aceh, Indonesia

ABSTRACT: **Purpose** – To develop a model, based on a farmer's characteristics, that can be used to minimize the risk of loan defaults for palm oil traders in Nagan Raya Regency of Aceh Province.

Design/methodology/approach – This research was conducted by interviewing palm oil traders in three sub-districts in Nagan Raya Regency in order to collect data regarding loan size and status, as well as the characteristics of the farmers who had been given the loan, and then analyzing the data with logistic regression.

Findings – Based on the Hosmer–Lemeshow test, logistic regression equations are able to explain the relationship between the 12 characteristics that were chosen for measurement and the probability of the farmers defaulting on the loan.

Research limitations/implications – An increase in the amount of data is expected to improve the model goodness of fit.

Practical implications – Palm oil traders are expected to use the model developed in this study to determine whether or not existing or prospective customers are eligible for a working capital loan.

Originality/value – Previously, default risk modeling research has generally been conducted on banking institutions, whereas this study focuses on SMEs that provide working capital loans as an incentive to farmers.

Keywords: Risk assessment, palm oil farmer, working capital, loan

1 INTRODUCTION

Indonesia is the world's largest producer of Crude Palm Oil (CPO). CPO is an oil that is obtained from the extraction of Fresh Fruit Bunches (FFBs) of palm oil. It is used for cooking oil, oleochemicals, and can even be used as fuel (biodiesel). Currently, palm oil plantations in Indonesia are situated on the islands of Sumatra (Aceh, North Sumatra, Riau, Jambi, West Sumatra and South Sumatra), Java (West Java), Kalimantan and Sulawesi.

In Aceh province, palm oil plantations are situated in almost every regency, from Aceh Tamiang, East Aceh and Great Aceh regency on the East Coast, to Aceh Jaya, West Aceh, Nagan Raya and South Aceh Regency on the West Coast. In Nagan Raya Regency, palm oil is a premium commodity. In 2016, palm oil plantations in Nagan Raya Regency occupied 49,401 hectares of land and produced 346,413 tons of FFBs (Central Bureau of Statistics, 2017).

In addition to large-scale palm oil plantations, such as PTPN I, PT Socfin Indonesia, PT Fajar Baizuri and PT Kalista Alam, there are also many smallholder plantations in Nagan Raya Regency. A smallholder palm oil plantation is a plantation owned by an individual with an area of less than 20 hectares, who is usually referred to as a palm oil farmer. Farmers with insufficient access to directly sell their FFB to Palm Oil Mills (POMs) present a new business opportunity: buying such farmers' FFBs and then selling them to POMs as a collecting merchant (commonly called a palm oil trader).

The business model of palm oil traders involves buying the FFBs from palm oil farmers and selling it to the POMs with a guaranteed profit margin. The profitability of this business means that a lot of people are interested in becoming a palm oil trader. Consequently, palm oil traders who wish to have a steady FFB supply from farmers must provide them with loans, either in the form of cash or production facilities, as a bond to ensure that the farmer's FFBs will be sold to the palm oil trader who has provided the loan.

However, the palm oil traders must be careful when assessing the personality or characteristics of the farmer who will be given a loan. This is because many palm oil traders have suffered losses, or even become bankrupt, because farmers do not repay the loans that they have been given.

Based on that background, the problem faced by palm oil traders is 'What model can be used to minimize the risk of loan defaults for palm oil traders in Nagan Raya Regency of Aceh Province, based on a farmer's characteristics?'

This study aims to create a model, based on a farmer's characteristics, that can be used to minimize the risk of loan defaults for palm oil traders in Nagan Raya Regency of Aceh Province. The palm oil traders are expected to use the model developed in this study to determine whether or not existing or prospective customers are eligible for a loan.

2 THE MODELS

The logistic regression model used in this research is:

$$g(p(x)) = \ln\left(\frac{p(x)}{1 - p(x)}\right) = \beta_0 + \beta_1 x_1 + \beta_2 x_2 + \cdots + + \beta_{12} x_{12}$$

or

$$p(x) = \frac{e^{\beta_0 + \beta_1 x_1 + \beta_2 x_2 + \cdots + + \beta_{12} x_{12}}}{\left(1 + e^{\beta_0 + \beta_1 x_1 + \beta_2 x_2 + \cdots + + \beta_{12} x_{12}}\right)}$$

where:
- $p(x)$: probability of loan default
- x_1 : sex (M/F)
- x_2 : age (years)
- x_3 : dependents (person)
- x_4 : other income (million rupiah)
- x_5 : land area (hectare)
- x_6 : production (100 kg FFB/month)
- x_7 : loan amount (million rupiah)
- x_8 : loan cycles (time)
- x_9 : level of education (years)
- x_{10} : experience as a palm oil farmer (years)
- x_{11} : family ties to the palm oil trader (Y/N)
- x_{12} : farmers loyalty (Y/N).

3 DATA AND SUMMARY STATISTICS

This research was conducted by interviewing palm oil traders in three sub-districts in Nagan Raya Regency—Kuala, Kuala Pesisir, and Darul Makmur—in order to collect information regarding loan size and status, as well as the characteristics of the farmers who were given the loan. The number of samples in this study was 72 farmers. Logistic regression was used to analyze the probability of default based on the characteristics of the farmers.

Based on IBM SPSS software output from the logistic regression analysis, the data shown in Table 1 was obtained.

The significance value of the omnibus tests was below 0.05, which illustrates the significant effect of the 12 farmer's characteristics in the model simultaneously affecting the probability of loan default.

Based on Nagelkerke's R-squared statistic (see Table 2), there is a 72.90% probability that loan defaults can be explained by these 12 characteristics, while other factors account for the remainder.

The significance value of the Hosmer–Lemeshow test of 0.503 (> 0.05) (see Table 3) illustrates that the logistic regression equation can explain the relationship between the 12 characteristics of the farmers and the probability of loan default.

It can be seen in Table 4 that the logistic regression model used is good enough because it is able to correctly predict 84.7% of the conditions that occur.

Table 5 shows that, of the 12 characteristics that were measured, only x_2 (age), x_6 (production), x_{10} (experience as a palm oil farmer), x_{11} (family ties to the palm oil trader), and x_{12} (farmers loyalty) have a significant influence on the probability of loan default.

Table 1. Omnibus tests of model coefficients.

		Chi-squared	df	Sig.
Step 1	Step	57.009	12	0.000
	Block	57.009	12	0.000
	Model	57.009	12	0.000

Table 2. Model summary.

	–2 Log likelihood	Cox & Snell R-squared	Nagelkerke R-squared
Step 1	42.804	0.547	0.729

Table 3. Hosmer– Lemeshow test.

	Chi-squared	df	Sig.
Step 1	7.315	8	0.503

Table 4. Classification table[a].

Observed			Predicted Y		% Correct
			Default	Smooth	
Step 1	Y	Default	12	6	66.7
		Smooth	5	49	90.7
		Overall percentage			84.7

a. The cut off value is 0.500.

Table 5. Variables in the equation.

		B	S.E.	Wald	df	Sig.	Exp(B)
Step 1[a]	$x_1(1)$	–0.012	1.411	0.000	1	0.993	0.988
	x_2	0.137	0.051	7.169	1	0.007	1.147
	x_3	–0.559	0.440	1.610	1	0.204	0.572
	x_4	–0.003	0.005	0.376	1	0.540	0.997
	x_5	0.009	0.005	2.499	1	0.114	1.009
	x_6	–0.052	0.026	4.043	1	0.044	0.949
	x_7	0.002	0.001	2.307	1	0.129	1.002
	x_8	0.205	0.231	0.788	1	0.375	1.228
	x_9	0.032	0.121	0.072	1	0.788	1.033
	x_{10}	–1.492	0.445	11.260	1	0.001	0.225
	$x_{11}(1)$	2.493	1.036	5.793	1	0.016	12.098
	$x_{12}(1)$	–4.209	1.497	7.900	1	0.005	0.015

a. Variable(s) entered on step 1: $x_1, x_2, x_3, x_4, x_5, x_6, x_7, x_8, x_9, x_{10}, x_{11}, x_{12}$.

4 RESULTS

$$p(x) = \frac{e^{0,137x_2-0,052x_6-1,492x_{10}+2,493x_{11}-4,209x_{12}}}{\left(1+ e^{0,137x_2-0,052x_6-1,492x_{10}+2,493x_{11}-4,209x_{12}}\right)}$$

If the value of p approaches 1, then the probability of non-default on the loan will be higher, and vice versa. If it is close to 0, then the probability of default on the loan by the farmers will be higher.

5 CONCLUSIONS

The model is expected to be used by palm oil traders before they decide to give a loan to farmers. The use of the model is expected to predict the probability of farmers being able to repay the loans that they are given.

Based on these statistics, it can be concluded that palm oil traders can use this model to minimize the probability of loan default.

REFERENCES

Caire, D., Barton, S., de Zubiria, A., Alexiev, Z., Dyer, J., Bundred, F. & Brislin, N. (2006). *A handbook for developing credit scoring systems in a microfinance context*. Washington, DC: US Agency for International Development.

Castro, C. & Garcia, K. (2014). Default risk in agricultural lending, the effects of commodity price volatility and climate. *Agricultural Finance Review, 74*(4), 501–521. doi:10.1108/AFR-10-2013-0036.

Central Bureau of Statistics. (2017). *Aceh in numbers*. Jakarta, Indonesia: Badan Pusat Statistik.

Dendramis, Y., Tzavalis, E. & Adraktas, G. (2017). Credit risk modelling under recessionary and financially distressed conditions. *Journal of Banking and Finance, 91*, 160–175. doi:10.1016/j.jbankfin.2017.03.020.

Hernandez-Trillo, F. (1995). A model-based estimation of the probability of default in sovereign credit markets. *Journal of Development Economics, 46*, 163–179.

Thomas, L.C. (2000). A survey of credit and behavioural scoring: Forecasting financial risk of lending to consumers. *International Journal of Forecasting, 16*, 149–172.

Business Innovation and Development in Emerging Economies – Trinugroho & Lau (Eds)
© *2019 Taylor & Francis Group, London, ISBN 978-1-138-35996-3*

Comparative financial risk of conventional and Islamic bank

O.S. Heningtyas, K.A. Rahayu & Payamta
Faculty of Economics, Sebelas Maret University, Surakarta, Indonesia

ABSTRACT: This research aims to analyze and compare the level of financial risk for both of conventional banks and Islamic banks for 2016. The data used in this study is secondary data in the form of company financial statements obtained from www.idx.co.id and webs of each bank. The method of analysis used in this research is Z-score discriminant analysis.

From the results of the data analysis in this research, it can be reported that conventional banks are in a high risk category because they have a mean Z score value of 0, 24 (0,24 <1,81) while Islamic banks also have high level of risk because the mean of their value of Z -score is also below 1.81 (0.96 <1.81). A comparison of the level of financial risk using the discriminant analysis results (Z-score) shows that both of banks are in a high risk position. However, conventional banks are greater risk than Islamic banks because they have a lower average Z-score.

The results of this study indicates that the banks are in a high risk financial position and poor financial management could lead to bankruptcy in the long run.

Keywords: Banking, Finance, Conventional bank, Islamic bank, Z-score

1 INTRODUCTION

The formation of Islamic economics began with the concept of economic and business *non-ribawi* (ban of usury in practice) as an effort to replace the conventional economic system that has an element of interest, and led to the formation of a rural financial institution with the name of Bank Mit Ghamr established by Ahmad El Najjar in 1963 (Karim, 2007). These efforts have had a positive impact on the collection of savings from the community, money deposit from *zakat, infaq* and *shadaqah,* and the lending of capital to people who have a low income, especially in agriculture (Ready, 1981 in Suyanto, 2006).

In addition to the ban on receiving interest (*riba*), gambling (*maisir*), excessive uncertainty (*gharar*) and investment restrictions in some sectors that produce products that are prohibited under Islamic law, what makes Islamic finance different is that financial assets should be based on the principle of Profit-Loss-Sharing (PLS). The requirements of this principles are practiced through contractual obligations built on purchases and sales (*murabahah*), leasing (*ijarah*) or partnerships in (*musyarakah/mudharabah*) in assets or asset portfolios (Iqbal & Mirakhor, 2011).

The development of sharia banking has been very rapid, both in Indonesia and in the world as a whole. A survey by the Bahrain Monetary Agency in 2004 showed that the number of sharia banking institutions jumped significantly from 176 in 1997 to 267 in 2004 operating in 60 countries around the world. With a growth rate of 15% this year, some parties claim that the sharia banking industry is the fastest growing sector in Muslim countries (Zaher & Hassan, in Fitria & Hartanti, 2010). Under normal circumstance, Karim Consulting Indonesia (KCI) predicts that asset returns (ROA) will reach 3,39 percent on assets of Rp 462.03 trillion. In an optimistic scenario, sharia banking ROA will reach 4,09 percent on assets of Rp 501.09 trillion. In 2018 the level of Non Performing Financing (NPF) expected to improve by between of 1,5 and 1.8 percent.

The main principle of Islamic finance is the prohibition of excessive risk taking. Regulator used to control the risk exposure of conventional banks. Unlike conventional banks, Islamic banks operate under the moral prohibitions set out in Islam in respect of excessive risk taking, interest-based transactions, and transactions with firms whose core business is considered outlawed in

Islam. these moral prohibitions along with the lack of prudential regulation of sharia banking, can lead to differences in risk taking between islamic and conventional banks. The rapid growth in the sharia banking industry has prompted scientific interest in investigating the risk issues for sharia banks in comparison with those for conventional banks (Cihák & Hesse,2010; Abedifar et al., 2013; Beck et al., 2013; Kabir et al., 2015; Mollah et al., 2017). Earlier empirical studies have examined accounting measures based on bankruptcy risk and credit risk for both islamic banks and conventional banks. Abedifar et al. (2013) and Mollah et al. (2017) state that there is no difference in risk between Islamic banks and conventional banks. In contrast, Cihák and Hesse (2010) and Beck et al. (2013) found that Islamic banks face a greater risk of bankruptcy. With reference to credit risk, Abedifar et al. (2013) reported that there is a lower credit risk, while Kabir et al. (2015) reported that islamic banks are at a greater credit risk than conventional banks.

Given these differing opinions in the literature, some research on the degree of risk in sharia banks and conventional banks is needed in Indonesia. In this study we tested the financial ratios to illustrate the level of risk and financial performance of sharia banks relative to conventional banks in Indonesia. This research was conducted in Indonesia because, on the one hand, majority of the population of Indonesia is Muslim and, on the other hand, because the magnitude of the proportion of muslim in Indonesia has an impact on the development of Islamic banks in Indonesia. We hope that this study may become one of the references for determining the strategy, evaluation and decision-making of managers and stakeholders regarding the risk assessment between Islamic banks and conventional banks.

2 LITERATURE REVIEW

2.1 Risks

According to Silalahi in Umar (2001), the definition of a risk is the chance of the occurrence of a loss, probability of loss, uncertainty, actual deviation from what is expected or probability that a result will differ from that is expected.

1. Risks in financial institutions

   a. Market risk
   Market risk is an inherent risk to the instruments and assets traded in the market and can arise in either micro or macro sources. On the one hand, systemic market risks are the result of overall price changes and policies in the economy. On the other hand, nonsystemic market risk arises when a specific asset or instrument price changes due to an event affecting the asset or instrument.
   b. Interest rate risk
   Interest rate risk is the exposure of a bank's financial conditions to changes in interest rates. Interest rate risk can arise from various sources.
   c. Credit risk
   Credit risk is the risk of the failure of creditors to fulfill their obligations in full and on time in accordance with their agreement. Credit risk can appear within either a trading book or banking book. In a banking book, credit risk arises when the customers in default fails to meet their debt obligations in full at the agreed time. Mean while the credit risk in a trading book, can also caused by incompetence or the ignorance of customers in fulfilling their obligations as contained in the contract.
   d. Liquidity risk
   Liquidity risk arises from a bank having inadequate liquidity to purchase its operational needs. Risk can also arise due to the difficulty of the bank in obtaining funds at a reasonable cost, either through loans or selling assets.

2.2 Conventional banks

According to Abdurahman in Umar (2001) the *Encyclopedia of Financial Economics and Trade,* state that 'the bank is a type of financial institution that performs various services, such as lending, currency exchange, currency control, acting as a store of valuable objects, finance business company business and others'. In Law Number 7 of 1992 of the Republic of Indonesia concerning banking. Article 1 Item 1stipulates that a bank is a business entity that collects funds from the community directly in the form of savings, and distributes it to the community in order to improve the standard of living of the people.

A conventional bank is a bank that conducts its business activities conventionally and provides payment service. The products of the accumulation of funds include demand deposits and time deposits. Disbursements can be formed as consumer loans, investment loans and working capital loans. Conventional banking services include services such as consulting services, handling export and import transactions, foreign exchange, and others. For banks that are based on conventional principles, the main advantage is obtained from the difference between interest on deposits granted to depositors and interest received from loans or disbursed loans. The advantage to the bank of this interest difference is referred as spread based. If a bank incurs a loss of interest difference, in which the deposit interest rate is greater than the lending rate, then it is known by the term negative spread.

2.3 *Islamic banks*

A sharia bank is a bank based on sharia principles with regards to a bank's business activities, sharia principles are rules of agreement under Islamic law between banks and other parties for the storage of funds and financing of business activities, or other activities declared to be in accordance with sharia. Rising funds in Islamic banks *wadi'ah* and applying the principle of *mudharabah*. The Al-Wadi'ah principle is relates to a pure deposit from one party to another party, where both individuals and legal entities must be maintained and returned to the cymbals.

The Al-Wadi'ah (trust depository) principle can be divided into Al-Wadi'ah Yad Amanah and Al-Wadi'ah Yad Dhamanah. The application of Al-Wadi'ah Yad Amanah concept in syariah bank is the party receiving the deposit should not use and utilize the money or goods that is entrusted, so it must be kept in accordance with the custom. In this case the recipient of the deposit can charge to the payee keeping fee. The concept of Al-Wadiah Yad Adh Dhamanah, in this concept the party receiving the deposit may use the money or goods deposited, of course the bank in this case get the profit sharing from the user.

In the *mudharabah* principle the depositor acts as the owner of the capital (*shahibul mall*) and the bank *as* the *mudharrib* (fund manager). The funds are used by the bank to conduct a *mudharabah*, whereby these two outcomes will be divided according to an agreed ratio. In the case of the bank using the funds to conduct a second *mudharabah*, the bank is fully responsible for any losses incurred. The *mudharabah* pillars applied *mudharrib*, who are own of funds. In this business they must share and took *Kabul permit*. This principle applies to time deposits and time deposits. The distribution of funds in Islamic banks is done through financing by the principle of sale and purchase, financing by the principle of lease, and financing by the principle of profit sharing. The principle of financing by sale and purchase is carried out in connection with the transfer of the ownership of goods or objects (transfer of property). The level of bank profit is determined in advance and becomes part of the price of the goods sold.

2.4 *Altman Z-score method*

a. Assess the Z-score risk level

The Altman Z-score is one method that is used to determine the level of corporate financial health and can be used to assess the success or failure of corporate management. The Altman Z-score formula that is used to predict bankruptcy is a multivariate formula used to measure the financial health of a company. Altman discovered five types of financial risks that can be combined in order to conclude the difference between a bankrupt and non-bankrupt company.

b. Predicted ratios of bank financial risk level

There are five financial ratios that are used to assess a bank's financial risks, namely:

1. Working Capital/Total Assets

Working capital here means the difference between current assets and current liabilities. Current assets in a banking company consist of cash, placements in other banks securities, accounts receivable, loans, and investments. Current liabilities consist of immediate liabilities, customer deposits, deposits from other banks, securities, derivative liabilities and acceptances, and taxes payable.

2. Retained Earnings/Total Assets

Retained here refers to retained earnings. Retained earning/total assets is a profitability ratio that can detect the ability of companies to generate profits within a certain period, which in terms of the company's ability to earn profits compared to the operating assets turnover rate as a measure business efficiency. This ratio regulates the accumulated profits during the company operate allows to facilitate the accumulated retained earnings.

3. Income before deducting the Interest Cost/Total Asset

The ratio of earning before interest and tax here refers to the operating profit. This ratio is the largest contributor to the model. Some indicators that can be used to detect problems with the company's profitability capability, inventory increases, decreased sales, delayed collection of accounts receivable, reduced corporate credibility and availability of creditors who can not pay at a given time.

4. Stock Market Price in Stock/Total Value of Debt

This ratio measures the ability of a company to provide assurance to every debt through its own capital. The market value equity ratio is the sum of the capital or equity value, while debt includes current liabilities and short-term debt.

5. Sales/Total Assets

This ratio measures the management's ability to use their assets to generate sales.

2.5 *Financial statement analysis*

According to Hery (2012): comparative financial statement analysis is done by reviewing the balance sheet, income statement, or cash flow in sequence from one period to the next. Analysis includes a review of the changes in the balance from one year to the next year, or over a few periods. By going the through comparative analysis of the financial reports, information can be obtained about the tendencies or trends of the account balance from year to year or over several periods. Through a comparative analysis, a bank can also assess the proliferation of relationships between one account balance and other interrelated account balances. Analysis of the comparative financial statements is also called horizontal analysis, which involves comparing the balances of existing accounts in the company's financial report over several different years.

3 FRAMEWORK

In order to provide a clear and systematic picture, the following figure presents a framework of the thinking behind the research and a guide to the overall research conducted. The theoretical framework in use is shown in Figure 1.

Information:

> In order to assess the level of financial risk, the required financial statement data consists of the income statement and the balance sheet. In order to calculate the Z-score, the variables of financial ratios must first be calculated, as shown in the above framework. From here, the level of business risk of a company can be seen.

A company is said to be healthy if it has a low risk level (Z > 2,99). If the value of the Z-score is between 1,81 and 2.99 this means that the company is considered to be in a *gray area*, and it is said to be unhealthy if it has a high risk level (Z < 1.81). Z-score is used as a tool to measure the level of financial risk of the banking industries in the period 2010–2012. As shown in the financial statements of the companies that were sampled in this research.

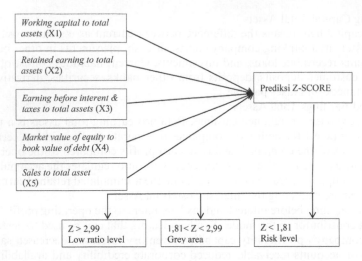

Working capital to total assets (X1)	
Retained earning to total assets (X2)	
Earning before interent & taxes to total assets (X3)	Prediksi Z-SCORE
Market value of equity to book value of debt (X4)	
Sales to total asset (X5)	

| Z > 2,99 Low ratio level | 1,81< Z < 2,99 Grey area | Z < 1,81 Risk level |

Figure 1.   Theoritical research framework.

4   RESEARCH METHODOLOGY

4.1   Objects and research subjects

The objects in this study are the annual financial statements of the sharia banks in Indonesia and the convention Banks listed on the Indonesian Stock Exchanges in 2016. The type of data used in this study is secondary data and quantitative data. The secondary data used in this study was taken from the annual financial statements that were obtained from official website www.idx.co.id 2016.

4.2   Sampling technique

The sampling technique used in this research is purposive sampling, which is based on certain specified considerations and criteria. The reason for using the purposive sampling method in this study was because the criteria for the sharia commercial banks and conventional commercial banks that will be make up the sample in this study were as follows:

1. Islamic banks in Indonesia and conventional banks that are listed the on BEI;
2. Sharia banks and conventional banks that publish annual financial report;
3. Published financial reports of sharia banks and conventional banks for 2016.

The number of conventional banks listed on the BEI in 2016 was 40 and the population of sharia banks in Indonesia amounted to 12. However, there were six conventional banks and one sharia banks that did not publish their financial statements, so that the final sample in this study was 34 conventional banks and 11 sharia banks.

4.3   Data collection technique

Data collected by using documentation method, that is done by tracing, collecting, recording and counting from data obtained. Documentation of data obtained from Statistics Islamic Banking and Conventional Banking published through the official website www.idx.co.id.

4.4   Data analysis method

In order to be able to perform the data analysis, data processing was initially undertaken by calculating the variables studied. The above variables described above were measured over a period of time and then an average each variable. Once the average of all these variables was known to they were included following in the formula (Supardi, 2003):

$$Z = 1{,}2\ (X1) + 1.4\ (X2) + 3.3\ (X3) + 0.6\ (X4) + 1.0\ (X5)$$

To find out which banks have a high or low risk level an assessment of the value of Z-score can be made, as follows:

1. If the value of the Z-score is smaller than or equal to 1,81 it means that the company is experiencing financial difficulties and has a high level of risk.
2. If the Z-score value is between 1,81 and 2.99 it means that the company is considered to be in a gray area. In this situation, the company has financial problems that must be addressed with proper management handling.
3. If the Z-score value is more than 2,99 it means that the company is in a healthy state and therefore has a low risk level.

5 RESULTS AND DISCUSSION

Bank's financial statements can indicate the level of financial risk of the company or predict the threat of bankruptcy. This can be achieved by calculating the financial ratios in order

Table 1. Result of conventional bank ratio calculation.

No	Conventional banks	X1	X2	X3	X4	X5	Z-score
1	PT Bank Artos Indonesia Tbk	0.071	−0.045	−0.014	0.190	0.112	0.202
2	PT Bank Rakyat Indonesia Agroniaga Tbk	0.856	0.021	0.011	0.170	0.085	1.280
3	PT Bank Agris Tbk	−0.145	0.006	0.002	0.141	0.094	0.018
4	PT Bank Capital Indonesia Tbk	0.097	0.035	0.009	0.093	0.094	0.344
5	PT Bank MNC International Tbk	0.116	−0.015	0.001	0.142	0.091	0.298
6	PT Bank Central Asia Tbk	−0.043	0.146	0.038	0.167	0.154	0.532
7	PT Bank Harda International Tbk	0.116	−0.012	0.005	0.185	0.110	0.361
8	PT Bank Negara Indonesia (Persero) Tbk	0.209	0.084	0.024	0.148	0.089	0.624
9	PT Mestika Benua Mas Tbk	0.212	2.262	0.029	0.255	0.073	3.745
10	PT Bank Yudha Bhakti Tbk	0.121	0.037	0.022	0.156	0.125	0.488
11	PT Bank Jtrust Indonesia Tbk	0.077	−0.695	−0.044	0.084	0.089	−0.888
12	PT Bank Nusantara Parahyangan Tbk	0.167	0.083	0.002	0.155	0.123	0.538
13	PT Bank Pembangunan Daerah Banten Tbk	0.029	−0,229	−0.097	0.165	0.101	−0.406
14	PT Bank Ganesha Tbk	0.219	−0.023	0.013	0.252	0.072	0.497
15	PT Bank Ina Perdana Tbk	−6.670	0.027	0.010	0.205	0.105	−7.706
16	PT Bank Pembangunan Daerah Jawa Timur Tbk	0.179	0.057	0.034	0.168	0.122	0.628
17	PT Bank QNB Indonesia Tbk	0.089	−0.015	−0.036	0.134	0.092	0.140
18	PT Bank Permata Tbk	0.252	0.117	−0.052	0.117	0.049	0.412
19	PT Bank CIMB Niaga Tbk	0.145	0.084	0.012	0.142	0.098	0.513
20	PT Bank Bumi Artha Tbk	0.080	0.066	0.015	0.182	0.104	0.451
21	PT Bank Maspion Indonesia Tbk	0.908	0.037	0.017	0.020	0.103	1.311
22	PT Bank Dinar Indonesia Tbk	0.093	0.057	0.007	0.193	0.086	0.418
23	PT Bank Tabungan Pensiunan Nasional Tbk	0.101	0.149	0.029	0.179	0.132	0.663
24	PT Bank of India Indonesia Tbk	0.145	−0.045	−0.134	0.257	0.097	−0.078
25	PT Bank Sinarmas Tbk	0.068	0.045	0.016	0.143	0.078	0.360
26	PT Bank China Contruction Bank Indonesia Tbk	0.845	0.033	0.006	0.195	0.089	1.288
27	PT Bank Mega Tbk	−0.189	0.040	0.022	0.174	0.112	0.117
28	PT Bank Mayapada International Tbk	0.828	0.043	0.018	0.116	0.100	1.282
29	PT Bank Artha Graha International Tbk	0.055	0.036	0.004	0.169	0.091	0.321
30	PT Bank Mitraniaga Tbk	−0.102	0.006	0.007	0.104	0.105	0.077
31	PT Bank Woori Saudara Indonesia 1906 Tbk	0.183	0.094	0.019	0.020	0.089	0.512
32	PT Bank PAN Indonesia Tbk	−0.454	0.096	0.017	0.172	0.094	−0.159
33	PT Bank National NOBU Tbk	−0.385	0.008	0.004	0.148	0.064	−0.283
34	PT Bank OCBC NISP Tbk	0.094	0.072	0.017	0.141	0.082	0.437
Average							0.245

Table 2. Result of calculation of islamic banks ratio.

No	Islamic banks	X1	X2	X3	X4	X5	Z-score
1	PT Bank Muamalat Indonesia Tbk	0.781	0.004	0.002	0.065	0.074	1.063
2	PT Bank Mybank Syariah Indonesia	0.644	−0.192	0.107	0.440	0.085	1.207
4	PT BankBCA Syariah	0.658	0.001	0.010	0.220	0.085	1.040
5	PT Bank BRI Syariah	0.557	0.018	0.009	0.090	0.095	0.871
6	PT Bank Pan Indonesia Tbk	0.149	0.028	0.017	0.172	0.094	0.471
7	PT Bank Jabar Banten Syariah	0.207	0.061	−0.073	0.118	0.271	0.433
8	PT Bank Syariah Bukopin	0.784	0.005	0.007	0.114	0.254	1.292
9	PT Bank BNI Syariah	0.814	0.030	0.013	0.088	0.116	1.231
10	PT bank Syariah Mandiri	1.482	0.005	0.006	0.081	0.152	2.004
11	PT Bank Tabungan Pensiunan Nasional Tbk	0.189	0.029	0.029	0.179	0.135	0.603
12	PTBank Victoria Syariah	0.248	−0.010	−0.017	0.120	0.099	0.399
Average							0.965

measure whether or not the company is healthy. To detect whether a corporation is financial distress or unable to trade, we use a Z-score analysis developed by Professor Edward Altman. A company needs to know its level of financial risk in order to be able to operate optimally. One of the factors that must be considered in order for a company to survive are the financial statements that are used to determine the financial risk of banking.

The results of the analysis of the level of financial risk of conventional banks in 2016 show that they have a Z-score value of below 1.81 (0.24 < 1.81) while the results for Islamic banks also show a high risk level due to also having a Z-score value below 1.81 (0.96 <1.81).

From the results of discriminant analysis, it can be seen that both conventional bank and sharia banks are in high risk position. However, the results of the Islamic banking analysis show that their Z-score value is higher than that derived from the analysis of the conventional bank analysis which means that the financial risk value of the result of the islamic bank analysis is lower than that of the conventional bank analysis. The low Z-score indicates that conventional banks still remain in a high financial risk positions but the increased in the Z-score value in the sharia banks suggests improvements in financial management handling while still remaining at high risk.

The Z-score calculation shown above is important because one of the most important aspects of the analysis of the financial statements of a banking company is its usefulness to assess the viability of banking. Banking survival is crucial for management to anticipate potential bankruptcy possibilities in order o ensure a bank's survival, as bankruptcy means payments of costs, both direct and indirect.

6 CONCLUSIONS, SUGGESTIONS, AND LIMITATIONS OF RESEARCH

6.1 Conclusion

Based on the results of the research on financial risk analysis in conventional banks and sharia banks using the Altman Z-score method the following conclusion can be drawn:

1. Altman Z-score analysis results for the financial performance of conventional banks in 2016 produced a Z-score value of 0.24. According to the Z-score, a value below 1.81 is categorized as a company that has very large financial difficulties and is at high risk. It can therefore be seen that conventional banks 2016 were experiencing a very large level of difficulty and were at high risk, so that the possibility of bankruptcy was very large.
2. Altman Z-Score analysis results for Sharia Bank financial performance in 2016 obtained Z-Score value of 0.96. Based on the Z-Score criteria <1.81 is categorized as a company

that has very big financial difficulties and high risk so that it can be seen that the Sharia Bank in 2016 has a very big difficulty and high risk so that the possibility of bankruptcy is very large.

6.2 Limitations of the research

Some limitations in this study are as follows:

1. The timeframe of this research is just 2016, so it represents a limited view of banking finance risk.
2. The variables used in this study are limited to financial risk variables.

6.3 Suggestion

Based on the above conclusions, the following suggestions can be made:

1. Only the financial risk variables are used in this study. Further research could be expected to add other variables that affect the risk level of both Islamic banks and Conventional banks such as adding disclosure as an additional variable.
2. It is also suggested that future researchers lengthen the study period so that a bigger sample is obtained, which should generate more accurate results.

REFERENCES

Abedifar, P., Molyneux, P., & Tarazi, A., (2013). Risk in Islamic banking. *Rev. Financ. 17*, 2035–2096.
Beck, T., Demirgüç-Kunt, A., & Merrouche, O., (2013). Islamic vs. conventional banking: Business model, efficiency and stability. *J. Bank. Financ. 37* (2), 433–447.
Čihák, M., & Hesse, H., (2010). Islamic banks and financial stability: An empirical analysis. *J. Financ. Serv. Res. 38*, 95–113.
Hery. (2012). *Financial Statement Analysis*. Jakarta, Indonesia: Bumi Aksara.
Iqbal, Z., Mirakhor, & A., (2011). *An introduction to islamic finance: Theory and practice*, 2nd ed. Singapore: John Wiley & Sons (Asia).
Kabir, M.N., Worthington, A., & Gupta, R., 2015. Comparative credit risk in Islamic and conventional bank. *Pac. Basin Financ. J. 34*, 327–353.
Karim, A.A. (2007). *Islamic micro economy*. Jakarta, Indonesia: PT Raja Grafindo Persada.
Mollah, S., & Zaman, M., (2015). Shari'ah supervision, corporate governance and performance: Conventional vs. Islamic banks. *J. Bank. Financ. 58*, 418–435.
Mollah, S., Hassan, M.K., Al Farooque, O., & Mobarek, A., (2017). The governance, risk-taking, and performance of Islamic banks. *J. Financ. Serv. Res. 51*, 195–219.
Suyanto, M. 2006. *The influence of the implementation of shariah principles on the performance and welfare of the community in the environment of Indonesia Islamic banking activities. OPTIMAL*, (*1*) 23–49.
Umar, (2001). *Manajemen Risiko Bisnis*. Jakarta, Indonesia: PT. Gramedia Pustaka Utama. (2018).

REFERENCES

Social sciences

Business Innovation and Development in Emerging Economies – Trinugroho & Lau (Eds)
© *2019 Taylor & Francis Group, London, ISBN 978-1-138-35996-3*

Social performance measurement through local culture in microfinance institutions

I.P. Astawa
Tourism Department, Politeknik Negeri Bali, Badung, Bali, Indonesia

I.M. Sudana
Accounting Department, Politeknik Negeri Bali, Badung, Bali, Indonesia

N.G.N.S. Murni
Tourism Department, Politeknik Negeri Bali, Badung, Bali, Indonesia

I.G.N. Sanjaya
Business Administration, Politeknik Negeri Bali, Badung, Bali, Indonesia

ABSTRACT: Currently there is no agreement regarding instrument used in social performance measurement although it is required by rural community-oriented microfinance institutions. This research aimed to explore microfinance institution programs related to social performance from the perspective of local culture implemented by the institutions in serving their customers. This paper analyzed the ways in which the LKM a (microfinance institutions) selected for this study measure and manage their social performance. Data were acquired from two sessions of focus group interviews, a series of semi-structured interviews, and extensive reviews of documentaries. Harmonious culture was used to explore the social performance of three microfinance institutions that had the largest assets and implemented a harmonious culture. The research results indicate that LKM programs consistently favored rural communities through credit services to improve the rural economy. Social activity that became its goal included cultural preservation and was used to evaluate social performance. The indicators of social performance were making a donation for the construction of places of worship; funding the spiritual journey activities; assisting ceremonies in a village; making a donations for religious ceremonies in a communal group environment; donation to make ceremonial facilities; donation of funds for ceremonies related to ancestors; improvement of community skills; assisting humanity ceremonies; providing entrepreneurship education; making a donation for funeral, health or; business assistance; improving education at the village level; assisting the construction of a meeting hall in the village; market, road, sport facilities; and garbage bins. This type of research was not conducted broadly; thus, it gave a contribution to the development of social performance. The implication of the research results is that microfinance institutions in Indonesia could include harmonious culture in their working programs when implementing a social mission.

Keywords: social performance, local culture, microfinance institutions

1 INTRODUCTION

The social performance issue is still debated among experts in microfinance institutions. The question arises whether microfinance institutions that provide social services for the poor in society have conducted a proper social performance measurement (Copestake, 2007). Social performance has different characters compared to the measurement of financial performance (Arena et al., 2015; Ebrahim & Rangan, 2014). Therefore, both performance measurements are important to develop organizational sustainability. The measurement of social performance has an impact on internal activities and social services, which are the goals of microfinance institutions.

The organizational mission could be better achieved if it is translated into measurable strategic activities (Bhimani & Willcocks, 2014). An understandable measurement could facilitate the achievement of company' vision and improve its activities (Copestake, 2007; Woller, 2007). Good and measurable mission translation could bring efficiency and effectiveness to work. It also facilitates the organization to objectively evaluate activities and performance. Clarity in performance measures helps in reducing miscommunication, conflict and misunderstanding among employees. It also gives clarity in optimizing the existing resources, tangible and intangible. Becker (1993), in Human Capital Theory, stressed that resources empowerment could give a significant contribution to the company in order to gain profit. Good resources management could improve employees' skills and the company's performance; thus, a clear standard of measure is needed in every activities (Newman et al., 2014; Ucbasaran et al., 2008; Baron & Ensley, 2006; Baum & Locke, 2004).

The measurement of social performance could provide recommendations to the company in providing better services to the customers as well as perform various internal policies effectively and efficiently. The management of social performance has a significant relationship with the achievement of social goals by microfinance institutions (Ledgerwood et al., 2013). Microfinance institutions should be able to measure their social performance in their own way or through a culture they implemented in order to achieve their mission. Measuring social performance using financial data is not recommended since it will give a different opinion and ruin the social mission (Copestake, 2007; Ebrahim et al., 2014). The management of social performance could be conducted in two forms: results achieved by a social mission and financial performance.

Regardless of the results of previous studies, there is still a gap in the measurement of social performance that is often measured by financial data. Internal process in social performance measurement suitable to institutional culture in conducting social mission has not conducted well. Most previous research results stress the impact of conducting a social program in order to achieve financial performance (Ebrahim & Rangan, 2014). This paper tries to fill the gap in previous research by conducting a social performance measurement implemented in the internal processes of microfinance institutions (Ebrahim & Rangan, 2014). The problem concerns the forms of local cultural-based social performance measurements conducted by the companies.

Regarding the research, microfinance institutions in Indonesia that use local culture are village credit institutions (LPD) (Astawa et al., 2013, 2016a, 2016b, 2018). Local culture implementation places emphasis on the balance of the harmonious relationship of the company with the employees (*pawongan*), working environment (*palemahan*) and God (*parahyangan*) (Astawa, 2013; Astawa & Sudika, 2014). The three relationships are called a harmonious culture or *tri hita karana* and are used as a foundation in implementing all LPD programs (Astawa et al., 2017).

2   THE ACTVITIES OF VILLAGE CREDIT INSTITUTIONS

Village credit institutions have an important role in supporting village economies (Baskara, 2013). There are 1,408 LPDs (Bank Pembangunan Daerah, 2015) that have a goal of providing credit service to rural communities in the hopes of improving the economy and preserving the culture (Astawa, 2013). The LPDs have provided assistance in the form of entrepreneurship training for debtors consisting of 1,658 small companies since 2012. The training was conducted for one, two, or three years period (Bank Pembangunan Daerah, 2015). The implementation of the value of a harmonious culture conducted by the LPDs that consisting of assistance in religious activities, public health improvements, and environmental improvements for the customers has increased the amount of savings (Astawa et al., 2016a). Various activities have been conducted for employees to improve services through skills training provided by the local government. *Bank Pembangunan Daerah* (Regional Development Bank) acts as an advisor for the LPDs and is assisted by the government. The types of financial services provided are savings and credit for rural communities. The scope of the services is in the rural communities where the LPD is established and they are not allowed to serve people outside the village.

3 LITERAURE REVIEW

Microfinance institutions are necessary in developing countries to provide credit service (Rashidah et al., 2016). Eighty percent of Indonesian people require a microfinance service (Karim et al., 2008). The government encourages the growth of microfinance institutions in various regions to give funding access for underprivileged rural communities to promote economic resilience. The services provided are financial and non-financial in nature (Ledgerwood, 1999). Service strategies should be considered in order to maintain business sustainability since there are social and financial performance measures and both are different (Woller, 2007). The sustainability of a microfinance institution can be seen from its social responsibility measured by financial, economic and social performances (Hubbard, 2009). Different opinions occur when the sustainability of microfinance institutions is viewed merely from income that is used as a measure for success since it could cover operational cost and make a profit (Bhanot & Bapat, 2015).

Another opinion is developed where the sustainability of a company is related to its financial performance. This opinion brings about two approaches namely, institutional and community welfare approaches. The institutional approach focuses on the number of borrowers and the range of services (Woller et al., 1999). The institutional approach is focused on financial performance; therefore, services related to the interest issue should be in accordance with the existing market conditions. Companies will give their employees trainings to provide better services in order to increase profit. A large profit is used to conduct a planned social mission. The community welfare approach has a different view, where the sustainability of a company can be assessed through the impact of lending, which is whether or not the customers have a better life after participating in the programs provided by the microfinance institutions (Olivares-Polanco, 2005). This approach does not view financial sustainability; instead, it merely focuses on the amount of subsidies or benefits given by the microfinance institutions to the rural communities (Bos & Millone, 2015).

The result of the literature review demonstrates that there are some problems for companies using profit or non-profit in measuring performance. Using profit as a measure of social performance is the best method (Durden, 2008; Norris & O'Dwyer, 2004). Most companies use a financial approach as a tool to measure social performance (Durden, 2008; Norris & O'Dwyer, 2004). This condition encourages the company to assess social performance subjectively (Stevens et al., 2014). Another impact of this method is the use of an informal approach to achieve a social mission (Norris & O'Dwyer, 2004). Only some literatures provides guidelines for assessing the social performance of microfinance institutions (Astawa et al., 2017; Astawa et al., 2016b; Copestake, 2007).

The informal approach uses in achieving a social mission can be conducted through the culture that exists in the microfinance institutions. One of the strategies is by building communication with the customers and paying attention to their culture (Astawa et al., 2016a; Astawa & Sudika, 2014; Astawa et al., 2013; Astawa, 2013; Astawa et al., 2012). The organizational culture followed by the microfinance institutions is sourced from the local culture in Indonesia, known as a harmonious culture. Harmonious culture emphasizes a harmonious relationship between microfinance institutions and God (*parahyangan*) as the creator of the universe through various religious and indigenous activities (Astawa et al., 2017; Sukawati & Astawa, 2017; Astawa & Sukawati, 2016; Astawa et al., 2016b). Other harmonious relationships are those between microfinance institutions and employees, customers, and the natural environment. The harmonious culture becomes the organizational culture in microfinance institutions and it has a significant impact on a company's performance (Astawa et al., 2017;

Sukawati & Astawa, 2017; Astawa & Sukawati, 2016; Astawa et al., 2016b) and is used as a strategy to win competition with commercial banks.

4 RESEARCH METHOD

Data collection was conducted during focus groups attended by three managers of the LPDs with the largest business activities (Bank Pembangunan Daerah, 2015). The invitation for focus group discussion (FGD) was submitted to the managers earlier and the purpose of the FGD for the measurement of social culture through the values of harmonious culture was explained. It also asked the managers to prepare key informants who understood the issue if they could not attend the FGD. In the initial stage, the researcher conducted a formal presentation before the key informants and continued with a debt interview for four hours. During the interview, documenting, recording, and taking pictures were conducted to facilitate data transcription. The results of the in-depth interview are presented in Table 1.

The interview was followed up by a phone call according to the agreement with the managers of the LPD.

To develop an understanding of LPDs' internal activities in social practices through harmonious culture, the data processing involved three stages. The first stage involved setting the results of interview transcripts in a chronological way, reviewing the documents, and conducting social performance analysis (Ahrens & Chapman, 2007). We further used formal documentary data as a means of triangulation to explain and confirm issues that emerged during our interviews and focus group sessions (Ahrens & Dent, 1998). The second stage involved understanding the transcripts that are significant to the mission of LPDs and using three analysis elements: intent, implementation, and outcome (Ledgerwood et al., 2013). The third stage involved combining the analysis result with the harmonious culture consisting of a harmonious culture with God, human beings, and the environment (Astawa et al., 2017; Sukawati & Astawa, 2017; Astawa & Sukawati, 2016; Astawa et al., 2016b).

5 FINDINGS

This section explains how LPDs measured the social performance of the programs implemented. The explanation consists of the LPD's intent and design, internal processes of LPDs' programs and evaluation of the implementation of LPDs' programs. Another explanation relates to the reasons why LPDs did not conduct a social measurement and how is the relationship with cultural values implemented by the LPDs.

5.1 Intent and design of the LPDs

The LPDs have been designed by the local government to help societies to improve their economies and preserve their cultures. The process of village economic improvement is conducted

Table 1. Focus group interview.

LPD	Title of participants	Tenure	Amount of credit distributed	The implementation of a harmonious culture
LPD 1	LPD's manager	16 years	On average 10.6 billion per year	Village fund, on average 500 million per year Assistance to the villagers, on average 345 million per year
LPD 2	LPD's manager	14 years	On average 10.1 billion per year	Village fund, on average 450 million per year
LPD 3	LPD's manager	13 years	On average 10 billion per year	Village fund, on average 450 million per year

Table 2. Result of debt interview.

Activities	LPD1	LPD2	LPD3
The emphasis of the LPD	Major financial improvement referring to the Regional Development Bank (BPD)	The balance of financial growth and social mission	Financial growth and serving the community
Organizational structure in service	The formation of community service division	Services are developed according to community demand	Services to the community are handled by one division
Business model implemented	Serving the rural community	Serving the rural community	Serving the rural community
Types of loan	Personal, group and company	Personal, group	Personal, group, company and village
Other services	Payment of electricity bill, Samsat (one roof administrative system), phone bill, tax and consumer cooperative	Services for electricity and state-owned water company (PDAM)	Services for electricity and phone bill payment
Services related to customary village	Give profits to the customary village	Give profits to the customary village	Give profits to the customary village
Type of cultural service related to God	Make a donation for the construction of worship places and the making of ceremonial means	Funding spiritual journey activities, the construction of worship places	Assist ceremonies in the village, donation for religious ceremonies in the community group environment
Type of cultural service related to human beings	Assist ceremonies related to the ancestors, improvement of community skills, assist humanity ceremonies	Give entrepreneurship education, ceremonies related to human, donation for funerals, health	Give business assistance, donation for religious activities in ceremonies for humans, improvement of village education
Type of cultural service related to environment	Assist the construction of village facilities and infrastructure	Assist the construction of village meeting hall, market, road and sports facilities	Assist the village road construction, the making of garbage bins, sports facilities renovation, market and meeting hall renovation.
The measurement of results and performance	Financial growth and credit service	Financial growth and credit service	Financial growth and credit service
Impact of evaluation	Internal company and customary village	Internal company and customary village	Internal company and customary village

Source: Processed data.

by lending, whereas cultural preservation is conducted by giving part of the profits to the head of a customary village. The services are limited to the members of the society in the village where the LPD is established. The results of in-depth interviews from three LPD managers obtained several statements related to the intent and design of the LPDs as follows;

> The societies need LPD and LPDs' mission has been set by the local government for economic and cultural purposes [LPD1, LPD2]. it is a government program for the

villages [LPD1] with the main missions include developing the village and preserving the culture [LPD3].

5.2 Internal processes

Social performance is implemented into the management programs existing in the LPDs. The summarized results related to the process of implementation of a social mission into programs and targets achieved in every activities of the three LPDs are shown in Table 2.

Based on Table 2, the three LPDs emphasized financial growth by referring to the regulations of the government represented by *Bank Pembangunan Daerah*. The financial strength became a base for carrying out social activities through various assistances for cultural activities. The programs of the LPDs were financial and social services-oriented but without an explanation of social performance measurement. The impact of the activities was evaluated and had internal and external roles (customary villages). The services developed were not merely in credit services but developed according to the needs of rural communities, such as electricity, tax, PDAM (water) and phone bill payments.

5.3 Program evaluation

Generally, the measurement of performance is conducted using financial growth and credit service. Whereas, performance measurement referring to banking regulations is conducted using CAMEL (capital, asset, management, earning, liquidity). The result of a performance measurement is explained as follows:

> LPD performs the measurement by referring to the regulations stated by Bank Pembangunan as our instructor [LPD1, LPD2, LPD3], our achievement is accounted for through village meeting in a simple way that consists of profit, credit service, problems and future plans [LPD1, LPD2, LPD3].

Performance evaluation was implemented internally through credit analysis and an internal regulatory body. The result of the evaluation was submitted to the head of the company to be used as a base for consideration or as a source in the company's policy making. The performance evaluations were mostly financially oriented, whereas the impact of lending had not been well conducted. It is stated by the three LPDs:

> we realize that evaluation on lending is needed; however, we have no standard rules about it and we are only based on how much profit we can give to the village and social donation related to culture [LPD1, LPD2, LPD3].

6 DISCUSSION

The summary of the results based on intent and design of the LPDs, internal processes and program evaluation of the three LPDs with the goal of improving the economy and culture are outlined in Figure 1.

In running their business, the LPDs are not merely economic or financially oriented as stated by banking regulations but they have a mission of cultural preservation and are expected to impact on society (Nabiha et al., 2018). Post-lending prosperity of the community is the main indicator in the mission of microfinance institutions in the world (Rashidah et al., 2016) and commercial banking is used as a reference in assessing performance. Microfinance institutions in Indonesia have a mission of preserving the culture and it is important to asses it. The measures of social performance could be assessed through cultural practices, as stated by the three LPDs, such as: making a donation for the construction of worship places, funding spiritual journey activities, assisting ceremonies in the village, donating for religious ceremonies in the community group environment, donating for the making of ceremonial means, donating funds for ceremonies related to the ancestors, improvement of community

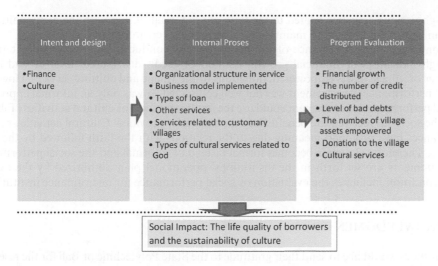

Figure 1.   Social performance of the LPDs.

skills, assisting humanity ceremonies, providing entrepreneurship education, donating for funeral, health and business assistance, improvement of education at village level, assisting the construction of meeting halls in the village, markets, roads, sport facilities and garbage bins. These indicators could be used to assess social performance. This view is strengthened by research results related to harmonious culture which state that culture has a significant impact on performance (Astawa et al., 2017; Astawa et al., 2016b; Copestake, 2007).

The social mission implemented by the LPDs develops from various perspectives of previous research results (Nabiha et al., 2018; Ebrahim & Rangan, 2014) with emphasis on the level of society's welfare through economy. Economic welfare is easy to quantify. In reality, certain properties in community's welfare are difficult to measure, namely the non-financial. LPDs, however, succeed in implementing non-financial measurements; thus financial and non-financial prosperities could work in harmony and strengthen the communities' economies (Astawa et al., 2017, Bos & Millone, 2015, Ledgerwood, 1999).

7   CONCLUSION

Village credit institutions focus on the service of rural communities in the economic sector as well as preserving the cultural values that exist and develop in every village. The obligation of financial and non-financial development is synergized between the LPD administrator and the head of the village. The government gives directions and guidance related to the governance of microfinance institutions on an ongoing basis to maintain the performance of the LPDs. LPD coaching has been set forth in a mutual agreement between the government and the village. At village level, agreement is also conducted between the administrator and the head of the village. This method helps accelerate the execution of the mission set forth in the work programs.

An emphasis on business is still dominant in the element of financial improvement along with the implementation of the social mission through cultural activities. The culture provides a different image to the microfinance institutions in the world. Microfinance institutions in the world provide details on non-financial performance; however there is no explanation about the cultural elements. To carry out the mission, a simple organizational structure is built that consists of a head, secretary and administration. The structure is given by the government and it could be developed accordingly. The business model is oriented toward serving the rural community to achieve a reliable and strong economy and a sustainable culture. Cultural practice becomes a significant differentiator from other microfinance institutions through various activities related to the faith in God, relationship with the communities or human beings, and related to maintaining a harmonious environment.

Monitoring the performance of an institution was conducted internally and externally through indicators of financial growth, amount of credit distributed, level of bad debt, number of village assets empowered, donations to the village and cultural services. However, social performance measurement was not well conducted. It was only an informal approach. Social performance measurement could use the implementation of cultural activities (Table 2) and the societies received the benefit from the assistance provided. Cultural activities-related assistances have profound meaning due to the relations with the faith followed by the borrowers. Therefore, closeness becomes formal instead of informal and the social performance measurements are set forth in the institutions' operational plan authorized by the owner. This condition facilitates the evaluation of social performance for microfinance institutions.

ACKNOWLEDGMENTS

The authors would like to send their gratitude to the State Polytechnic of Bali for the research funds and the head of the Center of Research and Community Service who provided the motivation to complete the research on time. We would also like to say thank you to the managers of the LPDs who participated in the research.

REFERENCES

Ahrens, T. & Chapman, C. (2007). "Management accounting as practice". *Accounting, Organizations and Society*, *32*(1/2), 1–27.

Ahrens, T. & Dent, J.F. (1998). "Accounting and organizations: Realizing the richness of field research." *Journal of Management Accounting Research*, *10*, 1–39.

Arena, M., Azzone, G. & Bengo, I. (2015). "Performance measurement for social enterprises VOLUN-TAS." *International Journal of Voluntary and Non profit Organizations*, *26*(2), 649–672.

Astawa, I.P. & Sudika, P. (2014). "The impact of local culture on financial performance in property firms in Bali." *Asia-Pacific Management and Business Application*, *3*(2), 106–115.

Astawa, I.P. & Sukawati, T.G.R. (2016). "Ubud gets the customer an ethnomethodology approach." *International Journal of Economic Research*, *13*(7), 2681–2692.

Astawa, I.P., Sukawati, T.G.R., Triyuni, N.N. & Abdi, I.N. (2016b). "Performance of microfinance institution in harmony cultural perspective in Bali." *Procedia-Social and Behavioral Sciences*, *219*, 113–120.

Astawa, I.P., Susyarini, N.P.W.A., & Ni, N.T. (2016a). "Information technology implementation on socialization of harmonious culture in Bali." In *MATEC Web of Conferences*, *58*, EDP Sciences.

Astawa, I.P., Triyuni, N.N., & Cipta, I.D.M. (2018). "Sustainable tourism and harmonious culture: Acase study of cultic model at village tourism." *Journal of Physics: Conference Series*, *953*, 1–9.

Astawa, P.I., (2013). "Ownership in the perspective of ethnomethodology at village credit institutional in Bali." *Research Journal of Finance and Accounting*, 4(8), 55–62.

Astawa, P.I., Sudana, I.M., & Murni, N.G.N.S. (2017). "Non-financial performance measures on local culture basis in assessing the health of microfinance institutions." *Journal of Finance and Banking Review*, *2*(3), 29–35.

Astawa, P.I., Sudarma, M., Aisjah, S., & Djumahir. (2013). "Institutional ownership and harmonious values in increasing financial performance of village credit institution (Lembaga Perkreditan Rakyat/ LPD) in Bali province," *Journal of Basic and Applied Scientific Research*, *3*(6), 813–824.

Bank Pembangunan Daerah Bali., (2015). "*Annual report*," Bali Province Press.

Baron, R.A., & Ensley, M.D. (2006). "Opportunity recognition as the detection of meaningful patterns: Evidence from comparisons of novice and experienced entrepreneurs," *Management Science*, *52*(9), 1331–1344.

Baskara, I.G.K. (2013). "Microfinance in Indonesia," *Journal Buletin Studi Ekonomi*, *18*(2), 114–124.

Baum, J. R., & Locke, E.A. (2004). "The relationship of entrepreneurial traits, skill, and motivation to subsequent venture growth," *Journal of Applied Psychology*, *89*(4), 587–598.

Becker, G.S. (1993). "Nobel lecture: The economic way of looking at behavior," *Journal of Political Economy*, 385–409.

Bhanot, D. & Bapat, V. (2015). "Sustainability index of micro finance institutions (MFIs) and contributory factors," *International Journal of Social Economics*, *42*(4), 387–403.

Bhimani, A. & Willcocks, L. (2014). "Digitisation, 'big data' and the transformation of accounting information," *Accounting and Business Research*, *44*(4), 469–490.

Bos, J.W.B. & Millone, M. (2015). "Practice what you preach: Microfinance business models and operational efficiency," *World Development*, *70*, 28–42.

Copestake, J. (2007). "Mainstreaming microfinance: Social performance management or mission drift?," *World Development*, *35*(10), 1721–1738.

Durden, C. (2008). "Towards a socially responsible management control system," *Accounting, Auditing and Accountability Journal*, *21*(5), 671–694.

Ebrahim, A. & Rangan, V.K. (2014). "What impact? A framework for measuring the scale and scope of social performance," *California Management Review*, *56*(3), 118–141.

Ebrahim, A., Battilana, J. & Mair, J. (2014). "The governance of social enterprises: Mission drift and accountability challenges in hybrid organizations," *Research in Organizational Behavior*, *34*, 81–100.

Hubbard, G. (2009). "Measuring organizational performance: Beyond the triple bottom line," *Business Strategy and the Environment*, *18*(3),177–191.

Karim, N., Tarazi, M., & Reille, X. (2008). *"Islamic microfinance: An emerging market niche," Focus Note 49.* Washington, D.C.: CGAP.

Ledgerwood, J. (1999). *Microfinance handbook: An institutional and financial perspective.* Washington, DC: World Bank.

Ledgerwood, J., Earne, J. & Nelson, C. (2013). *The new microfinance handbook.* Washington, DC: World Bank.

Nabiha, A.K.S., Azhar, Z., Isa, S.M. &, Nazariah, A.Z.S. (2018). "Measuring social performance: Reconciling the tension between commercial and social logics," *International Journal of Social Economics*, *45*(1), 205–222.

Newman, A., Schwarz, S., & Borgia, D. (2014). "How does microfinance enhance entrepreneurial outcomes in emerging economies? The mediating mechanisms of psychological and social capital," *International Small Business Journal*, *32*(2), 158–179.

Norris, G. & O'Dwyer, B. (2004). "Motivating socially responsive decision making: The operation of management controls in a socially responsive organization," *The British Accounting Review*, *36*(2),173–196.

Olivares-Polanco, F. (2005). "Commercializing microfinance and deepening outreach? Empirical evidence from Latin America," *Journal of Microfinance/ESR Review*, *7*(2), 47–69.

Rashidah, S.K., Zuraidah, A.R., Abideen, M.S. & Adeyemi, A. (2016). "Role of market orientation in sustainable performance: The case of a leading microfinance provider," *Humanomics*, *32*(3),76–87.

Stevens, R., Moray, N. & Bruneel, J. (2014). "The social and economic mission of social enterprises: Dimensions, measurement, validation, and relation," *Entrepreneurship: Theory and Practice*, *39*(5),1–32.

Sukawati, T.G.R & Astawa, I.P. (2017). "Improving performance by harmonious culture approach in internal marketing," *Polish Journal of Management Stu*dies, *6*(1), 226–233.

Ucbasaran, D., Westhead, P., & Wright, M. (2008). "Opportunity identification and pursuit: Does an entrepreneur's human capital matter?." *Small Business Economics*, *30*(2),153–173.

Woller, G. (2007). "Trade-offs between social & financial performance," *ESR Review*, *9*(2),14–19.

Woller, G., Dunford, C. & Woodworth, W. (1999). "Where to microfinance," *International Journal of Economic Development*, *1*(1), 29–64.

Business Innovation and Development in Emerging Economies – Trinugroho & Lau (Eds)
© 2019 Taylor & Francis Group, London, ISBN 978-1-138-35996-3

How to develop cultural capital in order to improve academic achievement from a gender perspective

Darma Rika Swaramarinda
Universitas Negeri Jakarta, Indonesia

ABSTRACT: This article illustrates how cultural capital can improve student academic achievement and discusses how it can be developed among students. Good academic achievement is supported by the cultural capital owned by the students. This capital is important because it contains value, ethics and empowerment. Cultural capital in this article is the involvement and participation of students in extracurricular and cultural activities and a reading climate, all of which can preventively overcome the social climate and strengthen the social interaction of the less able students with the academic community. The research methodology used in the article is descriptive and qualitative. In this way, analysis of culture capital building is made by exploring students' self potential in order to improve their academic achievement. Gender affects cultural capital and resulting enhancement of academic achievement. From the results of the calculations made, it can be concluded that female students are exposed to such capital from an early age, are used to discipline and more often spend their time involved in useful activities that add to their insight. They believe that extracurricular activities and student organizations can improve their academic performance, as well as allowing them to assess their friends who may be following negative currents of modernization, which may undermine their cultural capital and have an impact on any decline in their academic achievement. Another case, male students who recognize that the environment greatly affects the value of habituation that forms the cultural capital that affects their academic achievement.

Keywords: academic achievement, cultural capital, education, gender, social science

1 INTRODUCTION

Education is an activity that shapes human development and helps people become good and useful human beings. The measure of educational success is the academic achievement of students. Students in the context of this study are teenagers, so they are still in a period of transition and adjustment. Teenagers in general cannot be separated from the problems that may affect their academic achievement.

Academic achievement is procured by students in various ways; some students have high achievement, some moderate, and there are also students who have low achievement. This is of course influenced by various factors, both internal ones (derived from within students themselves) and external ones (derived from outside the students). Academic achievement can be seen at the end of each lecture. When students learn that their academic achievement is not good enough, then the consequence is that they will try to improve this. A factor that may affect academic achievement is cultural capital, which is the development of education by improving academic achievement in a humanist way, and can be enjoyed by all community groups or campus academic communities.

In addition to the leaders who must be aware of this, the academic community, or in this case the students, must be able to build awareness through the strengthening of cultural capital. It is important because it contains values, ethics and empowerment. Culture is the fruit of human thought through the process of creation, taste and intention and runs through everyday life; if optimized it can certainly encourage educational improvement and academic achievement.

The cultural capital that a person possesses can also differ because it can be influenced by gender. These differences can be affected, among other factors, by the hormonal conditions of each gender. Women are generally more calm than men and are more reactive in accepting new things.

At present, there are diverse conditions that come from various regions on campuses, which are necessary for the development of education-based cultural capital and multicultural capital. If carefully examined, culture-based and multicultural education are important for strengthening the sense of nationalism.

Building education with good academic achievement is supported by the cultural and multicultural capital of students, meaning that the values of local wisdom and diversity are highly regarded without discrediting minority groups, including increasing preventive activities to cope with student crime, such as character building, and the independence of the young generation. Good academic cultural capital is built through student participation in academic activities by upholding good values and ethics.

If we expect the existence of competent output and character, then the educational environment, in this case the campus, should also be able to provide a good academic culture. If a positive campus academic culture can be applied to maximum effect, it will be able to encourage the growth of the social climate and healthy interaction amongst the academic community. It can also help in exploring the potential of students themselves, allowing them to develop not only in the mind, but also from the heart, in sports, and in taste which can ultimately improve student academic achievement. Cultural capital within the campus involves participation in cultural activities, a reading climate, and extracurricular activities that can support improvement in student academic achievement.

Cultural activities in this regard include, for example, visiting historic places that can provide new knowledge; a reading climate that can also provide knowledge and insight that many students who have a cognitive ability that is directed to the intellectual abilities, ability to think and intelligence will be achieved beyond any other students. Extracurricular activities can provide targeted psychomotor skills that can encourage students to improve their academic performance. From previous studies, Eryanto and Swaramarinda (2013), it can be seen that the results and conclusions of the research can vary, which may be due to the varying characteristics of each country studied. Recognizing that cultural capital has an influence on student academic achievement, the researcher is interested in examining this phenomenon from a gender perspective with students of the Faculty of Economics, State University of Jakarta.

2 LITERATURE REVIEW

2.1 *Academic achievement*

Academic achievement is the result of lessons learned from learning activities in schools or colleges that are cognitive, and usually determined through measurement and assessment. The academic achievement of students during college education is measured by course assessment, semester assessment, final year academic assessment and the final assessment of the study program. The benchmark used in academic achievement is the Grade Point Average (GPA).

According to Sobur (2006), academic achievement is a change in terms of behavioral skills, or an ability that can increase over time not due to the growth process, but to the existence of learning situations. The embodiment of the learning process can be either oral or written, and the skills and problem solving can be directly measured or assessed by standardized tests.

Meanwhile, according to Chaplin, cited by Thantawy (2004), educational or academic achievement is a specified level of attainment or proficiency in academic work as evaluated by teachers by standardized tests, by teachers' tests, or by a combination of the two.

Furthermore, according to Suryabrata (2006), academic achievement is the final learning achievement achieved by students within a certain time period, which in the case of school students is usually expressed in the form of numbers or symbols. From these numbers or symbols, other people or students themselves will be able to know the extent of achievement. Thus, academic achievement in school is another form of the mastery of lesson material that has been achieved by students; report cards can be used as the last learning outcome of the mastery of the lesson.

The definition of academic achievement by Azwar (2002) is evidence of the improvement or achievement obtained by a student as a statement of whether or not there has been progress or success in education programs. In addition, Djamarah (2002) defines academic achievement as the result obtained in the form of impressions that result in changes in the individual as the end result of learning activities.

Based on this discussion, it can be concluded that academic achievement is the result of achievement, ability or skill that produces change over a certain time period because of the learning effort, which can be measured or assessed by teachers' evaluations, standardized tests, or by a combination of the two, expressed in terms of numbers or symbols.

2.2 Cultural capital

The term 'cultural capital' is not often discussed or heard by ordinary people, and little research on it is associated with the field of education, in this case the campus. Capital culture is the capital of a student able to act well and ethically. Associated actions are the habits or customs of students who are assessed in everyday life as individual beings and as social beings. Some notions of cultural capital have been expressed by the following researchers.

Lamount and Lareau (1988) argue that "At the most general level of cultural capital of science to knowledge of the dominant conceptual and normative codes inscribed in a culture. Cultural capital is used by individuals or groups positioned at different levels in social hierarchies as a means of promoting social relations. Consequently, cultural capital enables individuals and families with knowledge of institutionalized high-status cultural signals (attitudes, preferences, formal knowledge, behaviors, goods and credentials) to exclude others from advantages of social positions or high-status groups".

According to Bourdieu (1986), cultural capital is the skills, education, and advantages that people have, which give them a higher status in society. Parents provide their children with cultural capital by transmitting the attitudes and knowledge needed to succeed in the current educational system.

2.3 Types of cultural capital

Bourdieu distinguishes cultural capital into three types of capital, namely:

1. Tangible cultural capital; i.e. cultural capital embodied either consciously or acquired passively from "inheriting" the properties of oneself. It is used not in the genetic sense, but in the sense of acceptance from time to time, usually through the socialization, culture and traditions of the family.
2. Cultural capital of objectification; i.e. cultural capital that can be seen from the object. It consists of physical objects held, such as scientific instruments or artwork.
3. Cultural capital institutionalized; i.e. cultural capital consisting of institutional recognition, generally in the form of cultural capital qualifications owned by an individual. This is particularly prominent in the labour market, where it allows a diverse range of cultural capital to be presented in qualitative and quantitative measures and compared to other people's measured cultural capital.

2.4 Gender

Echols and Shadily (cited in Sutinah, 2004) state that gender means sex. Gender is a visible difference in men and women when viewed from the perspective of values and behaviour. However, gender is actually different from sex.

Gender is a group of attributes, behaviours, positions, and socially-culturally-shaped roles of men and women. For example, women are considered to be gentle, emotional, motherly, and so forth, while men are considered, for example, to be strong, rational and powerful. In fact, these properties are unnatural, because they do not last forever, and can still be changed. That is, there are men who are emotional and gentle, and women who are strong and rational (Minister of Women Empowerment, 2002).

Fakih (cited in Sutinah, 2004) argues that the various social injustices affecting women are due to the close relationship between gender differences and gender inequalities and the structure of society's widespread injustice.

2.5 Previous research findings

Two previous studies were reviewed, namely the results of the research conducted by Mads Meier Jaeger, (2010) and by the authors themselves are Henry Eryanto and Darma Rika Swaramarinda (2013). In this case, for this article, the authors only looked at one variable studied by previous researchers, i.e. cultural capital, but with additional assessment from a gender perspective. Research conducted by Mads Meier Jaeger (2010) provides a new estimate of the impact of cultural capital on academic achievement. The results of the study found that (1) cultural capital (measured by indicators of participation in cultural activities, reading climate, and extracurricular activities) had a positive effect on children's reading and the value of mathematics tests, (2) the influence of cultural capital in general weaker than previously reported, and (3) the effect of cultural capital varies across different SES groups. The model used by Jaeger was:

$$Y_{ijt} = \alpha + C_{ijt}\beta_1 + X_{it}\beta_2 + d_{ij}\beta_3 + k_i\beta_4 + \varepsilon_{ijt}$$

where i is the child (i = 1, ..., N) in family j (j = 1, ..., J) at time t (t = 1, ..., T). There are four types of explanatory variable in the model.

Variable C is a cultural capital variable that has ijt because they differ between individuals (different children have different values, so are given index i), between families (different siblings have different values, therefore index j) and between individuals over time (children have different values at different points in time, so are given index t).

Variable X is the income of parents who only have it because they vary from individual and time, but not in family.

Variable D is the child's gender and age, and has subscript ij because these vary between individuals and families, but not over time.

Variable K is parental education, which only has subscript i because this varies between individuals, but not between families or individuals. The term normal-distributed error summarizes the impact of all the observed variables that also affect academic achievement.

The second research, conducted by Eryanto and Swaramarinda (2013) used the following equation:

$$A_i = \alpha + \beta_1 C_i + \beta_2 PE_i + \varepsilon_i$$

where the variables used are academic achievement (A), influenced by cultural capital (C) and parent education level (PE).

The research was conducted using a survey method. The study used questionnaire-shaped instruments for cultural capital variables and secondary data from the campus to measure each variable of academic achievement and parental education level. The data for cultural capital consists of several indicators got from questionnaire, while other variables in the form of secondary data are in the form of ordinal data.

Academic achievement and parents education level are using ordinal data. According to Priyatno (2010), ordinal data is data categorization results that are not equivalent and can not be calculated arithmetically. Cultural capital uses questionnaires based on indicators of cultural capital variables expressed in 34 point statements. Parental education level is seen from the secondary data sourced by universities who viewed the information system used on the campus, that is SIAKAD UNJ, on education data of parents.

2.6 *Cultural capital and academic achievement from the gender perspective*

The cultural capital possessed by a student can have an impact on their academic achievement, as stated by some experts from their research results. Bourdieu (1973), The most dominant class culture functions as the culture that can dominate the different stages of the culture. Students who have been nurtured with cultural forms from childhood will have the greatest possibilities for achievement in academic life.

According to Jaeger (2010), cultural capital is a scarce resource which equips individuals with knowledge, practical skills and a sense of "the rules of the game" in the educational system, which is recognized and rewarded by institutional gatekeepers and peers.

De Graaf and Kraaykamp (2000) add that children who have more cultural capital are more comfortable at school, communicate more easily with teachers, and are therefore more likely to do well. Susan (2002) states that women are able to enlarge the modalities in the community, either in the form of social capital, economic and cultural.

From the studies mentioned above, it can be seen that students' gender can affect their cultural capital, which will have an impact on improving academic achievement. It can therefore be concluded that cultural capital can affect academic achievement.

3 RESEARCH METHOD

This research uses the descriptive quantitative survey method and the data is analysed by content analysis. The research uses primary data from observations of the respondents, who were students selected according to gender.

4 RESULTS AND DISCUSSION

4.1 *Discussion on developing cultural capital in order to improve academic achievement from a gender perspective*

The discussion in this article follows the previous research conducted by Jaeger (2010) with the addition of attention to gender. The research has proven the hypothesis that there is a positive influence of cultural capital on academic achievement, meaning the higher the cultural capital of a student, the higher his or her academic achievement. There are three types of cultural capital: tangible cultural capital; cultural capital of objectification; and capital of institutional culture.

There are basic personality differences between girls and boys; boys generally individualistic, aggressive, impatient, more assertive, more confident, more in control of the task, and more dominant. Girls are more 'warm', nurturing, more tolerant of others, and much praised. But this condition is not absolute everywhere.

The academic achievement of students can be influenced by cultural capital because it includes the values of character formation or the personality of each individual who has experienced it. If the individual has experienced habituation from when they were younger, in terms of receiving positive and constructive learning patterns, then automatically in adulthood it will be easy to become an individual with good academic value.

73% of female students state they have experienced such habituation since their early age, whereas only 42% of male students were habituated to character formation early on. This is because female students can address the existing learning environment to create comfortable learning conditions. Male students are more susceptible to environmental impact and achieve slower maturity than women.

Conversely, if at an early age individuals are in an environment that is less supportive of the application of patterns of positive character formation, then most likely it will be difficult for them to compete in academic terms than their peers, although it is possible there may be other individuals from poor environmental conditions who will achieve good academic grades.

The factors mentioned above are very influential on one's environment. Environmental conditions will form habits for the individuals who experience them; that is, the value of habituation that will be cultural capital for them in achieving their academic activities. 74% of male students recognize that the environment greatly affects the value of habituation which shapes their cultural capital.

The Faculty of Economics UNJ has diverse students from different backgrounds. These backgrounds are the cultural capital of these individuals in learning process on campus. 64% of female students and 51% of male students stated they came from areas outside Jakarta.

Previous research has explained that the cultural capital of FE UNJ students strongly influences academic value. If we examine the results of these studies, then this is indeed the case. The strong relationship between cultural capital and the academic value of FE UNJ students, can be the result of more overseas students (from outside Jakarta, others city in Indonesia) than students from within the city itself. Their sincerity and readiness to continue their education in college is a great asset to their success.

Students from outside Jakarta will work hard, hoping to obtain a scholarship or become outstanding student, that will facilitate their entry into the world of work. In addition, they are also used to the discipline of time, and accustomed to spending time on activities that will add insight. 78% of females are already accustomed to discipline and more often spend their time engaged in useful activities and add their insights, while the male student figure is only 36%.

Habits that they bring that can transmit other students to also spur the spirit of competing in a positive way. Open debate activities, faculty and campus extracurricular activities and student organizations are all related to the development of self potential and will certainly be cultivated by students because they will use the existing time to increase their potential.

Such habituation will directly increase the academic value of the student, and will even also indirectly increase the academic value of other students, because they will compete in a positive way.

Extracurricular activities and student organizations in campus institutions are also able to increase student academic value. they will gain experience directly, not only in theory. These activities can explore students' sensitivity in dealing with events significantly beyond the theoretical knowledge they gain in class. 83% of female students believe that extracurricular activities and student organizations can improve their academic achievement, while 67% of male students believe this.

The unification of theories that students learn in lectures with extracurricular activities or student organizations they follow is a cultural capital within the campus so that they are expected to be ready for their own future. On the other hand, there are also contradictory students who participate in the flow of modernization, which will undermine the cultural capital they already have. This will weaken the academic value they achieve. Conditions such as this that must be considered further; cultural capital in general will help students to obtain or improve their academic value. 87% of female students and 54% of male students believed that their friends who follow a negative stream of modernization will undermine their cultural capital, which will have an impact on the decline in their academic achievement.

Building the value of cultural capital is not as easy as thought, it should be planted early in students who are still in primary education. Character building formation is very beneficial in one's golden age. The hope of character formation at a young age is that children have embedded positive values not only from learning process activities, but also from the social norms that apply to the community. The result of the formation of habituation is that in the future it will be cultural capital for them to be able to develop into a whole person.

5 CONCLUSION

Gender affects cultural capital and thus enhances academic achievement. Female students experience such habituation from an early age, are used to discipline and more often spend their time involved in useful activities that add to their insight. They also believe that extracurricular activities and student organizations can improve their academic performance and that their friends who follow negative currents of modernization will undermine their cultural capital, which will have an impact on the decline of their academic achievement. Male students show that the environment greatly affects the value of habituation, which will form a cultural capital that affects their academic achievement

REFERENCES

Azwar, S., *Achievement Test: Development Function of Achievement of Learning Achievement* (Yogyakarta: Student Literature, 2002).

Bourdieu, P., Cultural reproduction and social reproduction in knowledge, education and cultural change (London: Tavistock, 1973).

De Graaf, Nan Dirk, Paul M. De Graaf and Gerbert Kraaykamp, *Parental Cultural Capital and Educational Attainment in the Netherlands: A Refinement of the Cultural Capital Perspective* (Sociology of Education Vol. 73, No. 2, Apr., 2000).

Djamarah, Psychology of learning. (Jakarta: PT Rineka Cipta, 2002).

Dumais, Susan A, *Cultural Capital, Gender, and School Success: The Role of Habitus* (Sociology of Education Vol.75: 44–68, 2002).

Eryanto, Henry and Darma Rika Swaramarinda, *Culture Capital Influence, Parent Education Level and Parent's Revenue Level on Academic Achievement at Student of Faculty of Economics, State University of Jakarta* (Journal of Economics and Business Education, 2013).

Lamont, Michele and Annette Lareau, *Cultural Capital: Allusions, Gaps and Glissandos in Recent Theoretical Developments* (Sociological Theory 6: 153–168, 1988).

Jæger, Mads Meier, *Does Cultural Capital Really Affect Academic Achievement?* (Danish School of Education, Aarhus University, Tuborgvej 164, DK-2400 Copenhagen NV, Denmark, 2010).

Minister of Women Empowerment, *Gender Guidance In Participatory Planning, Jakarta* (2002).

Sobur, *General Psychology* (Bandung: Loyal Library, 2006).

Suryabrata, S., *Educational Psychology* (Jakarta: PT Raja Grafindo Persada, 2006).

Sutinah, *Gender & Women's Studies, in Dwi Narwoko & Bagong Suyanto (ed) 2004.* (Sociology: Introduction & Applied Text, Jakarta: Prenada Media, 2004).

Thantawy R, *Characteristics of Learning Outstanding Student Culture* (Journal of Education Science Parameter, No.18 Year XXI, January, 2004).

Technology management and information system

Business Innovation and Development in Emerging Economies – Trinugroho & Lau (Eds)
© 2019 Taylor & Francis Group, London, ISBN 978-1-138-35996-3

The intention to use information technology system: Survey of hospital employees in Solo

D. Setyawan, D.A. Pitaloka, N.A. Budiadi & Y. Kristanto
Department of Management, Faculty of Economic, Setia Budi University of Surakarta, Surakarta, Indonesia

B. Setyanta
Faculty of Economic, Janabadra University of Yogyakarta, Yogyakarta, Indonesia

ABSTRACT: This study aims to identify the intention to use information technology systems for hospital employees in Solo. The acceptance model encourages the formation of individual behavioral intention in the use of information systems (Davis, 1989). In this study, there were added social influence and facilitating conditions as an implementation variable which refers to the model of Unified Theory of acceptance and Use of Technology (Venkatesh *et al.*, 2003). The number of samples in this study was 420 and used structural equation modeling (SEM). The results indicated that perceived usefulness, social influence, and facilitating conditions affect attitude to use information technology systems. Attitudes change a person's intentions to use information technology systems. This study indicates the critical role of attitude when evaluating the actual use of information technology systems.

Keywords: Perceived usefulness, perceived ease of use, attitude, social influence, facilitating conditions, and behavioral intention

1 INTRODUCTION

Research on the acceptance of information systems within an organization is important because of the diversity of results in various research objects (Aggelidis and Chatzoglou, 2008; Gagnon, 2011; Han, 2006). The behavior measured in the Technology Acceptance Model (TAM) should be the actual use of technology, but in fact, many research papers uses technology that does not necessarily reflect or measure actual usage in implementation. It seems to indicate that the TAM has a weakness in its application (Collerete et al., 2003; Gagnon, 2011; Yu et al., 2008).

Napitupulu and Sensuse (2014) found that 35% of information systems implemented in developing countries failed because they only used a technology approach and paid little attention to organizational approaches. The role of the organization is to improve the implementation of information systems by providing infrastructure and motivating employees (Ahmad et al., 2013; Al-Hadban et al., 2016).

The research of Aggelidis and Chatzoglou (2008) based on the Unified Theory of Acceptance and Utilization of Technology (UTAUT) developed by Venkatesh et al. (2003) indicate that the acceptance of information systems still has weaknesses. The UTAUT model can account for more than 70% of the variation in the formation of intentions to use technology by incorporating social influencing and facility conditions that are technological implementations (Aggelidis and Chatzoglou, 2008; Hoque and Sorwar, 2017; Liu et al., 2014;). Social influence, and pressure from superiors and co-workers, increases the use of information systems (Marumping et al., 2017). A condition of the facility–infrastructure and availability of technicians in charge of solving problems –increasesan employee's intention to use information systems (Nikou & Economides, 2017). Based on previous research, this study aims to examine the implementation of information system acceptance by hospital employees in Solo, Indonesia.

2 LITERATURE REVIEW

2.1 Behavioral intention

The intention to use information systems is an individual's desire to use technology (Davis, 1989). Behavioral intentions are formed through a process of belief in service, building a positive attitude toward service, and a desire to use a system (Lichtle & Plichon, 2008).

The study of intentions for receiving information systems often uses the theory of TAM. The variables contained in TAM are attitudes, perceptions of use, and ease of use of impressions (Davis et al., 1989). Because TAM still has weaknesses in its implementation, this study adds technology implementation variables; namely social influences and facilitating conditions (Aggelidis & Chatzoglou, 2008; Gagnon, 2011; Yu et al., 2008).

2.2 Attitude

Attitude is a positive or negative disposition to respond to things that are an evaluation of an object (Ajzen, 1991). Dimensions of belief in objects are obtained from knowledge and information, while evaluative aspects describe emotions and feelings toward object (See et al., 2008) which affects a person's speed in responding to an object; the higher the cognitive and affective aspects, the faster the response (Huskinson & Haddock, 2006).

Previous research has indicated the consistency of the effect of perspectives on behavior in various research backgrounds (Aggelidis & Chatzoglou, 2008; Gagnon, 2011; Yu et al., 2008;). In the context of online shopping, Warrington et al. (2001) found that by studying 15 major cities, those in the United States indicated a strong relationship between attitudes and intentions to buy online.

For research in the context of the use of online health services, Aggelidis and Chatzoglou (2008) conducted a study on the effect of attitudes on the intention to receive technology for health workers and hospital administration personnel in East Macedonia, Greece. They showed that positive attitude influenced the intention to accept technology in the hospital. Another study of the effect of attitudes on the intention to use information technology on medical personnel with different backgrounds, suggested that views have a positive and significant impact on the use of information technology (Gagnon, 2011; Yu et al., 2008).

Hypothesis 1: The higher the positive attitude toward information systems, the more elevated the intention of using information systems

2.3 Perceived Usefulness

The literature review shows consistency of results in the relationship of perceived usefulness to attitude in using information systems in various research backgrounds (Hartini, 2011; Hassanein & Head, 2007; Shih, 2004). Research by Shih (2004) and Hassanein and Head (2007) on the effect of perceived usefulness on the attitude of using information technology in different contexts and backgrounds indicates a consistency in research results. Research by Shih (2004), and Hassanein and Head (2007) identified that perceived usefulness positively affects the attitudes of using information technology. The results of this study indicated that if the information system's perceived usefulness by individuals can improve its performance, then the attitude of individuals toward information systems becomes more positive.

Hypothesis 2: The higher the perceived usefulness, the higher the positive attitude toward using information systems.

The TAM developed by Davis (1989) suggested that perceived usefulness has a substantial effect on the intention of using technology. The concept originated by Davis (1989) is supported by research conducted by Fang et al. (2009) who identified perceived usefulness as having a significant effect on the intention of using the technology, even in different contexts and backgrounds. The results of this study showed that the higher the perceived usefulness, the greater the plan to use the technology.

Hypothesis 3: The higher the perceived usefulness, the higher the intention of using information systems

2.4 *Perceived ease of use*

Perceived ease of use is a belief that an information system is easy to understand, easy to learn, easy to operate (Davis, 1989), is of minimal effort to use (Davis, 1989; Green & Pearson, 2011; Tong, 2010), easy to access, and flexible and reliable to improve performance (Lin et al., 2000).

Prior research on the acceptance of technology has consistent results in the influence of perceived ease of use on attitudes to use information systems in various contexts and research backgrounds (Daud et al., 2011; Ha & Stoel, 2008; Sanz-Blas et al., 2008). Perceived ease of use is a predictor of attitudes to using information technology systems (Daud et al., 2011; Hsieh and Liao, 2011; Lin, 2007). The study was supported by a survey by Daud et al. (2011) which researched the effect of e-retailer websites that look attractive and are easy to use. The results indicated that, if a website is more attractive and easy to use.

Other studies in the context of online shopping in Taiwan, indicated that perceived ease of use positively affects online shopping attitudes (Lin, 2007; Hsieh & Liao, 2011). Pavlou and Fygenson (2006) indicated that perceived ease of use obtains information, thus encouraging attitudes toward using an information system. Various studies show that if the perceived technology is easy to use, it will increase the positive attitude toward using technology.

Hypothesis 4: The higher the perceived ease of use, the higher the positive attitude toward using information systems.

There is a wide variety of research results on the perceived ease of use effect on the decision to receive information technology. Research by Chau and Hu (2002) identified that perceived ease of use can account for more than 50% of variations in intent to use information systems. In the context of the use of professional user information systems, perceived ease of use does not affect the intention to use *Personal Digital Assistant* (PDA) in hospitals (Probst et al., 2006). Research by Probst et al. (2006) is supported by Chau and Hu (2002) which identified that in the context of the use of telemedicine, perceived ease of use had no direct effect on the intention to use information technology. Previous research has shown that for professional users, it is easier to adopt new, more complex and sophisticated technologies (Probst et al., 2006). Celik (2011) proved that in the context of online shopping, perceived ease of use directly affects the intention to shop online.

Hypothesis 5: The higher perceived ease of use, the higher the intention of using information systems.

2.5 *Social influence*

Social influence plays a vital role in individual behavior and decision making. Social impact can be a perceived social pressure to perform or not to make a particular response (Aggelidis & Chatzoglou, 2008; Park, 2009). Han (2006) showed that the intention to use information systems is influenced by social influences in their environment and is stronger with the moderation role of innovation and age.

Another study stated that social influence is an implementation variable that has a significant effect on behavior (Aggeliddis & Chatzoglou, 2008). These findings are supported by Yu et al. (2008) which identified that social influence positively affects the intention to use information technology. Social pressure increases intention to use information systems

(Maillet et al., 2015; Marumping et al., 2017). The research of Maillet et al. (2015) and Marumping et al. (2017) showed that the higher the social impact of the use of information system technology, the plan to use information system technology is getting more positive.

Hypothesis 6: Higher social influence can increase the intention of using information systems.

2.6 *Facilitating conditions*

Facilitating conditions influences the adoption of information systems (Ahmadi et al., 2015; Nikou & Economides, 2017). They are a belief that organizations have adequate infrastructure and technology forusing information systems (Aggelidis & Chatzoglou, 2008; Marumping et al., 2017). Sufficient facilitating conditions increases the intention of nurses to use information systems (Maillet et al., 2015; Susanto & Aljoza, 2015). Providing facilities for the implementation process encourages the intention to use an information system (Witarsyah et al., 2017). Previous research on the different backgrounds of health workers, hospital administration and nurses indicate that facilitating conditions affect the intention to use information systems (Gagnon et al., 2011; Hoque & Sorwar, 2017). Based on these studies, the higher the conditions that facilitate, the higher the intention to use information systems.

Hypothesis 7: The higher the facilitating conditions, the higher the intention to use information systems.

3 RESEARCH MODEL

3.1 *Methodology and result*

The validity test in this research used confirmatory factor analysis which requires each question item to have factor loading of > 0.5. This assumption is one of the requirements to be able to undertake model analysis with structural equation modeling. The technique used was a rotated component matrix whose output must be perfectly extracted. The validity test results showed that 17 questionnaires had a validity of > 0.5, and perfectly extracted. Six questionnaire items were dropped because they were not perfectly extracted, thus indicating six items of questionnaires were invalid.

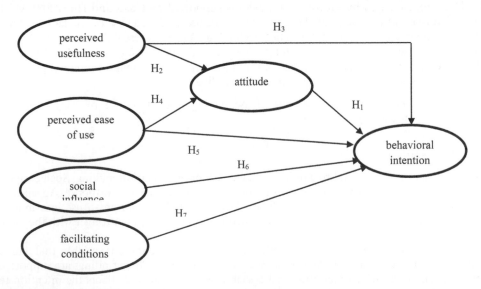

Figure 1. Research model.

Table 1. Goodness-of-fit test results.

Indeks	Cut-off	Result	Conclusion
Chi Square	Small	338.994	
Probability	≥ 0,05	0,995	Fit
CMIN/DF	≤ 2,00	0,821	Fit
Goodness-of-fit index	≥ 0,90	0,923	Fit
Adjusted goodness-of-fit index	≥ 0,90	0,886	Marginal
Comparative fit index	≥ 0,95	0,997	Fit
Tucker-Lewis Index	≥ 0,95	0,998	Fit
RMSEA	≤ 0,06	0	Fit
IFI	≥ 0,95	0,997	Fit

Table 2. Result of structural equation modeling.

	β	C.R	Conclusion
A → BI	0,383	3,427	Supported
PU → A	0,373	3,585	Supported
PU → BI	0,188	1,206	Not supported
EU → A	0,092	0,392	Not supported
EU → BI	0,045	0,082	Not supported
SI → BI	0,164	2,267	Supported
FC → BI	0,152	2,105	Supported

Level of significance: 5%.

The reliability test was performed to test the internal consistency of the research instrument. In this study, a reliability test was undertaken that used Cronbach's alpha. Based on the reliability test obtained, Cronbach's alpha value for all research variables was > 0.7. This meant that the research data was feasible to perform further statistical tests on.

The result of the goodness of fit test show that only Adjusted goodness of fit index (AGFI) is marginal, so the research model is indicated to be able to explain the phenomenon of research problems.

4 DISCUSSION

The results of this study indicated that attitude has a positive and significant impact on the intention to use information technology systems (β = 0.383, Critical Ratio = 3.427). This study supports the findings of Taylor and Todd (2001), Yu et al., (2008), Aggelidis and Chatzoglou (2008) and Gagnon (2011). These findings suggested that a positive attitude to using information technology can increase the intention to use information technology. To improve the generalization of concepts, this research model should be tested in different contexts and research backgrounds.

Perceived usefulness has a positive and significant impact on attitudes to use information technology systems (β = 0.373, Critical Ratio = 3.585). These findings indicated that information technology systems are perceived to provide benefits in completing work and so promoted positive attitude to using information technology systems. The results of this study support the research of Yu et al., (2008), and Aggelidis and Chatzoglou (2008), as well as supporting the concept developed by Davis (1989) which states that perceived usefulness has a significant effect on attitude. To improve the generalization of concepts, this research model should be tested in different contexts and research backgrounds.

This study showed that perceived usefulness has no significant effect on the intention to use information technology (β = 0.188, Critical Ratio = 1.206). Lin et al. (2010) revealed that the intention to use information technology is low in people who have poor experience in operating information system. This is contrary to previous research that indicated that perceived usefulness has a significant effect on behavioral intent (Ozer et al., 2013; Venkatesh & Davis, 2000; Yu et al., 2008).

This study found that ease of use does not affect attitudes toward using information technology systems ($\beta = 0.092$, CR = 0.392). This is contrary to research by Gagnon (2011), Yu et al. (2008), and Hsieh & Liao (2011) which indicated comfort in using information system technology increase a positive attitude.

This study supports previous research that indicated that ease of use does not help attitude toward using information systems (Ha & Stoel, 2008; Sentosa & Mat, 2012; Tong, 2010). The differences in the results of this study may be due to the sample in this study. The population studied were direct users of information technology systems, so had the skills and knowledge in using it. The diversity of research is caused by differences in experience and knowledge in operating of information technology systems. Needy individuals experience and knowledge feel more comfortable to operate manual information systems.

Perceived ease of use also does not affect the intention to use information technology systems ($\beta = 0.045$, CR = 0.082). Probst et al. (2005) identified that ease of use has no significant effect on the intention to use information technology systems. A different finding was put forward by Venkatesh and Davis (2000) who stated that ease of use influences the plan to use information technology systems.

Chau and Hu (2002) suggested that perceived ease of use had no significant effect on the attitude of using information systems in professional groups. These groups have strong competencies and highly skilled staff so that the ease of running an information system is not an important consideration. The results of the study support the research of Probst et al., (2006) which states that professional groups that have high competence are more accessible to adopting technologies that are perceived as complex and complicated. Differences in the characteristics of the sample as a direct user or not, influence the relationship of perceived ease of use and attitude (Wu & Wang, 2005).

Implementation of acceptance of information technology systems requires two variables that affect intentions for behavior: social influences and conditions that facilitate. The results indicated that social influences have a positive and significant impact on the intention to use information technology systems ($\beta = 0.164$, Critical Ratio = 2.267). The result support previous research that leaders and co-workers encourages the use of information technology system. (Han, 2006; Hoque & Sorwar, 2017; Maillet et al., 2015).

Facilitating conditions have a significant effect on intention to use information technology systems ($\beta = 0.152$, CR = 2.105). This study's results support the research of Aggelidis and Chatzoglou (2008), and Gagnon (2011) Work environment conditions support the intention to use information technology systems because it is an established system and is a routine work procedure. Employees have a positive perception of the availability of infrastructure and resources that support the implementation of information systems because it increases the use of information systems that improve performance (Nikou & Economides, 2017; Marumping et al., 2017).

5 CONCLUSION

The result of the study indicates that group of experienced and knowledgeable users form a social environment of workers who support the intention to use information systems. The intention to use information technology system is increasingly supported by complete facilities and pressure from leaders.

REFERENCES

Aggelidis, VP., Chatzoglou, PD. 2008. Using a Modified Technology Acceptance Model in Hospital. *International Journal of Medical Informatics.* Vol. 78. pp. 115–126.

Brooks, J.R., Ranganathan, S.K., Madupu, V., Sen, S. 2013. Affective and Cognitive Antacedents of Customer Loyalty Toward E-Mail Service Providers, *Journal of Service Marketing*, Vol. 27, No. 3, pp. 195–206.

Celik, H., 2011. Influence of Social Norms, Perceived Playfulness and Online Shopping Anxiety on Consumers' Adoption of Online Retail Shopping: An Empirical Study in The Turkish Context, *International Journal of Retail & Distribution Management*, Vol. 39, No. 6, pp. 390–413.

Chau, P.Y.K., Hu, P.J.H. 2002. Investigating Healtchcare Professionals' Decisions to Accept Telemedicine Technology, *Information & Management*, Vol. 39, No. 4, pp. 297–311.

Chrismar, W.G., Patton, S.W. 2002. Does the Extended Technology Acceptance Model Apply to Physicians. *Journal of Computer Society.*

Collerete, P., Legris, P., Ingham, J. 2003. Why Do People Use Information Technology? A Critical Review of Technology Acceptance Model, *Information & Management*, Vol. 40, pp. 191–204.

Davis, F.D., Bagozzi, R.P., Warshaw, P.R. 1989. User Acceptance of Computer Technology: A Comparison of Two Theoretical Model, *Management Science*, Vol. 35, No. 8, pp. 982–1003.

Davis, F.D. 1989. Perceived Usefullnes, Perceived Ease of Use, and User Acceptance of Information Technology. *MIS Quarterly.* Vol. 13. No. 3. pp. 319–340.

Daud, K.A.K., Yulihasri, Islam, M.A. 2011. Factors that Influence Customers' Buying Intention on Shopping Online, *International Journal of Marketing Studies*, Vol. 3, No. 1, February, pp. 128–139.

Fang, Y.H., Chiu, C.M., Chang, C.C., Cheng, H.L. 2009. Determinants of Customer Repurchase Intention in Online Shopping, *Online Information Review*, Vol. 33, No. 4, pp. 761–784.

Gagnon, M.P. 2011. Using a Modified Technology Acceptance Model to Evaluate Healthcare Professionals' Adoption of an New Telemonitoring System. *Telemedicine and E-Health.* Vol. 18. No. 1. pp. 54–59.

Han, S. 2006. Physicians' acceptance of mobile communication technology: an exploratory study. *International Journal of Mobile Communications.* Vol. 4. No. 2. pp. 210–230.

Hartini, S. 2011. Pengembangan Model TAM: Expertice dan Innovativeness sebagai Variabel Moderator Studi pada Penggunaan E-Banking, *Journal of Business and Banking*, Vol. 1, No. 2, November, pp. 155–164.

Ha, S., Stoel, L. 2008. Consumer E-Shopping Acceptance: Antacedents in a Technology Acceptance Model, *Journal of Business Research*, Vol. 62, Iss. 5, pp. 565–571.

Hassanein, K., Head, M. 2007. Manipulating Perceived Social Presence Through the Web Interface and its Impact on Attitude Toward Online Shopping, *International Journal Human-Computer Studies*, Vol. 65, pp. 687–708.

Hoque, R., Sorwar, G. 2017. Understanding Factors Influencing the Adoption of mHealth by The Elderly: An Extension of The UTAUT Model, *International Journal of Medical Informatics*, Vol. 101, pp. 75–84.

Hsieh, J.Y., Liao, P.W. 2011. Antecedent and Moderators of Online Shopping Behavior in Undergraduate Students, *Social Behavior and Personality*, Vol. 39, No. 9, pp. 1271–1280.

Hsu, C.L., Lin, J.C.C. 2008. Acceptance of Blog Usage: The Roles of Technology Acceptance, Social Influence and Knowledge Sharing Motivation, *Information & Management*, Vol. 45, pp. 65–74.

Huitt, W., Cain, S. 2005. An Overview of the Conative Domain, Educational Psychology Interactive, Valdosta, GA., Valdosta State University, diakses dari http://www.edpsycinteractive.org/brilstar/chapter/conative.pdf. tanggal 11 Desember 2014.

Huskinson, T.L.H., Haddock, G. 2006. Individual Differeces in Attitude Structure and The Accesibility of The Affective and Cognitive Component of Attitude. *Social Cognition*, 24(4): 453–468.

Kaufaris, M. 2002. Applying the Technology Acceptance Model and Flow Theory to Online Consumer Behavior, *Information Systems Research*, Vol. 13, No. 2, June, pp. 205–223.

Lichtle, M.C., Plichon, V. 2008. Understanding Better Consumer Loyalty, *Recherche et Applications en Marketing*, Vol. 23, No. 4, pp. 121–140.

Lin, H.F. 2007. Predicting Consumer Intentions to Shop Online: An Empirical Test of Competing Theories, *Electronic Consumer Research and Applications*, Vol. 6, pp. 433–442.

Lin Y.C., Fang, K., dan Tu, C.C. 2010. Predicting Consumer Repurchase Intentions to Shop Online. *Journal of Computers*, Vol. 5, No. 10, pp. 1527–1533.

Maillet, E., Mathieu, L., Sicotte, C. 2015. Modeling Factors Explaining the Acceptance, Actual Use, and Satisfaction of Nurse Using an Electronic Patient Record in Acute Care Settings: An Extension of the UTAUT, *International Journal od Medical Informatics*, Vol. 84, Iss. 1, pp. 36–47.

Marumping, L.M., Bala, H., Venkatesh, V., Brown, S.A. 2017. Going Beyond Intention: Integrating Behavioral Expectation Into the Unified Theory of Acceptance and Use of Technology, *Journal of The Association for Information Science and Technology*, Vol. 68, No. 3, pp. 623–637.

Nahl, D. 2001. A Conceptual Framework for Explaining Information Behavior, *Studies in Media & Information Literacy*, Vol. 1, Issue 2, May, pp. 1–15.

Napitupulu, D., Sensuse, D.I. 2014. The Critical Success Factor Study for e-Government Implementation, *International Journal of Computer Applications*, Vol. 89, No. 16, pp. 23–32.

Nikou, S.A., Economides, A.A. 2017. Mobile-Based Assessment: Investigating the Factors that Influence Behavioral Intention to Use, *Computer & Education*, Vol. 109, pp. 56–73.

Park, S.Y. 2009. An Analysis of the Technology Acceptance Model in Understanding University Students' Behavioral Intention to Use e-Learning. *Educational Technology & Society*. Vol. 12, No. 3, pp. 150–162.

Probst, J.C., Mun, Y.Y., Joyce, D.J., Jae, S.P. 2006. Understanding Information Technology Acceptance by Individual Professionals: Toward an Integrative View, *Information & Management*, Vol. 43, pp. 350–363.

Polychronopoulos, G., Giovanis, A.N., Binioris, S. 2012. An Extension of TAM Model with IDT and Security/Privacy Risk in the Adoption of Internet Banking Services in Greece, *EuroMed Journal of Business*, Vol. 7, No. 1, pp. 24–53.

Sanz-Blas, S., Bigne-Alcaniz, E., Ruiz-Mafe, C., Aldas-Manzano, J. 2008. Influence of Online Shopping Information Dependency and Innovativeness on Internet Shopping Adoption, *Online Information Review*, Vol. 32, No. 5, pp. 648–667.

See, Y.H.M., Petty, R.E. 2008. Affective and Cognitive Meta-Bases of Attitudes: Unique Effects on Information Interest and Persuasion. *Journal of Personality and Social Psychology*, 94(6): 938–955. DOI: 10.1037/0022-3514.94.6.938.

Shih, H.P. 2004. An Empirical Study on Predicting User Acceptance of E-Shopping on the Web, *Information & Management*, Vol. 41, pp. 351–368.

Sun, H., Zhang, P. 2006. Causal Relationship Between Perceived Enjoyment and Perceived Ease of Use: An Alternative Approach, *Journal of the Association for Information System*, Vol. 7, No. 9, September, pp. 618–645.

Taylor, S., and Petter A. Todd. 2001. Understanding Information Technology Ussage: A test of Competing Model. *Information System* Research. Vol 6. No 2. pp 144–176.

Wu, J.H., Wang, S.C. 2005. What Drives Mobile Commerce? An Empirical Evaluation of The Revised Technology Acceptance Model, *Information & Management*, Vol. 42, pp. 719–729.

Yu, P., Li, H., dan Gagnon, M.P. 2008. Health IT Acceptance Factors in Long-term Care Facilities: A Cross-Sectional Survey. *International Journal of Medical Informatics.* Vol 78. pp 219–229.

Venkatesh, V. 2000. Determinants of Perceived Ease of Use: Integrating Control, Intrinsic Motivation, and Emotion into the Technology Acceptance Model. *Information Systems Research*. Vol. 11, No. 4, pp. 342–365.

Venkatesh, V., Davis, F.D. 2000. Theoretical Extension of the Technology Acceptance Model: Four Longitudinal Field Studies. *Management Science*. Vol. 46, No. 2, pp. 186–204.

Venkatesh, V. Morris, M.G. Davis, G.B. Davis, F.D. 2003. User Acceptance of Information System: Toward a Unifield View. *MIS Quartely.* Vol. 27, No. 3, pp. 425–478.

Wintarsyah, D., Sjafrizal, T., Fudzee, M.F.MD., Salamat, M.A. 2017. The Critical Factors Affecting E-Government Adoption in Indonesia: A Conceptual Framework, *International Journal on Advanced Science Engineering Information Technology*, Vol. 7, No. 1, pp. 160–167.

Business Innovation and Development in Emerging Economies – Trinugroho & Lau (Eds)
© *2019 Taylor & Francis Group, London, ISBN 978-1-138-35996-3*

Factors affecting the quality of accounting information systems in Indonesian's higher education (research model)

C.D.K. Susilawati & Jerry
Lecturer of Accounting Program Study in Maranatha Christian University, Indonesia
Student of Accounting Science Doctoral at Padjajaran University, Indonesia

Y. Carolina & Rapina
Lecturer of Accounting Program Study in Maranatha Christian University, Indonesia

ABSTRACT: Accounting information system serves to generate accounting information that is useful in strategic decision making for companies including higher education/university. What is happening now is that many higher education with financial management experience problems because the utilization of existing resources is not optimal and income declines significantly resulting in the quality of higher education is also not optimal.

Researchers are interested to examine the factors that affect the quality of accounting information systems to produce quality accounting information in Higher Education, so that these factors can be managed properly in order to form a quality accounting information system to produce quality accounting information that is useful in the strategic decisions of Higher Education business. In this era of globalization, although a university is a type of business whose main purpose is not profitable, but Higher Education is also a business that must prioritize financial management in accounting information system quality to improve the quality of higher education in society. Factors identified by researchers include environmental uncertainty, user competence, organizational culture and information technology.

Keywords: Quality Accounting Information Systems, Higher Education

1 INTRODUCTION

1.1 *Background and problems of research*

There are various phenomena in Indonesia that occur related to the quality of accounting information systems in higher education revealed by the Chairperson of Suyatno ABPPTSI Center (2015) that there are 205 foundations that are problematic because financial conflicts, facilities, and Management authority related to financial delays occur in the preparation of financial statements, inappropriate use of accounting systems, and manual reporting is not based on information technology.

It can be said that the quality of accounting information systems in a number of private university foundations are still bad. Another phenomenon concerning the environmental uncertainty that influences the inflexible system is proposed by Minister of Research and Technology Nasir (2015) which states that the financial system in PTN-BH is currently lacking flexibility, such as difficulties in managing the budget given by Kemendikbud. This means that the budget absorption becomes low and the financial system applied by the Ministry of Finance (MoF) are complex and incompatible with dynamic academic activities, meaning that factors that have not been anticipated by this dynamic academic activity cause the financial system in PTN-BH is less flexible.

An Accounting information system is influenced by the use of information technology because through the use of information technology will improve the quality of accounting information systems (Wilkinson et al., 2000). The same thing was put forward by Romney & Steinbart (2015) who argued that information technology is one of the factors affecting accounting information systems. According to Thompson & Baril (2002) information technology is a combination of hardware and software. The same thing was put forward by Laudon & Laudon (2013) who argued that information technology consists of hardware and software that the organization needs to achieve organizational goals. According to Romney & Steinbart (2015) information technology is a computer and other electronic devices used to store, retrieve, transmit and manipulate data.

A well-designed accounting information system will provide added value to the company (Romney & Steinbart, 2015). So a quality accounting information system that can provide added value for the company. Qualified accounting information systems have integrated, flexible, accessibility, procedural and rich media characteristics (Heidmann 2008). The quality of accounting information systems is generally flexible, efficient, accesible and timely (Stair and Reynolds, 2010). A quality information systems must be useful and when the information system is used it can improve performance (Davis et al., 1989). The system is designed to produce a good quality information system design that is easy to use and will produce the correct function for the user. It is a fast in taking data and moving between data display, reliable, safe and well integrated with other systems Bocij, 2014). Quality systems are easy to use, easy to learn, accurate, flexible, satisfactory, integrated and customizable (Khosrow-Pour, 2011).

The result of a quality information system is quality accounting information in decision making. This is expressed by Laudon & Laudon (2012) who state that quality information systems will produce quality accounting information used for users in decision making. The same thing is expressed by Hall (2011) who states that the quality of accounting information is directly related to accounting information system activity. Similarly, Gelinas & Dull (2008) state thats the information system collects data to convert it into important and quality information. Also supported by Romney & Steibart (2015) who states that accounting information systems process data to generate information for decision making.

Information is organized and processed data to be more meaningful and improve decision making (Romney & Steinbart., 2015). The same thing is expressed by Wilkinson et al. (2000) who states that information as must be meaningful and useful data for those who need it. And Susanto (2013: 38) states that information is the result of data processing that gives meaning and benefits. According to Bocij (2015), information is data that has been processed to be more meaningful, for a purpose that it can be interpreted and understood by the user. Quality accounting information has the characteristics of completeness (scope), timeliness, easy to understand (format) and accuracy (Heidmann, 2008). The same thing is expressed by Romney & Steinbart (2015: 30) who state that useful information characteristics are relevant (for improved decision making), reliable (complete from bias or error), complete, timely) and accessibility. Further according to Stair et al. (2010), valuable information characteristics are accessible, accurate, complete, economical, flexible, relevant, secure and simple. According to Bocij (2015), the characteristics of quality information in terms of its contents are accurate, relevant, complete, concise and scope, and in terms of form clear, detailed, appropriate and presented in the correct form. According to Gelinas & Dull (2008), effective information is easy to understand, relevant, timely, predictive value, feedback value, verifiability and neutrality.

Rahayu (2012) found that the quality of data and quality of accounting information systems affect the quality of accounting information at the tax office in Bandung and in Jakarta (Indonesia). The same thing is raised Jun Shien (2015) in that quality accounting information system affect the quality of accounting information at the University in Bandung-Indonesia. According to Abdallah (2013), the use of information systems will improve the quality of tax information in Jordan.

From the description above the authors identify that there are several factors that affect the quality of accounting information systems that produce quality accounting information and the right decision-making such as environmental uncertainty, user competence, organizational culture and information technology in college. The author is interested to conduct research entitled Factors—Factors Affecting the Quality of Accounting Information Systems to produce Quality Accounting Information at Universities in Indonesia.

2 STUDY LITERATURE

2.1 *Relevant and current primary reference library and hypothesis development*

By prioritizing research results variable organization includes environmental uncertainty, user competence in the organization, organizational culture and information technology used by organizations in the Journal of Scientific and Development Hypothesis

Table 1. Effect of environmental uncertainty on the quality of SIA.

Year and Author	Result of environmental uncertainty study on quality of AIS
Hammad et al. (2012)	Environmental uncertainty becomes an important factor in designing an effective and efficient accounting system, and the results of the research shows the uncertainty of the environment affecting accounting systems that result in information for management
Gilbert & Singer (2011)	Environmental uncertainty shows a significant impact on strategy and information systems designed as a strategy.
Gull et al. (1993)	Conditions of perception of high environmental uncertainty improve the quality of information generated by information systems to produce decision-making that improves company performance.
Hwah & Huynh (2013)	Environmental uncertainties affect the accounting system and company performance as a moderating variable.

So it can be said from the results of previous studies formulated hypotheses:

H1: Environmental uncertainty affects the quality of accounting information systems.

Table 2. Effect of user competence on quality accounting information system.

Year and Author	User competence research on AIS quality
Nurhayati & Mulyani (2015)	User competence plays an important role in the implementation of quality AIS
Saleh (2013)	Personal competence improves the quality of AIS
Iskandar (2015)	User competence affects the quality of AIS.
Jun Shien (2015)	User competence affects the quality of AIS that can improve decision making.

H2: User competence affects the quality of AIS that can improve decision making.

Table 3. Influence of organization culture on quality accounting information system.

Years and Authors	The theory and results of organizational culture research on the quality of AIS
Susanto (2013)	Culture is an everyday internal environment that is visible to and felt by those who work within it. Organizational culture gives every organization its features and meaning and without the support of organizational culture all efforts will be in vain.
Ivancevich et al. (2008)	Argued that organizational culture is what employees perceive and how it creates patterns of beliefs, values and expectations.
Laudon & Laudon (2014)	Part of organizational culture can always be found embedded in information systems.
Romney & Steinbart (2015)	The design of accounting information systems is influenced by organizational culture.
Stair & Reynolds (2012:31)	Argued that organizational culture has a positive effect on the successful development of information systems.

(Continued)

Table 3. (*Continued*).

Years and Authors	The theory and results of organizational culture research on the quality of AIS
O'Brien & Marakas (2009)	The success of information systems is not only measured through efficiency in terms of minimizing cost, time and use of information resources, but the success of information systems must be measured through the organizational culture that supports it.
Napitupulu (2015)	Accounting information systems must be able to adjust to the organizational culture that exists within the organization.
Nusa (2015)	There is an influence of organizational culture on the quality of accounting information systems
Rapina (2014)	Organizational culture affects the quality of accounting information systems and the quality of accounting information.

Based on theories as the concepts in this study that have been described above and some recent research results that support the theory it can then be formulated:

H3: Organizational culture affects the quality of accounting information systems.

Table 4. Influence of information technology on quality accounting information system.

Year and Author	Theory and results of information technology research on quality of AIS
Susanto (2013)	Culture is an everyday internal environment that is visible to and felt by those who work within it, organizational culture gives every organization its features and meaning and without the support of organizational culture all efforts will be in vain.
Ivancevich et al. (2008)	Argued that organizational culture is what employees perceive and how it creates patterns of beliefs, values and expectations.
O'Brien & Marakas, (2009)	Success in today's dynamic business environment relies heavily on maximizing the use of Internet-based information technology and systems to meet customer needs in the global marketplace competition
Thompson & Baril (2002)	Information technology is a combination of hardware and software.
Laudon & Laudon (2013)	Information technology consists of the hardware and software organizations need to achieve organizational goals.
Wilkinson et al., (2000)	An accounting information system is influenced by the use of information technology because the use of information technology will improve the quality of accounting information systems.
Romney & Steinbart (2015)	Information technology is one of the factors that affect the accounting information system
O'Brien & Marakas (2009:17)	The success of information systems is not only measured through efficiency in terms of minimizing cost, time and use of information resources, but the success of information systems one of which must be measured through the organizational culture that supports it.
Sacer & Oluic (2013)	The use of information technology affect sthe accounting information system.

Based on theories as a concept in this study that have been described above and some recent research results that support the theory the following hypothesis can be formulate:

H4: Information technology affects the quality of accounting information systems.

Table 5. Influence of accounting information system quality to accounting information quality.

Year and Author	Theory and research results quality accounting information system of accounting information quality.
Susanto (2013)	Accounting information system are orientate to accounting information quality
Wilkinson et al. (2000)	Quality information should be relevant, accurate, timely, concise, clear, measurable and consistent.
Mancini et al. (2013)	A quality accounting information system can produce quality accounting information that can be used to meet control needs.
Susanto (2013:12)	Companies that use information effectively can benefit from opportunities to do things first (faster), better (more effectively) and cheaper (efficient) than their competitors.
Jun Shien (2014)	Top management support affects the quality of accounting information systems and their impact on the quality of accounting information generated by companies by improving the quality of financial reporting systems.
Rapina (2014)	Management commitment, organizational culture and organizational structure affect the quality of accounting information systems and their impact on the quality of accounting information.
Rahayu (2012)	The quality of accounting information systems have a significant effect on the quality of accounting information.

So from the description above the following hypothesis can be identified:

H5: The quality of accounting information system affects the quality of accounting information.

2.2 Preliminary study of researcher

A preliminary study that has been implemented is limited to literature study (textbook) and the results of research from the articles, and research that has been done by researchers related to the research that will be done:

1. Role of Uncertainty Environment on Management Information System—Literature approach. International Journal of Scientific & Technology Research Vol. 5 (6) 2016
2. How competence user impacts the quality management information system. Christine Dwi Karya Susilawati, Proceeding ICITB Darmajaya 2016, *https://jurnal.darmajaya.ac.id/index.php/icitb/article/view/571*
3. Application of Qualified Accounting Information System in Higher Education: Viewed From the Finance Section Perspective to Anticipate Environmental Uncertainty Political economy: government expenditures & related policies ejurnal 11(72). https://hq.ssrn.com/Journals/IssueProof.cfm?abstractid=3154194&journalid=1245083&issue_number=72&volume=11&journal_type=CMBO&function=showissue

2.3 Respondent's characteristic (Target)

Table 6. Characteristics of respondents in general.

Position	
Permanent lecturer	40%
Head of finance & budget	20%
Head of Program/Secretary of Program	20%
Finance Manager	5%
Staff accounting and reporting section	10%
Student	5%
	100%

(Continued)

Table 6. (*Continued*).

Gender		
Man		40%
Women		60%
No Charging		0%
		100%

Jumlah Perguruan Tinggi		66
Total Respondent		320

Characteristics of Higher Education

		Respondent	Percentage
Institute	5	23	7
Polytechnic	8	42	13
High School	6	24	8
University	47	231	72
	66	**320**	**100**

3 RESEARCH METHODS

3.1 *Statistical methods used for processing research results*

The data analysis method used in this research is the Structural Equation Model (SEM) method—Partial Least Square (PLS). According to Hair et al. (2013), PLS is an alternative method of SEM that can be used to overcome relationship problems between complex variables but the sample size of the data in a small range (30 to 100).

The SEM-PLS approach, is a component-based estimate that differs from covariance-based estimates that are generally resolved with software such as PLS Warp. The component-based estimation method in PLS is an iterative algorithm that separately breaks the model measurement block and then estimates the path coefficients in the model structural. Therefore, PLS is claimed to be able to explain the best residual variance of latent variables and manifest variables in any regression carried out in the model (Vinzi et al., 2010).

3.2 *Model and design research*
The research model used is described as follows:

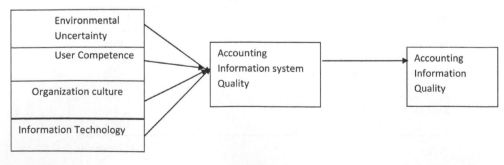

3.3 *Model and design research*
Variable operationalization

Table 7. Variable operationalization.

Variable	Dimensions	Indicator	Scale	Questionnaire
Environmental uncertainty is the individual's inability to accurately predict environmental conditions affecting the company. (Butler, 2001; Luhman & Cunliffe, 2012; Soliman, 2014; Kessler, 2013) As variable X1	1) **State uncertainty,** lack of information about the nature of the external environment leads to the uncertainty of perception so that members of the organization do not understand how things change or how things are related (Luhman & Cunliffe, 2012; Butler, 2001; Soliman, 2014; Kessler, 2013)	1) **Source of change:** source of external environmental factor change from price & demand, technology & sociological environment (Konig, 2009; Luhman & Cunliffe, 2012; Butler, 2001; Soliman, 2013; Kessler, 2013)	Ordinal	1
		2) **Level of predictability**: assumption divergent assumptions for rational management decision making **(Konig, 2009; Luhman & Cunliffe, 2012; Butler, 2001; Soliman, 2014; Kessler, 2013)**	Ordinal	2
	2) **Effect uncertainty** That is the lack of knowledge about environmental influences on the organization means that members can not predict the impact of external changes and have confidence to give a causal statement of environmental factors (Luhman & Cunliffe, 2012; Kessler, 2001; Soliman, 2014; Kessler, 2013)	1) **Downside risk/upside**: the risk of uncertainty on the organization's operational activities at the lower (execution) level and above (design/ planning) harmful or beneficial (Konig, 2009; Luhman & Cunliffe, 2012; Butler, 2001; Soliman, 2014: 214; Kessler, 2013)	Ordinal	3
		2) **The organization's performance measure (misaligned)** can not meet the needs of the environment, requiring additional costs for adaptation such as transaction costs and search costs (Konig, 2009;; Luhman & Cunliffe, 2012; Butler, 2001; Soliman, 2014; Kessler, 2013)	Ordinal	
	3) **Response uncertainty** lack of knowledge about the value or usefulness of any action leads to an inability to predict the consequences (Luhman & Cunliffe, 2012; Butler, 2001; Soliman, 2014; Kessler, 2013).	1) **Endogenous uncertainty**: action to overcome the inner uncertainty by proactively providing insetives (Konig, 2009; Luhman & Cunliffe, 2012; Butler, 2001; Soliman, 2013; Kessler, 2013)	Ordinal	4

(Continued)

Table 7. (*Continued*).

Variable	Dimensions	Indicator	Scale	Questionnaire
		2) **Exogenous uncertainty**: the organization has no direct influence, the organization can adopt an adaptive strategy (Konig, 2009; Luhman & Cunliffe, 2012; Butler, 2001; Soliman, 2014; Kessler, 2013)		5
User Competence is knowledge, skills as the best characteristic of a person's job (Cooper, 2000; Spencer & Spencer, 1993; Armstrong, 2011; Kessler, 2008; Gerber and Collin 2000). As a variable X2	1) **Motive**, something that is consistently about something desirable that results in action. (Spencer & Spencer, 1993).	1) **Impact and Influencing** (have influence);, Impact and influence from inside and outside/Intrinsically and extrinsically (Krausert, 2008; Mckee, 2012). 2) **Achievement Orientation** (achieve personal goal) (Krausert, 200:, Mckee, 2012)	Ordinal	6
	1) **Traits**, physical characteristics and a consistent response to the situation or information. (Spencer & Spencer, 1993; Armstrong, 2011)	1) **Analytical thinking** (conceptual thinking) (Krausert, 2008) 2) **Initiative and Persistence** (Krausert, 2008) 3) **Problem solving** (Krausert, 2008)	Ordinal	7
	2) **Self-Concept**, attitudes, values and self-image of a person (Spencer & Spencer, 1993; Armstrong, 2011)	1) **Flexibility** (Krausert, 2008) 2) **Self Confidence** (Krausert, 2008) 3) **Self Control** (Krausert, 2008)	Ordinal	8
	3) **Knowledge** is information of a person in a specific area (Spencer & Spencer, 1993; Gerber and Collin 2000).	1) **General Knowledge** (general information) (Spencer & Spencer, 1993; Gerber and Collin 2000) 2) **Specific knowledge** of information on specific areas—Levels of formal and informal education (Spencer & Spencer, 1993; Gerber and Collin 2000)	Ordinal	9
	4) **Skill**, ability that demonstrates ability in physical or mental task tasks (Spencer & Spencer, 1993; Gerber and Collin 2000).	1) **People Management Skills**: Socialized power, managing group process, positive regard and accurate self assessment (Spencer & Spencer, 1993; Mckee, 2012)	Ordinal	10
			Ordinal	11

666

Organizational culture				
		2) **Goal and Action Skills**: efficiency orientation, diagnostics use of concepts, proactivity (Spencer & Spencer, 1993; Mckee, 2012)	Ordinal	12
	1) **Orientation on detail** (Robbins, et al., 2013)	1) Employees are expected to show accuracy and analysis (Robbins, et al., 2013)	Ordinal	13
		2) Employees are expected to pay attention to details (Robbins, et al., 2013).	Ordinal	14
	2) **Orientation on results** (Robbins, et al., 2013)	1) The extent to which management is more focused on results (Robbins, et al., 2013)	Ordinal	15
		2) The extent to which management is more focused on the techniques and processes used to achieve results (Robbins, et al., 2013)		
	1) Orientation of the people (Robbins, et al., 2013)	1) To what extent management decisions consider what is produced in people in the organization (Robbins, et al., 2013)	Ordinal	16
		2) To what extent management decisions consider the effects of what is produced on the people within the organization (Robbins et al, 2013)		
	2)Team orientation (Robbins, et al., 2013)	1) To what extent management decisions consider what is produced in people in the organization (Robbins, et al., 2013)	Ordinal	17
		2) To what extent management decisions consider the effects of what is produced on the people within the organization (Robbins et al, 2013)		
	3) Aggressiveness (Robbins, et al., 2013)	1) To the extent to which people are aggressive (Robbins, et al., 2013)	Ordinal	18
		2) To the extent to which people are willing to compete (Robbins, et al., 2013)		
	4) Stability (Robbins, et al., 2013)	1) Organizational activities emphasize maintaining stability (Robbins, et al., 2013)	Ordinal	19
		2) Organizational activities lead to growth (Robbins, et al., 2013)		

(Continued)

Table 7. (*Continued*).

Variable	Dimensions	Indicator	Scale	Questionnaire
	6) Innovation and risk-taking (Robbins, et al., 2013)	Employees are encouraged to be innovative (Robbins, et al., 2013) 2) employees are encouraged to take risks (Robbins, et al., 2013)	Ordinal	20
Use of Information Technology (X4) (Thompson & Baril, 2002; Romney and Steinbart, 2015; Laudon and Laudon, 2013)	1) Based on function (Thompson & Baril, 2002)	1) Speed Thompson and Baril, 2002) 2) Reliability Thompson and Baril, 2002) 3) Usage Fee (Thompson and Baril, 2002) 4) Conditions of use (Thompson and Baril, 2002)	Ordinal	21
	2) Ease of use (Thompson & Baril, 2002)	1) Quality of user interface (Thompson and Baril, 2002) 2) Ease to become an expert (Thompson and Baril, 2002) 3) Portability (Thompson and Baril, 2002)	Ordinal	22
	3) Conformity (Thompson & Baril, 2002)	1) Conformity with standards (Thompson and Baril, 2002) 2) Operational conformance (Thompson and Baril, 2002)	Ordinal	23
	4) Maintenance (Thompson & Baril, 2002)	1) Modularity (Thompson and Baril, 2002) 2) Scalability (Thompson and Baril, 2002) 3) Flexibility (Thompson and Baril, 2002: 36) 1.1	Ordinal	24
Quality Accounting Information System (Y) (Susanto, 2013; Wilkinson et al., 2000; Bodnar & Hopwood, 2014; Romney & Steinbart, 2015; Bagranoff et al., 2010)	1. Integration (Heidmann, 2008; Susanto, 2013)	1) The system can facilitate the processing of information from various resources to support decision making (Heidmann, 2008) 2) Integration of all related elements and sub-elements in establishing an accounting information system (Susanto, 2013)	Ordinal	25
	2. Flexibility (Heidmann, 2008)	1) Information systems can adapt to the various users in need (Heidmann, 2008)	Ordinal	26

668

	2) Information systems can adapt to changing conditions (Heidmann, 2008)		
3. Accessibility (Heidmann, 2008)	1) Use of a computerized system (Heidmann, 2008)	Ordinal	27
	2) Ease in accessing information contained in information systems (Heidmann, 2008)	Ordinal	28
4. Formalization (Heidmann, 2008)	1) System contains rules or procedures (Heidmann, 2008)		
	2) Rules or procedures contained in the system are used to coordinate activities (Heidman, 2008)		
Quality of Accounting Information (Z) (Susanto, 2013; James A. Hall, 2011; Wilkinson et al., 2000; Romney and Steinbart, 2015)			
1. Accurate (Susanto, 2013; Hall, 2011)	1) Reflects the existing situation and conditions (Susanto, 2013)	Ordinal	29
	2) Information must be free from material fallacies (Hall, 2011)	Ordinal	30
2. Relevan (Susanto, 2013; James A.Hall, 2011)	1) Accounting information generated really fit the needs (Susanto, 2013)	Ordinal	31
	2) Accounting information in reports or documents shall be in accordance with the intended (Hall, 2011)		
3. Timeliness (Susanto, 2013; Hall, 2011)	1) Accounting information generated really fit the needs (Susanto, 2013)	Ordinal	32
	2) accounting information in reports or documents shall be in accordance with the intended (Hall, 2011)		
4. Complete (Susanto, 2013; Hall, 2011)	1) the resulting accounting information has been as complete as desired and needed (Susanto, 2013)	Ordinal	33
	2) no missing information (Hall, 2011)		

4   CONCLUSIONS IN LITERATURE

So the most must be considered by universities in Indonesia is information technology and organizational culture at the University in terms of detailed task orientation, teams, organizational orientation on more stable, aggressive and innovative work to improve the quality of accounting information systems. Supported by other factors that play an important role is the competence of users in higher education which is more dominant in knowledge and skill, traits, self concept and motive. Environmental uncertainty factor are important but difficult compared to other factors in this study, but it is important to note that this environment is always uncertain and should be anticipated.

This research will continue in the next article in Research Results and Discussion and Solution Suggestion.

ACKNOWLEDGEMENT

Thanks to Ristekdikti who has funded this applied and strategic research grant and the LPPM of Maranatha Christian University. And thanks also to the lecturer in the Doctoral Program of Accounting Science that provided science and knowledge at the time of study of doctoral program of accounting science for the process of our research work.

REFERENCES

Abdallah, A.A.J. (2013). The impact of using accounting information system on the quality of financial statement submitted to the income and sale tax department in Jordan. *European Scientific Journal.*

Bagranoff, N.A., Simkin, D.B.A., Norman, Mark G., Strand, C. (2010). *Core concept of accounting information systems.* (11th ed.). Printed in the United States of America.

Bocij, P. (2014). *Business information system: Technology, Development and Management for the E-Business.* (5th ed.). United Kingdom. Pearson

Bodnar, G.H. and Hopwood, W.S. (2014). *Accounting Information systems* (11th ed.). Pearson New International Edition. Always Learning.

Boockholdt., J.L. (1999). *Accounting Information Systems.* (5th ed.). Mc Graw Hill International Edition.

Davis, C.K. (1989). Technologies & methodologies for evaluating information technology in business. IRM Press Publisher.

Gelinas, U.J. and Dull, Richard B. (2008). *Accounting Information Systems.* Canada. Thomson South Western.

Gerber, R. dan Collin, L. (2000). *Training for a smart workforce.* London, England. Routledge.

Gilbert, A.H. and Singer, J.F. (2011). The strategic impact of environmental uncertainty and information system design. *The Review of Accounting Information System.*

Gull, F.A., Glen, W. & Hung, A.R. (1993). The effect of environmental uncertainty, computer usage and management accounting system in the small business. *The Journal of Entrepreneurial Finance.*

Hammad, S.A., Jusoh, R. Ghozali., I. (2012). Decentralization, perceived environmental uncertainty, managerial performance and management accounting system information in Egyptian hospitals. *International Journal of Information System.*

Hall, J.A. (2011). *Accounting Information Systems* (7th ed.). Cengage Learning.

Heidmann, M. (2008). *The role of management accounting systems in strategic sensemaking.* Springer Science Business and Media.

Holbeche, L. (2009). *Aligning Human Resources and Business Strategy* (2nd ed.). Butterworth-Heinemann is an imprint of Elsevier.

Iskandar, D. (2015). *Analysis Of Factors Affecting The Success Of The Application Of Accounting Information System.* International Journal of Scientific & Technology Research.

Ivancevich, J.M., Lyon, H.L. Adams, D.P. (2008.) Business in a dynamic environment. West.

Kessler, E.H. (2013). *Encyclopedia of management theory.* Sage Publication.

Khosrow-Pour, M. (2011). *Enterprise information systems: concepts, methodologies, tools and application information science Reference.* United Stated of America. IGI.

Konig, F. (2009). *The uncertainty-governance choice puzzled revisited.* Berlin. Gabler.

Krausert, A. (2008). *Performance management for different employee groups*. Phisyca Verlag. Springer Company.

Laudon, K.C. and Laudon, J.P. (2012). *Management information systems* – Managing The Digital Firm. (12th ed.). Pearson Prentice Hall.

Luhman, J.T,.Cunliffe, A.L. 2012. *Key concept in organization theory*. Sage.

Mancini, D., Vaassen, E.H.J. and Dameri, R.P. (2013). *Accounting information systems for decision making*. Springer.

Mckee, A. (2012). *Management a focus on a leader.* Prentice Hall. Pearson.

Meiryani, J.S. (2015). Influence of user ability and top management support on the quality of accounting information system and its impact on the quality of accounting information. *International Journal of Recent Advances in Multidisciplinary Research* 02, 03, 0277–0283.

Nasir, M. (2015). Menristekdikti: PTN-BH finance is less flexible. (Menristekdikti: keuangan PTN-BH kurang fleksibel). ihttp://www.antaranews.com/berita/517386/menristekdikti-keuangan-ptn-bh-kurang-fleksibel.

Napitupulu, I.H. (2015). Impact of organizational culture on the quality of management accounting information system: a theoretical approach. *Research Journal of Finance and Accounting, 6*(4).

Nurhayati, N. & Mulyani, S. (2015). User participation on system development, user competence and top management commitment and their effect on the success of the implementation of accounting information system (Empirical Study in Islamic Bank in Bandung). *European Journal of Business and Innovation Research.*

Nusa, I.B.S. (2015). Influence of organizational culture and structure on quality of accounting information system. *International Journal of Scientific & Technology Research.*

Robins, S.P. & Coulter, M. (2011). *Management.* (11th ed.). Prentice Hall.

Romney, M.B. & Steinbart, P.J. (2015). *Accounting information systems.* (13th ed.). Pearson Education.

Richardson, V., Chang, J. & Smith, R. (2014). *Accounting and information systems*. Mcgraw Hill Education.

Rahayu, S.K. (2012). The factors that support the implementation of accounting information system: a survey in Bandung and Jakarta taxpayer offices. *Journal of Global Management, 4*(1), 25–52.

Robbins, S.P., DeCenzo, D.A., Coulter, M. & Anderson, I. (2014). *Fundamentals of managements.* Canada: Pearson Canada Inc.

Saleh, F.M. (2013). Critical success factors and data quality in accounting information systems in Indonesian cooperative enterprises: An empirical examination. *Interdisciplinary Journal of Contemporary Research In Business Copy Right. Institute of Interdisiplinary Business Research.*

Sacer, I.M. & Oluic, A. (2013). Information technology and accounting information systems quality in Croatia middle and large companies. *Journal of Information and Organizational Sciences* 37(2).

Sekaran, U. & Bougie, R. (2013). *Research methods for business a skill-building approach.* (6th ed.). Wiley.

Spencer, Lyle & Spencer, Signe M. (1993). *Competence work models for superior performance.* Wiley.

Stair, R.M. & Reynolds, G.W. (2010). Principles of information system. (9th ed.). Management Approach.: Cengage Learning.

Susanto, Azhar. (2008). Accounting information systems development risk management structure *(Sistem Informasi Akuntansi Struktur Pengendalian Resiko* Pengembangan). Lingga Jaya.

Susilawati, C.D.K. (2016). Role of uncertainty environment on management information system—literature approach. *International Journal of Scientific & Technology Research,* 5(6),.

Susilawati, C.D.K. (2016). How Competence User Impact The Quality Management Information System. *Proceedings of International Conference on Information Technology and Business Darmajaya 2016,* https://jurnal.darmajaya.ac.id/index.php/icitb/article/view/571.

Susilawati, C.D.K. (2018). Application of qualified accounting information system in higher education: viewed from the finance section perspective to anticipate environmental uncertainty. *Political economy: Government Expenditures & Related Policies e-Journal* 11(72).

Thompson, R.W. Baril, W.C. (2002). Information technology and management. Irwin: McGraw-Hill.

Suyatno, T. 2015. PT Association Records 205 Problematic PTS Foundations.

(Asosiasi PT Catat 205 Yayasan PTS Bermasalah). http://www.beritasatu.com/pendidikan/302405-asosiasi-pt-catat-205-yayasan-pts-bermasalah.html.

Wilkinson, J.W. Cerullo, M.J, Raval, V., Wong-On-Wing, B. (2000). Accounting information systems. (4th ed.). United Stated of America. John Wiley and Sons.

Wang, David Han-Min, and Quang Linh Huynh. (2013). Effects of environmental uncertainty on computerized accounting system adoption and firm performance *International Journal of Humanities and Applied Sciences* 2(1).

Business Innovation and Development in Emerging Economies – Trinugroho & Lau (Eds)
© *2019 Taylor & Francis Group, London, ISBN 978-1-138-35996-3*

The role of management control systems in aspects of managerial entrepreneurship

S.A. Syahdan
Accounting Study Program, STIE Indonesia Banjarmasin, Indonesia
Doctorate Program of Economics Science, Faculty of Economics and Business,
Universitas Sebelas Maret, Indonesia

Rahmawati, Djuminah & E. Gantyowati
Department of Accounting, Faculty of Economics and Business, Universitas Sebelas Maret, Indonesia

ABSTRACT: This study examines the role of management control systems specifically social networking, innovative organizational culture, and formal control, in aspects of managerial entrepreneurship at local water companies in South Kalimantan, Indonesia. The research used a direct data collection method with purposive sampling, via a mail survey send to unit managers of local water componies. A total of 123 respondents from 12 local water componies in the South Kalimantan Regional Government area participated in the study. Research data were analyzed using multiple regression analysis to test the hypotheses developed. The results showed that social networking, innovative organizational culture, and formal control simultaneously positively influence managerial entrepreneurship with a significance value of 0.02. Mean while, social networking has a partial negative effect on managerial entrepreneurship with significance of 0.007. Innovative organization culture and formal control each have a positive effect on managerial entrepreneurship with significance of 0.030 and 0.049, respectively.

Keywords: social networking, innovative organizational culture, formal control, managerial entrepreneurship

1 INTRODUCTION

Some aspects of managerial behavior are problem issues for local water componies in Indonesia. To improve performance of such componies, managerial entrepreneurship is required for creation and enhancement af the innovative capacities of public managers (Peters and Savoie, 1995). To create order within managerial entrepreneurship activities, management control systems are needed to ensure that managers are directed and controlled in carrying out renewal or innovation (Lövstål, 2008)).

Based on data from the Drinking Water Provider System Management Improvement Agency (BPPSPAM) of the Ministry of Public Works and Public Housing, in 2015 and 2016 the performance appraisal level in poorly performing local water componies 47%, while in 2017 it was 45%. This shows that PDAM the performance of local water componies has bot been optimal three years to 2017. The BPPSPAM appraisal process assesses the four components of human resources (managerial) competency, and these cover many of the categories of poor performance in local water componies.

From the above data and the preliminary survey, it clear that the values of management entrepreneurship have not yet been introduced in these componies, This situation is based on the still very inherent, centralized business environment, in which regional heads (mayors and regent) still determine the rules for business both in technical and operational matters.

Low levels of entrepreneurship can be observed in these business as exemplified by their lack of risk taking behavior, and this is a barrier to public companies improving their performance (Syahdan & Santoso, 2004).

Moon's (1999) study defines one dimension of managerial entrepreneurship as being based on behavior (behavior-based entrepreneurship), that emphasizes the tendency for risk-taking. Risk-taking is defined here as a managerial form that addresses organizational change and innovative decision making. Organizational development that supports entrepreneurship is inseparable from the effective management both accounting and non-accounting information, allowing the decision-making process produced to be the basis for improving performance. Increased attention to the roles that accounting and control play in corporate entrepreneurship and innovation might enhance these two aspects of business activity. Davila et al. (2009) conclude that a new paradigm has emerged over the past decade, that highlights the relevance of formal accounting and control for innovation and entrepreneurship.

Management control system have evolved in recent years, focussing on formal and quantitative information to assist managerial decision-making with a broader range of information. This includes external information relating to the market, customers, and competitors, non-financial information related to the production process, predictive information and various decision mechanisms, and informal personal and social controls.

Conventionally, management control systems are considered as passive tools that provide information to help managers (Chenhall, 2003). However, Chenhall et al. (2011), develop this view by stating that a management control system is a set of controls consisting of social networks, innovative organizational culture, and formal controls. Furthermore, the management control system is a tool used by managers in a variety of control processes, such as those involved in planning and decision-making. However, the character of the is usage varies according to organizational settings and between the control process of unit managers. Therefore, to understand the relationship between management and entrepreneurship, control systems must be understood boot in context and in their implementation (Lövstål, 2008).

It is interesting to investigate when management control system are considered to inhibit managers in carrying out renewal or innovation and when, in contrast, they are seen as a form of managerial entrepreneurship to create and innovate the capacities of public managers and make them more responsible for decisions taken (Peter & Savoie, 1995).

A management control system can sometimes be interpreted as being a force contradictory to managerial entrepreneurship. Management control systems have the purpose of creating order, and making the processes more efficient, while in contrast, entrepreneurship is a process of innovation and creation. Many management control systems are based on ideas about stability and predictability, while entrepreneurship is located more in aspects of uncertainty and ambiguity (Lövstål, 2008).

Various opinions and suggestions about the relationship between management and managerial control systems and entrepreneurship have been raised, with some researchers showing that management control system have negative impacts on entrepreneurship. In contrast, Simons (1994, 1995) claims that management control systems encourage managerial entrepreneurship and can facilitate innovation and renewal. Furthermore, some researchers point out that whether the management control system has a positive or negative effect on entrepreneurship depend on the nature of the management control system it self. In other words, the emphasis is on how a system is interpreted and used in determining whether it is good or bad from the perspective of entrepreneurship (Lövstål, 2008).

This research is expected to provide contributions to the performance of local water componies making their management control system more efficient and competencies of human resources. These resources may have far demonstrated efficiency and bureaucratic practices that hinder innovation.

The weakness of managerial entrepreneurship aspects in local water componies and the main problems that will be examined in this study can be formulated by the following question: in local water componies can management control systems, including social networking, innovative organization culture, and formal control positively influence managerial entrepreneurship including the taking of risk?

2 LITERATURE REVIEW AND DEVELOPMENT OF HYPOTHESES.

2.1 *Contingency theory*

The contingency theory approach, as used as an analysis technique in management accounting, has long attracted the interest of researchers. The contingency approach used in management accounting is based on the premise that there is no one management accounting system that is universally appropriate to be applied to all organizations in every situation (Otley, 1980). The management accounting chosen system depends on situational factors. In management accounting studies, a contingency approach is needed to allow the evaluation of conditional factors that make a management control system more or less effective. Furthermore, Cadez & Guilding, (2008) state that the main proposition of contingency theory is that effective company performance will depend on the compatibility between its the contextual factors. The essence of contingency theory also says that organizations must adapt their structures to issues such as the environment, organizational size, and business strategy if they are to operate effectively and efficiently (Gerdin & Greve, 2008).

2.2 *Management control systems*

A management control system (MCS) is a management tool that directs the organization via competitive advantages toward its strategic goals (Anthony & Govindarajan, (2007). Ahrens and Chapman (2004) state that MCSs can play a dynamic role in helping managers to formulate new strategies and are therefore not just tools to support strategy implementation. Meanwhile, both the Institute of Management Accountants and the Chartered Institute of Management Accountants state that management accounting is an integral aspects of management control systems. Malmi and Brown (2008) put forward the general concept of a management control system as a package made up of a collection of controls and control systems comprising planning control, cybernetics control, award control and compensation, control administration and cultural control.

The concept of the control system is used in empirical research as an independent variable associated with competitive concepts, one of which is related to organizational entrepreneurship (Lövstål, (2008).

2.3 *Social networking*

Social networking refers to the way interorganizational exchanges are managed with an emphasis on personal and social relationships based on length of relationship and trust (Chenhall et al., 2011). Research into the effects of the socialization process that occurs in social networks shows a strong association between socialization and progress in sharing information, engaging in joint problem-solving, adapting to unexpected changes, and avoiding the abuse of power (Mahama, 2006), all of which are factors that can encourage innovation (Feldman, 1976).

The theory developed in this study is that social networking is increasingly important to organizations, although it is not commonly subject to control in Western companies (Anderson et al., 1994; Gulati et al., 2000). In contrast, social networking is used as a part of interorganizational control, especially for transactions such as joint ventures and for targeted regulation of buyers and suppliers (Hakansson & Lind, 2004).

2.4 *Innovative organizational culture*

The values and practices of the culture applied can be used by researchers in investigating the influence of culture in the organization. The relationship between organizational culture and various social functions in organizations has been a theme in social studies for some time (Denison & Mishra, 1995). Sociologists, social anthropologists and psychologists present culture and ideology as integral features of a social function. Hofstede (1980) developed a theory of cultural dimensions which was then widely studied in the fields of sociology, organizational theory and accounting.

Innovative organizational culture is the informal environment with in an organization that provides open, flexible and structured communication, while innovation is defined as the application of new ideas or behaviors by an organization (Zaltman et al., 1976). As in other MCS based research into innovation, It is necessary distinguish between innovation in general and product innovation in particular. While innovation is often considered to be closely related to an organization's need to become more competitive through broader implementation, administrative innovations are more complex to be implement and thus less profitable (Damanpour, 1990).

According to Holmes and Marsden (1996), the innovative culture of an organization has an influence on the behavior, work methods, and motivations of its managers and subordinates in achieving organizational performance. Lusch and Harvey (1994) argue that improving organizational performance can be influenced by intangible assets such as organizational culture, customer relationships, and corporate image (brand equity).

2.5 *Formal control*

Formal control is the control used for planning and for aspects as budget and analysis variants. The adoption of formal controls is encouraged, as it is seen to provide mechanisms to integrate diverse operations (Lawrence and Lorsch 1967; Merchant, 1981). The control in question comprises practices that can be used for planning and controlling such aspects as budgeting and analysis of variance, costs, and investment valuation techniques. Formal controls that are relevant to innovation include management accounting practices that assist in planning, such as investment valuation techniques, additional analysis, based on cost activities, budgeting, production scheduling, and controls such as quality control, inventory control, internal audit, and performance appraisal (Davila et al., 2006; Chenhall & Morris, 1995). From this review of the literature, it can be seen that there is sufficient support for the idea that formal control can help produce innovation.

2.6 *Managerial entrepreneurship (risk-taking).*

Peters and Savoie (1996) provide a summary of the power and context of managerial entrepreneurship developed to deal with reactions to conventional models that always reject risk and to high level of bureaucracy. Managerial entrepreneurship is also formed to create and innovate the capacities of public organizations and managers and also to make them more responsible for their decisions. Another idea of managerial entrepreneurship is that it can combine managerial understanding in the public sector with customer-driven government. Contextually, managerial entrepreneurship is the ability of the management of a company to continuously carry out renewal and innovation and constructively take risks in the field of operations (Miller, 1983).

Managerial entrepreneurship is defined as the ability of company management to continually renew, innovate and constructively take risks (Miller, 1983; Naman and Slevin, 1993). Drucker (1995) defines managerial entrepreneurship as the practice of making decision based on entrepreneurial knowledge and utilizing such decision to increase the effectiveness of new business ventures and small and medium businesses.

Entrepreneurship deals with innovation, risk-taking, and proactivity (Covin & Slevin, 1989). Stevenson et al. (as citied in Wood et al., 2000) suggest that managerial entrepreneurship is an organizational management's willingness to encourage and support creativity, flexibility, and risk-taking. Thus, according to Jarillo and Wood in Wood et al. (2000), organizational managerial entrepreneurship is related to three interrelated conceptual components, namely innovativeness, proactivity, and risk-taking. Innovation is the introduction of new products, services, technology, and markets, proactivity is the active seeking of ways to achieve organizational goals, while risk-taking is related to making the right decisions under conditions of environmental uncertainty.

The literature agrees that the concept construct and entrepreneurship theory are multidimensional (Hofer and Bygrave, 1992). Moon's (1999) research defines the managerial dimension of entrepreneurship as based on behavior (behavior-based entrepreneurship), and in this description, the dimension of behavior-based managerial entrepreneurship emphasizes the tendency for risk-taking. Risk-taking here is a managerial form that addresses trends in organizational change and innovative decision-making.

2.7 *Conceptual framework for research and hypothesis development*

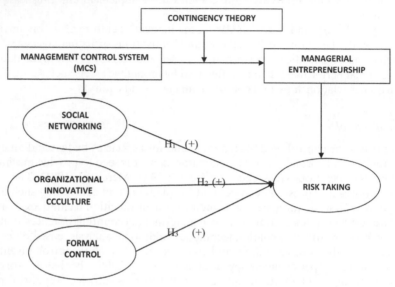

Figure 1.   Presents the research framework and hypotheses of the research.

2.8 *The relationship between social networking and risk-taking*

Empirical evidence about how social networking between organizations increases risk taking or decisions that innovative decisions is proposed by Goes and Park (1997) and Pennings and Harianto (1992) and confirms that innovative risk-taking abilities in organizations are greatly enhanced by interorganizational network links. In contrast, results of Chenhall et al. (2011) suggested that there was no significant direct relationship between innovation in decision-making and social networking.

Companies that use social networks are open to joint problem-solving, and such networks can assist in developing personality, trust, and reducing bureaucracy in relations between organizations. This interaction provides a climate that promotes information exchange that can foster the seeds of ideas for development. In this situation, organizations tend to be rich in opportunities for innovation and are stimulated to develop organizational innovations that can help enable acceptance of new ideas and free communication in the decision-making process. Thus the first hypothesis developed in this study is:

H1: Social networking has a positive influence on managerial entrepreneurship aspects of risk-taking.

2.9 *The relationship between Innovative organizational cultural and risk-taking*

Empirical research has show that innovative organizational culture as an organizational contextual factor influences management control systems. Additionally, organizational culture is also a moderating factor between management control systems and company performance (Fauzi & Hussain, 2008). The results of Chenhall et al., (2011) research found evidence that an innovative culture in organization has a significant relationship with decision-making innovation. Mean while, Moon (1999) stated that organizational culture has a significant influence on managerial entrepreneurship in the dimension of risk-taking. The second hypothesis developed in this study is therefore:

H2: Innovative organizational culture has a positive effect on managerial entrepreneurship aspects of risk-taking.

2.10 The relationship between formal control and risk-taking

Formal controls are relevant to innovation as management accounting practices that help in planning, such as investment valuation techniques, additional analysis based on cost activities, budgeting, production scheduling, and controls such as quality control, inventory control, internal audit, and performance appraisal (Davila et al., 2006; Chenhall and Morris, 1995). In the literature, there is sufficient support for the idea that formal control can help develop innovation.

The results of Chenhall et al. (2011) show that formal control has a significant relationship to innovation, while Moon (1999) stated in his research that the organizational structure associated with formalization has an influence on managerial entrepreneurship in term of risk-taking. Likewise the results of research by Caruana et al. (1998) state that formalization affects entrepreneurship. The third hypothesis developed is therefore:

H3: Formal control has a positive effect on managerial entrepreneurship for the aspects of risk-taking.

3   RESEARCH METHODS

3.1   Data selection and collection

The population and sample used in this study are managers of 12 local water companies in South Kalimantan District and City. The sample choice based on the managers in these being the decision-makers. The unit of analysis in the middle to upper managers in local water companies. In addition to the above considerations, the selection of samples also follows the Moon's (1999) research and is based on the job description in the field.

Data was collected using a questionnaire method. The questionnaire was effectively distributed directly or via email to 200 respondents expected to achieve a response rate of 60% of the distributed questionnaires. The primary data gathered in this study comes from respondents' answers to the questionnaire from unit managers.

3.2   Operational definitions and variable measurement

The dependent variables used in this study are managerial entrepreneurship aspects of risk reduction, while the independent variables in this study are social networks, innovative organizational culture, and formal control. To measure the variables in this study, an instrument which is the developed and modified from Moon (1999), as measured by a five-point Likert scale is employed. Temporary variables in this study were developed in line with Khandwalla (1972) and Chenhall and Morris (1995).

3.2.1   Dependent variables
Managerial entrepreneurship is reflected by risk-taking, their being the tendency of organizations to change and update their decision-making. Decision-making is a key part of a manager's activities and is a process in which a series of activities are chosen (Moon, 1999).

The questionnaire instrument for managerial entrepreneurship variables is developed from the research of Moon (1999) and uses four questions. The measurement scale used is a five-point Likert scale, where (1) states that the respondent strongly disagrees and (5) that he or she strongly agrees.

3.2.2 *Independent variable*

1. *Social networking (X1)*

   Social networks taking place in exchanges between organizations are measured by examining personal and social relationships based on the length of relationships and levels of trust (Chenhall et al., 2011).

   Questionnaire instruments for social networking variables are developed from research by Khandwalla (1972) and Chenhall and Morris (1995) use eight questions. The measurement scale used is a five-point Likert scale, where (1) is strongly disagree and (5) is strongly agree.

2. *Innovation organizational culture (X2)*

   An innovative culture comprises informal processes that provide for open, flexible, and structured communication. Innovation is defined as the application of new ideas (Zaltman et al., 1973).

   The questionnaire instrument for Organizational Innovation Culture variables is developed from the research of Khandwalla (1972) and Chenhall and Morris (1995), and uses eight questions. The measurement scale used is a five-point Likert scale, where (1) is strongly disagree and (5) is strongly agree.

3. *Formal control (X3)*

   Formal controls are defined that are relevant to innovation such as management audits, additional analysis based on costs, budgeting, calculation, internal audit, and performance (Davila et al., 2009; Chenhall and Morris, 1995).

   The questionnaire instrument for Organizational Innovation Culture variables is developed from the research of Khandwalla (1972) and Chenhall and Morris (1995), and uses eight questions. The measurement scale used is a five-point Likert scale, where (1) is strongly disagree and (5) is strongly agree.

3.3 *Regression model*

The data analysis method used in this study is multiple linear regression. In general, the multiple regression analysis equation can be formulated as follows:

$$ME = B0 + B1.JS + B2.BIO + B3.PF + e \tag{1}$$

where
ME   = managerial entrepreneurship
JS   = social network
BIO   = organizational Innovative Culture
PF   = formal control
B0   = constants
B1.B3 = regression coefficient X1, X2, X3
e   = error.

4 RESEARCH RESULTS AND DISCUSSION

4.1 *Reliability and validity test*

Tests are used to determine the consistency and accuracy of the data collected. The procedures performed to test the quality of the data in the study are as follows:

Test of internal consistency (reliability), intended to test the consistency of questionnaires in measuring the same construct (Sekaran, 2016), and if measurements are taken from time to time by others (Ghozali, 2011). This testing is performance by calculating the Cronbach alpha coefficient of each instrument in one variable. The closer to 1 the Cronbach alpha coefficient, the higher the internal coefficient of reliability of instrument (Sekaran, 2016).

Table 1. Reliability and validity test results.

Variable	Cronbach alpha	Validity of correlation*
Managerial entrepreneurship	0,557	Valid
Social networking	0.721	Valid
Innovative organizational culture	0,776	Valid
Formal control	0,655	Valid

Source: Primary data processed, 2018.

Validity testing shows that the correlation between each indicator against the total construct score has a value > 0.6. From this, it can be concluded that each question indicator is valid because it is correlated and significant. The results of the validity test are summarized in Table 1.

4.2 Classic assumption tests

1. Multicollinearity tests
   The R-squared summary data model output is quite low, at only 0.118, indicating that 11.8% of the above models do not indicate the occurrence of multicollinearity. The second indication is seen from the variance inflation factor (VIF) and tolerance values for the three independent variables (social networking, innovative organizational culture, and formal control: VIF values of around 1 social network, 1.034, organizational innovative culture, 1.048; and formal control 1,015 and tolerance values that are also close to 1 social networking, 0.967, organizational innovation culture, 0.954 and formal control, 0.985. This means that the three independent variables do not have multicollinearity symptoms with other variables.

2. Heterocedasticity tests
   Testing of the presence or absence of heteroscedasticity is achieved in this study by looking at the graph of the plot of the dependent variable predictive value (ZPRED) and the residuals (SRESID). If there are patterns, such as the points forming regular pattern that is wavy, widened and then narrowed, then there is heteroscedasticity. If there is no pattern, and the points are spread above and below the number 0 on the Y axis, then there is no heteroscedasticity.
   Based on the scatterplot graph of (SRESID) and ZPRED where the Y axis is Y which has been predicted and the X axis is the residuals (Y predictions with true Y) that have been studentized, the points are spread randomly, without t forming a clear pattern, and are well spread above or below the number 0 on the Y axis. From this, It can be concluded that there is no heteroscedasticity in the regression model, so the appropriate regression model can be used to predict managerial entrepreneurship based on the input of the three independent variables (social networking, innovative organizational culture, and formal control).

3. Normality tests
   The normality test was conducted to find out whether, the dependent variable and the independent variables in the regression model have normal distribution This was performed using graph analysis and the Sample Kolmogorov-Smirnov test. Chart analysis to test the normality of can be carried out via a histogram that compares the observation data with a distribution that approaches the normal distribution or by looking at the normal probability plot that represent at normal distribution. In this, if the data is spread around diagonal line and follows the direction of the diagonal line or if the histogram line shows a normal distribution pattern, the regression model meets the assumption of normality. In contrast, if the data spreads far from the diagonal line or does not follow the direction of the diagonal line, the regression model does not meet the assumption of normality.

From graph analysis can be concluded that the histogram provides a distribution pattern that is close to normal, and the normal graph shows points spread around the diagonal line. This means that, because the assumption of normality has been fulfilled, the regression model is suitable to be used to predict the interest of managerial entrepreneurship, based on inputs from independent variables. The one sample Kolmogorov-Smirnov also indicates that the data distribution is normal.

4.3 *Analysis and discussion*

To assess the accuracy of the sample regression function estimating the actual value can be measured from the goodness of fit. Statistically, this can be measured from the value of the test statistic of individual parameters, the statistical value of F and the determinant coefficient. The results of data analysis carried out using ANOVA testing and F-test obtained F calculated at 5.317 with a probability level of 0.002. Because the probability is much smaller than 0.05, the regression model can be used to predict managerial entrepreneurship by stating that social networking, an innovative organizational culture, and formal control together influence managerial entrepreneurship.

The adjusted R-squared value of 0.96 or 9.6% variations in managerial entrepreneurship can be explained by variations in the three independent variables of social networks, innovative organizational culture, and formal control. While the remaining 90.4% is explained by other factors outside the model. The Standard error of the estimate is 1.47518.

4.4 *First hypothesis testing (H_1)*

The first hypothesis tested in this study (H_1) is that social networks influence managerial entrepreneurship from the aspects of risk-taking. From regression results for, it can be seen that social networks have a significant negative effect on managerial entrepreneurship, with a significance level of 0.007, (that less than 0.05). Thus, regression results the hypothesis, indicate that social networking has a negative effect on managerial entrepreneurship in terms risk-taking.

The results of this study are not in line with research conducted by Chenhall et al. (2011) which shows no direct relationship between social networking and innovation. However, the results of this study are in line with Goes and Park (1997), who suggest that social networking between organizations increases innovation, as well as wwith the research of Pennings and Harianto (1992) which confirms that innovative abilities and the application of innovation in organizations are greatly enhanced by interorganizational network links.

4.5 *Second hypothesis testing (H_2)*

The second hypothesis tested in this study (H_2) is that innovative organizational culture influencing entrepreneurship. Regression results for the second hypothesis, indicate that the innovative organizational culture of influences managerial entrepreneurship, with a significance level of 0.03 (that is, greater than 0.05). Thus regression results for the hypothesis, indicate that the innovative culture an the organization influences managerial entrepreneurship for the dimension of risk-taking.

The results of this study are in line with the research of Chenhall et al. (2011), which states that an organization's innovative culture is positively related to innovation. Similarly, the study by Moon (1999) indicates that organizational culture influences managerial entrepreneurship in term risk-taking, meaning that risk-taking behavior tends to be influenced by organizational culture and structural characteristics.

Organizational research has widely stated that an innovative culture in organizations will support innovation because such business are more adaptive and responsive, have open communication, free flow of information, and involve employees in developing new ideas (Burns & Stalker 1961; Mintzberg & Waters, 1985; Quinn 1980). Mintzberg

(1994) summarizes the idea that organizations generate new ideas, notes the important roles that support creativity but sometimes the innovation process is not comfortable and that a more open, informal process and support culture is needed to support innovation.

4.6   Third hypothesis testing (H_3)

The third hypothesis tested in this study (H_3) is that formal control influences entrepreneurship. Regression results for indicate that formal control has a significant effect on entrepreneurship, with significance level of 0.04 (that it, smaller than 0.05). Regression results suggest that formal control affects managerial entrepreneurship for the dimensions of risk-taking. The results of this study are in line with the research of Chenhall et al. (2011), which shows a positive relationship between formal control and risk-taking innovation. However, it is different from the research of Moon (1999) which shows that formalization has no effect on managerial entrepreneurship in term risk-taking.

4.7   Discussion

The results of this study indicate that taken together, social networking variables, innovative organizational culture, and formal control affect managerial entrepreneurship from aspects risk-taking. This shows that there is a relationship between management control systems and managerial entrepreneurship in the risk-taking innovation model. The management control system is described as a set of controls relating to social networking, innovative culture of organization, and formal control. The management control system has evolved in recent years, to focus on formal and quantitative information that can assist managerial decision-making with a broader range of information. This includes external information relating to the market, customers, competitors, non-financial information related to the production process, predictive information and various decision mechanisms, and informal personal and social controls. More conventionally, management control systems are considered as being passive tools that provide information to help managers (Chenhall, 2003).

It would appear that social networking is partially a reflection of the creation of entrepreneurial behavior, as reflected in the regression results with a significance value of 0.007 (tahat is, smaller than alpha 0.05). This shows that wider social networks in regional companies, will support managers in making innovations via risk taking, even though a negative relationship indicates in expanding social networks, public sector organizations collide with political aspects of local governments as their largest shareholders. It is we known that social networking is a way of managing interorganizational exchanges with an emphasis on personal and social relationships based on the length of the relationship and trust. Theoretical support for the role of social networks in improve innovation can be found in the organizational studies literature, and Noteboom (1999) daimed that social networking can help in improving managerial entrepreneurship, especially in decision-making for innovation, because to be innovative, companies need outside source cognition and competence to enable to complete.

Research into the effects of the socialization process that usually occurs in social networks shows a strong association between socialization and willingness to share information, engage in joint problem-solving, adapt to unexpected changes, and avoid the ab-use of power (Mahama, 2006), all of which encourage increased managerial entrepreneurship, especially in term of innovation (Feldman, 1976). Empirical evidence about how social networking between organizations (managerial entrepreneurship) will increase innovation is expressed by (Goes & Park, 1997 and Penings & Harianto (1992), who confirm that the manager's ability to innovate and implement innovation in organizations is greatly enhanced by interorganizational networking links. The existence of relationships and social contacts for managers is believed to be able to contribute to risk-taking innovation.

The results of this study show that an this variable has an influence on managerial entrepreneurship. The regression results provide support for the significance value is 0.03 (that is, smaller than 0,05). This shows that more innovative culture in an organization will improve manager's make innovations in decision-making.

Organizational culture is considered to have an important role in an organization, with cultural characteristics being considered as key elements that increases entrepreneurship (Covin and Slevin, 1991; Hornsby et al. 1993; Zahra, 1993). Others (Doney et al., 1998 argue that organizational culture at both the personal and organizational levels can provide a competitive advantage and needs to be considered in term of strengthening the relationship between subordinates and superiors. Therefore, the role of organizational culture has importance in the development of an organization and needs to be considered as an organizational strategy. In addition, Holmes and Marsden (1996) state that corporate or organizational culture has an influence on the behavior, work methods and motivation of managers and subordinates in achieving organizational performance can be influenced by. Lusch and Harvey (1994) argue that improving organizational performance can be influenced by intangible assets such as organizational culture, customer relationships and corporate image (brand equity).

The formal control variable also has an influence on managerial entrepreneurship, with a significance value of 0.0049 (that is, smaller than 0.05). It can be interpreted that higher level oftr formal control will improve management behavior in term of making more innovative decisions. Recently it has been argued that formal control has a role in a control package that aims to encourage innovation (Chenhall & Morris 1995; Simons 1995; Henri 2006b; Widener 2007).

Caruana et al., (1998) research provides some results that indicate formal control ensures organizations are able to maintain individual creativity in solving organizational goals without there being a problem of centralization policies crippling entrepreneurship in term of innovation, risk-taking, and proactivity. Organizations can achieve formal control in several ways, including those that helps to ensure that individuals are both team player and innovative, and do not pursue random or excessive opportunities that are inconsistent with the company's mission and strategic direction.

5 CONCLUSIONS, LIMITATIONS, SUGGESTIONS AND RECOMMENDATIONS

5.1 *Conclusion*

Based on the results of the research and discussion, it can be concluded thats:

1. H_1 stating that social networking variables negatively affect managerial entrepreneurship is proved.
2. H_2 stating that the organization's innovative culture variables have a positive effect on managerial entrepreneurship is proved.
3. H_3 stating that formal control variables have a positive effect on managerial entrepreneurship is proved.

5.2 *Limitations and suggestions*

The researcher is aware of some limitations that might affect the results, even though overall these results provide support for previous research, These limitation include 1) the object of research is limited, so generalization cannot be made from the actual results, and 2) the coefficient of determinant R_2 is low, so the three variables cannot explain overall managerial entrepreneurship. In line with Chenhall's research (2011), it possible that there are still other variables that will strengthen managerial entrepreneurship.

Some suggestions that may be considered are as follows:

1. It is expected that future research can include other variables in order to increase the determination coefficient R_2 and that a more accurate survey method could be used.

2. The results of this study are expected to be used as a reference for further research, in line with the research of Moon (1999) in that while there is extensive interdisciplinary research into managerial entrepreneurship and an attempt to introduce entrepreneurship to the public sector, there is still a lack of empirical research about entrepreneurship.

This research is also expected to provide a contribution to developing information about how entrepreneurship that relates to decision-making behavior or risk is influenced by three independent variables such as social networking, innovative organization culture, and formal control (management control systems), especially in regional companies and regional public service bodies. Such organizations are regional assets that have not received much attention and which still lag behind by private companies in terms of human resources, technology and innovation.

REFERENCES

Abernethy, M.A. & Brownell, P. (1997). Management control systems in research and development organizations: The role of accounting, behavior and personnel controls. *Accounting, Organizations and Society*. 22 (3/4), 233–248.

Abernethy, M.A. & Lillis, A.M. (1995). The impact of manufacturing flexibility on management control system design. *Accounting, Organizations and Society*. 20, (4), 241–258.

Ahrens, T. & Chapman, C. (2004). Accounting for flexibility and efficiency: A field study of management control systems in a restaurant chain, *Contemporary Accounting Research*, 21 (2), 271–301.

Anderson, J.C., Ha°kansson, H & Johanson, J. (1994). Dyadic business relationships within a business network context. *Journal of Marketing, 58: 1–15.*

Anthony, R.N., Dearden, J. & Bedford, N.M. (1998). *Managerial control system*, Homeword Illinois, Irwin.

Anthony, R.N & Govindarajan, V. (2012). Management control system. Jakarta: Salemba Empat.

Atkinson, A.A, Banker, R.J., Kaplan, R.S. & Young, S.M. (1995), *Management accounting*, Engle Wood Cliifts, NJ; Prentice Hall.

Buku Kinerja PDAM, (2017). BPPSPAM Kementerain Pekerjaan Umum dan Perumahan Rakyat, Jakarta.

Burns, T. & Stalker, G.M. (1961). The management of innovation. London, UK: Tavistock.

Cadez, S. & Guilding, C. (2008). An explanatory investigation of an integrated contingency model of strategic management accounting. *Accounting, Organization and Society,* 33(4), 836–863.

Caruana, A., Morris, M.H. & Valla, A.J. (1998), The effect of centralization and formalization on entrepreneurship in export firm, *Journal of Small Business Management*, 36 (1), 16–29.

Chenhall, R.H. & Morris, D. (1995). Organic decision and communication processes and management accounting systems in entrepreneurial and conservative business organizations. *Omega* 23: 485–497.

Chenhall, R.H. (2003). Management control system design within Its orgaizational context: findings from contingency-based research and directions for the future. *Accounting, Organizations and Society*, 28 (1), 127–168.

Chenhall, R.H., Kallunki, J.P. & Silvola H. 2011, Exploring the relationships between strategy, innovation, and management control systems: The roles of social networking, organic innovative culture, and formal controls, *Journal of Management Accounting Research*, 23 (1), 99–128.

Covin, J.G. & Slevin, D.P. (1991), A conceptual model of entrepreneurship as firm behavior. *Entrepreneurship Theory and Practice,* 16 (1), 7–25.

Damanpour, F. (1990). *Innovation effectiveness, adoption and organizational performance,* In M.A. West & J.L. Farr (Eds), *In Innovation and Creativity at Work*, (125–141). Chichester, UK: John Wiley.

Davila, A. Foster, G & Oyon, D, (2009). *Accounting and control, entrepreneurship and innovation: venturing Into new research opportunities. European Accounting Review,* 18 (2), 281–311.

Denison, D.R. & Mishra, A.K. (1995). Toward a theory of organizational culture and effectiveness. *Organization Science*, 6(2), 204–223.

Dent, J.F. (1990). Strategy, Organization and control: Some possibilities for accounting research. *Accounting, Organizations and Society*, 15(1–2), 3–25.

Dilulio, J.J., Garvey, G & Kettl, D. (1993). *Improving government performance: An owner's manual.* Washington, DC: Brookings Institution.

Doney, M., Cannon, J.P. & Mullen, M.R. (1998). Understanding the influence of national culture on the development of trust, *Academy of Management Review*, 23(3), 601–620.

Drucker, P. (1995). *Innovation in entrepreneurship*, London, UK: Pan Books.

Fauzi, H. & Hussain, M.M. (2008). Relationship between contextual variables and management control systems: experience with Indonesian hospitality industry, (*Working paper*, pp. 1–34).

Gerdin, J. & Greve, J. (2008). The appropriateness of statistical methods for testing contingency hypotheses in management accounting research. *Accounting, Organizations and Society*, 33(5), 995–1009.

Ghozali, I. (2011), *Aplikasi Multivariate dengan SPSS*, Badan Penerbit Universitas Diponegoro.

Goes, J.B. & Park, S.H. (1997). Interorganizational links and innovation: The case of hospital services. Academy of Management Journal, 40 (3), 673–696.

Greve, A. & Salaff, J.W. (2003). Social networks and entrepreneurship *Journal Entrepreneurship Theory and Practice*, 28 (1), 1–22.

Gulati, R. Nohria, N. & Zaheer, A. (2000). Strategic networks, *Strategic Management Journal*, 21(3), 203–216.

Hakansson, H. & Lind, J. (2004). Accounting and network coordination. Accounting, *Organizations and Society*, 29 (1), 1–93.

Han, J.K., Kim, N., & Srivastava, R.K. (1998). Market orientation and organizational performance: Is innovation a missing Link? *Journal of Marketing*, 62 (4), 30–45.

Henri, J.F. (2006). Management control systems and strategy: A resource-based perspective. *Accounting, Organization and Society,* 31(6), 529–558.

Hofer, C.W. & Bygrave, W.D. (1992). Researching entrepreneurship. *Entrepreneurship Theory and Practice, 16*(3), 91–100.

Hofstede, G. (1980). *Culture's consequences: International differences in work-related values.* London UK: Sage Publications.

Holmes, S. & Marsden, S. (1996). An exploration of espoused organizational culture of public accounting firms. *Accounting Horizons*, 10(3), 26–53.

Hornsby, J.S., Naffziger, D.W., Kuratko, D.F. & Montagno, R,V. (1993). An interactive model of the corporate entrepreneurship process, *Entrepreneurship Theory and Practice,* 17(2), 29–37.

Ingraham, P. & Romzek, B.S. (1994). *New paradigms for government: Issues for The changing public service.* San Francisco: McGill-Queen's University Press.

Kearney, C., Hisrich R. & Roche, R. (2007). Facilitating public sector corporate entrepreneurship process: A conceptual model. *Journal of Enterprising Culture*, Vol. 15(3), 275–299.

Khandawalla, P. (1972). The effect of different types of competition on the use of management controls. *Journal of Accounting Research*, 10(2), 275–285.

Kloot, L. (1997). Organizational learning and management control systems: Responding to environmental change. *Management Accounting Research*, 8(1), 47–73.

Langfield-Smith, K. (1997). Management control systems and strategy: A critical review. *Accounting, Organizations and Society*, 22(2), 207–232.

Lövstål, E. (2008), *Management control systems in entrepreneurial organisations – A balancing challenge*. Jönköping International Business School, Jönköping University, JIBS Dissertation Series No. 045.

Mahama, H. (2006). Management control Systems, cooperation and performance in strategic supply relationships: A survey in the mines. *Management Accounting Research*, 17(2), 315–339.

Malmi, T. & Brown, D.A. (2008). Management control systems as a package: Opportunities, challenges and research directions. *Management Accounting Research,* 19(2), 287 –300.

Merchant, K.A. (1985). Organizational control and discretionary program decision making: A field study. *Accounting, Organizations and Society*, 10(1) 67–85.

Merchant, K.A. & Van der Stede, W.A. (2007). *Management control systems: Performance measurement, evaluation and incentives* (2nd). UK: Prentice Hall.

Miller, D. & Friesen, P.H. (1982). Innovation in conservative and entrepreneurial firms: Two Models of strategic momentum. *Strategic Management Journal*, 3(1), 1–25.

Miller, D. (1983). The correlates of entrepreneurship in three types of firms. *Management Science*, 29(7), 770–791.

Mintzberg, H. (1994). *The rise and fall of strategic planning*. New York, NY: The Free Press.

Mintzberg, H. & Waters, J.A. (1985). Of strategies, deliberate and emergent. *Strategic Management Journal*, 6, 257–272.

Moon, M.J. (1999), The pursuit of managerial entrepreneurship: Does organization matter? *Public Administration Review*, 59(1), 31–43.

Naman, J.L. & Slevin, D.P. (1993). Entrepreneurship and the concept of fit: A model and empirical test. *Strategic Management Journal*, 14(2), 137–153.

Noteboom, B. (1999). Innovation and the inter-firm linkages: New implications for policy. *Research Policy*, 28: 793–805.

Otley, D.T. (1980), The contingency theory of management accounting: Achievement and prognosis, *Accounting Organization and Society*, 5: 413–428.

Pennings, J.M. and Harianto, F. (1992). The diffusion of technological innovation in the commercial banking industry. *Strategic Management Journal*, 13(1), 29–46.

Peters, B.G. & Savoie, D.J. (1995), *Governance in a new environment, Montreal, Canada*: McGill/Queens University Press.

Quinn, J.B. (1980). Strategies for change, logical incrementalism. Homewood, IL: Irwin.

Sanger, M.B. & Levin, M.A. (1992), Using old stuff in new ways: Innovation as a case of evolutionary tinkering, *Journal of Policy Analysis and Management*, 1, 88–115.

Sekaran, U. 2016. *Research methods for business: A skill-building approach*. (3rd Ed.). Chichester, UK: John Wiley and Sons.

Simons, R. (1994). How New top managers use control systems as levers of strategic renewal. *Strategic Management Journal*, 15(5), 46–62.

Simons, R. (1995). Levers of control. Boston, MA: Harvard Business School Press.

Syahdan, S.A, & Santoso, P.B. (2004), The effect of organizational structure, culture, and environment on managerial entrepreneurship dimensions in regional companies of South East Kalimantan. *Journal Business and Strategy*, 13(2), 190–204.

Widener, S.K. (2007). An empirical analysis of the levers of control framework. Accounting, *Organizations and Society, 32(7/8)*, 757–788.

Wood, V.R., Bhuin, S., & Kiecker, P. (2000). Market orientation and organizational performance in not-for-profit hospitals. *Journal of Business Research*, 48(2), 213–226.

APPENDIX 1. MULTIPLE REGRESSION OUTPUT

Variables entered/removed[a]

Model	Variables entered	Variables removed	Method
1	X3, X1, X2[b]		Enter

a. Dependent variable: Y.
b. All requested variables entered.

Model summary[b]

Model	R	R-squared	Adjusted R-squared	Std. error of the estimate	Durbin-Watson
1	0.344[a]	0.118	0.096	1.47518	2.146

a. Predictors: (constant), X3, X1, X2.
b. Dependent variable: Y.

ANOVA[a]

Model		Sum of squares	Df	Mean square	F	Sig.
1	Regression	34.712	3	11.571	5.317	0.002[b]
	Residual	258.962	119	2.176		
	Total	293.675	122			

a. Dependent variable: Y.
b. Predictors: (constant), X3, X1, X2.

Coefficients[a]

Model		Unstandardized coefficients		Standardized coefficients	t	Sig.	Collinearity statistics	
		B	Std. error	Beta			Tolerance	VIF
1	(Constant)	12.344	2.089		5.910	0.000		
	X1	−0.084	0.030	−0.241	−2.758	0.007	0.967	1.034
	X2	0.094	0.043	0.193	2.192	0.030	0.954	1.048
	X3	0.084	0.042	0.173	1.989	0.049	0.985	1.015

a. Dependent variable: Y.

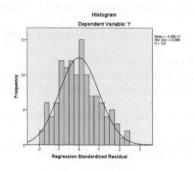

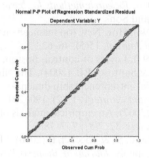

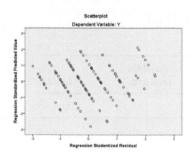

Tourism management

Business Innovation and Development in Emerging Economies – Trinugroho & Lau (Eds)
© *2019 Taylor & Francis Group, London, ISBN 978-1-138-35996-3*

Determinants of health-visit appeal in terms of medical tourism

R.T. Ratnasari, Sedianingsih & A. Prasetyo
Faculty of Economics and Business, Universitas Airlangga, Surabaya, Indonesia

ABSTRACT: The purpose of this study is to determine the factors that influence the attractiveness of a health visit to Surabaya, as a destination for city-based *medical tourism*. The research design used in this study was an exploratory quantitative approach. This quantitative approach used a factor analysis method (analyzer) and used a questionnaire as a tool to determine the respondent's perception. There are two groups. The first group is the renderer of services (the management of the health service institution, the management and employees of hospitals, health centers, and health clinics in Surabaya). The second group is the service user (the consumers' hospital, health center, and health clinic).

This method was carried out in three stages. The first stage determined indicators of factors affecting the appeal of a visit to Surabaya for city-based medical tourism, obtained from the process of extracting data through in-depth interviews with both groups of informants in a *preliminary test.* The second stage involved a pre-survey of 40 respondents to test if the material questionnaire was understood. The third phase was a structured *close-ended question* to 400 respondents, consisting of 100 outpatients at the hospital in Surabaya, 100 consumers in Surabaya, and 200 consumers outside Surabaya. The sampling technique was conducted with a *purposive non-random* sampling.

The results of the research show that there are six factors influencing the attractiveness of a health visit to Surabaya, as a medical tourism-based city. The six factors are: credibility of medical team; quality of service to patient; technology of medical equipment; hospital management; price comparison of benefit; communication with patient and family. Of these, the credibility of the medical team is the highest factor in influencing the appeal of health visits to Surabaya as a medical tourism-based city, and communication with patient and family is the lowest.

The implications include the need for the medical team to make improvements in terms of developing a good communication relationship pattern with patients and their families, developing better ethical and empathic attitudes, and prioritizing patient safety so that people are satisfied. The government also needs to involve relevant industries, such as the health industry, the insurance industry, the pharmaceutical industry and the regulators, to ensure integration with local/national tourism development programs.

Keywords: Factors, attraction, health visiting, *medical tourism*

1 INTRODUCTION

The tourism industry has grown to become the largest service industry in the world and it is increasingly difficult for the government to manage. Most developed or developing countries take into account the tourism industry as one of their main economic priorities. According to Rosy Mary (2014) in Ganguli (2017), Asian countries Thailand, Singapore and India are recognized as the three main destinations on this continent and are projected to account for more than 80% of the Asian market in the future. This study provides a framework to identify the factors (that need to be assessed to determine if a place has the potential for medical tourism and how this potential can be realized. Clearly, we need to take a holistic approach in assessing the suitability of a place, and to help identify performance deficiencies of public and private stakeholders that are relevant. This approach combines with feedback from relevant stakeholders. The importance of tourism is not limited to creating career opportunities and generating revenue. The tourism and travel industry has undergone tremendous changes since the 1950s, a characteristic of the period that is a revelation. These changes are improving tourism services and in the near future we will benefit from high-quality tourism services. Currently, tourism is seen as a strong and professional activity in the world, and is considered to be one of the most important economic resources (Hallmann et al., 2012; Constantin, 2015, pp. 13–21).

Medical tourism is one of the fastest growing industrial sectors (Wang, 2012). Tourism refers to the activity of the people who are traveling to places outside their usual environment for the purpose of recreation, relaxation and pleasure. However, in the new era of globalization, human lifestyles continue to change rapidly with the development of the international aviation industry. Where once there were clear regional characteristics, there is now developing a service mode competing with the traditions. Regarding medical tourism, various definitions have been proposed. Hunter-Jones (2005) defines medical tourism as the way to recovery, instead of going directly to get treatment. In addition, Hall (2011) suggests that, until recently, medical tourism has been included in the overall context of health tourism. Glinos and Baeten (2006) describe medical tourism as an activity of patients who go abroad to seek medical care because of some relative weakness in the national health care system of their own community. Medical tourism activity is not only for traveling but also for recovery and treatment. Named philosophy was developed from the concept of *medical tourism*. Even in Asian countries the concept is already widely developed. China became a destination for organ transplants, Korea as the superior in plastic surgery and Thailand is popular as a place to repair teeth and tighten facial skin. At this time, at least more than 600,000 Indonesian patients are seeking treatment abroad at an annual cost of at least 20 trillion rupiah. This number is simply fantastic in the middle of the current economic difficulties (Zuardin, 2015). According to Nagar (2011), for example, many hospitals in Thailand and Malaysia have a specific prayer room and halal food for their Muslim patients. Therefore, it is necessary for the Surabaya city government to develop medical tourism or medical travel.

There are several definitions of medical tourism. One definition considers medical tourism or medical travel as a trip abroad for a detox diet, dental care or surgery. This trip should involve at least a chance to stay in a location where treatment takes (Sadrmomtaz & Agharahimi, 2010, pp. 516–524). According to Edelheit (2009), said that medical tourism refers to people traveling in countries other than the home country of the consumer to receive medical treatment.

Medical tourism is one of the most important indicators of the tourism industry to the economy. The social benefits are significant, known as international trips where someone makes use of treatment costing less compared to the same treatment in their home country (Edelheit, 2008; Constantin, 2015, pp. 9–1, pp. 9–10). Medical tourism requires infrastructure support, one element of which is the hospital. A hospital is built to provide health services to the community. The American Hospital Association (1996; Tarin, 2009, pp. 19) provides a definition of the hospital as being an organization where a trained professional employed by the hospital will give medical action, continuous nursing care, diagnosis and treatment of diseases suffered by patients (Ratnasari & Masmira, 2016).

However, the role of the hospital is not only as a management function, implemented and devoted to providing health services to the community (Ratnasari & Kurniawati, 2016). Currently, the hospital can play a supportive role in health-based tourism. The healthcare industry has the potential to be developed into a tourist health service. Surabaya, known for its trade and services, should optimize its service sector, especially its health services. Hopefully, Surabaya will have the appeal for a visit to its health sector, known as *medical tourism*, by consumers in Surabaya (who do not need medical treatment abroad), by consumers in Eastern Indonesia, and even consumers from foreign countries. Hence, Surabaya is expected as the world faces a health referral center of ASEAN Economic Community. The new effect has been evident since December 2015.

Surabaya mayor, Tri Rismaharini (www.tempo.com, 2015) stated that the actual quality of hospitals and doctors in Indonesia, especially in Surabaya, is very good compared to the hospitals in Singapore, Malaysia and Thailand. Furthermore, Surabaya can at least take back citizens who would otherwise choose to seek treatment in Penang, Malaysia or Singapore. This becomes very important, because according to the Surabaya Central Statistics Agency (BPS, 2015) the potential consumers who live in the Surabaya region amount to 3.2 million people; this is a *market* that cannot be removed just like that. In addition, the need for health services in East Java has experienced a rapid increase. Based on the information from BPS, the number of healthcare facilities available in Surabaya amounted to 1,043, in the form of hospitals, health centers, pharmacies, and other health facilities (http://surabayakota.bps.go.id).

Health tourism or medical tourism can provide quality health services, as well as several other facilities, such as better accommodation, shopping and arrangements for consumers to conduct recreational activities (Sultana et al., 2014). Therefore, it becomes very necessary for hospital management to pay attention to good healthcare, and the need to implement a quality service. Quality of care is particularly important in customer satisfaction (Cronin & Taylor, 1992). Quality of service is recorded as a main prerequisite for establishing and maintaining satisfactory relationships with consumers (Lassar et al., 2000). Quality of service is the result of long-term cognitive evaluation generated by the consumers, with the services delivered by marketers (Lovelock & Wright, 2007, pp. 96). Superior service performance results when the service is at the level of service expected by the consumer, including, as in this case, healthcare. According to research, Gratzer and Ljungbo (2014), in Smith (2016) stated that the consequences of increased competition in the global medical tourism sector are a necessity in order to increase specialization in hospitals, clinics and medical research institutes.

Therefore, Surabaya has a lot of potential, having a number of the largest hospitals in Indonesia: it has 62 type-D hospitals, 60 health centers, five major laboratories and five health clinics (yield interview with the management of Health Office of East Java Province, December 2015). This suggests that the government makes the Surabaya region an appealing destination for health visits, known as *medical tourism*, in addition to the Surabaya that is known for its *business and entertainment tourism* and educational tourism. Smith (2016) stated within the tourism sector, healthcare and medical means it is important to consider innovative service. Johnstom (2015) directs the research results on the opinion of prospective medical tourists. Information was obtained that prospective medical tourists will participate in medical tourism; they will adjust their travel distance (Buzinde & Yarnal, 2015). However, it is different in the public health and medical discourse. As medical tourists, people traveling across the border for medical treatment have high levels of strength and relatively high freedom to choose medical treatment. It is expected that, in the future, the hospitals in Surabaya will be able to offer a package of health services and provide a superior service. Likewise, the citizens of Surabaya who prefer medical treatment abroad will return again for treatment to a hospital in Surabaya. Therefore, it becomes very important to do research under the heading of 'Factors Affecting Appeal of Health Visiting to Surabaya in City-Based *Medical Tourism*'.

2 LITERATURE REVIEW

2.1 *Tourism destination competitiveness*

A desirable destination for medical tourism should have the types of products that involve the core rewards, such as commercial infrastructure and environmental factors. It should have comparative advantages in terms of climate, environment, flora and fauna, while also having the competitive advantage associated with items produced, such as in the health and medical care areas, historic sites, events, site transportation, government policies, the actual quality of management and the skills of workers (Sultana et al., 2014).

One of the most important factors in the medical tourism industry is the quality of service provided by the destination country. Medical tourists are very focused on this problem. Thus the country of destination must meet the expectations of the medical tourists in terms of the quality of its service and performance. In service provider organizations, service quality is proven to be an important determinant of competitiveness (Sultana et al., 2014).

According to the study by Ryan (1995) in Chiu (2016) satisfaction could be one of the most researched variables in tourism literature. Satisfaction can be regarded as post-purchase evaluation or a visit. In tourism research, Hunt (1983) in Chiu (2016) suggested that satisfaction is not just about a pleasurable travel experience but also involves an evaluation. So satisfaction will come when consumers compare their original expectations with their actual perceptions. Once perceived experiences surpass expectations, consumers are satisfied (Yüksel & Yüksel, 2001).

2.2 *Medical tourism*

As we have already seen, there are several definitions of medical tourism. Mukherjee (2016) states that medical tourism is composed of three forms, namely health tourism, travel and curative health care (Iordache et al., 2013). Health tourism relates to health promotion, such as ecotourism and nature-based travel. Health tourism consists of two aspects: spas (spa, yoga, aromatherapy, herbal treatments, Ayurveda treatments) and rehabilitation (such as hemodialysis treatment or therapy). Contantin (2015) explain the main features of medical tourism in carrying out activities based on health services *medical tourism*, namely:

a. A large number of people traveling for treatment;
b. Patients traveling to treatment in developing rather than developed countries, due to high-quality care offered at a low price, and a variety of resources available on the Internet, through which the patient can calculate in advance the approximate price of healthcare services, so that consumers will benefit from savings on travel costs;
c. Industrial development: both in the private and public health sector of developed countries, but also to develop sustainable investment in promoting medical tourism, which is a source of potential increase in external revenue. The care tourism industry dynamic is reinforced by a number of factors, including the economic climate, changes in domestic policy, political instability, travel restrictions, advertising practices, changes in the geopolitical situation, and other forms of innovative and pioneering treatment that can help change consumption and production patterns in the health services, both domestically and abroad.

2.3 *Quality of service*

Quality is determined as being an important issue for competitive success. Improving the quality and potential in the travel and tourism sector is very difficult to implement because it involves people from different countries with different cultural backgrounds and demands. However, companies can improve service quality by lowering distribution costs and improving the services provided (Sultana et al., 2016). Quality is a word that should be applied to all providers of services to customers. Gietsch and Davis, in Tjiptono (2007, pp. 4), define define the quality of a dynamic condition associated with products, services, people, processes and environments that meet or exceed expectations. Kotler and Keller (2016, pp. 310) question whether all the features and quality of a product or service that affects the ability to satisfy are stated or implied, while the definition of quality in Lupiyoadi (2008, pp. 175) is a blend of characteristics that determine the extent to which *output* can meet customer expectations. Based on the above notions, a successful supplier is providing good quality when the product or service provided can meet or exceed customer expectations.

Quality of care is particularly important in customer satisfaction (Cronin & Taylor, 1992). Quality of service is recorded as being a main prerequisite for establishing and maintaining satisfactory relationships with consumers (Lassar et al., 2000). Numerous recent studies show that by improving the quality of services, a company is able to satisfy its customers and retain their loyalty (Lee & Murphy, 2008). Quality of service is in the form of long-term cognitive evaluation generated by consumers, with the services delivered by marketers

(Lovelock & Wright, 2007, pp. 96). Quality of service has become the biggest differentiator, and the most powerful competitive weapon for service organizations. In this context, an understanding of the interaction between factors such as quality of health services, outcomes and patient satisfaction, has been a very valuable input for designing, managing and *benchmarking* health systems. Therefore, the concept of service quality in the health context is a necessary one. As a result, we expect that the results will be used to guide the development of a competitive health travel policy (Chang et al., 2013).

2.4 *Excellent service quality*

Ratnasari and Aksa (2011, pp. 129–131) state that one of the excellent service quality approaches popularly used as a reference in marketing research is the SERVQUAL (*Service Quality*) model, developed by Parasuraman et al. (1985). SERVQUAL is built on their comparison of two main factors, namely the perception of the real customers for the services they receive (*Perceived Service*) and the actual services expected/desired (*Expected Service*). If the reality is better than expected, then the service can be said to have qualified, while if the reality is poorer than expected, then the service is said to not have qualified. And if the results are the same as expected, it can also be called satisfactory. Thus the *service quality* can be defined as being the degree of difference between reality and the expectations of customers for services they receive/acquire.

3 METHOD

The research design used in this study is exploratory quantitative approach. In this quantitative approach the method (analyzer) used is factor analysis. This method aims to reduce the number of indicators of research, still using as much information as possible. The measurement method of research will be more easily understood through quantitative methods using a questionnaire as a tool to determine the respondent's perception. There are two groups: the first group is the renderer of services (management of the health service institution, the management and employees of hospitals, health centers, and health clinics in Surabaya, the second biggest city in Indonesia); the second is the service user (consumer's hospital and health center and health clinic). The method is applied in three stages. The first stage is to determine indicators of factors affecting the appeal of a health visit to Surabaya for city-based medical tourism, obtained from the process of extracting data through in-depth interviews with both sets of informants in a preliminary test. The second stage is to perform a pre-survey of 40 respondents to test if the material questionnaire is understood. The third phase follows with a structured close-ended question to 400 respondents.

4 RESULTS

The results in Table 1 of the factor rotation calculation show that there are six factors formed (highlighted in color). A rotation factor can be considered if it has a value loading factor of more than 0.5:

1. The first factor has a value of more than 0.5 for indicators X_9, X_{10}, X_{11}, and X_{13};
2. The second factor has a value of more than 0.5 for indicators X_3 and X_4;
3. The third factor has a value of more than 0.5 for indicators X_1, X_2, and X_6;
4. The fourth factor has a value of more than 0.5 for indicators X_{12}, X_{17}, and X_8;
5. The fifth factor has a value of more than 0.5 for indicators X_7 and X_{14};
6. The sixth factor has a value of more than 0.5 for indicators X_{15}, X_6, X_5, and X_{16}.

After the factor rotation step, the next step is to perform factor interpretation. The purpose of this step is to determine which indicators can fit in a factor and which are not included in a factor. The naming of each factor in this research uses a surrogate method, that is, a

Table 1. Factor rotation results.

	Component					
	1	2	3	4	5	6
X_1	0.128	0.208	0.708	−0.230	0.073	0.089
X_2	0.164	−0.088	0.813	0.157	0.075	0.105
X_3	0.423	0.654	0.000	0.156	−0.080	−0.366
X_4	−0.028	0.732	0.062	0.141	0.027	0.082
X_5	0.222	0.069	0.102	−0.024	0.210	0.682
X_6	−0.149	0.276	0.581	0.314	−0.044	0.621
X_7	0.027	0.231	0.220	0.453	0.651	0.112
X_8	0.058	0.487	0.179	0.500	0.228	0.232
X_9	0.670	0.127	−0.110	0.198	0.039	0.150
X_{10}	0.535	0.053	0.208	0.399	−0.113	−0.221
X_{11}	0.575	−0.042	0.189	0.352	0.173	−0.123
X_{12}	0.223	0.138	−0.257	0.705	−0.240	0.105
X_{13}	0.643	0.410	0.068	0.038	0.053	−0.061
X_{14}	0.149	0.136	−0.023	−0.048	0.827	0.086
X_{15}	−0.016	0.094	0.101	−0.006	0.117	0.879
X_{16}	0.086	0.064	0.102	0.039	0.099	0.692
X_{17}	0.466	0.405	0.021	0.677	−0.403	0.215

Source: Field work research, 2017.

method named factor based on the value of loading the highest factor on each factor formed (Simamora, 2008, in Tetuko, 2015, pp. 75).

The first factor has a value of more than 0.5 on the following indicators: X_9 performs the best health services; X_{10} provides after-hours service from the hospital, such as timed control; X_{11} is the availability of the training center; X_{13} is the hygiene and the comfort of the waiting room and the check room. Based on these indicators, the factor is named the **Quality of medical service to the patient**.

The second factor has a value of more than 0.5 on the following indicators: X_3 is sophisticated medical equipment; X_4 is that the distribution of competent medical personnel at several hospitals in Surabaya is quite good. Based on these indicators, the factor is named the **Medical equipment technology**.

The third factor has a value of more than 0.5 on the following indicators: X_1 is expertise is not much different from doctors who are in high-grade hospitals abroad; X_2 is the credibility of the doctor according to his specialist expertise; X_6 is that many doctors in Surabaya have certain specializations recognized by physician associations abroad. Based on these indicators, the factor is named the **Credibility of the medical team**.

The fourth factor has a value of more than 0.5 on the following indicators: X_{12} promotes patient safety so that people are satisfied; X_{17} involves four industries, namely the healthcare industry, insurance finance industry, drug supply industry and the government regulator; X_8 is more professional. Based on these indicators, the factor is named the **Management of the hospital**.

The fifth factor has a value of more than 0.5 on the following indicators: X_7 is that the price of services in the hospital is on a more expensive premium scale, but the package is more complete; X_{14} is that the cost of handling doctors and drugs tends to be more expensive. Based on these indicators, the factor is called the **Comparison of prices with benefits**.

The sixth factor has a value of more than 0.5 on the following indicators: X_{15} develops a good pattern of communication relationships with patients and their families; X_6 develops a better ethical attitude to patients; X_5 develops better empathy for patients; X_{16} is that time consultation with the doctor is not too short. Based on these indicators, the factor is named the **Communication with patient and family**.

Table 2. Results of confirmatory factor analysis.

Factor	Indicator	Corrected item – total correlation	Cronbach's alpha	Note
1 (Credibility of medical team)	X_9	0.568	0.765	Reliable
	X_{10}	0.567		
	X_{11}	0.529		
	X_{13}	0.508		
2 (Quality of service to patient)	X_3	0.518	0.675	Reliable
	X_4	0.544		
3 (Technology of medical equipment)	X_1	0.506	0.680	Reliable
	X_2	0.571		
	X_6	0.655		
4 (Hospital management)	X_{12}	0.583	0.618	Reliable
	X_{17}	0.491		
	X_8	0.514		
5 (Price comparison with benefit)	X_7	0.486	0.654	Reliable
	X_{14}	0.486		
6 (Communication with patient and family)	X_{15}	0.414	0.661	Reliable
	X_6	0.514		
	X_5	0.472		
	X_{16}	0.532		

Source: Field work research, 2017.

4.1 Validity and reliability test

Based on the new dimensions formed after exploratory factor analysis, the validity and reliability of these factor results were then tested. The resulting calculations show that the six factors that have been formed fulfill the validity and reliability requirements. These six factors are the Credibility of the medical team, Quality of service to patient, Technology of medical equipment, Hospital management, Price comparison with benefits, and Communication with patient and family.

According to Sugiyono (2008) reliability involves a series of measurements or a series of measuring tools that have consistency when measurements are made repeatedly, while Sudjana (2014) aid that the reliability of a test is the accuracy or consistency of the test in assessing what it's for, meaning that whenever the test is done it will give the same or relatively similar results. According to Suryabrata (2004, pp. 28) reliability shows the extent to which the measurement results with the tool can be trusted. Measurement results must be reliable in terms of having consistency and stability.

5 DISCUSSION

The research results show that there are six factors influencing the attractiveness of a health visit to Surabaya for city-based *medical tourism*. The six factors are:

5.1 Credibility of medical team

The research results show that the credibility of the medical team has influence on the attractiveness of a health visit to Surabaya, with alpha value of 0.765 (> 0.6). This shows that the factor has a reliable internal consistency of construct validity.

There are three indicators of credibility of the medical team, which are that expertise is little different from doctors in high-grade hospitals abroad, the credibility of the doctor in terms of their specialist expertise, and that many doctors in Surabaya have particular specializations recognized by physician associations abroad.

It shows that the more credible the medical team, the more attracted citizens are to having medical treatment in Surabaya. Credibility of the medical team has high influence on the attractiveness of a health visit to Surabaya. Thus, the medical team should pay more attention to this matter to make sure that citizens will prefer medical treatment in Surabaya than anywhere else. In other words, patients believe that the more credible the medical team, the better the result they will have in their medical treatment.

5.2 *Quality of service to patient*

The research results show that quality of service to patient has influence on the attractiveness of a health visit to Surabaya, with alpha value 0.675 (> 0.6). This shows that the factor has a reliable internal consistency of construct validity.

There are four indicators of quality medical service to the patient. There are performance of the best health services, provision of after-hours service from the hospital, the availability of the training center, and the hygiene and comfort of the waiting room and the consultation room.

This indicates that quality of service to the patient in Surabaya for city-based medical tourism affects its attractiveness to citizens when they need medical treatment. Thus the medical team need to make improvements in service quality so that patients are more satisfied, and then the patients who prefer medical treatment abroad will return again for treatment to a hospital in Surabaya.

5.3 *Technology of medical equipment*

The research results show that technology of medical equipment has an influence on the attractiveness of a health visit to Surabaya, with alpha value 0.680 (> 0.6). This shows that the factor has a reliable internal consistency of construct validity.

There are two indicators that affect the technology of medical equipment factor. These are the sophistication of the medical equipment, and quite good distribution of competent medical personnel at several hospitals in Surabaya.

This shows that citizens prefer a place for medical treatment that has medical equipment of high technology to a place that does not. To have modern medical equipment technology, the medical team needs support from many sides to help them realize their development programs in technology. This is needed because citizens will assume that the higher the technology of medical equipment in use, the better the chance that patients will recover.

5.4 *Hospital management*

The research results show that hospital management has an influence on the attractiveness of a health visit to Surabaya, with alpha value 0.618 (> 0.6). This shows that the factor has a reliable internal consistency of construct validity.

There are three indicators of hospital management, which are promotion of patient safety so that people are satisfied, involvement of the four industries of healthcare, insurance finance, drug supply and government regulation, and being more professional.

This indicates that the better the hospital management, the more attracted citizens are to having medical treatment in Surabaya. Hospital management has a high influence on the attractiveness of a health visit to Surabaya. It also involves operating with many industries. So, the medical team must develop programs that will make it work, while satisfying patients with their service and hospitality.

5.5 *Price comparison with benefit*

The research results show that price comparison with benefit has an influence on the attractiveness of a health visit to Surabaya, with alpha value 0.654 (> 0.6). This shows that the factor has a reliable internal consistency of construct validity.

There are two elements of the price comparison with benefit factor. These are the price of services in the hospital is on a more expensive, premium scale (but the package is more complete), and that the cost of handling doctors and drugs tends to be more expensive.

It cannot be denied that citizens always compare medical treatment in many hospitals so that they can choose the best one for them. They will prefer an affordable price to get a more complete package of medical treatment. So, the better the medical treatment that they receive, the more satisfied they get, and the price paid is comparable with the benefits.

5.6 Communication with patient and family

The research results show that communication with the patient and family has an influence on the attractiveness of a health visit to Surabaya, with alpha value 0.661 (> 0.6). This shows that the factor has a reliable internal consistency of construct validity.

There are four indicators that affect the communication with the patient and family factor. These are development of a good pattern of communication relationships with patients and their families, development of a better ethical attitude to patients, development of better empathy for patients, and consultation time with the doctor is not too short.

It can be seen that the medical team needs to make improvements in terms of developing a good communication relationship pattern with patients and their families, developing better ethical and empathic attitudes, and prioritizing patient safety so that people are satisfied with their services. The more satisfied they get, the more attracted people are to using medical treatment in Surabaya again, rather than going anywhere else for treatment.

6 CONCLUSION

The conclusion of the research shows that there are six factors influencing the attractiveness of a health visit to Surabaya as a medical tourism-based city. The six factors are the credibility of the medical team, the quality of service to the patient, the technology of medical equipment, hospital management, price comparison with benefit, and communication with patient and family. Of these, the credibility of the medical team is the highest factor in influencing the appeal of health visits to Surabaya as a medical tourism-based city, and communication with patients and families is the lowest.

The implications are that the medical team needs to make improvements in terms of developing a good communication relationship pattern with patients and their families, develop better ethical and empathic attitudes, and prioritize patient safety so that people are satisfied. The government also needs to involve relevant industries, such as the health industry, the insurance industry, the pharmaceutical industry and the regulators, which must be integrated with local/national tourism development programs.

REFERENCES

Anshori, M. & Iswati, S. (2009). *Metodologi Penelitian Kuantitatif (Bahan Ajar)* (Unpublished thesis, Faculty of Economics, Airlangga University, Surabaya, Indonesia).

Arikunto, S. (2006). *Prosedur Penelitian: Suatu Pendekatan Praktik*. Jakarta, Indonesia: Penerbit Rineka Cipta.

Baker, D.M. (2015). Medical tourism development, challenges and opportunities for Asia. *Journal of Tourism, Culture and Territorial Development*, 6(12), 193–210.

Edelheit, J. 2009, "The Effects of the World Economic Recession On Medical Tourism: How will it Affect you?", Medical Tourism Magazine, vol. 09, pp. 22–24. Retrieve d from http://www.medicaltourismmag. com/issue-detail.php?item= 06&issue=9.

Glinos, Irene A. And Rita Baeten. 2006. A Literature Review of Cross-Border Patient Mobility in the European Union. Europe for Patients. September.

Kotler, P. & Keller, K.L. (2016). *Marketing management* (11th ed.). New York, NY: Prentice Hall.

Parasuraman, A., Zeithaml, V.A. & Berry, L.L. (1985). A conceptual model of service quality and its implications for future research. *Journal of Marketing*, 49(4), 41–50.

Ratnasari, R.T. & Aksa, M.H. (2011). *Manajemen Pemasaran Jasa*. Bogor, Indonesia: Ghalia Indonesia.

Ratnasari, R.T. & Kurniawati, M. (2016). Excellent service based on the concept of corporate entrepreneurship in hospital. *International Journal of Business Management & Research*, 6(2), 69–78.

Sugiyono, D. (2008). *Metode Penelitian Bisnis*. Bandung, Indonesia: Alfabeta.

Tjiptono, F. (2007). *Pemasaran Jasa* (1st ed.). Yogyakarta, Indonesia: Banyu Media.

Business Innovation and Development in Emerging Economies – Trinugroho & Lau (Eds)
© 2019 Taylor & Francis Group, London, ISBN 978-1-138-35996-3

Author index